Fourth edition

Geography

An Integrated Approach

David Waugh

Nelson Thornes

Contents

Introduction

● ●

Geography: An Integrated Approach (affectionately referred to as *GAIA*) has been written as much for those students who have an interest in Geography, an enquiring mind and a concern for the future of the planet upon which they live, as for those specialising in the subject. The text has been written as concisely as seemed practical in order to minimise the time needed for reading and note-taking, and to maximise the time available for discussion, individual enquiry and wider reading. Photographs, sketches and maps are used throughout to illustrate the wide range of natural and human-created environments. Annotated diagrams are included to show interrelationships and to help explain the more difficult concepts and theories. A wide range of graphical skills has been used to handle geographical data – data that are as up to date as possible at the time of writing and which you can continue to update for yourself by referring to relevant websites and other sources.

It is because Geography is concerned with interrelationships that this book has included, and aims to integrate, several fields of study. These involve physical environments (atmosphere, lithosphere and hydrosphere) and the living world (biosphere); economic development (or lack of it); the frequent misuse of the environment, the long-overdue concern over the resultant consequences, and the need for careful management and sustainable development; together with the application, where appropriate, of a modern scientific approach using statistical methods in investigations.

It is intended that this single book will:

■ satisfy the requirements of the latest Advanced Subsidiary (AS), A2, Advanced GCE, IB and other main Geography specifications
■ allow you to read more widely in Geography than just to be limited to the core and option modules in your examination specifications.

What it is not intended to do is to match the specifications, or methods of assessment, of individual syllabuses, as these are subject to change over periods of time. Rather the book aims to show the scope, width and everyday relevance of Geography in an ever-changing world.

By coincidence, the initial letters of the title of this book form the word *GAIA*. In Ancient Greece, Gaia was the goddess of the Earth. Today the term has been reintroduced to mean 'a new look at life on Earth', an approach that looks at the Earth in its entirety as a living organism. It is hoped that this book reflects aspects of this approach.

There is no rigid or prescribed sequence in the order either of the chapters themselves or in their structure. Each is open to several routes of enquiry. Terminology can be a major problem, as geographers may use several terms, some borrowed from other disciplines, to describe the same phenomenon. When a term is introduced for the first time it is shown in **bold** type,

has a list of alternatives (one of which is subsequently retained for consistency), and is defined. Alternative terms and specific examples often appear in brackets in order to save space. The detailed index, to allow you to cross-reference, has the key page reference for each entry in **bold** type.

The book sets out to provide an easily accessible store of information which will help you understand basic processes and concepts, to enter discussions and to develop your own informed, rather than subjective, values and attitudes. Theory is, whenever possible, supported by specific examples, which have been highlighted in the text as **Places**. Although there are over one hundred Places, limited space means these may be shorter than is ideal. Nevertheless they should enable you either to build upon your earlier knowledge or to stimulate you into reading more widely. At the end of each chapter is a more detailed **Case Study**. These include natural hazards, problems created by population growth, and by the misuse of the natural environment, and the attempts – or lack of – to manage the environment and the Earth's resources. **Further references** given at the conclusion of each chapter are those to which the author has himself referred, but they are not intended to be a comprehensive bibliographical list. In this edition, they include suggested reliable and useful **websites**.

As the reader, it is essential you appreciate that Geography is a dynamic subject with data, views, policies and terms which change constantly. Consequently, your own research must not be limited to textbooks, which in any case are out of date even before their publication, but should be widened to include the use of the Internet, CD-ROMs, newspapers, journals, television, radio and many 'non-academic' media.

GAIA also includes 19 **Frameworks** whose function is to stimulate discussion on methodological and theoretical issues. They illustrate some of the skills required, and the problems involved, in geographical enquiry, e.g. the uses, limitations and reliability of models; quantitative techniques; the collection of data, including using the Internet; Geographical Information Systems (GIS); maps; making classifications; and the dangers of stereotyping and of making broad generalisations. Geography is also concerned with the development of graphical skills. The media show an increasing amount of data in a graphical form, and this is likely to grow as Geographical Information Systems develop. It is assumed that the reader already understands those skills covered by current GCSE and Standard Grade examination specifications and therefore only new skills are explained in this book. Quantitative and statistical techniques are incorporated at appropriate points, although each may be relevant elsewhere in many of the physical and human/economic chapters. Following an explanation of each technique, there is a worked example.

The questions at the end of each chapter have been revised for this new edition. They are not written to be 'in the style of' any one specification or awarding body; rather they aim to provide all students, irrespective of the exam that they will be sitting, with graded practice, working towards the general style of questions that they might expect to face in their AS, A2, A Level, Baccalaureate or other exam being sat at the end of their course. The questions are arranged into four sections, which are graded in difficulty as students move towards structuring and planning their own answers. These sections are:

- **Activities** – highly structured sets of questions, designed mainly to test comprehension of key ideas and to be answered mainly by extracting relevant material from the text.
- **Exam practice: basic structured questions** – contain fewer sub-sections than the previous Activities, and are designed to be similar to the type of structured questions to be found on some AS papers.
- **Exam practice: structured questions** – contain fewer sub-sections than the basic structured questions and generally move on to test more complex and sophisticated knowledge and understanding.
- **Exam practice: essays** – designed to provide the 'stretch and challenge' that is such an important feature of the latest revised A Level specifications, following the agenda set by QCA. A minimum of structure is provided here, as A2 candidates are expected to plan extended essays on their own and show their ability to bring together knowledge and understanding from different areas of their study of Geography. It is hoped that these essay questions will provide opportunities for students of average ability to show evidence that they have learned good geographical skills. However, the essays are also intended to allow higher-ability students to demonstrate what they know and understand from their studies of Geography, and these students are expected to respond to the stretch and challenge provided by producing excellent answers.

This, the Fourth edition of *GAIA*, was written when advances in space-shrinking technologies and the speed of globalisation processes mean that events taking place in one part of the world can either be seen by people across the planet almost as they occur (the earthquake in south-east China or sporting events such as the Olympic Games) or have an immediate impact on every country (changing oil prices, climate change or the collapse of world banking). During the writing of the previous (Third) edition of this book in 1998–99, the most up-to-date data I could find was often for two or three years earlier and was, at best, updated annually. At that time, only 13 per cent of the world's population had access to landline telephones and 1.4 per cent to the Internet, while 2.5 per cent had a mobile phone. In 2009, data is now readily available not only for the current year but is often updated monthly or even more frequently. Over 50 per cent of people now have access to landlines and the Internet is now available to nearly 60 per cent in developed and over 10 per cent in developing countries. Over 90 per cent of the population in developed and 30 per cent in developing countries have a mobile phone (or something far more advanced!).

Apart from adding new, more relevant and appropriate Places, Case Studies and topics (Goa and Dubai, Fairtrade and WaterAid) and giving more depth to the emerging countries of China and India, this edition also introduces new terms (such as globalisation, ecological footprint, carbon credit and value-added chain), and updates information (often using 2007 or 2008 data), definitions and Places and Case Studies (climate change, coastal management, types of energy, famine, transnational corporations and HIV/AIDS).

Best wishes with your studies

David Waugh [signature]

David Waugh

Author's acknowledgements

To help with the writing of this Fourth edition of *Geography: An Integrated Approach*, several leading geographers were asked to comment on the current accuracy and relevance of the Third edition, and to advise on recent changes in terminology, concepts and approach. I am, therefore, most grateful to the following for their advice on the content of specific chapters in this book: **Dr David Chester** (*University of Liverpool*) and **Professor Angus Duncan** (*University of Bedfordshire*) for 'Plate tectonics, earthquakes and volcanoes'; **Professor Andrew Goudie** (*University of Oxford*) for 'Weathering and slopes', 'Periglaciation', 'Deserts' and 'Rock types and landforms'; **Dr Mike Bentley** (*University of Durham*) for 'Glaciation'; **Mr Nick Gee** (*UEA*) for 'Coasts', 'Farming and food supplies' and 'Rural land use'; **Dr Antoinette Mannion** (*University of Reading*) for 'Drainage basins and rivers', 'Biogeography' and 'Population'; **Dr Grant Bigg** (*University of Sheffield*) for 'Weather and climate'; **Dr Steven Trudgill** (*University of Cambridge*) for 'Soils'; **Bob Digby** for 'Urbanisation'; **Dr Nick Middleton** (*University of Oxford*) for 'Energy resources'; **Dr Louise Crewe** (*University of Nottingham*) for 'Manufacturing industries'; **Dr Jane Dove** (*St Paul's Girls' School*) for 'Tourism'; **Dr Alisdair Rogers** (*University of Oxford*) and **Dr Richard Knowles** (*University of Salford*) for 'Development and globalisation'.

My thanks also to the following contributors: **Pete Murray** for questions in Chapters 1–12, written originally for the Third edition, some of which have been re-used in this new edition; **John Smith** for the revision, updating and restructuring of the questions throughout the book, and for the Issues Analysis on the Serengeti in Chapter 11 (pages 311–312); **Mike Brown** for local knowledge, information and photographs for the Goa tourism Case Study (pages 600–601) and Places 88 on Pune (page 574); **Bob Digby** for the Issues Analysis on the Westfield Centre (pages 458–459); **Roger Jeans** (*Education, OS*) for advice and assistance on the updating and revision of Framework 9 on GIS (pages 277–278); **Alison Rae** for the Issues Analysis on population policies in India and China (pages 386–387); **Simon Ross** for Framework 1 (pages 22–23) on the use of the internet in study and research, and for the new feature on mapping (pages 98–99); **John Rutter** for updating Framework 9 on GIS (pages 277–278).

My special thanks go to the following who have helped with the production of this new edition: **Barry Page**, who has no equal as a project manager; **Katherine James**, who must have corrected thousands of my mistakes over almost 20 years of editing my books; **Sue Sharp**, for finding so many stunning photos; **Lynne Adams**, for her hours researching both new and updated material; **Melanie Grey**, for her help with the proofreading of such a long book; my very good friend **John Smith**, for again writing and revising the many questions; my wife **Judith**, who had to put up with my absence at the computer. Without them, a book as big and detailed as *GAIA* could never have been produced.

Plate tectonics, earthquakes and volcanoes

1

'. . . how does a supercontinent begin to rift and how do the pieces move apart? What effects do such movements have on the shaping of the continental landscapes, on hot climates and ice ages, on the evolution of life in general and on humanity's relationship with the upper crust of the Earth in particular?'

R. Redfern, *The Making of a Continent*, 1983

Figure 1.1

The geological timescale

The history of the Earth

It is estimated that the Earth was formed about 4 600 000 000 years ago. Even if this figure is simplified to 4600 million years, it still presents a timescale far beyond our understanding. Nigel Calder, in his book *The Restless Earth*, made a more comprehensible analogy by reducing the timespan to 46 years. He ignored the eight noughts and compared the 46 years with a human lifetime (Places 1).

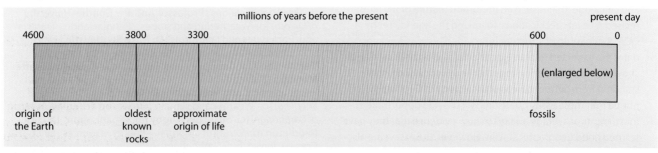

Era	Geological period	Epoch	Millions of years before present	Conditions and rocks in Britain	Major world events
Cenozoic	Quaternary	Holocene	0.01	Post ice age. Alluvium deposited, peat formed	Early civilisations
		Pleistocene	1.8	Ice age, with warm periods	Emergence of the human
	Tertiary	Pliocene	5	Warm climate: Crag rocks in East Anglia	
		Miocene	24	No deposits in Britain	Formation of the Alps
		Oligocene	33	Warm shallow seas in south of England	Rockies and Himalayas begin to form
		Eocene	54	Nearly tropical: London clay	Volcanic activity in Scotland
Mesozoic	Cretaceous		136	Chalk deposited: Atlantic ridge opens	End of the dinosaurs/Age of the dinosaurs
	Jurassic		195	Oxford clays and limestones: warm	Pangaea breaks up
	Triassic		225	Desert: sandstones	First mammals
Palaeozoic	Permian		280	Desert: New Red Sandstones, limestones	Formation of Pangaea
	Carboniferous		345	Tropical coast with swamps: coal	First amphibians and insects
	Devonian		395	Warm desert coastline: sandstones	First land animals
	Silurian		440	Warm seas with coral: limestones	First land plants
	Ordovician		500	Warm seas: volcanoes (Snowdonia) sandstones, shales	First vertebrates
	Cambrian		570	Cold at times: sea conditions	Abundant fossils begin
	Pre-Cambrian			Igneous and sedimentary rocks	

'…Or we can depict Mother Earth as a lady of 46, if her "years" are megacenturies. The first seven of those years are wholly lost to the biographer, but the deeds of her later childhood are to be seen in old rocks in Greenland and South Africa. Like the human memory, the surface of our planet distorts the record, emphasising more recent events and letting the rest pass into vagueness – or at least into unimpressive joints in worn down mountain chains.

Most of what we recognise on Earth, including all substantial animal life, is the product of the past six years of the lady's life. She flowered, literally, in her middle age. Her continents were quite bare of life until she was getting on for 42 and flowering plants did not appear until she was 45 – just one year ago. At that time, the great reptiles, including the dinosaurs, were her pets and the break-up of the last supercontinent was in progress.

The dinosaurs passed away eight months ago and the upstart mammals replaced them. In the middle of last week, in Africa, some man-like apes turned into ape-like men and, at the weekend, Mother Earth began shivering with the latest series of ice ages. Just over four hours have elapsed since a new species calling itself *Homo sapiens* started chasing the other animals and in the last hour it has invented agriculture and settled down. A quarter of an hour ago, Moses led his people to safety across a crack in the Earth's shell, and about five minutes later Jesus was preaching on a hill farther along the fault line. Just one minute has passed, out of Mother Earth's 46 "years", since man began his industrial revolution, three human lifetimes ago. During that time he has multiplied his numbers and skills prodigiously and ransacked the planet for metal and fuel.'

N. Calder, *The Restless Earth,* 1972

Geologists have been able to study rocks and fossils formed during the last 600 million years, equivalent to the last 'six years of the lady's life', and have produced a time chart, or **geological timescale**. Not only have they been able to add dates with increasing confidence, but they have made progress in describing and accounting for the major changes in the Earth's surface, e.g. sea-level fluctuations and landform development, and in its climate. The timescale, shown in Figure 1.1, should be a useful reference for later parts of this book.

Earthquakes

Even the earliest civilisations were aware that the crust of the Earth is not rigid and immobile. The first major European civilisation, the Minoan, based in Crete, constructed buildings such as the Royal Palace at Knossos which withstood a succession of earthquakes. However, this civilisation may have been destroyed by the effects of a huge volcanic eruption on the nearby island of Thera (Santorini). Later, inhabitants of places as far apart as Lisbon (1755), San Francisco (1906), Tokyo (1923), Mexico City (1985), Los Angeles (1994 – Case Study 15A), Kobe (1995), Sri Lanka and Sumatra (2004 – Places 4) and China (2008 – Places 2) were to suffer from the effects of major earth movements.

It was by studying earthquakes that geologists were first able to determine the structure of the Earth (Figure 1.2). At the **Mohorovičić** or **'Moho' discontinuity**, it was found that shock waves begin to travel faster, indicating a change of structure – in this case, the junction of the Earth's **crust** and **mantle** (Figure 1.2). The 'Moho' discontinuity is the junction between the Earth's crust and the mantle where seismic waves are modified. The Moho is at about 35–40 km beneath continents (reaching 70 km under mountain chains) and at 6–10 km below the oceans.

Earthquakes result from a slow build-up of pressure within crustal rocks. If this pressure is suddenly released then parts of the surface may experience a jerking movement. Within the crust, the point at which the release in pressure occurs is known as the **focus**. Above this, on the surface and usually receiving the worst of the shock or **seismic waves**, is the **epicentre**. Unfortunately, it is not only the immediate or primary effects of the earthquake that may cause loss of life and property; often the secondary or after-effects are even more serious (Places 2). These may include fires from broken gas pipes, disruption of transport and other services, exposure caused by a lack of shelter, a shortage of food, clean water and medical equipment, and disease caused by polluted water supplies. These problems may be exacerbated by after-shocks which often follow the main earthquake.

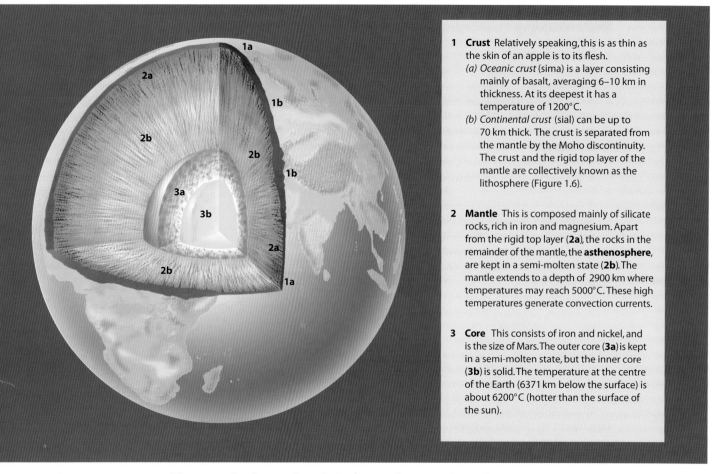

1 **Crust** Relatively speaking, this is as thin as the skin of an apple is to its flesh.
   (a) *Oceanic crust* (sima) is a layer consisting mainly of basalt, averaging 6–10 km in thickness. At its deepest it has a temperature of 1200°C.
   (b) *Continental crust* (sial) can be up to 70 km thick. The crust is separated from the mantle by the Moho discontinuity. The crust and the rigid top layer of the mantle are collectively known as the lithosphere (Figure 1.6).

2 **Mantle** This is composed mainly of silicate rocks, rich in iron and magnesium. Apart from the rigid top layer (**2a**), the rocks in the remainder of the mantle, the **asthenosphere**, are kept in a semi-molten state (**2b**). The mantle extends to a depth of 2900 km where temperatures may reach 5000°C. These high temperatures generate convection currents.

3 **Core** This consists of iron and nickel, and is the size of Mars. The outer core (**3a**) is kept in a semi-molten state, but the inner core (**3b**) is solid. The temperature at the centre of the Earth (6371 km below the surface) is about 6200°C (hotter than the surface of the sun).

Figure 1.2

The internal structure of the Earth

The strength of an earthquake is measured on the Richter scale (Figure 1.3). To cover the huge range of earthquakes, the magnitude of the scale is logarithmic, each unit representing a tenfold increase in strength and around a 30-fold increase in energy. This means that the 1755 Lisbon earthquake was 10 times stronger and released 30 times more energy than the 1985 Mexico City earthquake, and was nearly 100 times stronger and released almost 900 times more energy than the 1989 San Francisco earthquake (Figure 1.3).

Figure 1.3

The Richter scale

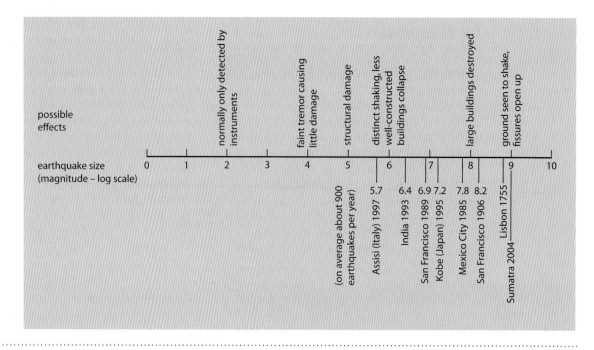

12 May 2008

Just before 1430 hours local time, an earthquake measuring 7.9 on the Richter scale hit Sichuan province in south-west China. It was the worst to affect the country since the city of Tangshan was levelled with the loss of over 220 000 lives in 1976. The epicentre was at Wenchuan, 80 km north of Chengdu. Sichuan, known as the 'rice-bowl of China', is one of the most densely populated and poorest of the country's provinces. The earthquake, which lasted 20 seconds, occurred in a region where the fertile plains of Sichuan give way to high cliffs, steep gorges and forests of pine and bamboo – the last being the sole remaining natural habitat for the giant panda – near to the Tibetan Plateau.

Apart from the collapse of thousands of buildings, giant landslides of mud and rubble blocked roads and rivers. A huge emergency plan was immediately put into effect, including the marching of parts of the army for 30 hours over the mountains to try to help survivors. At least six schools were destroyed, their teachers and students buried under the rubble – indeed it was in schools where poor-quality cement had been used that a high proportion of deaths occurred. Telephone links, including those for mobiles, were lost and people were left without power, fresh water and shelter. Two large dams developed wide cracks and the seemingly endless heavy rain following the quake caused further landslides that killed several relief workers and created over 40 new lakes.

26 May 2008

Whereas after most earthquakes the aftershocks decrease in magnitude and frequency fairly rapidly, in Sichuan they continued. Two weeks after the main event, during which there had been several hundred shockwaves, a tremor of 5.9 magnitude killed six people, injured over a thousand more, and demolished many of the already severely damaged buildings. By this time the official death toll had been put at 67 000 people with another 20 000 still missing. Some 5 million people – equivalent to the combined populations of Manchester and Birmingham – were reported to be homeless. The newly created 'quake' or 'barrier' lakes, together with up to 400 purpose-built reservoirs, became a major concern as they continued to fill following the heavy rains. The talk was of completely abandoning towns such as Wenchuan and Beichuan, where virtually all of the buildings had been destroyed, and creating new settlements.

Earthquakes, volcanoes and young fold mountains

These do not occur at random over the Earth's surface but have a clearly identifiable pattern. This can be seen by working through the following activities.

1 On an outline map of the world, **mark by a dot** (there is no need to name the places) the location of the following earthquakes:

1924	Philippines
1925	California
1926	Rhodes
1927	Japan
1928	Chile
1929	Aleutians, Japan
1931	New Zealand
1932	Mexico
1933	California
1935	Sumatra
1938	Java
1939	Chile, Turkey
1940	Burma, Peru
1941	Ecuador, Guatemala
1943	Philippines, Java
1944	Japan
1946	West Indies, Japan
1949	Alaska
1950	Japan, Assam
1953	Turkey, Japan
1956	California
1957	Mexico
1958	Alaska
1960	Chile, Morocco
1962	Iran
1963	Yugoslavia
1964	Alaska, Turkey, Mexico, Japan, Taiwan
1965	El Salvador, Greece
1966	Chile, Peru, Turkey
1967	Colombia, Yugoslavia, Java, Japan
1968	Iran
1970	Peru
1971	New Guinea, California
1972	Nicaragua
1976	Guatemala, Italy, China, Philippines, Turkey
1978	Japan
1980	Italy
1985	Mexico, Colombia
1988	Armenia
1989	San Francisco, Iran
1993	Java, Japan, India, Egypt
1994	Los Angeles
1995	Japan, Greece
1996	China, Indonesia
1997	Afghanistan, Italy, Iran
1998	Iraq, Afghanistan
1999	Turkey, Taiwan
2001	India, El Salvador
2002	Alaska, Mexico
2003	Japan, Iran
2004	Morocco, Sumatra
2005	Pakistan
2006	Java
2007	Peru
2008	China

2 On a tracing overlay, mark and **name** the following **volcanoes**: Aconcagua, Chimborazo, Cotopaxi, Nevado del Ruiz, Paricutín, Popocatepetl, Mount St Helens, Fuji, Mount Pinatubo, Mayon, Krakatoa, Merapi, Ruapehu, Erebus, Helgafell, Surtsey, Azores archipelago, Ascension, St Helena, Tristan da Cunha, Vesuvius, Etna, Pelée, Montserrat, Mauna Loa, Kilauea.

3 On a second overlay, **mark and name** the following **fold mountains**: Andes, Rockies, Atlas, Pyrenees, Alps, Caucasus, Hindu Kush, Himalayas, Southern Alps.

4 Use the Internet (see Framework 1, page 22) to find the names of more earthquakes and volcanic eruptions, after 2008.

Plate tectonics

As early as 1620, Francis Bacon noted the jigsaw-like fit between the east coast of South America and the west coast of Africa. Others were later to point out similarities between the shapes of coastlines of several adjacent continents.

In 1912, a German meteorologist, Alfred Wegener, published his theory that all the continents were once joined together in one large supercontinent which he named **Pangaea**. Later, this landmass somehow split up and the various continents, as we know them, drifted apart. Wegener collated evidence from several sciences:

■ **Biology** Mesosaurus was a small reptile living in Permian times (Figure 1.1); its remains have been found only in South Africa and Brazil. A plant which existed when coal was being formed has only been located in India and Antarctica.

■ **Geology** Rocks of similar type, age, formation and structure occur in south-east Brazil and South Africa, and the Appalachian Mountains of the eastern USA correspond geologically with mountains in north-west Europe.

■ **Climatology** Coal, formed under warm, wet conditions, is found beneath the Antarctic ice-cap, and evidence of glaciation had been noted in tropical Brazil and central India. Coal, sandstone and limestone could not have formed in Britain with its present climate.

Wegener's theory of **continental drift** combined information from several subject areas, but his ideas were rejected by specialists in those disciplines, partly because he was not regarded as an expert himself but perhaps mainly because he could not explain how solid continents had changed their positions. He was unable to suggest a mechanism for drift.

Figure 1.4a shows Wegener's Pangaea and how it began to divide up into two large continents, which he named **Laurasia** and **Gondwanaland**; it also suggests how the world may look in the future if the continents continue to drift.

Figure 1.4

The wandering continents

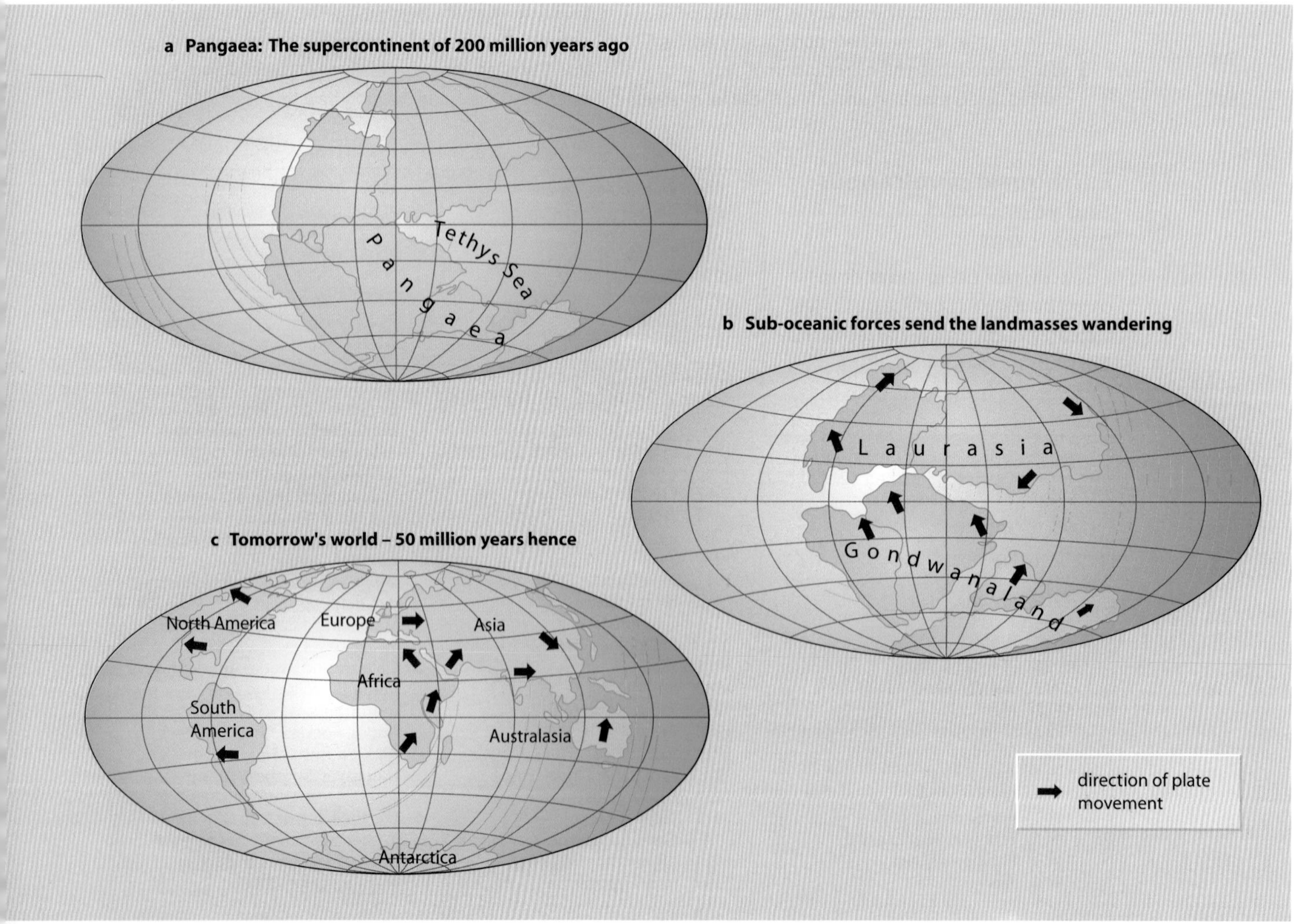

a Pangaea: The supercontinent of 200 million years ago

b Sub-oceanic forces send the landmasses wandering

c Tomorrow's world – 50 million years hence

direction of plate movement

Since Wegener first put forward this theory, three groups of new evidence have become available to support his ideas.

1   **The discovery and study of the Mid-Atlantic Ridge** While investigating islands in the Atlantic in 1948, Maurice Ewing noted the presence of a continuous mountain range extending the whole length of the ocean bed. This mountain range, named the Mid-Atlantic Ridge, is about 1000 km wide and rises to 2500 m in height. Ewing also noted that the rocks of this range were volcanic and recent in origin – not ancient as previously assumed was the case in mid-oceans. Later investigations show similar ranges on other ocean floors, the one in the eastern Pacific extending for nearly 5000 km (Figure 1.8).

2   **Studies of palaeomagnetism in the 1950s** During underwater volcanic eruptions, basaltic magma is intruded into the crust and cools (Figure 1.31). During the cooling process, individual minerals, especially iron oxides, align themselves along the Earth's magnetic field, i.e. in the direction of the magnetic pole. Recent refinements in dating techniques enable the time at which rocks were formed to be accurately calculated. It was known before the 1950s that the Earth's magnetic pole varied a little from year to year, but only then was it discovered that the magnetic field reverses periodically, i.e. the magnetic pole is in the south for a period of time and then in the north for a further period, and so on. It

is claimed that there have been 171 reversals over 76 million years. If formed when the magnetic pole was in the north, new basalt would be aligned to the north. After a reversal in the magnetic poles, newer lava would be oriented to the south. After a further reversal, the alignment would again be to the north. Subsequent investigations have shown that these alternations in alignment are almost symmetrical in rocks on either side of the Mid-Atlantic Ridge (Figure 1.5).

3   **Sea floor spreading** In 1962, Harry Hess studied the age of rocks from the middle of the Atlantic outwards to the coast of North America. He confirmed that the newest rocks were in the centre of the ocean, and were still being formed in Iceland, and that the oldest rocks were those nearest to the USA and the Caribbean. He also suggested that the Atlantic could be widening by up to 5 cm a year.

One major difficulty resulting from this concept of sea floor spreading was the implication that the Earth must be increasing in size. Since this is not so, evidence was needed to show that else-where parts of the crust were being destroyed. Such areas were found to correspond to the fringes of the Pacific Ocean – the region where you plotted some major earthquakes and volcanic eruptions (page 11). These discoveries led to the development of the theory of **plate tectonics** which is now virtually universally accepted, but which may still be modified following further investigation and study.

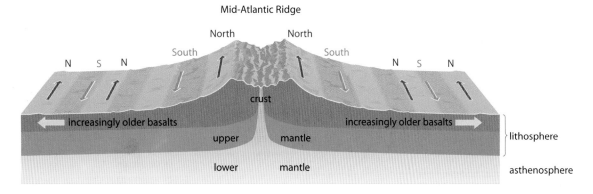

Figure 1.5

The repeated reversal of the Earth's magnetic field – the timings are irregular but show a mirror image

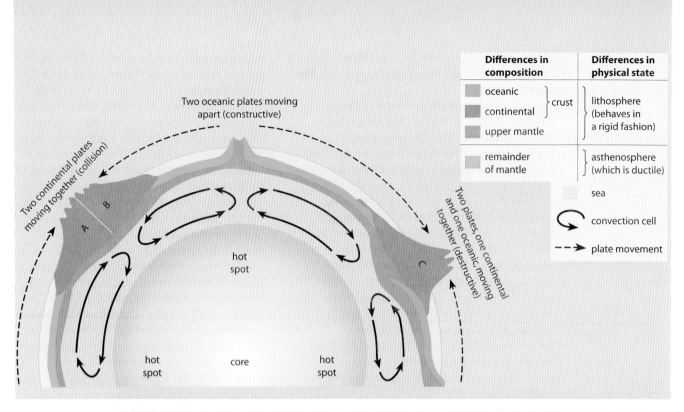

	Differences in composition		Differences in physical state
▨	oceanic	} crust	lithosphere (behaves in a rigid fashion)
▨	continental		
▨	upper mantle		
▨	remainder of mantle		asthenosphere (which is ductile)
	sea		
↻	convection cell		
⇢	plate movement		

Figure 1.6

How plates move

Figure 1.7

Differences between continental and oceanic crust

	Continental crust (sial)	Oceanic crust (sima)
Thickness	35–40 km on average, reaching 60–70 km under mountain chains	6–10 km on average
Age of rocks	very old, mainly over 1500 million years	very young, mainly under 200 million years
Weight of rocks	lighter, with an average density of 2.6	heavier, with an average density of 3.0
Nature of rocks	light in colour; many contain silica and aluminium; numerous types, granite is the most common	dark in colour; many contain silica and magnesium; few types, mainly basalt

The theory of plate tectonics

The **lithosphere** (the Earth's crust and the rigid upper part of the mantle) is divided into seven large and several smaller **plates**. The plates, which are rigid, float like rafts on the underlying semi-molten mantle (the **asthenosphere**) and are moved by currents which form **convection cells** (Figure 1.6). Plate tectonics is the study of the movement of these plates and their resultant landforms.

There are two types of plate material: **continental** and **oceanic**. Continental crust is composed of older, lighter rock of granitic type. Oceanic crust consists of much younger, denser rock of basaltic composition. However, as most plates consist of areas of both continental and oceanic crust, it is important to realise that the two terms do not refer to our named continents and oceans. The major differences between the two types of crust are summarised in Figure 1.7.

Plate movement

As a result of the convection cells generated by heat from the centre of the Earth, plates may move towards, away from or sideways along adjacent plates. It is at plate boundaries that most of the world's major landforms occur, and where earthquake, volcanic and mountain-building zones are located (Figure 1.8). However, before trying to account for the formation of these landforms, several points should be noted.

1   Due to its relatively low density, continental crust does not sink and so is permanent; being denser, oceanic crust can sink. Oceanic crust is being formed and destroyed continuously.

2   Continental plates, such as the Eurasian Plate, may consist of both continental and oceanic crust.

3   Continental crust may extend far beyond the margins of the landmass.

4   Plates cannot overlap. This means that either they must be pushed upwards on impact to form mountains (AB on Figure 1.6) or one plate must be forced downwards into the mantle and destroyed (C on Figure 1.6).

5   No 'gaps' may occur on the Earth's surface so, if two plates are moving apart, new oceanic crust originating from the mantle must be being formed.

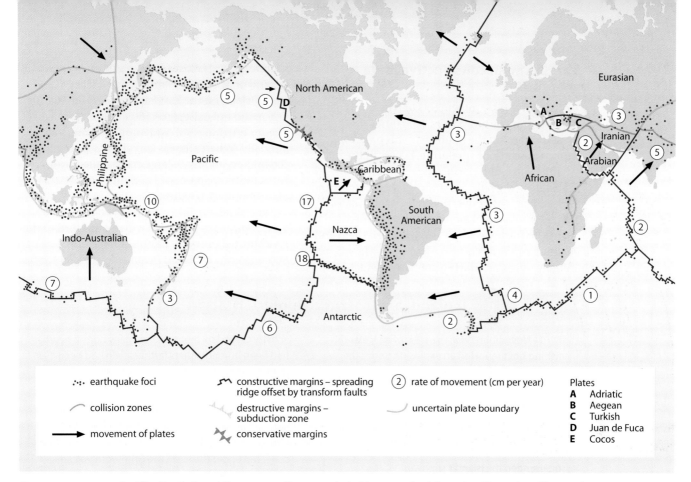

Figure 1.8

Plate boundaries and active zones of the Earth's crust

6 The Earth is neither expanding nor shrinking in size. Thus when new oceanic crust is being formed in one place, older oceanic crust must be being destroyed in another.

7 Plate movement is slow (though not in geological terms) and is usually continuous. Sudden movements are detected as earthquakes.

8 Most significant landforms (fold mountains, volcanoes, island arcs, deep-sea trenches, and batholith intrusions) are found at plate boundaries. Very little change occurs in plate centres (shield lands). Figure 1.9 summarises the major landforms resulting from different types of plate movement.

Figure 1.9

The major landforms resulting from plate movements

Type of plate boundary	Description of changes	Examples
A Constructive margins (spreading or divergent plates)	two plates move away from each other; new oceanic crust appears forming mid-ocean ridges with volcanoes	Mid-Atlantic Ridge (Americas moving away from Eurasian and African Plates) East Pacific Rise (Nazca and Pacific Plates moving apart)
B Destructive margins (subduction zones)	oceanic crust moves towards continental crust but, being heavier, sinks and is destroyed forming deep-sea trenches and island arcs with volcanoes	Nazca sinks under South American Plate (Andes) Juan de Fuca sinks under North American Plate (Rockies) Island arcs of the West Indies and Aleutians
Collision zones	two continental crusts collide and, as neither can sink, are forced up into fold mountains	Indian Plate collided with Eurasian Plate, forming Himalayas African Plate collided with Eurasian Plate, forming Alps
C Conservative or passive margins (transform faults)	two plates move sideways past each other – land is neither formed nor destroyed	San Andreas Fault in California
Note: centres of plates are rigid...	rigid plate centres form a shields lands (cratons) of ancient worn-down rocks b depressions on edges of the shield which develop into large river basins	Canadian (Laurentian) Shield, Brazilian Shield Mississippi–Missouri, Amazon
...with one main exception	Africa dividing to form a rift valley and possibly a new sea	African Rift Valley and the Red Sea

Landforms at constructive plate margins

Constructive plate margins occur where two plates diverge, or move away, from each other and new crust is created at the boundary. This process, known as **sea-floor spreading**, occurs in the mid-Atlantic where the North and South American Plates are being pulled apart from the Eurasian and African Plates by convection cells. As the plates diverge, molten rock or **magma** rises from the mantle to fill any possible gaps between them and, in doing so, creates new oceanic crust. The magma initially forms **submarine volcanoes** which may in time grow above sea-level, e.g. Surtsey, south of Iceland on the Mid-Atlantic Ridge (Places 3) and Easter Island on the East Pacific Rise. The Atlantic Ocean did not exist some 150 million years ago (Figure 1.4) and is still widening by some 2–5 cm annually. Where there is lateral movement along the mid-ocean ridges, large cracks called **transform faults** are produced at right-angles to the plate boundary (Figure 1.8).

The largest visible product of constructive divergent plates is Iceland where one-third of the lava emitted onto the Earth's surface in the last 500 years can be found (Figures 1.10b and 1.26).

Places 3 · Iceland: a constructive plate margin

On 14 November 1963, the crew of an Icelandic fishing boat reported an explosion under the sea south-west of the Westman Islands. This was followed by smoke, steam and emissions of pumice stone. Having built up an ash cone of 130 m from the seabed, the island of Surtsey emerged above the waves. On 4 April 1964, a lava flow covered the unconsolidated ash and guaranteed the island's survival.

Just before 0200 hours on 23 January 1973, an earth tremor stopped the clock in the main street of Heimaey, Iceland's main fishing port. Once again the North American and Eurasian Plates were moving apart (Figure 1.10b). Fishermen at sea witnessed the crust of the Earth break open and lava and ash pour out of a fissure 2 km in length (page 25). Eventually the activity became concentrated on the volcanic cone of Helgafell and the inhabitants of Heimaey were evacuated to safety. By the time volcanic activity ceased six months later, many homes nearby had been burned; others farther afield had been buried under 5 m of ash; and the entrance to the harbour had been all but blocked.

A large volcanic eruption in a fissure under the Vatnajökull icecap melted 3000 m³ of the glacier above it in October 1996. The resultant meltwater collected under the ice in the Grimsvötn volcanic crater (caldera) until, in November, an eruption spewed a 4270 m high column of ash into the air and released the trapped water. The subsequent torrent, which contained house-sized blocks of ice and black sulphurous water, demolished three of Iceland's largest bridges and several kilometres of the south coast ring road (Figure 1.25). A further event in December 1998 resulted in five craters within the caldera becoming active along a 1300 m long fissure and the creation of an eruption plume 10 km in height.

Figure 1.10

A constructive plate margin: Iceland

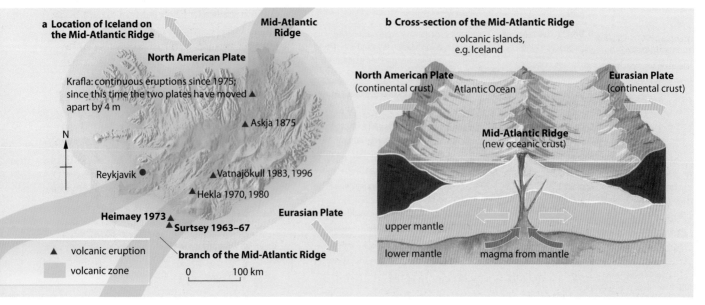

a Location of Iceland on the Mid-Atlantic Ridge

Mid-Atlantic Ridge

North American Plate

Krafla: continuous eruptions since 1975; since this time the two plates have moved apart by 4 m

N

▲ Askja 1875

Reykjavik ●

▲Vatnajökull 1983, 1996

▲Hekla 1970, 1980

Heimaey 1973 ▲

Eurasian Plate

▲ Surtsey 1963–67

▲ volcanic eruption

volcanic zone

branch of the Mid-Atlantic Ridge

0   100 km

b Cross-section of the Mid-Atlantic Ridge

volcanic islands, e.g. Iceland

North American Plate (continental crust)   Atlantic Ocean

Eurasian Plate (continental crust)

Mid-Atlantic Ridge (new oceanic crust)

upper mantle

lower mantle   magma from mantle

Figure 1.11

The African Rift Valley

a Location

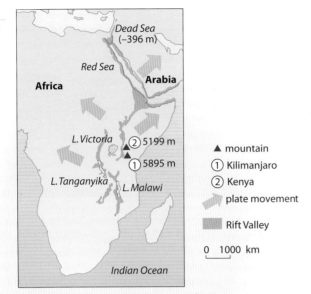

- ▲ mountain
- ① Kilimanjaro
- ② Kenya
- ⟶ plate movement
- ▨ Rift Valley

0   1000   km

b Idealised cross-section

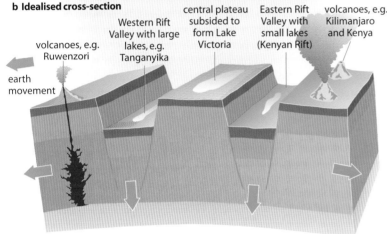

volcanoes, e.g. Ruwenzori

earth movement

Western Rift Valley with large lakes, e.g. Tanganyika

central plateau subsided to form Lake Victoria

Eastern Rift Valley with small lakes (Kenyan Rift)

volcanoes, e.g. Kilimanjaro and Kenya

Figure 1.12

A destructive plate margin – the Nazca and South American Plate boundary

(ii) some lava reaches surface to form volcanoes 6000 m high, e.g. Chimborazo and Cotopaxi

Peru–Chile deep sea trench

Coastal Range

Pacific Ocean

sea level

Nazca Plate (oceanic crust)

earthquake foci

subduction zone, oceanic plate breaks up producing earthquake foci

saline lakes (e.g. Titicaca), remnants of disappearing former oceans

Western Cordillera   Eastern Cordillera

young fold mountains of the Andes, separated by the Altiplano (High Plateau)

Amazon and Parana lowlands (sedimentary rocks)

Atlantic Ocean

(i) lava rises, but cools within the crust forming the granite Andean batholith

Brazilian Plateau, an ancient shield having always been part of a stable continental plate

American Plate (continental crust)

friction from the subduction zone gives extra heat producing either (i) or (ii) above

The Atlantic Ocean was formed as the continent of Laurasia split into two, a process that may be repeating itself today in East Africa. Here the brittle crust has fractured and, as sections moved apart, the central portion dropped to form the Great African Rift Valley (Figure 1.11) with its associated volcanic activity. In Africa the rift valley extends for 4000 km from Mozambique to the Red Sea. In places its sides are over 600 m in height while its width varies between 10 and 50 km. Where the land has been pulled apart and dropped sufficiently, it has been invaded by the sea. It has been suggested that the Red Sea is a newly forming ocean. Looking 50 million years into the future (Figure 1.4c), it is possible that Africa will have moved further away from Arabia.

Landforms at destructive plate margins

Destructive margins occur where continental and oceanic plates converge. The Pacific Ocean, which extends over five oceanic plates, is surrounded by continental plates (Figure 1.8). The Pacific Plate, the largest of the oceanic plates, and the Philippines Plate move north-west to collide with eastern Asia. In contrast, the smaller Nazca, Cocos and Juan de Fuca Plates travel eastwards towards South America, Central America and North America respectively. Figure 1.12 shows how the Nazca Plate, made of oceanic crust which cannot override continental crust, is forced to dip downwards at an angle to form a **subduction zone** with its associated **deep-sea trench**. As oceanic lithosphere descends, the increase in pressure can trigger major earthquakes, while dehydration of the subducted oceanic crust, caused by the increase in pressure, results in the release of water into the overlying mantle which promotes partial melting and the generation of magma. Being less dense than the mantle, the newly formed magma will try to rise to the Earth's surface. Where it does reach the surface, volcanoes will occur. These volcanoes are likely to form either a long chain of **fold mountains** (e.g. the Andes) or, if the eruptions take place offshore, an **island arc** (e.g. Japan, Caribbean). Estimates claim that 80 per cent of the world's present active volcanoes are located above subduction zones. As the rising magma at destructive margins is more acidic than the lava of constructive margins (page 24), it is more viscous and flows less easily. It may solidify within the mountain mass to form large **intrusive** features called **batholiths** (Figure 1.31).

Figure 1.13

The 2004 tsunami as it hit the coast of Thailand

Tsunamis are giant waves, often generated at destructive plate margins, that can cross oceans – indeed the four tsunamis that followed the eruption of Krakatoa in 1883 travelled three times around the world. Tsunamis are rare events, but they can cause enormous damage and considerable loss of life. They occur when a sudden, large-scale change in the area of an ocean bed leads to the displacement of a large volume of water and the subsequent formation of one or more huge waves. Although tsunamis can result from a major coastal landslide (e.g. Alaska 1958), their origin is more likely to be seismic – either following a volcanic eruption (Krakatoa 1883 – Places 35, page 289) or a shallow submarine earthquake (Indian Ocean 2004 – Places 4).

Tsunamis have exceptionally long wave-lengths of up to 100 km, unlike wind-driven waves where the distance between consecutive wave crests is only a few metres (page 141). Tsunamis can cross oceans at speeds of up to 700 km/hr yet their small height, perhaps only half a metre, makes them almost imperceptible. On approaching a coastline, their speed may rapidly decrease to only 30 km/hr (still faster than people can run) while their height can increase to 20 m or more.

Places 4 Indian Ocean: the 2004 tsunami

Andaman Islands
1 hour
South China Sea
$1\frac{1}{2}$ hours

tsunami waves travelling outwards and time taken

India
$\frac{1}{2}$ hour
Thailand
Phuket and Phi Phi Island

Sri Lanka
Galle
Indian Ocean
Banda Aceh
Malaysia

$2\frac{1}{2}$ hours
2 hours
$1\frac{1}{2}$ hours
1 hour
$\frac{1}{2}$ hour
Sumatra
epicentre
Eurasian Plate (mainly continental crust)

0 500 km
focus

Indian Plate (mainly oceanic crust)

seabed forced to rise, pushing water upwards and outwards forming a giant wave

tsunamis
seawater displaced
seawater displaced
fault
earthquake focus
seabed distorted

Figure 1.14

Track of the Indian Ocean tsunami

The Indian Ocean tsunami of 26 December 2004 was caused by a horizontal movement of some 15 m along a 1200 km section of fault line where the mainly oceanic crust of the north-eastwards moving Indian Plate is subducted under the mainly continental crust of the Eurasian Plate. The magnitude of the earthquake that triggered this movement was measured as 9.0 on the Richter scale and had its epicentre just off the west coast of Sumatra in Indonesia. As part of the seabed directly above the epicentre was forced to rise locally, water above it was pushed upwards and outwards forming the tsunami.

Part of the resultant wave travelled eastwards to devastate, first, those parts of Banda Aceh in Sumatra that had not been destroyed a few minutes earlier by the earthquake, and, later, several coastal resorts in Thailand (Figure 1.13). The remainder of the wave travelled westwards across the Indian Ocean where it affected, without warning, the Andaman Islands, Sri Lanka and southern India (Figure 1.14). In all those places the 15 m wave, preceded by a retreat of the sea, raced inland carrying people and property with it and then rushing back to the ocean dragging bodies and debris. In all 300 000 people died, including 200 000 in Indonesia and 30 000 in Sri Lanka, and nearly 2 million were left homeless. Hundreds of kilometres of roads were destroyed, as were many schools and hospitals, fishing boats and coastal crops (Places 103, page 633).

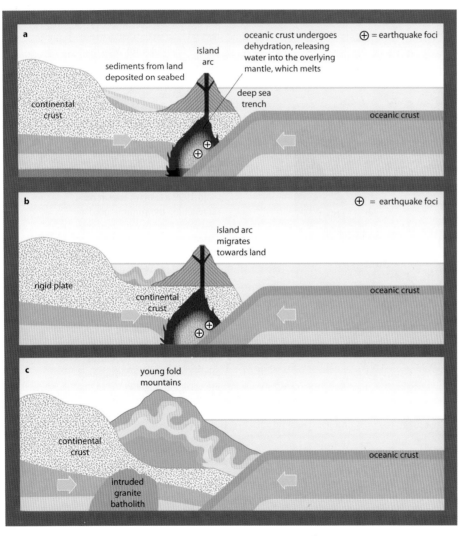

Figure 1.15

A collision plate margin – the formation of fold mountains (orogenesis)

Landforms at collision plate margins

The formation of fold mountains is often extremely complex. As has already been explained in the context of the Pacific, fold mountains often occur where oceanic crust is subducted by continental crust (Figure 1.15). A second, though less frequent, occurrence is when two plates composed of continental crust move together. In Places 5 the Indian subcontinent, forming part of the Indo-Australian Plate, is shown to have moved north-eastwards and to have collided with the Eurasian Plate. Because continental crust cannot sink, the subsequent collision caused the intervening sediments, which contained seashells, to be pushed upwards to form the Himalayas – an uplift that is still continuing. It is where these continental collisions occur that fold mountains form and the Earth's crust is at its thickest (Figures 1.6 and 1.7).

Measurements of current convergence rates suggest that the Indo-Australian Plate is moving towards the Eurasian Plate at a rate of 5.8 cm/year. Although the convergence of two plates of continental crust has pushed up the Himalayas and caused the formation of the Tibetan Plateau, in parts the Indian Plate is being pushed under Tibet to form the mountain roots up to 70 km deep shown on Figure 1.16.

This movement causes great stresses which are released by periodic, often extremely violent and destructive, earthquakes. Earthquakes this century have included:

- Gujarat in northern India in 2001 when over 30 000 people were killed in an earthquake lasting 45 seconds

- northern Pakistan in 2005 when more than 78 000 died in a quake that measured 7.8 and lasted 32 seconds

- Sichuan in south-west China in 2008 when the death toll in an earthquake of 7.9 was in excess of 80 000 (Places 2).

Recent measurements have led scientists to believe that this plate movement is causing Mount Everest to rise by up to 3 cm a year (Figure 1.17). The Himalayas are not only the world's highest mountain range, they are also one of the youngest.

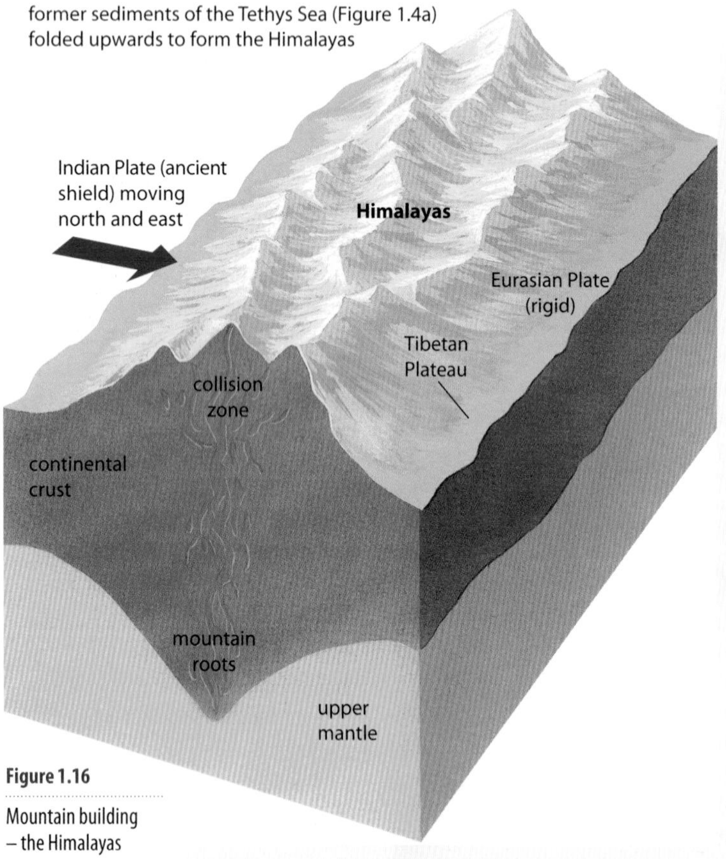

In the 1950s, the height of Mount Everest was given as 29 002 feet (8840 m) but this was revised later in the century to 29 029 feet (8848 m). Was this difference in height due to the uncertainty of the rock summit which was covered in ice and snow to a then estimated depth of 20 feet (6 m); to plate movement having caused the mountain to be pushed up higher during that time; or to the fact that earlier measurements were inaccurate?

In 1999 a team of researchers, on reaching the summit, used an ice-coring drill to reach down to solid rock, and the global positioning system (GPS) to help fix the height, which was given as 29 035 feet (8850 m). Apart from suggesting that Everest is rising by up to 1.2 inches (3 cm) a year, the team hoped that, by monitoring the position of the summit, they might be able to predict when future earthquakes in the region might occur.

Figure 1.17

Is Mount Everest still rising?

former sediments of the Tethys Sea (Figure 1.4a) folded upwards to form the Himalayas

Figure 1.16

Mountain building – the Himalayas

Landforms at conservative plate margins

Conservative margins occur where two plates move parallel or nearly parallel to each other. Although frequent small earth tremors and occasional severe earthquakes may occur as a consequence of the plates trying to slide past each other, the margin between the plates is said to be **conservative** because crustal rocks are being neither created nor destroyed here. The boundary between the two plates is characterised by pronounced transform faults (Figure 1.18a). The San Andreas Fault is the most notorious of several hundred known transform faults in California (Places 6 and Case Study 15A).

The San Andreas Fault forms a junction between the North American and Pacific Plates. Although both plates are moving north-west, the Pacific Plate moves faster giving the illusion that they are moving in opposite directions. The Pacific Plate moves about 6 cm a year, but sometimes it sticks (like a machine without oil) until pressure builds up enabling it to jerk forwards as it did in San Francisco in 1906 and 1989 and is predicted to do again before 2032. Should these plates continue to slide past each other, it is likely that Los Angeles will eventually be on an island off the Canadian coast.

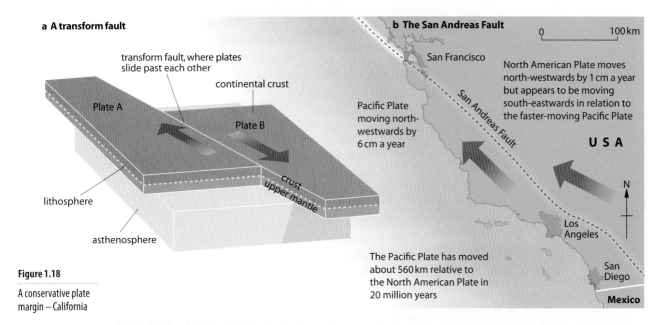

a A transform fault

transform fault, where plates slide past each other

continental crust

Plate A

Plate B

lithosphere

crust
upper mantle

asthenosphere

b The San Andreas Fault

0   100 km

San Francisco

North American Plate moves north-westwards by 1 cm a year but appears to be moving south-eastwards in relation to the faster-moving Pacific Plate

Pacific Plate moving north-westwards by 6 cm a year

San Andreas Fault

USA

N

Los Angeles

San Diego

Mexico

The Pacific Plate has moved about 560 km relative to the North American Plate in 20 million years

Figure 1.18

A conservative plate margin – California

San Francisco: earthquakes

1906

At 0512 hours on the morning of 18 April, the ground began to shake. There were three tremors, each one increasingly more severe. The ground moved by over 6 m in an earthquake which measured 8.2 on the Richter scale. Many apartment buildings collapsed, bridges were destroyed – the Golden Gate had not then been built – and water pipes fractured. The worst damage was 'downtown' where the housing density was greatest. Although many people were trapped within collapsed buildings there were relatively few deaths.

Then came the fire! It started in numerous places resulting from overturned stoves or sparked by electricity or the ignition of gas escaping from the broken mains. As the water pipes had been fractured, it hardly mattered that there were only 38 horse-drawn fire engines to cope with 52 fires. As the fire spread, houses were blown up with dynamite to try to create gaps to thwart the flames, but the explosions only caused further fires. It took over three days to put out the fires, by which time over 450 people (mainly those previously trapped) had died, 28 000 buildings in 500 blocks had been destroyed, and an area six times greater than that destroyed by the Great Fire of London had been ravaged.

1989

During the early evening rush-hour on 17 October 1989, an earthquake measuring 6.9 on the Richter scale shook the city for 15 seconds. The early-warning system had given no clues. Skyscrapers swayed 3 m, fractured gas pipes caused fires in one residential area, and parts of a downtown shopping centre collapsed. The greatest loss of life occurred when 1.5 km of the upper section of the two-tier Interstate Highway 880 collapsed onto the lower portion, killing people in their vehicles.

The final casualty figures of 67 dead and 2000 homeless were, however, low compared with an earthquake of similar magnitude in Armenia, a less developed country, which had killed 55 000 people 11 months earlier. San Francisco has the money and technology to enable it to take precautions to reduce the effects of an earthquake and to train and fully equip emergency services. Armenia lacks these resources, which is why the death toll and the damage incurred there were so much greater.

Plate tectonics and the British Isles

During the Cambrian period (Figure 1.1), northern Scotland lay on the American Plate while the rest of Britain was on the Eurasian Plate, as it is today. Both plates are thought then to have been in the latitude of present-day South Africa. In the Ordovician and Silurian periods, the two plates began to converge causing volcanic activity and the formation of mountains in Snowdonia and the Lake District (a collision zone). Being continental crust, sediment between the plates was pushed up to form the Caledonian Mountains which linked Scotland to the rest of Britain. During the Devonian period, the locked plates drifted northwards through a desert environment (the present Kalahari Desert) when the Old Red Sandstones were deposited (page 201). This northward movement continued in Carboniferous times, accompanied by a sinking of the land which allowed the limestones of that period to form in warm, clear seas (page 196).

As the land began to emerge from these seas, millstone grit was formed from sediments in a shallow sea, and then coal measures were laid down under the hot, wet, swampy conditions usually associated with equatorial areas. It was during the Permian and Triassic periods that the continents collided to form Pangaea (Figure 1.4a). Africa moved towards Europe, and Britain's New Red Sandstones (page 201) were laid down under dry, hot desert conditions (in the position of the present-day Sahara Desert). A further submergence during Jurassic/Cretaceous times enabled the Cotswold limestones and then the chalk of the Downs to form – again in warm, clear seas (page 196).

During the Tertiary era, the North American and Eurasian Plates split apart forming a constructive boundary and the volcanoes of north-western Scotland (page 29). At the same time, the African Plate moved further north pushing up the Alps and the hills of southern England. Subsequently, although Britain has been located away from the volcanoes and severe earthquakes associated with various plate margins, its landscape has been modified both during and since the ice ages. These, however, have been due to climatic change rather than plate movement.

Framework 1 — Using the Internet for studying geography

The Internet is a rich global resource base. For geographers it offers enormous potential but it does need to be treated with caution.

A source of facts and figures

Geography is full of facts and figures and the Internet is a good resource for such information. Several encyclopaedias, such as Wikipedia (www.wikipedia.org), offer information on a range of topics and issues. Government agencies, such as the Environment Agency (www.environment-agency.gov.uk) and the Office for National Statistics (www.statistics.gov.uk), are now more likely to provide information online than as hard copy. Through such sites, geography students have easy and immediate access to huge quantities of information.

Providing up-to-date case studies

Geographers are interested in studying places, which is why you are asked to support your work with case studies. Here the Internet offers many opportunities. Global media organisations such as the BBC (www.bbc.co.uk) use the immediacy of their websites to post up-to-date information on events such as earthquakes, pollution incidents and extreme weather events.

Photo library

The Internet enables you to bring your studies to life by including photographs. Most photo-journalists and agencies make their material available online and a carefully directed search (be specific and include '+ photo' in your search) will reveal a wide range of possible illustrations.

GIS

Geographic Information Systems (GIS) is a relatively recent innovation but you are expected to understand what it is and how it can be used in

Figure 1.19

The website for the British Geological Survey has useful worldwide information

geography (Framework 9, page 277). Put simply, GIS is the integrated use of digital information in the form of statistical data, maps and photos. Digital technology enables data to be presented spatially using a series of 'layers'. The operator (user) has considerable control over the use and interpretation of this information. Some useful sites provide a portal on GIS information and applications, such as the Royal Geographical Society (www.gis.rgs.org) and the Staffordshire Learning Net (www.sln.org.uk/geography/gis.htm).

Internet issues

Authenticity

Geography books (such as this one!) take very many months to write. Experienced geographers write them and they undergo all sorts of editorial checks before being published. For the most part you can be assured of their accuracy. This assurance does not necessarily apply to the Internet and you need to exercise care when using sites. You should always refer to the source of information (give its web address) and be aware of possible bias. Follow recommendations from your teacher or from other trustworthy sources such as the *Geography Review*.

As a general rule, government sites (which have 'gov' in the URL) and universities (with 'ac' in the URL) are likely to provide authentic information. The same is true of major media websites such as the BBC and newspapers such as The Independent (www.independent.co.uk), Guardian (www.guardian.co.uk) and The Times (www.timesonline.co.uk). Including reference to a known authentic site in a search (e.g. 'global warming + bbc') can streamline a search and ensure quality of information.

Time consuming

Searching the Internet can be rewarding but takes up time, too. The key is to make specific searches, narrowing down your field by using, for example, 'and' or '+' and adding specific aspects such as dates, locations or websites. So, for example, when looking for information about the 2008 Chinese earthquake, a search on 'earthquakes' will be much less productive than 'chinese earthquake 2008 + bbc', which will take you straight to a special report published by a recognised authentic source.

Information overload

Even a fairly specific search such as 'chinese earthquake 2008 + bbc' reveals nearly 1 million sites. How often do you look at more than the first two or three sites let alone the second page?

Figure 1.20

The Met Office: a valuable source of climatic information

Streamline your search as much as possible and skip sites to look at recognised URLs (see above).

Using the Internet

Having found some potentially useful information, you need to decide how to use it. All too often students rely on 'copy, paste'. This is inappropriate and to be discouraged unless you wish to capture information to work on at a later stage. The 'copy, paste' function will not help you to learn material and may well result in inappropriate information being retained.

Use the Internet as a source of information rather than as the end product of your research:

- Select only that which is of direct relevance to your research. Selectivity is a key geographical skill at AS/A level.
- Only include detail you can understand. Academic sites and even Wikipedia often contain information that is of a much higher level than AS/A level. There is no benefit to you in including terminology or concepts that you do not understand.
- Re-write text in your own words.
- Add labels or annotations to diagrams and photos. By doing this you are showing initiative, which will be rewarded.

Do use the Internet to support your studies – but do not rely on it.

Social networking sites

You might expect social networking sites to be frowned upon. Far from it – they have much to offer in geographical research. The opinions of individuals are increasingly important in academic geography. Hearing first hand from people affected by an earthquake or a hurricane is valuable. Videos on YouTube can capture events and provide interesting portrayals of people's sense of place. Just bear in mind the issue of authenticity.

Volcanology

The term **volcanology** includes all the processes by which solid, liquid or gaseous materials are forced into the Earth's crust or are ejected onto the surface. Although material in the mantle has a high temperature, it is kept in a semi-solid state because of the great pressure exerted upon it. However, if this pressure is released locally by folding, faulting or other movements at plate boundaries, some of the semi-solid material becomes molten and rises, forcing its way into weaknesses in the crust, or onto the surface, where it cools, crystallises and solidifies.

The molten rock is called **magma** when it is below the surface and **lava** when on the surface. When lava and other materials reach the surface they are called **extrusive**. The resulting landforms vary in size from tiny cones to widespread lava flows. Materials injected into the crust are referred to as **intrusive**. These may later be exposed at the surface by erosion of the overlying rocks. Both extrusive and intrusive materials cooled from magma are known as **igneous rocks**.

Extrusive landforms

There are several types of extrusive landform whose nature depends on how gaseous and/or viscous the lava is when it reaches the Earth's surface (Figure 1.21).

- Lava produced by the upward movement of material from the mantle is **basaltic** and tends to be located along mid-ocean ridges, over hot spots and alongside rift valleys.
- Lava that results from the process of subduction is described as **andesitic** (after the Andes) and occurs as island arcs or at destructive plate boundaries where oceanic crust is being destroyed.
- **Pyroclastic material** (meaning 'fire broken') is material ejected by volcanoes in a fragmented form. Tephra, fragments of different sizes, include ash, lapilli (small stones) and bombs (larger material) which are thrown into the air before falling back to earth. Pyroclastic flows move down the side of a volcano as a fast-moving cloud (Figure 1.46). Subsequent heavy rainfall, e.g. Mount Pinatubo (Case Study 1) or the melting of ice and snow, e.g. Nevado del Ruiz (Case Study 2A) can rework the fragmented pyroclastic material to form mudflows (or lahars).

How can volcanoes be classified?

Because of the large number of volcanoes and wide variety of eruptions, it is convenient to group together those with similar characteristics (Framework 7, page 167). Unfortunately, there is no universally accepted method of classification. One of the two most quoted groupings is according to the **shape** of the volcano and its vent which, because it describes landforms, is arguably of more value to the geographer (page 25). The other is the **nature of the eruption**, which has traditionally been the method used by volcanologists (page 28).

Figure 1.21

Basic and acid lava

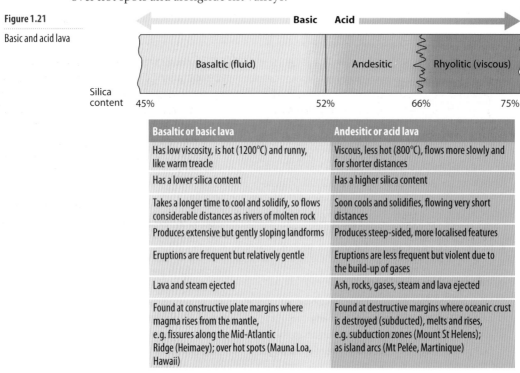

Basaltic or basic lava	Andesitic or acid lava
Has low viscosity, is hot (1200°C) and runny, like warm treacle	Viscous, less hot (800°C), flows more slowly and for shorter distances
Has a lower silica content	Has a higher silica content
Takes a longer time to cool and solidify, so flows considerable distances as rivers of molten rock	Soon cools and solidifies, flowing very short distances
Produces extensive but gently sloping landforms	Produces steep-sided, more localised features
Eruptions are frequent but relatively gentle	Eruptions are less frequent but violent due to the build-up of gases
Lava and steam ejected	Ash, rocks, gases, steam and lava ejected
Found at constructive plate margins where magma rises from the mantle, e.g. fissures along the Mid-Atlantic Ridge (Heimaey); over hot spots (Mauna Loa, Hawaii)	Found at destructive margins where oceanic crust is destroyed (subducted), melts and rises, e.g. subduction zones (Mount St Helens); as island arcs (Mt Pelée, Martinique)

The shape of the volcano and its vent

1 **Fissure eruptions** When two plates move apart, lava may be ejected through fissures rather than via a central vent (Figure 1.22a). The Heimaey eruption of 1973 (Places 3, page 16) began with a fissure 2 km in length. This was small in comparison with that at Laki, also in Iceland, where in 1783 a fissure exceeding 30 km opened up. The basalt may form large plateaus, filling in hollows rather than building up into the more typical cone-shaped volcanic peak. The remains of one such lava flow, formed when the Eurasian and North American Plates began to move apart, can be seen in Northern Ireland, north-west Scotland, Iceland and Greenland. The columnar jointing produced by the slow cooling of the lava provides tourist attractions at the Giant's Causeway in Northern Ireland (Figure 1.27) and Fingal's Cave on the Isle of Staffa.

2 **Basic or shield volcanoes** In volcanoes such as Mauna Loa on Hawaii, lava flows out of a central vent and can spread over wide areas before solidifying. The result is a 'cone' with long, gentle sides made up of many layers of lava from repeated flows (Figure 1.22b).

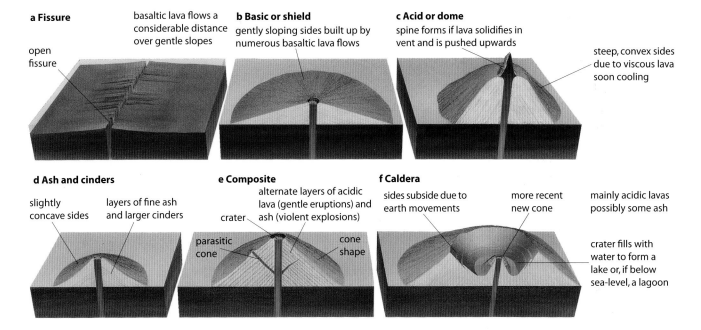

a Fissure
basaltic lava flows a considerable distance over gentle slopes
open fissure

b Basic or shield
gently sloping sides built up by numerous basaltic lava flows

c Acid or dome
spine forms if lava solidifies in vent and is pushed upwards
steep, convex sides due to viscous lava soon cooling

d Ash and cinders
slightly concave sides
layers of fine ash and larger cinders

e Composite
alternate layers of acidic lava (gentle eruptions) and ash (violent explosions)
crater
parasitic cone
cone shape

f Caldera
sides subside due to earth movements
more recent new cone
mainly acidic lavas possibly some ash
crater fills with water to form a lake or, if below sea-level, a lagoon

Figure 1.22

Classification of volcanoes based on their shape (not to scale)

3 **Acid or dome volcanoes** Acid lava quickly solidifies on exposure to the air. This produces a steep-sided, convex cone as in most cases the lava solidifies near to the crater (Figure 1.22c). In one extreme instance, that of Mt Pelée, the lava actually solidified as it came up the vent and produced a spine rather than flowing down the sides.

4 **Ash and cinder cones** (Figure 1.22d) Paricutín, for example, was formed in the 1940s by ash and cinders building up into a symmetrical cone.

5 **Composite cones** Many of the larger, classically shaped volcanoes result from alternating types of eruption in which first ash and then lava (usually acidic) are ejected, e.g. Mt Etna and Fujiyama (Figure 1.22e).

6 **Calderas** When the build-up of gases becomes extreme, huge explosions may clear the magma chamber beneath the volcano and remove the summit of the cone. This causes the sides of the crater to subside, thus widening the opening to several kilometres in diameter. In the cases of both Thera (Santorini) and Krakatoa, the enlarged craters or calderas have been flooded by the sea and later eruptions have formed smaller cones within the resultant lagoons (Figures 1.22f and 1.29).

Minor extrusive features

These are often associated with, but are not exclusive to, areas of declining volcanic activity. They include solfatara, fumaroles, geysers and mud volcanoes (Figure 1.24, Places 7 and Figure 17.16).

Figure 1.23

Minor extrusive landforms

a **Mud volcano**: hot water mixes with mud and surface deposits

b **Solfatara**: created when gases, mainly sulphurous, escape onto the surface

c **Geyser**: water in the lower crust is heated by rocks and turns to steam; pressure increases and the steam and water explode onto the surface

d **Fumaroles**: superheated water turns to steam as its pressure drops when it emerges from the ground

magma chamber
(probably solid by this stage)

Places 7 Solfatara, Italy: an area of declining volcanic activity

Solfatara is a small volcano on the outskirts of Naples. Its crater is 2 km in diameter, making it larger than that of nearby Vesuvius, but there is no volcanic cone. Solfatara takes its name from the gases which escape to the surface; they are mainly sulphurous and can be smelled from a considerable distance. Many rocks are coated with sulphur. Solfatara has given its name to all similar features of this type. **Fumaroles**, resulting from superheated water being turned to steam as it cools on its ejection through the thin crust, are numerous in the area (Figure 1.24). Evidence of the thinness of the crust (magma is only 3 m below the surface) is provided by a guide who throws a boulder onto the surface and makes groups of tourists jump in harmony to hear the hollowness of the ground. The guide, who is needed to keep visitors safely away from bubbling mud volcanoes and areas too hot to walk on, also shows volcanic activity by lighting twigs and stirring loose material to cause a miniature eruption.

The only minor feature missing is the **geyser**, an intermittent fountain of hot water (e.g. Old Faithful, Yellowstone National Park, USA, Figure 17.16).

During the mid-1980s the temperature (160°C), pressure and surface of Solfatara all increased, giving rise to fears of a new eruption – the last was in 1198. Despite the appearance of a small fissure near to the observatory, which led to its abandonment, activity appears to have stabilised.

Figure 1.24

Inside the Solfatara crater, near Naples, Italy

Figure 1.25

Results of the 1996 Grimsvötn eruption, Iceland

Figure 1.26

The boundary between the North American and Eurasian Plates in Iceland, showing the split and a volcano along the boundary margin

Figure 1.27

Giant's Causeway, Northern Ireland

Figure 1.28

Vesuvius: notice the new cone within the old crater of Monte Somma

Figure 1.29

Anak Krakatoa (meaning 'child of Krakatoa') is now a small island volcano that has risen from the centre of the original, much larger volcano which erupted in 1883

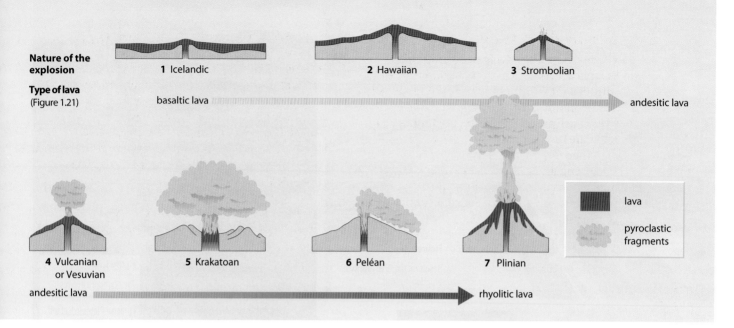

Nature of the explosion			
1 Icelandic	2 Hawaiian	3 Strombolian	

Type of lava
(Figure 1.21)

basaltic lava ➞ andesitic lava

| lava |
| pyroclastic fragments |

4 Vulcanian or Vesuvian   5 Krakatoan   6 Peléan   7 Plinian

andesitic lava ➞ rhyolitic lava

Figure 1.30

Classification of volcanoes according to the nature of the explosion

The nature of the eruption

This classification of volcanoes is based on the degree of violence of the explosion which is a consequence of the pressure and amount of gas in the magma (Figure 1.30). Its categories may be summarised as follows:

1. Icelandic, where lava flows gently from a fissure
2. Hawaiian, where lava is emitted gently but from a vent
3. Strombolian, where small but very frequent eruptions occur
4. Vulcanian, or Vesuvian, which is more violent and less frequent (Figure 1.28)
5. Krakatoan, which has an exceptionally violent explosion that may remove much of the original cone (Figure 1.29)
6. Peléan, where a violent eruption is accompanied by pyroclastic flows that may include a nuée ardente ('glowing cloud')
7. Plinian, where large amounts of lava and pyroclastic material are ejected.

Hydromagmatic refers to any eruptive process in which magma and lava interact with external water. According to Parfitt and Wilson (2008), such interactions can take place in a wide range of environments, including:

- deep marine locations where volcanoes grow on the ocean floor, mainly at mid-ocean ridges where the pressure of the overlying water helps to suppress the explosivity of the eruption and the lava undergoes rapid cooling to form pillow lavas
- where lava flows into the sea either with or without an explosion, e.g. Kilauea in Hawaii
- shallow marine locations (e.g. Surtsey – Places 3) or a crater lake (e.g. Taal in the Philippines) where the eruptions may be highly dramatic
- subglacial locations where an eruption occurs under an ice cap or a glacier, e.g. Vatnajökull in Iceland
- where magma comes into contact with groundwater before erupting, e.g. Ukinrek, Alaska.

The most explosive interactions occur when water makes up about 25–30 per cent of the volume of the exploding mixture.

Figure 1.31

Diagrammatic model showing intrusive landforms: batholith, dyke and sills

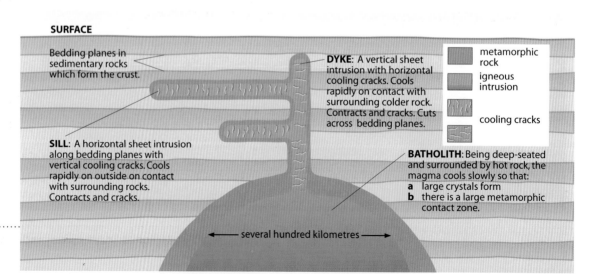

SURFACE

Bedding planes in sedimentary rocks which form the crust.

DYKE: A vertical sheet intrusion with horizontal cooling cracks. Cools rapidly on contact with surrounding colder rock. Contracts and cracks. Cuts across bedding planes.

| metamorphic rock |
| igneous intrusion |
| cooling cracks |

SILL: A horizontal sheet intrusion along bedding planes with vertical cooling cracks. Cools rapidly on outside on contact with surrounding rocks. Contracts and cracks.

BATHOLITH: Being deep-seated and surrounded by hot rock, the magma cools slowly so that:
a large crystals form
b there is a large metamorphic contact zone.

← several hundred kilometres →

Intrusive landforms

Usually, only a relatively small amount of magma actually reaches the surface as most is intruded into the crust, where it solidifies. Such intrusions may initially have little impact upon the surface **geomorphology**, but if the overlying rocks are later worn away, distinctive landforms may then develop (Figure 1.32).

During the Tertiary era, an upthrust of magma was **intruded** into the sedimentary rocks of Arran to form the Northern Granite. As the magma slowly cooled, it formed large crystals (unlike on the surface where rapid cooling forms fine crystals), contracted and cracked resulting in a series of joints. The magma also produced a large, deep-seated, dome-shaped **batholith** as it solidified.

Surrounding the batholith is a **metamorphic aureole** where the original sedimentary rocks have been changed (metamorphosed) by the heat and pressure of the intrusion from sandstones into **schists**. Since then, the overlying rocks have been removed by water, ice and even the sea to leave the granite batholith with its jointing exposed (Figure 1.32). These joints have been widened by chemical weathering (pages 42–44) to form the large granite slabs and tors surrounding Goatfell (compare Figure 8.14).

Figure 1.32

Idealised transect through northern Arran

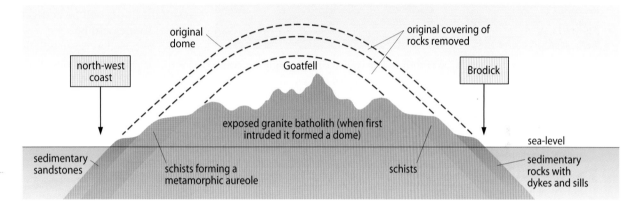

Figure 1.33

Fieldsketch of a dyke at Kildonan, Arran

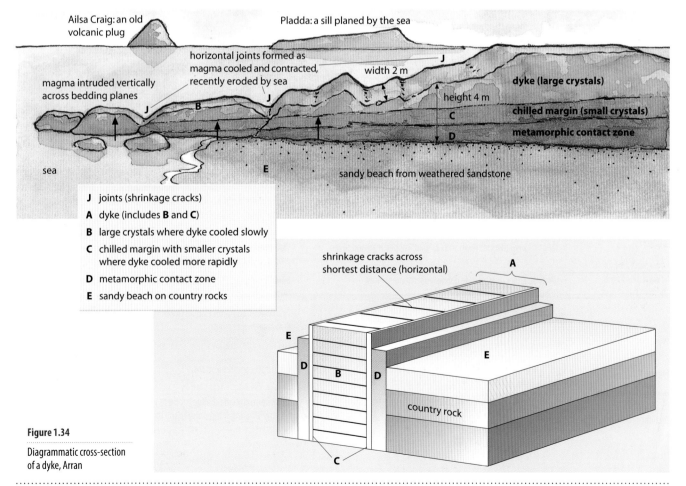

J joints (shrinkage cracks)

A dyke (includes **B** and **C**)

B large crystals where dyke cooled slowly

C chilled margin with smaller crystals where dyke cooled more rapidly

D metamorphic contact zone

E sandy beach on country rocks

Figure 1.34

Diagrammatic cross-section of a dyke, Arran

Figure 1.35

Dyke at Kildonan, Arran

although those parts that come into contact with the surrounding rock will cool more rapidly to produce a chilled margin (Figure 1.34). Most dykes on Arran were formed after, and radiate from, the batholith intrusion; they are so numerous that they have been termed a 'dyke swarm'. Most of the dykes are more resistant to erosion than the surrounding sandstones and so where they cross the island's beaches they stand up like groynes (Figure 1.35). Although averaging 3 m, these dykes vary from 1 to 15 m in width.

A **sill** is formed when the igneous rock is intruded along the bedding planes between the existing sedimentary rocks (Figure 1.31). The magma cools and contracts but this time the resultant joints will be vertical and their hexagonal shapes can be seen when the landform is later exposed as on headlands such as that at Drumadoon on the west coast of Arran (Figures 1.36 and 1.37) and the Giant's Causeway in Northern Ireland (Figure 1.27). The sill at Drumadoon is 50 m thick.

Figure 1.36

Fieldsketch of a sill exposed at Drumadoon, Arran

If, in trying to rise to the surface, magma cuts across the bedding planes of the sedimentary rock, it is called a **dyke** (Figures 1.31 and 1.33). The material which forms the dyke cools slowly

original covering of sandstone removed

vertical joints (columnar) jointing) formed as magma cooled and contracted

magma intruded horiztally between bedding planes

metamorphic contact zone under a chilled margin

50 m

talus (scree) covering sandstone (11 m)

30 m

raised beach

sea

Figure 1.37

Sill at Drumadoon, Arran

Benefits	Hazards
Ash weathers into a fertile soil ideal for farming. Basic lava may also produce fertile soils (the region surrounding Mount Etna) but needs very careful management. The fertility of acid lava is low.	Earthquakes destroy buildings and result in loss of life.
Igneous rock contains minerals such as gold, copper, lead and silver.	Violent eruptions with blast waves and gas may destroy life and property (Mt Pelée, Mount St Helens).
Extinct volcanoes may provide defensive settlement sites (Edinburgh).	Mudflows/lahars may be caused by heavy rain and melting snow (Armero in Colombia and Pinatubo in the Philippines).
Igneous rock is used for building purposes (Naples, Aberdeen).	Tidal waves/tsunamis (Indian Ocean tsunami and following the eruption of Krakatoa).
Geothermal power is being developed (Iceland, New Zealand).	Ejection of ash and lava ruins crops and kills animals.
Geysers and volcanoes are tourist attractions (Yellowstone National Park), generating revenue for local communities.	Interrupts communications.
Volcanic eruptions may produce spectacular sunsets (Krakatoa).	Short-term climatic changes occur as volcanic dust absorbs solar energy, lowering temperatures and increasing rainfall.

Figure 1.38

Benefits and hazards resulting from tectonic processes

What are natural hazards?

Natural hazards, which include earthquakes, volcanic eruptions, floods, drought and storms, result from natural processes within the environment (Figure 1.39). They are, therefore, different from environmental disasters, such as desertification, ozone depletion and acid rain, which are caused by human activity and the mismanagement of the environment. It is important, however, to stress the difference between a **natural hazard** and a **hazard event**. Natural hazards have the potential to affect people and the environment; it is the hazard event that causes the damage. An event only becomes a hazard if it affects, or threatens, people and property. For example, the submarine volcanic eruption which created the new island of Surtsey (Places 3, page 16) was hardly a hazard event, whereas the China earthquake of 2008 killed over 80 000 people, destroyed towns and for a time ended normal human activities. The impact of a hazard event may be felt over a wide area; the effects may be long-term as well as immediate; and the event can be costly to property and dangerous to people.

Figure 1.39

Types of natural hazard (*after* Burton and Kates)

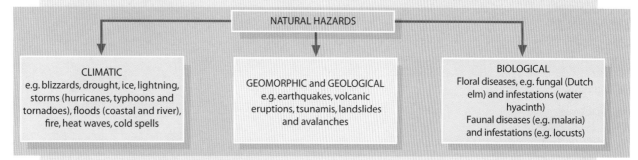

NATURAL HAZARDS

CLIMATIC
e.g. blizzards, drought, ice, lightning, storms (hurricanes, typhoons and tornadoes), floods (coastal and river), fire, heat waves, cold spells

GEOMORPHIC and GEOLOGICAL
e.g. earthquakes, volcanic eruptions, tsunamis, landslides and avalanches

BIOLOGICAL
Floral diseases, e.g. fungal (Dutch elm) and infestations (water hyacinth)
Faunal diseases (e.g. malaria) and infestations (e.g. locusts)

The International Strategy for Disaster Reduction (ISDR)

The United Nations, through the ISDR, tries to reduce loss of life, property damage and social and economic destruction caused by natural disasters, especially those occurring in less well-off developing countries. There is, however, a problem in classifying the type of hazard and in quantifying data after the event (Figure 1.40).

Data provided by the ISDR suggests that about 60 per cent of natural disasters and over 80 per cent of deaths occur in developing countries, especially those in the South-east Asia/Pacific Rim region. Developing countries are less likely to have the equipment needed to predict the occurrence of a hazard and less money either for planning how to reduce its impact or for organising a rapid and effective response after it. Figures 1.41 and 1.42 show that despite the incidence of occasional severe earthquakes/tsunamis, and even with the introduction of early storm- and flood-warning systems in places like Bangladesh and the Caribbean, globally over 80 per cent of deaths are still caused by tropical storms and flooding.

How may people react to natural hazards?

Geographers need to ask the following questions when studying either the risk of a potential natural hazard or a specific hazard event.

1 Many natural disasters result from a combination of events, meaning that it becomes impossible to attribute the losses to a single cause. For example, in Sumatra following the Indian Ocean tsunami of 2004 (Places 4), how many deaths were due to the initial earthquake and how many to drowning caused by the subsequent flooding? Again, how many deaths in New Orleans in 2005 were the result of tropical storm Katrina or the flooding that followed? Classifying events under specific headings can lead to double counting after extreme events.

2 Even direct deaths and damage may be difficult to quantify accurately in some developing countries due to a lack of reliable census data or population registers. Hence initial reports of 'hundreds killed' or 'damage estimated in millions of dollars' may be grossly exaggerated, while those recorded as 'missing' or who die later from disease or malnutrition caused by the disaster may be under-estimated.

Adapted from an article by Keith Smith in *Teaching Geography*, Sept 1996

Figure 1.40

The problems of defining natural disasters

1 What are people's perception of the natural hazard? Perception is how individuals or groups of people view the hazard risk. This often depends on their knowledge and experience of the potential event. The inhabitants of Pompeii, prior to the eruption of Vesuvius in AD 79, had not realised that the

Figure 1.41

Number of natural disasters by type, 1970–2005

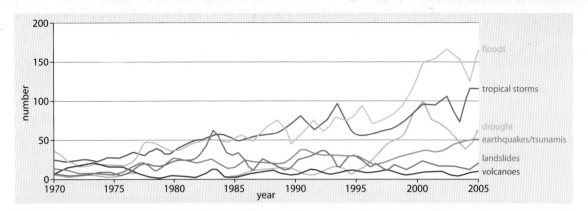

Figure 1.42

Natural disasters

a Type	Events			Deaths		
	Per year 1967–91	2004	2005	Per year 1967–91	2004	2005
Earthquakes	32	29	21	27 000	882	76 241
Volcanic eruptions	4	5	7	117	2	3
Tsunamis	1	2	0	267	226 435	0
Tropical storms (hurricanes/typhoons)	37	81	69	37 400	6 513	4 672
Floods (rivers/coasts)	57	107	168	12 750	6 957	6 135
Storms (depressions/tornadoes/lightning)	34	27	17	2 300	827	269
Cold wave (blizzards/heatwaves)	5	7	15	204	239	923
Drought	18	15	22	55 570	149	11 100
Avalanches	1	2	1	54	42	12
Landslides	10	16	12	1 750	357	649
Fires (bush)	30	8	10	420	14	47

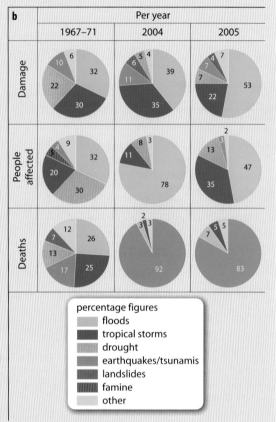

percentage figures
- floods
- tropical storms
- drought
- earthquakes/tsunamis
- landslides
- famine
- other

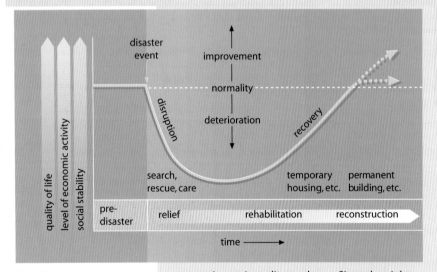

Figure 1.43

The responses to a hazard event (*after* Chris Park)

mountain was in reality a volcano. Since then it has erupted on numerous occasions. The question is, Why do people continue to live in this and other hazardous areas? It may be because they:

- perceive the area as providing the best of opportunities to earn a living
- are too concerned with day-to-day problems to consider the hazard risk

- have the capital and technology to cope with the hazard event.

2 *What are the immediate and long-term effects of the event?*

3 *How do people respond to the event (Figure 1.43)?*

4 *How might people adjust to and plan for a future event?* It has been suggested that people have six options. They may try to: prevent the event; modify the hazard; lessen the possible amount of damage; spread the losses caused by the event; claim for losses; or do nothing but pray that the event will not occur again (at least not in their own lifetime).

5 *Can a future event be predicted?* This involves predicting where the next event will take place, when it is likely to occur and how big it is likely to be.

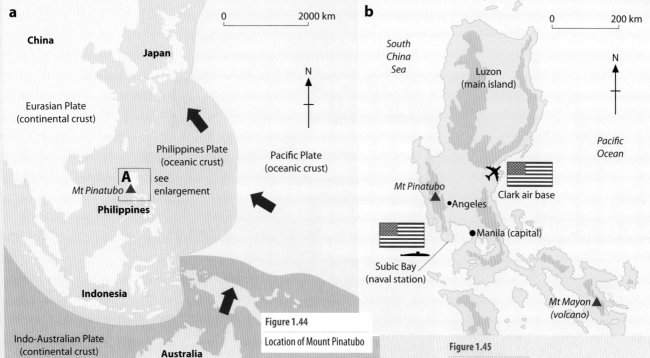

Figure 1.44
Location of Mount Pinatubo

Figure 1.45
Eyewitness account of the eruption

Why is Mount Pinatubo in a hazard risk area?

Mount Pinatubo is located in the Philippines (Figure 1.44). The Philippines lie on a destructive plate margin where the Philippines Plate, composed of oceanic crust, moves towards and is subducted by the Eurasian Plate, which consists of continental crust. As the oceanic plate is subducted, it is converted into magma which rises to the surface and forms volcanoes. The Philippines owe their existence to the frequent ejection of lava over a period of several million years. Even before Pinatubo erupted in 1991, there were over 30 active volcanoes in the Philippines.

Why did people live in this hazard risk area?

As Mount Pinatubo had not erupted since 1380, people living in the area no longer considered it to be a hazard. During that time, ash and lava from earlier eruptions had weathered into a fertile soil, ideal for rice growing. By 1991, people no longer perceived Pinatubo to be a danger. On the lower slopes of the mountain, the Aeta, recognised as the aboriginal inhabitants of

the islands, practised subsistence farming (slash and burn agriculture, Places 66, page 480). Near the foothills was the rapidly growing city of Angeles, together with an American air base and a naval station (Figure 1.44b).

What were the nature, effects and consequences of the eruption?

1 Immediate effects

The volcano began to show signs of erupting in early June 1991. Fortunately, there were several advance-warning signs which allowed time for the evacuation of thousands of people from Angeles and the 15 000 personnel from the American air base. The number and size of eruptions increased after 9 June. On 12 June, an explosion sent a cloud of steam and ash 30 km into the atmosphere – the third-largest eruption experienced anywhere in the world this century (Figure 1.45). Up to 50 cm of ash fell nearby, and over 10 cm within a 600 km radius. The eruptions were, characteristically, accompanied by earthquakes and torrential rain – except that the rain, combining with the ash, fell as thick

Seismologists said a mixture of searing gas, ash and molten rock quickly raced down the mountain's west and northern flanks and into the Marella, Maraunot and O'Donnel rivers [Figure 1.46]. Ash also rained down on seven towns in the region and traces of ash were detected near the Subic Bay naval base, 50 miles [80 km] to the south-west. Pumice fragments measuring up to 1.2 inches [3 cm] long fell on villages south-west of the volcano. At a refugee centre in Olongapo, about 35 miles [56 km] south-west of the volcano, survivors said they saw the sky grow dark and then heard a tremendous explosion followed by a rain of ash and stones as big as a man's head.

Other reporters described panic as people fled with their belongings and livestock over roads made slippery by the falling ash. Refugees wore cardboard boxes with air and peepholes to protect themselves from the ash. The ash was so thick in the air that at noon motorists were driving with their headlights on and wipers operating to clear the debris.

Adapted from *The Independent*, 13 June 1991

mud. The ash destroyed all crops on adjacent farmland and its weight caused buildings to collapse, including 200 000 homes, a local hospital and many factories. Power supplies were cut off for three weeks and water supplies became contaminated. Relief operations were hindered as many roads became impassable and bridges were destroyed.

2 Longer-term effects

The thick fall of ash not only ruined the harvest of 1991, but made planting impossible for 1992. Over one million farm animals died, many through starvation due to the lack of grass. Several thousand farmers and their families had to take refuge in large cities. The majority were forced to seek food and shelter in shanty-type refugee camps. Disease, especially malaria, chickenpox and diarrhoea, spread rapidly and doctors had to treat hundreds of people for respiratory and stomach disorders. Soon after the event, and again in 1993, typhoons brought heavy rainfall which caused flooding and lahars (mud-flows). Lahars form when surface water picks up large amounts of volcanic ash in mountainous areas and deposits it as mud over lower-lying areas (Figure 1.47). The ash that was ejected into the atmosphere is believed to have caused changes in the Earth's climate, including the lowering of world temperatures and ozone depletion (Figure 1.48).

The eruption and its after-effects were blamed for about 700 deaths. Of these, only six were believed to have been a direct result of the eruption itself. Over 600 people were to die from disease and a further 70 from suffocation by lahars.

Figure 1.46

A pyroclastic cloud produced by the eruption, June 1991

Figure 1.47

A lahar at Angeles, near Mount Pinatubo

IT HAS been described as the world's greatest climatic experiment, but unlike most scientific endeavours it was unplanned. When the tropical tranquillity of the Philippines was shattered last June by a volcanic explosion, Mount Pinatubo was a relatively obscure volcano, known in the scientific community only to a handful of geologists. Having sent more than 20 million tonnes of dust and ash into the atmosphere, altering its heat balance and accelerating ozone depletion over a large part of the globe, Pinatubo has become the focus of several far-reaching studies.

Climatologists now use the term 'Pinatubo effect' to describe how volcanic ash and debris, if sent high enough into the atmosphere, can influence temperature and weather for several years afterwards. The dust from Pinatubo was ejected as high as 20 miles [32 km] above the Earth. From the haven of Earth orbit, satellites observed the plume of volcanic ash as it girdled the globe at speeds approaching 75 miles [120 km] per hour. A month after the eruption which killed 350 people, a 3000 mile [4800 km] cloud of ash and sulphur compounds circled the Earth.

Satellite temperature measurements confirmed that the dust had effectively shaded the surface of the Earth from the sun's rays, resulting in a lowering of the average global temperature. A NASA team at the Goddard Institute for Space Studies in New York, led by James Hansen, tried to assess what effect the cooling caused by the dust of Mount Pinatubo would have on global warming caused by man-made emissions of carbon dioxide. They concluded that Pinatubo would in effect delay global warming by several years. While global warming experts argue about the effect of Pinatubo's eruption on average temperatures, ozone specialists are interested in the effect the volcano has had and will have on the ozone layer. The volcano has spewed out huge quantities of sulphate aerosols, particles containing sulphur that remain suspended in the atmosphere for several years. These sulphate particles are important in the chemistry of ozone destruction for two reasons: first, they act as sites where ozone-destroying reactions take place; and secondly, they mop up nitrogen-containing compounds that help to prevent ozone destruction. This winter American and European scientists undertook the most intensive investigation of ozone depletion over the northern hemisphere, including Europe and North America. More than 300 scientists from 17 countries were involved and their work has shown that ozone levels fell by 10 to 20 per cent more than expected. 'The eruption of Mount Pinatubo has increased the abundance of natural sulphate particles, potentially enhancing ozone losses due to chemical reactions that occur on particle surfaces,' the NASA ozone monitoring team said earlier this month.

Adapted from *The Independent on Sunday*, 10 May 1992

Figure 1.48

The climatic effects of the eruption

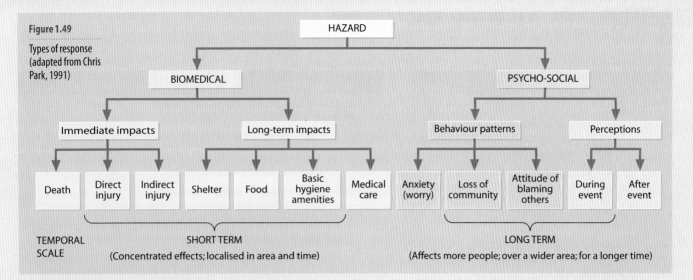

Figure 1.49

Types of response (adapted from Chris Park, 1991)

| | | TEMPORAL SCALE | | SHORT TERM (Concentrated effects; localised in area and time) | | | | | LONG TERM (Affects more people; over a wider area; for a longer time) | | |

HAZARD
- BIOMEDICAL
  - Immediate impacts
    - Death
    - Direct injury
    - Indirect injury
  - Long-term impacts
    - Shelter
    - Food
    - Basic hygiene amenities
    - Medical care
- PSYCHO-SOCIAL
  - Behaviour patterns
    - Anxiety (worry)
    - Loss of community
    - Attitude of blaming others
  - Perceptions
    - During event
    - After event

How did people respond to the hazard event?

Chris Park has divided human responses during and after any hazard event into two categories (Figure 1.49).

Within a few weeks of the major Pinatubo eruption, groups of evacuees from the affected area began to consider their future options and their next move. Their range of responses included the following:

1   Some members of the Aeta tribe (Figure 1.50) decided not to return to their former homes. As a spokesperson explained: 'Everything we have planted has been destroyed. There is no point in going back. The government will have to put us somewhere else.'

2   In contrast, the majority of the Aeta tribe decided to return. To them, the mountain slopes, although vastly changed, were still their home and the hard way of life in the hills was preferable to the foreign habits of the lowlanders and to living in urban areas.

3   Most of the people who fled from the city of Angeles have, so far, opted against returning home. To them, life in the shanty refugee camps is safer than returning to an area where eruptions and earthquakes are still occurring and where the heavy rain is likely to cause lahars for several years until the regrowth of vegetation stabilises the slopes.

Can future eruptions be predicted?

At present, although it may be possible to predict fairly accurately where volcanic eruptions are likely to occur (i.e. at constructive and destructive plate margins, Figure 1.8), there is less prospect of scientists being able to predict accurately either the precise time or the scale of a specific event. Prediction is easier in places where volcanoes erupt regularly, as they will be better monitored (Figure 1.51), than in places where eruptions have not occurred for several centuries (Mount St Helens and Mount Pinatubo) and where people's perception of the hazard risk is less. Monitoring potential eruptions is also more likely in an economically more developed country with its greater wealth and technology, or in places where a high population density is a risk.

Figure 1.50

Members of the Aeta tribe

a Colour-coded alert levels

Colour	Implication
Green	No eruption. Volcano is quiet/dormant.
Yellow	Eruption possible in next few weeks (with little or no additional warning). Local earth tremors and/or increased levels of volcanic gas emissions.
Orange	Explosive eruption possible within a few days (with little or no additional warning). Increased number/strength of local earth tremors/quakes. Non-explosive extrusion of a lava dome and/or lava flows. Any ash plume will be under 9 km.
Red	Major explosive eruption expected within a day. Strong earthquake activity detected even at a distance. Ash plume exceeds 9 km.

b Numerical alert levels

	Indicative phenomena	Volcano status
0	Typical background surface activity; low levels of seismic deformation and heat flows	Dormant or quiescent
1	Apparent seismic, geodic, thermal and other unrest indicators	No eruption threat
2	Increase in number/intensity of unrest indicators including heat flows, seismicity and deformation	Eruption threat
3	Minor steam eruptions; high/increasing trends in indicators of unrest; significant effects on volcano and possibly beyond	Minor eruption started; real threat of a major event
4	Eruption of new magma; sustained high levels of unrest indicators on both the volcano and beyond	Hazardous local eruption; threat of a serious event
5	Destruction/major damage beyond the volcano; significant hazard risk over a wide area	Large hazardous eruption in progress

Figure 1.51

Two volcano alert systems (abridged from Parfitt and Wilson)

These procedures are easier to adopt in volcanoes that erupt frequently as they are monitored partly to learn more about their internal structure and partly for signs of activity. Continuous monitoring instruments are both expensive and vulnerable. Data can be collected:

- on the volcano using seismometers to record minor seismic tremors, any inflation or tilt, an increase in pressure or the release of volcanic gases
- using satellites that can detect changes in temperature, vegetation (caused by the release of gases) and the local magnetic field
- by studying previous timescales of cycles of eruptions and maps showing paths taken by **earlier lava or pyroclastic flows.**

Predicting and planning for earthquakes

Scientists can use sensitive instruments to measure increases in earth movements and a build-up of pressure. They can also map the epicentres and frequency of previous earthquakes to see if there is either a repeat location or a time-interval pattern. In Kanto, the region surrounding Tokyo, there has been a severe earthquake, on average, every 70–80 years for the last five centuries. As the last event was in 1923, with an estimated 14 000 deaths, then an equally severe earthquake might be expected to occur early in the 21st century. Even so, such methods can predict neither the precise timing nor the exact location of the earthquake. A less scientific method, but successfully used in China, has been the observation of unusual animal behaviour shortly before a major earth movement

(though not before the 2008 event), e.g. mice have fled houses, dogs have howled, fish have jumped out of water and the giant panda has moaned.

In earthquake-prone areas, especially in more wealthy countries, buildings can be constructed to withstand earthquakes. They are built with steel (which can sway during earth movement) and fire-resistant materials – never with bricks or reinforced concrete blocks. Foundations are sunk deep into bedrock and are separated from the superstructure by shock-absorbers. Open spaces should be provided for people to assemble, and roads made sufficiently wide to allow rapid access by emergency services. The emergency services themselves need to be trained and well-equipped, while local residents need to be made aware as to how they should respond both during and after the event.

Further reference

Alexander, D. (2001) *Confronting Catastrophe*, Terra Publishing.

Buranakul, S. (2005) 'Asian Tsunami: the Aftermath' in *Geography Review* Vol 19 No 1 (September).

Calder, N. (1973) *The Restless Earth*, BBC Publications.

Chester, D. (1993) *Volcanoes and Society*, Hodder Arnold.

Francis, P. and Oppenheimer, C. (2003) *Volcanoes*, Oxford University Press.

Goudie, A.S. (2001) *The Nature of the Environment*, WileyBlackwell.

Keller, E.A. and Pinter, N. (1995) *Active Tectonics: Earthquakes, uplift and landscape*, Prentice Hall.

Marti, J. and Ernst, G.G.J. (2008) *Volcanoes and the Environment*, Cambridge University Press.

Parfitt, L. and Wilson, L. (2008) *Fundamentals of Physical Volcanology*, Blackwell.

Park, C.C. (1991) *Environmental Hazards*, Nelson Thornes.

Petley, D. (2005) 'Tsunami' in *Geography Review* Vol 18 No 5 (May).

Earthquake information:
http://quake.wr.usgs.gov/
www.rcep.dpri.kyoto-u.ac.jp/~sato/tottori/index.html

Plate tectonics:
http://vulcan.wr.usgs.gov/Glossary/PlateTectonics/framework.html
http://eos.higp.hawaii.edu/volcanolis.html
http://pubs.usgs.gov/gip/dynamic/understanding.html
http://eos.higp.hawaii.edu/volcanolist.html

Further links:
www.physicalgeography.net/fundamentals/10i.html
http://vulcan.wr.usgs.gov/Servers/earth_servers.html

Activities

1 Study Figure 1.3 (page 10).

a What is an earthquake? *(3 marks)*

b Why is an earthquake that measures 7.0 on the Richter scale 100 times more severe than one that measures 5.0? *(3 marks)*

c How severe was the earthquake in San Francisco in 1989? *(1 mark)*

d How much bigger was the earthquake in San Francisco in 1906 than the one in Kobe in 1972? *(3 marks)*

e Describe one way in which buildings may be made 'earthquake proof'. *(4 marks)*

f List **two** rules that you would need to follow if your home was in an earthquake area. Explain why they would be important. *(4 marks)*

g How do local and national authorities try to prepare for earthquakes in areas where they may occur? *(7 marks)*

2 Create a table using the headings in the left column of the table below. Use it to provide details of a volcanic or earthquake event you have studied. *(marks as shown)*

Heading	Description from case study	
Location	Identify where the disaster occurred	(2 marks)
Pre-disaster potential	Description of geology of the area to identify the reason for an event to occur	(3 marks)
Disaster event	Timing, size and nature of the disaster	(3 marks)
Disruption	Details of immediate damage	(3 marks)
Relief	Types of immediate relief needed	(3 marks)
Recovery	Nature of the required recovery programme	(4 marks)
Time	Timescale of the continuing impact of the event	(3 marks)
Reconstruction	Type and amount of long-term aid required	(4 marks)

Exam practice: basic structured questions

4 a Study Figure 1.52 and identify the internal structure of the Earth by naming A, B, C and D. *(4 marks)*

b Identify the **two** types of crust of the Earth and describe the differences between them. *(4 marks)*

c Explain why crustal plates move. *(5 marks)*

d Choose **one** of the following types of plate margin:
• constructive (spreading) margin
• destructive (subduction) margin
• conservative (slip) margin.
Describe the distinctive landforms that develop there, and explain their development. *(12 marks)*

3 For **either** a volcanic eruption **or** an earthquake you have studied:

a Draw a sketch map to show the location of the area where it occurred. *(3 marks)*

b Describe the hazard event. *(3 marks)*

c Explain, with the aid of a diagram, the causes of the event. *(4 marks)*

d How big was the event? *(2 marks)*

e How frequently do such events occur in this area? *(2 marks)*

f How large an area was affected by the event? *(3 marks)*

g Describe the effects of the event on the area. *(4 marks)*

h What lessons for the future were learned from this event? *(4 marks)*

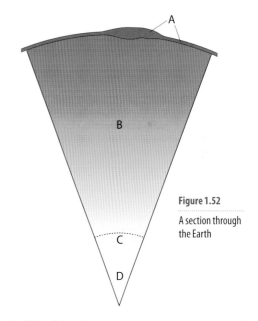

Figure 1.52
A section through the Earth

5 a i What is lava? *(2 marks)*

ii What happens to lava when it is exposed on the ground surface? *(1 mark)*

iii Why does some lava flow quickly and some flow more slowly? *(4 marks)*

b Making use of annotated diagrams, describe **two** different kinds of volcano. Name an example of each of your kinds of volcano. *(8 marks)*

c With reference to one or more areas that you have studied, explain why people continue to live close to active volcanoes. *(10 marks)*

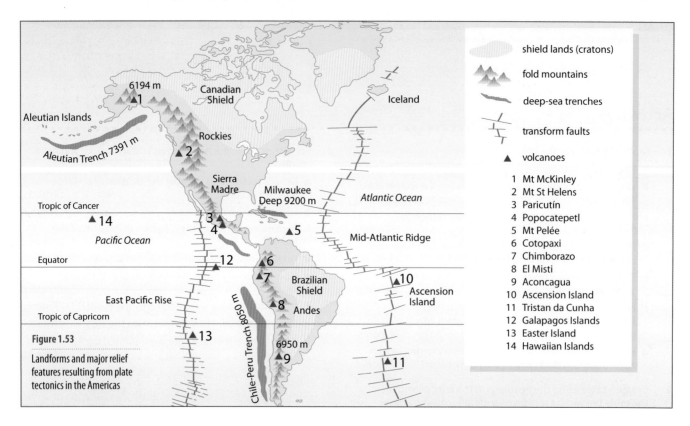

Figure 1.53

Landforms and major relief features resulting from plate tectonics in the Americas

Key (map legend):
- shield lands (cratons)
- fold mountains
- deep-sea trenches
- transform faults
- ▲ volcanoes

1 Mt McKinley
2 Mt St Helens
3 Paricutín
4 Popocatepetl
5 Mt Pelée
6 Cotopaxi
7 Chimborazo
8 El Misti
9 Aconcagua
10 Ascension Island
11 Tristan da Cunha
12 Galapagos Islands
13 Easter Island
14 Hawaiian Islands

6 Study Figure 1.53 and answer the following questions.

a i Name an example of **each** of the following from the map: shield lands (cratons); fold mountains; deep-sea trenches. *(3 marks)*

ii Explain the meaning of **each** of these terms: shield lands (cratons); fold mountains; deep-sea trenches. *(6 marks)*

b i Identify the **compass direction** for the movement of the Earth's crust at each of Ascension Island (number 10) and Easter Island (number 13). *(13 marks)*

ii For **each** of these places, explain why you think the crust moves in that direction. *(4 marks)*

c Choose **one** volcano marked on the map and, referring to plate movements, explain why it occurs there. *(10 marks)*

7 a Draw a labelled diagram to show the features of a composite volcano. *(4 marks)*

b Name **one** intrusive landform and explain how it was formed. *(5 marks)*

c With reference to the photographs on page 34 (Figures 1.46 and 1.47):

i describe a pyroclastic cloud and explain why it is a threat to people living nearby *(8 marks)*

ii describe a lahar and explain why it is a threat to people living nearby. *(8 marks)*

Exam practice: structured questions

8 a In areas where there are volcanic eruptions, earthquakes also occur. Suggest how volcanoes and earthquakes are linked to each other. *(5 marks)*

b Earthquakes occur in areas where there is no evidence of volcanic eruption. For **one** area where there are earthquakes but no volcanoes, explain the causes of earthquake activity. *(10 marks)*

c Name an area where earthquakes have occurred.

Describe **one** landscape feature found in that area that was formed by earthquake activity.

Explain how it was formed. *(10 marks)*

9 a i Draw an annotated diagram and describe the features which may be found associated with a **constructive plate margin**. *(8 marks)*

ii For **one** of these features, explain the processes that have led to its formation. *(8 marks)*

b i Explain **one** way in which areas close to a constructive plate margin may be of economic value.

ii Suggest how people can exploit the economic resource you have identified. *(9 marks)*

c With reference to one or more areas that you have studied, explain how people can exploit the economic resources that can be found at constructive plate margins. *(9 marks)*

10 a i Draw an annotated diagram and describe the features associated with a **destructive plate margin**.
(8 marks)

   ii For one of these features, explain the processes that have led to its formation. *(8 marks)*

   b i Explain one way in which an area close to a destructive plate margin may be of economic value.

   ii Suggest how people can exploit the economic resource you have identified. *(9 marks)*

   c With reference to one or more areas that you have studied, explain how people can exploit the economic resources that can be found at destructive plate margins. *(9 marks)*

11 Look at Figure 1.53 and make use of Figure 1.8 (page 15).

   a Describe the distribution of **cratons** (shield areas) and of young fold mountain ranges in the Americas. *(8 marks)*

   b Explain, with the use of diagrams, the origins of:

   i Ascension Island (number 10 on Figure 1.53) *(7 marks)*

   ii the Chile–Peru trench and the volcanic mountains (numbered 7, 8 and 9 on Figure 1.53). *(10 marks)*

12 a Identify **two different ways** in which volcanoes may be classified. For **one** of the ways you have identified, explain how one type of volcano fits into the classification. *(8 marks)*

   b Why do people continue to live close to active volcanoes? *(7 marks)*

   c Using an example of a real upland area, explain what happens to a volcanic area once volcanic activity ceases. *(10 marks)*

13 Study Figure 1.54.

   a Choose **one** geological factor from the table. Explain how that geological factor influences the assessment of the danger from a volcanic hazard in an area. *(7 marks)*

   b Explain **two** ways in which a volcanic eruption could affect an urban area outside the zone of direct lava and pyroclastic outfall. *(6 marks)*

   c With reference to examples that you have studied, explain how people in areas at different stages development can prepare for the hazards of volcanic eruptions. *(12 marks)*

Figure 1.54
Range of factors affecting volcanic hazards

Danger factor	Assessment of danger
Geological factors	
Plate margin type	There will be more explosive activity on a destructive margin than on a constructive margin.
Volcano type	A shield volcano will be less explosive than a stratovolcano.
Extruded material	A lava eruption is less dangerous than a pyroclastic eruption.
Silica content	Silica-rich magmas produce more explosive eruptions than silica-poor magmas.
Dormancy period	Volcanoes with longer periods of dormancy tend to be more explosive than those with shorter dormancy periods.
Environmental and topographical factors	
Wind direction	Pyroclastic flows are thicker downwind from an active vent.
Topography	Valleys funnel pyroclastic and other flows. Ridges across the route of flows can shelter areas within a blast zone.
Social and economic factors	
Settlement density	More densely settled areas will be at greater risk of immediate damage.
Economic status	Total cost will be greater in more economically developed areas but response will be faster and more effective. Loss of life will be lower. In less developed areas, loss of life will be greater and economic damage will be greater in proportion to the total.

14 a i What is a natural hazard? *(2 marks)*

   ii Under what circumstances can a volcanic eruption be described as a hazard event? *(4 marks)*

   b For any volcanic event that you have studied:

   i identify the causes of the volcanic event *(7 marks)*

   ii evaluate the severity of the effects of the event on the surrounding area and its inhabitants. *(12 marks)*

Exam practice: essays

15 Describe the theory of plate tectonics and explain **three** pieces of evidence which provide support for the theory. *(25 marks)*

16 For any **one** area that experiences volcanic and/or earthquake hazards that you have studied, explain how people perceive and manage the hazard. *(25 marks)*

17 'The extent to which earthquakes represent hazards depends on where they occur.' Discuss this statement. *(25 marks)*

18 'In the last 30 years or so natural hazards caused by tectonic pressures have led to an increased death rate around the world. This increase is due more to an increase in world population than to an increased frequency of tectonic events.' Discuss this statement. *(25 marks)*

2

Weathering and slopes

• •

'Every valley shall be exalted, and every mountain and hill shall be made low: and the crooked shall be made straight, and the rough places plain.'

The Bible, Isaiah 40:4

Weathering

The majority of rocks have been formed at high temperatures (igneous and many metamorphic rocks) and/or under great pressure (igneous, metamorphic and sedimentary rocks), but in the absence of oxygen and water. If, later, these rocks become exposed on the Earth's surface, they will experience a release of pressure, be subjected to fluctuating temperatures, and be exposed to oxygen in the air and to water. They are therefore vulnerable to **weathering**, which is the disintegration and decomposition of rock *in situ* – i.e. in its original position. Weathering is, therefore, the natural breakdown of rock and can be distinguished from erosion because it need not involve any movement of material. Weathering is the first stage in the **denudation** or wearing down of the landscape; it loosens material which can subsequently be transported by such agents of erosion as running water (Chapter 3), ice (Chapter 4), the sea (Chapter 6) and the wind (Chapter 7). The degree of weathering depends upon the structure and mineral composition of the rocks, local climate and vegetation, and the length of time during which the weathering processes operate.

There are two main types of weathering:

1 **Mechanical** (or **physical**) **weathering** is the disintegration of rock into smaller particles by mechanical processes but without any change in the chemical composition of that rock. It is more likely to occur in areas devoid of vegetation, such as deserts, high mountains and arctic regions. Physical weathering usually produces sands.

2 **Chemical weathering** is the decomposition of rock resulting from a chemical change. It produces changed substances and solubles, and usually forms clays. Chemical weathering is more likely to take place in warmer, more moist climates where there is an associated vegetation cover.

It should be appreciated that although in any given area either mechanical or chemical weathering may be locally dominant, both processes usually operate together rather than in isolation.

Mechanical weathering
Frost shattering

This is the most widespread form of mechanical weathering. It occurs in rocks that contain crevices and joints (e.g. joints formed in granite as it cooled, bedding planes found in sedimentary rocks, and pore spaces in porous rocks), where there is limited vegetation cover and where temperatures fluctuate around 0°C (page 134). In the daytime, when it is warmer, water enters the joints, but during cold nights it freezes. Frost leads to mechanical breakdown in two ways:

1 As ice occupies 9 per cent more volume than water, it exerts pressure within the joints.
2 When water freezes within the rock it attracts small particles of water, creating increasingly large ice crystals.

In each case the alternating **freeze–thaw process**, or **frost shattering**, slowly widens the joints and, in time, causes pieces of rock to shatter from the main body. Where this **block disintegration** occurs on steep slopes, large angular rocks collect at the foot of the slope as **scree** or **talus** (Figure 2.1); if the slopes are gentle, however, large **blockfields** (felsenmeer) tend to develop. Frost shattering is more common in upland regions of Britain where temperatures fluctuate around freezing point for several months in winter, than in polar areas where temperatures rarely rise above 0°C.

Salt crystallisation

If water entering the pore spaces in rocks is slightly saline then, as it evaporates, salt crystals are likely to form. As the crystals become larger, they exert stresses upon the rock, causing it to disintegrate. This process occurs in hot deserts where capillary action draws water to the surface and where the

Figure 2.1

The formation of screes resulting from frost shattering: Moraine Lake, Banff National Park, Canada

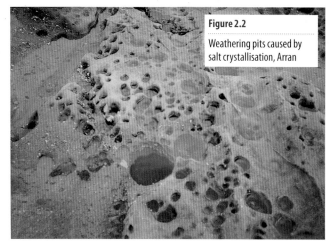

Figure 2.2

Weathering pits caused by salt crystallisation, Arran

rock is sandstone (page 182). Individual grains of sand are broken off by **granular disintegration**. Salt crystallisation also occurs on coasts where the constant supply of salt can lead to the development of weathering pits (Figure 2.2).

Pressure release

As stated earlier, many rocks, especially intrusive jointed granites, have developed under considerable pressure. The confining pressure increases the strength of the rocks. If these rocks, at a later date, are exposed to the atmosphere, then there will be a substantial release of pressure. (If you had 10 m of bedrock sitting on top of you, you would be considerably relieved were it to be removed!) The release of pressure weakens the rock allowing other agents to enter it and other processes to develop. Where cracks develop parallel to the surface, a process called **sheeting** causes the outer layers of rock to peel away. This process is now believed to be responsible for the formation of large, rounded rocks called **exfoliation domes** (Figures 2.3 and

7.6) and, in part, for the granite tors of Dartmoor and the Isle of Arran (Figures 8.14 and 8.15). Jointing, caused by pressure release, has also accentuated the characteristic shapes of glacial cirques and troughs (Figures 2.4, 4.14 and 4.15).

Thermal expansion or insolation weathering

Like all solids, rocks expand when heated and contract when cooled. In deserts, where cloud and vegetation cover are minimal, the diurnal range of temperature can exceed 50°C. It was believed that, because the outer layers of rock warm up faster and cool more rapidly than the inner ones, stresses were set up that would cause the outer thickness to peel off like the layers of an onion – the process of **exfoliation** (page 181). Initially, it was thought that it was this expansion–contraction process which produced exfoliation domes. Changes in temperature will also cause different minerals within a rock to expand and contract at different rates. It has been suggested that this causes **granular disintegration** in rocks composed of several minerals (e.g. granite which consists of quartz, feldspar and mica), whereas in homogeneous rocks it is more likely to cause block disintegration.

Laboratory experiments (e.g. by Griggs in 1936 and Goudie in 1974) have, however, cast doubt on the effectiveness of insolation weathering (page 181).

Biological weathering

Tree roots may grow along bedding planes or extend into joints, widening them until blocks of rock become detached (Figure 2.5). It is also claimed that burrowing creatures, such as worms and rabbits, may play a minor role in the excavation of partially weathered rocks.

Figure 2.3

An exfoliation dome: Sugar Loaf Mountain in Rio de Janeiro, Brazil

Figure 2.4

The process of pressure release tends to perpetuate landforms: as new surfaces are exposed, the reduction in pressure causes further jointing parallel to the surface

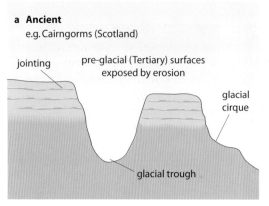

a Ancient
e.g. Cairngorms (Scotland)

jointing

pre-glacial (Tertiary) surfaces exposed by erosion

glacial cirque

glacial trough

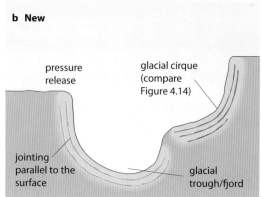

b New

pressure release

glacial cirque (compare Figure 4.14)

jointing parallel to the surface

glacial trough/fjord

Chemical weathering

Chemical weathering tends to:
- attack certain minerals selectively
- occur in zones of alternate wetting and drying, e.g. where the level of the water table fluctuates

Figure 2.5

Mechanical (biological) weathering caused by expanding tree roots in Geltsdale, Cumbria

Figure 2.6

Oxidation in Geltsdale, Cumbria

- occur mostly at the base of slopes where it is likely to be wetter and warmer.

This type of weathering involves a number of specific processes which may operate in isolation but which are more likely to be found in conjunction with one another. Formulae for the various chemical reactions are listed at the end of the chapter, page 57.

Oxidation

This occurs when rocks are exposed to oxygen in the air or water. The simplest and most easily recognised example is when iron in a **ferrous** state is changed by the addition of oxygen into a **ferric** state. The rock or soil, which may have been blue or grey in colour (characteristic of a lack of oxygen), is discoloured into a reddish-brown – a process better known as **rusting** (Figure 2.6). Oxidation causes rocks to crumble more easily.

In waterlogged areas, oxidation may operate in reverse and is known as **reduction**. Here, the amount of oxygen is reduced and the soils take on a blue/green/grey tinge (see **gleying**, page 272).

Hydration

Certain rocks, especially those containing salt minerals, are capable of absorbing water into their structure, causing them to swell and to become vulnerable to future breakdown. For example, gypsum is the result of water having been added to anhydrite ($CaSO_4$). This process appears to be most active following successive periods of wet and dry weather and is important in forming clay particles. Hydration is in fact a physio-chemical process as the rocks may swell and exert pressure as well as changing their chemical structure.

Hydrolysis

This is possibly the most significant chemical process in the decomposition of rocks and formation of clays. Hydrogen in water reacts with minerals in the rock or, more specifically, there is a combination of the H+ and OH– ions in the water and the ions of the mineral (i.e. the water combines with the mineral rather than dissolving it).

The rate of hydrolysis depends on the amount of H+ ions, which in turn depends on the composition of air and water in the soil (Figure 10.4), the activity of organisms (page 268), the presence of organic acids (page 271) and the cation exchange (page 269). An example of hydrolysis is the breakdown of feldspar (Figure 2.7), a mineral found in igneous rocks such as granite, into a residual clay deposit known as kaolinite (china clay). Granite consists of three minerals – quartz, mica and feldspar (Figure 8.2c) – and, as the table below shows, each reacts at a different rate with water.

Quartz	Mica	Feldspar
Not affected by water, remains unchanged as sand (Figure 2.7)	May be affected by water under more acid conditions releasing aluminium and iron	Readily attracts water producing a chemical change which turns the feldspar into clay (kaolin or china clay)

Figure 2.7

Decomposition of granite by hydrolysis on Goatfell, Arran

Figure 2.8

Carbonation of limestone near Ingleton, North Yorkshire

Carbonation

Rainwater contains carbon dioxide in solution which produces carbonic acid (H_2CO_3). This weak acid reacts with rocks that are composed of calcium carbonate, such as limestone. The limestone dissolves and is removed in solution (calcium bicarbonate) by running water. Carboniferous limestone is well-jointed and bedded (Chapter 8), which results in the development of a distinctive group of landforms (Figure 2.8).

Solution

Some minerals, e.g. rock salt, are soluble in water and simply dissolve *in situ*. The rate of solution can be affected by acidity since many minerals become more soluble as the pH of the solvent increases (page 269).

Organic weathering

Humic acid, derived from the decomposition of vegetation (humus), contains important elements such as calcium, magnesium and iron. These are released by a process known as **chelation** (page 271). The action of bacteria and the respiration of plant roots tends to increase carbon dioxide levels which helps accelerate solution processes, especially carbonation. Lichen can also extract iron from certain rocks through the process of reduction. Recent research suggests that lichen and blue-green algae, which form the pioneer community in the development of a lithosere (page 288), play a far greater weathering role than was previously thought. However, it should be remembered that the presence of a vegetation cover dramatically reduces the extent of mechanical weathering.

Acid rain

Human economic activities (such as power generation and transport) release increasingly more carbon dioxide, sulphur dioxide and nitrogen oxide into the atmosphere. These gases then form acids in solution in rainwater (page 222). Acid rain readily attacks limestones and, to a lesser extent, sandstones, as shown by crumbling buildings and statues (Figure 2.9). The increased level of acidity in water passing through the soil tends to release more hydrogen and so speeds up the process of hydrolysis. An indirect consequence of acid rain is the release from certain rocks of toxic metals, such as aluminium, cadmium, copper and zinc, which can be harmful to plants and soil biota (page 268).

Some authorities, including Andrew Goudie, prefer to divide weathering into three categories rather than the two described here. Their alternative classification includes, as a third category, **biological weathering**. Instead of including 'biological' under mechanical weathering and 'organic' under chemical weathering, they would group these two types together under the heading 'biological weathering'.

Climatic controls on weathering
Mechanical weathering

Frost shattering is important if temperatures fluctuate around 0°C, but will not operate if the climate is too cold (permanently frozen), too warm (no freezing), too dry (no moisture to freeze), or too wet (covered by vegetation). Mechanical weathering will not take place at **X** on Figure 2.10a where it is too warm and there is insufficient moisture, while at **Y**, the high temperature and heavy rainfall will give a thick protective vegetation cover against insolation.

Chemical weathering

This increases as temperatures and rainfall totals increase. It has been claimed that the rate of chemical weathering doubles with every 10°C temperature increase. Recent theories suggest that, in humid tropical areas, direct removal by solution may be the major factor in the lowering of the landscape, due to the continuous flow of water through the soil. Chemical weathering will be rapid at **S** (Figure 2.10b) due to humic acid from the vegetation. It will be limited at **P**, because temperatures are low, and at **R**, where there is insufficient moisture for the chemical decomposition of rocks. Carbon dioxide is an exception in that, being more soluble at lower as opposed to higher temperatures, it can accelerate rates of solution in cold climates.

Weathering regions

Peltier, an American physicist and climatologist, attempted to predict the type and rate of weathering at any given place in the world from its mean annual temperature and mean annual rainfall (Figure 2.10c). It should be realised that mechanical and chemical weathering usually operate together at the same time and at the same place, but it is likely that in each situation one type or the other will be the more significant.

Figure 2.9

Acid rain damage to stone statues, Exeter Cathedral

Figure 2.10

Climatic controls on weathering (*after* Peltier)

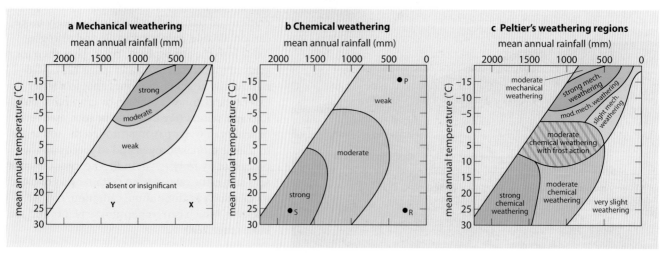

One type of model (Framework 12, page 352) widely adopted by geographers to help explain phenomena is the **system**. The system is a method of analysing relationships within a unit and consists of a number of components between which there are linkages. The model is usually illustrated schematically as a flow diagram.

Systems may be described in three ways:

- **Isolated:** there is no input or output of energy or matter. Some suggest the universe is the sole example of this type; others claim the idea is not applicable in geography.

- **Closed:** there is input, transfer and output of energy but not of matter (or mass).

- **Open:** most environmental systems are open and there are inputs and outputs of both energy and matter.

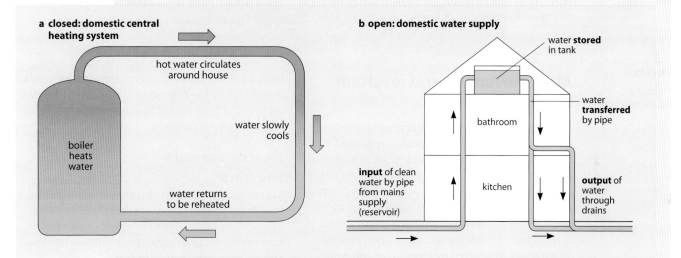

a closed: domestic central heating system

hot water circulates around house

boiler heats water

water slowly cools

water returns to be reheated

b open: domestic water supply

water **stored** in tank

water **transferred** by pipe

bathroom

input of clean water by pipe from mains supply (reservoir)

kitchen

output of water through drains

Figure 2.11

Closed and open systems in the house

Examples of the systems approach used and referred to in this book (chapter number is given in brackets):

Geomorphological	Climate, soils and vegetation	Human and economic
Slopes (2)	Atmosphere energy budget (9)	Population change (13)
Drainage basins (3)	Hydrological cycle (9)	Farming (16)
Glaciers (4)	Soils (10)	Industry (19)
	Ecosystems (11)	
	Nutrient cycle (12)	

When opposing forces, or inputs and outputs, are balanced, the system is said to be in a state of **dynamic equilibrium**. If one element in the system changes because of some outside influence, then it upsets this equilibrium and affects the other components. For example, equilibrium is upset when:

- prolonged heavy rainfall causes an increase in the discharge and velocity of a river or a lowering of base level (page 81), both of which lead to an increase in the rate of erosion

- an increase in carbon dioxide into the atmosphere causes global temperatures to rise (global warming, Case Study 9)

- drought affects the carrying capacity of animals (or people) grazing (living) in an area as the water shortage reduces the availability of grass (food supplies) (page 378)

- an increase in the number of tourists to places of scenic attraction harms the environment (especially where it is fragile) that was the original source of the attraction (page 591).

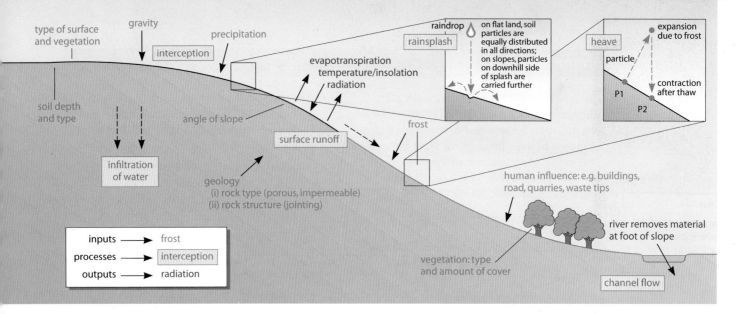

inputs	→	frost
processes	→	interception
outputs	→	radiation

Figure 2.12

The slope as a
dynamic open system

Mass movement and resultant landforms

The term **mass movement** describes all downhill movements of weathered material (**regolith**), including soil, loose stones and rocks, in response to gravity. However, it excludes movements where the material is carried by ice, water or wind. When gravitational forces exceed forces of resistance, slope failure occurs and material starts to move downwards. A slope is a **dynamic open system** (Framework 3) affected by biotic, climatic, gravitational, groundwater and tectonic inputs which vary in scale and time. The amount, rate and type of movement depend upon the degree of slope failure (Figure 2.12).

Although by definition mass movement refers only to the movement downhill of material under the force of gravity, in reality water is usually present and assists the process. When Carson and Kirkby (1972) attempted to group mass movements, they used the speed of movement and the amount of moisture present as a basis for distinguishing between the various types (Figure 2.13). The following classification is based on speed of flow related to moisture content and angle of slope (Framework 7, page 167).

Slow movements
Soil creep

This is the slowest of downhill movements and is difficult to measure as it takes place at a rate of less than 1 cm a year. However, unlike faster movements, it is an almost continuous process. Soil creep occurs mainly in humid climates where there is a vegetation cover. There are two major causes of creep, both resulting from repeated expansion and contraction.

1 **Wet–dry periods** During times of heavy rainfall, moisture increases the volume and weight of the soil, causing expansion and allowing the regolith to move downhill under gravity. In a subsequent dry period, the soil will dry out and then contract, especially if it is clay. An extreme case of contraction in clays occurred in south-east England during the 1976 drought when buildings sited on almost imperceptible slopes suffered major structural damage.

2 **Freeze–thaw** When the regolith freezes, the presence of ice crystals increases the volume of the soil by 9 per cent. As the soil expands, particles are lifted at right-angles to the slope in a process called **heave** (Figure 2.12 and page 132). When the ground later thaws and the regolith contracts, these particles fall back vertically under the influence of gravity and so move downslope.

Figure 2.13

A classification of
mass movement
processes (*after*
Carson and Kirkby,
1972)

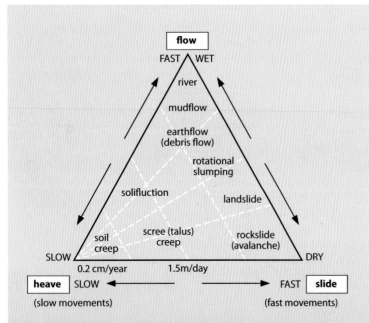

Figure 2.14

Terracettes in Wharfedale, Yorkshire Dales

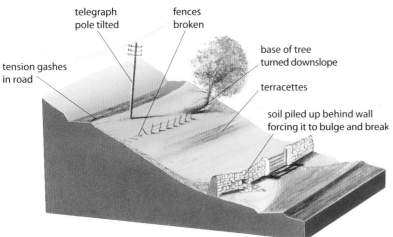

Figure 2.15

The effects of soil creep

telegraph pole tilted

fences broken

base of tree turned downslope

tension gashes in road

terracettes

soil piled up behind wall forcing it to bulge and break

periglacial conditions (Chapter 5) where vegetation cover is limited. During the winter season, both the bedrock and regolith are frozen. In summer, the surface layer thaws but the underlying layer remains frozen and acts like impermeable rock. Because surface meltwater cannot infiltrate downwards and temperatures are too low for much effective evaporation, any topsoil will soon become saturated and will flow as an **active layer** over the frozen subsoil and rock (page 131). This process produces **solifluction sheet** or **lobes** (Figure 5.12), rounded, tongue-like features reaching up to 50 m in width, and **head**, a mixture of sand and clay formed in valleys and at the foot of sea cliffs (Figure 5.13). Solifluction was widespread in southern Britain during the Pleistocene ice age; covered most of Britain following the Pleistocene; and continues to take place in the Scottish Highlands today.

Flow movements

Earthflows

When the regolith on slopes of 5–15° becomes saturated with water, it begins to flow downhill at a rate varying between 1 and 15 km per year. The movement of material may produce short **flow tracks** and small bulging lobes or tongues, yet may not be fast enough to break the vegetation.

Mudflows

These are more rapid movements, occurring on steeper slopes, and exceeding 1 km/hr. When Nevado del Ruiz erupted in Colombia in 1985, the resultant mudflow reached the town of Armero at an estimated speed of over 40 km/hr (Case Study 2A). Mudflows are most likely to occur following periods of intensive rainfall, when both volume and weight are added to the soil giving it a higher water content than an earthflow. Mudflows may result from a combination of several factors (Figure 2.16).

Soil creep usually occurs on slopes of about 5° and produces **terracettes** (Figure 2.14). These are step-like features, often 20–50 cm in height, which develop as the vegetation is stretched and torn: they are often used and accentuated by grazing animals, especially sheep. The effects of soil creep are shown in Figure 2.15.

Solifluction

This process, meaning 'soil flow', is a slightly faster movement usually averaging between 5 cm and 1 m a year. It often takes place under

Figure 2.16

Fieldsketch showing the causes of a mudflow, Glen Rosa, Arran

granite slabs (impermeable): bare rock results in rapid runoff (influence of geology and vegetation)

BEINN A CHLIABHAIN

unusually heavy rainfall for 48 hours: extra moisture caused hillside to swell, and added extra weight (influence of climate)

scar of an earlier mudflow

scar

steep valley sides of over 40° resulting from a valley glacier (influence of slope)

thin soils overlying impermeable bedrock – easily saturated (influence of soils)

slope decreasing to 10°

flowtrack over 100 m in length: material presumably flowed downhill at over 10 m/sec

GLEN ROSA

lobe or debris fan: soil, loose rock, large boulders deposited when mud lost its momentum

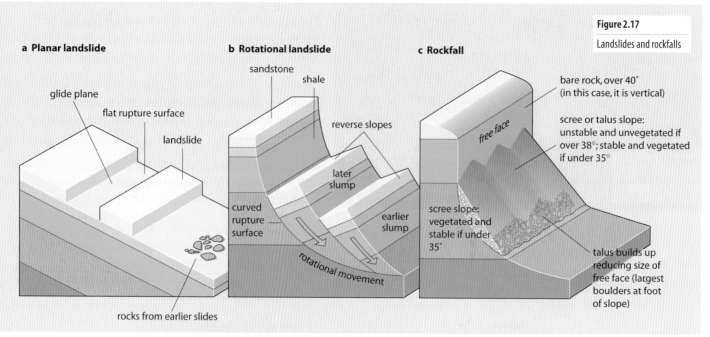

Figure 2.17

Landslides and rockfalls

a Planar landslide

glide plane

flat rupture surface

landslide

rocks from earlier slides

b Rotational landslide

sandstone

shale

reverse slopes

curved rupture surface

later slump

earlier slump

rotational movement

c Rockfall

bare rock, over 40°
(in this case, it is vertical)

scree or talus slope:
unstable and unvegetated if
over 38°; stable and vegetated
if under 35°

free face

scree slope:
vegetated and
stable if under
35°

talus builds up
reducing size of
free face (largest
boulders at foot
of slope)

Rapid movements
Slides

The fundamental difference between slides and flows is that flows suffer internal derangement whilst, in contrast, slides move 'en masse' and are not affected by internal derangement. Rocks that are jointed or have bedding planes roughly parallel to the angle of slope are particularly susceptible to landslides. Slides may be planar or rotational (Figure 2.17a and b). In a planar slide, the weathered rock moves downhill leaving behind it a flat rupture surface (Figure 2.17a). Where rotational movement occurs, a process sometimes referred to as **slumping**, a curved rupture surface is produced (Figure 2.17b). Rotational movement can occur in areas of homogeneous rock, but is more likely where softer materials (clay or sands) overlie more resistant or impermeable rock (limestone or granite). Slides are common in many coastal areas of southern and eastern England. In Figure 2.18, the cliffs, composed of glacial deposits, are retreating rapidly due to frequent slides. The slumped material can be seen at the foot of the cliff.

Very rapid movements
Rockfalls

These are spontaneous, though relatively rare, debris movements on slopes that exceed 40°. They may result from extreme physical or chemical weathering in mountains, pressure release, storm-wave action on sea cliffs, or earthquakes. Material, once broken from the surface, will either bounce or fall vertically to form scree, or talus, at the foot of a slope (Figures 2.17c and 2.19).

Figure 2.18

Landslides on the Norfolk coast

Figure 2.19

Rockfalls in the crater of Vesuvius, Italy

Petropolis

The town of Petropolis, named after a former king of Brazil, lies in the Serro do Mar Mountains some 60 km north of Rio de Janeiro (Figure 2.21). Today, with a population of 300 000, it is one of Rio's two main mountain resorts to which people escape in summer to avoid the heat and humidity of the coast. But the steep-sided mountains can also prove to be a hazard, as in 2001 when 50 people were killed in a series of landslides (Figure 2.20).

As shown below, December of that year was an exceptionally wet month for Petropolis. The result was a series of more than 20 significant landsides, 14 of which were between them responsible for the 50 fatalities.

Period	Rainfall
1 to 16 December	up to 250 mm
17 to 23 December	up to 125 mm
24 December (in 12 hours)	up to 200 mm

The area, with its steep hillsides and heavy seasonal rainfall, is prone to natural landslides but investigations following this event suggested that the two main causes, on this occasion at least, resulted from human activity:

1   The construction of poor-quality, unauthorised building: many of the shanty settlements had been built on steep hillsides, often where the slope was over 45° and in places even up to 80°.
2   The failure to provide rainwater drainage channels: such drains could have taken away some of the excess surface water and so reduced the hazard risk.

Of the 50 deaths, 24 were attributed to unauthorised settlements and 22 to the lack of drainage channels.

Rio de Janeiro

Rio de Janeiro experiences the same problems of mass movement, but on an even larger scale, as Petropolis. Figure 15.34 shows one of Rio's many favelas (shanty settlements) that have been built on the steep hillsides. One flash flood in 1988 led to mudslides which carried away many of the flimsy houses that had probably been built from waste materials such as wood, corrugated iron and broken bricks. The mudslides were responsible for the deaths of more than 200 people.

Figure 2.21

The town of Petropolis has spread up steep hillsides from the valley bottom

Figure 2.20

A landslide in Petropolis, 2001

Development of slopes

Slope development is the result of the interaction of several factors. Rock structure and lithology, soil, climate, vegetation and human activity are probably the most significant. All are influenced by the time over which the processes operate. Slopes are an integral part of the drainage basin system (Chapter 3) as they provide water and sediment for the river channel.

The effects of rock structure and lithology

- Areas of bare rock are vulnerable to mechanical weathering (e.g. frost shattering) and some chemical weathering processes.
- Areas of alternating harder/more resistant rocks and softer/less resistant rocks are more likely to experience movement, e.g. clays on limestones (Vaiont Dam, Case Study 2B).

- An impervious underlying rock will cause the topsoil to become saturated more quickly, e.g. glacial deposits overlying granite.
- Steep gradients are more likely to suffer slope failure than gentler ones. In Britain, especially in lowland areas, most slopes are under 5° and few are over 40°.
- Failure is also likely on slopes where the equilibrium (balance) of the system (Framework 3, page 45), has been disturbed, e.g. a glaciated valley.
- The presence of joints, cracks and bedding planes can allow increased water content and so lead to sliding (Vaiont Dam, Case Study 2B).
- Earthquakes (Mount Huascaran in Peru) and volcanic eruptions (Nevado del Ruiz in Colombia) can cause extreme slope movements (Case Study 2A).

Figure 2.22

The effect of pore-water pressure and capillary action on soil movement

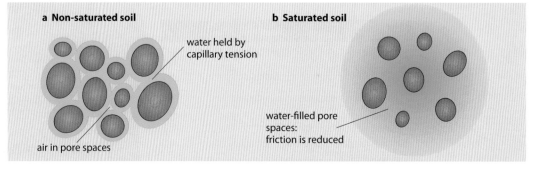

a Non-saturated soil

water held by capillary tension

air in pore spaces

b Saturated soil

water-filled pore spaces: friction is reduced

Soil

- Thin soils tend to be more unstable. As they can support only limited vegetation, there are fewer roots to bind the soil together.
- Unconsolidated sands have lower internal cohesion than clays.
- A porous soil, e.g. sand, is less likely to become saturated than one that is impermeable, e.g. clay.
- In a non-saturated soil (Figure 2.22a), the surface tension of the water tends to draw particles together. This increases cohesion and reduces soil movement. In a saturated soil (Figure 2.22b), the pore water pressure (page 267) forces the particles apart, reducing friction and causing soil movement.

Climate

- Heavy rain and meltwater both add volume and weight to the soil.
- Heavy rain increases the erosive power of any river at the base of a slope and so, by removing material, makes that slope less stable.
- Areas with freeze–thaw or wet–dry periods are subjected to alternating expansion–contraction of the soil.

- Heavy snowfall adds weight and is thus conducive to rapid movements, e.g. avalanches, Case Study 4a.

Vegetation

- A lack of vegetation means that there are fewer roots to bind the soil together.
- Sparse vegetation cover will encourage surface runoff as precipitation is not intercepted (page 59).

Human influence

- Deforestation increases (afforestation decreases) the rate of slope movement.
- Road construction or quarrying at the foot of slopes upsets the equilibrium, e.g. during the building of the M5 in the Bristol area.
- Slope development processes may be accentuated either by building on steep slopes (Hong Kong and Rio de Janeiro, Case Study 2B) or by using them to deposit industrial or mining waste (Aberfan, Case Study 2B).
- The vibration caused by heavy traffic can destabilise slopes (Mam Tor, Derbyshire).
- The grazing of animals and ploughing help loosen soil and remove the protective vegetation cover.

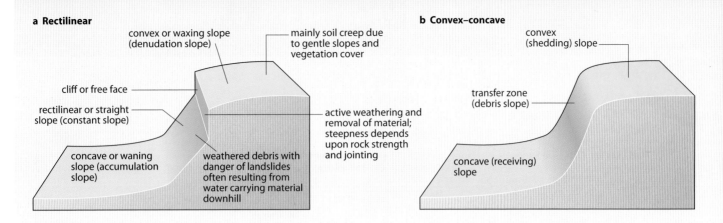

a Rectilinear

convex or waxing slope (denudation slope)

mainly soil creep due to gentle slopes and vegetation cover

cliff or free face

rectilinear or straight slope (constant slope)

concave or waning slope (accumulation slope)

weathered debris with danger of landslides often resulting from water carrying material downhill

active weathering and removal of material; steepness depends upon rock strength and jointing

b Convex–concave

convex (shedding) slope

transfer zone (debris slope)

concave (receiving) slope

Figure 2.23

Slope element models

Slope elements

Two models try to show the shape and form of a typical slope. The first, Figure 2.23a, is more widely used than the second (Figure 2.23b) – although, in this author's view, the first is less easily seen in the British landscape. Regardless of which model is used, confusion unfortunately arises because of the variation in nomenclature used to describe the different facets of the slope.

In reality, few slopes are likely to match up perfectly with either model, and each individual slope is likely to show more elements than those in Figure 2.23.

Slope development through time

Figure 2.24

Slope development theories

How slopes have developed over time is one of the more controversial topics in geomorphology.

This is partly due to the time needed for slopes to evolve and partly due to the variety of combinations of processes acting upon slopes in various parts of the world. Slope development in different environments has led to three divergent theories being proposed: **slope decline**, **slope replacement** and **parallel retreat**. Figure 2.24 is a summary of these theories.

None of the theories of slope development can be universally accepted, although each may have local relevance in the context of the climate and geology (structure) of a specific area. At the same time, two different climates or processes may produce the same type of slope, e.g. cliff retreat due to sea action in a humid climate or to weathering in a semi-arid climate.

	Slope decline (W.M. Davis, 1899)	**Slope replacement** (W. Penck, 1924)	**Parallel retreat** (L.C. King, 1948, 1957)
Region of study	Theory based on slopes in what was to Davis a normal climate, north-west Europe and north-east USA.	Conclusions drawn from evidence of slopes in the Alps and Andes.	Based on slopes in South Africa.
Climate	Humid climates.	Tectonic areas.	Semi-arid landscapes. Sea cliffs with wave-cut platforms.
Description of slope	Steepest slopes at beginning of process with a progressively decreasing angle in time to give a convex upper slope and a concave lower slope.	The maximum angle decreases as the gentler lower slopes erode back to replace the steeper ones giving a concave central portion to the slope.	The maximum angle remains constant as do all slope facets apart from the lower one which increases in concavity.
	slope decline stage 4 / stage 3 / stage 2 / stage 1 / concave curve / watershed worn down / convex curve / peneplain By stage 4 land has been worn down into a convex-concave slope	**slope replacement** stage 3 / stage 2 / stage 1 / A / A / A / C / B / C / B / C / B talus-scree slope B will replace slope A; slope C will eventually replace slope B	**slope retreat** stage 4 / stage 2 / stage 3 / stage 1 / convex / free face concave debris slope pediment (can be removed by flash floods)
Changes over time	Assumed a rapid uplift of land with an immediate onset of denudation. The uplifted land would undergo a cycle of erosion where slopes were initially made steeper by vertical erosion by rivers but later became less steep (slope decline) until the land was almost flat (peneplain).	Assumed landscape started with a straight rock slope with equal weathering overall. As scree (talus) collected at the foot of the cliff it gave a gentler slope which, as the scree grew, replaced the original one.	Assumed that slopes had two facets – a gently concave lower slope or pediment and a steeper upper slope (scarp). Weathering caused the parallel retreat of the scarp slope allowing the pediment to extend in size.

A Natural causes

All slopes are affected by gravity and, consequently, by one or more of the several mass movement processes by which weathered material is transported downhill (pages 46–48). Where slopes are gentle (about 5°), the movement of material is slow and has relatively little effect on property, life or human activity. As slope angles increase, however, so too do the rate and frequency of slope movement and the risk of sudden slope failure. Slope failure, occurring in the form of either mudflows or landslides, is a natural event. When this failure occurs in densely populated areas, it becomes a potentially dangerous natural hazard (Framework 2, page 31). Three examples of how slope failure caused by natural events can cause serious loss of property and life (Figure 2.25) are:

  (i) earthquakes
  (ii) volcanic activity
  (iii) excessive rainfall.

(i) Earthquakes – avalanches and rockfalls (Peru 1970)

In 1970 an offshore earthquake measuring 7.7 on the Richter scale shook parts of Peru to the north of its capital, Lima. The shock waves loosened a mass of unstable ice and snow near the summit of Huascaran, the country's highest peak (6768 m). The falling ice and snow formed a huge avalanche which rushed downhill, falling 3000 m into the Rio Santo Valley, collecting rocks and boulders en route. In its path stood the town of Yungay with a population of 20 000.

Estimates suggest that the avalanche was travelling at a speed of 480 km/hr when it hit the settlement. It took rescue workers three days to reach the town. Once there, they found very few survivors and only the tops of several 30 m palm trees, which marked the location of the former town square (Figures 2.25 and 2.26).

Figure 2.25

Sites of some recent hazardous events in South and Central America

Figure 2.26

The site of Yungay after the avalanche

(ii) Volcanic eruptions – mudflows (Colombia 1985)

The Colombian volcano of Nevado del Ruiz had not erupted since 1595 until, in November 1985, it showed signs of activity by emitting gas and steam. As an increasing amount of magma welled upwards towards the crater, the whole peak must have become warmer, as was made evident by the increased melting of ice and snow around its summit. A mudflow, 20 m in height, which travelled 27 km down the Lagunillas Valley, proved an advance warning that went unheeded. Ice and snow continued to melt until, on 13 November, there was a major eruption. Although this eruption was small in comparison with other eruptions such as Mount St Helens, the lava, ash and hot rocks ejected were sufficient to melt the remaining ice and snow, releasing a tremendous volume of meltwater. This meltwater, swelled by torrential rain (often associated with volcanic eruptions), raced down the Lagunillas Valley collecting with it large amounts of ash deposited from previous eruptions. The resultant mud tidal wave (a lahar), estimated to have been 30 m in height, travelled down the valley at over 80 km/hr.

Some 50 km from the crater, the mudflow emerged onto more open ground on which was situated the town of Armero. The time was 2300 hours when the mudflow struck, and most of the 22 000 inhabitants had already gone to bed. The few survivors claimed that the first onrush of muddy water was ice-cold, but became increasingly warmer. By morning a layer of mud, up to 8 m deep, covered Armero and the surrounding area (Figures 2.25 and 2.27). The death toll was put at 21 000, making this the worst single natural disaster ever to have affected people in the western hemisphere.

(iii) Heavy rainfall – Hurricane Stan (Guatemala 2005)

Hurricane Stan swept across Central America during September 2005. Although by hurricane standards it was not the strongest, it proved particularly lethal because it struck a region where most people lived in flimsy shanty dwellings constructed around, or at the foot of, steep mountainsides. As is often the case with hurricanes, it was not the strength of the winds that was to cause

Figure 2.27
Armero, Colombia

so many deaths, but rather the effects of the torrential rainfall.

The rainwater collected soil and other material as it rushed down the mountain slopes creating a mudflow 15 m deep that engulfed the town of Panabaj (Figures 2.25 and 2.28). The devastation was so complete that the authorities and relief workers soon abandoned efforts to retrieve survivors, or even bodies, and declared the area a mass grave. In all, 1400 people disappeared and all that was left of the town were the tops of the taller trees. The handful of lucky survivors described how they were awoken by rumblings from the mountainsides, and managed to escape because they were nearer to the edges of the mudflow.

Raging rivers destroyed bridges and made roads impassable, so the hard-pressed authorities had to struggle to airlift in food, drinking water and emergency supplies.

Figure 2.28

The view across Lake Atitlan in Guatemala to the volcanic peaks on the far shore. Beside the lake, which is a caldera (page 25), are several long-established Mayan settlements and a few modern tourist resorts. One Mayan town was Panabaj

B Human mismanagement

The probability of slope failure in populated areas is often increased by thoughtless planning, or a total lack of it, or where human activity exerts too much pressure upon the land available. Three examples of how slope instability and the risk of slope failure may be increased by human activity are when land is used for:

(i) building dams to create reservoirs
(ii) the extraction of a natural resource or the dumping of waste material
(iii) rapid urbanisation.

(i) Building dams to create reservoirs (Italy 1963)

The Vaiont Dam, built in the Italian Alps, was completed in 1960. The dam, the third highest in the world at that time, was built in a narrow valley with steep sides consisting of alternate layers of clays and limestone (Figure 2.29), and where landslides were not uncommon. Down the valley were several hamlets and the small town of Longarone.

Heavy rain in October 1963 saturated the clay. Just before midnight on 9 October, a landslide of rocks, clay, mud and vegetation slid over the harder beds of limestone and into the reservoir. The dam itself stood, but a wave of water spilled over the lip creating a towering wall of water which swept down the valley. Longarone was virtually destroyed. The final death toll was put at almost 1900, although several bodies were never recovered. Debris from the landslide filled in almost two-thirds of the lake. A court of enquiry concluded that the site was geologically unstable and that even during construction many smaller landslides had occurred. The dam was closed.

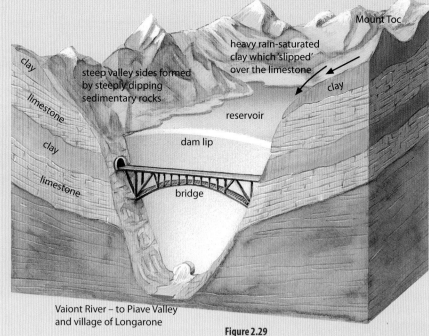

Vaiont River – to Piave Valley and village of Longarone

Figure 2.29

The Vaiont Dam

Figure 2.30

Aberfan immediately after the mudflow

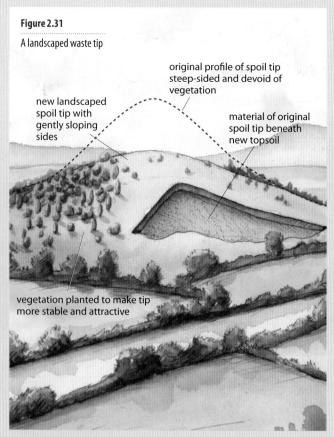

Figure 2.31

A landscaped waste tip

original profile of spoil tip steep-sided and devoid of vegetation

new landscaped spoil tip with gently sloping sides

material of original spoil tip beneath new topsoil

vegetation planted to make tip more stable and attractive

(ii) Dumping waste material (Aberfan 1966)

Aberfan, like many other settlements in the South Wales valleys, grew up around its colliery. However, the valley floors were rarely wide enough to store the coal waste and so it became common practice to tip it high above the towns on the steep valley sides. At Aberfan, the spoil tips were on slopes of 25°, over 200 m above the town and, unknowingly, on a line of springs. Water from these springs added weight to the waste heaps, which reduced their internal cohesion. Following a wet October in 1966 and a night of heavy rain, slope failure resulted in the waste material suddenly and rapidly moving downhill. The resultant mudflow, estimated to contain over 100 000 m³ of material, engulfed part of the town which included the local junior school (Figure 2.30). The time was just after 0900 hours on 21 October, soon after lessons in the school had begun. Of the 147 deaths in Aberfan that morning, 116 were children and five their teachers.

Since then, the colliery has closed and, as elsewhere in the former coal-mining valleys, the potentially dangerous waste tips have been lowered, regraded and landscaped to try to prevent any occurrence of a similar event (Figure 2.31).

(iii) Urbanisation (Hong Kong 1957 to 2007)

Many parts of the world, especially in economically less developed countries, are experiencing rapid urbanisation (page 418). As most of the best sites for residential development have long since been used, it means that newcomers to a city are forced to live on land previously considered unusable (e.g. flood-prone valleys in Nairobi – Places 58, page 444), or unsafe (e.g. steep hillsides in Caracas 1999, and Rio de Janeiro – Places 57, page 443).

In Hong Kong, landslips have been responsible for 430 deaths since 1957 (Figure 2.32). Most landslips during this time have been attributed to two factors: the inadequacies of hillside construction works in the last 50 years, and deficiencies in maintaining slopes once they are utilised (Figure 2.33). In 1966, torrential rainstorms triggered massive landslides which killed 64 people, made 2500 homeless and

caused 8000 to be evacuated. In 1976, a major landslide led to 22 deaths. The consequence of this was the setting up, in 1977, of the Geotechnical Engineering Office (GEO). GEO's main functions were:

- to investigate slopes for potential risk and to take preventive measures
- to control geotechnic aspects of new buildings and roads
- to promote slope maintenance by owners
- to undertake landslide warning and emergency services
- to advise on land-use plans to minimise public risk.

In 1997, most of Hong Kong experienced over 300 mm of rain in 24 hours. At the centre of the storm, 110 mm fell in one hour and 800 mm in the day. Resultant landslips caused the death of one person, injuries to eight people, the disruption of the Kowloon–Guangzhou railway (Places 106, page 640) and the closure of a six-lane highway for several hours. These losses and disruptions were, however, relatively minor, because the community had learned to cope better with the landslip hazard. Indeed the Hong Kong authorities now collaborate with their counterparts in other cities in Asia and South America with similar climatic and topographic characteristics, and where economic and social development is creating an unacceptable level of landslip risk.

The success of GEO can be seen by the decrease in the number of deaths (Figure 2.32).

Figure 2.32

Number of landslip fatalities in Hong Kong, 1957–2007

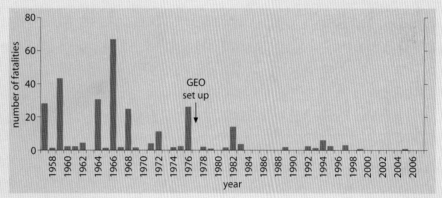

Figure 2.33

Consequences of a landslip in Hong Kong

Further reference

Carson, M.A. and Kirby, N.J. (1972) *Hillslope Form and Process,* Cambridge University Press.

Goudie, A.S. (2001) *The Nature of the Environment,* WileyBlackwell.

Guerra, T. *et al* (2007) 'Mass movement in Petropolis, Brazil' in *Geography Review* Vol 20 No 4 (March).

Trudgill, S.T. (1986) *Weathering and Erosion,* Heinemann.

Geoweb, landslides:
www.georesources.co.uk/edexunit6.htm

Glossary of related terminology:
www.scottishgeology.com/glossary/glossary.html

Slope weathering:
www.bgrg.org – search for 'slope weathering'
http://earthsci.org/Flooding/unit3/u3-02-03.html
www.georesources.co.uk/edexunit6.htm

Questions & Activities

Activities

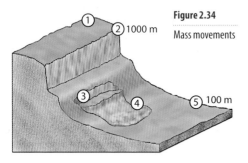

Figure 2.34

Mass movements

1000 m

100 m

1   a What is meant by the following terms?

    i   slope element   *(1 mark)*

    ii   mass wasting   *(2 marks)*

    iii scree   *(2 marks)*

    iv   terracette   *(2 marks)*

  b Choose **three** of the features marked 1 to 5 on Figure 2.34. Describe the appearance of each of the features you have chosen.   *(6 marks)*

  c For each of your chosen features, explain the role of mass wasting in its formation.   *(12 marks)*

2 Study Figure 2.35 and answer the following questions.

  a i Explain the meaning of each of the following slope movement terms:

       earth flow; mud flow; slide; rock fall.   *(6 marks)*

    ii Name **two** types of slope movement it is possible to see in the photograph. State where they can be found.   *(4 marks)*

    iii Identify **two** ways in which people have tried to protect slopes in this photograph. For each one suggest how it is intended to work.   *(6 marks)*

  b Had the slope movement finished when this photograph was taken? Suggest reasons for your answer.   *(4 marks)*

  c Should cliffs, such as the one in the photograph, be protected? Give reasons for your answer.   *(5 marks)*

Figure 2.35

Holbeck Hall Hotel, Scarborough

3 Use Case Study 2B (iii) on Hong Kong (page 55) to answer the following questions.

  a Describe the physical features of the hillside shown in the photograph.   *(3 marks)*

  b Why have people settled on this hillside?   *(3 marks)*

  c Why is a hillside, such as the one in the photograph, in danger of rapid mass movement even without human activities?   *(7 marks)*

  d Give **two** examples of human activities which increase the danger of rapid mass movement on such slopes. Explain how they increase the danger.   *(6 marks)*

  e The heavy rainfall in 1997 was an extreme climatic event but it created relatively little damage. Explain **one** way in which authorities such as those in Hong Kong are trying to manage the problems caused by the physical environment in which they operate.   *(6 marks)*

Exam practice: basic structured questions

4 **a** Define the term 'weathering'. *(2 marks)*

b Choose one type of **mechanical weathering**.

   i Making use of diagrams, explain the processes involved in the type of weathering. *(4 marks)*

   ii Describe the landscape features which result from the weathering type you have chosen. *(4 marks)*

c Choose any **one** climatic region and identify the type of **chemical weathering** that will dominate the area. Explain why this type of chemical weathering will be dominant. *(8 marks)*

d Human activity can influence the rate of weathering that occurs in an area. With the aid of specific examples, explain how human activity influences the rate of weathering. *(7 marks)*

Exam practice: structured questions

5 **a** **i** Study Figure 2.36. Match each of the following types of slope movement with one of the labels on the graph numbered 1 to 5: earth/mudflow; solifluction; rockfall; slide; soil creep. *(5 marks)*

   ii For any **two** of the flow movements above, explain how the process occurs and describe the landform shape that results. *(10 marks)*

b Use examples of **two** types of rural land use you have studied to explain how people in rural areas try to manage slopes to reduce the downslope movement of soil. *(10 marks)*

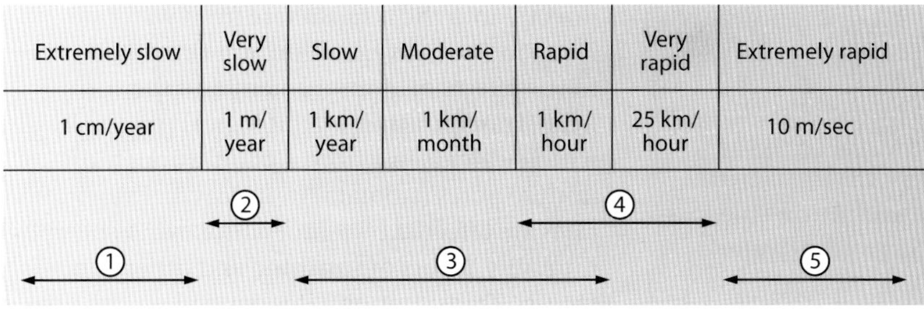

6 **a** Study the photograph of Holbeck Hall Hotel (Figure 2.35).

   i Draw an annotated diagram or sketch map **only** to illustrate the landscape features of the slopes. *(8 marks)*

   ii Explain what has happened to these slopes and suggest why it has occurred. *(8 marks)*

b Making good use of examples, explain how human activities can increase the stability of some slopes and destabilise other slopes. *(9 marks)*

7 Choose a drainage basin that you have studied.

a Describe and suggest reasons for the variation in slope types that exist within the drainage basin. *(10 marks)*

b For any **one** slope, identify and explain changes that are likely to affect the slope in the future. *(8 marks)*

c Suggest how human activity can influence the rate of change and shape of slopes. *(7 marks)*

Figure 2.36

Speed of movement of mass movements

Extremely slow	Very slow	Slow	Moderate	Rapid	Very rapid	Extremely rapid
1 cm/year	1 m/year	1 km/year	1 km/month	1 km/hour	25 km/hour	10 m/sec

Exam practice: essays

8 'A range of processes, which differ in contrasting environments, affect slope shapes.' Discuss this statement with reference to slopes you have studied. In your answer you should refer to:
- the variation of slope elements in different environments
- the variation in importance of types of weathering process in different environments
- the interaction of factors within environments to create slopes. *(25 marks)*

9 With reference to case studies from a range of environments, explain how an understanding of natural slope processes can be used in planning urban developments. *(25 marks)*

Formulae for chemical weathering processes	
Oxidation	$4\,FeO + O_2 \rightarrow 2Fe_2O_3$ (ferrous oxide + oxygen → ferric oxide)
Hydrolosis	Formula varies depending on rock type involved. For the hydrolosis of feldspar/granite to kaolin, this is a common example: $K_2O, Al_2O_3, 6SiO_2 + H_2O \rightarrow Al_2O_3, 2SiO_2, 2H_2O$ (feldspar + water → kaolin)
Hydration	$CaSO_4 + 2H_2O \rightarrow CaSO_4 2H_2O$ (anhydrite + water → gypsum)
Carbonation	This process is in two stages: $H_2O + CO_2 \rightarrow H_2CO_3$ (water + carbon dioxide → carbonic acid) $CaCO_3 + H_2CO_3 \rightarrow Ca(HCO_3)_2$ (calcium carbonate + carbonic acid → calcium bicarbonate)
Acid rain	$2SO_2 + O_2 + 2H_2O \rightarrow 2H_2SO_4$ (sulphur dioxide + oxygen + water → weak sulphuric acid)

Drainage basins and rivers

'All the rivers run into the sea; yet the sea is not full; unto the place from whence the rivers come, thither they return again.'

The Bible, Ecclesiastes 1:7

A **drainage basin** is an area of land drained by a river and its tributaries. Its boundary is marked by a ridge of high land beyond which any precipitation will drain into adjacent basins. This boundary is called a **watershed**.

A drainage basin may be described as an **open system** and it forms part of the hydrological or water cycle. If a drainage basin is viewed as a system (Framework 3, page 45) then its characteristics are:

- **inputs** in the form of precipitation (rain and snow)
- **outputs** where the water is lost from the system either by the river carrying it to the sea or through **evapotranspiration** (the loss of water directly from the ground, water surfaces and vegetation).

Within this system, some of the water:

- is **stored** in lakes and/or in the soil, or
- passes through a series of **transfers** or flows, e.g. infiltration, percolation, throughflow.

Elements of the drainage basin system

Figure 3.1

The drainage basin as an open system

Figure 3.1 shows the drainage basin system as it is likely to operate in a temperate humid region such as the British Isles.

Precipitation

This forms the major input into the system, though amounts vary over time and space. As a rule, the greater the intensity of a storm, the shorter its duration. Convectional thunderstorms are short, heavy and may be confined to small areas, whereas the passing of a warm front of a depression (page 231) will give a longer period of more steady rainfall extending over the entire basin.

Evapotranspiration

The two components of evapotranspiration are outputs from the system. **Evaporation** is the physical process by which moisture is lost directly into the atmosphere from water surfaces, including vegetation and the soil, due to the effects of air movement and the sun's heat. **Transpiration** is a biological process by which water is lost from a plant through the minute pores (stomata) in its leaves. Evaporation rates are affected by temperature, wind speed, humidity, hours of sunshine and other climatic factors. Transpiration rates depend on the time of year, the type and amount of vegetation, the availability of moisture and the length of the growing season. It is also possible to distinguish between the potential and the actual evapotranspiration of an area. For example, in deserts there is a high **potential evapotranspiration** because the amount of moisture that could be lost is greater than the amount of water actually available. On the other hand, in Britain the amount

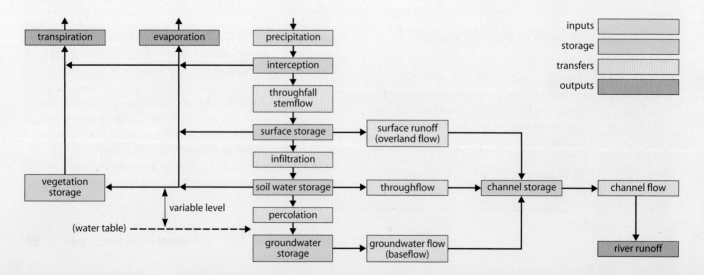

of water available for evapotranspiration nearly always exceeds the amount which actually takes place, hence the term **actual evapotranspiration**. In other words, transpiration is limited by the availability of water in the soil.

Interception

The first raindrops of a rainfall event will fall on vegetation which shelters the underlying ground. This is called **interception storage**. It is greater in a woodland area or where tree crops are grown than on grass or arable land. If the precipitation is light and of short duration, much of the water may never reach the ground and it may be quickly lost from the system through evaporation. Estimates suggest that in a woodland area up to 30 per cent of the precipitation may be lost through interception, which helps to explain why soil erosion is limited in forests. According to Newson (1975), 'Interception is a dynamic process of filling and emptying a shallow store (about 2 mm in most UK trees). The emptying occurs because evaporation is very efficient for small raindrops held on tree surfaces.' In an area of deciduous trees, both interception and evapotranspiration rates will be higher in summer, although the two processes do not occur simultaneously.

If a rainfall event persists, then water begins to reach the ground by three possible routes: dropping off the leaves, or **throughfall**; flowing down the trunk, or **stemflow**; and by undergoing **secondary interception** by undergrowth. Following a warm, dry spell in summer, the ground may be hard; at the start of a rainfall event water will then lie on the surface (**surface storage**) until the upper layers become sufficiently moistened to allow it to soak slowly downwards. If precipitation is very heavy initially, or if the soil becomes saturated, then excess water will flow over the surface, a transfer known as **surface runoff** (or, in Horton's term, **overland flow**) (Figure 3.2).

Infiltration

In most environments, overland flow is relatively rare except in urban areas – which have impermeable coverings of tarmac and concrete – or during exceptionally heavy storms. Soil will gradually admit water from the surface, if the supply rate is moderate, allowing it slowly to infiltrate vertically through the pores in the soil. The maximum rate at which water can pass through the soil is called its **infiltration capacity** and is expressed in mm/hr. The rate of infiltration depends upon the amount of water already in the soil (**antecedent precipitation**), the **porosity** (Figure 8.2) and structure of the soil, the nature of the soil surface (e.g. crusted, cracked, ploughed), and the type, amount and seasonal changes in vegetation cover. Some of the water will flow laterally as **throughflow**. During drier periods, some water may be drawn up towards the surface by **capillary action**.

Percolation

As water reaches the underlying soil or rock layers, which tend to be more compact, its progress is slowed. This constant movement, called percolation, creates **groundwater storage**. Water eventually collects above an impermeable rock layer, or it may fill all pore spaces, creating a **zone of saturation**. The upper boundary of the saturated material, i.e. the upper surface of the groundwater layer, is known as the **water table**. Water may then be slowly transferred laterally as **groundwater flow** or **baseflow**. Except in areas of Carboniferous limestone, groundwater levels usually respond slowly to surface storms or short periods of drought (Figure 3.5). During a lengthy dry period, some of the groundwater store will be utilised as river levels fall. In a subsequent wetter period, groundwater must be replaced before the level of the river can rise appreciably (Figure 3.3). If the water table reaches the surface, it means that the ground is saturated; excess water will then form a marsh where the land is flat, or will become surface runoff if the ground is sloping.

Channel flow

Although some rain does fall directly into the channel of a river (**channel precipitation**), most water reaches it by a combination of three transfer processes: surface runoff (overland flow), throughflow, or groundwater flow (baseflow). Once in the river, as **channel storage**, water flows towards the sea and is lost from the drainage basin system.

Figure 3.2

Surface runoff (overland flow), Blyford, Suffolk

The water balance

This shows the state of equilibrium in the drainage basin between the inputs and outputs. It can be expressed as:

$$P = Q + E \pm \text{change in storage}$$

where:

$P =$ precipitation (measured using rain gauges)

$Q =$ runoff (measured by discharge flumes in the river channel), and

$E =$ evapotranspiration. (This is far more difficult to measure – how can you measure accurately transpiration from a forest?)

In Britain, the annual precipitation nearly always, in most years and in most places, exceeds evapotranspiration. As, therefore, precipitation input exceeds evapotranspiration loss, then there is **positive water balance** (or water budget). However in some years, e.g 1974 and 1975, and 1995 and 1996, the long, dry summers, especially in the south and east of the country, resulted in evapotranspiration exceeding precipitation to give a temporary **negative water balance**. Changes in **storage** in the water balance reflect the amount of moisture in the soil. The **soil moisture budget** is, according to Newson, a sub-system of the catchment water balance.

Figure 3.3 is a graph showing the soil moisture balance for an area in south-east England. During winter, precipitation exceeds evapotranspiration creating a **soil moisture surplus** which results in considerable surface runoff and a rise in river levels. In summer, evapotranspiration exceeds precipitation and so plants and humans have to **utilise** water from the soil store leaving it depleted and causing river levels to fall. By autumn, when precipitation again exceeds evapotranspiration, the first of the surplus water has to be used to **recharge** the soil until it reaches its **field capacity** (page 267). At no time in Figure 3.3 was the utilisation of water sufficient to create a **soil moisture deficit** (as in Figure 3.4b).

Figure 3.3

A model illustrating soil moisture budget

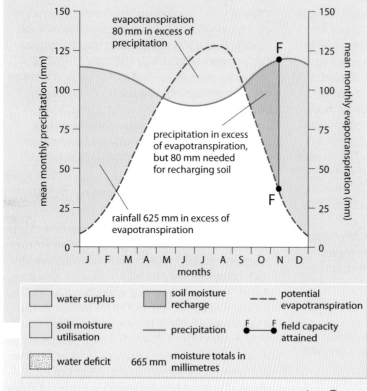

Figure 3.4

Soil moisture budget for two towns in the USA

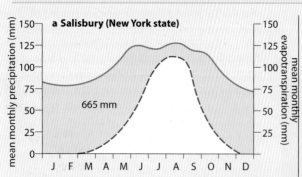

As precipitation is above potential evapotranspiration throughout the year then there is, in an average year, neither a water shortage nor a need to utilise moisture from the soil.

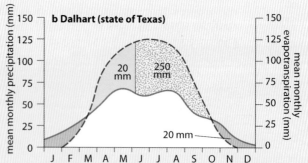

By spring, potential evapotranspiration is greater than precipitation. As there is no water surplus, then plants have to utilise moisture from the soil. By midsummer, water in the soil has been used up and there is a water deficit – meaning that plants can only survive if they are either drought-resistant or if they can obtain water through irrigation. When precipitation does exceed potential evapotranspiration, in winter, the rain is needed to replace (recharge) that taken from the soil earlier in the year, and amounts are insufficient to give a water surplus.

Figure 3.5

The storm hydrograph

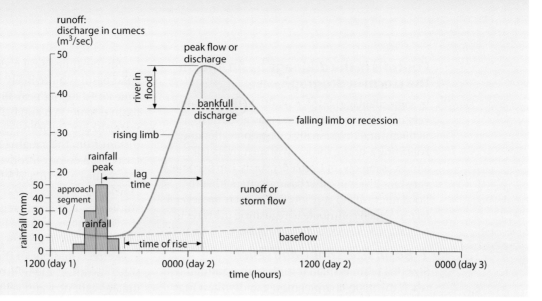

The storm hydrograph

An important aspect of **hydrology** (the study of water, **precipitation**, **runoff** and **evaporation/transpiration processes**) is how a drainage basin reacts to a period of rain. This is important because it can be used in predicting the flood risk and in making the necessary precautions to avoid damage to property and loss of life. The response of a river can be studied by using the **storm** or **flood hydrograph**. The hydrograph is a means of showing the discharge of a river at a given point over a short period of time. **Discharge** is the amount of water originating as precipitation which reaches the channel by surface runoff, throughflow and baseflow. Discharge is therefore the water *not* stored in the drainage basin by interception, as surface storage, soil moisture storage or groundwater storage or lost through evapotranspiration (Figure 3.1). The model of a storm hydrograph, Figure 3.5, shows how the discharge of a river responds to an individual rainfall event.

Measuring discharge

Discharge is the velocity (speed) of the river, measured in metres (m) per second, multiplied by the cross-sectional area of the river, measured in m^2. This gives the volume in m^3/sec or **cumecs**. It can be expressed as:

$$Q = A \times V$$

where:

Q = discharge
A = cross-sectional area
V = velocity.

Interpreting the hydrograph

Refer to the hydrograph in Figure 3.5. The graph includes the **approach segment** which shows the discharge of the river before the storm (the antecedent flow rate). When the storm begins, the river's response is negligible for although some of the rain does fall directly into the channel,

most falls elsewhere in the basin and takes time to reach the channel. However, when the initial surface runoff and, later, the throughflow eventually reach the river there is a rapid increase in discharge as indicated by the **rising limb**. The steeper the rising limb, the faster the response to rainfall – i.e. water reaches the channel more quickly. The **peak discharge** (peak flow) occurs when the river reaches its highest level. The period between maximum precipitation and peak discharge is referred to as the **lag time**. The lag time varies according to conditions within the drainage basin, e.g. soil and rock type, slope and size of the basin, drainage density, type and amount of vegetation and water already in storage. Rivers with a short lag time tend to experience a higher peak discharge and are more prone to flooding than rivers with a long lag time. The **falling** or **recession limb** is the segment of the graph where discharge is decreasing and river levels are falling. This segment is usually less steep than the rising limb because throughflow is being released relatively slowly into the channel. By the time all the water from the storm has passed through the channel at a given location, the river will have returned to its baseflow level – unless there has been another storm within the basin. **Stormflow** is the discharge, both surface and subsurface flow, attributed to a single storm. **Baseflow** is very slow to respond to a storm, but by continually releasing groundwater it maintains the river's flow during periods of low precipitation. Indeed, baseflow is more significant over a longer period of time than an individual storm and reflects seasonal changes in precipitation, snowmelt, vegetation and evapotranspiration. Finally, on the graph, **bankfull discharge** occurs when a river's water level reaches the top of its channel; any further increase in discharge will result in flooding of the surrounding land. This happens, on average, once every year or two.

Controls in the drainage basin and on the storm hydrograph

In some drainage basins, river discharge increases very quickly after a storm and may give rise to frequent, and occasionally catastrophic, flooding. Following a storm, the levels of such rivers fall almost as rapidly and, after dry spells, can become very low. Rivers in other basins seem neither to flood nor to fall to very low levels. There are several factors which contribute to regulating the ways in which a river responds to precipitation.

1 Basin size, shape and relief

Size If a basin is small it is likely that rainfall will reach the main channel more rapidly than in a larger basin where the water has much further to travel. Lag time will therefore be shorter in the smaller basin.

Shape It has long been accepted that a circular basin is more likely to have a shorter lag time and a higher peak flow than an elongated basin (Figure 3.6a and b). All the points on the watershed of the former are approximately equidistant from the gauging station, whereas in the latter it takes longer for water from the extremities of the basin to reach the gauging station. However, Newson (1994) has pointed out that studies made in many regions of the world have shown that basin shape is less reliable as a flood indicator than basin size and slope.

Relief The slope of the basin and its valley sides also affect the hydrograph. In steep-sided upland valleys, water is likely to reach the river more quickly than in gently sloping lowland areas (Figure 3.6c).

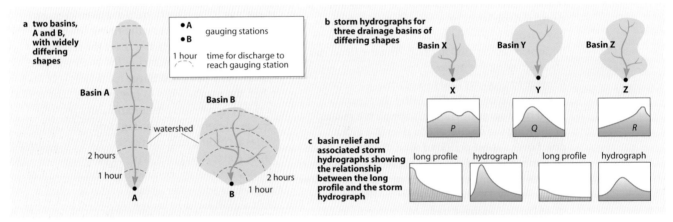

Figure 3.6

Drainage basin shape

2 Types of precipitation

Prolonged rainfall Flooding most frequently occurs following a long period of heavy rainfall when the ground has become saturated and infiltration has been replaced by surface runoff (overland flow).

Intense storms (e.g. convectional thunderstorms) When heavy rain occurs, the rainfall intensity may be greater than the infiltration capacity of the soil (e.g. in summer in Britain, when the ground may be harder). The resulting surface runoff is likely to produce a rapid rise in river levels (flash floods) – Boscastle, Cornwall, Places 12, page 80.

Snowfall Heavy snowfall means that water is held in surface storage and river levels drop. When temperatures rise rapidly (in Britain, this may be with the passage of a warm front and its associated rainfall, page 231), meltwater soon reaches the main river. It is possible that the ground will remain frozen for some time, in which case infiltration will be impeded.

3 Temperature

Extremes of temperature can restrict infiltration (very cold in winter, very hot and dry in summer) and so increase surface runoff. If evapotranspiration rates are high, then there will be less water available to flow into the main river.

4 Land use

Vegetation Vegetation may help to prevent flooding by intercepting rainfall (storing moisture on its leaves before it evaporates back into the atmosphere – page 59). Estimates suggest that tropical rainforests intercept up to 80 per cent of rainfall (30 per cent of which may later evaporate) whereas arable land may intercept only 10 per cent. Interception is less during the winter in Britain when deciduous trees have shed their leaves and crops have been harvested to expose bare earth. Plant roots, especially those of trees, reduce throughflow by taking up water from the soil.

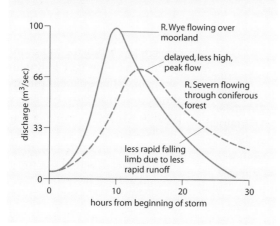

Figure 3.7

The effect of vegetation on the storm hydrographs of the Rivers Wye and Severn (geology and precipitation are the same in both basins)

Flooding is more likely to occur in deforested areas, e.g. the increasingly frequent and serious flooding in Bangladesh is attributed to the removal of trees in Nepal and other Himalayan areas. In areas of afforestation, flooding may initially increase as the land is cleared of old vegetation and drained, but later decrease as the planted trees mature. Newson (1994) points out that, after 20 years of data collecting, the evidence suggests that the canopy has more effect on medium flows than on high flows, as the main ditches remain active.

Figure 3.7 contrasts the storm hydrographs of two rivers. Although they rise very close together, the River Wye flows over moors and grassland, whereas the River Severn flows through an area of coniferous forest.

Urbanisation Urbanisation has increased flood risk. Water cannot infiltrate through tarmac and concrete, and gutters and drains carry water more quickly to the nearest river. Small streams may be either canalised so that

(with friction reduced) the water flows away more quickly, or culverted, which allows only a limited amount of water to pass through at one time (Figure 3.8).

5 Rock type (geology)
Rocks that allow water to pass through them are said to be **permeable**. There are two types of permeable rock:

- **Porous**, e.g. sandstone and chalk, which contain numerous pores able to fill with and store water (Figure 8.2).
- **Pervious**, e.g. Carboniferous limestone, which allow water to flow along bedding planes and down joints within the rock, although the rock itself is impervious (Figure 8.1).

As both types permit rapid infiltration, there is little surface runoff and only a limited number of surface streams. In contrast **impermeable rocks**, such as granite, do not allow water to pass through them and so they are characterised by more surface runoff and a greater number of streams.

6 Soil type
This controls the rate and volume of infiltration, the amount of soil moisture storage and the rate of throughflow (page 265). Sandy soils, with large pore spaces, allow rapid infiltration and do not encourage flooding. Clays have much smaller pore spaces and they are less well connected; this reduces infiltration and throughflow, but encourages surface runoff and increases the risk of flooding.

7 Drainage density
This refers to the number of surface streams in a given area (page 67). The density is higher on impermeable rocks and clays, and lower on permeable rocks and sands. The higher the density, the greater is the probability of flash floods. A **flash flood** is a sudden rise of water in a river, shown on the hydrograph as a shorter lag time and a higher peak flow in relation to normal discharge.

8 Tides and storm surges
High spring tides tend to prevent river floodwater from escaping into the sea. Floodwater therefore builds up in the lower part of the valley. If high tides coincide with gale-force winds blowing onshore and a narrowing estuary, the result may be a **storm surge** (Places 19, page 148). This happened in south-east England and in the Netherlands in 1953 and prompted the construction of the Thames Barrier and the implementation of the Dutch Delta Plan.

Figure 3.8

An urban river

Drainage basins and rivers   **63**

River regimes

The regime of a river is the term used to describe the annual variation in discharge. The average regime, which can be shown by either the mean daily or the mean monthly figures, is determined primarily by the climate of the area, e.g. the amount and distribution of rainfall, together with the rates of evapotranspiration and snow-melt. Local geology may also be significant. There are few rivers flowing today under wholly natural conditions, especially in Britain. Most are managed, regulated systems which result from human activity, e.g. reservoirs and flood protection schemes.

Regimes of rivers, which are used to demonstrate seasonal variations, may be either simple, with one peak period of flow, or complex with several peaks (Places 9).

Places 9 — **River Don, Yorkshire and River Torridge, Devon: river discharge**

Figure 3.9 shows the rainfall and runoff figures for the River Don (South Yorkshire) for one year. Discharge is usually at its highest in winter when Britain receives most of its depressions and when evapotranspiration is limited due to the low temperatures. Early spring may also show a peak if the source of the river is in an upland area liable to heavy winter snowfalls – in this case, the Pennines. It is possible for runoff to exceed precipitation, e.g. when heavy snowfall at the end of a month melts during a milder, drier period at the beginning of the next month. In contrast, river levels are lowest in summer when most of Britain receives less rainfall and when evapotranspiration rates are at their highest. There is often a correlation, or relationship, between the two variables of rainfall and runoff. This relationship can be shown by means of a scattergraph (Framework 19, page 612). Rainfall is plotted along the base (the x axis) because it is the independent variable, i.e. it does not depend on the amount of runoff. Runoff is plotted on the vertical or y axis because it is the dependent variable, i.e. runoff does depend upon the amount of rainfall.

The Environment Agency (EA) also produces hydrographs covering longer periods of time than for a single storm (Figure 3.5) but with far greater, and more useful, data than that given for the annual regime of a river (Figure 3.9).

Figure 3.10 gives rainfall and discharge for a wet month in late 1992 for the River Torridge in Devon. It shows that:

a as most of the peak discharges occur within a day of peak rainfall then the river must respond quickly to rainfall and, therefore, is likely to pose a flood risk

b the highest discharge (on the 30th) came after several very wet days during which river levels had no time to drop, rather than after a very wet day (the 17th) which followed a relatively dry spell of weather.

Figure 3.9

Rainfall and runoff for the River Don, Yorkshire

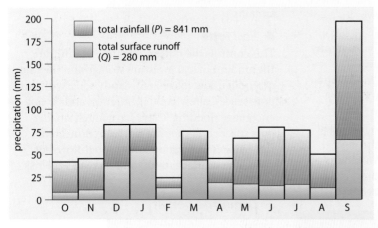

Figure 3.10

Hydrograph for the River Torridge at Torrington, Devon, late 1992

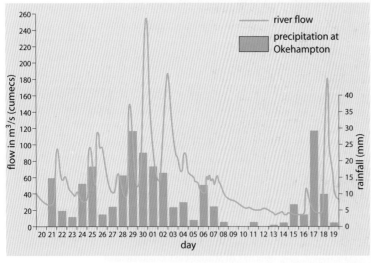

Morphometry of drainage basins

Morphometry means 'the measurement of shape or form'. The development of morphometric techniques was a major advance in the quantitative (as opposed to the qualitative) description of drainage basins (Framework 4). Instead of studies being purely subjective, it became possible to compare and contrast different basins with precision. Much of the early work in this field was by R.E. Horton. In the mid-1940s he devised the 'Laws of drainage composition' which established a hierarchy of streams ranked according to 'order'. One of these laws, the **law of stream number**, states that within a drainage basin a constant geometric relationship exists between stream order and stream number (Figure 3.12a).

Figure 3.11 shows how one of Horton's successors, A.N. Strahler, defined streams of different order. All the initial, unbranched source tributaries he called **first order** streams. When two first order streams join they form a **second order**; when two second order streams merge they form a **third order**; and so on. Notice that it needs two stream segments of equal order to join to produce a segment of a higher order, while the order remains unchanged if a lower order segment joins a higher order segment. For example, a second order plus a second order gives a third order but if a second order stream joins a third order, the resultant stream remains as a third order. A basin may therefore be described in terms of the highest order stream within it, e.g. a 'third order basin' or a 'fourth order basin'.

If the number of segments in a stream order is plotted on a semi-log graph against the stream order, then the resultant best-fit line will be straight (Figure 3.12a). On a semi-log graph, the vertical scale, showing the dependent variable (Framework 19, page 612), is divided into cycles, each of which begins and ends ten times greater than the previous cycle, e.g. a range of 1 to 10, 10 to 100, 100 to 1000, and so on. (If the horizontal scale, showing the independent variable, had also been divided into cycles instead of having an arithmetic scale, then Figure 3.12 would have been referred to as a log-log graph (Figure 18.25).) Logarithmic graphs are valuable when:

- the rate of change is of more interest than the amount of change: the steeper the line the greater the rate of change
- there is a greater range in the data than there is space to express on an arithmetic scale (a log scale compresses values)
- there are considerably more data at one end of the range than the other.

Figure 3.12a shows a perfect negative correlation (Figure 21.14): as the independent variable (in this case the stream order) increases, then the dependent variable (the number of streams) decreases. Studies of stream ordering for most rivers in the world produce a similar straight-line relationship. For any exceptions to Horton's law of stream ordering, further studies can be made to determine which local factors alter the relationship. Relationships also exist between stream order and the mean length of streams (Figure 3.12b), and stream order and mean drainage basin area (Figure 3.12c).

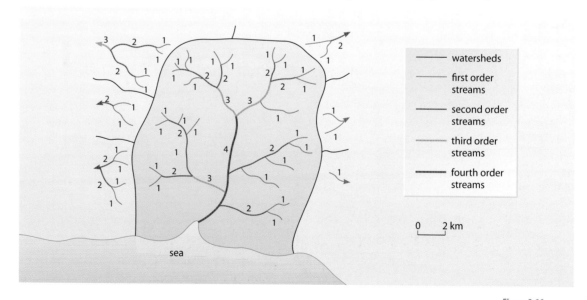

Figure 3.11

Strahler's method of stream ordering

Figure 3.12

Relationships
between stream order
and other variables

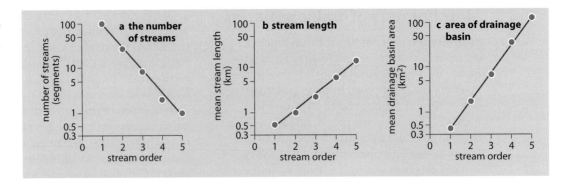

Comparing drainage basins

Horton's work has made it possible to compare different drainage basins scientifically (quantitatively) rather than relying on subjective (qualitative) descriptions by individuals. It also allows studies of drainage basin morphometry in different parts of the world to use the same standards, measurements and 'language'.

Figure 3.13 shows two imaginary and adjacent basins. These can be compared in several different ways, including:

- the bifurcation ratio, and
- drainage density.

The bifurcation ratio

This is the relationship between the number of streams of one order and those of the next highest order. It is obtained by dividing the number of streams in one order by the number in the next highest order, e.g. for basin **A** in Figure 3.13:

$$\frac{N1}{N2} = \frac{\text{(number of first order streams)}}{\text{(number of second order streams)}} = \frac{26}{6} = 4.33$$

$$\frac{N2}{N3} = \frac{\text{(number of second order streams)}}{\text{(number of third order streams)}} = \frac{6}{2} = 3.00$$

$$\frac{N3}{N4} = \frac{\text{(number of third order streams)}}{\text{(number of fourth order streams)}} = \frac{2}{1} = 2.00$$

and then finding the mean of all the ratios in the basin being studied, i.e.

$$\frac{4.33 + 3.00 + 2.00}{3} = 3.11 = \begin{array}{l}\text{bifurcation}\\\text{ratio for}\\\text{basin } \textbf{A}\end{array}$$

The human significance of the bifurcation ratio is that as the ratio is reduced so the risk of flooding within the basin increases. It also indicates the flood risk for parts, rather than all, of the basin. Most British rivers have a bifurcation ratio of between 3 and 5.

Figure 3.13

A comparison between
two adjacent drainage
basins on clays and
sands

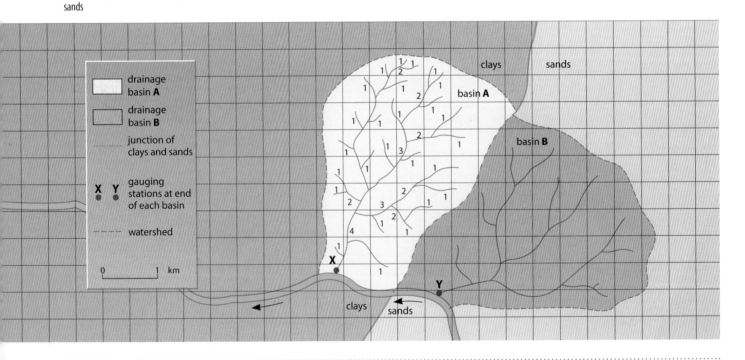

Drainage density

This is calculated by measuring the total length of all the streams within the basin (L) and dividing by the area of the whole basin (A). It is therefore the average length of stream within each unit area. For basin **A** in Figure 3.13, this will be:

$$\frac{L}{A} = \frac{22.65}{12.50} = 1.81 \text{ km per km}^2$$

In Britain most drainage densities lie between 2 and 4 km per km^2 but this varies considerably according to local conditions. A number of factors influence drainage density. It tends to be highest in areas where the land surface is impermeable, where slopes are steep, where rainfall is heavy and prolonged, and where vegetation cover is lacking.

a **Geology and soils** On very permeable rocks or soils (e.g. chalk, sands) drainage densities may be under 1 km per km^2, whereas this increases to over 5 km per km^2 on highly impermeable surfaces (e.g. granite, clays).

In Figure 3.13 with two adjacent drainage basins of approximately equal size, shape and probably rainfall, the difference in drainage density is likely to be due to basin **A** being on clays and basin **B** on sands.

b **Land use** The drainage density, especially of first order streams, is much greater in areas with little vegetation cover. The density decreases, as does the number of first order streams, if the area becomes afforested. Deserts tend to have the highest densities of first order channels, even if the channels are dry for most of the time.

c **Time** As a river pattern develops over a period of time, the number of tributaries will decrease, as will the drainage density.

d **Precipitation** Densities are usually highest in areas where rainfall totals and intensity are also high.

e **Relief** Density is usually greater on steeper slopes than on more gentle slopes.

Framework 4 — Quantitative techniques and statistical methods of data interpretation

As geography adjusted to a more scientific approach in the 1960s, a series of statistical techniques were adopted which could be used to quantify field data and add objectivity to the testing of hypotheses and theories. This period is often referred to as the 'Quantitative Revolution'.

At first it seemed to many, the author included, that mathematics had taken over the subject, but it is now accepted that these techniques are a useful aid provided they are not seen as an end in themselves. They provide a tool which, if carefully handled and understood, gives greater precision to arguments, helps in the identification of patterns and may contribute to the discovery of relationships and possible cause–effect links. In short, by providing greater accuracy in handling data they reduce the reliance upon subjective conclusions.

It is essential to select the most appropriate techniques for the data and for the job in hand. Therefore some understanding of the statistical methods involved is important.

Statistical methods may be profitably employed in these areas.

1 **Sampling** (Framework 6, page 159) Rapid collection of the data is made possible.

2 **Correlation and regression** (Framework 19, page 612) This not only shows possible relationships between two variables but quantifies or measures the strength of those relationships.

3 **Spatial distributions** (Framework 19, page 612) Not only may this approach be used to identify patterns, but it may also demonstrate how likely it is that the resultant distributions occurred by chance.

When these new techniques first appeared in schools in the 1970s, they appeared extremely daunting until it was realised that often the difficulty of the worked examples detracted from the usefulness of the technique itself. Where such techniques appear in this book, the mathematics have been simplified to show more clearly how methods may be used and to what effect. With the wider availability of calculators and computers it has become easier to take advantage of more complex calculations to test geographical hypotheses (Framework 10, page 299). Much of the 'number crunching' has now been removed by the increasing availability of statistical packages for computers.

Figure 3.14

Turbulence in a river: the confluence of the Rio Amazon (red with silt from the Andes) and the Rio Negro (black with plant acids)

horizontal movement of water so rarely experienced in rivers that it is usually discounted. Such a form of flow, if it existed, would travel over sediment on the river bed without disturbing it. Turbulent flow, the dominant mechanism, consists of a series of erratic eddies, both vertical and horizontal, in a downstream direction (Figures 3.14 and 3.15b). Turbulence varies with the velocity of the river which, in turn, depends upon the amount of energy available after friction has been overcome. It is estimated that under 'normal' conditions about 95 per cent of a river's energy is expended in order to overcome friction.

Influence of velocity on turbulence

- If the velocity is high, the amount of energy still available after friction has been overcome will be greater and so turbulence increases. This results in sediment on the bed being disturbed and carried downstream. The faster the flow of the river, the larger the quantity and size of particles which can be transported. The transported material is referred to as the river's **load**.
- When the velocity is low, there is less energy to overcome friction. Turbulence decreases and may not be visible to the human eye. Sediment on the river bed remains undisturbed. Indeed, as turbulence maintains the transport of the load, a reduction in turbulence may lead to deposition of sediment.

The velocity of a river is influenced by three main factors:

1 channel shape in cross-section
2 roughness of the channel's bed and banks, and
3 channel slope.

River form and velocity

A river will try to adopt a channel shape that best fulfils its two main functions: transporting water and sediment. It is important to understand the significance of channel shape in order to identify the controls on the flow of a river.

Types of flow

As water flows downhill under gravity, it seeks the path of least resistance – i.e. a river possesses potential energy and follows a route that will maximise the rate of flow (velocity) and minimise the loss of this energy caused by friction. Most friction occurs along the banks and bed of the river, but the internal friction of the water and air resistance on the surface are also significant.

There are two patterns of flow, **laminar** and **turbulent**. Laminar flow (Figure 3.15a) is a

Figure 3.15

Types of flow in a river

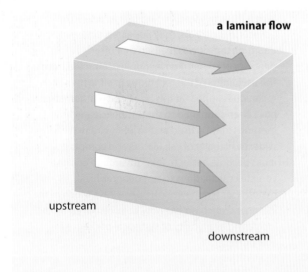

a laminar flow

upstream

downstream

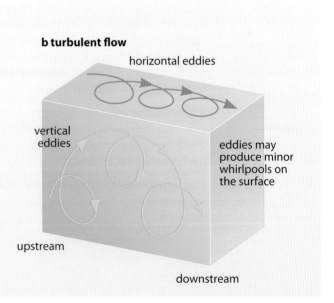

b turbulent flow

horizontal eddies

vertical eddies

eddies may produce minor whirlpools on the surface

upstream

downstream

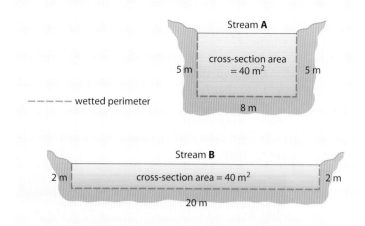

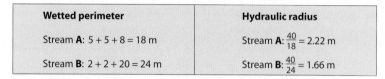

Wetted perimeter	Hydraulic radius
Stream **A**: 5 + 5 + 8 = 18 m	Stream **A**: $\frac{40}{18}$ = 2.22 m
Stream **B**: 2 + 2 + 20 = 24 m	Stream **B**: $\frac{40}{24}$ = 1.66 m

Figure 3.16

The wetted perimeter, hydraulic radius and efficiency of two different-shaped channels with equal area

1 Channel shape

This is best described by the term **hydraulic radius**, i.e. the ratio between the area of the cross-section of a river channel and the length of its wetted perimeter. The cross-section area is obtained by measuring the width and the mean depth of the channel. The **wetted perimeter** is the total length of the bed and bank sides in contact with the water in the channel. Figure 3.16 shows two channels with the same cross-section area but with different shapes and hydraulic radii.

Stream **A** has a larger hydraulic radius, meaning that it has a smaller amount of water in its cross-section in contact with the wetted perimeter. This creates less friction which in turn reduces energy loss and allows greater velocity. Stream **A** is said to be the more efficient of the two rivers.

Stream **B** has a smaller hydraulic radius, meaning that a larger amount of water is in contact with the wetted perimeter. This results in greater friction, more energy loss and reduced velocity. Stream **B** is less efficient than stream **A**.

The shape of the cross-section controls the point of maximum velocity in a river's channel. The point of maximum velocity is different in a river with a straight course where the channel is likely to be approximately symmetrical (Figure 3.17a) compared with a meandering channel where the shape is asymmetrical (Figure 3.17b).

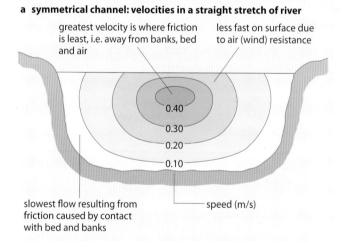

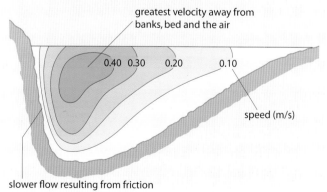

Figure 3.17

Cross-sections of a symmetrical and an asymmetrical stream channel

Figure 3.18

Tiger Leaping Gorge on the River Yangtze, China. This gorge has been suggested as a site for a future hydro-electric power station. It is nearly 1500 km upstream from the Three Gorges Dam

2 Roughness of channel bed and banks

A river flowing between banks composed of coarse material with numerous protrusions and over a bed of large, angular rocks (Figure 3.18) meets with more resistance than a river with cohesive clays and silts forming its bed and banks.

Figure 3.19 shows why the velocity of a mountain stream is less than that of a lowland river. As bank and bed roughness increase, so does turbulence. Therefore a mountain stream is likely to pick up loose material and carry it downstream.

Roughness is difficult to measure, but Manning, an engineer, calculated a **roughness coefficient** by which he interrelated the three factors affecting the velocity of a river. In his formula, known as 'Manning's N':

$$v = \frac{R^{0.67} \; S^{0.5}}{n}$$

where:

v = mean velocity of flow
R = hydraulic radius
S = channel slope
n = boundary roughness.

The formula gives a useful approximation: the higher the value, the rougher the bed and banks. For example:

Bed profile	Sand and gravel	Coarse gravel	Boulders
Uniform	0.02	0.03	0.05
Undulating	0.05	0.06	0.07
Highly irregular	0.08	0.09	0.10

Figure 3.19

Why a river increases in velocity towards its mouth

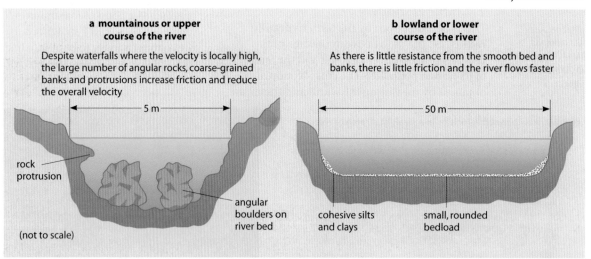

a mountainous or upper course of the river

Despite waterfalls where the velocity is locally high, the large number of angular rocks, coarse-grained banks and protrusions increase friction and reduce the overall velocity

5 m

rock protrusion

angular boulders on river bed

(not to scale)

b lowland or lower course of the river

As there is little resistance from the smooth bed and banks, there is little friction and the river flows faster

50 m

cohesive silts and clays

small, rounded bedload

Figure 3.20

The characteristic long profile of a river

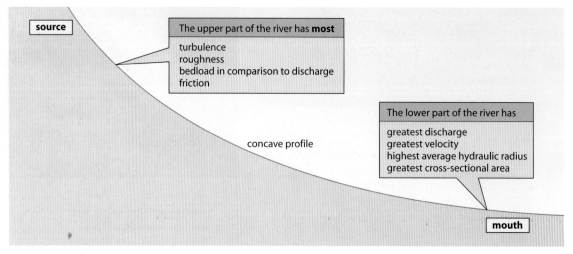

source

The upper part of the river has **most**

turbulence
roughness
bedload in comparison to discharge
friction

concave profile

The lower part of the river has

greatest discharge
greatest velocity
highest average hydraulic radius
greatest cross-sectional area

mouth

3 Channel slope

As more tributaries and water from surface runoff, throughflow and groundwater flow join the main river, the discharge, the channel cross-section area and the hydraulic radius will all increase. At the same time, less energy will be lost through friction and the erosive power of bedload material will decrease. As a result, the river flows over a gradually decreasing gradient – the characteristic concave **long profile (thalweg)** as shown in Figure 3.20.

In summarising this section it should be noted that:

- a river in a deep, broad channel, often with a gentle gradient and a small bedload, will have a greater velocity than a river in a shallow, narrow, rock-filled channel – even if the gradient of the latter is steeper
- the velocity of a river increases as it nears the sea – unless, like the Colorado and the Nile (Places 73, page 490), it flows through deserts where water is lost through evaporation or by human extraction for water supply
- the velocity increases as the depth, width and discharge of a river all increase
- as roughness increases, so too does turbulence and the ability of the river to pick up and transport sediment.

Transportation

Any energy remaining after the river has overcome friction can be used to transport sediment. The amount of energy available increases rapidly as the discharge, velocity and turbulence increase, until the river reaches flood levels. A river in flood has a large wetted perimeter and the extra friction is likely to cause deposition on the floodplain. A river at bankfull stage can move large quantities of soil and rock – its load – along its channel. In Britain, most material carried by a river is either sediment being redistributed from its banks, or material reaching the river from mass movement on its valley sides.

The load is transported by three main processes: **suspension**, **solution** and as **bedload** (Figure 3.21 and Places 10, page 73).

Suspended load

Very fine particles of clay and silt are dislodged and carried by turbulence in a fast-flowing river. The greater the turbulence and velocity, the larger the quantity and size of particles which can be picked up. The material held in suspension usually forms the greatest part of the total load; it increases in amount towards the river's mouth, giving the water its brown or black colour.

Dissolved or solution load

If the bedrock of a river is readily soluble, like limestone, it is constantly dissolved in flowing water and removed in solution. Except in limestone areas, the material in solution forms only a relatively small proportion of the total load.

Figure 3.21

Transportation processes in a river or stream

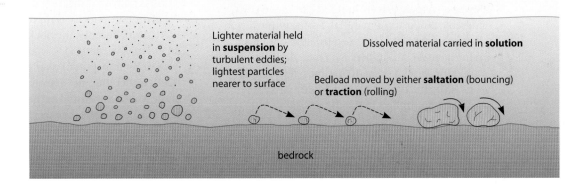

Bedload

Larger particles which cannot be picked up by the current may be moved along the bed of the river in one of two ways. **Saltation** occurs when pebbles, sand and gravel are temporarily lifted up by the current and bounced along the bed in a hopping motion (compare saltation in deserts, page 183). **Traction** occurs when the largest cobbles and boulders roll or slide along the bed. The largest of these may only be moved during times of extreme flood.

It is much more difficult to measure the bedload than the suspended or dissolved load. Its contribution to the total load may be small unless the river is in flood. It has been suggested that the proportion of material carried in one year by the River Tyne is 57 per cent in suspension, 35 per cent in solution and 8 per cent as bedload. This is the equivalent of a 10-tonne lorry tipping its load into the river every 20 minutes throughout the year. In comparison, the Amazon's load is equivalent to four such lorries tipping every minute of the year!

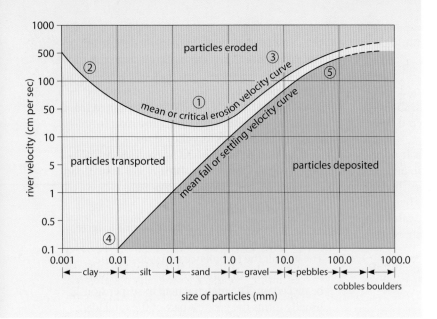

Figure 3.22

The Hjulström graph, showing the relationship between velocity and particle size. It shows the velocities necessary ('critical') for the initiation of movement (erosion); for deposition (sedimentation); and the area where transportation will continue to occur once movement has been initiated

Figure 3.23

Potholes in the bed of the Afon Glaslyn, Snowdonia

Competence and capacity

Two further terms should be noted at this point: the competence and capacity of a river. **Competence** is the maximum size of material which a river is capable of transporting. **Capacity** is the total load actually transported. When the velocity is low, only small particles such as clay, silt and fine sand can be picked up (Figure 3.22). As the velocity increases, larger material can be moved. Because the maximum particle mass which can be moved increases with the sixth power of velocity, rivers in flood can move considerable amounts of material. For example, if the stream velocity increased by a factor of four, then the mass of boulders which could be moved would increase by 4^6 or 4096 times; if by a factor of five, the maximum mass it could transport would be multiplied 15 625 times.

The relationship between particle size (competence) and water velocity is shown on the Hjulström graph (Figure 3.22). The **mean**, or **critical**, **erosion velocity** curve gives the approximate velocity needed to pick up and transport,

in suspension, particles of various sizes. The material carried by the river (capacity) is responsible for most of the subsequent erosion. The **mean fall** or **settling velocity** curve shows the velocities at which particles of a given size become too heavy to be transported and so will fall out of suspension and be deposited.

The graph shows two important points:
1 Sand can be transported at lower velocities than either finer or coarser particles. Particles of about 0.2 mm diameter can be picked up by a velocity of 20 cm per second (labelled **1** on the graph) whereas finer clay particles (**2**), because of their cohesive properties, need a velocity similar to that of pebbles (**3**) to be dislodged. During times of high discharge and velocity, the size and amount of the river's load will increase considerably, causing increased erosion within the channel.
2 The velocity required to maintain particles in suspension is less than the velocity needed to pick them up. For very fine clays (**4**) the velocity required to maintain them is virtually nil – at which point the river must almost have stopped flowing! This means that material picked up by turbulent tributaries and lower order streams can be kept in suspension by a less turbulent, higher order main river. For coarser particles (**5**), the boundary between transportation and deposition is narrow, indicating that only a relatively small drop in velocity is needed to cause sedimentation. Recently, Keylock has argued that an alternative method to that of Hjulström for measuring transport of river sediment is by flow depth rather than flow velocity. He suggests that shear stress – a measure of the force per unit area that the flow exerts on a particle on the river bed – can cause particles to roll out of their riverbed location.

Erosion

The material carried by a river can contribute to the wearing away of its banks and, to a lesser extent and mainly in the upper course, its bed. There are four main processes of erosion.

Corrasion

Corrasion occurs when the river picks up material and rubs it along its bed and banks, wearing them away by **abrasion**, rather like sandpaper. This process is most effective during times of flood and is the major method by which the river erodes both vertically and horizontally. If there are hollows in the river bed, pebbles are likely to become trapped. Turbulent eddies in the current can swirl pebbles around to form **potholes** (Figure 3.23).

Attrition

As the bedload is moved downstream, boulders collide with other material and the impact may break the rock into smaller pieces. In time, angular rocks become increasingly rounded in appearance.

Hydraulic action

The sheer force of the water as the turbulent current hits river banks (on the outside of a meander), pushes water into cracks. The air in the cracks is compressed, pressure is increased and, in time, the bank will collapse. **Cavitation** is a form of hydraulic action caused by bubbles of air collapsing. The resultant shock waves hit and slowly weaken the banks. This is the slowest and least effective erosion process.

Solution, or corrosion

This occurs continuously and is independent of river discharge or velocity. It is related to the chemical composition of the water, e.g. the concentration of carbonic acid and humic acid.

Deposition

When the velocity of a river begins to fall, it has less energy and so no longer has the competence or capacity to carry all its load. So, starting with the largest particles, material begins to be deposited (Figure 3.22). Deposition occurs when:

- discharge is reduced following a period of low precipitation
- velocity is lessened on entering the sea or a lake (resulting in a delta)
- shallower water occurs on the inside of a meander (Figure 3.25)
- the load is suddenly increased (caused by debris from a landslide)
- the river overflows its banks so that the velocity outside the channel is reduced (resulting in a floodplain).

As the river loses energy, the following changes are likely:

- The heaviest or bedload material is deposited first. It is for this reason that the channels of mountain streams are often filled with large boulders (Figures 3.18 and 3.27). Large boulders increase the size of the wetted perimeter.
- Gravel, sand and silt – transported either as bedload or in suspension – will be carried further, to be deposited over the floodplain (Figure 3.31) or in the channel of the river as it nears its mouth (Figure 3.32).
- The finest particles of silt and clay, which are carried in suspension, may be deposited where the river meets the sea – either to infill an estuary or to form a delta (Figure 3.33).
- The dissolved load will not be deposited, but will be carried out to sea where it will help to maintain the saltiness of the oceans.

Places 10 Afon Glaslyn, North Wales: river processes

Figure 3.24

The Glaslyn Valley, North Wales

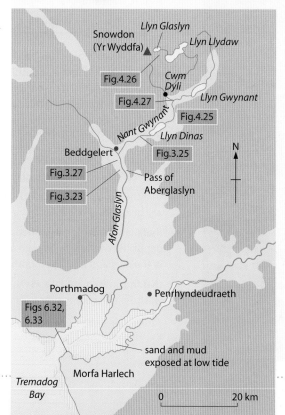

The Afon Glaslyn rises near the centre of the Snowdon massif and flows in a general southerly direction towards Tremadog Bay (Figure 3.24).

Figure 3.25

Erosion and deposition in the middle Afon Glaslyn

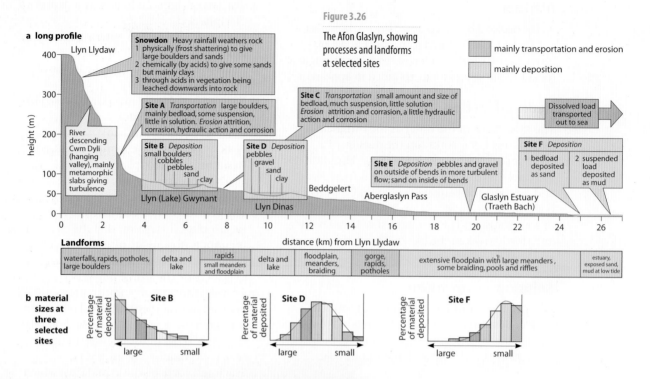

Figure 3.26

The Afon Glaslyn, showing processes and landforms at selected sites

a long profile

▨ mainly transportation and erosion

▨ mainly deposition

Llyn Llydaw

Snowdon Heavy rainfall weathers rock
1 physically (frost shattering) to give large boulders and sands
2 chemically (by acids) to give some sands but mainly clays
3 through acids in vegetation being leached downwards into rock

Site C *Transportation* small amount and size of bedload, much suspension, little in solution
Erosion attrition and corrasion, a little hydraulic action and corrosion

Dissolved load transported out to sea

Site A *Transportation* large boulders, mainly bedload, some suspension, little in solution. *Erosion* attrition, corrasion, hydraulic action and corrosion

Site F *Deposition*
1 bedload deposited as sand
2 suspended load deposited as mud

River descending Cwm Dyli (hanging valley), mainly metamorphic slabs giving turbulence

Site B *Deposition*
small boulders
cobbles
pebbles
sand
clay

Site D *Deposition*
pebbles
gravel
sand
clay

Site E *Deposition* pebbles and gravel on outside of bends in more turbulent flow; sand on inside of bends

Llyn (Lake) Gwynant

Llyn Dinas

Beddgelert

Aberglaslyn Pass

Glaslyn Estuary (Traeth Bach)

distance (km) from Llyn Llydaw

Landforms

| waterfalls, rapids, potholes, large boulders | delta and lake | rapids small meanders and floodplain | delta and lake | floodplain, meanders, braiding | gorge, rapids, potholes | extensive floodplain with large meanders, some braiding, pools and riffles | estuary, exposed sand, mud at low tide |

b material sizes at three selected sites

Percentage of material deposited — **Site B** — large → small

Percentage of material deposited — **Site D** — large → small

Percentage of material deposited — **Site F** — large → small

Figure 3.27

The boulder-strewn river bed of the upper Afon Glaslyn

The long profile of the Glaslyn, as shown in Figure 3.26, does not, however, match the smooth curve of the model shown in Figure 3.20. This is partly because of:

• the effect of glaciation in the upper course (Figure 4.25) and

• differences in rock structure in the middle course (the Aberglaslyn Pass in Figure 3.27).

Figure 3.26 (a summary of an Open University programme) shows the relationships between the processes of fluvial transportation, erosion and deposition. By studying this diagram, how likely are the following hypotheses (Framework 10, page 299):

• that as the competence of the river decreases, material is likely to be carried greater distances

• that the largest material, carried as the bedload, will be deposited first

• that material carried in suspension will be deposited over the floodplain or in the channel of the river as it nears its mouth

• that the finest material and the dissolved load will be carried out to sea?

Figure 3.28

V-shaped valley with interlocking spurs, small rapids and no floodplain: Peak District National Park

Fluvial landforms

As the velocity of a river increases, surplus energy becomes available which may be harnessed to transport material and cause erosion. Where the velocity decreases, an energy deficit is likely to result in depositional features.

Effects of fluvial erosion
V-shaped valleys and interlocking spurs

As shown in Figure 3.27, the channel of a river in its upper course is often choked with large, angular boulders. This bedload produces a large wetted perimeter which uses up much of the river's energy. Erosion is minimal because little energy is left to pick up and transport material. However, following periods of heavy rainfall or after rapid snowmelt, the discharge of a river may rise rapidly. As the water flows between boulders, turbulence increases and may result either in the bedload being taken up into suspension or, as is more usual because of its size, in its being rolled or bounced along the river bed. The result is intensive **vertical erosion** which enables the river to create a steep-sided valley with a characteristic V shape (Figure 3.28). The steepness of the valley sides depends upon several factors.

- **Climate** Valleys are steeper where there is sufficient rainfall:
  - a to instigate mass movement on the valley sides and
  - b to create sufficient discharge to allow the river to create enough energy to move its bedload and, therefore, to erode vertically, or
  - c for rivers to cross desert areas which have little rain to wash down the valley sides, e.g. the Grand Canyon (Figure 7.19).
- **Rock structure** Resistant, permeable rocks like Carboniferous limestone (Figure 8.5) often produce almost vertical sides in contrast to less resistant, impermeable rocks such as clay which are likely to produce more gentle slopes.
- **Vegetation** Vegetation may help to bind the soil together and thus keep the hillslope more stable.

Interlocking spurs form because the river is forced to follow a winding course around the protrusions of the surrounding highland. As the resultant spurs interlock, the view up or down the valley is restricted (Figure 3.28).

A process characteristic at the source of a river is **headward erosion**, or **spring sapping**. Here, where throughflow reaches the surface, the river may erode back towards its watershed as it undercuts the rock, soil and vegetation. Given time this can lead to river capture or piracy (page 85).

Waterfalls

A waterfall forms when a river, after flowing over relatively hard rock, meets a band of less resistant rock or, as is common in South America and Africa, where it flows over the edge of a plateau. As the water approaches the brink of the falls, velocity increases because the water in front of it loses contact with its bed and so is unhampered by friction (Figure 3.29). The underlying softer rock is worn away as water falls onto it. In time, the harder rock may become undercut and unstable and may eventually collapse. Some of this collapsed rock may be swirled around at the foot of the falls by turbulence, usually at times of high discharge, to create a deep **plunge pool**. As this process is repeated, the waterfall retreats upstream leaving a deep, steep-sided **gorge** (Places 11). At Niagara, where a hard band of limestone overlies softer shales and sandstone, the Niagara River plunges 50 m causing the falls to retreat by 1 m a year and so creating the Niagara Gorge.

Rapids

Rapids develop where the gradient of the river bed increases without a sudden break of slope (as in a waterfall) or where the stream flows over a series of gently dipping bands of harder rock. Rapids increase the turbulence of a river and hence its erosive power (Figure 3.27).

Places 11 Iguaçu Falls, Brazil: a waterfall

The Iguaçu River, a tributary of the Parana, forms part of the border between Brazil and Argentina. At one point along its course, the Iguaçu plunges 80 m over a 3 km wide, crescent-shaped precipice (Figure 3.30). The Iguaçu Falls occur where the river leaves the resistant basaltic lava which forms the southern edge of the Brazilian plateau and flows onto less resistant rock, while their crescent shape results from the retreat of the falls upstream (Figure 3.29).

By the end of the rainy season (January/February) up to 4 million litres of water a day can pour over the individual cascades – numbering up to 275 – which combine to form the falls. The main attraction is the Devil's Throat where 14 separate falls unite to create a deafening noise, volumes of spray, foaming water and a large rainbow. In contrast, by the end of the dry season (June/July), river levels may be very low – indeed, for one month in 1978 it actually dried up.

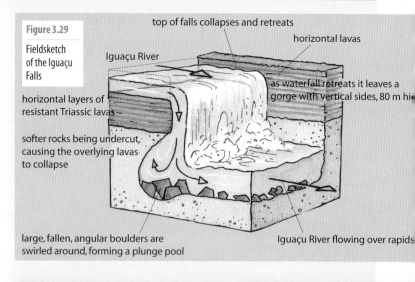

Figure 3.29

Fieldsketch of the Iguaçu Falls

top of falls collapses and retreats

horizontal lavas

Iguaçu River

horizontal layers of resistant Triassic lavas

as waterfall retreats it leaves a gorge with vertical sides, 80 m high

softer rocks being undercut, causing the overlying lavas to collapse

large, fallen, angular boulders are swirled around, forming a plunge pool

Iguaçu River flowing over rapids

Figure 3.30

The Iguaçu Falls

Effects of fluvial deposition

Deposition of sediment takes place when there is a decrease in energy or an increase in capacity which makes the river less competent to transport its load. This can occur anywhere from the upper course, where large boulders may be left, to the mouth, where fine clays may be deposited.

Floodplains

Rivers have most energy when at their bankfull stage. Should the river continue to rise, then the water will cover any adjacent flat land. The land susceptible to flooding in this way is known as the **floodplain** (Figure 3.31 and Places 10, page 74). As the river spreads over its floodplain, there will be a sudden increase in both the wetted perimeter and the hydraulic radius. This results in an increase in friction, a corresponding decrease in velocity and the deposition of material previously held in suspension. The thin veneer of silt, deposited by each flood, increases the fertility of the land, while the successive flooding causes the floodplain to build up in height (as yet it has proved impossible to bore down to bedrock in the lower Nile valley). The floodplain may also be made up of material deposited as point bars on the inside of meanders (Figure 3.38) and can be widened by the **lateral erosion** of the meanders. The edge of the floodplain is often marked by a prominent slope known as the **bluff line** (Figure 3.31).

Levées

When a river overflows its banks, the increase in friction produced by the contact with the floodplain causes material to be deposited. The coarsest material is dropped first to form a small, natural embankment (or levée) alongside the channel (Figure 3.31). During subsequent periods of low discharge, further deposition will occur within the main channel causing the bed of the river to rise and the risk of flooding to increase. To try to contain the river, the embankments are sometimes artificially strengthened and heightened (the levée protecting St Louis from the Mississippi is 15.8 m higher than the floodplain which it is meant to protect). Some rivers, such as the Mississippi and Yangtze, flow above the level of their floodplains which means that if the levées collapse there can be serious damage to property, and loss of life (Case Study 3A).

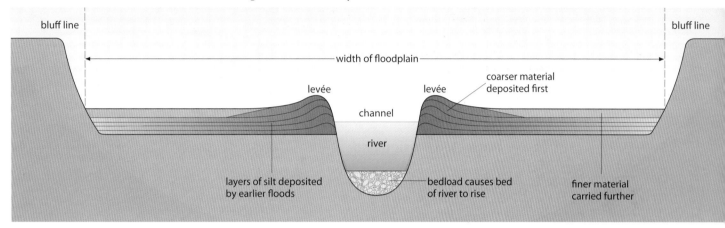

Figure 3.31

Cross-section of a floodplain showing levées and bluffs

Braiding

For short periods of the year, some rivers carry a very high load in relation to their velocity, e.g. during snowmelt periods in Alpine or Arctic areas. When a river's level falls rapidly, competence and capacity are reduced, and the channel may become choked with material, causing the river to braid – that is, to divide into a series of diverging and converging segments (Figures 3.32 and 5.16).

Figure 3.32

A braided river, South Island, New Zealand

Deltas

A delta is usually composed of fine sediment which is deposited when a river loses energy and competence as it flows into an area of slow-moving water such as a lake (Figure 4.22) or the sea. When rivers like the Mississippi or the Nile reach the sea, the meeting of fresh and salt water produces an electric charge which causes clay particles to coagulate and to settle on the seabed, a process called **flocculation**.

Deltas are so called because it was thought that their shape resembled that of delta, the fourth letter of the Greek alphabet (Δ). In fact, deltas vary greatly in shape but geomorphologists have grouped them into three basic forms:

- **arcuate:** having a rounded, convex outer margin, e.g. the Nile
- **cuspate:** where the material brought down by a river is spread out evenly on either side of its channel, e.g. the Tiber
- **bird's foot:** where the river has many distributaries bounded by sediment and which extend out to sea like the claws of a bird's foot, e.g. the Mississippi (Figure 3.33).

Although deltas provide some of the world's most fertile land, their flatness makes them high flood-risk areas, while the shallow and frequently changing river channels hinder navigation.

Figure 3.35

A pool and riffles in the
River Gelt, Cumbria

Effects of combined erosion and deposition

Pools, riffles and meanders

Rivers rarely flow in a straight line. Indeed, testing under laboratory conditions suggests that a straight course is abnormal and unstable. How meanders begin to form is uncertain, but they appear to have their origins during times of flood and in relatively straight sections where pools and riffles develop (Figure 3.34). The usual spacing between **pools**, areas of deeper water, and **riffles**, areas of shallower water, is usually very regular, being five to six times that of the bed width. The pool is an area of greater erosion where the available energy in the river builds up due to a reduction in friction. Energy is dissipated across the riffle area. As a higher proportion of the total energy is then needed to overcome friction, the erosive capacity is decreased and, except at times of high discharge, material is deposited (Figure 3.35). The regular spacings of pools and riffles, spacings which are almost perfect in an alluvial stretch of river, are believed to result from a series of secondary flows which exist within the main flow. Secondary flows include **helicoidal flow**, a corkscrew movement, as shown in Figure 3.15b, and a series of converging and diverging lateral rotations. Helicoidal flow is believed to be responsible for moving material from the outside of one meander bend and then depositing much of it on the inside of the next bend. It is thought, therefore, that it is the secondary flows that increase the sinuosity (the curving nature) of the meander (Figure 3.36), producing a regular meander wavelength which is about ten times that of the bed width. Sinuosity is described as:

$$\frac{\text{actual channel length}}{\text{straight-line distance}}$$

Figure 3.34

A possible sequence in the development of a meander

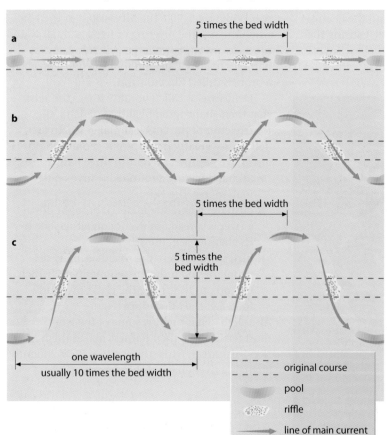

a

5 times the bed width

b

c

5 times the bed width

5 times the bed width

one wavelength
usually 10 times the bed width

- - - - original course
pool
riffle
→ line of main current

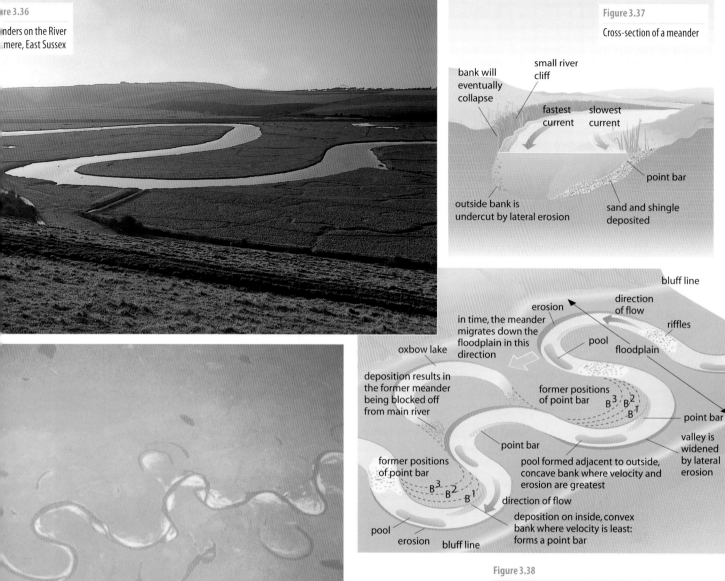

Figure 3.37

Cross-section of a meander

bank will
eventually
collapse

small river
cliff

fastest
current

slowest
current

point bar

outside bank is
undercut by lateral erosion

sand and shingle
deposited

bluff line

erosion

direction
of flow

in time, the meander
migrates down the
floodplain in this
direction

riffles

oxbow lake

pool

floodplain

deposition results in
the former meander
being blocked off
from main river

former positions
of point bar

B^3 B^2
B^1

point bar

point bar

valley is
widened
by lateral
erosion

former positions
of point bar

B^3 B^2
B^1

pool formed adjacent to outside,
concave bank where velocity and
erosion are greatest

direction of flow

pool

erosion

bluff line

deposition on inside, convex
bank where velocity is least:
forms a point bar

Figure 3.38

Meanders, point bars and oxbow lakes, showing migration of
meanders and changing positions of point bars over time

Figure 3.39

Meanders and
oxbow lakes,
Alaska, USA

Meanders, point bars and oxbow lakes

A meander has an asymmetrical cross-section
(Figure 3.37) formed by erosion on the outside
bend, where discharge and velocity are greatest
and friction is at a minimum, and deposition on
the inside, where discharge and velocity are at
a minimum and friction is at its greatest (Figure
3.25). Material deposited on the convex inside of
the bend may take the form of a curving **point
bar** (Figure 3.38). The particles are usually graded
in size, with the largest material being found on
the upstream side of the feature (there is rarely

any gradation up the slope itself). As erosion
continues on the outer bend, the whole meander
tends to migrate slowly downstream. Material
forming the point bar becomes a contributory
factor in the formation of the floodplain. Over
time, the sinuosity of the meander may become
so pronounced that, during a flood, the river
cuts through the narrow neck of land in order to
shorten its course. Having achieved a temporary
straightening of its channel, the main current
will then flow in mid-channel. Deposition can
now take place next to the banks and so, eventually, the old curve of the river will be abandoned,
leaving a crescent-shaped feature known as an
oxbow lake or **cutoff** (Figures 3.38 and 3.39).

On the afternoon of 16 August 2004, 200.2 mm of rainfall – the equivalent of three normal months – was recorded in only four hours on Bodmin Moor, an upland area lying behind the Cornish village of Boscastle. As the ground was already saturated, most of this water swept downhill and through two narrow, steep-sided valleys which converged on the village itself (Figure 3.40). Added to this volume of water was an estimated further 50 mm of rain that fell between 1300 and 1500 hours that same afternoon on Boscastle itself. The result was a wall of water over 3 m in height that swept through the village (Figure 3.41).

The floodwater carried with it cars, tree branches and other debris which became trapped behind the two bridges in the village, which then acted as dams. As the volume of water increased the bridges were swept away, causing further surges in the height of the River Valency. Residents and tourists alike were forced to flee. Although some managed to reach higher ground, the only means of escape for most people was to clamber upstairs and to await eventual rescue by helicopter from either upper-storey windows or rooftops.

Six helicopters (1 in Figure 3.42) rescued 120 people from rooftops and upper-storey windows (buildings 4, 5, 6, 7 and 8), while two lifeboats searched the harbour fearing people might have been swept out to sea. The car park (2) and two bridges (9 and 16) were destroyed. Vehicles were carried through the village by the torrent, some being deposited en route (12 and Figure 3.41) and over 30 in the harbour. Two shops (10 and 17) and four houses were destroyed while other buildings were badly damaged including the Visitor Centre (3) and two tourist shops (11 and 15). Among buildings flooded was a restaurant (13) and the village store (4), museum (14) and Youth Hostel (18). Power had to be switched off to protect rescuers and survivors from electrocution. When the floodwater receded, the village was left under a carpet of thick brown mud.

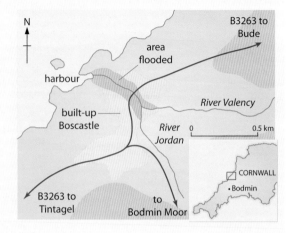

Figure 3.40

The flood at Boscastle

Figure 3.42

Annotated photo from the *Daily Telegraph*, Tuesday 17 August 2004

Figure 3.41

Water rages through the village of Boscastle carrying cars with it

Base level and the graded river

Base level

This is the lowest level to which erosion by running water can take place. In the case of rivers, this theoretical limit is sea-level. Exceptions occur when a river flows into an inland sea (e.g. the River Jordan into the Dead Sea) and if there happens to be a temporary **local base level**, such as where a river flows into a lake, where a tributary joins a main river, or where there is a resistant band of rock crossing a valley.

Grade

The concept of **grade** is one of a river forming an open system (Framework 3, page 45) in a state of dynamic equilibrium where there is a balance between the rate of erosion and the rate of deposition. In its simplest interpretation, a graded river has a gently sloping long profile with the gradient decreasing towards its mouth (Figure 3.43a). This balance is always transitory as the slope (profile) has to adjust constantly to changes in discharge and sediment load. These can cause short-term increases in either the rate of erosion or deposition until the state of equilibrium has again been reached. This may be illustrated by two situations:

- The long profile of a river happens to contain a waterfall and a lake (Figure 3.43b). Erosion is likely to be greatest at the waterfall, while deposition occurs in the lake. In time, both features will be eliminated.
- There is a lengthy period of heavy rainfall within a river basin. As the volume of water rises and consequently the velocity and load of the river increase, so too will the rate of erosion. Ultimately, the extra load carried by the river leads to extra deposition further down the valley or out at sea.

In a wider interpretation, grade is a balance not only in the long profile, but also in the river's cross-profile and in the roughness of its channel. In this sense, balance or grade is when all aspects of the river's channel (width, depth and gradient) are adjusted to the discharge and load of the river at a given point in time. If the volume and load change, then the river's channel morphology must adjust accordingly. Such changes, where and when they do occur, are likely to take lengthy periods of geological time.

Changes in base level

There are three groups of factors which influence changes in base level:
- **Climatic:** the effects of glaciation and/or changes in rainfall.
- **Tectonic:** crustal uplift, following plate movement, and local volcanic activity.
- **Eustatic and isostatic adjustment:** caused by the expansion and contraction of ice sheets (page 123).

As will be seen in Chapter 6, changes in base level affect coasts as well as rivers. There are two types of base level movement: positive and negative.
- **Positive change** occurs when sea-level rises in relation to the land (or the land sinks in relation to the sea). This results in a decrease in the gradient of the river with a corresponding increase in deposition and potential flooding of coastal areas.
- **Negative change** occurs when sea-level falls in relation to the land (or the land rises in relation to the sea). This movement causes land to emerge from the sea, steepening the gradient of the river and therefore increasing the rate of fluvial erosion. This process is called **rejuvenation**.

Figure 3.43

River profiles

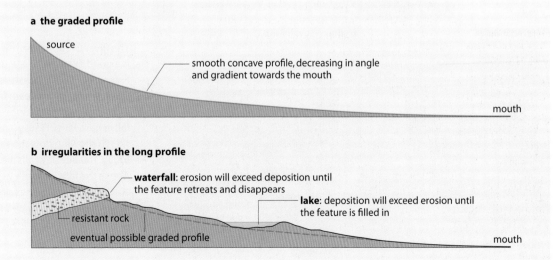

a the graded profile
source
smooth concave profile, decreasing in angle and gradient towards the mouth
mouth

b irregularities in the long profile
waterfall: erosion will exceed deposition until the feature retreats and disappears
lake: deposition will exceed erosion until the feature is filled in
resistant rock
eventual possible graded profile
mouth

Figure 3.44

The effect of rejuvenation on the long profile

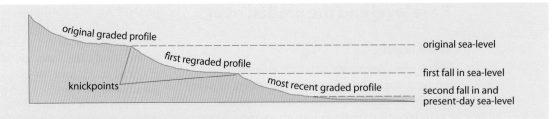

several partly graded profiles (Figure 3.44). Where the rise in the land (or drop in sea-level) is too rapid to allow a river sufficient time to erode vertically to the new sea-level, it may have to descend as a waterfall over recently emerged sea cliffs (Figure 3.45). In time, the river will cut downwards and backwards and the waterfall will retreat upstream. The **knickpoint**, usually indicated by the presence of a waterfall, marks the maximum extent of the newly graded profile (Places 13). Should a river become completely regraded, which is unlikely because of the timescale involved, the knickpoint and all of the original graded profile will disappear.

River terraces and incised meanders

River terraces are remnants of former floodplains which, following vertical erosion caused by rejuvenation, have been left high and dry above the maximum level of present-day flooding. They offer excellent sites for the location of towns (e.g. London, Figures 3.47 and 14.9). Above the present floodplain of the Thames at London are two earlier ones forming the Taplow and Boyn Hill terraces. If a river cuts rapidly into its floodplain, a pair of terraces of equal height may be seen flanking the river and creating a **valley-in-valley** feature. However, more often than not, the river cuts down relatively slowly, enabling it to meander at the same time. The result is that the terrace to one side of the river

Figure 3.45

A rejuvenated river, Antalya, Turkey: the land has only recently experienced tectonic uplift and the river has had insufficient time to re-adjust to the new sea-level

Rejuvenation

A negative change in base level increases the potential energy of a river, enabling it to revive its erosive activity; in doing so, it upsets any possible graded long profile. Beginning in its lowest reaches, next to the sea, the river will try to regrade itself.

During the Pleistocene glacial period, Britain was depressed by the weight of ice. Following deglaciation, the land slowly and intermittently rose again (**isostatic uplift**, page 123). Thus rejuvenation took place on more than one occasion, with the result that many rivers today show

Figure 3.46

The River Greta (*after* D.S. Walker)

Places 13 River Greta, Yorkshire Dales National Park: a rejuvenated river

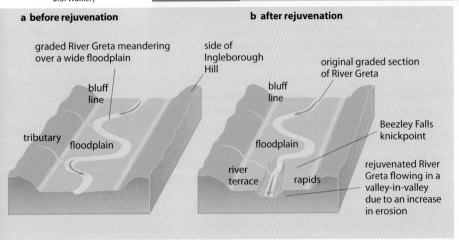

The River Greta, in north-west Yorkshire, is a good example of a rejuvenated river. Figure 3.46a is a reconstruction to show what its valley (upstream from the village of Ingleton) might have looked like before the fall in base level. Figure 3.46b is a simplified sketch showing how the same area appears today. The Beezley Falls are a knickpoint. Above the falls, the valley has a wide, open appearance. Below the falls, the river flows over a series of rapids and smaller falls in a deep, steep-sided 'valley-in-valley'.

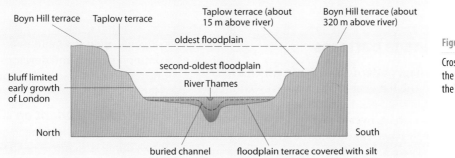

may be removed as the meanders migrate downstream. Figure 3.49 shows terraces, not paired, on a small stream crossing a beach on southern Arran. In this case, rejuvenation takes place twice daily as the tide ebbs and sea-level falls.

If the uplift of land (or fall in sea-level) continues for a lengthy period, the river may cut downwards to form incised meanders. There are two types of incised meander. **Entrenched meander**s have a symmetrical cross-section and result from either a very rapid incision by the river, or the valley sides being resistant to erosion (the River Wear at Durham, Figures 3.48a and 14.6). **Ingrown meanders** occur when the uplift of the land, or incision by the river, is less rapid, allowing the river time to shift laterally and to produce an asymmetrical cross-valley shape (the River Wye at Tintern Abbey, Figure 3.48b). As with meanders in the lower course of a normal river, incised meanders can also change their channels to leave an abandoned meander with a central **meander core** (Figure 3.48b).

Figure 3.48

Incised meanders and associated cross-valley profiles

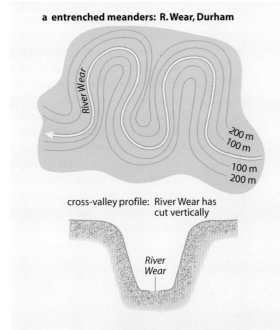

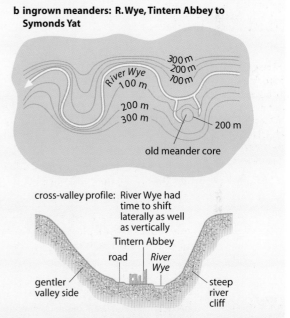

Figure 3.49

Rejuvenation on a micro scale: a small stream crossing a beach at Kildonan, Arran, has cut downwards to the level of the falling tide – note the ingrown meander, river terraces and valley-in-valley features

Drainage patterns

A **drainage pattern** is the way in which a river and its tributaries arrange themselves within their drainage basin (see Horton's Laws, page 65). Most patterns evolve over a lengthy period of time and usually become adjusted to the structure of the basin. There is no widely accepted classification, partly because most patterns are descriptive.

Patterns independent of structure

Parallel This, the simplest pattern, occurs on newly uplifted land or other uniformly sloping surfaces which allow rivers and tributaries to flow downhill more or less parallel with each other, e.g. rivers flowing south-eastwards from the Aberdare Mountains in Kenya (Figure 3.50a).

Dendritic Deriving its name from the Greek word dendron, meaning a tree, this is a tree-like pattern in which the many tributaries (branches) converge upon the main river (trunk). It is a common pattern and develops in basins having one rock type with no variations in structure (Figure 3.50b).

Patterns dependent on structure

Radial In areas where the rocks have been lifted into a dome structure (e.g. the batholiths of Dartmoor and Arran) or where a conical volcanic cone has formed (e.g. Mount Etna), rivers radiate outwards from a central point like the spokes of a wheel (Figure 3.50c).

Trellised or rectangular In areas of alternating resistant and less resistant rock, tributaries will form and join the main river at right-angles (Figure 3.50d). Sometimes each individual segment is of approximately equal length. The main river, called a **consequent river** because it is a consequence of the initial uplift or slope (compare parallel drainage), flows in the same direction as the dip of the rocks (Figure 3.51a). The tributaries which develop, mainly by headward erosion along areas of weaker rocks, are called **subsequent streams** because they form at a later date than the consequents. In time, these subsequents create wide valleys or vales (Figure 3.51b). **Obsequent streams** flow in the opposite direction from the consequent streams, i.e. down the steep scarp slope of the escarpment (Figure 3.51b). It is these obsequents that often provide the sources of water for scarp-foot springline settlements (Figure 14.4). The development of this drainage pattern is also responsible for the formation of the **scarp and vale topography** of south-east England (Figure 8.9).

Figure 3.50

Drainage patterns

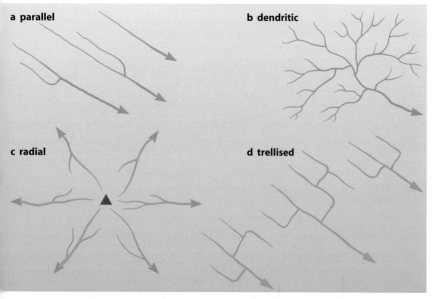

a parallel

b dendritic

c radial

d trellised

Figure 3.51

Development of a trellised drainage pattern

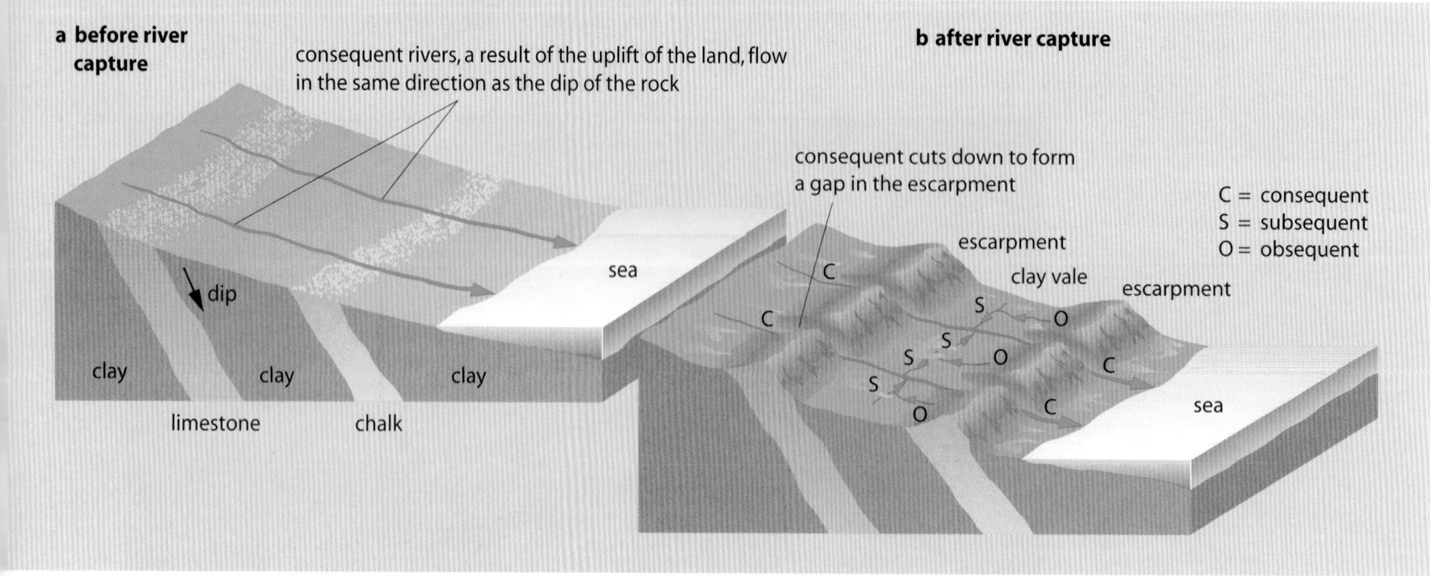

a before river capture

consequent rivers, a result of the uplift of the land, flow in the same direction as the dip of the rock

dip

clay

clay

clay

limestone

chalk

sea

b after river capture

consequent cuts down to form a gap in the escarpment

escarpment

clay vale

escarpment

C = consequent
S = subsequent
O = obsequent

sea

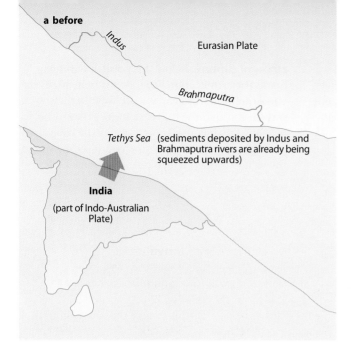

Figure 3.52

Antecedent drainage, Himalayas

Patterns apparently unrelated to structure

Antecedent Antecedence is when the drainage pattern developed before such structural movements as the uplift or folding of the land, and where vertical erosion by the river was able to keep pace with the later uplift. The Brahmaputra River rises in Tibet, but turns southwards to flow through a series of deep gorges in the Himalayas before reaching the Bay of Bengal (Figure 3.52). It must at one stage have flowed southwards into the Tethys Sea (Figure 1.4) which had existed before the Indo-Australian Plate moved northwards and collided with the Eurasian Plate forming the Himalayas (pages 19 and 20). The Brahmaputra, with an increasing gradient and load, was able to cut downwards through the rising Himalayas to maintain its original course.

Superimposed In several parts of the world, including the English Lake District, the drainage pattern seems to have no relationship to the present-day surface rocks. When the Lake District was uplifted into a dome, the newly-formed volcanic rocks were covered by sedimentary limestones and sandstones. The radial drainage pattern which developed, together with later glacial processes, cut through and ultimately removed the surface layers of sedimentary rock to superimpose itself upon the underlying volcanic rocks.

River capture

Rivers, in attempting to adjust to structure, may capture the headwaters of their neighbours. For example, most eastward-flowing English rivers between the Humber and central Northumberland have had their courses altered by **river capture** or **piracy** (Figure 3.53).

Figure 3.54a shows a case where there are two consequent rivers with one having a greater discharge and higher erosional activity than the other. Each has a tributary (subsequents X and Y) flowing along a valley of weaker rock, but subsequent X (the tributary of the master, or larger, consequent) is likely to be the more vigorous. Subsequent X will, therefore, cut backwards by headward erosion until it reaches subsequent Y (the tributary of the weaker consequent); then, by a process known as **watershed migration**

Figure 3.53

River capture, Northumberland

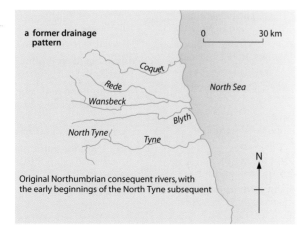

a former drainage pattern

0   30 km

Coquet

Rede

Wansbeck

North Sea

Blyth

North Tyne

Tyne

N

Original Northumbrian consequent rivers, with the early beginnings of the North Tyne subsequent

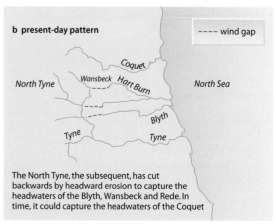

b present-day pattern

- - - - wind gap

Coquet

North Tyne   Wansbeck   Hart Burn

North Sea

Blyth

Tyne   Tyne

The North Tyne, the subsequent, has cut backwards by headward erosion to capture the headwaters of the Blyth, Wansbeck and Rede. In time, it could capture the headwaters of the Coquet

Drainage basins and rivers   **85**

(Figure 3.54b), it will begin to enlarge its own drainage basin at the expense of the smaller river. In time, the headwaters of the minor consequent will be captured and diverted into the drainage basin of the major consequent (Figure 3.54c).

The point at which the headwaters of the minor river change direction is known as the **elbow of capture**. Below this point, a **wind gap** marks the former course of the now **beheaded consequent** (a wind gap is a dry valley which was cut through the hills by a former river). The beheaded river is also known as a **misfit stream**, as its discharge is far too low to account for the size of the valley through which it flows (Figure 3.54c).

Figure 3.54

Stages in river capture shown in plan and cross-profile

a before capture (piracy) occurs

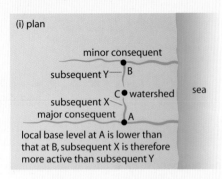

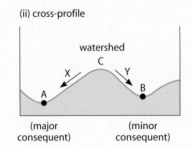

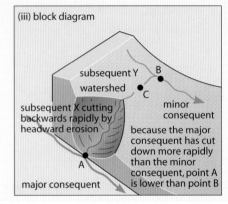

b watershed migration (recession)

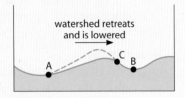

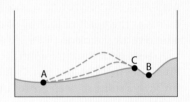

c after capture has taken place

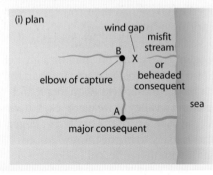

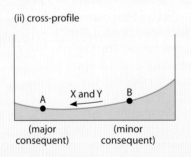

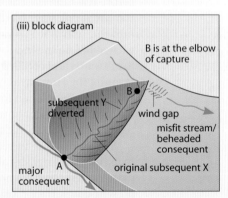

Further reference

Charlton, R. (2007) *Fundamentals of Fluvial Geomorphology*, Routledge.

Environment Agency (2000) *River Rehabilitation – Practical Aspects from 16 Case Studies*, Environment Agency Publications.

Gregory, K.J. and Walling, D.E. (2005) *Drainage Basin: Form, Process and Management*, WileyBlackwell.

Leopold, L. (2006) *A View of the River*, Harvard University Press.

Newson, M. (1994) *Hydrology and the River Environment*, Oxford University Press.

Oakes, S. (2006) 'Hi-tech flood warnings' in *Geography Review* Vol 20 No 1 (September).

Robert, A. (2003) *River Processes: An introduction to fluvial dynamics*, Hodder Arnold.

Weyman, D.R. (1975) *Runoff Processes and Streamflow Modelling*, Oxford University Press.

River management:

http://earthsci.org/Flooding/unit3/u3-01-06.html

www.broads-authority.gov.uk/managing/rivers-and-broads.html

Environment Agency, environmental information index (UK rivers, floods):

www.environment-agency.gov.uk/?lang=_e

www.floodarchive.co.uk

Minnesota River Basin:

www.soils.umn.edu/research/mn-river/

Newfoundland and Labrador site (examples of drainage basins and flood-risk zones):

www.heritage.nf.ca/sitemap.html

Norfolk Broads Authority:

www.broads-authority.gov.uk/broads/pages/river4.html

Yellow River, China:

www.cis.umassd.edu/~gleung/

The need for river management

Case Study 3

A River flooding: the Mississippi, 1993

Flooding by rivers is a natural event which, because people often choose to live in flood-risk areas, becomes a hazard (page 31). To people living in the Mississippi valley, 'that their river should flood is as natural as sunshine in Florida or snowfall in the Rockies'. Without human intervention, the Mississippi would flood virtually every year. Indeed, it has been this frequency of flooding which has, over many centuries, allowed today's river to flow for much of its course over a wide, fertile, flat, alluvial floodplain (Figures 3.55 and 3.56).

Figure 3.55

The flood hazard and the Mississippi River

Enquiry route on river flooding	Application to a specific event: the Mississippi, 1993
1 Where is the river/drainage basin located?	The Mississippi – together with its main tributaries, the Missouri and the Ohio – drains one-third of the USA and a small part of Canada (Figure 3.56).
2 What is the frequency of flooding?	Left to its own devices, flooding would be an almost annual event with late spring being the peak period.
3 What is the magnitude of flooding?	Until recently, major floods occurred every 5–10 years (there were six in the 1880s) and a serious/extreme flood occurred approximately once every 40 years.
4 What are the natural causes of flooding?	Usually it results from heavy rainfall (January–May) in the Appalachian Mountains, especially if this coincides with snowmelt (Figure 3.56).
5 What are the consequences of flooding?	Initially, it was to develop the wide, alluvial floodplain. The 1927 flood caused 217 deaths; 700 000 people were evacuated; the river became up to 150 km wide (usual width 1 km); livestock and crops were lost; services were destroyed.
6 What attempts can be made to reduce the flood hazard?	Until the 1927 flood, the main policy was 'hold by levées' – by 1993, some levées were 15 m high (Figure 3.57). After 1927, new schemes included building dams and storage reservoirs (6 huge dams and 105 reservoirs on Missouri); afforestation to reduce/delay runoff; creating diversion spillways (e.g. Bonnet Carré floodway diverts floodwater into Lake Pontchartrain and the sea); cutting through meanders to straighten and shorten the course (Figure 3.57).
7 How successful have the attempts to reduce flooding been?	In 1883, Mark Twain claimed that 'You cannot tame that lawless stream'. By 1973, it appeared that the river had been tamed: there was no further flooding … until 1993. Has human intervention made the danger worse? (page 96)

Usually, of course, the great floods occur in the lower river, in the last 1600 km below Cairo, Illinois. This is where the plain flattens out (the river drops less than 120 m from here to its mouth) and where the Ohio and Tennessee flow into the Mississippi.

Of the water that flows past Memphis, only about 38 per cent comes from the Missouri–Mississippi network. The bulk comes from the Ohio and Tennessee, from the lush Appalachians, rather than the dry Mid-West. 'We don't mind too much about the Missouri,' says Donna Willett, speaking for the US Army Corps of Engineers (who have the responsibility of flood prevention). 'It can rain there for weeks, and we wouldn't mind. We can handle three times the water coming down in those floods. But the Ohio, well, that's another story. When that starts rising, we start watching …'

Figure 3.56

Flooding in the Mississippi Basin

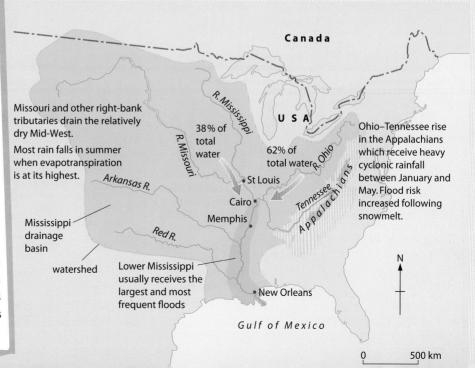

Missouri and other right-bank tributaries drain the relatively dry Mid-West.

Most rain falls in summer when evapotranspiration is at its highest.

Ohio–Tennessee rise in the Appalachians which receive heavy cyclonic rainfall between January and May. Flood risk increased following snowmelt.

Mississippi drainage basin

watershed

Lower Mississippi usually receives the largest and most frequent floods

38% of total water

62% of total water

Canada

USA

R. Mississippi

R. Missouri

Arkansas R.

Red R.

St Louis

Cairo

Memphis

R. Ohio

Tennessee

Appalachians

New Orleans

Gulf of Mexico

N

0   500 km

1a Height (metres) of levées at Memphis

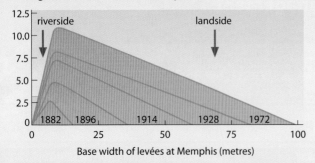

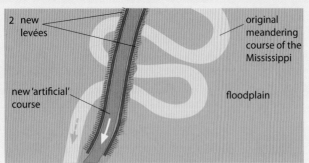

2 new levées

original meandering course of the Mississippi

new 'artificial' course

floodplain

b The 1993 flood at St Louis

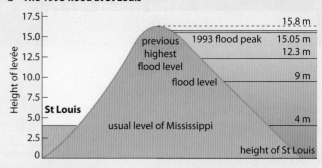

By making the course straighter and shorter, floodwater could be removed from the river basin as quickly as possible. It was achieved by cutting through the narrow necks of large meanders. Between 1934 and 1945 one stretch of the river alone was reduced from 530 km to almost 230 km. By shortening the distance, the gradient and therefore the velocity of the river increases. (But rivers try to create meanders rather than flow naturally in straight courses.)

Engineering/planning schemes in the Mississippi basin

Prior to the 1993 flood, it was perceived that the flow of the Mississippi had been controlled. This had been achieved through a variety of flood prevention schemes (Figure 3.57).

- Levées had been heightened, in places to over 15 m, and strengthened. There were almost 3000 km of levées along the main river and its tributaries.
- By cutting through meanders, the Mississippi had been straightened and shortened: for 1750 km, it flows in artificial channels.
- Large spillways had been built to take excess water during times of flood.
- The flow of the major tributaries (Missouri, Ohio and Tennessee) had been controlled by a series of dams.

Why did the Mississippi flood in 1993?

The Mid-West was already having a wet year when record-setting spring and summer rains hit. The rain ran off the soggy ground and into rapidly rising rivers. Several parts of the central USA had over 200 per cent more rain than was usual for the time of year (Figure 3.58). It was the ferocity, location and timing of the flood that took everyone by surprise. Normally, river levels are falling in midsummer, the upper Mississippi was not perceived to be the major flood-risk area, and people believed that flooding in the basin had been controlled. Floodwater at St Louis reached an all-time high (Figure 3.58). Satellite photographs showed the extent of the flooding (Figure 3.59). Figure 3.60 describes some of its effects.

Figure 3.57

Two engineering schemes to try to control flooding

After the flood: should rivers run freer?

Since the first levée was built on the Mississippi in 1718, engineers have been channelling the river to protect farmland and towns from floodwaters. But have the levées, dams and diversion channels actually aggravated the flooding? There are two schools of thought. One advocates accepting that rivers are part of a complex ecological balance and that flooding should be allowed as a natural event (Figure 3.71). The other argues for better defences and a more effective control of rivers (Figure 3.70).

Figure 3.58

Extract from *US Today*, a daily newspaper

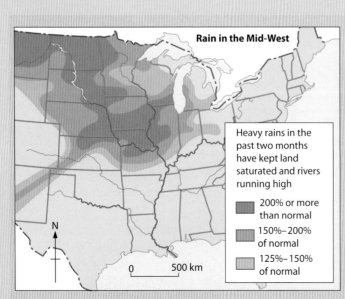

Rain in the Mid-West

Heavy rains in the past two months have kept land saturated and rivers running high

- 200% or more than normal
- 150%–200% of normal
- 125%–150% of normal

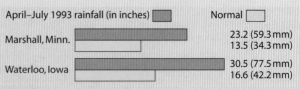

April–July 1993 rainfall (in inches)	Normal
Marshall, Minn.	23.2 (59.3 mm) / 13.5 (34.3 mm)
Waterloo, Iowa	30.5 (77.5 mm) / 16.6 (42.2 mm)

St Louis

Although there were some nervous moments, the city's massive 11-mile long, 52-foot floodwall protected the downtown from flooding. The river crested here August 1 at a record 49.4 feet, and the amount of water flowing past the Gateway Arch surpassed a record 1 million cubic feet per second.

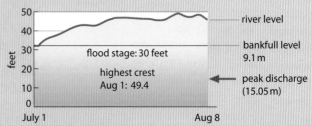

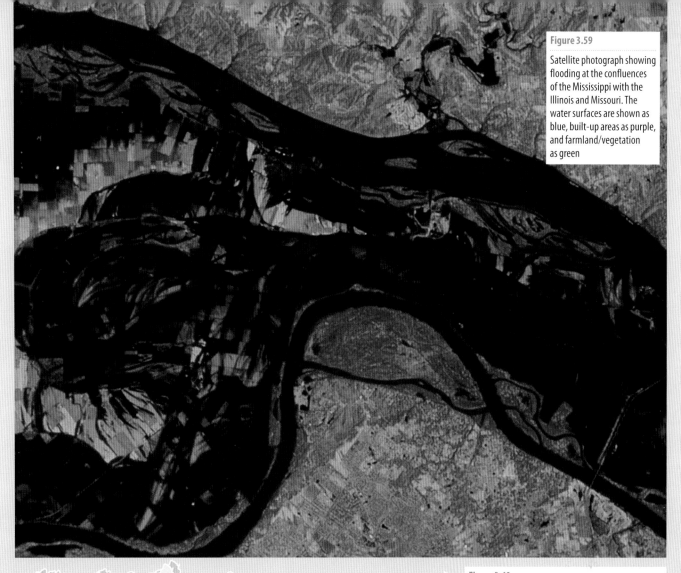

Figure 3.59

Satellite photograph showing flooding at the confluences of the Mississippi with the Illinois and Missouri. The water surfaces are shown as blue, built-up areas as purple, and farmland/vegetation as green

Figure 3.60

The consequences of flooding in the St Louis area

US Today, 9 August 1993

Flood of '93

Damage $10.5 billion	Deaths 45	Evacuated 74 000	Houses 45 000	Crops $6.5 billion

Nearly half of the counties in nine states bordering the upper reaches of the Mississippi and Missouri rivers have been declared federal disaster areas. This is the first step in becoming eligible for federal aid, including direct grants from Congress, Federal Emergency Management Agency and many other groups:

Declared disaster areas

Peak discharge: 26 June

Illinois: In the fight against flooding rivers, 17 levées were breached, including one that flooded the town of Valmeyer and 70 000 acres of surrounding farmland. One flood-related death was reported.

In Alton, the treatment plant was flooded Aug 1, cutting off water to the town's 33 000 residents. "Our levee did not breach, but the water came in through the street, the drains, anywhere there was a hole, at such a rate that pumps couldn't keep up," says Mayor Bob Towse.

Statewide property losses may top $365 million, including damage to 140 miles of roads and eight bridges. Agricultural damage is estimated at more than $610 million. An estimated 4% of the state's cropland—900 000 acres—was flooded. In addition, 15 727 people were displaced, 860 businesses closed and nearly 9000 jobs lost.

Missouri: The highest death toll—25—and the greatest property damage—$1.3 billion—of all flooded states were reported here. Statewide, 13 airports have been closed, and 25 000 residents evacuated. Flooding on 1.8 million acres of farmland has caused about $1.7 billion in crop losses.

Heroic efforts apparently saved historic Ste Genevieve, which has been battling rising waters since the start of July.

B River flooding: Mozambique

Mozambique has a pronounced single wet season followed by a lengthy dry season. As shown in Figure 3.61, both Maputo, the capital city, and Beira, the second city, receive almost 75 per cent of their annual rainfall during the five or six summer months when the sun is almost overhead (Figure 12.12) and when the south-east trades, blowing over the warm offshore Mozambique Current, are at their strongest (page 319). This rainfall pattern is repeated in the countries to the west and where Mozambique's three main rivers, the Zambezi, Save and Limpopo, have their headwaters.

The people of Mozambique are accustomed to the threat of seasonal flooding. In 2000 the country experienced its worst floods for over 50 years, an event that, in the following years, seemed to become an almost annual occurrence until 2008 when the government introduced its 'prevention-focused rather than response-oriented' policy.

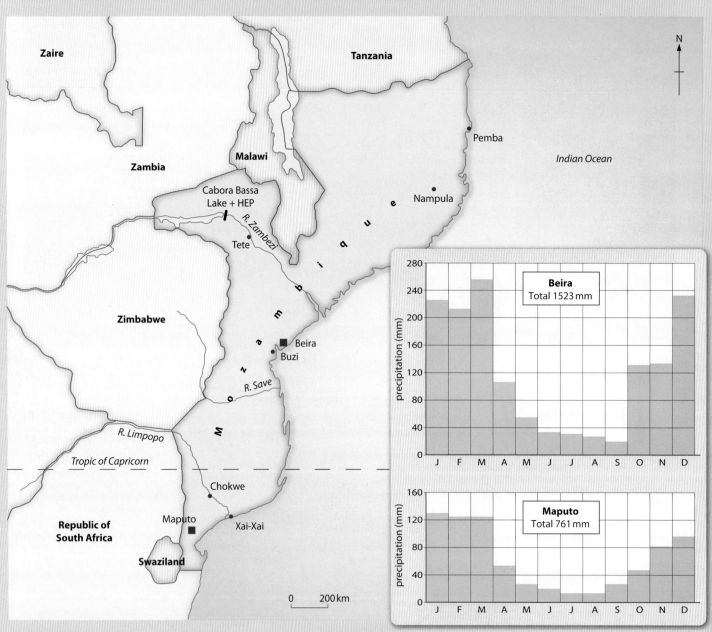

2000

Rivers, especially the Limpopo, began to overflow their banks in early February after several days of heavy rain, with the extreme south of the country the most severely affected. In Maputo, tens of thousands of people were forced to leave their homes, the worst-hit being those living in flimsy shanty settlements located on the edges of the city. Houses, roads, bridges and crops were destroyed, electricity supplies were disrupted and towns were left without a clean water supply after pumping stations were either inundated or swept away.

Figure 3.61

Mozambique, with rainfall graphs for Maputo and Beira

On 22 February the coastal region near Beira received the full impact of tropical storm Eline – a relatively rare hazard event in Mozambique. Winds of up to 260 km/hr hit a coastal area just north of the still-affected flooded regions. By 24 February, further heavy rainfall over much of southern Africa had swollen Mozambique's rivers by up to 8 m above their normal level (Figure 3.62). On 27 February, flash floods inundated more areas near to Chokwe and Xai-Xai. Estimates suggested that up to 7000 people, without food and water for several days, were surviving in the tops of trees or on small islands of high ground (Figure 3.63). International relief aid, when it eventually arrived, was to last for several months.

Final figures stated that 7000 people died, half a million were left homeless, 2 million had their lives affected, 11 per cent of farmland was ruined, 20 000 cattle were drowned and local industries in Maputo were forced to close.

2001

Over a month of heavy rain caused rivers in central areas, including the Zambezi near to Chokwe, to overflow. These floods led to 41 deaths, made 750 000 people homeless and affected half a million people in total. Roads and bridges, some only just repaired from the previous year, were swept away.

2006 and 2007

Following droughts in 2004 and 2005, heavy rainfall at the end of December 2005 and through early 2006 again affected thousands of people, although this time the death toll was down to 21. However, in 2007, several weeks of heavy rain resulted in the worst Zambezi floods since 2000. Fears that the huge Cabora Bassa dam (Figure 3.61) might overflow led to water being released from the lake behind it. This resulted in the level of the Zambezi rising even higher, and increased flooding in the lower basin. As a result 30 people died and 70 000 people were forced to leave their homes.

2008

Although an estimated 115 000 people were affected by the 2008 flood, the death toll was limited to 20. This was, according to UN aid workers, due to Mozambique's success in preparing for the flood event (Figure 3.64).

Figure 3.62

Aerial photo showing the extent of the 2000 flood

Figure 3.63

People awaiting rescue from tree tops (2000)

There has been, this year, a significant improvement in the government's disaster management. During the previous year the government had revamped its policies, making them prevention-focused rather than response-oriented. Realising that floods (and droughts) are going to happen, then the best approach is to try to minimise their impact. The Disaster Agency opened regional branches and began monitoring weather forecasts, upstream dam capacities and rainfall in neighbouring countries. It also set up an early-warning system and moved boats, together with reserves of food and medical supplies, to places with a high flood risk. Finally it drew up contingency plans aimed at evacuating low-lying villages should the need arise.

Figure 3.64

Extract from a 2008 UN report (UN/BBC News Africa)

C Flooding: the Severn in England, 2007

For many parts of England and Wales, 2007 was the wettest year, and certainly the wettest summer, ever recorded. The main reason was a failure by the polar front jet stream to move northwards as it usually does at this time of year (Figure 9.37). This meant that instead of the drier, more settled weather associated with a British summer, winds still came from the now warm Atlantic Ocean. Being warm, these winds were able to collect more moisture than was usual as they crossed the sea, resulting in heavy rainfall as they reached the British Isles. Torrential rain during June caused severe flooding in Hull, Doncaster and Sheffield that was to leave some properties uninhabitable for over a year.

20 July

Although forecasters had warned of heavy rain for up to a week beforehand and the Met Office had issued a severe weather warning two days before, no one quite expected the downpours of 20 July. Two months of rain fell in two hours, and three times July's normal total in 24 hours in parts of the Midlands where the soil was already saturated and many rivers were close to their bankfull level. Pershore, in Worcestershire, received 145 mm in that one day. Flash flooding immediately affected several towns in the Avon and lower Severn valleys (Figure 3.65). By early evening much of Evesham and parts of Stratford-upon-Avon were under water, 1 billion litres of water was pouring through Gloucester where up to 2000 people were to spend the night in emergency shelters, and residents in Tewkesbury, at the confluence of

the Severn and Avon, had begun to leave their homes. The flooding and the volume of traffic caused gridlock on major roads in the area, with an estimated 10 000 motorists left stranded for up to 10 hours on the M5 between Worcester and Gloucester (Figure 3.66). This gridlock prevented the emergency services moving equipment such as portable steel flood barriers to places like Upton-upon-Severn which were threatened by flooding, and hampered their attempts to rescue people already trapped. The result was the largest deployment of rescue helicopters and the biggest peacetime emergency ever in the United Kingdom.

Figure 3.65

Lower Severn valley

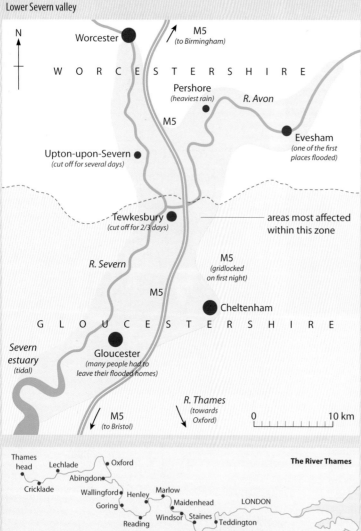

Figure 3.66

Gridlocked traffic on a flooded road near Tewkesbury

Figure 3.67

Flooded Tewkesbury, at the confluence of the Severn and Avon

22 July

More rain, together with runoff arriving from the headwaters of the River Severn, made the situation even worse. Helicopters were still rescuing people from Tewkesbury where 75 000 residents were completely cut off (Figure 3.67). Nearby, the Avon began to flood a water treatment works at the Mythe, forcing it to close down and leaving 350 000 people without water for washing, cooking or sewerage. Some 20 km to the south, a major crisis arose as floodwater began seeping into an electricity sub-power station, threatening to cut off supplies to 600 000 people. This led to the military being called in to help construct a 1 km embankment around the station to prevent further flooding and then to pump out water that was already

in it. This was achieved despite having only six hours before a high tide at nearby Gloucester would cause the level of the Severn to peak at almost 8 m above its usual level. Meanwhile further heavy rain was beginning to cause major disruptions to places further east in the Thames Valley.

23 July

Half of Gloucestershire was now without water and people were told that it might be two weeks before supplies could be restored, and 50 000 homes were without electricity. Freshwater tankers and bottled water suppliers were struggling to reach places still cut off, while supermarkets were experiencing panic buying. Of the thousands of people who had had to evacuate their homes in the region, some

were warned it would be over a year before they could return. While the Severn was still over its banks in several places and severe flood warnings remained in place between Tewkesbury and Gloucester, it was now people living close to the Thames in Oxfordshire who were faced with a real threat from flooding.

24 July

Floodwater had by now receded from most places in the Severn valley apart from properties adjacent to the river itself. Mopping up could begin but the real clean-up was expected to take months. Initial estimates of flood damage were put at over £2 billion.

D Flood and river management

Economically more developed countries such as the United Kingdom have the capital and technology that enable them to better predict, plan for, manage and respond to the flood risk than do less economically developed countries such as Mozambique.

Flood management in the UK is the responsibility of the Environment Agency (EA). The EA has the powers to set measures in place to reduce the risk of flooding on rivers and tidal waters. It also has the lead role in providing flood warnings and, wherever possible, to protect people and property at risk. Dynamic issues such as climate change, floodplain development and evolving technology mean that the EA has to frequently update its flood warning service and advice. The EA aims to reduce the impacts of flooding by:

- strategic and development planning
- investment in planning and managing flood defences
- mapping areas at risk of flooding and managing flooding information
- managing floods and providing the flood warning service.

Flood incidents vary in scale and impact, from low impact of unpopulated floodplains to severe flooding in large towns and cities which can disrupt key parts of the urban, and even regional, infrastructure. According to the EA, a flood incident involves planning for floods, communicating the risk of flooding, detecting and forecasting flooding, issuing flood warnings, providing information on flooding and responding to flooding (Figures 3.68 and 3.69).

Figure 3.68

How the EA prepares for and manages a flood event

	Role of the EA	Organisations involved
Planning for flooding	We constantly plan for flooding and organise how we will respond to each incident. We regularly meet with our professional partners to create multi-agency response plans and major incident plans for flooding. These detail how each organisation will respond to flooding in specific locations.	Police, ambulance, fire and rescue services. Local authorities, utility companies and community groups
Communicating flood risk	We talk to the public throughout the year about all aspects of our flood risk management work. We focus on flood awareness, our flood warning service (Figure 3.69) and providing information about what to do before, during and after the event.	Residents and property owners living or working in the area
Detecting flooding	We monitor rivers and sea conditions, 24 hours a day, 365 days a year, so we are prepared for potential flooding. We use remote detection systems to measure rainfall, wind speeds and direction, water levels and water flows in rivers and seas.	Met Office
Forecasting flooding	We use flood forecasting so that we know when and where to issue flood warnings and when to operate our flood defences. We share this with our professional partners so that they can also respond to flooding.	Met Office, emergency services, utility companies, local authorities
Issuing flood warnings	We send warnings by automated voice messages to land-line and mobile phones, and by fax, pager, SMS text, email, static sirens, public address loudhailers and broadcasts by radio and television.	General public, professional partners, the media
Providing information on flooding	If the public have not received flood warnings or want confirmation of the warnings issued, they can view warnings in force by: visiting our website at www.environment-agency.gov.uk/floodline, viewing Teletext (page 154) and Ceefax (page 149), or contacting Floodline on 0845 988 1188.	Website, the media, telephone
Responding to flooding	During a flood our priority is to issue flood warnings and make sure that our flood defences are working properly.	Emergency services, local authorities

Figure 3.69

Guide to the EA's flood warning codes

Flood Watch

Flooding of low-lying land and roads is expected. Be aware, be prepared, watch out.

Flood Warning

Flooding of homes and businesses is expected. Act now!

Severe Flood Warning

Severe flooding is expected. There is extreme danger to life and property. Act now!

Triggers

• Recorded rainfall that will cause flooding
• Recorded or forecast water levels that will cause flooding
• Snowmelt forecast

Triggers

• Heavy rainfall that could cause flash flooding
• Snowmelt
• Observed rising level – critical trigger point reached
• Forecast level or flow – trigger point for Flood Warning forecast
• Site observations, e.g. blockages or defence failures
• Actual flooding

Triggers

As for Flood Warning plus:
• Site observations of severe flooding or major problems with infrastructure and services
• Forecasts predict a worsening situation and severe flooding likely
• Actual flooding
• Professional judgement, including consultation with professional partners

Impact on the ground

• Fast-flowing rivers
• Bankfull rivers
• Flooding of fields and recreation land
• Minor road flooding
• Car park flooding
• Farmland flooding
• Surface water flooding (linked to river flooding)
• Overland flow from rivers and streams
• Localised flooding due to heavy storms

Impact on the ground

• Flooding of homes
• Flooding of businesses
• Flooding of cellars and basements
• Underground rail stations and lines vulnerable
• Flooding of major road infrastructure
• Flooding of rail infrastructure
• Significant floodplain inundation (high risk to caravan parks or campsites)
• Flooding of major tourist/recreational attractions
• Damage to flood defences

Impact on the ground

• Large numbers (at least 100) of homes/businesses expected to flood
• Large numbers of people are likely to be affected by flooding
• Highest risk to life
• Severe adverse impact on local infrastructure anticipated, e.g. transport, hospitals, utilities
• Significant impact on the capacity of professional partners, organisations and the public (e.g. vulnerable groups) to respond effectively
• Flood defence failures or overtopping which could result in extreme flooding

All Clear

We also use an 'All Clear' message to indicate receding floodwaters and a settled outlook.

Management in the future

Climate modellers are now predicting that by 2080, due to climate change, floods like those experienced in England in 2007, which have previously only happened once in every 150 years, could happen every 20 to 30 years. Insurers expect that by that time annual losses will be £21 billion – five times greater than in 2007. Since the floods, environmental risk consultants have been urging the British government to take urgent steps to deal with the increased risk of heavy rainfall events and that, instead of trying to control and contain the flow of rivers as in the past, space should be found for the excess water to go. A government report of 2004, 'Making Space for Water', came to the same conclusion, proposing the sacrificing of farmland, meadows and other areas of open space as a way of ensuring least damage to property and disruption to human activity (although this seemed at odds with government plans to build thousands of new homes in flood-risk areas – page 400).

Flood experts have begun detailed mapping of large urban areas in Britain. They hope, by using three-dimensional maps, not only to show which places are at greatest risk, but also to predict how deep the water might get and how long it might take before draining away.

Others are pointing out that by constructing hard defences and flood walls (Figure 3.70) all that is achieved is to push the problem further downstream. They suggest that there needs to be a major upgrading of the sewerage network and drainage systems to cope with more severe storm events, that houses should be built with the ground floor used for car parking and living space above it, and the use of stone and concrete for flooring would enable a flooded house to be hosed down and dried out more quickly than at present.

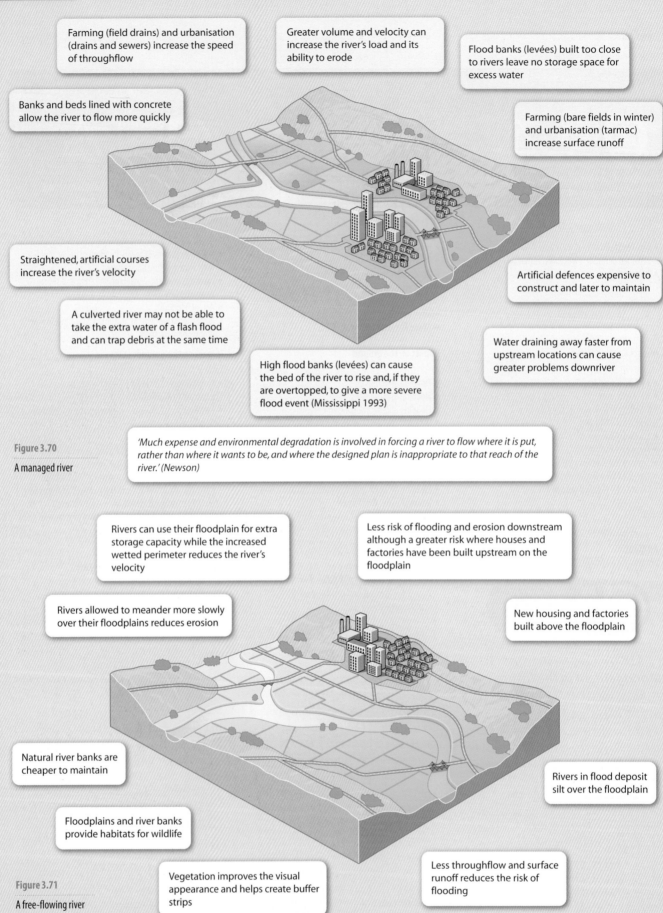

Farming (field drains) and urbanisation (drains and sewers) increase the speed of throughflow

Greater volume and velocity can increase the river's load and its ability to erode

Flood banks (levées) built too close to rivers leave no storage space for excess water

Banks and beds lined with concrete allow the river to flow more quickly

Farming (bare fields in winter) and urbanisation (tarmac) increase surface runoff

Straightened, artificial courses increase the river's velocity

Artificial defences expensive to construct and later to maintain

A culverted river may not be able to take the extra water of a flash flood and can trap debris at the same time

Water draining away faster from upstream locations can cause greater problems downriver

High flood banks (levées) can cause the bed of the river to rise and, if they are overtopped, to give a more severe flood event (Mississippi 1993)

Figure 3.70

A managed river

'Much expense and environmental degradation is involved in forcing a river to flow where it is put, rather than where it wants to be, and where the designed plan is inappropriate to that reach of the river.' (Newson)

Rivers can use their floodplain for extra storage capacity while the increased wetted perimeter reduces the river's velocity

Less risk of flooding and erosion downstream although a greater risk where houses and factories have been built upstream on the floodplain

Rivers allowed to meander more slowly over their floodplains reduces erosion

New housing and factories built above the floodplain

Natural river banks are cheaper to maintain

Rivers in flood deposit silt over the floodplain

Floodplains and river banks provide habitats for wildlife

Figure 3.71

A free-flowing river

Vegetation improves the visual appearance and helps create buffer strips

Less throughflow and surface runoff reduces the risk of flooding

Should rivers be managed or not?

People living and working in flood-risk areas naturally want their lives, property and way of life protecting yet increasingly this can only be done at greater financial and environmental costs. Some of the problems created by trying to control rivers are shown in Figure 3.70. Yet as flood events increase in frequency and severity, there may come a time when it is impossible to finance new defences or maintain existing ones. Figure 3.71 shows some of the ways by which the EA has, in a published pack of 16 schemes, tried to rehabilitate both rivers and their floodplains in an attempt to allow people to live with, rather than trying to control, them.

The River Skerne, near Darlington in County Durham, had, over 200 years, been progressively straightened for flood control, drainage, housing and industrial development (Figure 3.72). The floodplain had been a place for tipping contaminated waste while the river itself had become polluted, unsightly and, in places, inaccessible. Towards the end of the 20th century various organisations, including the EA, Northumbrian Water, English Nature, the Countryside Commission and Darlington Borough Council, worked together, with considerable effect, to rehabilitate the river (Figure 3.73). This has been achieved without compromising flood protection standards.
Rivers may be rehabilitated by:

- creating new habitats for wildlife (otters, birds, fish)
- reshaping river banks and channels and replacing artificial beds and banks ('hard' engineering) with natural materials
- recreating meanders and riffles
- reopening culverts.

Floodplains may be rehabilitated by:

- restoring former ponds and wetland areas or establishing new ones
- raising water tables and allowing increased flooding on floodplains
- planting trees and shrubs and creating buffer strips
- creating recreation areas.

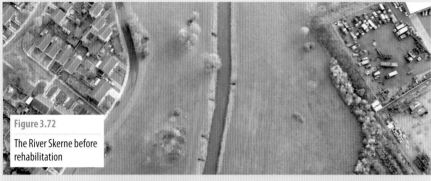

Figure 3.72

The River Skerne before rehabilitation

Figure 3.73

The River Skerne after rehabilitation

Sources of maps

(see pages 98–99)

Textbooks
Ross, S. (2002) *Essential Mapwork Skills*, Nelson Thornes, ISBN 978-0-7487-6461-7
Ross, S. (2006) *Essential Mapwork Skills 2*, Nelson Thornes, ISBN 978-0-7487-8436-3

Shops
In the UK, Stanfords (branches in London and Bristol) carries an astonishing range of maps and is well worth a visit (website address below).

Online
British Geological Survey
www.bgs.ac.uk/enquiries/rocks_beneath.html
Caribbean Disaster Emergency Response Agency (CDERA)
www.cdera.org
Cassini Historical Maps
www.cassinimaps.co.uk
China (topographic maps)
http://cartographic.com

Environment Agency
www.enviroment-agency.gov.uk/maps
Geological Survey of India
www.gsi.gov.in
Get Mapping
www.getmapping.com
GOAD maps available through Experian at
www.business-strategies.co.uk/sitecore/content/Products%20and%20services/Goad.aspx
Google maps
www.maps.google.co.uk
Land use maps Brighton and Hove
www.sussex.ac.uk/geography/1-2-4-1-2.html
Florida
www.mapwise.com/maps/florida/land-use-zoning.html
Map Action
www.mapaction.org
Met Office
www.metoffice.gov.uk
Multimap
www.multimap.com

National Hurricane Center
www.nhc.noaa.gov
Omnimap.com
www.omnimap.com
Ordnance Survey
www.ordnancesurvey.co.uk/oswebsite
www.ordnancesurvey.co.uk/oswebsite/getmap/
Ordnance Survey of Northern Ireland
www.osni.gov.uk
Population Reference Bureau
www.prb.org/Publications/GraphicsBank/PopulationTrends.aspx
School for Disaster Geo-Information Management
www.itc.nl/unu/dgim/diag/pakistan.asp
Soil Survey Maps
www.cranfield.ac.uk/sas/nsri/index.jsp
Stanfords Maps
www.stanfords.co.uk
Streetmap
www.streetmap.co.uk
US Geological Survey
www.usgs.gov

Maps provide a rich source of information for geographical study. There are many different types, including the traditional topographic Ordnance Survey (OS) maps, and specialist ones such as soil maps, geology maps and historical maps. Detailed maps exist for many parts of the world, providing a huge amount of information on land use, tourism and communications. The Internet is a great source of maps, enabling the user to have control over scale and coverage. See page 97 for some useful sources of maps, including those described below.

Paper maps

In the UK the maps most commonly used by geography students are the topographic OS maps. These are widely available and cover England, Wales and Scotland. Maps of Northern Ireland (produced by the OS of Northern Ireland) are slightly different, although there is widespread coverage. The most commonly used OS maps are the Landranger 1:50 000 maps and the Explorer 1:25 000 maps. Now that all the cartographic details are stored digitally it is possible to obtain site-centred maps at a great variety of scales, including 1:10 000, 1:5000 and even 1:1250, which give detailed layouts of houses and gardens.

Across the world, topographic maps similar to the UK's OS maps have been produced mostly using satellite information and exploiting GIS. Recently 1:50 000 topographic maps of China have been produced and these are now widely used to support economic development.

Many specialist paper maps are available for geographical study:

- The National Soil Resources Institute at the UK's Cranfield University publishes extremely detailed soil maps.

- The British Geological Survey has produced similarly detailed geological maps identifying rock types and geological features (Figure 3.74). These have many applications, for example in studying the location of landslides or the distribution of farms.

    - The Geological Survey of India publishes geology maps at various scales. These show details of geology as well as hazards and earth resources.

    - Historical maps are now available for many parts of the UK and these are an excellent resource when investigating changes over time, for example for an inner city area such as London Docklands or on a rural–urban fringe.

- Land use maps provide a further useful historical record for geographical study. Two sets of such maps cover the UK. These were drawn up in the 1930s and 1960s. More recently in 1996, the UK Geographical Association conducted a land use survey of 1000 × 1 km^2 squares – 500 rural and 500 urban – to enable comparisons to be made with the historical land use maps. Similar maps are available for other parts of the world.

- In South Africa a large range of city maps is available from Omnimap.com, together with a selection of topographic maps at different scales and thematic maps covering land uses, resources and geology. Omnimap.com also sells a range of maps of Malaysia, including land use maps and detailed geology/mineral maps.

- International Travel Maps (printed in Canada) give an excellent coverage of South America including the Amazon rainforest. These maps can be obtained from Stanfords bookshop (see 'Sources of maps' on page 97). Similar maps published by Globetrotter give good coverage of the Middle East, and are also available from Stanfords.

- In the UK, students may come across GOAD maps at GCSE. Essentially these plot commercial land uses in towns and cities. Buildings are drawn to scale and the nature of the building use is described; individual shops and stores are named. GOAD maps provide wonderful historical records and can be used to demonstrate changing urban land use (particularly retailing). While these maps are only available for the UK, they are a useful source of information for anyone studying geography.

Maps on the Internet

Today when asked for a map, most students automatically turn to the Internet. There are several Internet map providers, including Google Maps, Multimap, Get Mapping and Streetmap. The Ordnance Survey also provides maps online, and has a service Get-a-Map by which it is possible to find a map for a named place and print it, subject to certain conditions.

The Internet gives access to maps of all kinds, quickly and cheaply (often free of charge), and usually offers interactivity, with zoom and navigation facilities. Increasingly GIS enables the user to select particular

Figure 3.74

Extract from a geology map. Notice how rock types (coloured) are superimposed onto a traditional OS map
Source: www.bgs.ac.uk

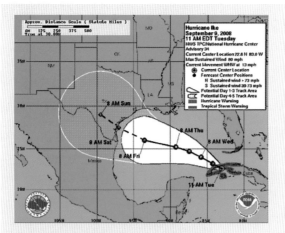

information to include on a map. Aerial photographs and so-called 'hybrid' maps (traditional maps superimposed over aerial photos) provide a further dimension for the geography student.

Many organisations provide specialist maps. For example:

- Map Action produces maps of areas hit by natural disasters such as volcanic eruptions, earthquakes or hurricanes. These maps are produced very quickly following an event to support relief agencies in their work.

- The School for Disaster Geo-Information Management has a tremendous selection of maps relating to the 2005 Pakistan earthquake; some 40 maps have been produced at a scale of 1:50 000 to assist aid workers in the region.

- Maps plotting hurricanes can be found at the National Hurricane Center (Figure 3.75).

- A huge variety of maps to support the study of tectonics, water resources and geology can be found at the US Geological Survey.

- For disasters in the Caribbean, such as earthquakes, volcanic eruptions, hurricanes and landslides, the Caribbean Disaster Emergency Response Agency provides excellent information including maps.

- Up-to-date and archive weather maps can be found at the Met Office and a range of UK postcode-related environmental maps can be found at the Environment Agency's website.

- A great site providing population maps is the Population Reference Bureau.

Using maps in geographical research

Maps are an essential part of study at AS/A level and you should make use of them when conducting your own individual research. At the most basic level a map identifies the location of a study area. It also helps to provide context, for example where a place is in relation to other places, or important features of the landscape. Geography is about interrelationships and connections and maps are often invaluable in this respect.

Information on maps can be directly relevant to geographical study, providing an alternative source of information about an area. In physical geography, for example, maps can be used to identify features such as corries, raised beaches and sea stacks. In human geography they provide information about services, patterns of roads and settlements, and land uses.

Sketch maps

Topographic maps are wonderfully detailed but sometimes they contain too much information so that it is difficult to see the overall picture. A sketch map enables a geographer to be more focused by making a careful copy of just a few selected pieces of information. Sketch maps are invaluable when researching case studies, for example in identifying landforms along a stretch of coastline. When drawing a sketch map you must be clear about its purpose and avoid adding irrelevant detail. Ensure that your map is as accurate as possible and remember to always include a scale and a north arrow. Use labels or annotations to provide interpretation of your map.

Using maps in exams

There is a strong chance that you will be given a map extract in one or more of your exam modules; so you do need to prepare yourself thoroughly as part of your revision. Practise the essential mapwork skills such as using grid references, measuring distance, describing orientation and drawing simple sketch maps. Make sure you know most of the symbols so that you can 'read' a map without having to keep referring to the key.

Take time to learn how to interpret a map in different geographical contexts. For example, be clear what different types of housing look like in an urban area, and make sure that you can identify a high tide line when examining a stretch of coastline.

Figure 3.76

Detailed topographic map of Singapore
Source: www.omnimap.com

Activities

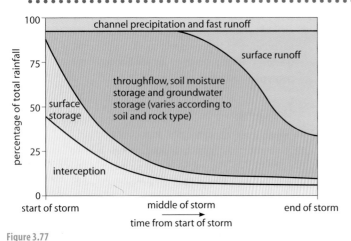

Figure 3.77

The relationship between rainfall and runoff in the course of a typical storm

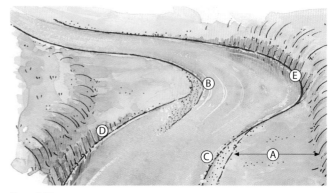

Figure 3.78

Fieldsketch of a meander

1 Study Figure 3.77.

  a i What is surface storage? *(2 marks)*

    ii Why does interception decrease during a storm? *(3 marks)*

    iii What happens to surface runoff during the storm? *(4 marks)*

  b What would happen to a river at the following stages:

    i at the start of this storm

    ii at the middle of the storm

    iii at the end of the storm? *(8 marks)*

  c The figure shows the reaction of a vegetated area to a heavy rainstorm. Describe and explain which parts of the model would change if the area were covered in concrete paving and drains. *(8 marks)*

2 **a** Study Figure 3.3 (page 60) and answer the following questions:

    i What is a 'soil moisture budget'? *(2 marks)*

    ii Explain each of the following terms used in the description of a soil moisture (water) budget: field capacity; water balance; soil moisture utilisation. *(7 marks)*

    iii Why is there no soil moisture deficit shown in Figure 3.3? *(4 marks)*

  b Why would a farmer need to understand the water balance of farmland? *(6 marks)*

  c Why do water companies in Britain depend on winter rainfall to maintain reservoirs? *(6 marks)*

3 **a i** Study the diagram of a meander (Figure 3.78) and identify the location of the following landforms:

    inside of the bend; outside of the bend; floodplain; slip-off slope; river cliff. *(5 marks)*

    ii Describe the features of the **channel cross-section** of a typical river meander. *(5 marks)*

  b Choose **one** of the following features of a river: waterfall; cascade; rapids. Using one or more sketches/diagrams, describe the features of your chosen landform and explain how it is eroded by a river. *(7 marks)*

  c i How does a meandering river form an oxbow lake? *(6 marks)*

    ii How could the formation of an oxbow lake lead to management problems on the floodplain of a river? *(4 marks)*

Exam practice: basic structured questions

4 **a i** What is a 'storm hydrograph'? *(3 marks)*

    ii What is meant by each of the following terms used in relation to a storm hydrograph: lag time; peak discharge; recession (falling) limb? *(6 marks)*

  b i Identify two drainage basin characteristics that make a river react quickly to a rainstorm (have a 'flashy' regime). For **each one** explain why it has this effect. *(7 marks)*

    ii With reference to specific example/s, suggest how river management strategies may be used to alleviate the problems caused by a 'flashy' regime. *(9 marks)*

5 **a i** Study Figure 3.27 (page 74). Describe the river bed shown in the photograph. *(3 marks)*

    ii Suggest where the loose boulders shown beside the river have come from. *(4 marks)*

iii How does a river erode a river bed such as the one in the photograph? *(6 marks)*

b Explain **two** ways in which you would know that loose rocks found on a field trip had been worn away by a river. *(6 marks)*

c With the aid of diagrams of a waterfall, show how it is being changed over time by river processes. *(6 marks)*

6 a i Describe the characteristic features of a **dendritic drainage pattern**. *(3 marks)*

ii Making good use of annotated diagrams, explain the development of a trellis drainage pattern. *(8 marks)*

b i Study Figure 3.53 (page 85). Describe the valley shape you would see if you were walking from the River Wansbeck to the Hart Burn. *(2 marks)*

ii Explain how the present drainage pattern evolved from the former drainage pattern. *(6 marks)*

c Choose and name an example of a drainage pattern other than a trellis pattern. Describe it and explain how it has been formed. *(6 marks)*

Exam practice: structured questions

7 a Using annotated diagram/s to help your answer, illustrate the components of a *storm hydrograph*. *(5 marks)*

b Explain how it is possible to measure the discharge of a stream in the field and how the results collected will be processed. *(10 marks)*

c Why do lag times differ on the same stream at different times? *(10 marks)*

8 When a housing estate is built on the rural/urban fringe, pre-existing drainage patterns are changed and river systems respond in a different way to storm events.

a Study of such changes must start before building to establish a 'baseline' for change. Briefly describe one technique you could use to measure the discharge of a stream in a rural catchment. *(5 marks)*

b Describe and account for **two** changes to discharge which may occur once the housing estate is built *(10 marks)*

c Describe **two** problems that could occur in the area due to the altered discharge pattern. *(10 marks)*

9 a Using annotated diagram/s **only**, show how the velocity of a typical river varies across its cross-section. *(5 marks)*

b i Describe the processes by which the load of a river is transported. *(8 marks)*

ii What factors affect the size of the particles eroded, transported and deposited by a river? *(12 marks)*

10 a Describe and suggest reasons for the cross-section shape of a river:

  i near the source of the river

  ii close to the mouth of the river. *(12 marks)*

b Identify and suggest reasons for **two** variations in the **long profile** of a river. *(13 marks)*

11 a i What is the difference between general base level and local base level? *(6 marks)*

ii Explain what happens to base level in a river system if sea-level falls. *(4 marks)*

b Choose **two** landforms formed in a river valley by a change in base level. Identify the direction of change involved and describe and explain the formation of each landform. *(15 marks)*

12 a Under what circumstances do rivers deposit material? *(12 marks)*

b i Explain how levées form as a result of natural river processes. *(5 marks)*

ii How do levées affect rivers and their tributaries? *(8 marks)*

13 Study Case Study 3B on pages 90 and 91.

a Describe the seasonal rainfall pattern in Mozambique and explain why this distribution of rainfall makes flooding common in the country's major river basins. *(7 marks)*

b Population densities are increasing in both the rural and urban areas of Mozambique. Suggest how this increases the flood hazard in the country. *(8 marks)*

c '… the government introduced its prevention-focused rather than its response-focused policy.'

Suggest what these policy changes might have meant in different parts of Mozambique. *(10 marks)*

Exam practice: essays

14 With reference to one or more river basins that you have studied, describe and evaluate river rehabilitation schemes. *(25 marks)*

15 Explain how changes in the base level of a river can affect the valley cross-section and the river's long profile. *(25 marks)*

16 'Flood hazards, resulting from a combination of physical and human influences, are increasing in many parts of the world.'

Discuss this statement with reference to rivers in countries at different stages of economic development. *(25 marks)*

4

Glaciation

● ●

'Great God! this is an awful place.'
The South Pole, **Robert Falcon Scott**, *Journal*, 1912

Ice ages

It appears that roughly every 200–250 million years in the Earth's history there have been major periods of ice activity (Figure 4.1). Of these, the most recent and significant occurred during

Figure 4.1

A chronology of ice ages (in bold)

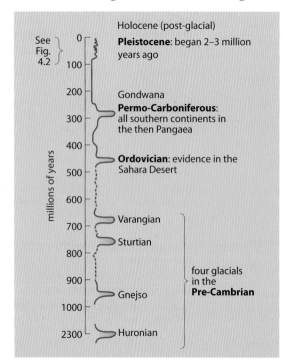

Figure 4.2

Generalised trends in mean global temperatures during the past 1 million years

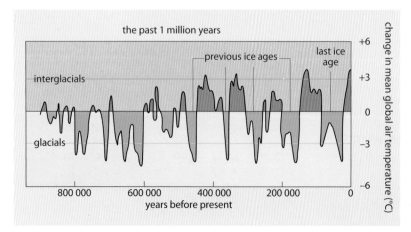

the Pleistocene epoch of the Quaternary period (Figure 1.1). In the 2 million years since the onset of the Quaternary, the time subject to most public interest and scientific research, there have been fluctuations in global temperature of up to 10°C which have led to cold phases (**glacials**) and warm phases (**interglacials**). Recent analyses of both ocean floor and Antarctic ice cores (Places 14) confirm that over the last 750 000 years the Earth has experienced eight **ice ages** (glacials) separated by eight interglacials (Figure 4.2).

When the ice reached its maximum extent, it is estimated that it covered 30 per cent of the Earth's land surface (compared with some 10 per cent today). However, its effect was not only felt in polar latitudes and mountainous areas, for each time the ice advanced there was a change in the global climatic belts (Figure 4.3). Only 18 000 years ago, at the time of the maximum advance within the last glacial, ice covered Britain as far south as the Bristol Channel, the Midlands and Norfolk. The southern part of Britain experienced **tundra conditions** (page 333), as did most of France.

Climatic change

Although it is accepted that climatic fluctuations occur on a variety of timescales, as yet there is no single explanation for the onset of major ice ages or for fluctuations within each ice age. The most feasible of theories to date is that of Milutin Milankovitch, mathematician/astronomer. Between 1912 and 1941, he performed exhaustive calculations which show that the Earth's position in space, its tilt and its orbit around the Sun all change. These changes, he claimed, affect incoming radiation from the Sun and produce three main cycles of 100 000, 40 000 and 21 000 thousand years (Figure 4.6). His theory, and the timescale of each cycle, has been given considerable support by evidence gained, since the mid-1970s, from ocean floor cores. As yet, although the relationship appears to have been established, it is not known precisely how these celestial cycles relate to climatic change.

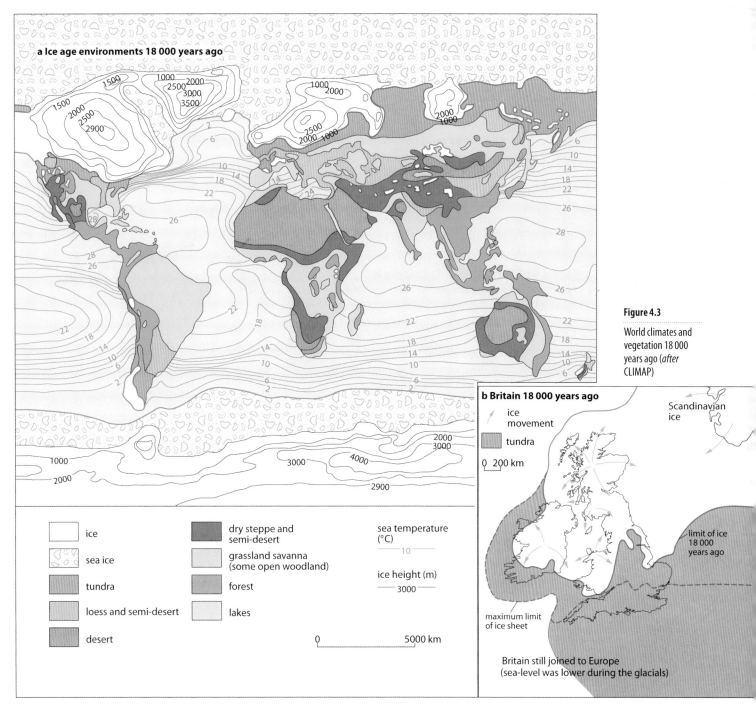

a Ice age environments 18 000 years ago

Figure 4.3

World climates and vegetation 18 000 years ago (*after* CLIMAP)

b Britain 18 000 years ago

ice movement

tundra

0   200 km

Scandinavian ice

limit of ice 18 000 years ago

maximum limit of ice sheet

Britain still joined to Europe (sea-level was lower during the glacials)

ice

sea ice

tundra

loess and semi-desert

desert

dry steppe and semi-desert

grassland savanna (some open woodland)

forest

lakes

sea temperature (°C)
— 10 —

ice height (m)
— 3000 —

0   5000 km

Other suggestions have been made as to the causes of ice ages. Some of these processes are likely to act in combination (Places 14) and may well amplify Milankovitch's variations.

- Variations in sunspot activity may increase or decrease the amount of radiation received by the Earth.
- Injections of volcanic dust into the atmosphere can reflect and absorb radiation from the Sun (page 207 and Figure 1.48).
- Changes in atmospheric carbon dioxide gas could accentuate the greenhouse effect (Case Study 9B). Initially extra CO_2 traps heat in the atmosphere, possibly raising world temperatures by an estimated 3°C. In time, some of this CO_2 will be absorbed by the seas, reducing the amount remaining in the atmosphere and causing a drop in world temperatures and the onset of another ice age (Figure 4.5).
- The movement of plates – either into colder latitudes or at constructive margins, where there is an increase in altitude – could lead to an overall drop in world land temperatures.
- Changes in ocean currents (page 211) or jet streams (page 227).

Antarctica

In **1988**, the Russians announced the first results of a five-year drilling experiment in Antarctica in which they extracted ice cores descending downwards through the ice sheet for nearly 2 km. Each core is a cylinder of ice 10 cm in diameter and about 3 m in length. The cores show a succession of rings, each of which is equivalent to the accumulation of one year of snow (Figure 4.4). From this, it was estimated that the ice at the bottom of the core had been formed 160 000 ago.

In 2004, the European Project for Ice Coring in Antarctica (EPICA) went deeper. The team, from ten countries and including members of the former British Antarctic Survey, produced a 3 km deep ice core that contained, at it lowest point, snowfall from 740 000 years ago. The consortium are still drilling and hope, by 2010, to reach base rock under the ice sheet and to recover ice that fell as snow over 900 000 years ago.

Analysis of the core showed how temperature has changed in the past and how the concentration of gases, mainly CO_2 and methane, and particles in the atmosphere, have varied. Results confirmed that:

- there have been eight glacials in the last 750 000 years and our present warm period is part of an interglacial that could last for at least another 15 000 years (although this could, without evidence, be longer if global warming continues)
- there is a close link between temperature change and the content of CO_2 in the atmosphere (Figure 4.5) and the last glacial began when the CO_2 content was very low
- there have been several previous periods of considerable global volcanic activity
- there is a likelihood of the Earth wobbling on its axis causing Milankovitch's 21 million year cycle.

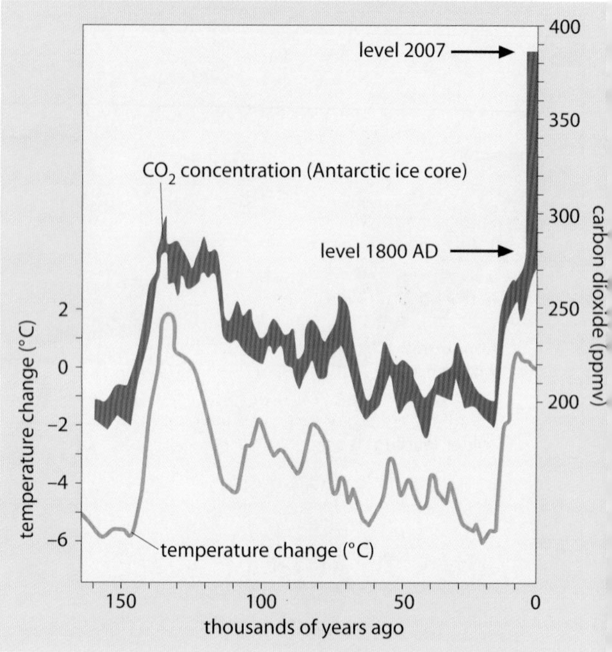

Figure 4.4

Dirt bands (englacial debris) in an Icelandic glacier: the amount of ice between each dirt band represents one year's accumulation of snow

Figure 4.5

Atmospheric CO_2 concentration and temperature change

Greenland, 1998

Two projects conducted from 1989 to 1993 collected parallel cores of ice from two places 30 km apart in the central part of the Greenland ice sheet. Each core was over 2 km deep and has been shown to extend back 110 000 years. During that period snowfall averaged 15–20 cm a year. At the same time as the snow was being compressed into ice (page 105), volcanic dust, wind-blown dust, sea salt, gases and chemicals which were present in the atmosphere, were trapped within the ice. The gases included two types of oxygen isotope, O-16 and O-18 (page 248). The ratio between these two isotopes changes as the proportion of global water bound up in the ice changes (the amount of O-18 in the atmosphere increases as air temperature falls, and decreases as air temperature rises). The changing ratios from the Greenland cores showed short-term and long-term changes in temperature, and that rapid global change is more the norm for the Earth's climate than the stability and gradual adjustment that was previously assumed. The recent ice core from Antarctica directly correlates 'with an astounding regularity' with the abrupt climate changes in both polar areas. However, findings also suggest that as Antarctica warms up, Greenland cools and, likewise, when temperatures rise in Greenland, they fall in Antarctica. This link suggests that the two icy regions are connected by ocean currents in a bipolar seesaw (Case Study 4).

a the 100 000 year eccentricity

The Earth's orbit stretches from being nearly circular to an elliptical shape and back again in a cycle of about 95 000 years. During the Quaternary, the major glacial–interglacial cycle was almost 100 000 years. Glacials occur when the orbit is almost circular and interglacials when it is a more elliptical shape.

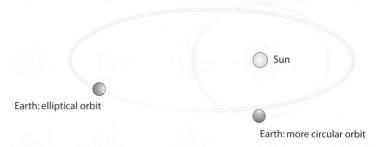

b the 40 000 year obliquity

Although the tropics are set at 23.5°N and 23.5°S to equate with the angle of the Earth's tilt, in reality the Earth's axis varies from its plane of orbit by between 21.5° and 24.5°. When the tilt increases, summers will become hotter and winters colder, leading to conditions favouring interglacials.

a = 21.5°

b = 24.5°

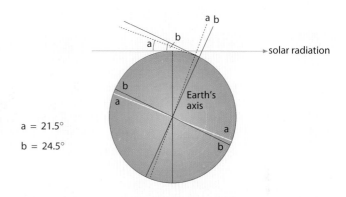

c the 21 000 year precession

As the Earth slowly wobbles in space, its axis describes a circle once in every 21 000 years.
1 At present, the orbit places the Earth closest to the Sun in the northern hemisphere's winter and furthest away in summer. This tends to make winters mild and summers cool. These are ideal conditions for glacials to develop.
2 The position was in reverse 12 000 years ago, and this has contributed to the onset of our current interglacial.

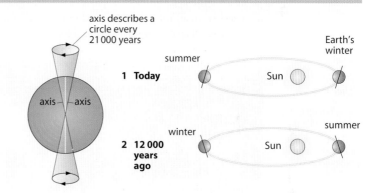

Figure 4.6

The orbital forcing mechanisms of Milankovitch's climatic change theory

Snow accumulation and ice formation

As the climate gets colder, more precipitation is likely to be in the form of snow in winter and there is less time for that snow to melt in the shorter summer. If the climate continues to deteriorate, snow will lie throughout the year forming a permanent **snow line** – the level above which snow will lie all year. In the northern hemisphere, the snow line is at a lower altitude on north-facing slopes, as these receive less insolation than south-facing slopes. The snow line is also lower nearer to the poles and higher nearer to the Equator: it is at sea-level in northern Greenland; at about 1500 m in southern Norway; at 3000 m in the Alps; and at 6000 m at the Equator. It is estimated that the Cairngorms in Scotland would be snow-covered all year had they been 200 m higher. In 2003 when Sir Edmund Hillary revisited the base camp for his 1953 ascent of Mount Everest, he found the snow-line had retreated uphill by 8 km in 50 years.

When snowflakes fall they have an open, feathery appearance, trap air and have a low density. Where snow collects in hollows, it becomes compressed by the weight of subsequent falls and gradually develops into a more compact, dense form called **firn** or **névé**. Firn is compacted snow which has experienced one winter's freezing and survived a summer's melting. It is composed of randomly oriented ice crystals separated by air passages. In temperate latitudes, such as in the Alps, summer meltwater percolates into the firn only to freeze either at night or during the following winter, thus forming an increasingly dense mass. Air is progressively squeezed out and after 20–40 years the firn will have turned into solid ice. This same process may take several hundred years in Antarctica and Greenland where there is no summer melting. Once ice has formed, it may begin to flow downhill, under the force of gravity, as a **glacier**.

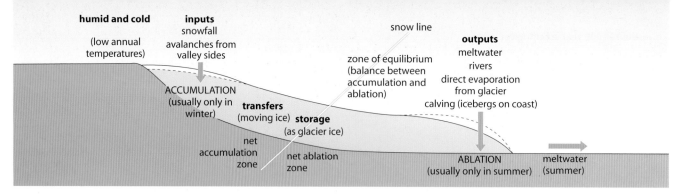

humid and cold
(low annual temperatures)

inputs
snowfall
avalanches from valley sides

snow line

outputs
meltwater
rivers
direct evaporation from glacier
calving (icebergs on coast)

ACCUMULATION
(usually only in winter)

zone of equilibrium
(balance between accumulation and ablation)

transfers
(moving ice) **storage**
(as glacier ice)

net accumulation zone

net ablation zone

ABLATION
(usually only in summer)

meltwater
(summer)

Figure 4.7

The glacial system showing inputs, stores, transfers and outputs

Glaciers and ice masses

Glaciers may be classified (Framework 7, page 167) according to size and shape – characteristics that are relatively easy to identify by field observation.

1 **Corrie** or **cirque glaciers** are small masses of ice occupying armchair-shaped hollows in mountains (Figure 4.14). They often overspill from their hollows to feed valley glaciers.

2 **Valley glaciers** are larger masses of ice which move down from either an icefield or a cirque basin source (Figure 4.8). They usually follow former river courses and are bounded by steep sides.

3 **Piedmont glaciers** are formed when valley glaciers extend onto lowland areas, spread out and merge.

Figure 4.8

The Gigjökull glacier, Iceland, showing the zones of accumulation, equilibrium (snow line) and ablation

Figure 4.9

The glacial budget or net balance (northern hemisphere)

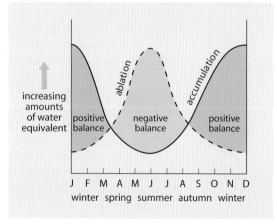

increasing amounts of water equivalent

positive balance | negative balance | positive balance

ablation

accumulation

J F M A M J J A S O N D
winter spring summer autumn winter

4 **Icecaps** and **ice sheets** are huge areas of ice which spread outwards from central domes. Apart from exposed summits of high mountains, called **nunataks**, the whole landscape is buried. Ice sheets, which once covered much of northern Europe and North America (Figure 4.3) are now confined to Antarctica (86 per cent of present-day world ice) and Greenland (11 per cent).

5 **Ice shelves** form when ice sheets reach the sea and begin to float. **Icebergs** form when ice breaks away, a process known as **calving**.

Glacial systems and budgets

A glacier behaves as a system (Framework 3, page 45), with inputs, stores, transfers and outputs (Figure 4.7). Inputs are derived from snow falling directly onto the glacier or from **avalanches** along valley sides (Case Study 4). The glacier itself is water in storage and transfer. Outputs from the glacier system include evaporation, calving (the formation of icebergs), and meltwater streams which flow either on top of or under the ice during the summer months.

The upper part of the glacier, where inputs exceed outputs, is known as the **zone of accumulation**; the lower part, where outputs exceed inputs, is called the **zone of ablation**. The **zone of equilibrium** is where the rates of accumulation and ablation are equal, and it corresponds with the snow line (Figures 4.7 and 4.8).

The **glacier budget**, or **net balance**, is the difference between the total accumulation and the total ablation for one year. In temperate glaciers (page 108), there is likely to be a negative balance in summer when ablation exceeds accumulation, and a positive balance in winter when the reverse occurs (Figure 4.9). If the summer and winter budgets cancel each other out, the glacier appears to be stationary. It appears stationary because the **snout** – i.e. the end of the glacier – is neither advancing nor retreating, although ice from the accumulation zone is still moving down-valley into the ablation zone. Because glaciers are acutely affected by changes to inputs and outputs, they are sensitive indicators of climatic change, both short term and long term.

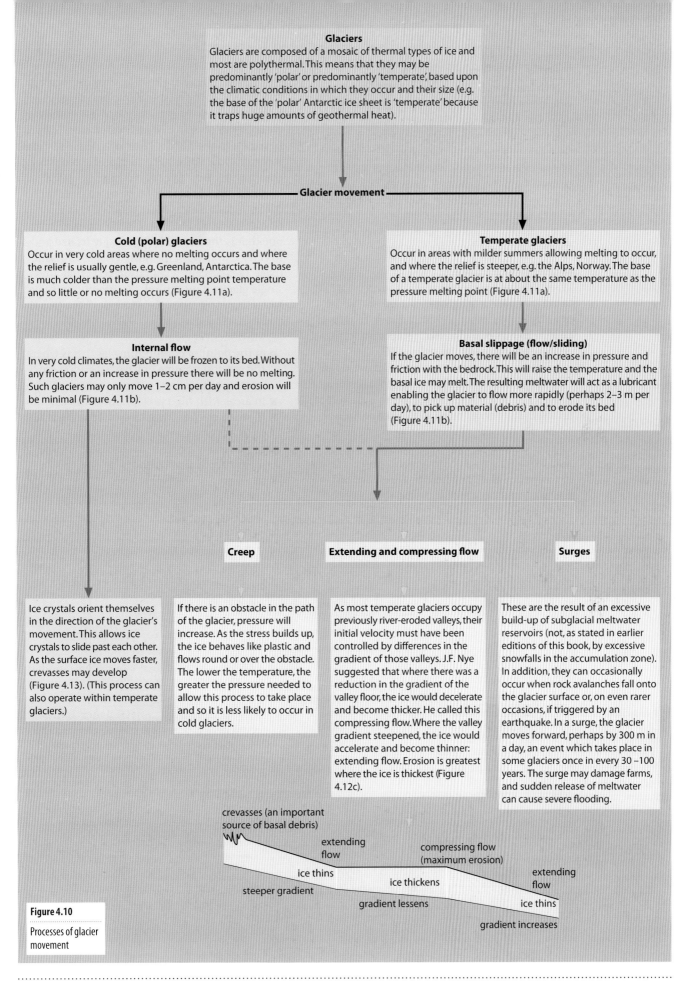

Glaciers

Glaciers are composed of a mosaic of thermal types of ice and most are polythermal. This means that they may be predominantly 'polar' or predominantly 'temperate', based upon the climatic conditions in which they occur and their size (e.g. the base of the 'polar' Antarctic ice sheet is 'temperate' because it traps huge amounts of geothermal heat).

Glacier movement

Cold (polar) glaciers

Occur in very cold areas where no melting occurs and where the relief is usually gentle, e.g. Greenland, Antarctica. The base is much colder than the pressure melting point temperature and so little or no melting occurs (Figure 4.11a).

Temperate glaciers

Occur in areas with milder summers allowing melting to occur, and where the relief is steeper, e.g. the Alps, Norway. The base of a temperate glacier is at about the same temperature as the pressure melting point (Figure 4.11a).

Internal flow

In very cold climates, the glacier will be frozen to its bed. Without any friction or an increase in pressure there will be no melting. Such glaciers may only move 1–2 cm per day and erosion will be minimal (Figure 4.11b).

Basal slippage (flow/sliding)

If the glacier moves, there will be an increase in pressure and friction with the bedrock. This will raise the temperature and the basal ice may melt. The resulting meltwater will act as a lubricant enabling the glacier to flow more rapidly (perhaps 2–3 m per day), to pick up material (debris) and to erode its bed (Figure 4.11b).

Creep

Extending and compressing flow

Surges

Ice crystals orient themselves in the direction of the glacier's movement. This allows ice crystals to slide past each other. As the surface ice moves faster, crevasses may develop (Figure 4.13). (This process can also operate within temperate glaciers.)

If there is an obstacle in the path of the glacier, pressure will increase. As the stress builds up, the ice behaves like plastic and flows round or over the obstacle. The lower the temperature, the greater the pressure needed to allow this process to take place and so it is less likely to occur in cold glaciers.

As most temperate glaciers occupy previously river-eroded valleys, their initial velocity must have been controlled by differences in the gradient of those valleys. J.F. Nye suggested that where there was a reduction in the gradient of the valley floor, the ice would decelerate and become thicker. He called this compressing flow. Where the valley gradient steepened, the ice would accelerate and become thinner: extending flow. Erosion is greatest where the ice is thickest (Figure 4.12c).

These are the result of an excessive build-up of subglacial meltwater reservoirs (not, as stated in earlier editions of this book, by excessive snowfalls in the accumulation zone). In addition, they can occasionally occur when rock avalanches fall onto the glacier surface or, on even rarer occasions, if triggered by an earthquake. In a surge, the glacier moves forward, perhaps by 300 m in a day, an event which takes place in some glaciers once in every 30–100 years. The surge may damage farms, and sudden release of meltwater can cause severe flooding.

crevasses (an important source of basal debris)

extending flow

compressing flow (maximum erosion)

extending flow

ice thins

ice thickens

ice thins

steeper gradient

gradient lessens

gradient increases

Figure 4.10

Processes of glacier movement

Glacier movement and temperature

The character and movement of ice depend upon whether it is warm or cold, which in turn depends upon the **pressure melting point (PMP)**. The pressure melting point is the temperature at which ice is on the verge of melting. A small increase in pressure can therefore cause melting. PMP is normally 0°C on the surface of a glacier, but it can be lower within a glacier (due to an increase in pressure caused by either the weight or the movement of ice). In other words, as pressure increases, then the freezing point for water falls below 0°C.

Warm and cold ice

Warm ice has a temperature of around 0°C (PMP) throughout its depth (Figure 4.11a) and consequently is able, especially in summer, to release large amounts of meltwater. Temperatures in cold ice are permanently below 0°C (PMP) and so there is virtually no meltwater (Figure 4.11a). It is the presence of meltwater that facilitates the movement of a glacier. Temperature is therefore an alternative criterion to size or shape for use when categorising glaciers – they may be either **temperate** (mainly warm ice) or **polar** (mainly cold ice) – Figure 4.10. Movement is much faster in temperate glaciers where the presence of meltwater acts as a lubricant and reduces fric-

tion (Figure 4.11b). It can take place by one of four processes: **basal flow** (or **slipping**); **creep**; **extending–compressing flow**; and **surges** (Figure 4.10). Polar glaciers move less quickly as, without the presence of meltwater, they tend to be frozen to their beds. The main process here is **internal flow**, although creep and extending–compressing flow may also occur.

Both types of glacier move more rapidly on the surface and away from their valley sides (Figure 4.12a and b), but it is the temperate one that is the more likely to erode its bed and to carry and deposit most material as **moraine** (page 117). Recent research suggests that any single glacier may exhibit, at different points along its profile, the characteristics of both polar and temperate glaciers.

Movement is greatest:
- at the point of equilibrium – as this is where the greatest volume of ice passes and consequently where there is most energy available
- in areas with high precipitation and ablation
- in small glaciers, which respond more readily to short-term climatic fluctuations
- in temperate glaciers, where there is more meltwater available, and
- in areas with steep gradients.

Figure 4.11

Comparison of temperature and velocity profiles in polar and temperate glaciers

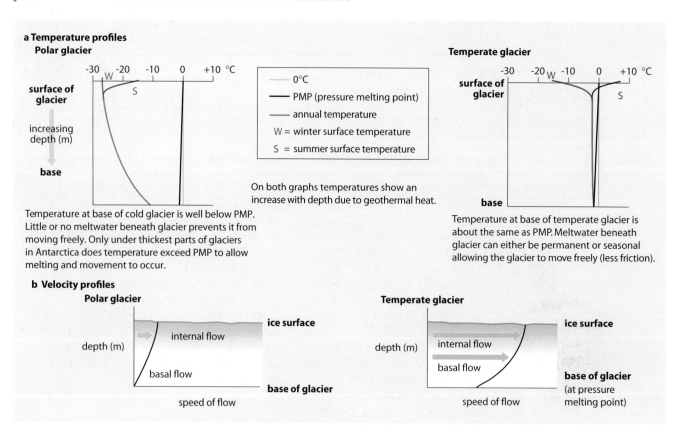

a Temperature profiles

Polar glacier

surface of glacier

increasing depth (m)

base

	0°C
	PMP (pressure melting point)
	annual temperature
W =	winter surface temperature
S =	summer surface temperature

On both graphs temperatures show an increase with depth due to geothermal heat.

Temperature at base of cold glacier is well below PMP. Little or no meltwater beneath glacier prevents it from moving freely. Only under thickest parts of glaciers in Antarctica does temperature exceed PMP to allow melting and movement to occur.

Temperate glacier

surface of glacier

base

Temperature at base of temperate glacier is about the same as PMP. Meltwater beneath glacier can either be permanent or seasonal allowing the glacier to move freely (less friction).

b Velocity profiles

Polar glacier

ice surface

internal flow

depth (m)

basal flow

base of glacier

speed of flow

Temperate glacier

ice surface

internal flow

depth (m)

basal flow

base of glacier (at pressure melting point)

speed of flow

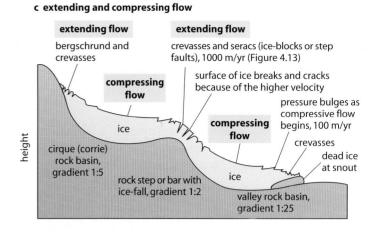

a surface velocity of a glacier

valley wall

glacier movement

valley wall

centre of glacier

vertical section across glacier

0 25 50 75 100
m per year

○ original position of marker poles
● position of marker poles after one year

b changes in velocity with depth

0

vertical section through glacier

glacier surface

200m

valley floor

0 10 20 30
m per year

c extending and compressing flow

extending flow
bergschrund and crevasses

extending flow
crevasses and seracs (ice-blocks or step faults), 1000 m/yr (Figure 4.13)

compressing flow

surface of ice breaks and cracks because of the higher velocity

compressing flow

pressure bulges as compressive flow begins, 100 m/yr

crevasses

dead ice at snout

height

ice

cirque (corrie) rock basin, gradient 1:5

rock step or bar with ice-fall, gradient 1:2

ice

valley rock basin, gradient 1:25

Figure 4.12

Plan view to show **a** and **b** velocity **c** flow of a glacier

Figure 4.13

Crevasses on an icefall, Skafta glacier, Iceland

Transportation by ice

Glaciers are capable of moving large quantities of debris. This rock debris may be transported in one of three ways:

1 **Supraglacial debris** is carried on the surface of the glacier as lateral and medial moraine (page 117). It consists of material that has fallen onto the glacier from the surrounding valley sides. In summer, the relatively small load carried by surface meltwater streams often disappears down crevasses.

2 **Englacial debris** is material carried within the body of the glacier. It may once have been on the surface, only to be buried by later snow-falls or to fall into crevasses (Figure 4.4).

3 **Subglacial debris** is moved along the floor of the valley either by the ice or by melt-water streams formed by pressure melting (page 108).

Glacial erosion

Ice that is stationary or contains little debris has limited erosive power, whereas moving ice carrying with it much debris can drastically alter the landscape. Although ice lacks the turbulence and velocity of water in a river, it has the 'advantage' of being able to melt and refreeze in order to overcome obstacles in its path (Figure 4.10) and consequently has the ability to lower (i.e. erode) the landscape more quickly than can running water. Virtually all the glacial processes of erosion are physical, as the climate tends to be too cold for chemical reactions to operate (Figure 2.10).

Processes of glacial erosion

The processes associated with glacial erosion are: frost shattering, abrasion, plucking, rotational movement, and extending and compressing flow.

Frost shattering

This process (page 40) produces much loose material which may fall from the valley sides onto the edges of the glacier to form **lateral moraine**, be covered by later snowfall, or plunge down crevasses to be transported as **englacial debris**. Some of this material may be added to rock loosened by frost action as the climate deteriorated (but before glaciers formed) to form **basal debris** (page 117).

Abrasion

This is the sandpapering effect of angular material embedded in the glacier as it rubs against the valley sides and floor. It usually produces smoothened, gently sloping landforms.

Plucking

At its simplest, this process involves the glacier freezing onto rock outcrops, after which ice movement pulls away masses of rock. In reality, as the strength of the bedrock is greater than that of the ice, it would seem that only previously loosened material can be removed. Material may be continually loosened by one of three processes:

1 The relationship between local pressure and temperature (the PMP) produces sufficient meltwater for freeze–thaw activity to break up the ice-contact rock.
2 Water flowing down a **bergschrund** (a large, crevasse-like feature found near the head of some glaciers – Figure 4.14b) or smaller crevasses will later freeze onto rock surfaces.
3 Removal of layers of bedrock by the glacier causes a release in pressure and an enlarging of joints in the underlying rocks (pressure release, page 41).

Plucking generally creates a jagged-featured landscape.

Figure 4.14

Processes in the formation of a cirque

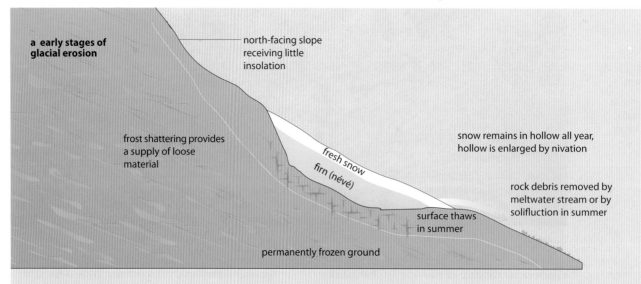

a early stages of glacial erosion

north-facing slope receiving little insolation

frost shattering provides a supply of loose material

fresh snow

firn (névé)

snow remains in hollow all year, hollow is enlarged by nivation

rock debris removed by meltwater stream or by solifluction in summer

surface thaws in summer

permanently frozen ground

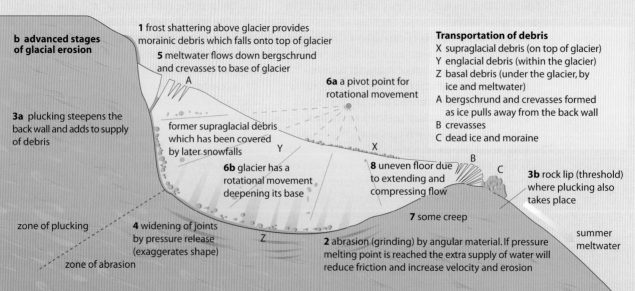

b advanced stages of glacial erosion

1 frost shattering above glacier provides morainic debris which falls onto top of glacier

5 meltwater flows down bergschrund and crevasses to base of glacier

A

6a a pivot point for rotational movement

3a plucking steepens the back wall and adds to supply of debris

former supraglacial debris which has been covered by later snowfalls

Y

6b glacier has a rotational movement deepening its base

8 uneven floor due to extending and compressing flow

X

B

C

3b rock lip (threshold) where plucking also takes place

Transportation of debris
X supraglacial debris (on top of glacier)
Y englacial debris (within the glacier)
Z basal debris (under the glacier, by ice and meltwater)
A bergschrund and crevasses formed as ice pulls away from the back wall
B crevasses
C dead ice and moraine

zone of plucking

4 widening of joints by pressure release (exaggerates shape)

Z

7 some creep

2 abrasion (grinding) by angular material. If pressure melting point is reached the extra supply of water will reduce friction and increase velocity and erosion

summer meltwater

zone of abrasion

(Numbers refer to different glacial processes)

Figure 4.15

A cirque in West Wales (Cader Idris). The steep wall maintains its shape as freeze–thaw still operates. Broken-off material forms scree which is beginning to infill the lake, which itself has been dammed behind a natural rock lip

Rotational movement

This is a downhill movement of ice which, like a landslide (Figure 2.17), pivots about a point. The increase in pressure is responsible for the over-deepening of a cirque floor (Figure 4.14b).

Extending and compressing flow

Figures 4.10 and 4.12c show how this process causes differences in the rate of erosion at the base of a glacier.

Maximum erosion occurs:

- where temperatures fluctuate around 0°C, allowing frequent freeze–thaw to operate
- in areas of jointed rocks which can be more easily frost shattered
- where two tributary glaciers join, or the valley narrows, giving an increased depth of ice, and
- in steep mountainous regions in temperate latitudes, where the velocity of the glacier is greatest.

Landforms produced by glacial erosion
Cirques

These are amphitheatre or armchair-shaped hollows with a steep back wall and a rock basin (Figure 4.15). They are also known as **corries** (Scotland) and **cwms** (Wales – Figures 4.25 and 4.26).

During periglacial times (Chapter 5), before the last glacial, snow collected in hollows, especially on north-facing slopes. A series of processes, collectively known as **nivation** and which included freeze–thaw, solifluction and possibly chemical weathering, operated under and around the snow patch (Figure 4.14a). These processes caused the underlying rocks to disintegrate. The resultant debris was then removed by summer meltwater streams to leave, in the enlarged hollow, an embryo cirque. It has been suggested that the overdeepening process might need several periglacials or interglacials and

glacials in which to form. As the snow patch grew, its layers became increasingly compressed to form firn and, eventually, ice (page 105).

It is accepted that several processes interact to form a fully developed cirque (Figure 4.14b). Plucking is one process responsible for steepening the back wall, but this partly relies upon a supply of water for freeze–thaw and partly upon pressure release in well-jointed rocks. A rotational movement, aided by water from pressure point melting and angular subglacial debris from frost shattering, enables abrasion to over-deepen the floor of the cirque. A **rock lip** develops where erosion decreases. This may be increased in height by the deposition of morainic debris at the glacier's snout. When the climate begins to get warmer, the ice remaining in the hollow melts to leave a deep, rounded lake or tarn (Figures 4.15 and 4.26).

In Britain, as elsewhere in the northern hemisphere, cirques are nearly always oriented between the north-west (315°), through the north-east (where the frequency peaks) to the south-east (135°). This is because in the UK:

- northern slopes receive least **insolation** and so glaciers remained there much longer than those facing in more southerly directions (less melting on north-facing slopes)
- western slopes face the sea and, although still cold, the relatively warmer winds which blew from that direction were more likely to melt the snow and ice (more snow accumulated on east-facing slopes)
- the prevailing westerly winds cause snow to drift into east-facing hollows.

Lip orientation is the direction of an imaginary line from the centre of the back wall of the cirque to its lip. Of 56 cirques identified in the Snowdon area, 51 have a lip orientation of between 310° and 120°, and of 15 on Arran, 14 have an orientation between 5° and 115°.

Mean, median and mode are all types of average (measures of dispersion, Framework 8, page 246).

1 The **mean** (or arithmetic average) is obtained by totalling the values in a set of data and dividing by the number of values in that set. It is expressed by the formula:

$$\bar{x} = \frac{\Sigma x}{n}$$

where:

$\bar{x}$ = mean, Σ = the sum of, x = the value of the variable, n = the number of values in the set

The mean is reliable when the number of values in the sample is high and their range, i.e. the difference between the highest and lowest values, is low, but it becomes less reliable as the number in the sample decreases, as it is then influenced by extreme values.

2 The **median** is the mid-point value of a set of data. For example, you have to find the median height of students in your class. To do this you will have to rank each person in descending order of height. If there were 15 students then the mid-point would be the eighth student as there will be seven taller and seven shorter. Had there been an even number in the sample, such as 16, then the median would have been the mean of the two middle values. The median is a less accurate measure of dispersion than the mean because widely differing sets of data can return the same median, but it is less distorted by extreme values.

3 The **mode** is the value or class that occurs most frequently in the data. In the set of values 4, 6, 4, 2, 4 the mode would be 4. Although it is the easiest of the three 'averages' to obtain, it has limited value. Some data may not have two values in the same class (e.g. 1, 2, 3, 4, 5), while others may have more than one modal value (e.g. 1, 1, 2, 4, 4).

Relationships between mean, median and mode

When data is plotted on a graph we can often make useful observations about the shape of the curve. For example, we would expect A-level results nationally to show a few top grades, a smaller number of 'unclassifieds' and a large number of average passes. Graphically this would show a normal distribution, with all three averages at the peak. If the distribution is skewed, then by definition only the mode will lie at the peak (Figure 4.16).

Figure 4.16

Normal and skewed distributions

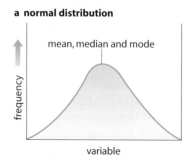

a normal distribution

mean, median and mode

frequency

variable

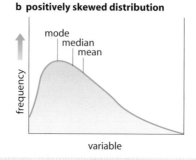

b positively skewed distribution

mode
median
mean

frequency

variable

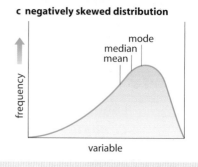

c negatively skewed distribution

mode
median
mean

frequency

variable

Arêtes and pyramidal peaks

When two adjacent cirques erode backwards or sideways towards each other, the previously rounded landscape is transformed into a narrow, rocky, steep-sided ridge called an **arête**, as at Striding Edge in the Lake District (Figure 4.17) and Crib Goch on Snowdon (Figure 4.25). If three or more cirques develop on all sides of a mountain, a **pyramidal peak**, or horn, may be formed. This feature has steep sides and several arêtes radiating from the central peak (Figures 4.18 and 4.19), e.g. the Matterhorn.

Figure 4.17

An arête: Striding Edge on Helvellyn in the Lake District

Figure 4.18

Arêtes in the Karakoram
Mountains, northern Pakistan

Figure 4.19

A pyramidal peak:
Machhapuchare, Nepal

Glacial troughs, rock steps, truncated spurs and hanging valleys

These features are interrelated in their formation. Valley glaciers straighten, widen and deepen
preglacial valleys, turning the original V-shaped,
river-formed feature into the characteristic U
shape typical of glacial erosion, e.g. Wast Water in
the Lake District (Figure 4.20). These steep-sided,
flat-floored valleys are known as **glacial troughs**.
The overdeepening of the valleys is credited to the
movement of ice which, aided by large volumes
of meltwater and subglacial debris, has a greater
erosive power than that of rivers. Extending and
compressing flow may overdeepen parts of the
trough floor, which later may be occupied by long,
narrow **ribbon lakes**, such as Wast Water, or may
leave less eroded, more resistant **rock steps**.

Theories to explain pronounced overdeepening of valley floors are debated amongst glaciologists and geomorphologists. Suggested causes
include: extra erosion following the
confluence of two glaciers; the presence of
weaker rocks; an area of rock deeply weathered
in preglacial times; or a zone of well-jointed rock.
Should the deepening of the trough continue
below the former sea-level, then during deglaciation and subsequent rises in sea-level the valley
may become submerged to form a **fiord** (Figures
4.21 and 6.48).

Abrasion by englacial and subglacial debris
and plucking along the valley sides remove the
tips of preglacial interlocking spurs leaving cliff-
like **truncated spurs** (Figure 4.20, and to the left
of Figure 4.27).

Figure 4.20

Glacial trough with ribbon lake:
Wast Water in the Lake District

Figure 4.21

A fiord: Milford Sound, New Zealand

Figure 4.22

Hanging valley:
Lake Bigden,
Norway

Hanging valleys result from differential erosion between a main glacier and its tributary glaciers. The floor of any tributary glacier is deepened at a slower rate so that when the glaciers melt it is left hanging high above the main valley and its river has to descend by a single waterfall or a series of waterfalls, e.g. Lake Bigden, Norway (Figure 4.22) and Cwm Dyli, Snowdonia (Figure 4.25).

Striations, roches moutonnées, rock drumlins and crag and tail

These are all smaller erosion features which help to indicate the direction of ice movement. As a glacier moves across areas of exposed rock, larger fragments of angular debris embedded in the ice tend to leave a series of parallel scratches and grooves called **striations** (e.g. Central Park in New York).

A **roche moutonnée** is a mass of more resistant rock. It has a smooth, rounded upvalley or stoss slope facing the direction of ice flow, formed by abrasion, and a steep, jagged, down-valley or lee slope resulting from plucking (Figures 4.23 and 4.24).

Rock drumlins are more streamlined bedrock which lack the quarried lee face of the roche moutonnée. They are sometimes referred to as **whalebacks** as they resemble the backs of whales breaking the ocean surface.

A **crag and tail** consists of a larger mass of resistant rock or crag (e.g. the basaltic crag upon which Edinburgh Castle has been built) which protected the lee-side rocks from erosion, thus forming a gently sloping tail of deposited material (e.g. the tail down which the Royal Mile extends).

It should be remembered that while many of these erosional landforms may be found together in most glaciated uplands, their arrangement, frequency and presence is likely to change from one area to another. Places 15 describes some of these glacial features as found in one part of Snowdonia.

Figure 4.23

A roche moutonnée:
Yosemite National
Park, California

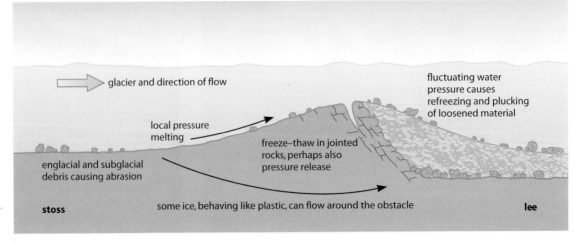

glacier and direction of flow

fluctuating water
pressure causes
refreezing and plucking
of loosened material

local pressure
melting

freeze–thaw in jointed
rocks, perhaps also
pressure release

englacial and subglacial
debris causing abrasion

some ice, behaving like plastic, can flow around the obstacle

stoss

lee

Figure 4.24

The formation of a
roche moutonnée

Llyn Peris – ribbon lake

Nant Llanberis – glacial trough

Snowdon – pyramidal peak

Crib y Ddysgl – arête

Cwmbrwynog – arête

Y Lliwedd – arête

Bwlch Main – arête

Crib Goch – arête

Glaslyn – corrie

hanging valley

Llyn Llydaw – corrie

Cwm Dyli – hanging valley

truncated spur

waterfall

Llyn Gwynant – ribbon lake

Nant Gwynant – glacial trough

Figure 4.25

Landsketch of glacial features in Snowdonia (looking west)

Snowdonia is an example of a glaciated upland area. Although Snowdon itself has the characteristics of a pyramidal peak, the ice age was too short (by several thousand years) for the completed development of the classic pyramidal shape which makes the appearance of the Matterhorn so spectacular (compare Figure 4.19). What *are* well developed are the arêtes, such as Crib Goch and Bwlch Main, which radiate from the central peak. Between these arêtes are up to half a dozen cirques (cwms, as this is Wales), including the eastward-facing Glaslyn and the north-eastward-oriented (page 111) Llyn (lake) Llydaw. Glaslyn, which is trapped by a rock lip, is 170 m higher than Llyn Llydaw (Figure 4.26). Striations and roches moutonnées can be found in several places where the rocks are exposed on the surface. To the north and south-east of Snowdon are the glacial troughs of Nant (valley) Llanberis, Nant Ffrancon and Nant Gwynant. These valleys have the characteristic U shape, with steep valley sides, truncated spurs and a flat valley floor (Figure 4.27). Located on the valley floors are ribbon lakes, including Llyn Peris and Llyn Gwynant (Figure 3.24). Numerous small rivers, with their sources in hanging valleys, descend by waterfalls, as at Cwm Dyli, into the two main valleys. Although the ice has long since gone, the actions of frost and snow, together with that of rain and more recently people, continue to modify the landscape – remember that rarely does a landscape exhibit stereotyped 'textbook' features (see Figure 4.25)!

Figure 4.26

Llyn Glaslyn: a cirque lake formed behind a rock lip, and a hanging valley into Llyn Llydaw

Figure 4.27

Nant Ffrancon: a glacial trough with, at the sides, truncated spurs and hanging valleys

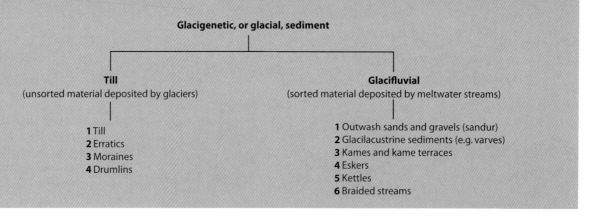

```
                    Glacigenetic, or glacial, sediment
                                    |
         ┌──────────────────────────┴──────────────────────────┐
         |                                                       |
        Till                                               Glacifluvial
(unsorted material deposited by glaciers)        (sorted material deposited by meltwater streams)
         |                                                       |
    1 Till                                          1 Outwash sands and gravels (sandur)
    2 Erratics                                      2 Glacilacustrine sediments (e.g. varves)
    3 Moraines                                      3 Kames and kame terraces
    4 Drumlins                                      4 Eskers
                                                    5 Kettles
                                                    6 Braided streams
```

Figure 4.28

Landforms resulting from glacial deposition

Glacial deposition

Glacigenetic sediment (or **glacial sediments**) has replaced 'drift' as the term which was used historically by British geologists and glaciologists when referring collectively to all glacial deposits (Figure 4.28). These deposits, which include boulders, gravels, sands and clays, may be subdivided into **till**, which includes all material deposited directly by the ice, and **glacifluvial material**, which is the debris deposited by meltwater streams. Glacifluvial material includes deposits which may have been deposited initially by the ice and which were later picked up and redeposited by meltwater – either during or after the ice age. Till consists of largely unsorted material, whereas glacifluvial deposits have been sorted. Deposition occurs in upland valleys and across lowland areas. A study of glacigenetic deposits helps to explain the:

■ nature and extent of an ice advance
■ frequency of ice advances
■ sources and directions of ice movement, and
■ postglacial chronology (including climatic changes, page 294).

Till deposits

Although the term **till** is often applied today to all materials deposited by ice, it is more accurately used to mean an unsorted mixture of rocks, clays and sands. This material was largely transported as supraglacial debris and later deposited to form moraine – either during periods of active ice movement, or at times when the glacier was in retreat. In Britain, till was commonly called **boulder clay** but – since some deposits may contain neither boulders nor clay – this term is now obsolete. Individual stones are sub-angular – that is, they are not rounded like river or beach material but neither do they possess the sharp edges of rocks that have recently been broken up by frost shattering. The composition of till reflects the character of the rocks over which it has passed; East Anglia, for example, is covered by chalky till because the ice passed over a chalk escarpment, i.e. the East Anglian Heights.

Till fabric analysis is a fieldwork technique used to determine the direction and source of glacial deposits. Stones and pebbles carried by a glacier tend to become aligned with their long axes parallel to the direction of ice flow, as this offers least resistance to the ice. For example, a small sample of 50 stones was taken from a moraine in Glen Rosa, Arran. As each stone was removed, its geology was examined and its orientation was carefully measured using a compass. The results allowed two conclusions to be reached:

1 The pebbles were grouped into classes of 20° and plotted onto a rose diagram (Figure 4.29). The classes were plotted as respective radii from the midpoint of the diagram and then the ends of the radii were joined up to form a star-like polygonal graph. As each stone has two orientations which must be opposites (e.g. 10° and 190°), the graph will be symmetrical. The results show that the ice must have come from the north-north-west or the south-south-east.

2 Although most of the pebbles taken in the sample were composed of local rock, some were of material not found on the island (erratics). This suggests that some of the ice must have come from the Scottish mainland.

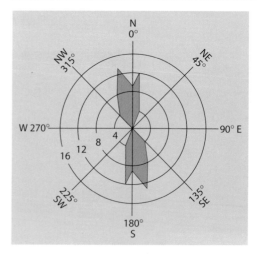

Figure 4.29

Till fabric analysis: orientation of a sample of stones taken from a moraine in Glen Rosa, Arran

Landforms characteristic of glacial deposition

Erratics

These are boulders picked up and carried by ice, often for many kilometres, to be deposited in areas of completely different lithology (Figure 4.30). **Lithology** is the study of the nature and composition of rocks. By determining where the boulders originally came from, it is possible to track ice movements. For example, volcanic material from Ailsa Craig in the Firth of Clyde has been found 250 km to the south on the Lancashire plain, while some deposits on the north Norfolk coast originated in southern Norway.

Figure 4.30

An erratic near Ingleborough in the Yorkshire Dales: Silurian rock lying on top of Carboniferous limestone (Figure 1.1)

Moraine

Moraine is a type of landform that develops when the debris carried by a glacier is deposited. It is *not*, therefore, the actual material that is being transported by the glacier – with the exception of the medial moraine, which is a term that refers to a landform both on the glacier and in the valley after glacial recession. It is possible to recognise at least five types of moraine (Figure 4.31):

- **Lateral moraine** is formed from debris derived from frost shattering of valley sides and carried along the edges of the glacier (Figure 4.32). When the glacier melts, it leaves an embankment of material along the valley side.
- **Medial moraine** is found in the centre of a valley and results from the merging of two lateral moraines where two glaciers joined (Figure 4.32).
- **Terminal** or **end moraine** is often a high mound (or series of mounds) of material extending across a valley, or lowland area, at right-angles to and marking the maximum advance of the glacier or ice sheet.
- **Recessional moraines** mark interruptions in the retreat of the ice when the glacier or ice sheet remained stationary long enough for a mound to build up. Recessional moraines are usually parallel to the terminal moraine.
- **Push moraines** may develop if the climate deteriorates sufficiently for the ice temporarily to advance again. Previously deposited moraine may be shunted up into a mound. It can be recognised by individual stones which have been pushed upwards from their original horizontal positions, or even large blocks of sediment that have been bulldozed whole, while frozen.

Figure 4.31

Types of moraine

1 cirque glacier
2 lateral moraine
3 medial moraine
4 valley glacier
5 frost shattering
6 meltwater streams
7 recessional moraine
8 push moraine
9 terminal moraine

Figure 4.32

Medial and lateral moraines, Meade Glacier, Alaska

Figure 4.33

Morainic mounds above Haweswater, Cumbria

Drumlins

These are smooth, elongated mounds of till with their long axis parallel to the direction of ice movement. Drumlins may be over 50 m in height, over 1 km in length and nearly 0.5 km in width. The steep stoss end faces the direction from which the ice came, while the lee side has a more gentle, streamlined appearance. The highest point of the feature is near to the stoss end (Figure 4.34). The shape of drumlins can be described by using the **elongation ratio**:

$$E = \frac{I}{W}$$

where I is the maximum bedform length, and W is the maximum bedform width. Drumlins are always longer than they are wide, and they are usually found in **swarms** or *en echelon*.

There is much disagreement as to how drumlins are formed. Theories suggest they may be an erosion feature, or formed by deposition around a central rock. However, neither of these accounts for the fact that the majority of drumlins are composed of till which, lacking a central core of rock and consisting of unsorted material, would be totally eroded by moving ice. The most widely accepted view is that they were formed when the ice became overloaded with material, thus reducing the capacity of the glacier. The reduced competence may have been due to the melting of the glacier or to changes in velocity related to the pattern of extending–compressing flow. Once the material had been deposited, it may then have been moulded and streamlined by later ice movement. The most recent theory (1987) is based on evidence that drumlins can be composed of both till and glacifluvial sediments. The most widely accepted view now is that 'they are subglacially deformed masses of pre-existing sediment to which more sediment may be added by the melting out of debris from the glacier base' (D. Evans, 1999).

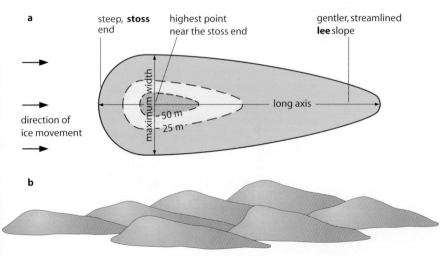

Figure 4.34

Drumlins
a plan showing typical dimensions
b swarm – *en echelon*

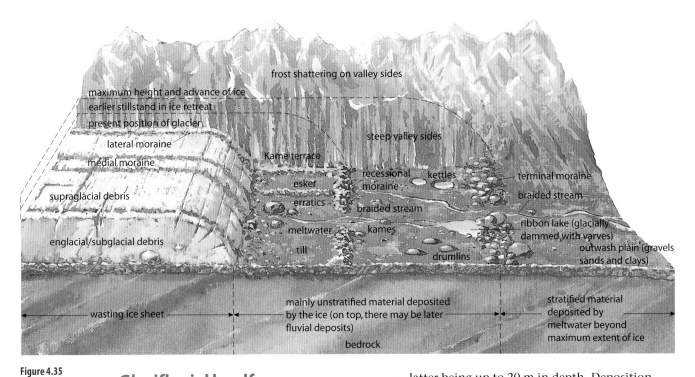

frost shattering on valley sides

maximum height and advance of ice
earlier stillstand in ice retreat
present position of glacier

lateral moraine

medial moraine

supraglacial debris

englacial/subglacial debris

Kame terrace

esker

erratics

meltwater

till

steep valley sides

recessional moraine

braided stream

kames

kettles

drumlins

terminal moraine

braided stream

ribbon lake (glacially dammed, with varves)

outwash plain (gravels sands and clays)

wasting ice sheet

mainly unstratified material deposited by the ice (on top, there may be later fluvial deposits)

bedrock

stratified material deposited by meltwater beyond maximum extent of ice

Figure 4.35

Features of lowland glaciation

Glacifluvial landforms

Glacifluvial landforms are those moulded by glacial meltwater and have, in the past, been considered to be mainly depositional. More recently it has been realised that meltwater plays a far more important role in the glacial system than was previously thought, especially in temperate glaciers and in creating erosion features as well as depositional landforms. Most meltwater is derived from ablation. The discharge of glacial streams, both supraglacial and subglacial, is high during the warmer, if not warm, summer months. As the water often flows under considerable pressure, it has a high velocity and is very turbulent. It is therefore able to pick up and transport a larger amount of material than a normal river of similar size. This material can erode vertically, mainly through abrasion but partly by solution, to create sub-glacial valleys and large potholes, some of the

latter being up to 20 m in depth. Deposition occurs whenever there is a decrease in discharge, and it is responsible for a group of landforms (Figures 4.35 and 4.37).

Outwash plains (sandur)

These are composed of gravels, sands and, uppermost and furthest from the snout, clays. They are deposited by meltwater streams issuing from the ice either during summer or when the glacier melts. The material may originally have been deposited by the glacier and later picked up, sorted and dropped by running water beyond the maximum extent of the ice sheets. In parts of the North German Plain, deposits are up to 75 m deep. Outwash material may also be deposited on top of till following the retreat of the ice (Figure 4.35).

Glacilacustrine sediments (varves)

A varve is a distinct layer of silt lying on top of a layer of sand, deposited annually in lakes found near to glacial margins. The coarser, lighter-coloured sand is deposited during late spring when meltwater streams have their peak discharge and are carrying their maximum load. As discharge decreases towards autumn when temperatures begin to drop, the finer, darker-coloured silt settles. Each band of light and dark materials represents one year's accumulation (Figure 4.36). By counting the number of varves, it is possible to date the origin of the lake; variations in the thickness of each varve indicate warmer and colder periods (e.g. greater melting causing increased deposition).

Figure 4.36

The formation of varves in a postglacial lake

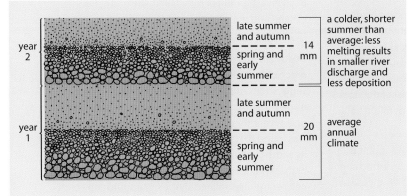

year 2

late summer and autumn

spring and early summer

14 mm

a colder, shorter summer than average: less melting results in smaller river discharge and less deposition

year 1

late summer and autumn

spring and early summer

20 mm

average annual climate

Kames and kame terraces

Kames are undulating mounds of sand and gravel deposited unevenly by meltwater, similar to a series of deltas, along the front of a stationary or slowly melting ice sheet (Figure 4.35). As the ice retreats, the unsupported kame often collapses. Kame terraces, also of sand and gravel, are flat areas found along the sides of valleys. They are deposited by meltwater streams flowing in the trough between the glacier and the valley wall. Troughs occur here because, in summer, the valley side heats up faster than the glacier ice and so the ice in contact with it melts. Kame terraces are distinguishable from lateral moraines by their sorted deposits.

Eskers

These are very long, narrow, sinuous ridges composed of sorted coarse sands and gravel. It is thought that eskers are the fossilised courses of subglacial meltwater streams. As the channel is restricted by ice walls, the hydrostatic pressure and the transported load are both considerable. As the bed of the channel builds up (there is no floodplain), material is left above the surrounding land following the retreat of the ice. Like kames, eskers usually form during times of deglaciation (Figure 4.35).

Kettles

These form from detached blocks of ice, left by the glacier as it retreats, and then partially buried by the glacifluvial deposits left by meltwater streams. When the ice blocks melt, they leave enclosed depressions which often fill with water to form kettle-hole lakes and 'kame and kettle' topography (Figure 4.35).

Braided streams

Channels of meltwater rivers often become choked with coarse material as a result of the marked seasonal variations in discharge (compare Figures 3.32 and 5.16).

Places 16 Arran: glacial landforms

Using fieldwork to answer an Advanced GCE question: 'Describe the landforms found near the snout of a former glacier.'

Figure 4.28 lists the types of feature formed by glacial deposition, subdividing them into those composed of unsorted material, left by the glacier, and sorted material deposited by glacifluvial activity. If the snout of a glacier had remained stationary for some time, indicating a balance between accumulation and ablation, and had then slowly retreated, several of these landforms might be visible following deglaciation. One such site studied by a sixth form was the lower Glen Rosa valley on the Isle of Arran (Figure 4.37).

The dominant feature was a mound **A**, 14 m high, into which the Rosa Water had cut, giving a fine exposed section of the deposited material. As the mound was a long, narrow, ridge-like feature extending across the valley, it was suggested that it might be either a terminal or a recessional moraine. It was concluded that the feature was ice-deposited because the material was unsorted: many of the largest boulders were high up in the exposure; also, most of the stones were sub-angular (not more rounded as might be expected in glacifluvial deposits).

However, an observation downstream at point **B** revealed that material there was also unsorted and this, together with some large granite erratics seen earlier nearer the coast, seemed to indicate that the mound could not be a terminal moraine as it did not mark the maximum advance of the ice. When a till fabric analysis was carried out, it was noted that the average dip of the stones was about 25°, suggesting that the feature might instead have been a push moraine resulting from a minor re-advance during deglaciation. The orientation of 50 sample stones (Figure 4.29) showed that the ice must have come either from the north-north-west (probable, as this was the highland) or the south-south-east (unlikely, as the lower ground would not be the source of a glacier). An examination of the geology of the stones showed that 80 per cent were granite, and therefore were erratics carried from the upper Rosa valley; 15 per cent were schists (the local rock); and 5 per cent were other igneous rocks not found on the island. It was inferred from the presence of these other rocks that some of the ice must have originated on the Scottish mainland. Also at point **B**, an investigation of river banks showed a mass of sand and gravel with some level of sorting – as might be expected in an outwash area.

Upstream from **A** was a second mound, **C**, filling much of the valley floor (Figure 4.38). Student suggestions as to the nature of the feature included its being a drumlin, a lateral, a medial, a recessional or even another push moraine. When measured, it was found that its length was slightly greater than its width (an elongation ratio of 1.25:1) and the highest point was nearest the up-valley end; it had neither the streamlined shape nor a sufficiently high

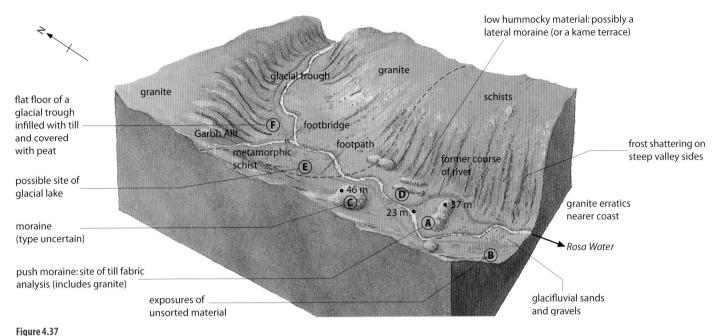

Figure 4.37

Sketch to show features of glacial deposition in the lower Glen Rosa valley, Arran

Labels on Figure 4.37:

flat floor of a glacial trough infilled with till and covered with peat

possible site of glacial lake

moraine (type uncertain)

push moraine: site of till fabric analysis (includes granite)

exposures of unsorted material

granite

glacial trough

granite

low hummocky material: possibly a lateral moraine (or a kame terrace)

schists

Garbh Allt

footbridge

footpath

metamorphic schist

F

E

C • 46 m

D

23 m •

A

B

• 37 m

former course of river

frost shattering on steep valley sides

granite erratics nearer coast

Rosa Water

glacifluvial sands and gravels

---

elongation ratio to be a drumlin (and there were no signs of a swarm!). It appeared to be too far from the valley side to be a lateral moraine; and as two glaciers could not have met here, neither could it have been a medial moraine. It was concluded that it *was* another moraine – perhaps formed during an intermediate stillstand in the glacier's retreat, or if the glacier lost momentum after having negotiated a bend in the glacial trough.

Across the river (**D**), was an area of low hummocky material winding along the foot of the valley side to as far as **A**. It was speculated that the feature may have been formed in one of three ways: meltwater depositing sands and gravel between the valley side and the former glacier as a kame terrace; a lateral moraine from frost shattering on the valley sides; or solifluction deposits (page 47) formed as the climate grew milder and the glacier retreated (the feature was not flat enough for a river terrace to be seriously considered).

Upstream, the valley floor was extremely flat (**E**). This could be the remains of a former glacial lake, formed when meltwater from the retreating glacier had become trapped behind the moraine at **C** and before it had had time to cut through the deposits. It was impossible to gain a profile to prove or disprove the existence of a lake.

After crossing the Garbh Allt (a hanging valley), the steep-sided, flat-floored U-shape of the glacial trough through which the Rosa Water flows was visible. The flatness of the floor was probably due to

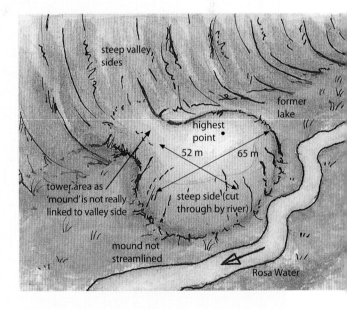

Figure 4.38

Fieldsketch of landform at **C** in Figure 4.37

Labels on Figure 4.38:

steep valley sides

former lake

highest point

52 m   65 m

tower area as 'mound' is not really linked to valley side

steep side (cut through by river)

mound not streamlined

Rosa Water

---

the deposition of subglacial debris – although the till has since been covered by peat, a symptom of the cold, wet conditions.

Although not every feature of glacial deposition was present – there was no evidence of eskers or kettles – this small area did contain several of the landforms and deposits that might be expected at, or near to, the snout of a former glacier.

Other effects of glaciation

Drainage diversion and proglacial lakes

Where ice sheets expand, they may divert the courses of rivers. For example, the preglacial River Thames flowed in a north-easterly direction. It was progressively diverted southwards by advancing ice (Figure 4.40).

Where ice sheets expand and dam rivers, proglacial lakes are created (Figure 4.39), e.g. Lakes Lapworth and Harrison (Figure 4.40). Before the ice age, the River Severn flowed northwards into the River Dee, but this route became blocked during the Pleistocene by Irish Sea ice. A large lake, Lapworth, was impounded against the edge of the ice until the waters rose high enough to breach the lowest point in the southern watershed. As the water overflowed through an **overspill channel**, there was rapid vertical erosion which formed what is now the Ironbridge Gorge. When the ice had completely melted, the level of this new route was lower than the original course (which was also blocked by drift), forcing the present-day River Severn to flow southwards.

Other rivers, e.g. the Warwickshire Avon (Figure 4.40) and the Yorkshire Derwent (Places 17), have also been diverted as a consequence of glacial activity. Sometimes the glacial overspill channels have been abandoned, e.g. at Fenny Compton, where the Warwickshire Avon temporarily flowed south-east into the Thames (O^1 in Figure 4.40). Proglacial lakes are also found behind eskers and recessional moraines.

Figure 4.40

Glacial diversion of drainage and proglacial lakes in England and Wales

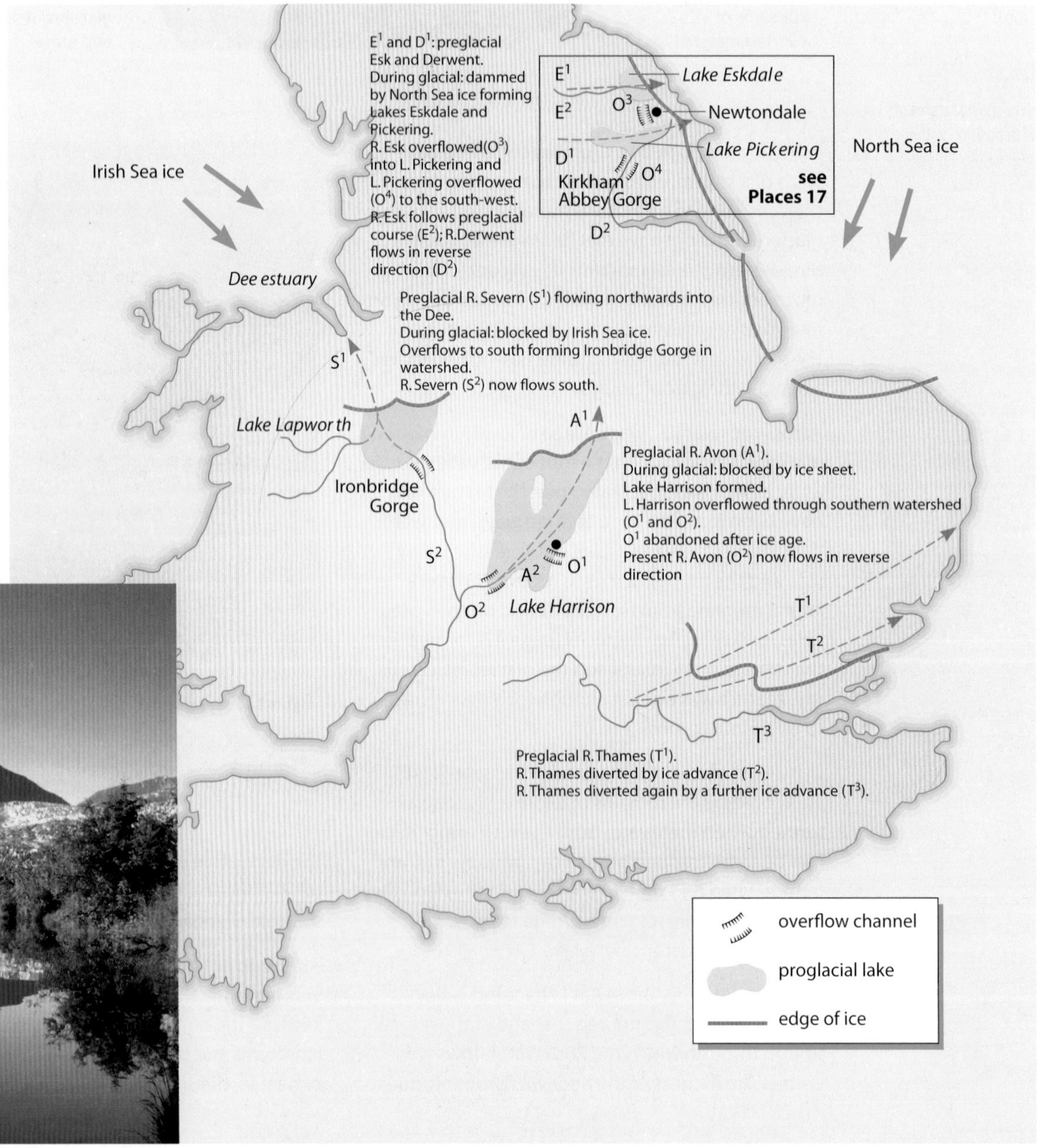

Figure 4.39

Ice-dammed lake: Mendenhall Glacier, Alaska

E^1 and D^1: preglacial Esk and Derwent. During glacial: dammed by North Sea ice forming Lakes Eskdale and Pickering. R. Esk overflowed(O^3) into L. Pickering and L. Pickering overflowed (O^4) to the south-west. R. Esk follows preglacial course (E^2); R.Derwent flows in reverse direction (D^2)

Preglacial R. Severn (S^1) flowing northwards into the Dee. During glacial: blocked by Irish Sea ice. Overflows to south forming Ironbridge Gorge in watershed. R. Severn (S^2) now flows south.

Preglacial R. Avon (A^1). During glacial: blocked by ice sheet. Lake Harrison formed. L. Harrison overflowed through southern watershed (O^1 and O^2). O^1 abandoned after ice age. Present R. Avon (O^2) now flows in reverse direction

Preglacial R. Thames (T^1). R.Thames diverted by ice advance (T^2). R.Thames diverted again by a further ice advance (T^3).

Irish Sea ice
North Sea ice
Dee estuary
Lake Eskdale
Newtondale
Lake Pickering
Kirkham Abbey Gorge
see Places 17
Lake Lapworth
Ironbridge Gorge
Lake Harrison

overflow channel
proglacial lake
edge of ice

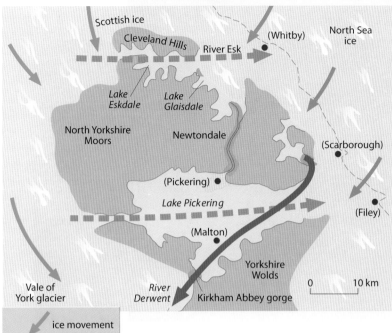

Figure 4.41

Proglacial lakes and overflow channels in North Yorkshire

Lake Eskdale, a proglacial lake, formed when the North Sea ice sheet blocked the mouth of the River Esk. The level of the lake rose until its water found a new route over a low point in its southern watershed on the North Yorkshire Moors. The overflow river flowed through Lake Glaisdale before cutting the deep, narrow, steep-sided, flat-floored Newtondale valley. At the end of this valley, the river formed a delta where it flowed into another proglacial lake – Lake Pickering. Lake Pickering, also dammed by North Sea ice, found an outlet to the south-west where it formed an overflow channel – the present-day Kirkham Gorge. After the ice melted, the Esk reverted to its original course, entering the sea near Whitby; Newtondale became virtually a dry valley; and the River Derwent, its eastward exit from Lake Pickering blocked by glacial deposits, continued to follow its new south-westerly course. Today, the site of Lake Pickering forms the fertile, flat-floored Vale of Pickering.

Changes in sea-level

The expansion and contraction of ice sheets affected sea-level in two different ways. **Eustatic** (also now called **glacio-eustatic**) refers to a world-wide fall (or rise) in sea-level due to changes in the hydrological cycle caused by water being held in storage on land in ice sheets (or released following the melting of ice sheets). **Isostatic** (or **glacio-isostatic**) adjustment is a more local change in sea-level resulting from the depression (or uplift) of the Earth's crust by the increased (or decreased) weight imposed upon it by a growing (or a declining) ice sheet. Evans (1991) claims that 'Because of their great weight, ice sheets depress the Earth's crust below them by approximately 0.3 times their thickness. So, at the centre of an ice sheet 700 m thick, there will be a maximum of 210 m of depression.' The history of sea-level depends on the location. For example, an equatorial site will experience the rise and fall of the sea solely associated with eustatic changes. In contrast, a site close to, or under, a glacier will have a history dominated by the isostatic rebound of the crust after glacial retreat. The sequence of events resulting from eustatic and isostatic changes during and after the last glacial can be summarised as follows:

1 At the beginning of the glacial, water in the hydrological cycle was stored as ice on the land instead of returning to the sea. There was a universal (eustatic) fall in sea-level, giving a negative change in base level (page 81).
2 As the glacial continued towards its peak, the weight of ice increased and depressed the Earth's crust beneath it. This led to a local (isostatic) rise in sea-level relative to the land and a positive change in base level.
3 As the ice sheets began to melt, large quantities of water, previously held in storage, were returned to the sea causing a worldwide (eustatic) rise in sea-level (a positive change in base level). This formed fiords, rias and drowned estuaries (page 163 and Places 22, page 164).
4 Finally, and still continuing in several places today, there was a local (isostatic) uplift of the land as the weight of the ice sheets decreased (a negative change in base level). This change created raised beaches (Places 23, page 166) and caused rejuvenation of rivers (page 82).

Looking into the future:

■ If the ice sheets continue to melt at their present rate, caused by global warming (Case Study 9B) or a milder climate, sea-levels could rise by 60 cm by the end of the century, with 1 m probably a reasonable high-end (and pessimistic?) estimate.
■ If isostatic uplift continues in Britain, it will increase the tilt that has already resulted in north-west Scotland rising by an estimated 10 m in the last 9000 years, and south-east England sinking. Tides in London are now more than 4 m higher than they were in Roman times – hence the need for the Thames Barrier (and its proposed replacement) – due to a combination of south-east England sinking and modern sea-level rise.

4a Case Study

Figure 4.42

An avalanche

An avalanche is a sudden downhill movement of snow, ice and/or rock (Case Study 2A). It occurs, like a landslide, when the weight (mass) of material is sufficient to overcome friction (Figure 4.42). This allows the debris to descend at a considerable speed under the force of gravity (mass movement). The average speed of descent is 40–60 km/hr, but video-recordings have shown extreme speeds in excess of 200 km/hr.

There are several different types of avalanche, which makes a simple classification difficult. Figure 4.43 gives a mainly descriptive classification put forward in the 19th century, while Figure 4.44 gives a modern classification based more on genetic and morphological characteristics.

Figure 4.43

A late 19th-century classification of avalanches

a Staublawinen (airborne powder snow)	Pure (completely airborne)
	Common (some contact with the ground)
b Grundlawinen (ground-hugging)	Rolling
	Sliding

a Avalanche break-away point	single point – loose snow avalanche	easier (not easy) to predict and manage; originates from a single point, usually soon after the snow falls
	large area, or 'slab'	often localised, hardest to predict, greatest threat to off-piste skiers; originates from a wider area and after the snow has had time to develop cohesion
b Depth	total snow depth	total mass of snow moves
	top layers of snow move over lower layers	alpine inhabitants regard this as the most dangerous
c Channel (track) width	unconfined – no channel	wide area, hard to manage
	gulley – confined to narrow track	dangerous, as it can reach higher speeds, but easier to manage
d Nature of snow (water content)	dry snow – mainly rolling	above ground-level so friction is reduced; can reach speeds of 200 km/hr – very destructive
	wet snow – mainly sliding	follows ground topography, occurs under föhn conditions (page 241), limited protection, much damage

Figure 4.44

A more recent classification of avalanches (1979)

Causes

- Heavy snowfall compressing and adding weight to earlier falls, especially on windward slopes.
- Steep slopes of over 25° where stability is reduced and friction is more easily overcome.
- A sudden increase in temperature, especially on south-facing slopes and, in the Alps, under föhn wind conditions (page 241).
- Heavy rain falling upon snow (more likely in Scotland than the Alps).
- Deforestation, partly for new ski-runs, which reduces slope stability.
- Vibrations triggered by off-piste skiers, any nearby traffic and, more dangerously, earth movements (Case Study 2A).
- Very long, cold, dry winters followed by heavy snowfalls in spring. Under these conditions, earlier falls of snow will turn into ice over which later falls will slide (some local people perceive this to pose the greatest avalanche risk).

Consequences

Avalanches can block roads and railways, cut off power supplies and telecommunications and, under extreme conditions, destroy buildings and cause loss of life. Between 1980 and 1991 there were, in Alpine Europe alone, 1210 recorded avalanche deaths, of whom nearly half were skiers – virtually all in off-piste areas. This death rate is increasing as the popularity of skiing grows and alpine weather becomes less predictable (a record total of 145 deaths in 1998–99).

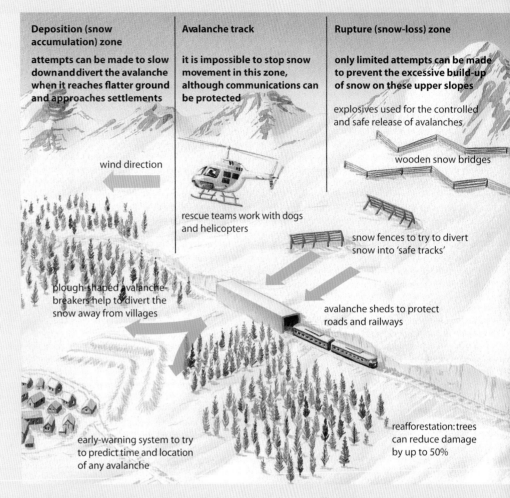

Deposition (snow accumulation) zone

attempts can be made to slow down and divert the avalanche when it reaches flatter ground and approaches settlements

wind direction

plough-shaped avalanche-breakers help to divert the snow away from villages

early-warning system to try to predict time and location of any avalanche

Avalanche track

it is impossible to stop snow movement in this zone, although communications can be protected

rescue teams work with dogs and helicopters

Rupture (snow-loss) zone

only limited attempts can be made to prevent the excessive build-up of snow on these upper slopes

explosives used for the controlled and safe release of avalanches

wooden snow bridges

snow fences to try to divert snow into 'safe tracks'

avalanche sheds to protect roads and railways

reafforestation: trees can reduce damage by up to 50%

Figure 4.45

Avalanche management schemes

Management

There is a close link between avalanches and:

- time of year – almost 80 per cent of avalanches in the French Alps occur between January and March, the 'avalanche season'
- altitude – over 90 per cent occur between 1500 and 3000 m.

Although it is possible to predict *when* and in which regions avalanches are most likely to occur, it is less easy to predict exactly *where* an event is likely to happen. It is this unpredictability that makes avalanches a major environmental hazard in alpine areas. However, despite this uncertainty, many avalanches do tend to follow certain 'tracks'. Consequently, as well as setting up early-warning systems and training rescue teams (Figure 4.46), it is possible to take some measures to try to protect life and property (Figure 4.45).

Figure 4.46

Avalanche protection and rescue schemes

Changed rates of melting ice and subsequent potential rises in sea-level are the main reasons why most scientists are working on glaciers at the present time, and why it should interest so many other people.

Ice helps to stabilise the world's climate by insulating large areas of ocean in summer and preventing heat loss in winter. Ice and snow also have a higher reflectivity, or **albedo** (page 207), than any other surface, reflecting 80 per cent of incoming solar radiation back into the atmosphere. As ice melts then the albedo will be reduced, less solar radiation will be reflected back and the Earth's temperature will rise.

(i) Ice shelves: Antarctica

Antarctica is covered by two huge ice sheets: the larger East Antarctic Ice Sheet (EAIS), which is bigger than the USA and holds most of the world's fresh water in storage; and the smaller West Antarctic Ice Sheet (WAIS). Scientists predict that even if only the EAIS melted, the world's sea-level would rise by 61 m. On the edges of the two ice sheets, and extending from them, are several ice shelves, the two largest being the Ross and Ronne (Figure 4.47). As global temperatures rise, especially around the Antarctic peninsula which extends beyond the Antarctic Circle, these ice shelves are becoming less stable and parts are collapsing.

The collapse of the Larsen B ice shelf in 2002 was the latest and most spectacular (it was the size of East Anglia) of ten collapses that have occurred off the coast of the Antarctic Peninsula since the mid-1980s (Figure 4.48). In 2008, part of the nearby Wilkins ice shelf was said to be 'hanging on by a thread'. The ice, following its collapse, drifts away from the polar region, often as huge icebergs, into warmer water where it melts. Being fresh water in a frozen state, its melting adds to the volume of the ocean, causing a global rise in sea-level. As ice shelves collapse, glaciers moving behind them on the ice sheet are accelerating by

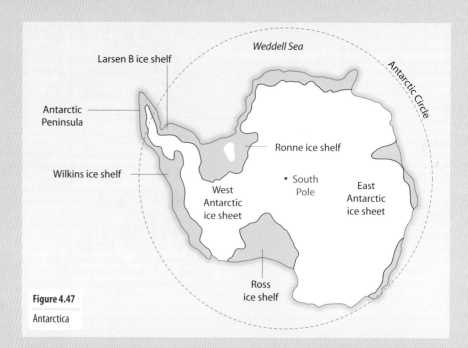

Figure 4.47

Antarctica

1 per cent a year, the fastest now travelling at 3.5 km/yr.

The collapses are credited to global warming, the average annual temperature in the Antarctic having risen by 2.5°C in the last 50 years compared with 0.5°C globally. According to Bentley in a series of articles in *Geography Review*, 'the key to the collapse is the formation of pools of meltwater on the surface of the ice shelf during the Antarctic

summer. In some places, the meltwater begins to fill crevasses in the ice shelf. Normally, crevasses are only tens of metres deep, but as the meltwater progressively fills them the weight of water forces the lowermost tip of the crevasse to crack even more deeply into the ice. Eventually the crevasses may penetrate through the full thickness of the ice shelf and a chunk of ice will break off.'

Figure 4.48

The collapse of the Larsen B ice shelf

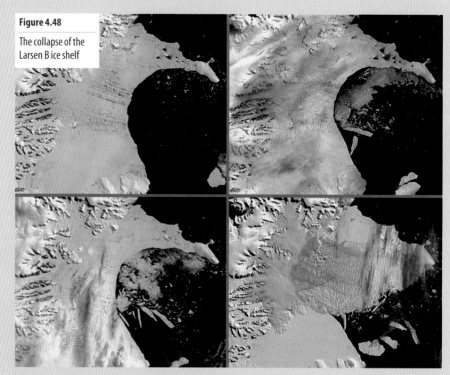

(ii) Ice sheets: Greenland

The average thickness of the Greenland ice sheet has been calculated to be 1800 m. However, while this thickness was believed to have decreased by an average of 1 m/yr throughout the last century, satellite imagery suggests that the rate of decrease had accelerated to 5 m/yr in 2000 and 10 m/yr by 2007. The increase in surface melting is creating more meltwater which sinks down crevasses to the bedrock where it acts as a lubricant accelerating basal flow (pages 107–108). This in turn causes glaciers leading from the ice sheet to flow faster. One of these, the Jokobshavn, reaches a speed of 1 m/hr as it nears the coast, making it the fastest-flowing glacier in the world.

As in Antarctica, Greenland's ice is fresh water in frozen storage. It is believed that should the whole ice sheet totally melt then the global sea-level would rise by 6.7 m.

(iii) Sea ice: the Arctic

Sea ice is frozen salt water and forms when temperatures remain for some time below −1.5°C. Recent satellite images have shown that the area covered by sea ice is now decreasing by 8 per cent annually. More significantly, nuclear submarines, operating under the ice for over half a century, have indicated that the thickness of the ice has decreased in that time from 4 m to 1.3 m. As the ice thins, the remaining ice will melt more quickly, speeding up the process. In the 19th century, explorers tried unsuccessfully to find a sea route around the north of Canada – the so-called North West Passage – and in the early 20th century the first explorers claiming to have reached the North Pole only did so after several weeks' travelling over sea ice. Some scientists are now predicting that, due to global warming, all the polar sea ice will have disappeared within 30 years (Figure 4.49).

As it is frozen seawater that is melting, then the effect on global sea-level will be minimal. Figure 4.50 shows some of the advantages and disadvantages that will result from an ice-free Arctic.

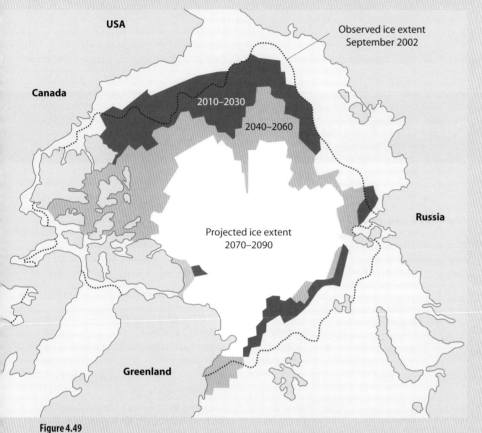

Figure 4.49

Present and predicted coverage of sea ice in the Arctic

Figure 4.50

Advantages and disadvantages of an ice-free Arctic

Advantages	Disadvantages
Easier to exploit resources such as oil and natural gas found under the seabed.	Less ice will mean a reduced albedo and an increase in global warming.
Improved navigation will reduce distances and travel time, e.g. i Tokyo to New York – distance reduced from 18 000 km to 14 000 km via the North West Passage (Canada) which in 2007 was open apart from 100 km of scattered ice floes ii Tokyo to London – distance reduced from 21 000 km to 13 000 km via the North East Passage (Russia) which in 2007 was open for six weeks.	An increase in the number of icebergs from surrounding ice shelves could make navigation more dangerous. An increased threat to wildlife – polar bears and other species threatened with extinction.

Further reference

Benn, D. and Evans, D.J.A. (1998) *Glaciers and Glaciation*, Hodder Arnold.

Bentley, M. (2004) 'Antarctic ice shelf collapse' in *Geography Review* Vol 18 No 2 (Novermber).

Bentley, M. (2005) 'Is the East Antarctic ice sheet stable?' in *Geography Review* Vol 19 No 2 (November).

Bentley, M. (2007) 'Where has all the sea ice gone?' in *Geography Review* Vol 20 No 5 (May).

Bentley, M. (2008) 'Climate warming on the Antarctic Peninsula' in *Geography Review* Vol 21 No 4 (April).

Dawson, A.G. (1992) *Ice Age Earth*, Routledge.

Hambrey, M. (1994) *Glacial Environments*, Routledge.

Knight, P.G. (2006) *Glacier Science and Environmental Change*, WileyBlackwell.

Mitchell, W. (2008) 'The Ribblehead drumlins' in *Geography Review* Vol 21 No 3 (February).

Alaska Science Forum – Water, Snow and Ice Index:
http://dogbert.gi.alaska.edu/ScienceForum/water.html

Cyberspace Snow and Avalanche Center (CSAC):
www.csac.org/

Glacial landforms:
www.bgrg,org/pages/education/alevel/coldenvirons/Lesson%2015.htm

Glacier Project:
http://glacier.rice.edu

Questions & Activities

Activities

1 **a** Define the terms 'interglacial' and 'interstadial'. *(4 marks)*

   b Describe the extent of ice across the British Isles at the height of the last ice advance 18 000 years ago. *(4 marks)*

   c Suggest and explain **one** theory for the cause of ice ages. *(4 marks)*

   d How is glacier ice formed? *(6 marks)*

   e Explain the **difference** in movement processes between **temperate** and **polar glaciers.** *(7 marks)*

2 Choose **one** of the features named in Figure 4.25 (page 115) and give its name.

   a i With the aid of a labelled diagram, describe the feature. *(5 marks)*

     ii Explain how a glacier created the feature you have chosen. *(5 marks)*

     iii Describe and explain **one** change in the feature, probably since the last ice age. *(4 marks)*

   b Many hollows in a glaciated upland are filled with water. Where does the water come from? *(2 marks)*

   c Suggest **two** pieces of evidence you would look for to suggest the direction of movement of a glacier if you were to carry out a study of a glaciated valley. *(4 marks)*

   d Describe and explain **one** difference between a glaciated upland area and an unglaciated one. *(5 marks)*

3 A glacier erodes, transports and deposits material using a range of methods.

   a i Name **two** types of glacial erosion. *(2 marks)*

     ii For **one** of the types of erosion in **a i**, explain how the glacier erodes. *(4 marks)*

   b Some loose material is carried on top of the glacier. Making good use of diagrams, show where, on the surface, this material is carried. *(4 marks)*

   c Where else is material carried by a glacier? *(2 marks)*

   d Choose **one** of the following landforms created by glacial deposition: drumlin; end moraine; kame terrace.

     i Describe its shape, size and composition. *(6 marks)*

     ii Explain how it was created by the glacier. *(7 marks)*

4 The area in front of a glacier is a **glacifluvial** landform often called a **sandur** or an **outwash plain**.

   a i Describe the characteristic deposits (shape and composition) of this area. *(4 marks)*

     ii Explain how glacifluvial processes helped to create the characteristics you have identified. *(4 marks)*

   b Choose one of the following features of a sandur: lakebed deposits; esker; kame; braided stream. Describe the shape and characteristics of the feature. *(4 marks)*

   c i What is a **kettle lake**? *(2 marks)*

     ii How is a kettle lake formed? *(5 marks)*

     iii Suggest how a kettle lake may disappear after the glacial period. *(6 marks)*

5 **a** What is a valley glacier? *(2 marks)*

   b Describe and explain the origins of **two** surface features of a moving glacier. *(6 marks)*

   c Explain how you could measure the movement of a valley glacier. *(4 marks)*

   d Why does the snout of a glacier sometimes retreat even though the ice always moves forward? *(6 marks)*

   e What feature may mark where the snout of a retreating glacier was in the past? Describe the shape and composition of the feature. *(7 marks)*

6 Ice movement during the last ice age had **indirect** as well as **direct** effects on the landscape. Indirect effects occur where the ice itself was not involved in the effect.

   a i Explain what is meant by the term 'drainage diversion'. *(2 marks)*

ii Choose one example of drainage diversion. Draw a sketch map to show the diversion and explain the role of glacier ice in the cause of the diversion. *(6 marks)*

b Why did the land experience an isostatic change of sea-level during the ice age? *(4 marks)*

c Why are 'raised beaches' found in coastal areas where glacial ice caused an isostatic change in sea-level? *(6 marks)*

d Choose **one** landform (other than a raised beach) which has been affected by sea-level change associated with glaciation. Describe the feature and explain how it was formed. *(7 marks)*

7 In a field survey (till fabric analysis) the orientation of clasts (stones) showed the data given in the table on the right. Orientation shows **two** possible directions (e.g. NW/SE).

a i Draw a graph to illustrate the data. *(6 marks)*

ii Using the data, suggest an interpretation of the ice movement in this area. *(7 marks)*

b Why do glacial deposits have a particular orientation? *(7 marks)*

c Suggest **two** other sources of data to indicate the direction of ice movement in an area. For one of these sources, explain how it shows the direction of ice movement. *(5 marks)*

Degrees	No. of clasts	Degrees	No. of clasts	Degrees	No. of clasts
0	0	120	2	240	8
15	0	135	3	255	3
30	10	150	1	270	1
45	12	165	1	285	1
60	8	180	0	300	2
75	3	195	0	315	3
90	1	210	10	330	1
105	1	225	12	345	1

Exam practice: basic structured questions

8 a Describe how ice can erode the rocks of upland areas by:
   i frost shattering
   ii plucking
   iii abrasion. *(9 marks)*

b Explain how these processes combine to produce cirques (also known as corries or cwms). *(6 marks)*

c With reference to one or more areas that you have studied, explain why upland glaciated areas are often difficult for human settlement. *(10 marks)*

9 Study Figure 4.25 on page 115. Select and name any **two** features of glacial erosion shown on the diagram.

a Describe **each** of your chosen features. *(5 + 5 marks)*

b Explain how **each** of these features was formed. *(15 marks)*

Exam practice: structured questions

10 a Identify **two** pieces of evidence to suggest that climatic change in an area has included at least **one** glacial period. For one of these pieces of evidence, show how it suggests a past glacial period. *(5 marks)*

b i Describe how a glacier operates as an 'open system'. *(8 marks)*

ii How and why does a glacier budget vary between winter and summer seasons? *(12 marks)*

11 a Geographers often classify glaciers into different types. Describe **one** system of classification. *(5 marks)*

b Why does movement of glacier ice vary across and within the glacier? *(12 marks)*

c Explain the difference in movement between glaciers in polar and temperate latitudes. *(8 marks)*

12 a i How has glacial ice affected sea-level in the past, and how might it affect sea-level in the next century or so? *(9 marks)*

ii How is glacial ice involved in sea-level change? *(9 marks)*

b i Describe the shape and scale of a fiord.

ii Explain the roles of glacial processes and sea level change in the formation of a fiord. *(12 marks)*

Exam practice: essays

13 Describe and evaluate the evidence **(including geomorphological evidence)** that there has been a series of ice ages in the northern hemisphere during the last million years. *(25 marks)*

14 For any one drainage diversion system you have studied, discuss the role of glacial ice and other factors in its formation. *(25 marks)*

15 Describe the features of glacial and glacifluvial deposition that might be found on a lowland plain from which an ice sheet had

recently melted, and explain how you would recognise the difference between selected features of glacial origin and selected features of glacifluvial origin. *(25 marks)*

16 Scientists have suggested that there is evidence from the Arctic and Antarctic ice sheets that global warming is happening. Describe and evaluate this evidence, and suggest how melting of the ice might affect the Earth's future geography. *(25 marks)*

5

Periglaciation

· ·

'Perennially frozen material lurks beneath at least one-fifth, and perhaps as much as one-fourth, of the Earth's land surface.'

Frederick Nelson, 1999

The term **periglacial**, strictly speaking, means 'near to or at the fringe of an ice sheet', where frost and snow have a major impact upon the landscape. However, the term is often more widely used to include any area that has a cold climate – e.g. mountains in temperate latitudes such as the Alps and the Plateau of Tibet – or which has experienced severe frost action in the past – e.g. southern England during the

Quaternary ice age (Figure 4.3b). Today, the most extensive periglacial areas lie in the Arctic regions of Canada, Alaska and Russia. These areas, which have a tundra climate, soils and vegetation (pages 333–334), exhibit their own characteristic landforms.

Permafrost

Permafrost is permanently frozen ground. It occurs where soil temperatures remain below 0°C for at least two consecutive years. Permafrost covers almost 25 per cent of the Earth's land surface (Figure 5.1) although its extent changes over periods of time. Its depth and continuity also vary (Figure 5.2).

Figure 5.1

Permafrost zones of the Arctic

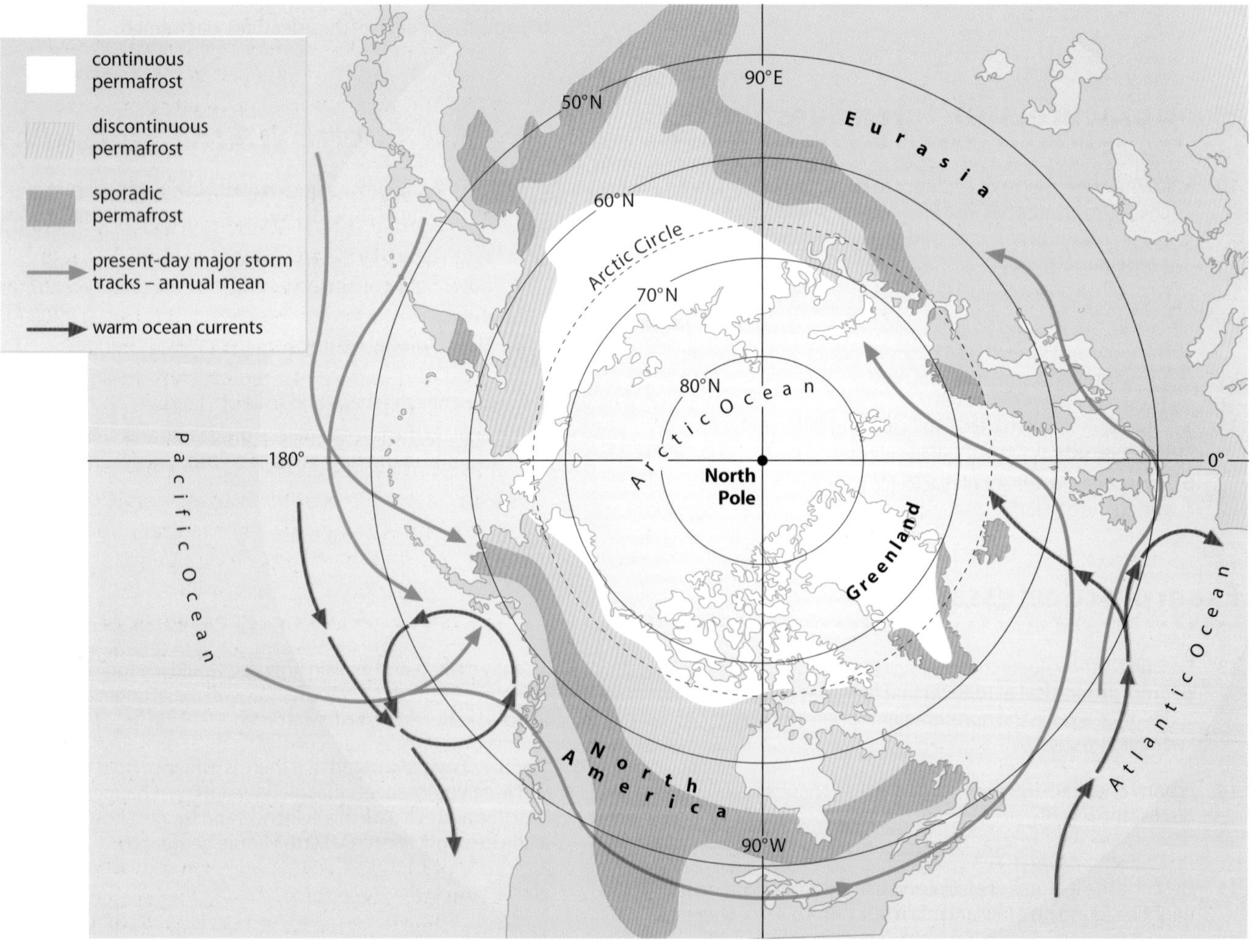

continuous permafrost

discontinuous permafrost

sporadic permafrost

present-day major storm tracks – annual mean

warm ocean currents

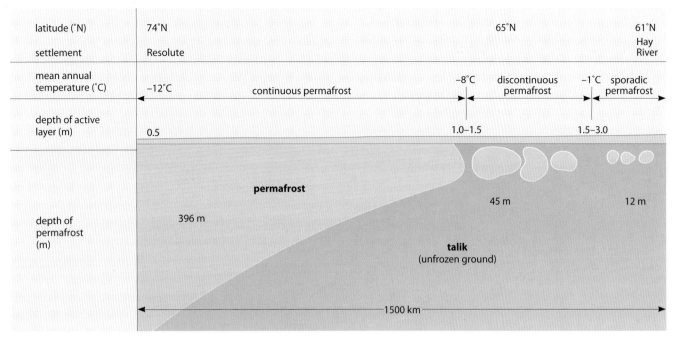

latitude (°N)	74°N		65°N	61°N
settlement	Resolute			Hay River
mean annual temperature (°C)	–12°C	continuous permafrost	–8°C discontinuous permafrost	–1°C sporadic permafrost
depth of active layer (m)	0.5		1.0–1.5	1.5–3.0
depth of permafrost (m)	permafrost 396 m		45 m talik (unfrozen ground)	12 m

1500 km

Figure 5.2

Transect through part of the permafrost zone in northern Canada

Continuous permafrost is found mainly within the Arctic Circle where the mean annual air temperature is below –5°C. Here winter temperatures may fall to –50°C and summers are too cold and too short to allow anything but a superficial melting of the ground. The permafrost has been estimated to reach a depth of 700 m in northern Canada and 1500 m in Siberia. As Figure 5.1 shows, continuous permafrost extends further south in continental interiors than in coastal areas which are subject to the warming influence of the sea, e.g. the North Atlantic Drift in north-west Europe.

Discontinuous permafrost lies further south in the northern hemisphere, reaching 50°N in central Russia, and corresponds to those areas with a mean annual temperature of between –1°C and –5°C. As is shown in Figure 5.2, discontinuous permafrost consists of islands of permanently frozen ground, separated by less cold areas which lie near to rivers, lakes and the sea.

Sporadic permafrost is found where mean annual temperatures are just below freezing point and summers are several degrees above 0°C. This results in isolated areas of frozen ground (Figure 5.2).

In areas where summer temperatures rise above freezing point, the surface layer thaws to form the **active layer**. This zone, which under some local conditions can become very mobile for a few months before freezing again, can vary in depth from a few centimetres (where peat or vegetation cover protects the ground from insolation) to 5 m. The active layer is often saturated because meltwater cannot infiltrate downwards through the impermeable permafrost. Meltwater is unlikely to evaporate in the low summer temperatures or to drain downhill since most of the slopes are very gentle. The result is that permafrost regions contain many of the world's few remaining wetland environments.

The unfrozen layer beneath, or indeed any unfrozen material within, the permafrost is known as talik. The lower limit of the permafrost is determined by geothermal heat which causes temperatures to rise above 0°C (Figure 5.3).

Temperatures taken over a period of years in the discontinuous and continuous permafrost suggest that, in Canada, Alaska and Russia, there is a general thawing of the frozen ground, an event accredited to global warming (Case Study 5).

Figure 5.3

Soil temperatures in permafrost at Yakutsk, Siberia

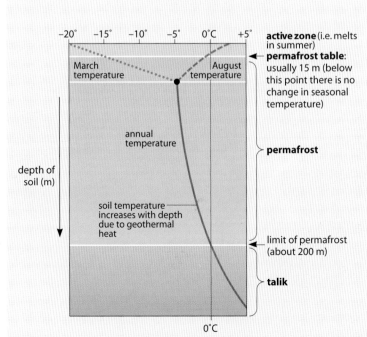

Periglacial processes and landforms

Most periglacial regions are sparsely populated and underdeveloped. Until the search for oil and gas in the 1960s, there had been little need to study or understand the geomorphological processes which operate in these areas. Although significant strides have been made in the last 30 years, there is still uncertainty as to how certain features have developed and, indeed, whether such features are still being formed today or are a legacy of a previous, even colder climate – i.e. a fossil or relict landscape. Figure 5.4 gives a classification of the various processes which operate, and the landforms which develop, in periglacial areas.

	Processes	Landforms
Ground ice	Ice crystals and lenses (frost-heave)	Sorted stone polygons (stone circles and stripes: patterned ground)
	Ground contraction	Ice wedges with unsorted polygons: patterned ground
	Freezing of groundwater	Pingos
Frost weathering	Frost shattering/Freeze–thaw	Blockfields, talus (scree), tors (Chapter 8)
Snow	Nivation	Nivation hollows
Meltwater	Solifluction	Solifluction sheets, rock streams
	Streams	Braiding, dry valleys in chalk (Chapter 8)
Wind	Windblown	Loess (limon), dunes

Ground ice
Frost-heave: ice crystals and lenses

Frost-heave includes several processes which cause either fine-grained soils such as silts and clays to expand to form small domes, or individual stones within the soil to be moved to the surface (Figure 5.5). It results from the direct formation of ice – either as crystals or as lenses. The **thermal conductivity** of stones is greater than that of soil. As a result, the area under a stone becomes colder than the surrounding soil, and ice crystals form. Further expansion by the ice widens the capillaries in the soil, allowing more moisture to rise and to freeze. The crystals, or the larger ice lenses which form at a greater depth, force the stones above them to rise until eventually they reach the surface. (Ask a gardener in northern Britain to explain why a plot that was left stoneless in the autumn has become stone-covered by the spring, following a cold winter.)

During periods of thaw, meltwater leaves fine material under the uplifted stones, preventing them from falling back into their original positions. In areas of repeated freezing (ideally where temperatures fall to between –4°C and –6°C) and thawing, frost-heave both lifts and sorts material to form **patterned ground** on the surface (Figure 5.6). The larger stones, with their extra weight, move outwards to form, on almost flat areas, stone circles or, more accurately, **stone polygons**. Where this process occurs on slopes with a gradient in excess of 2°, the stones will slowly move downhill under gravity to form elongated **stone stripes**.

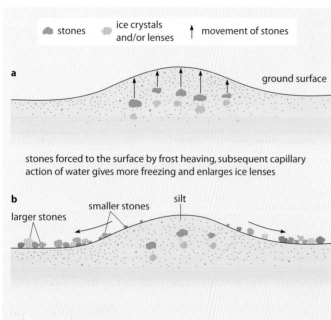

stones

ice crystals and/or lenses

↑ movement of stones

a

ground surface

stones forced to the surface by frost heaving, subsequent capillary action of water gives more freezing and enlarges ice lenses

b

larger stones

smaller stones

silt

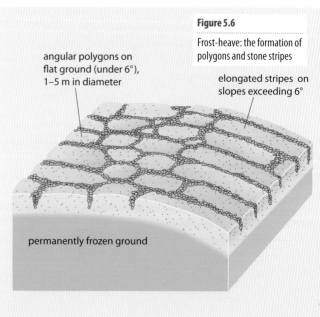

angular polygons on flat ground (under 6°), 1–5 m in diameter

elongated stripes on slopes exceeding 6°

permanently frozen ground

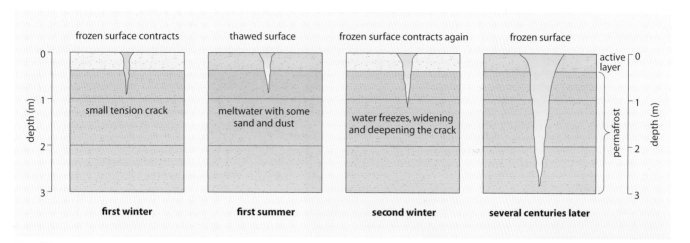

frozen surface contracts | thawed surface | frozen surface contracts again | frozen surface

small tension crack | meltwater with some sand and dust | water freezes, widening and deepening the crack

active layer
permafrost

depth (m) 0 1 2 3

first winter | first summer | second winter | several centuries later

Figure 5.7

The formation of ice wedges

Ground contraction

The refreezing of the active layer during the severe winter cold causes the soil to contract. Cracks open up which are similar in appearance to the irregularly shaped polygons found on the bed of a dried-up lake. During the following summer, these cracks open, close or fill with melt-water and, sometimes, also with water and wind-blown deposits. When the water refreezes, during the following winter the cracks widen and deepen to form **ice wedges** (Figure 5.7). This process is repeated annually until the wedges, which underlie the perimeters of the polygons, grow to as much as 1 m in width and 3 m in depth. **Fossil ice wedges**, i.e. cracks filled with sands and silt left by meltwater, are a sign of earlier periglacial conditions (Figure 5.9).

Patterned ground (Figure 5.8) can, therefore, be produced by two processes: frost-heaving (Figure 5.6) and ground contraction (Figure 5.7). Frost-heaving results in small dome-shaped polygons with larger stones found to the outside

Figure 5.8

Ice wedge polygons near Barrow, Alaska. The patterned ground is formed by polygons up to 30 m in diameter. The polygon boundaries mark the position of the ice wedges

Figure 5.9

Fossil ice wedge

of the circles, whereas ice contraction produces larger polygons with the centre of the circles depressed in height and containing the bigger stones. The diameter of an individual polygon can reach over 30 m.

Freezing of groundwater

Pingos are dome-shaped, isolated hills which interrupt the flat tundra plains (Figure 5.10). They can have a diameter of up to 500 m and may rise 50 m in height to a summit that is sometimes ruptured to expose an icy core. As they occur mainly in sand, they are not susceptible to frost-heaving. American geographers recognise two types of pingo (Figure 5.11a and b), although recent investigations have led to the suggestion of a third type: **polygenetic** (or mixed) pingos.

Figure 5.10

A pingo, Mackenzie
Delta, Canada

the permafrost is continuous. They often form on the sites of small lakes where water is trapped (**enclosed**) by freezing from above and by the advance of the permafrost inwards from the lake margins. As the water freezes it will expand, forcing the ground above it to rise upwards into a dome shape. This type of pingo is known as the **Mackenzie type** as over 1400 have been recorded in the delta region of the River Mackenzie. It results from the downward growth of the permafrost (Figure 5.11b).

As the surface of a pingo is stretched, the summit may rupture and crack. Where the ice-core melts, the hill may collapse leaving a melt-water-filled hollow (Figure 5.11c). Later, a new pingo may form on the same site, and there may be a repeated cycle of formation and collapse.

Open-system (hydraulic) pingos occur in valley bottoms and in areas of thin or discontinuous permafrost. Surface water is able to infiltrate into the upper layers of the ground where it can circulate in the unfrozen sediments before freezing. As the water freezes, it expands and forms localised masses of ice. The ice forces any overlying sediment upwards into a dome-shaped feature, in the same way that frozen milk lifts the cap off its bottle. This type of pingo, referred to as the **East Greenland type**, grows from below (Figure 5.11a).

Closed-system (hydrostatic) pingos are more characteristic of flat, low-lying areas where

Frost weathering

Mechanical weathering is far more significant in periglacial areas than is chemical weathering, with freeze–thaw being the dominant process (Figure 2.10). On relatively flat upland surfaces, e.g. the Scafell range in the Lake District and the Glyders in Snowdonia, the extensive spreads of large, angular boulders, formed *in situ* by frost action, are known as **blockfields** or **felsenmeer** (literally, a 'rock sea').

Scree, or **talus**, develops at the foot of steep slopes, especially those composed of well-jointed

Figure 5.11

Formation of pingos

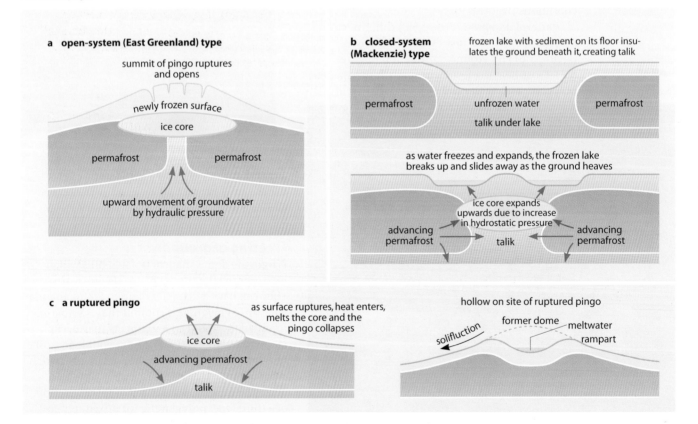

a **open-system (East Greenland) type**
summit of pingo ruptures and opens
newly frozen surface
ice core
permafrost
permafrost
upward movement of groundwater by hydraulic pressure

b **closed-system (Mackenzie) type**
frozen lake with sediment on its floor insulates the ground beneath it, creating talik
permafrost
unfrozen water
permafrost
talik under lake

as water freezes and expands, the frozen lake breaks up and slides away as the ground heaves
ice core expands upwards due to increase in hydrostatic pressure
advancing permafrost
talik
advancing permafrost

c **a ruptured pingo**
as surface ruptures, heat enters, melts the core and the pingo collapses
ice core
advancing permafrost
talik

hollow on site of ruptured pingo
former dome
meltwater
solifluction
rampart

rocks prone to frost action. Freeze–thaw may also turn well-jointed rocks, such as granite, into **tors** (page 202). One school of thought on tor formation suggests that these landforms result from frost shattering, with the weathered debris later having been removed by solifluction. If this is the case, tors are therefore a relict (fossil) of periglacial times.

Snow

Snow is the agent of several processes which collectively are known as **nivation** (page 111). These nivation processes, sometimes referred to as 'snowpatch erosion', are believed to be responsible for enlarging hollows on hillsides. Nivation hollows are still actively forming in places like Iceland, but are relict features in southern England (as on the scarp slope of the South Downs behind Eastbourne).

Meltwater

During periods of thaw, the upper zone (active layer) melts, becomes saturated and, if on a slope, begins to move downhill under gravity by the process of solifluction (page 47). Solifluction leads to the infilling of valleys and hollows by sands and clays to form **solifluction sheets** (Figures 5.12 and 5.13a) or, if the source of the flow was a nivation hollow, a rock stream (Figure 5.21). Solifluction deposits, whether they have in-filled valleys or have flowed over cliffs, as in southern England, are also known as **head** or, in chalky areas, **coombe** (Figure 5.13b).

The chalklands of southern England are characterised by numerous dry valleys (Figure 8.11). The most favoured of several hypotheses put forward to explain their origin suggests that the valleys were carved out under periglacial conditions. Any water in the porous chalk at this time would have frozen, to produce permafrost, leaving the surface impermeable. Later, meltwater rivers would have flowed over this frozen ground to form V-shaped valleys (page 200).

Rivers in periglacial areas have a different regime from those flowing in warmer climates. Many may stop flowing altogether during the long and very cold winter (Figure 5.14) and have a peak discharge in late spring or early summer when melting is at its maximum (Places 18). With their high velocity, these rivers are capable of transporting large amounts of material when at their peak flow. Later in the year, when river levels fall rapidly, much of this material will be deposited, leaving a braided channel (Figures 3.32 and 5.16).

Figure 5.12

Solifluction sheet in the Ogilvie Mountains, Yukon, Canada

Figure 5.13

Formation of solifluction sheet and head

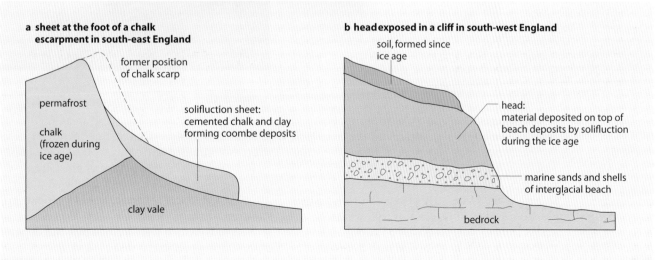

a sheet at the foot of a chalk escarpment in south-east England

former position of chalk scarp

permafrost

chalk (frozen during ice age)

solifluction sheet: cemented chalk and clay forming coombe deposits

clay vale

b head exposed in a cliff in south-west England

soil, formed since ice age

head: material deposited on top of beach deposits by solifluction during the ice age

marine sands and shells of interglacial beach

bedrock

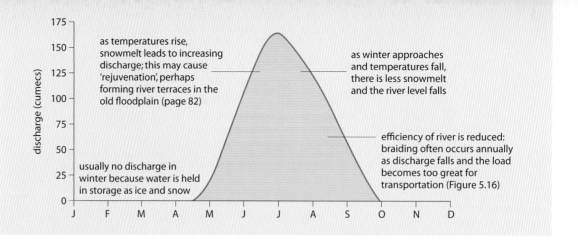

Figure 5.14

Model of a river regime in a periglacial area

as temperatures rise, snowmelt leads to increasing discharge; this may cause 'rejuvenation', perhaps forming river terraces in the old floodplain (page 82)

as winter approaches and temperatures fall, there is less snowmelt and the river level falls

usually no discharge in winter because water is held in storage as ice and snow

efficiency of river is reduced: braiding often occurs annually as discharge falls and the load becomes too great for transportation (Figure 5.16)

Places 18 Alaska: periglacial river regimes

Permafrost also affects the hydrological regimes of subarctic rivers. Figure 5.15 shows the regime of two Alaskan rivers, both of which flow in first order drainage basins (page 65). One river, however, is located in northern Alaska where over 50 per cent of the basin is underlain with continuous permafrost. The other river, in contrast, is located further south where most of the basin consists of discontinuous permafrost and only 3 per cent is continuous permafrost. The northern river, flowing over more impermeable ground (more permafrost giving increased surface runoff and reduced throughflow)

responds much more readily to changes in both temperature (increased snowmelt or freezing) and rainfall (amounts and seasonal distribution). It has a more extreme regime showing that it is more likely to flood in summer and to have a higher peak discharge and then to dry up sooner, and for a longer period, in winter or during dry spells. Figure 5.16 was taken on 7 August 1996 in the Dynali National Park. The river level had already fallen (as had the first snow of winter!), and the large load carried by the early summer meltwaters had already been deposited.

Figure 5.15

Contrasting regimes of rivers flowing over continuous and discontinuous permafrost

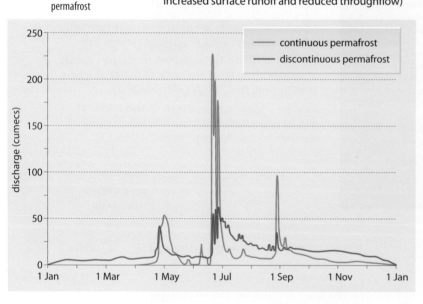

continuous permafrost
discontinuous permafrost

Figure 5.16

A river in the Dynali National Park

Wind

A lack of vegetation and a plentiful supply of fine, loose material (i.e. silt) found in glacial environments enabled strong, cold, out-blowing winds to pick up large amounts of dust and to redeposit it as **loess** in areas far beyond its source. Loess covers large areas in the Mississippi–Missouri valley in the USA. It also occurs across France (where it is called **limon**) and the North European Plain and into north-west China (where in places it exceeds 300 m in

depth and forms the yellow soils of the Huang He valley – Case Study 10). In all areas, it gives an agriculturally productive, fine-textured, deep, well-drained and easily worked soil which is, however, susceptible to further erosion by water and wind if not carefully managed (Figure 10.35). Large tracts of central Europe, other than those consisting of loess, are covered in dunes (coversands) which were formed by wind deposition during periglacial times.

In 2008, Dr Mike Bentley claimed in *Geography Review* that one of the most important, yet least publicised, effects of global warming is the melting of the permafrost (Figure 5.19). Measurements taken along a north–south transect adjacent to the Alaskan pipeline suggest that the depth of the active layer is increasing and the depth of the permafrost table is getting lower (Figure 5.3).

Causes

- Global warming is causing temperatures to rise more quickly in arctic areas, where the permafrost is located, than in more temperate regions. As the air temperature rises, the frozen ground beneath it warms up. In northern Canada, where there has been an increase in temperature of just over 1°C since 1990, the rate of thaw has trebled. However, although global warming is the main and obvious cause for the melting of the permafrost, there are other contributory reasons.

- The removal of mosses and other tundra vegetation (page 333) for construction purposes means that in summer more heat penetrates the soil, increasing the depth of thaw.
- The construction of centrally heated buildings warms the ground beneath them, while the laying of pipes in the active zone, for heating oil, sewerage and water, increases the rate of thaw (Figure 5.17).
- Heat produced by drilling for oil and natural gas in both Alaska and Russia melts the surrounding permafrost.

Effects

- There is a reduction in the polar extent of the permafrost in arctic areas and an increase in the frequency of landslips and slope failure in more temperate, mountainous regions.
- There is evidence that the tree line (page 331) is beginning to extend further northwards and that the length of the growing season has increased by three days in Canada and Alaska and by one day in Russia.
- There is an increase in the extent of **thermokarst**, which is a landscape that develops where masses of ground ice melts. As the depth of the active layer increases, parts of the land surface subside. Thermokarst is, therefore, the general name given to irregular, hummocky terrain with marshy or lake-filled hollows created by the disruption of the thermal equilibrium of the permafrost (Figures 5.18 and 12.43). This development also increases the risk of local flooding.
- Houses and other buildings tilt as their foundations subside and sink into the ground (Figure 5.20).
- Earth movements can alter the position of the supports for oil pipelines, threatening to fracture the pipes. Roads and railways can lose alignment, and dams and bridges may develop cracks.
- A new railway across the permafrost that makes up much of the Tibetan Plateau has had to be built on crushed rock as this reduces temperatures and consequently the rate of thaw.

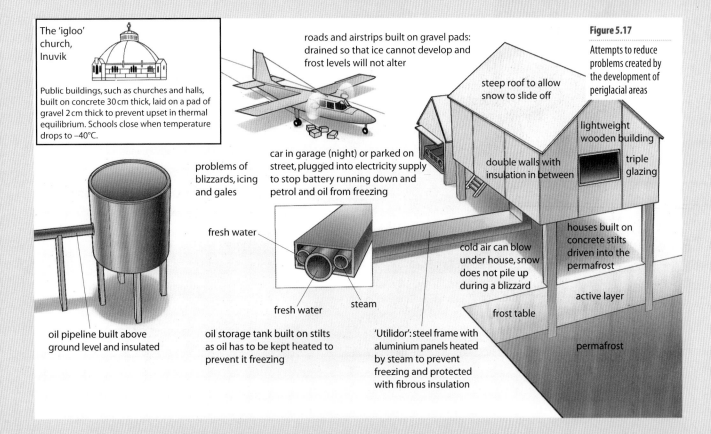

The 'igloo' church, Inuvik

Public buildings, such as churches and halls, built on concrete 30 cm thick, laid on a pad of gravel 2 cm thick to prevent upset in thermal equilibrium. Schools close when temperature drops to –40°C.

roads and airstrips built on gravel pads: drained so that ice cannot develop and frost levels will not alter

Figure 5.17

Attempts to reduce problems created by the development of periglacial areas

steep roof to allow snow to slide off

lightweight wooden building

double walls with insulation in between

triple glazing

problems of blizzards, icing and gales

car in garage (night) or parked on street, plugged into electricity supply to stop battery running down and petrol and oil from freezing

fresh water

cold air can blow under house, snow does not pile up during a blizzard

houses built on concrete stilts driven into the permafrost

fresh water   steam

active layer

frost table

oil pipeline built above ground level and insulated

oil storage tank built on stilts as oil has to be kept heated to prevent it freezing

'Utilidor': steel frame with aluminium panels heated by steam to prevent freezing and protected with fibrous insulation

permafrost

- Of all the effects resulting from the melting of the permafrost, it is the release of organic matter from permafrost soils as they thaw that is causing scientists the most concern (Figure 5.19). This organic matter contains large amounts of carbon in storage. As temperatures rise due to global warming, this carbon is released as one of two greenhouse gases – either CO_2 in drier areas or methane in wetter places (Figure 9.78). The release of these gases will increase the speed of global warming which in turn will accelerate the rate of melting in the permafrost, creating a vicious cycle.

Figure 5.18

Thermokarst scenery

Conclusion

Latest estimates suggest that the depth of the active layer could increase by 20 to 30 per cent by 2050, and that between 60 per cent (the most conservative figure) and 90 per cent (the worst-case scenario) of the permafrost could disappear by 2100. As Dr Bentley suggests: 'Permafrost may seem like a remote irrelevance to us in the temperate mid-latitudes, but it has the potential to affect every one of us through its impact on greenhouse gas emissions.'

Figure 5.19

Extract from an article in *Geography Review* February 2008, by Dr Mike Bentley

Figure 5.20

Buildings in Yukon, Canada, whose footings have sunk into the permafrost

Normally, the soils of permafrost areas are crammed with undegraded, well-preserved organic matter in the form of leaves, roots, twigs and so on. This is an enormous store of carbon, kept inert by being frozen in the ground. But if that ground begins to melt and the organic material can start rotting, it will release its carbon as carbon dioxide or methane, both greenhouse gases.

In other words, the newly thawed soils may release vast amounts of greenhouse gases into the atmosphere, which will of course give a further 'kick' to global warming. This will melt more permafrost and so on, in a worsening positive feedback cycle. This process is an example of biogeochemical feedback which could influence global climate change. The alarming thing about it is the amount of carbon contained in the Arctic, and the speed at which warming is occurring. The combined effect could be catastrophic.

To illustrate this, consider that the Arctic is estimated to contain about 900 gigatonnes (Gt) of carbon. Humans emit about 9 Gt of carbon from fossil fuels and deforestation every year. So it would only take the release of 1% of carbon in Arctic permafrost soils to effectively double our emissions of greenhouse gases.

Further reference

Bentley, M. (2008) 'On shaky ground' in *Geography Review* Vol 21 No 3 (February).

French, H.M. (2007) *The Periglacial Environment*, WileyBlackwell.

Goudie, A.S. (2001) *The Nature of the Environment*, WileyBlackwell.

Middleton, N. (2008) 'Arctic warming' in *Geography Review* Vol 21 No 4 (April).

Periglacial processes and landforms:
www.bgrg.org/pages/education/alevel/coldenvirons/Lesson%2019.htm
www.fettes.com/Cairngorms/periglacial.htm

Activities

1 Study Figure 5.1 (page 130), which shows where there is permafrost in the northern hemisphere, and Figure 5.2 (page 131).

  a i Where is the place closest to the North Pole where there is no permafrost?

    ii How close to the North Pole is this place? *(2 marks)*

  b i From Figure 5.1 suggest **two** reasons why there is no permafrost in some places while there is in other places. Give examples from the map to support your answer. *(6 marks)*

    iii Identify the cause/s of the 'pocket' of permafrost in north-west Scandinavia. *(2 marks)*

  c What is the 'active layer' in permafrost like? *(3 marks)*

  d i What is meant by the term 'mean annual temperature'? *(3 marks)*

    ii How deep is **a** the active layer and **b** the permafrost at Resolute Bay? *(2 marks)*

    iii Use data from Figure 5.2 to suggest the relationship between depth of permafrost and latitude. *(2 marks)*

  e Why does the permafrost not occur throughout the crustal rocks? *(5 marks)*

2 Study Figure 5.14 (page 136) which shows the flow of a river (its regime) in a periglacial area.

  a i When does water not flow in this river? *(2 marks)*

    ii Why does water not flow during this time? *(3 marks)*

    iii How would you recognise 'river terraces in the old floodplain' cut by such a river? *(5 marks)*

  b Using diagrams in your answer, explain the meaning of the term 'braiding' as used in the diagram. *(5 marks)*

  c Give **two** reasons why the wind has a greater erosional effect in periglacial environments than in most other areas. *(5 marks)*

  d How could you recognise that the wind had:

    i removed material from one area and

    ii deposited the material elsewhere? *(5 marks)*

Exam practice: basic structured questions

3 **a** Describe the shape and scale of **two** of the following periglacial landforms: ice wedge polygons; scree; nivation hollow; solifluction terracettes. *(6 marks)*

  b For **one** of the landforms you have described in **a**, explain how periglacial processes have led to its formation. *(6 marks)*

  c Figure 5.10 (page 134) shows a pingo in northern Canada. Write a description of the pingo from the photograph, including the area around it and its scale. *(6 marks)*

  d How is a pingo formed? *(7 marks)*

Exam practice: structured questions

4 Study Figure 5.21 which shows a range of periglacial landforms and their locations.

  a Choose **one** of the landforms labelled B to H. Describe its size and location in the field and suggest how it has been formed. *(8 marks)*

  b Explain the processes that are operating in the snow patch (A). *(5 marks)*

  c Explain the role of **i** wind and **ii** meltwater in the formation of landforms in areas of periglacial landscape. *(12 marks)*

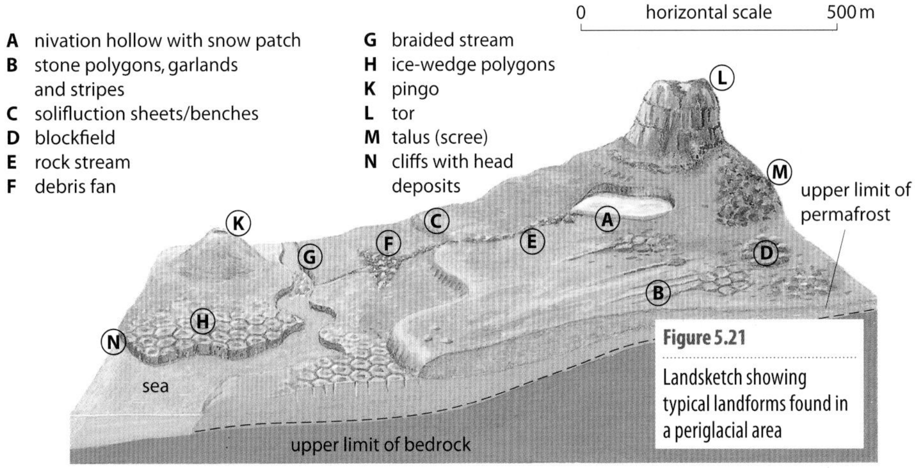

A nivation hollow with snow patch	**G** braided stream
B stone polygons, garlands and stripes	**H** ice-wedge polygons
C solifluction sheets/benches	**K** pingo
D blockfield	**L** tor
E rock stream	**M** talus (scree)
F debris fan	**N** cliffs with head deposits

Figure 5.21

Landsketch showing typical landforms found in a periglacial area

Exam practice: essays

5 'Changes to soil stability due to frost are a major problem for development in regions where there is a periglacial climate.'

Using examples you have studied, explain why this could be the case, and describe methods people use to overcome the problems of living in such areas. *(25 marks)*

6 'Permafrost may seem like a remote irrelevance to us in the temperate mid-latitudes, but its destruction could have big implications both locally and globally.'

Discuss this statement. *(25 marks)*

6 Coasts

'A recent estimate of the coastline of England and Wales is 2750 miles and it is very rare to find the same kind of coastal scenery for more than 10 to 15 miles together.'

J.A. Steers, *The Coastline of England and Wales*, 1960

'I do not know what I may appear to the world; but to myself I seem to have been only a boy playing on the sea-shore, and diverting myself in now and then finding a smoother pebble or a prettier shell than ordinary, while the great ocean of truth lay all undiscovered before me.'

Isaac Newton, *Philosophiae Naturalis Principia Mathematica*, 1687

The coast is a narrow zone where the land and the sea overlap and directly interact. Its development is affected by terrestrial, atmospheric, marine and human processes (Figure 6.1) and their interrelationships. The coast is the most varied and rapidly changing of all landforms and ecosystems.

Waves

Waves are created by the transfer of energy from the wind blowing over the surface of the sea. (An exception to this definition is those waves – **tsunamis** – that result from submarine shock waves generated by earthquake or volcanic activity.) As the strength of the wind increases, so too does **frictional drag** and the size of the waves. Waves that result from local winds and travel only short distances are known as **sea**, whereas those waves formed by distant storms and travelling large distances are referred to as **swell**.

The energy acquired by waves depends upon three factors: the wind velocity, the period of time during which the wind has blown, and the length of the fetch. The **fetch** is the maximum distance of open water over which the wind can blow, and so places with the greatest fetch potentially receive the highest-energy waves. Parts of south-west England are exposed to the Atlantic Ocean and when the south-westerly winds blow it is possible that some waves may have originated several thousand kilometres away. The Thames estuary, by comparison, has less open water between it and the Continent and consequently receives lower-energy waves.

Figure 6.1

Factors affecting coasts

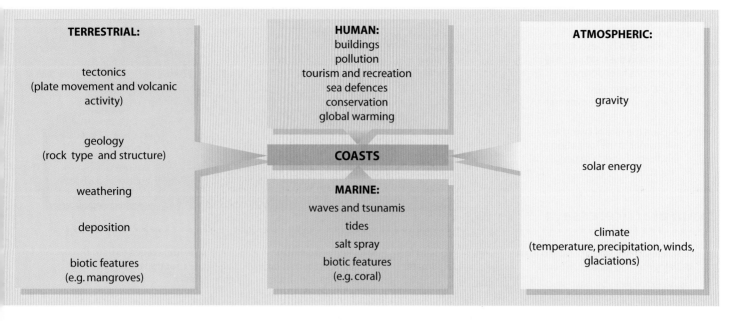

Wave terminology

The **crest** and the **trough** are respectively the highest and lowest points of a wave (Figure 6.2).

Wave height (*H*) is the distance between the crest and the trough. The height has to be estimated when in deep water. Wave height rarely exceeds 6 m although freak waves of 15 m have been reported by offshore oil-rigs, and 25 m by a wave-tracking satellite. Such waves can be a serious hazard to shipping.

Wave period (*T*) is the time taken for a wave to travel through one wave length. This can be timed either by counting the number of crests per minute or by timing 11 waves and dividing by 10 – i.e. the number of intervals.

Wave length (*L*) is the distance between two successive crests. It can be determined by the formula:

$$L = 1.56\ T^2$$

Wave velocity (*C*) is the speed of movement of a crest in a given period of time.

Wave steepness (*H ÷ L*) is the ratio of the wave height to the wave length. This ratio cannot exceed 1:7 (0.14) because at that point the wave will break. Steepness determines whether waves will build up or degrade beaches. Most waves have a steepness of between 0.005 and 0.05.

The **energy** (*E*) of a wave in deep water is expressed by the formula:

$$E \propto \text{(is proportional to) } LH^2$$

This means that even a slight increase in wave height can generate large increases in energy. It is estimated that the average pressure of a wave in winter is 11 tonnes per m², but this may be three times greater during a storm – it is little wonder that under such conditions sea defences may be destroyed and that wave power is a potential source of renewable energy (page 541).

Swell is characterised by waves of low height, gentle steepness, long wave length and a long period. **Sea**, with opposite characteristics, usually has higher-energy waves.

Waves in deep water

Deep water is when the depth of water is greater than one-quarter of the wave length:

$$(D = > \tfrac{L}{4})$$

The drag of the wind over the sea surface causes water and floating objects to move in an **orbital motion** (Figure 6.3). Waves are surface features (submerged submarines are unaffected by storms) and therefore the sizes of the orbits decrease rapidly with depth. Any floating object in the sea has a small net horizontal movement but a much larger vertical motion.

Waves in shallow water

As waves approach shallow water, i.e. when their depth is less than one-quarter of the wave length,

$$(D = < \tfrac{L}{4})$$

friction with the seabed increases. As the base of the wave begins to slow down, the circular oscillation becomes more elliptical (Figure 6.4). As the water depth continues to decrease, so does the wave length.

Meanwhile the height and steepness of the wave increase until the upper part spills or plunges over. The point at which the wave breaks is known as the **plunge line**. The body of foaming water which then rushes up the beach is called the **swash**, while any water returning down to the sea is the **backwash**.

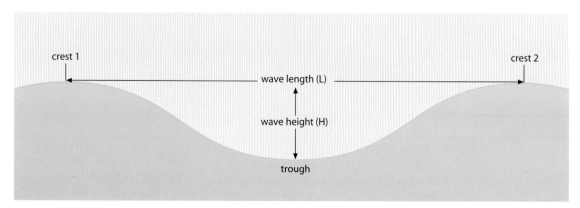

Figure 6.2

Wave terminology

Figure 6.3

Movement of an object in deep water: the diagrams show the circular movement of a ball or piece of driftwood through five stages in the passage of one wave length (crest 1 to crest 2); although the ball moves vertically up and down and the wave moves forward horizontally, there is very little horizontal movement of the ball until the wave breaks; the movement is orbital and the size of the orbit decreases with depth

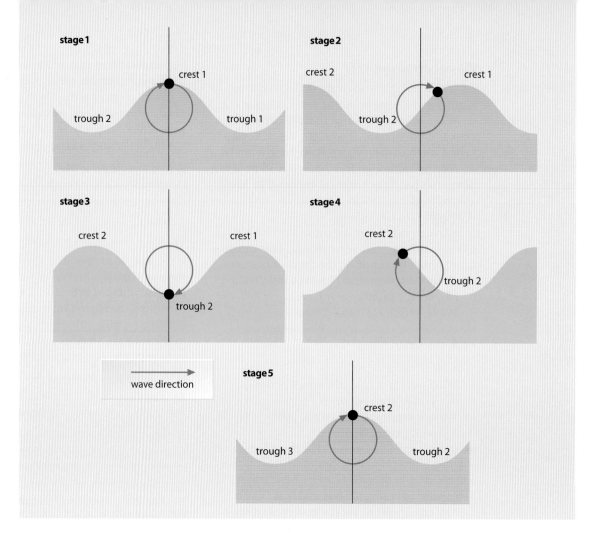

Figure 6.4

Why a wave breaks

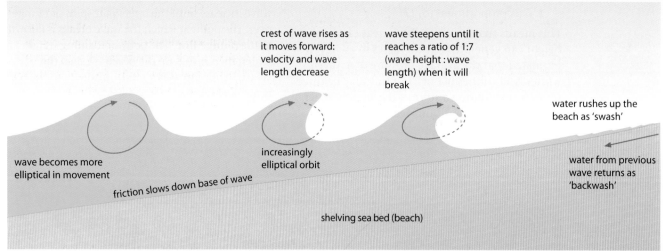

Wave refraction

Where waves approach an irregular coastline, they are refracted, i.e. they become increasingly parallel to the coastline. This is best illustrated where a headland separates two bays (Figure 6.5). As each wave crest nears the coast, it tends to drag in the shallow water near to a headland, or indeed any shallow water, so that the portion of the crest in deeper water moves forward while that in shallow water is retarded (by frictional drag), causing the wave to bend. The **orthogonals** (lines drawn at right-angles to wave crests) in Figure 6.5 represent four stages in the advance of a particular wave crest. It is apparent from the convergence of lines S^1, S^2, S^3 and S^4 that wave energy becomes concentrated upon, and so accentuates erosion at, the headland. The diagram also shows the formation of **longshore** (littoral) **currents**, which carry sediment away from the headland.

Figure 6.5

Wave refraction at a headland

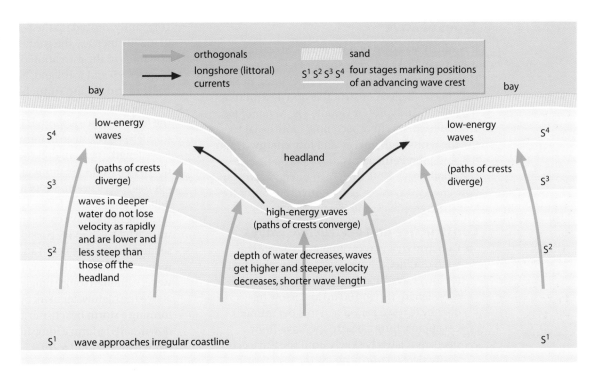

Beaches

Beaches may be divided into three sections – **backshore** (upper), **foreshore** (lower) and **nearshore** – based on the influence of waves (Figure 6.6). A beach forms a buffer zone between the waves and the coast. If the beach proves to be an effective buffer, it will dissipate wave energy without experiencing any net change itself. Because it is composed of loose material, a beach can rapidly adapt its shape to changes in wave energy. It is, therefore, in dynamic equilibrium with its environment (Framework 3, page 45).

Beach profiles fall between two extremes: those that are wide and relatively flat; and those that are narrow and steep. The gradient of natural beaches is dependent upon the interrelationship between two main variables:

- **Wave energy** Field studies have shown a close relationship between the profile of a beach and the action of two types of wave: constructive and destructive (page 144). However, the effect of wave steepness on beach profiles is complicated by the second variable.
- **Particle size** There is also, due to differences in the relative dissipation of wave energy, a distinct relationship between beach slope and particle size. This relationship is partly due to grain size and partly to percolation rates, both of which are greater on shingle beaches than on sand (pages 145–146). Consequently, shingle beaches are steeper than sand beaches (Figure 6.6).

Figure 6.6

Wave zones and beach morphology (*after* King, 1980)

143

Types of wave

It is widely accepted that there are two extreme wave types that affect the shape of a beach. However, whereas the extreme types have, in the past, been labelled **constructive** and **destructive** (Figure 6.7, and Andrew Goudie *The Nature of the Environment*), it is now becoming more usual to use the terms **high energy** and **low energy** (Figure 6.8, and John Pethick *An Introduction to Coastal Geomorphology*). Note that 'high-energy waves' and 'low-energy waves' are *not* synonymous terms for 'constructive waves' and 'destructive waves'.

Constructive and destructive waves

- **Constructive waves** often form where the fetch distance is long. They are usually small (or low) waves, flat in form and with a long wave length (up to 100 m) and a low frequency (a wave period of 6 to 8 per minute). On approaching a beach, the wave front steepens relatively slowly until the wave gently 'spills' over (Figure 6.7a). As the resultant swash moves up the beach, it rapidly loses volume and energy due to water percolating through the beach material. The result is that the backwash, despite the addition of gravity, is weak and has insufficient energy either to transport sediment back down the beach or to impede the swash from the following wave. Consequently sand and shingle is slowly, but constantly, moved up the beach. This will gradually increase the gradient of the beach and leads to the formation of **berms** at its crest (Figures 6.9 and 6.10) and, especially on sandy beaches, **ridges** and **runnels** (Figure 6.6).

- **Destructive waves** are more common where the fetch distance is shorter. They are often large (or high) waves, steep in form and with a short wave length (perhaps only 20 m) and a high frequency (10 to 14 per minute). These waves, on approaching a beach, steepen rapidly until they 'plunge' over (Figure 6.7b). The near-vertical breaking of the wave creates a powerful backwash which can move considerable amounts of sediment down the beach and, at the same time, reduce the effect of the swash from the following wave. Although some shingle may be thrown up above the high-water mark by very large waves, forming a **storm beach**, most material is moved downwards to form a **longshore (breakpoint) bar** (Figures 6.6 and 6.7b).

High-energy waves and low-energy waves

Recent opinion appears to support the view that beach shape is more dependent on, and linked to, wave energy. The correlation between the two types of wave energy and beach profile is given in Figure 6.8.

Figure 6.7

Constructive and destructive waves

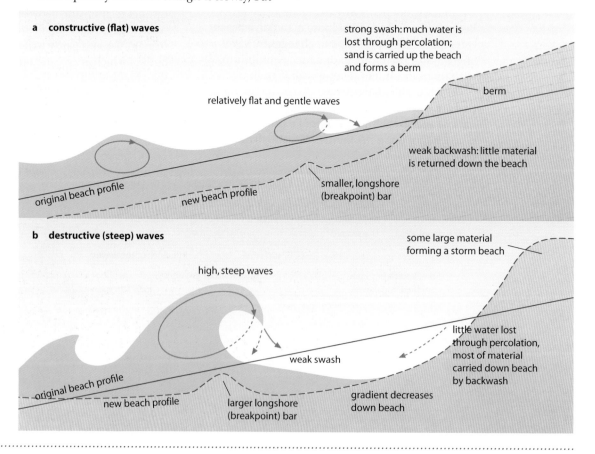

a **constructive (flat) waves**

strong swash: much water is lost through percolation; sand is carried up the beach and forms a berm

relatively flat and gentle waves

berm

weak backwash: little material is returned down the beach

original beach profile

new beach profile

smaller, longshore (breakpoint) bar

b **destructive (steep) waves**

some large material forming a storm beach

high, steep waves

weak swash

little water lost through percolation, most of material carried down beach by backwash

original beach profile

new beach profile

larger longshore (breakpoint) bar

gradient decreases down beach

Figure 6.8

High-energy and
low-energy waves
(*after* J. Pethick)

High-energy waves		Low-energy waves
Produced by distant storms	**Source**	Formed more locally
Large	**Fetch distance**	Short
Long (up to 100 m)	**Wave length**	Short (perhaps only 20 m)
High and short	**Wave height**	Low and flat
Move quickly and so lose little energy	**Speed of wave movement**	Move less quickly and so lose more energy
Spilling	**Type of breaker**	Surging
Long	**Dissipation distance**	Shorter
Flat and wide	**Beach shape**	Steeper and narrower

Particle size

This factor complicates the influence of wave steepness on the morphology of a beach. The fact that shingle beaches have a steeper gradient than sandy beaches is due mainly to differences in percolation rates resulting from differences in particle size – i.e. water will pass through coarse-grained shingle more rapidly than through fine-grained sand (Figure 8.2).

Figure 6.9

Storm beaches and
berms: berms mark
the limits of
successively lower
high tides

S + 1 = high tide after the spring high tide
S + 2 = second high tide after the spring high tide
S + 3 = third high tide after spring high tide

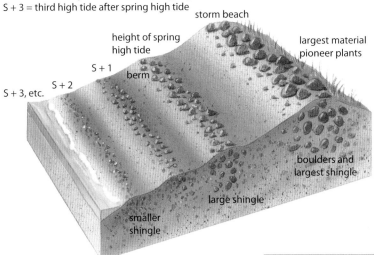

Shingle beaches

Shingle may make up the whole, or just the upper part, of the beach and, like sand, it will have been sorted by wave action. Usually, the larger the size of the shingle, the steeper the gradient of the beach, i.e. the gradient is in direct proportion to shingle size. This is an interesting hypothesis to test by experiment in the field (Framework 10, page 299).

Regardless of whether waves on shingle beaches are constructive or destructive, most of the swash rapidly percolates downwards leaving limited surface backwash. This, together with the loss of energy resulting from friction caused by the uneven surface of the shingle (compare this with the effects of bed roughness of a stream, page 70), means that under normal conditions, very little shingle is moved back down the beach. Indeed, the strong swash will probably transport material up the beach forming a berm at the spring high-tide level. Above the berm there is often a storm beach, composed of even bigger boulders thrown there by the largest of waves, while below may be several smaller ridges, each marking the height of the successively lower high tides which follow the maximum spring tide (Figures 6.9 and 6.10).

Figure 6.10

Berms and storm
beaches in north-east
Anglesey, Wales

Sand beaches

Sand usually produces beaches with a gentle gradient. This is because the small particle size allows the sand to become compact when wet, severely restricting the rate of percolation. Percolation is also hindered by the storage of water in pore spaces in sand which enables most of the swash from both constructive and destructive waves to return as backwash. Relatively little energy is lost by friction (sand presents a smoother surface than shingle) so material will be carried down the beach. The material will build up to form a longshore bar at the low-tide mark (Figure 6.6). This will cause waves to break further from the shore, giving them a wider beach over which to dissipate their energy. The lower parts of sand beaches are sometimes crossed by shore-parallel ridges and runnels (Figure 6.6). The ridges may be broken by channels which drain the runnels at low tide.

The interrelationship between wave energy, beach material and beach profiles may be summarised by the following generalisations which refer to net movements:

- Destructive waves carry material down the beach.
- Constructive waves carry material up the beach.
- Material is carried upwards on shingle beaches.
- Material is carried downwards on sandy beaches.

Tides

The position at which waves break over the beach, and their range, are determined by the state of the tide. It has already been seen that the levels of high tides vary (berms are formed at progressively lower levels following spring high tides; Figure 6.9). Tides are controlled by gravitational effects, mainly of the moon but partly of the sun, together with the rotation of the Earth and, more locally, the geomorphology of sea basins.

The moon has the greatest influence. Although its mass is much smaller than that of the sun, this is more than compensated for by its closer proximity to the Earth. The moon attracts, or pulls, water to the side of the Earth nearest to it. This creates a bulge or **high tide** (Figure 6.11a), with a complementary bulge on the opposite side of the Earth. This bulge is compensated for by the intervening areas where water is repelled and which experience a **low tide**. As the moon orbits the Earth, the high tides follow it.

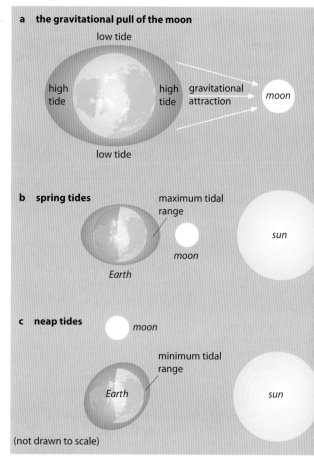

Figure 6.11

Causes of tides

a the gravitational pull of the moon

low tide
high tide · high tide · gravitational attraction · moon
low tide

b spring tides

maximum tidal range
Earth · moon · sun

c neap tides

moon
minimum tidal range
Earth · sun

(not drawn to scale)

A lunar month (the time it takes the moon to orbit the Earth) is 29 days and the tidal cycle (the time between two successive high tides) is 12 hours and 25 minutes, giving two high tides, near enough, per day. The sun, with its smaller gravitational attraction, is the cause of the difference in tidal range rather than of the tides themselves. Once every 14/15 days (i.e. twice in a lunar month), the moon and sun are in alignment on the same side of the Earth (Figure 6.11b). The increase in gravitational attraction generates the **spring tide** which produces the highest high tide, the lowest low tide and the maximum tidal range.

Midway between the spring tides are the **neap tides**, which occur when the sun, Earth and moon form a right-angle, with the Earth at the apex (Figure 6.11c). As the sun's attraction partly counterbalances that of the moon, the tidal range is at a minimum with the lowest of high tides and the highest of low tides (Figure 6.12). Spring and neap tides vary by approximately 20 per cent above and below the mean high-tide and low-tide levels.

So far, we have seen how tides might change on a uniform or totally sea-covered Earth. In practice, the tides may differ considerably from the above scenario due to such factors as: the Earth's rotation (and the effect of the Coriolis force, page 224); the distribution of land masses; and the size, depth and configuration of ocean and sea basins.

Figure 6.12

Tidal cycles during the lunar month

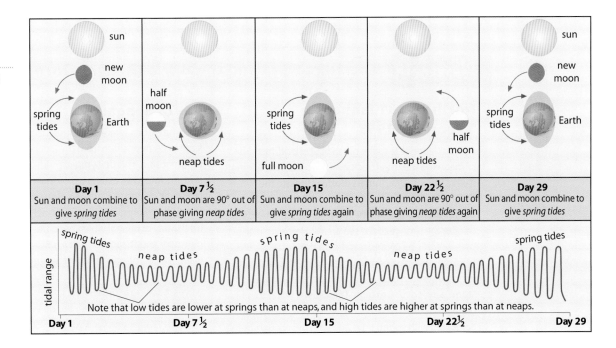

Day 1	Day 7½	Day 15	Day 22½	Day 29
Sun and moon combine to give *spring tides*	Sun and moon are 90° out of phase giving *neap tides*	Sun and moon combine to give *spring tides* again	Sun and moon are 90° out of phase giving *neap tides* again	Sun and moon combine to give *spring tides*

Note that low tides are lower at springs than at neaps, and high tides are higher at springs than at neaps.

Figure 6.13

Tidal range and difference in times of high tide in the North Sea

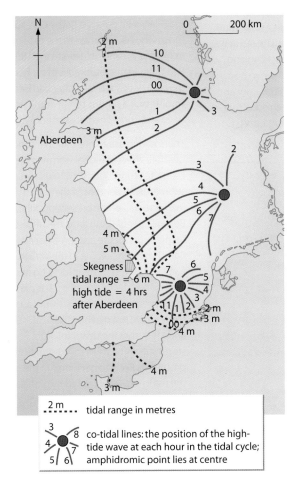

Skegness tidal range = 6 m high tide = 4 hrs after Aberdeen

2 m ---- tidal range in metres

co-tidal lines: the position of the high-tide wave at each hour in the tidal cycle; amphidromic point lies at centre

The morphology of the seabed and coastline affects tidal range. In the example of the North Sea, as the tidal wave travels south it moves into an area where both the width and the depth of the sea decrease. This results in a rapid accumulation, or funnelling, of water to give an increasingly higher tidal range – the range at Dover is several metres greater than in northern Scotland (Figure 6.13). Estuaries where incoming tides are forced into rapidly narrowing valleys also have considerable tidal ranges, e.g. the Severn estuary with 13 m, the Rance (Brittany) with 11.6 m and the Bay of Fundy (Canada) with 15 m. It is due to these extreme tidal ranges that the Rance has the world's first tidal power station, while the Bay of Fundy and the Severn have, respectively, experimental and proposed schemes for electricity generation (page 542). Extreme narrowing of estuaries can concentrate the tidal rise so rapidly that an advancing wall of water, or **tidal bore**, may travel upriver, e.g. the Rivers Severn and Amazon. In contrast, small enclosed seas have only minimal tidal ranges, e.g. the Mediterranean with 0.01 m.

Storm surges

Storm surges are rapid rises in sea-level caused by intense areas of low pressure, i.e. depressions (page 230) and tropical cyclones (page 235). For every drop in air pressure of 10 mb (page 224), sea-level can rise 10 cm. In tropical cyclones, pressure can fall by 100 mb causing the sea-level to rise by 1 m. Areas at greatest risk are those where sea basins become narrower and more shallow (e.g. southern North Sea and the Bay of Bengal) and where tropical cyclones move from the sea and cross low-lying areas (e.g. Bangladesh and Florida). When these storms coincide with hurricane-force winds and high tides, the surge can be topped by waves reaching 8 m in height. Where such events occur in densely populated areas, they pose a major natural hazard as they can cause considerable loss of life and damage to property (Places 19 and 31, page 238).

North Sea, 31 January – 1 February 1953

A deep depression to the north of Scotland, instead of following the usual track which would have taken it over Scandinavia, turned southwards into the North Sea (Figure 6.14). As air is forced to rise in a depression (page 230), the reduced pressure tends to raise the surface of the sea area underneath it. If pressure falls by 56 mb, as it did on this occasion, the level of the sea may rise by up to 0.5 m. The gale-force winds, travelling over the maximum fetch, produced storm waves over 6 m high. This caused water to pile up in the southern part of the North Sea. This event coincided with spring tides and with rivers discharging into the sea at flood levels. The result was a high tide, excluding the extra height of the waves, of over 2 m in Lincolnshire, over 2.5 m in the Thames estuary and over 3 m in the Netherlands. The immediate result was the drowning of 264 people in south-east England and 1835 people in the Netherlands. To prevent such devastation by future surges, the Thames Barrier and the Dutch Delta Scheme have since been constructed. Both schemes needed considerable capital and technology to implement.

Figure 6.14

The North Sea storm surge of 1 February 1953

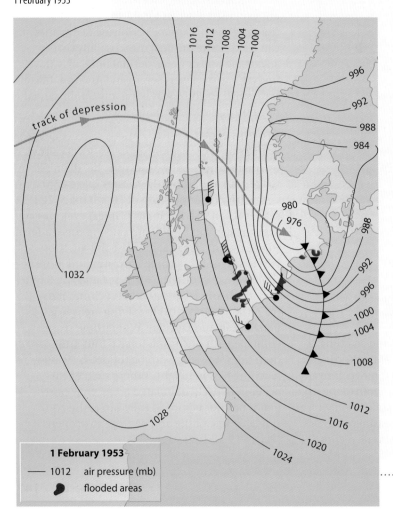

1 February 1953

— 1012   air pressure (mb)

🌊 flooded areas

Bay of Bengal

The south of Bangladesh includes many flat islands formed by deposition from the Rivers Ganges and Brahmaputra. This delta region is ideal for rice growing and is home to an estimated 40 million people. However, during the autumn, tropical cyclones (tropical low pressure storms) funnel water northwards up the Bay of Bengal which becomes increasingly narrower and shallower towards Bangladesh. The water sometimes builds up into a surge which may exceed 4 m in height and which may be capped by waves reaching a further 4 m. The result can be a wall of water which sweeps over the defenceless islands. Three days after one such surge in 1994, the Red Cross suggested that over 40 000 people had probably been drowned, many having been washed out to sea (Places 31, page 238). The only survivors were those who had climbed to the tops of palm trees and managed to cling on despite the 180 km/hr winds. The Red Cross feared outbreaks of typhoid and cholera in the area because fresh water had been contaminated. Famine was a serious threat as the rice harvest had been lost under the salty waters.

There is increasing international concern about the possible effect of global warming on Bangladesh. Estimates suggest that a 1 m rise in sea-level could submerge 25 per cent of the country, affecting over one-half of the present population (page 169). Because Bangladesh lacks the necessary capital and technology, for the last three decades the World Bank has been helping in the construction of cyclone early warning systems, providing flood shelters and improving coastal defences. It is partly because of these precautions, and partly because recent storm surges have not reached the peak heights of 1990 and 1991, that the death toll from flooding caused by storm surges has decreased significantly. However, the problem is likely to get worse in the near future due to the rising sea-level caused by global warming, and the lowering in height of the delta region resulting from the extraction of groundwater for agriculture.

Year	Height of storm surge	Death toll (estimated)
1966	6.1	80 000
1985	5.7	40 000
1988	4.8	25 000
1990	6.3	140 000
1991	6.1	150 000
1994	5.8	40 000
2007	5.1	2 300

Figure 6.15

Waves breaking on Filey Brigg, Yorkshire: wave energy is absorbed by a band of residual rock and so the cliff behind is protected

Processes of coastal erosion

Subaerial According to J. Pethick, 'Cliff recession is primarily the result of mass failure.' Mass failure may be caused by such non-marine processes as: rain falling directly onto the cliff face; by throughflow or, under extreme conditions, surface runoff of water from the land; and the effects of weathering by the wind and frost. These processes, individually or in combination, can cause mass movement either as soil creep on gentle slopes or as slumping and landslides on steeper cliffs (Figures 2.17 and 2.18).

Wave pounding Steep waves have considerable energy. When they break as they hit the foot of cliffs or sea walls, they may generate shockwaves of up to 30 tonnes per m². Some sea walls in parts of eastern England need replacing within 25 years of being built, due to wave pounding (Case Study 6).

Hydraulic pressure When a parcel of air is trapped and compressed, either in a joint in a cliff or between a breaking wave and a cliff, then the resultant increase in pressure may, over a period of time, weaken and break off pieces of rock or damage sea defences.

Abrasion/corrasion This is the wearing away of the cliffs by sand, shingle and boulders hurled against them by the waves. It is the most effective method of erosion and is most rapid on coasts exposed to storm waves.

Attrition Rocks and boulders already eroded from the cliffs are broken down into smaller and more rounded particles.

Corrosion/solution This includes the dissolving of limestones by carbonic acid in sea water (compare Figure 2.8), and the evaporation of salts to produce crystals which expand as they form and cause the rock to disintegrate (Figure 2.2). Salt from sea water or spray is capable of corroding several rock types.

Factors affecting the rate of erosion

Breaking point of the wave A wave that breaks as it hits the foot of a cliff releases most energy and causes maximum erosion. If the wave hits the cliff before it breaks, then much less energy is transmitted, whereas a wave breaking further offshore will have had its energy dissipated as it travelled across the beach (Figure 6.15).

Wave steepness Highest-energy waves, associated with longer fetch distances, have a high, steep appearance. They have greater erosive power than low-energy waves, which are generated where the fetch is shorter and have a lower and flatter form (Figure 6.8).

Depth of sea, length and direction of fetch, configuration of coastline A steeply shelving beach creates higher and steeper waves than one with a more gentle gradient. The longer the fetch, the greater the time available for waves to collect energy from the wind. The existence of headlands with vertical cliffs tends to concentrate energy by wave refraction (page 142).

Supply of beach material Beaches, by absorbing wave energy, provide a major protection against coastal erosion.

Beach morphology Beaches, by dissipating wave energy, act as a buffer between waves and the land. As they receive high-energy inputs at a rapid rate from steep waves, and low-energy inputs at a slower rate from flat waves, they must adopt a morphology (shape) to counteract the different energy inputs. High, rapid energy inputs are best dissipated by wide, flat beaches which spread out the oncoming wave energy. In contrast, the lower-energy inputs of flatter waves can easily be dissipated by narrow, steep beaches which act rather like a wall against which the waves flounder. An exception is when steep waves break onto a shingle beach. As energy is rapidly dissipated through friction and percolation, then a wide, flat beach profile is unnecessary (page 145).

Rock resistance, structure and dip The strength of coastal rocks influences the rate of erosion (Figure 6.16). In Britain, it is coastal areas where glacial till was deposited that are being worn back most rapidly (Places 20). When Surtsey first arose out of the sea off the southwest coast of Iceland in 1963 (Places 3, page 16), it consisted of unconsolidated volcanic ash. It was only when the ash was covered and protected by a lava flow the following year that the island's survival was seemingly guaranteed.

Rocks that are well-jointed (Figure 8.1) or have been subject to faulting have an increased vulnerability to erosion. The steepest cliffs are usually where the rock's structure is horizontal or vertical and the gentlest where the rock dips upwards away from the sea. In the latter case, blocks may break off and slide downwards (Figure 2.17). Erosion is also rapid where rocks of different resistance overlie one another, e.g. chalk and Gault clay in Kent.

Figure 6.16

Rock type and average rates of cliff recession

Rock type	Location	Rate of erosion (m/yr)
Volcanic ash	Krakatoa	40
Glacial till	Holderness	2
Glacial till	Norfolk	1
Chalk	South-east England	0.3
Shale	North Yorkshire	0.09
Granite	South-west England	0.001

Human activity The increase in pressure resulting from building on cliff tops and the removal of beach material which may otherwise have protected the base of the cliff both contribute to more rapid coastal erosion.

Although rates of erosion may be reduced locally by the construction of sea defences, such defences often lead to increased rates of erosion in adjacent areas. Human activity therefore has the effect of disturbing the equilibrium of the coast system (Case Study 6).

Places 20 Holderness: coastal processes

The coastline at Holderness is retreating by an average of 1.8 m a year. Since Roman times, the sea has encroached by nearly 3 km, and some 50 villages mentioned in the Domesday Book of 1086 have disappeared.

The following extract was taken from a management report, 'Humber Estuary & Coast' (1994) prepared by Professor J.S. Pethick (then of the University of Hull and now at the University of Newcastle) for Humberside County Council.

'The soft glacial till cliffs of Holderness are eroding at a rapid rate. The reasons for such erosion are, however, less to do with the soft sediment of the cliff than with the lack of beach material and the poorly developed nearshore zone [Figure 6.6]. Retreat of the cliff line here is matched by progressive lowering of the seabed to give a wide shallow platform stretching several kilometres seaward. Eventually this platform will be so extensive that most of the incident wave energy will be expended here rather than at the cliff so that erosion rates will decrease or even halt. Since this may take several thousand years, it cannot form part of any management plan for this coast – yet it

is important to recognise that the natural erosional processes here are neither random nor pernicious.

The process of cliff retreat along the Holderness coast is more complex than appears at first sight. Mass failures of the cliff are triggered by wave action at the cliff toe. Such failures may be 50 to 100 m wide and up to 30 m deep giving a scalloped edge to the cliff. The retreat rate varies temporarily; a large failure may produce a 10 m retreat in one year but no further retreat will then occur for 3 or 4 years – giving a periodicity of 4 or 5 years in total. This means that attempts to measure erosion rates over periods of less than 10 years, that is over 2 cycles, can be extremely misleading, resulting in massive over- or under-estimates of the long-term retreat rate which is remarkably constant at 1.8 m per year [Figure 6.17]. Three issues may be highlighted here.

- The beaches of Holderness are thin veneers covering the underlying glacial tills. The beaches do not increase in volume since, south of Hornsea, a balance exists between the input of sand by erosion and the removal of the sand by wave action, principally from the north-east, which drives sands south.

- The sediment balance on the Holderness coast is maintained by the action of storm waves from the north-east. These waves approach the coast obliquely, the angle between wave crest and shore being critical for the sediment transport rate. A clockwise movement would increase the transport and erosion rate while an anti-clockwise swing would decrease both of these. Random changes in the orientation of the shore are quickly eradicated by changes in the sediment balance, but any permanent change in the orientation of the coastline, such as that caused by the introduction of hard sea defences as at Hornsea, Mappleton and Withernsea, means that the sediment balance is disturbed.

Figure 6.17

Houses collapsing into the sea, Holderness

- Hard defences [Case Study 6A] can have two long-term effects: first, although erosion is halted at the defence itself, several kilometres to the north erosion continues as before. This causes an anti-clockwise re-orientation of the coast, sand transport is reduced and sand accumulates immediately north of the defences – as can be seen north of Hornsea. Second, the accumulation of sand north of the defences starves the beaches to the south causing an increase in erosion there. The fine-grained sediments from the Holderness cliff and seabed erosion are not transported along the beaches as are the sands and shingle but are moved in suspension. Research is presently under way which is intended to chart the precise movement of this material but it is clear that its dominant movement is south towards the Humber. A large proportion may enter the estuary and become deposited there. The remainder is moved south and east into the North Sea where the transport pathway is towards the Dutch and German coast.'

Erosion landforms

Headlands and bays

These are most likely to be found in areas of alternating resistant and less resistant rock. Initially, the less resistant rock experiences most erosion and develops into **bays**, leaving the more resistant outcrops as **headlands**. Later, the headlands receive the highest-energy waves and so become more vulnerable to erosion than the sheltered bays (Figure 6.5). The latter now experience low-energy breakers which allow sand to accumulate and so help to protect that part of the coastline.

Abrasion or wave-cut platforms

Wave energy is at its maximum when a high, steep wave breaks at the foot of a cliff. This results in undercutting of the cliff to form a **wave-cut notch** (Figure 6.18). The continual undercutting causes increased stress and tension in the cliff until eventually it collapses. As these processes are repeated, the cliff retreats leaving, at its base, a gently sloping **abrasion** or **wave-cut platform** which has a slope angle of less than 4° (Figure 6.19). The platform, which appears relatively even when viewed from a distance, cuts across rocks regardless of their type and structure. A closer inspection of this inter-tidal feature usually reveals that it is deeply dissected by abrasion, resulting from material carried across it by tidal movements, and corrosion. As the cliff continues to retreat, the widening of the platform means that incoming waves break further out to sea and have to travel over a wider area of beach. This dissipates their energy, reduces the rate of erosion of the headland, and limits the further extension of the platform. It has been hypothesised that wave-cut platforms cannot exceed 0.5 km in width.

Where there has been negative change in sea-level (page 81), former wave-cut platforms remain as **raised beaches** above the present influence of the sea (Figure 6.51).

Figure 6.18

Wave-cut notch at Coromandel Peninsula, New Zealand

Figure 6.19

Abrasion or wave-cut platform at Flamborough Head, Yorkshire

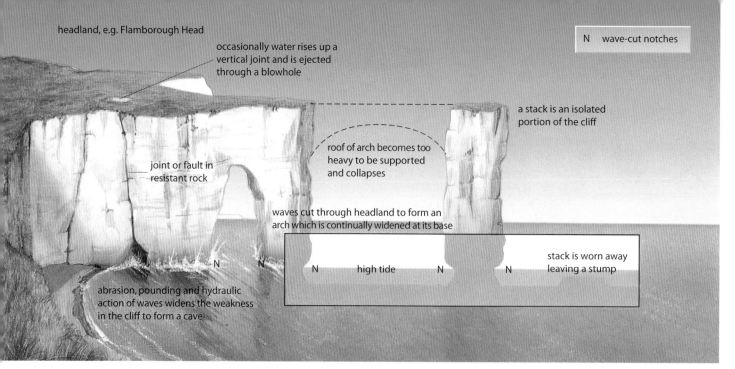

headland, e.g. Flamborough Head

occasionally water rises up a vertical joint and is ejected through a blowhole

N wave-cut notches

joint or fault in resistant rock

roof of arch becomes too heavy to be supported and collapses

a stack is an isolated portion of the cliff

waves cut through headland to form an arch which is continually widened at its base

abrasion, pounding and hydraulic action of waves widens the weakness in the cliff to form a cave

N   N   N   high tide   N   N

stack is worn away leaving a stump

Figure 6.20

The formation of caves, blowholes, arches and stacks

Caves, blowholes, arches and stacks

Where cliffs are of resistant rock, wave action attacks any line of weakness such as a joint or a fault. Sometimes the sea cuts inland, along a joint, to form a narrow, steep-sided inlet called a **geo**, or at other times it can undercut part of the cliff to form a **cave**. As shown in Figure 6.20, caves are often enlarged by several combined processes of marine erosion. Erosion may be vertical, to form **blowholes**, but is more typically backwards through a headland to form **arches** and **stacks** (Figures 6.20 and 6.21).

These landforms, which often prove to be attractions to sightseers and mountaineers, can be found at The Needles (Isle of Wight), Old Harry (near Swanage) and Flamborough Head (Yorkshire, Figure 6.19), which are all cut into chalk, and at The Old Man of Hoy (Orkneys) which is Old Red Sandstone (Figure 8.12).

Figure 6.21

Icelandic coastline

Transportation of beach material

Up and down the beach

As we have already seen, flat, constructive waves tend to move sand and shingle up the beach, whereas the net effect of steep, destructive waves is to comb the material downwards.

Longshore (littoral) drift

Usually wave crests are not parallel to the shore, but arrive at a slight angle. Only rarely do waves approach a beach at right-angles. The wave angle is determined by wind direction, the local configuration of the coastline, and refraction at headlands and in shallow water. The oblique wave angle creates a nearshore current known as **longshore (or littoral) drift** which is capable of moving large quantities of material in a down-drift direction (Figure 6.22). On many coasts, longshore drift is predominantly in one direction; for example, on the south coast of England, where the maximum fetch and prevailing wind are both from the south-west, there is a predominantly eastward movement of beach material. However, brief changes in wind – and therefore wave – direction can cause the movement of material to be reversed.

Of lesser importance, but more interesting and easier to observe, is the movement of material along the shore in a zigzag pattern. This is because when a wave breaks, the swash carries material up the beach at the same angle as that at which the wave approached the shore. As the swash dies away, the backwash and any material carried by it returns straight down the beach, at right-angles to the waterline, under the influence of gravity. If beach material is carried a considerable distance, it becomes smaller, more rounded and better sorted.

Where beach material is being lost through longshore drift, the coastline in that locality is likely to be worn back more quickly because the buffering effect of the beach is lessened. To counteract this process, wooden breakwaters or **groynes** may be built (Figure 6.23). Groynes encourage the local accumulation of sand (important in tourist resorts) but can result in a depletion of material, and therefore an increase in erosion, further along the coast (Case Study 6A).

Figure 6.22

The effects of longshore drift

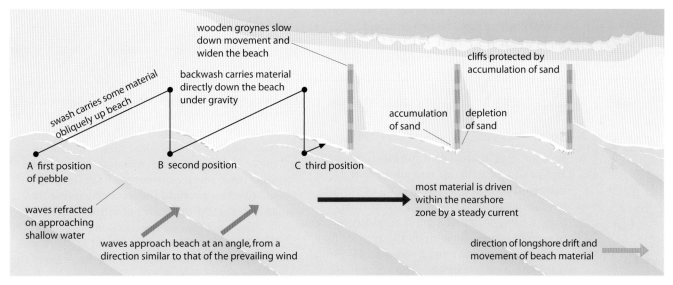

Figure 6.23

The effect of groynes on longshore drift, Southwold, Suffolk: this type of coastal management is usually undertaken at holiday resorts where sandy beaches are a major tourist attraction

Figure 6.24

A spit: Dawlish Warren at the mouth of the River Exe, Devon

Coastal deposition

Deposition occurs where the accumulation of sand and shingle exceeds its depletion. This may take place in sheltered areas with low-energy waves or where rapid coastal erosion further along the coast provides an abundant supply of material. In terms of the coastal system, deposition takes place as inputs exceed outputs, and the beach can be regarded as a store of eroded material.

Spits

Spits are long, narrow accumulations of sand and/or shingle with one end joined to the mainland and the other projecting out to sea or extending part way across a river estuary (Figure 6.24). Whether a spit is mainly composed of sand or shingle depends on the availability of sediment and wave energy (pages 145–146). Composite spits occur when the larger-sized shingle is deposited before the finer sands.

Figure 6.25

Stages in the formation of a spit

In Figure 6.25, the line **X–Y** marks the position of the original coastline. At point **A**, because the prevailing winds and maximum fetch are from the south-west, material is carried eastwards by longshore drift. When the orientation of the old coastline began to change at **B**, some of the larger shingle and pebbles were deposited in the slacker water in the lee of the headland. As the spit continued to grow, storm waves threw some larger material above the high-water mark (**C**), making the feature more permanent; while, under normal conditions, the finer sand was carried towards the end of the spit at **D**. Many spits develop a hooked or curved end. This may be for two reasons: a change in the prevailing wind to coincide with the second-most-dominant wave direction and second-longest fetch, or wave refraction at the end of the spit carrying some material into more sheltered water.

Eventually the seaward side of the spit will retreat, while longshore drift continues to extend the feature eastwards. A series of recurved ends may form (**E**) each time there is a series of storms from the south-east giving a lengthy period of altered wind direction. Having reached its present-day position (**F**), the spit is unlikely to grow any further – partly because the faster current of the river will carry material out to sea and partly because the depth of water becomes too great for the spit to build upwards above sea-level. Meanwhile, the prevailing south-westerly wind will pick up sand from the beach as it dries out at low tide and carry it inland to form **dunes** (**G**). The stability of the spit may be increased by the anchoring qualities of marram grass. At the same time, gentle, low-energy waves entering the sheltered area behind the spit deposit fine silt and mud, creating an area of **saltmarsh** (**H**).

Figure 6.28 shows the location of some of the larger spits around the coast of England and Wales. How do these relate to the direction of the maximum fetch and of the prevailing and dominant winds?

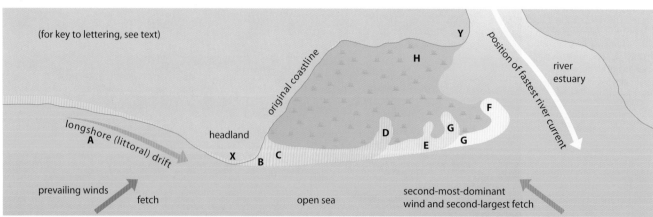

Figure 6.27

A bar: Slapton Ley, Devon

Figure 6.26

A tombolo: Loch Eriboll, Highland, Scotland

Tombolos, bars and barrier islands

A **tombolo** is a beach that extends outwards to join with an offshore island (Figure 6.26). Chesil Beach, in Dorset, links the Isle of Portland to the mainland. Some 30 km long and up to 14 m high, it presents a gently smoothed face to the prevailing winds in the English Channel.

If a spit develops in a bay into which no major river flows, it may be able to build across that bay, linking two headlands, to form a **bar**. Bars straighten coastlines and trap water in lagoons on the landward side. Bars, such as that at Slapton Ley, in Devon (Figure 6.27), may also result in places where constructive waves lead to the landward migration of offshore, seabed material.

Barrier islands are a series of sandy islands totally detached from, but running almost parallel to, the mainland. Between the islands, which may extend for several hundred kilometres, and the mainland is a tidal lagoon (Figure 6.29). Although relatively uncommon in Britain, they are widespread globally, accounting for 13 per cent of the world's coastlines. They are easily recognisable on maps of the eastern USA (Places 21), the Gulf of Mexico, the northern Netherlands, West Africa and southern and western Australia. Although their origin is uncertain, they tend to develop on coasts with relatively high-energy waves and a low tidal range. One theory suggests that they formed, below the low-tide mark, as offshore bars of sand and have moved progressively landwards. An alternative theory suggests that rises in post-glacial sea-level may have partly submerged older beach ridges. In either case, the breaches between the islands seem likely to have been caused by storm waves.

Figure 6.28

Location of some major spits, tombolos and bars in England and Wales

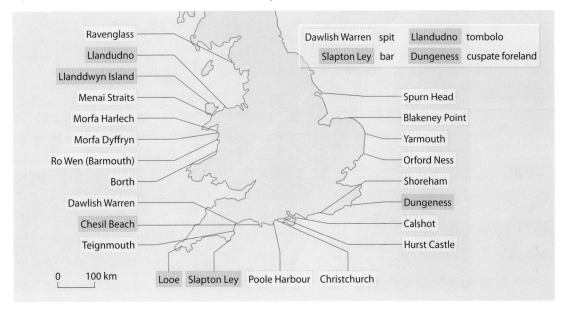

Ravenglass
Llandudno
Llanddwyn Island
Menai Straits
Morfa Harlech
Morfa Dyffryn
Ro Wen (Barmouth)
Borth
Dawlish Warren
Chesil Beach
Teignmouth

| Dawlish Warren | spit | Llandudno | tombolo |
| Slapton Ley | bar | Dungeness | cuspate foreland |

Spurn Head
Blakeney Point
Yarmouth
Orford Ness
Shoreham
Dungeness
Calshot
Hurst Castle

0   100 km

Looe   Slapton Ley   Poole Harbour   Christchurch

Figure 6.29

Barrier islands off North Carolina, USA, taken from the Apollo space-craft (X = position of Figure 6.30)

Barrier islands have a unique morphology, flora and fauna. The smooth, straight, ocean edge is characterised by wide, sandy beaches which slope gently upwards to sand dunes which are anchored by high grasses (Figure 6.30). Behind the dunes, the 'island' interior may contain shrubs and woods, deer and snakes, insects and birds. The landward side is punctuated by sheltered bays, quiet tidal lagoons, saltmarshes and, towards the tropics, mangrove swamps. These wetlands provide a natural habitat for oysters, fish and birds. Although barrier islands form the interface between the land and the ocean, they seem fragile in comparison with the power that the wind and sea brings to them. It is virtually impossible for a tropical storm or hurricane to move ashore without first crossing either of the two longest stretches of barrier islands in the world: either that which extends for 2500 km from New Jersey to the southern tip of Florida (Figure 6.29); or the one stretching for 2100 km along the Gulf Coast states to Mexico.

Barrier islands are subject to a process called 'wash over'. This process, which might occur up to 40 times in some years, is when storm waves carry large quantities of sand over the island from the seaward face to the landward side. This results in the seaward side being eroded and pushed backwards. The landward marshes and mangrove swamps become suffocated, and the tidal lagoons are narrowed. From a human viewpoint, barrier islands form an essential natural defence against hurricanes and their storm-force waves.

Figure 6.30

Barrier island on Core Banks, looking south (see X on Figure 6.29 for location)

Sand dunes

Sand dunes are a dynamic landform whose equilibrium depends on the interrelationship between mineral content (sand) and vegetation.

Longshore drift may deposit sand in the intertidal zone. As the tide ebbs, the sand will dry out allowing winds from the sea to move material up the beach by saltation (page 183). This process is most likely to occur when the prevailing winds come from the sea and where there is a large tidal range which exposes large expanses of sand at low tide. Sand may become trapped by seaweed and driftwood on berms or at the point of the highest spring tides. Plants begin to colonise the area (Figure 11.10), stabilising the sand and encouraging further accumulation. The regolith has a high pH value due to calcium carbonate from seashells.

Embryo dunes are the first to develop (Figure 6.31). They become stabilised by the growth of lyme and marram grasses. As these grasses trap more sand, the dunes build up and, due to the high rate of percolation, become increasingly arid. Plants need either succulent leaves to store water (sand couch), or thorn-like leaves to reduce transpiration in the strong winds (prickly saltwort), or long tap-roots to reach the water table (marram grass). As more sand accumulates, the embryo dunes join to form **foredunes** which can attain a height of 5 m (Figures 6.31 and 6.32). Due to a lack of humus, their colour gives them the name **yellow dunes**. The dunes become increasingly grey as humus and bacteria from plants and animals are added, and they gradually become more vegetation-covered and acidic. These **grey (mature) dunes** may reach a height of 10–30 m before the supply of fresh sand is cut off by their increasing distance from the beach (Figure 11.11). There may be several parallel ridges of old dunes (as at Morfa Harlech, Figure 6.33), separated by low-lying, damp slacks. Heath plants begin to dominate the area as acidity, humus and moisture content all increase (Figure 11.9). Paths cut by humans and animals expose areas of sand. As the wind funnels along these tracks, **blowouts** may form in the now **wasting dunes**. To combat further erosion at Morfa Harlech, parts of the dunes have been fenced off and marram grass has been planted to try to re-stabilise the area and to prevent any inland migration of the dunes.

The above idealised scheme can be interrupted at any stage by storms or human use. If the supply of sand is cut off, then new embryo dunes cannot form and yellow dunes may be degraded so that it is the older, grey dunes that line the beach.

Figure 6.31

A transect across sand dunes, based on fieldwork at Morfa Harlech, North Wales

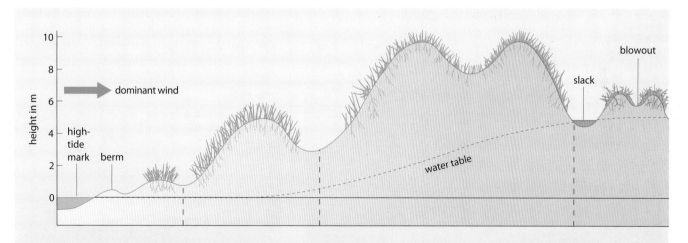

	Embryo	Fore or yellow dunes	Grey dunes and dune ridges	Wasting dunes with blowouts
Dune height (m)	1	5	8–10	6–8
Percentage of exposed sand	80	20	less than 10	over 40 on dunes
Humus and moisture content	very little humus, mixed salt and fresh water	some humus, very little moisture, fresh water	humus increases inland, water content still low, fresh water	high humus, brackish water in slacks
pH	over 8	slightly alkaline	increasingly acid inland: pH 6.5–7	acid: pH 5–6
Plant types	sand couch, lyme grass	marram, xerophytic species	creeping fescue, sea spurge, some marram, cotton grass, heather	heather, gorse on dunes, Juncus in slacks

Figure 6.33

Morfa Harlech from Harlech Castle showing foredunes, grey or wasting dunes, old cliff-line and, in the distance, saltmarsh

Figure 6.32

Embryo and foredunes at Morfa Harlech, North Wales (refer also to Figures 11.10 and 11.11)

Saltmarshes

Where there is sheltered water in river estuaries or behind spits, silt and mud will be deposited either by the gently rising and falling tide or by the river, thus forming a zone of **inter-tidal mudflats**. Initially, the area may only be uncovered by the sea for less than 1 hour in every 12-hour tidal cycle. Plants such as algae and *Salicornia* can tolerate this lengthy submergence and the high levels of salinity. They are able to trap more mud around them, creating a surface that remains exposed for increasingly longer periods between tides (Figure 6.34). *Spartina* grows throughout the year and since its introduction into Britain has colonised, and become dominant in, many estuaries. The landward side

of the inter-tidal mudflats is marked by a small cliff (Figure 11.12), above which is the flat **sward zone**. This zone may only be covered by the sea for less than 1 hour in each tidal cycle (Figure 6.12). Seawater collects in hollows which become increasingly saline as the water evaporates. The hollows often enlarge into **saltpans** (Figure 11.13) which are devoid of vegetation except for certain algae and the occasional halophyte (page 291). As each tide retreats, water drains into **creeks** which are then eroded rapidly both laterally and vertically (Figure 6.35). The upper sward zone may only be inundated by the highest of spring tides.

Figure 6.34

Llanrhidian saltmarsh, Gower peninsula, South Wales (refer also to Figures 11.13 and 11.14)

Figure 6.35

Llanrhidian saltmarsh showing the sward zone, creeks and saltpan

Figure 6.36

A sample population in relation to the total population

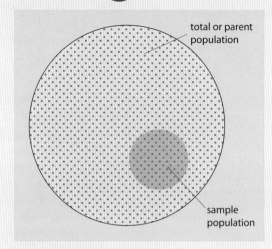

Why sample?

Geographers are part of a growing number of people who find it increasingly useful and/or necessary to use data to quantify the results of their research. The problem with this trend is that the amount of data may be very expensive, too time-consuming, or just impracticable to collect – as it would be, for example, to investigate everybody's shopping patterns in a large city, to find the number of stones on a spit, or to map the land use of all the farms in Britain.

Sampling is the method used to make statistically valid inferences when it is impossible to measure the **total population** (Figure 6.36). It is essential, therefore, to find the most accurate and practical method of obtaining a **representative sample**. If that sample can be made with the minimum of bias, then statistically significant conclusions may be drawn. However, even if every effort is made to achieve precision, it must be remembered that any sample can only be a close estimate.

Sampling basics

Most sampling procedures assume that the total population has a **normal distribution** (Figure 4.16a) which, when plotted on a graph, produces a symmetrical curve on either side of the mean value. This shows that a large proportion of the values are close to the average, with few extremes. Figure 6.37 shows a normal distribution curve and the **standard deviation** (page 247) – the measure of dispersion from the mean. Where most of the values are clustered near to the mean, the standard deviation is low.

The larger the sample, the more accurate it is likely to be, and the more likely it is to resemble the parent population; it is also more likely to conform to the normal distribution curve. While the generally accepted minimum size for a sample is 30, there is no upper limit – although there is a point beyond which the extra time and cost involved in increasing the sample size do not give a significant improvement in accuracy (an example of the law of diminishing returns, page 462).

Figure 6.37 shows that, in a normal distribution, 68.27 per cent of the values in the sample occur within a range of ±1 standard deviations (SDs) from the mean; 95 per cent of the values fall within ±2 SDs; and 99 per cent within ±3 SDs. These percentages are known as **confidence limits**, or **probability levels**. Geographers usually accept the 95 per cent probability level when sampling. This means that they accept the chance that, in 5 cases out of every 100, the true mean will lie outside 2 SDs to either side of their sample mean.

Figure 6.37

A normal distribution showing standard deviations from the mean

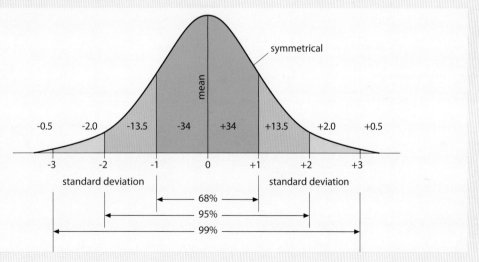

Sampling techniques

Several different methods may be used according to the demands of the required sample and the nature of the parent population. There are two major types, with one refinement:

- **Random sampling** This is the most accurate method as it has no bias.
- **Systematic sampling** This method is often quicker and easier to use, although some bias or selection is involved.
- **Stratified sampling** This method is often a very useful refinement for geographers; it can be used with either a random or a systematic sample.

Random sampling

Under normal circumstances, this is the ideal type of sample because it shows no bias. Every member of the total population has an equal chance of being selected, and the selection of one member does not affect the probability of selection of another member. The ideal random sample may be obtained using **random numbers**. These are often generated by computer and are available in the form of printed tables of random numbers, but if necessary they can be obtained by drawing numbers out of a hat. Random number tables usually consist of columns of pairs of digits. Numbers can be chosen by reading either along the rows or down the columns, provided only one method is used. Similarly, any number of figures may be selected – six for a grid reference, four for a grid square, three for house numbers in a long street, etc. Using the grid shown in Figure 6.38, the random number table given above yields eight 6-figure grid references: 927114; (986691 has to be excluded because the grid does not contain these numbers); 906126; etc.

One feature of a genuine random sample is that the same number can be selected more than once – so remember that if you are pulling numbers from a hat, they should be replaced immediately after they have been read and recorded.

There are three alternative ways of using random numbers to sample areal distributions (patterns over space) (Figure 6.38).

1   **Random point** A grid is superimposed over the area of the map to be sampled. Points, or map references, are then identified using random number tables, and plotted on the map. The eight points identified earlier (in the random number table) have been plotted on Figure 6.38a. A large number of points may be needed to ensure coverage of the whole area – see Figure 6.40.

2   **Random line** Random numbers are used to obtain two end points which are then joined by a line, as in Figure 6.38b which uses the same eight random points, in the order in which they occurred in the table. Several random lines are needed to get a representative sample (e.g. lines across a city to show transects of variation in land use).

3   **Random area** Areas of constant size, e.g. grid squares or quadrats, are obtained using random numbers. By convention, the number always identifies the south-west corner of a grid square. If sample squares one-quarter the size of a grid square are used, together with the same sample points, their locations are as shown on Figure 6.38c – note that the point in the north-east cannot be used because part of the sample square lies outside the study area. This method can be used to sample land-use areas or the distribution of plant communities over space.

Part of a random number table			
9271	0143	2141	9381
1498	3796	4413	1405
6691	4294	6077	9091
9061	1148	9493	1940
2660	7126	7126	4591
3459	7585	4897	8138
6090	7962	5766	7228
2191	9271	9042	5884

Figure 6.38

Random sampling using point, line and area techniques

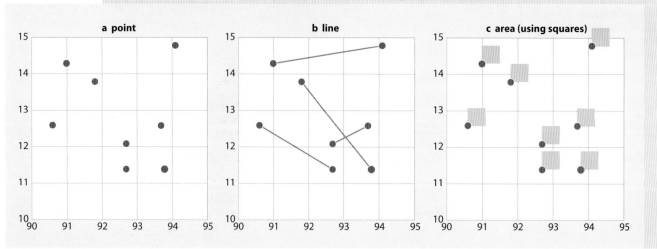

The advantages of random sampling include its ability to be used with large populations and its avoidance of bias. Careful **sample design** is needed, however, to avoid the possibility of achieving misleading results, for example when sampling small populations, and when sampling over a large area. Also, when used in the field, it may involve considerable time and energy in visiting every point.

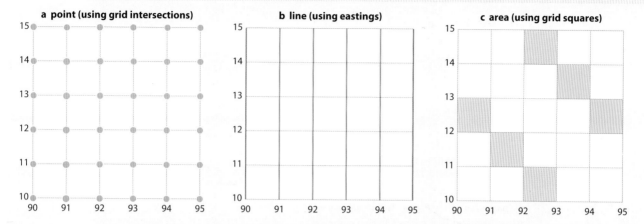

Figure 6.39

Systematic sampling using point, line and area techniques

Systematic sampling

A systematic sample is one in which values are selected in a regular way, e.g. choosing every 10th person on a list, or every 20th house in a street. This can be an easier method in terms of time and effort than random sampling. Like random sampling, it can be operated using individual points, lines or areas (Figure 6.39).

1 **Systematic point** This can show changes over distance, e.g. by sampling the land use every 100 m. It can also show change through time, e.g. by sampling from the population censuses (taken every 10 years).

2 **Systematic line** This may be used to choose a series of equally spaced transects across an area of land, e.g. a shingle spit.

3 **Systematic area** This is often used for land-use sampling, to show change with distance or through time (if old maps or air photographs are available). Quadrats, positioned at equal intervals, are used for assessing plant distributions.

The main advantage of systematic sampling lies in its ease of use. However, its main disadvantage is that all points do not have an equal chance of selection – it may either overstress or miss an underlying pattern (Figure 6.40).

Stratified sampling

When there are significant groups of known size within the parent population, in order to ensure adequate coverage of all the sub-groups it may be advisable to stratify the sample, i.e. to divide the population into categories and sample within each. Although categorising into groups (layers or strata) may be a subjective decision, the practical application of this technique has considerable advantages for the geographer. Once the groups have been decided, they can be sampled either systematically or randomly, using point, line or area techniques.

1 **Stratified systematic sampling** This method can be useful in many situations – when interviewing people, sampling from maps, and during fieldwork. For example, in political opinion polls, the total population to be sampled can be divided (stratified) into equal age and/or socio-economic groups, e.g. 10–19, 20–29, etc. The number interviewed in each category should be in proportion to its known size in the parent population. This is most easily achieved by sampling at a regular interval (systematically) throughout the entire population, so that the required total sample size is obtained. For example, if a sample size of 800 is required from a total population of 8000 (i.e. a 10 per cent sample), every 10th person would be interviewed.

Figure 6.40

Poor sample design and selection can lead to inaccurate results: an area of woodland is completely missed in this example

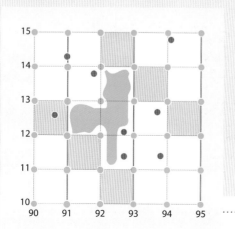

- • random point
- • systematic point
- | systematic line
- ▨ systematic area
- ⬛ woodland

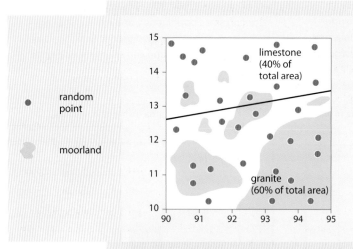

Figure 6.41

A random point sample, stratified by area

random point

moorland

limestone (40% of total area)

granite (60% of total area)

rock types: granite occupies 60% of the total area and limestone 40%. To discover whether the proportion of moorland cover varies with rock type, the sampling must be in proportion to their relative extents. Thus, if a sample size of 30 points is derived using random numbers, 18 are needed within the granite area (18 is 60 per cent of 30) and 12 within the limestone area (12 is 40 per cent of 30). If it was decided to area sample, 18 quadrats would have to fall within the granite area, and 12 in the limestone.

The advantages of stratified sampling include its potential to be used either randomly or systematically, and in conjunction with point, line or area techniques. This makes it very flexible and useful, as many populations have geographical sub-groups. Care must be taken, however, to select appropriate strata.

2 **Stratified random sampling** This method can be used to cover a wide range of data, both in interviewing and in geographical fieldwork and map work. For example, Figure 6.41 shows the distribution of moorland on two contrasting

Changes in sea-level

Although the daily movement of the tide alters the level at which waves break onto the foreshore, the average position of sea-level in relation to the land has remained relatively constant for nearly 6000 years (Figure 6.42). Before that time there had been several major changes in this mean level, the most dramatic being a result of the Quaternary ice age and of plate movements.

During times of maximum glaciation, large volumes of water were stored on the land as ice – probably three times more than today. This modification of the **hydrological cycle** meant that there was a worldwide, or **eustatic** (glacio-eustatic, page 123), fall in sea-level of an estimated 100–150 m.

As ice accumulated, its weight began to depress those parts of the crust lying beneath it. This caused a local, or **isostatic** (glacio-isostatic, page 123), change in sea-level.

Figure 6.42

Eustatic changes in sea-level since 18 000 BC

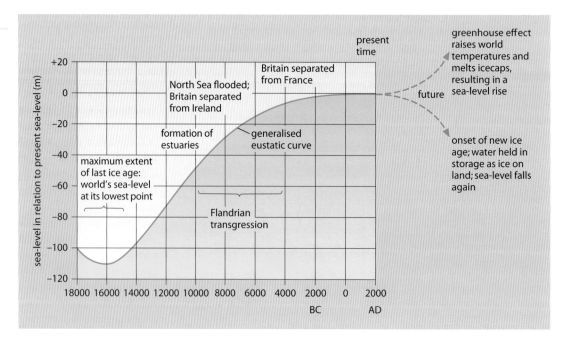

The world's sea-level was at its minimum 18 000 years ago when the ice was at its maximum (Figure 6.42). Later, as temperatures began to rise and icecaps melted, there was first a eustatic rise in sea-level followed by a slower iso-static uplift which is still operative in parts of the world today. This sequence of sea-level changes may be summarised as follows:

1 Formation of glaciers and ice sheets. Eustatic fall in sea-level gives rise to a negative change in base level (page 81).
2 Continued growth of ice sheets. Isostatic depression of the land under the ice produces a positive change in base level.
3 Ice sheets begin to melt. Eustatic rise in sea-level with a positive change in base level.
4 Continued decline of ice sheets and gla-ciers. Isostatic uplift of the land under former ice sheets results in a negative change in base level.

During this deglaciation, there may have been a continuing, albeit small, eustatic rise in sea-level but this has been less rapid than the isostatic uplift so that base level appears to be falling. Measurements suggest that parts of north-west Scotland are still rising by 4 mm a year and some northern areas of the Gulf of Bothnia (Scandinavia) by 20 mm a year (Places 23, page 166). The uplift in northern Britain is causing the British Isles to tilt and the land in south-east England to be depressed. This process is of utmost importance to the future natural development and human management of British coasts (Figure 6.56).

Tectonic changes have resulted in:
■ the uplift (**orogeny**) of new mountain ranges, especially at destructive and collision plate margins (pages 17 and 19)
■ local tilting (**epeirogeny**) of the land, as in south-east England, which has increased the flood risk, and in parts of the Mediterranean, leading to the submergence of several ancient ports and leaving others stranded above the present-day sea-level
■ local volcanic and earthquake activity, as in Iceland.

Landforms created by sea-level changes

Changes in sea-level have affected:
■ the shape of coastlines and the formation of new features by increased erosion or deposition
■ the balance between erosion and deposition by rivers (page 81) resulting in the drowning of lower sections of valleys or in the rejuvena-tion of rivers, and
■ the migration of plants, animals and people.

Landforms resulting from submergence
Eustatic rises in sea-level following the decay of the ice sheets led to the drowning of many low-lying coastal areas.

Estuaries are the tidal mouths of rivers, most of which have inherited the shape of the former river valley (Figure 6.45). In many cases, estuaries have resulted from the lower parts of the valleys being drowned by the post-glacial rise of sea-level. Being tidal, estuaries are subject to the ebbs and flows of the tide, and usually large expanses of mud are revealed at low tide (Figure 6.43). Many estuaries widen towards the sea and narrow to a meandering section inland (Figure 6.44).

Estuaries are affected by processes that are very different from those at work along rivers and coasts, because of particular features.

■ **Residual currents** are created by the mixing of fresh water (from rivers) and saline water (sea water brought in by the tides). Mixing tends to take place only when discharge and velocities are high; otherwise the fresh river water, being less dense, tends to rise and flow over the saline water.
■ **Tidal currents** have a two-way flow associ-ated with the incoming (flood) and outgoing (ebb) tide.
■ Continuous variations in both **discharge** and **velocity** resulting from the tidal cycle. Tidal velocities are highest at mid-tide and reduce to zero around high and low water. Times of zero velocity result in the deposition of fine-grained sediments, especially in upper estuary channels, which form mudflats and saltmarsh.

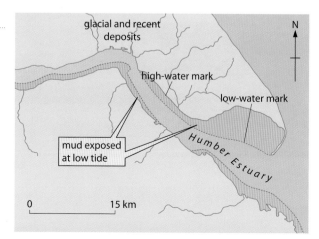

Figure 6.43

The Humber estuary

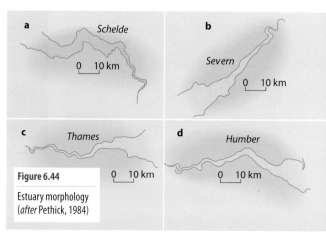

Figure 6.44

Estuary morphology (*after* Pethick, 1984)

Classification of estuaries

a According to origin This traditional method divides estuaries into different shapes but on the basis of their river valley origins.

- **Drowned river valleys**, resulting from post-glacial rises in sea-level, includes most estuaries.
- **Rias**, formed when valleys in a dissected upland are submerged, are one type of drowned river valley (Places 22).
- **Dalmatian coasts** are similar to rias except that their rivers flow almost parallel to the coast, in contrast to rias where they flow more at right-angles, e.g. Croatia.
- **Fiords**, formed by the drowning of glacial troughs (page 113), are extremely deep and steep-sided estuaries (Places 22).
- **Fiards** are drowned, glaciated lowland areas, e.g. Strangford Lough, Northern Ireland.

b According to tidal process and estuary shape This modern approach, supported by Pethick, acknowledges that it is tidal range that determines the tidal current, the residual current velocities and, therefore, the amount and source of sediment.

- **Micro-tidal** estuaries, which have a tidal range of less than 2 m, are dominated by freshwater river discharge and wind-driven waves from the sea. They tend to be long, wide and shallow, often with a fluvial delta or coastal spits and bars.
- **Meso-tidal** estuaries have a tidal range of between 2 m and 4 m. This fairly limited range means that, although fresh water has less influence, the tidal flow does not extend far upstream and the resultant shape is said to be stubby, with the presence of tidal meanders in the landward section.
- **Macro-tidal** estuaries have a tidal range in excess of 4 m and a tidal influence that extends far inland. They have a characteristic trumpet shape (Figure 6.44) and long, linear sand bars formed parallel to the tidal flow.

Places 22 Devon and Norway: a ria and a fiord

Kingsbridge estuary

During the last ice age, rivers in south-west England were often able to flow during the warmer summer months (compare Figure 5.14), cutting their valleys downwards to the then lower sea-level (page 163). When, following the ice age, sea-levels rose (Figure 6.42), the lower parts of many main rivers and their tributaries were drowned to form sheltered, winding inlets called **rias**. The Kingsbridge estuary (Figures 6.45 and 6.46) is a natural harbour produced by the drowning of a dendritic drainage system (Figure 3.50b). The deepest water is at the estuary mouth, a characteristic of a ria, with depth decreasing inland. The result is a fine natural harbour with an irregular shoreline and, at low tide, 800 hectares of tidal creeks and mudflats.

Apart from south-west England, rias are also found in south-west Wales, south-west Ireland, western Brittany and north-west Spain.

Figure 6.45

Kingsbridge estuary

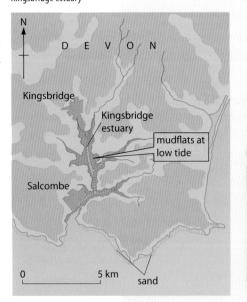

Figure 6.46

Kingsbridge estuary, looking north

Sognefjorden

Fiords (fjords) such as Sognefjorden (the Sogne Fiord) were formed by glaciers eroding their valleys to form deep glacial troughs (page 113). When the ice melted, the glacial troughs were flooded by a eustatic rise in sea-level (page 163) to form long, deep, narrow inlets with precipitous sides, a U-shaped cross-section, and hanging valleys (Figure 4.21). Glaciers seem to have followed lines of weakness, such as a pre-glacial river valley or, as suggested by their rectangular pattern, a major fault line (Figure 6.47). Unlike rias, fiords are deeper inland and have a pronounced shallowing towards their seaward end. The shallow entrance, comprising a rock bar, is known as a **threshold**.

The Sognefjorden extends 195 km inland and, at its deepest, has a depth of 1308 m (Figure 6.48). One description of the Sognefjorden is given in Figure 6.49.

Apart from Norway, fiords are also found on the west coasts of the South Island of New Zealand, British Columbia, Alaska, Greenland and southern Chile.

Figure 6.47

Location of Sognefjorden

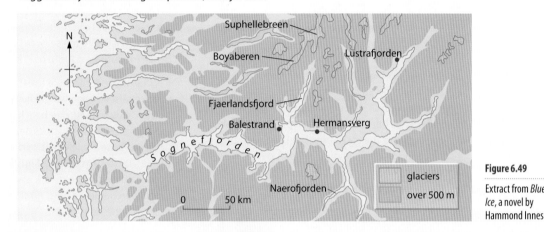

Figure 6.48

Sognefjorden

Figure 6.50

Erosion surfaces (marine peneplanation) at St David's, Dyfed, South Wales

Figure 6.49

Extract from *Blue Ice*, a novel by Hammond Innes

> As we sailed up the fjord, the wind died away leaving the water as flat as glass. The view was breathtakingly beautiful. Mountains rose to snow-covered, jagged peaks. The dark green of the pines covered the lower slopes, but higher up the vegetation vanished leaving sheer cliffs of bare rock which seems to rise to the blue sky. In the distance, on a piece of flat land, was Balestrand, with a steamer moving to the quay. Beyond was the hotel on a delta of green and fertile land.
>
> 'Isn't it lovely?' Dahler said. 'It is the sunniest place in all the Sogne Fjord. The big hotel you see is built completely of wood. Here the fjord is friendly, but when you reach Fjaerlandsfjord you will find the water like ice, the mountains dark and terrible, rising to 1300 metres in precipitous cliffs. High above you will see the Boya and Suphelle glaciers, and from these rivers from the melting snow plunge as giant waterfalls into the calm, cold, green coloured fjord.'

Landforms resulting from emergence

Following the global rise in sea-level, and still occurring in several parts of the world today, came the isostatic uplift of land as the weight of the ice sheets decreased. Landforms created as a result of land rising relative to the sea include erosion surfaces and raised beaches.

Erosion surfaces In Dyfed, the Gower peninsula (South Wales) and Cornwall, flat planation surfaces dominate the scenery. Where their general level is between 45 m and 200 m, the surfaces are thought to have been cut during the Pleistocene period when sea-levels were higher – hence the alternative name of **marine platforms** (Figure 6.50).

Coasts **165**

Raised beaches As the land rose, former wave-cut platforms and their beaches were raised above the reach of the waves. Raised beaches are characteristic of the west coast of Scotland (Figure 6.51). They are recognised by a line of degraded cliffs fronted by what was originally a wave-cut platform. Within the old cliff-line may be relict landforms such as wave-cut notches, caves, arches and stacks (Figure 6.52). The presence of such features indicates that isostatic uplift could not have been constant. It has been estimated that it would have taken an unchanging sea-level up to 2000 years to cut each wave-cut platform. (This evidence has been used to show that the climate did not ameliorate steadily following the ice age.)

Places 23 Arran: raised beaches

The Isle of Arran is one of many places in western Scotland where raised beaches are clearly visible. Early workers in the field claimed that there were three levels of raised beach on the west coast of Scotland, found at 25, 50 and 100 feet above the present sea-level. These are now referred to as the 8 m, 15 m and 30 m raised beaches. However, this description is now considered too simplistic, since it has been accepted that places nearest to the centre of the ice depression have risen the most and that the amount of uplift decreases with distance from that point. Thus, for example, the much-quoted '8 m raised beach' on Arran in fact lies at heights of 4–6 m. Where the raised beach is extensive, there is a considerable difference in height between the old cliff on its landward side and the more recent cliff to the seaward side, e.g. the 30 m beach in south-east Arran rises from 24 to 38 m.

It is now more acceptable to estimate the time at which a raised beach was formed by carbon-dating seashells found in former beach deposits, rather than by referring solely to its height above sea-level (i.e. to indicate a 'late glacial raised beach' rather than a '100 ft/30 m beach'). Figure 6.53 is a labelled transect, based on fieldwork, showing the two raised beaches in western Arran.

Figure 6.51

Raised beaches on the Isle of Arran: the lower one relates to the younger '8 m beach'; the upper one to the older '30 m beach'

Figure 6.52

The abandoned cliff-line at King's Cave, Arran, with its '8 m raised beach' (see Figure 6.53)

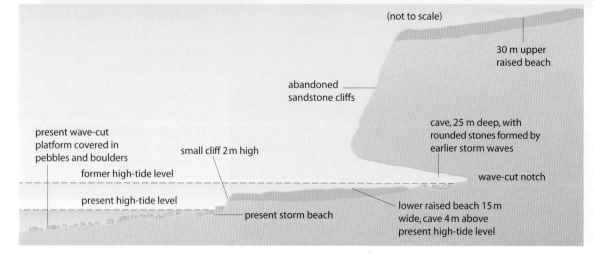

(not to scale)

30 m upper raised beach

abandoned sandstone cliffs

cave, 25 m deep, with rounded stones formed by earlier storm waves

present wave-cut platform covered in pebbles and boulders

former high-tide level

small cliff 2 m high

wave-cut notch

present high-tide level

present storm beach

lower raised beach 15 m wide, cave 4 m above present high-tide level

Figure 6.53

Diagrammatic transect across raised beaches of Arran

Figure 6.54

A concordant (Pacific) coastline: Lulworth Cove, Dorset

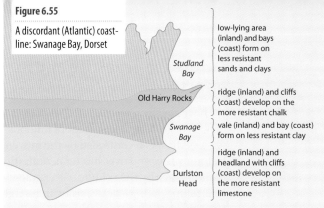

Figure 6.55

A discordant (Atlantic) coastline: Swanage Bay, Dorset

Studland Bay — low-lying area (inland) and bays (coast) form on less resistant sands and clays

Old Harry Rocks — ridge (inland) and cliffs (coast) develop on the more resistant chalk

Swanage Bay — vale (inland) and bay (coast) form on less resistant clay

Durlston Head — ridge (inland) and headland with cliffs (coast) develop on the more resistant limestone

Rock structure

Concordant coasts and **discordant coasts** are located where the natural relief is determined by rock structure (geology). They form where the geology consists of alternate bands of resistant and less resistant rock which form hill ridges and valleys (page 199). Concordant coasts occur where the rock structure is parallel to the coast, as at Lulworth Cove, Dorset (Figure 6.54). Should there be local tectonic movements, a eustatic rise in sea-level, or a breaching of the coastal ridge, then summits of the ridge may be left as islands and separated from the mainland by drowned valleys. These can be seen on atlas maps showing Croatia/the former Yugoslavia (Dalmatian coast) or San Francisco and southern Chile (Pacific coasts). Discordant coasts occur where the coast 'cuts across' the rock structure, as in Swanage Bay, Dorset (Figure 6.55). Here the ridges end as cliffs at headlands, while the valleys form bays.

Framework 7 Classification

Why classify?

Geographers frequently utilise classifications, e.g. types of climate, soil and vegetation, forms and hierarchy of settlement, and types of landform. This is done to try to create a sense of order by grouping together into classes features that have similar, if not identical, characteristics into identifiable categories. For example, no two stretches of coastline will be exactly the same, yet by describing Kingsbridge estuary as a ria, and Sognefjorden as a fiord (Places 22), it may be assumed that their appearance and the processes leading to their formation are similar to those of other rias and fiords, even if there are local differences in detail.

How to classify

When determining the basis for any classification, care must be taken to ensure that:

- only meaningful data and measures are used
- within each group or category, there is the maximum number of similarities
- between each group, there is the maximum number of differences
- there are no exceptions, i.e. all the features should fit into one group or another, and
- there is no duplication, i.e. each feature should fit into one category only.

As classifications are used for convenience and to assist understanding, they should be easy to use. They should not be oversimplified (too generalised), or too complex (unwieldy); but they should be appropriate to the purpose for which they are to be used.

No classification is likely to be perfect, and several approaches may be possible.

An example

The following landforms have already been referred to in this book:

arch; braided river; corrie; delta; esker; hanging valley; knickpoint; moraine; raised beach; rapids; spit; wave-cut platform.

Can you think of at least three different ways in which they may be categorised? The following are some possibilities:

a Perhaps the simplest classification is a two-fold division based on whether they result from erosion or from deposition.

b They could be reclassified into two different categories: those formed under a previous climate (i.e. relict features) and those still being formed today.

c The most obvious may be a three-fold division into coastal, glacial and fluvial landforms.

d A more complex classification would result from combining either **a** and **b**, or **a** and **c**, to give six groups.

Future sea-level rise and its effects

We have already seen (page 162) that over long periods of geological time (tens of millions of years) sea-level has been controlled by the movement of tectonic plates and over shorter periods (the last million years) by the volume of ice on the land (sea-level falling during glacials, rising in interglacials). Since the 'Little Ice Age' in the 17th century, when glaciers in alpine and arctic regions advanced, the world has slowly been warming. This warming helps to explain why global sea-levels are now some 20 cm higher than they were a century ago and why they are rising by 2 mm a year.

The fact that sea-level is continuing to rise, and at an accelerating rate, is due almost entirely to two factors:

1 **Thermal expansion** Since 1961, the average temperature of the global ocean has increased to depths of over 3000 m and the sea is now absorbing more than 80 per cent of the heat added to the climatic system through global warming. Such warming causes seawater to expand, contributing significantly to sea-level rise.

2 **Melting ice** A less significant, but increasing, contribution is from melting ice – mainly alpine glaciers, including the 1500 or so in the Himalayas – and, to a lesser extent as yet, polar ice sheets and ice caps.

Global sea-level rose at a rate of 1.8 mm/yr between 1965 and 2005 and by 3.1 mm/yr between 1993 and 2005. Some computer models are suggesting that between 1990 and 2090 it could be as high

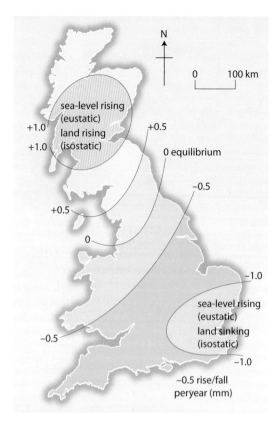

Figure 6.56

Relative sea-level (RSL): the combined net effect of sea and land surface changes

as 3.7 mm/yr, increasing to 5 mm/yr by 2100 (Figure 6.57). Other models have suggested a greater 'Doomsday' scenario with sea-levels rising by 8 mm/yr by the end of this century (one has even suggested 16 mm/yr). Whichever prediction eventually proves to be the most accurate, sea-level rise will have serious consequences:

Figure 6.57

Projections of future sea-level rise resulting from global warming: the extreme values cover the 95 per cent probability range (*after* Clayton, 1992)

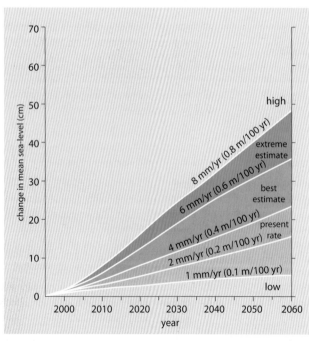

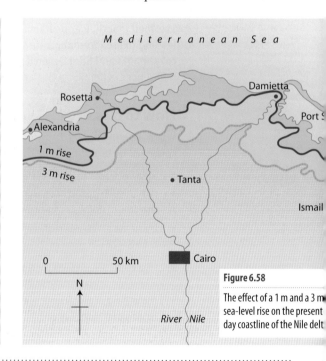

Figure 6.58

The effect of a 1 m and a 3 m sea-level rise on the present day coastline of the Nile delta

- Storm surges, tsunamis, higher tides and larger waves will cause more damage.
- An increase in the frequency and severity of coastal flooding would inundate numerous coastal settlements such as Tokyo, Shanghai, Lagos, London, Bangkok, Kolkata, Hong Kong and Miami, causing the displacement of large centres of populations as well as destroying industry and farmland (Figure 6.58). At present over 65 million people live in annual flood-risk areas, 50 million of those in danger of storm surges. A rise of 1 m in the next 100 years would inundate one-quarter of the land area of Bangladesh, affecting nearly 70 per cent of its population.
- Several low-lying ocean states such as the Maldives in the Indian Ocean and Tuvalu and the Marshall Islands in the Pacific are likely to be inundated.
- There will be an increase in coastal erosion and expensive coastal defences will need to be built and maintained.
- Various coastal ecosystems will be threatened, including sand dunes, saltmarshes, mangrove swamps, coral reefs and coral islands, which may not be able to adapt quickly enough if the rise is too rapid.
- Some sea-life species will migrate to cooler waters.

Larger waves

Mid-Atlantic waves that eventually pound the western coasts of the British Isles have increased in height over the last 30 years. Oceanographers have found that the mean height of these waves in winter has risen from 4 m to 5.3 m. Added to that, the mean height of the largest and most destructive type of wave has risen from 8 m to 11 m. This suggests that waves now have far more energy than they did in 1980 and while they may be a potential form of renewable energy, at present they undermine cliffs, strip sand from beaches and threaten coastal defences (Figure 6.59).

Freak waves of 15 m and over in height were in the past considered to be a marine myth. Opinions began to change when workers on off-shore oil-rigs reported that waves of that height occurred fairly frequently. Two orbiting satellites launched by the European Space Agency in 2000 were given the task of recording and plotting these so-called freak waves. Radar sensors on the satellites soon showed that freak waves were relatively common and, within one period of three weeks, a team of land-based observers noted the existence of more than ten waves of over 25 m spread across the various oceans. Freak waves may explain the sudden disappearance of ships, some as large as oil-tankers.

Figure 6.59

Some impressive waves
a North Cape, Norway – note the relative size of the people
b A wave breaking over a lighthouse, Seaford, Sussex

A The need for management

Although Britain's coasts are rarely affected by extreme events such as the Indian Ocean tsunami (Places 4), storm surges as in the Bay of Bengal (Places 19) or the tropical storms in Central America and Florida (Places 31, page 238), large stretches are under threat from one or more sources (Figure 6.60). Much of Britain's coastline is used for human activity and although in some more remote places there is often a demand from only one or two main land users, in many others there is competition for, and conflict over, land use (Figure 6.61). Combining the threats posed by:

- natural events such as flooding and erosion, and
- human demands that include settlement, economic activities and recreational use

there is a continuing need for a national, sustainable management plan. Such a plan has to consider on the one hand the rapidly increasing costs of providing new defences and maintaining both new and existing defences, and on the other hand the need to protect people and property.

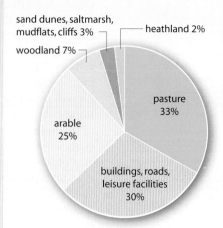

- 23% of the UK lies within 10 km of the coast.
- 17.2 million people live within this coastal zone.
- 35% of UK manufacturing and electricity production is close to the coast.
- Most of the coastline is used for recreational purposes, especially walking.
- Coasts attract larger number of specialist groups (ornithologists, geologists, school parties).

Figure 6.61

Coastal land use

Figure 6.60

Threats to Britain's coasts

Threat	Examples
Increased risk of flooding	
rising sea-level linked to global warming	estuaries, south-east England
higher high tides	Thames estuary
risk of increased number of storm surges	southern North Sea
Increased risk of erosion	
larger waves (generating more energy)	western Britain
human activity (use of footpaths, building on cliff-tops)	Yorkshire, East Anglia
Overuse and/or misuse	
settlements and economic development	estuaries
leisure and tourism (caravan and car parks, golf courses)	close to large urban areas

Who is responsible for coastal management?

The Department for Environment, Food and Rural Affairs (DEFRA) has overall responsibility for coastal defences in England, although the Environment Agency has powers to reduce flooding in tidal waters. In order to protect the coast, DEFRA has to produce a **shoreline management plan (SMP)**. To do this, it is necessary to understand coastal processes in any given stretch of coastline. It would be impossible to achieve this for the whole British coastline, so it has been divided into a number of separate units referred to as 'coastal cells' (Figure 6.62); there are eleven for England and Wales. The location and size of each of these cells is defined so that coastal processes within each individual cell are totally self-contained, and changes

Figure 6.62

Coastal cells around the coasts of England and Wales

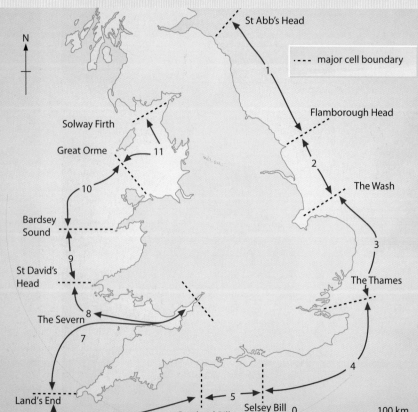

taking place within that cell do not significantly affect the coastline of adjacent cells.

Two basic principles in SMP production are that:
- natural processes should not be interfered with unless it is necessary to protect life or property
- all schemes must be economically viable and undergo a cost–benefit analysis to ensure that they make good use of public money.

What are the options?

A shoreline management plan has, for each coastal cell, four defence options:
- Do nothing, other than monitor and review.
- Hold the existing defence line by maintaining or changing the standard of protection.
- Advance the existing defence line.
- Retreat the existing defence line by realigning the coast, i.e. **managed retreat**.

SMPs are developed by groups of people that include planners, engineers, geomorphologists and others with special local knowledge.

How has the coast been protected in the past?

Traditional sea defences, now referred to as **hard defences** (Figure 6.63), involved the construction of distinctive features:
- Concrete sea walls were often built, in the 19th century, at holiday resorts. They created more space for promenades and leisure amenities and protected hotels from storm waves.
- Groynes, usually of wood, were constructed at right-angles to the coastline. They helped to reduce the force of the waves and trapped material being moved along the coast by longshore drift (Figure 6.23). This helped to widen beaches and to reduce the removal of beach material.
- Concrete breakwaters protected small harbours from strong wave action.

More recently it has been realised that:
- concrete sea walls absorb, rather than reflect, wave energy and so now they are often curved at the top (bullnose) to divert waves
- groynes, by trapping sand, cause the loss of replacement material further along the coast, increasing the problem elsewhere.

More recent hard defences include:
- wooden slatted revetments, constructed parallel to the coast, which dissipate the force of waves
- concrete blocks, known as rip-rap, which also absorb the power of waves
- offshore breakwaters and reefs which reduce wave energy but still allow some longshore drift (Figure 6.70).

Most of the earlier schemes, apart from being unsustainable, were not environmentally friendly, either visually or in relation to local habitats (ecosystems), and were expensive to build and to maintain. Wherever possible they are being replaced or supplemented by **soft defences**. Soft defences include:
- the use of beach replenishment at the base of cliffs and sea walls where lost sand and shingle is replaced (although such replacement is expensive and needs to be maintained for long periods)
- cliff stabilisation, either by inserting pipes to remove excess water or by planting vegetation to reduce mass movement.

Figure 6.63

Coastal defences
a Rip-rap
b Groynes and a bullnose sea wall
c Revetments

B Coastal management schemes in East Anglia

Erosion has always been a major problem along much of the coast of Norfolk (Figure 6.64) while further south flooding is the major hazard in Suffolk and Essex (Places 19). Present-day shoreline management plans (SMPs, page 170) must aim to strike the seemingly impossible balance between protecting the coastline at a viable cost and minimising the disruption of natural processes and nearby defence schemes. In north Norfolk, hard engineering solutions are now less in favour than softer options. In Suffolk and Essex controversy has arisen over SMP proposals to re-align parts of the coastline in a 'managed retreat'. This case study considers several specific places and their problems.

Aldeburgh and East Lane Point, Suffolk

Aldeburgh, in Suffolk, at the northern end of Orford Ness (Figures 6.28 and 6.65), was protected by a sea wall and timber groynes to reduce the loss of beach material. Six streets to the east of the town have been lost to the sea since the 16th century, and the only visible remains of the former village of Slaughden, 1 km to the south, are a Martello tower and what is now a marina.

Following the partial failure of the sea wall in 1988, Anglian Water and the National Rivers Authority (now the Environment Agency) devised a £4.9 million plan to provide sea defences that would also protect the tidal freshwater areas of the River Alde immediately to the west of the town. The existing sea wall was extended at its base in the section considered most threatened. Several 10-tonne rock blocks were placed in front of the sea wall to absorb the wave energy; 200 m of wall originally protecting the northern end of Orford Ness was demolished, and a rock armour bank put in its place. A total of 24 new groynes were built, stretching south beyond the Martello tower (Figure 6.66), and 75 000 m³ of shingle were deposited as beach replenishment. More rocks were brought in to make a 400 m bank between the existing sea wall to the south and the shingle bank. The scheme was completed in 1992. It took into account the risk that storm damage could cause to an important natural area.

In 2004 there were increasing fears that Aldeburgh could become an island and that the Suffolk coastline as far south as Felixstowe could change if the sea broke through obsolete defences during the next winter's storms. At greatest immediate risk is East Lane Point, near Bawdsey, south of Aldeburgh (Figure 6.67). Much of the land behind the Point is considered by the government to be a 'non-viable flood defence area' as it does not reach the requisite number of points required for funding under the new DEFRA scoring system mainly because the area is sparsely populated. A spokesperson for DEFRA stated that 'there will never be sufficient money available for every coastal defence need and so priority must go to protecting people and their property'.

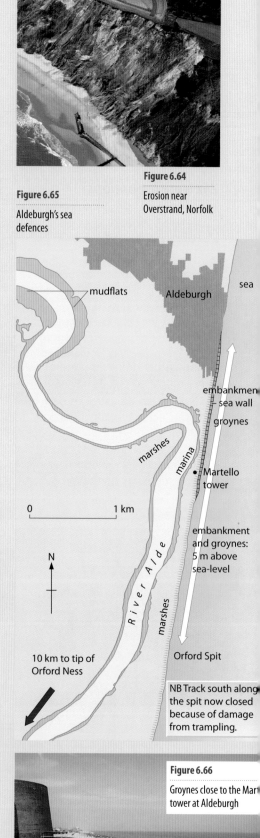

Figure 6.64

Erosion near Overstrand, Norfolk

Figure 6.65

Aldeburgh's sea defences

10 km to tip of Orford Ness

NB Track south along the spit now closed because of damage from trampling.

Figure 6.66

Groynes close to the Martello tower at Aldeburgh

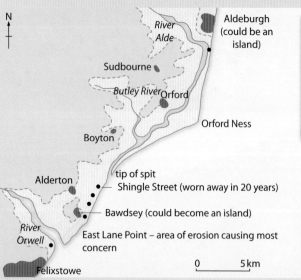

Legend	
——	present coastline
▬▬	present rivers
- - -	predicted new coastline – flood zone to 4 m
	land
	area likely to flood
	settlement
•	Martello towers

Figure 6.67

How Suffolk may flood if defences at East Lane Point are breached

0   5 km

Sea Palling, Norfolk

Much of the Norfolk coastline from Cromer southwards to Great Yarmouth is protected by expensive coastal defences. At Sea Palling the beach is backed by sand dunes which, in earlier times, helped provide a natural defence. Behind these are 6000 ha of land used for settlement, farming and (this area being part of the Norfolk Broads) tourism and wildlife. In 1953 a storm surge (Places 19) broke through the coastal defences, flooding large areas and, at Sea Palling itself, washing away houses and drowning seven people.

Following the flood, a sea wall was constructed in front of the dunes (Figure 6.68) and there was some replenishment of beach material. However, by the 1990s the beach in front of the sea wall had narrowed due to the removal of material southwards by longshore drift during times of northerly and easterly gales, a process that led to an increase in wave energy. Following the severe winter storms of 1991, rip-rap was positioned against the sea wall as a temporary measure.

In 1992 a beach management strategy was introduced with the conditions that it would not significantly affect adjacent coastal areas, it would have minimal environmental impacts and it would be cost-effective. Over 150 000 tonnes of rock were placed in front of the wall to prevent further undermining and 1.4 million m³ of replenishment sand were added in front of the rock. The major part of the scheme was the construction of four offshore reefs designed to reduce incoming wave energy and to protect the beach while at the same time allowing some longshore drift so as not to deplete the supply of sand to beaches further along the coast (Figure 6.69). These reefs were completed in 1995 but almost immediately presented a problem that had not been predicted: sand began to accumulate in the sheltered lee of the reefs, leading to the formation of tombolos (page 155 and Figure 6.70) which in turn interrupted the process of longshore drift. To try to overcome this problem, the next five reefs to be built were shorter (to reduce areas of shelter behind them), lower (to allow more overtopping waves) and closer together (to prevent erosion in the gaps). A further five are planned 3 km to the south.

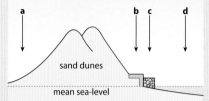

a houses, farmland, SSSIs and nature reserves just above sea-level
b 1.6 m high sea wall built in 1954
c rip-rap added in 1992
d beach material replenished as needed since 1992

Figure 6.68

Sea defences 1954–92

Figure 6.69

Artificial reefs at Sea Palling

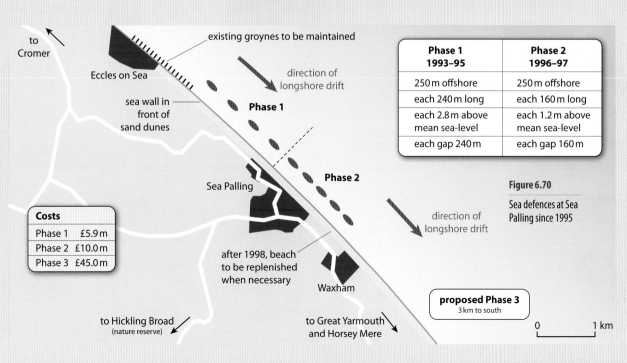

	Phase 1 1993–95	Phase 2 1996–97
	250 m offshore	250 m offshore
	each 240 m long	each 160 m long
	each 2.8 m above mean sea-level	each 1.2 m above mean sea-level
	each gap 240 m	each gap 160 m

Costs

Phase 1	£5.9 m
Phase 2	£10.0 m
Phase 3	£45.0 m

Figure 6.70

Sea defences at Sea Palling since 1995

Proposed 'managed retreat' in Norfolk

Controversial plans by Natural England to flood parts of Norfolk emerged in early 2008. The proposal, if accepted, would see Britain for the first time admitting defeat in the battle to maintain all of its coastal defences. Experts doubt if the present defences can cope with the rising sea-level resulting from global warming and the sinking of south-east England, and the plan to 'realign the coast' in a 'managed retreat' is the less expensive and more practical option. This would involve building a new sea wall further back from the present coastline, at a cost of a fraction of that of trying to maintain the existing defences. The Environment Agency, in response, stated that it is committed to 'holding the present line' of sea defences for the next 50 years, although it admitted that that option was becoming increasingly difficult and more expensive, while DEFRA said it was committed to the sustainable protection of people and property here in Norfolk and elsewhere.

Should the scheme go ahead, it would mean allowing the sea, over a period of time, to breach 25 km of the north Norfolk coast between Eccles on Sea and Winterton-on-Sea. In time the sea would create an area of saltwater lake and salt-marsh covering 65 km² (Figure 6.71). Over the next 50 years or so this lake would eliminate six villages: four on the coast (Eccles on Sea, Sea Palling, Waxham and Horsey) and two inland (Hickling and Potter Heigham). The lake would also inundate about 600 houses, many hectares of good-quality arable farmland and five fresh-water lakes that currently form part of the Norfolk Broads, including the tourist area of Hickling Broad (Figure 6.72) and the rare fauna and flora of Horsey Mere.

Opponents to the plan claim that it would mean in the short term making their properties unsaleable and, in the long term, relocating hundreds of people and paying them compensation. A millennium of history would vanish under the waves and with it villages like Hickling, which is mentioned in the Domesday Book, and Sea Palling, which the sea failed to destroy in the 1953 flood. Churches and other buildings listed by English Heritage would also be lost.

Proposers suggest that the plan is more economically sustainable than present policies and that a newly created saltmarsh could be used by farmers for cattle grazing, it could act as a buffer zone helping to dissipate wave energy, it would provide storage for excess water during times of storm surges, and provide a welcome haven for wildlife when little of Britain's original salt-marsh ecosystem remains (page 175). They also claim that experiments have shown that a sea wall can costs £5000 a metre to build and maintain, whereas an inland retreat of 80 m, allowing a saltmarsh to form a buffer against tides and waves, only costs £400 a metre to build and maintain.

Natural England claim that the 'surrender' option is only one of several possibilities, but it considers the issue to be so important that it is time to open discussions and to encourage debate. No final decision has been made about the plan.

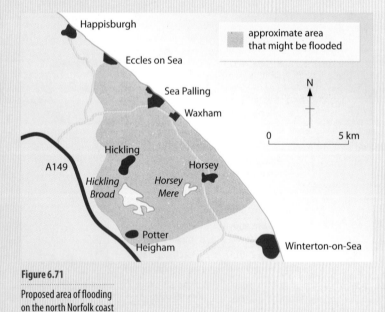

Figure 6.71

Proposed area of flooding on the north Norfolk coast

Figure 6.72

Hickling Broad

Sand dunes and saltmarsh

Large tracts of the coast of East Anglia consist either of sand dunes (pages 157 and 290) or saltmarsh (pages 158 and 291). Both are fragile ecosystems that are under threat and receive less attention and management than they deserve and need.

As we have seen, sand dunes fringe much of the Norfolk coast, either backing sandy beaches (Figure 11.10) or stabilising spits such as that at Blakeney Point. Sand dunes are under threat from:

- the rising sea-level which attacks the embryo and foredunes (Figure 6.32), narrowing beaches and thus depriving them of their source material
- excavation for sand by construction companies

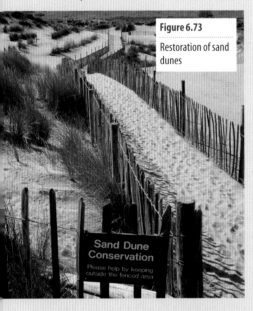

Figure 6.73

Restoration of sand dunes

- people either walking along paths within them, especially where they form part of a coastal footpath, or playing (or sheltering from the wind) in blow-outs.

Where human influence is limited, the ecosystem can repair itself, but where it is severe the damage may be irreversible. One solution is to fence off selected areas to allow time for recovery (Figure 6.73).

Saltmarsh develops behind coastal spits as at Blakeney Point (Figure 11.14) but is most extensive in the river estuaries of Suffolk and Essex (Figure 11.13). Saltmarsh has been under threat since Saxon times when parts were drained around the present-day Norfolk Broads. Essex was said to have 30 000 ha of saltmarsh in 1600, yet 400 years later only 2500 ha remain. This remaining saltmarsh supports around two million wildfowl and wading birds in winter and is a habitat for rare species of plants, birds and insects. Currently another 100 ha/yr of saltmarsh is being lost across England alone due to the rising sea-level and human activity. However, there are several plans in Essex to recreate more saltmarsh to provide alternative habitats for wildlife, to act as a buffer zone against the larger waves, and as storage for surplus water during storm surges or as the mean high-tide level rises. The most ambitious and expensive project (£12 million) is being undertaken by the RSPB, which intends to break the sea walls (Figure 6.74) around Wallasea Island, near Southend, changing 730 ha of farmland back into a mosaic of saltmarsh, creeks and mud-flats – although these will only be covered by 50 cm of water at high tide.

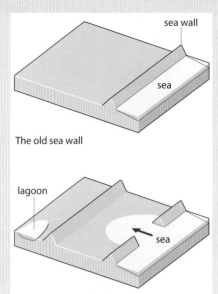

The old sea wall

A new bank is built well back using soil dug out to create lagoons.
A hole is made in the old wall, allowing the sea in.

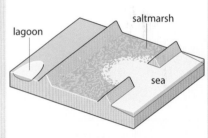

Saltmarsh grows in between the banks, soaking up wave energy and creating a habitat for wildlife.

Figure 6.74

Breaching of an old sea wall to create a saltmarsh

Further reference

Chapman, R. (2005) 'Changing attitudes to coastal protection' in *Geography Review* Vol 18 No 4 (March).

Dove, J. (2000) 'Plant distribution in the Sonoran Desert' in *Geography Review* Vol 14 No 2 (November).

Gee, N. (2005) 'Coastal management: Sea Palling, Norfolk' in *Geography Review* Vol 18 No 3 (January).

Gee, N. (2008) 'Management of the East Anglian coast' in *Geography Review* Vol 21 No 3 (February).

Goudie, A.S. (2001) *The Nature of the Environment*, WileyBlackwell.

Holmes, D. (2003) 'Investigating coastal sand dunes' in *Geography Review* Vol 16 No 3 (January).

Horton, B. (2005) 'Climate and sea-level change' in *Geography Review* Vol 18 No 4 (March).

Marshak, S. (2007) *Earth – Portrait of a Planet*, W.W. Norton & Co.

Pethick, J. (1984) *An Introduction to Coastal Geomorphology*, Hodder Arnold.

Skinner, B.J. and Porter, S.C. (2003) *The Dynamic Earth*, Wiley.

Coastal erosion:
www.walrus.wr.usgs.gov/hazards/erosion.html

Coastal management case studies:
www.westdorset-dc.gov.uk/westbay

Holderness coastline:
www.hull.ac.uk/coastalobs/general/erosionandflooding/erosion.html

Land Ocean Interaction Study:
www.nerc.ac.uk/research/programmes/lois/

Sea-level changes (Antrim coast):
www.ehsni.gov.uk/natural/earth/geology.shtml

Activities

1 **a** Study the photograph in Figure 6.75 and answer the following questions.

    i Describe the material found between the two stacks.

        (3 marks)

    ii Describe the beach material found in the foreground of the photograph.   *(3 marks)*

    iii Describe the main stack.   *(4 marks)*

  b How is a feature like this stack formed?   *(6 marks)*

  c Some cliff coastlines, such as Old Harry Rock near Swanage (Figure 6.21, page 152), have no beach while others, such as Marsden Rock (Figure 6.75), have.

    Suggest a reason for this difference.   *(4 marks)*

Figure 6.75
Marsden Rock

  d Marine erosion is concentrated at the base of a cliff.

    Suggest two ways in which the rest of the cliff is eroded.

        (5 marks)

2 **a** Making good use of diagrams, describe **two** landforms that may be found on a beach.   *(6 marks)*

  b Why are large stones and boulders found at the **back** of a beach?   *(4 marks)*

  c Making good use of diagrams, explain how sand and other material is moved along a beach by the action of waves.   *(5 marks)*

  d Why are shingle beaches steeper, on average, than sandy beaches?   *(5 marks)*

  e How and why may human activity change this marine transport process?   *(5 marks)*

3 **a** Making good use of annotated diagrams, explain the process of longshore drift.   *(5 marks)*

  b i Study Figure 6.23 (page 153). Suggest, with reasons, the direction of longshore drift on this coastline.

        (3 marks)

    ii Why were the sea defences put along this shoreline?

        (6 marks)

    iii What effect would you expect there to be further down the coast as a result of the building of these sea defences? Explain your answer.   *(6 marks)*

  c Choose **one** landform created by marine deposition.

    Describe the size and shape of the landform and suggest how marine deposition has helped to create it.   *(5 marks)*

Exam practice: basic structured questions

4 **a** What is meant by each of the following terms used in relation to the effects of waves on a coastline:

    i abrasion (sometimes called corrasion)

    ii attrition

    iii hydraulic action?   *(6 marks)*

  b Explain how the processes identified in **a** cause a cliff to change its shape.   *(6 marks)*

  c Study Figure 6.17 (page 150).

    i Describe and suggest reasons for the shape of the cliff shown in the photograph.   *(6 marks)*

    ii Although there are houses on top of this cliff it has been decided not to attempt to protect this coastline. Suggest **two** reasons for this decision.   *(7 marks)*

5 **a** Explain the terms 'eustatic' and 'isostatic' used when studying sea-level change.   *(4 marks)*

  b Explain how:

    i an ice age

    ii one other mechanism could cause sea-level change.

        (7 marks)

  c Choose **one** landform that has been created by or significantly changed by a fall in sea-level.

    Describe the landform and explain the role of sea-level change in its formation.   *(7 marks)*

  d Choose **one** landform that has been created or changed significantly by a rise in sea-level. Describe the landform and explain the role of sea-level change in its formation.   *(7 marks)*

6 **a** Study Figure 6.25 on page 154.

    Why has saltmarsh formed at H?   *(6 marks)*

  b Explain the meaning of:

    i dominant wind

    ii embryo dune.   *(4 marks)*

  c Explain how sand dunes go through a series of stages from the appearance of berms to the formation of grey (or mature) dunes.   *(15 marks)*

Exam practice: structured questions

7   a   On a coastline with cliffs, deposition can cause the shape of the coastline to change. Suggest where there will be deposition on such a coastline and the reasons for deposition there.   *(10 marks)*

   b i   Study Figure 6.75. Draw an annotated diagram to identify the main features of the landform in the photograph.   *(5 marks)*

      ii   With reference to evidence from the photograph, explain how marine processes may have created this landform   *(10 marks)*

8   a   With reference to one or more examples of cliff coastlines, explain how marine and sub-aerial processes have combined to shape the cliffs.   *(12 marks)*

   b i   Identify and describe **two** ways in which people can manage the erosion of a cliff foot.   *(6 marks)*

      ii   Evaluate the success of one of these management strategies.   *(7 marks)*

9   a   Choose two of the following micro-morphological features of a beach: berm; beach cusps; ridge and runnels; longshore bar.

      For each feature that you have chosen:

      i   Making use of annotated diagrams, describe its shape and location on a beach.   *(6 marks)*

      ii   Explain how it is formed.   *(10 marks)*

   b   What effect do storm waves have on a beach profile? *(9 marks)*

   c   Describe **one** method you could use to survey the profile of a beach.   *(5 marks)*

10 a   Using an annotated diagram only, explain the process by which beach material is moved along the coastline. *(5 marks)*

   b   Choose **one** landform that is created when beach material is deposited. Name and describe the landform.

      Explain the processes by which the landform is created.   *(10 marks)*

   c   Why do people try to reduce the movement of beach material on some coastlines? Suggest and explain **two** methods for reducing such movement.   *(10 marks)*

11 a   Using your own case studies, choose **two** examples of hazards that occur on marine coasts. For each hazard:

      i   Identify the hazard and its location.   *(2 marks)*

      ii   Explain how the action of the sea leads to danger on the coast.   *(12 marks)*

   b   Describe **one** way in which the people prepare to face marine hazards and evaluate their success when the danger occurs   *(11 marks)*

12 a   Using an example from your studies, explain why a particular coastal management scheme was felt to be necessary.

                                                   (6 marks)

   b   Describe the planning and decision-making process involved in the creation of the management plan for the area.   *(6 marks)*

   c   Outline the plan and suggest why the changes outlined should overcome the identified problem/s.   *(6 marks)*

   d   Evaluate the success of the project.   *(7 marks)*

13   Study the sand dune area in Figure 6.76.

   a i   Identify and locate **one** feature of the photograph which indicates that this area is popular with people.

         Explain how it shows the presence of people.   *(4 marks)*

      ii   Explain **one** piece of evidence from the photograph which shows that this popularity is causing damage to the environment.   *(4 marks)*

   b i   Suggest **one** possible effect of the environmental damage caused in this area.   *(7 marks)*

      ii   Explain how conservation work could overcome the damage done to this sand dune belt.   *(10 marks)*

Figure 6.76

Damaged sand dunes at Gower, Wales

Exam practice: essays

14   'The interface between the sea and the land is an area of conflict in nature and for people.' Using examples, explain this statement.
                                                   (25 marks)

15   Discuss possible causes of future changes in sea-level and explain how these changes might produce both short-term and long-term effects on the physical and human environment.   *(25 marks)*

16   Choose **one** system of coastal classification. Describe and explain the principles on which it is based and, making use of examples, describe some of the problems of applying your classification system to cover all coastal areas.
                                                   (25 marks)

17   Discuss the arguments for and against the managed retreat of parts of the coastline in the UK. Evaluate the strength of these arguments as they apply to **one or more** areas that you have studied.   *(25 marks)*

18   'Coastal sand dunes form some of the most important defences against the sea, so every effort should be made to conserve and strengthen our dune systems.'

   Evaluate this statement.   *(25 marks)*

7

Deserts

• •

'Now the wind grew strong and hard and it worked at the rain crust in the cornfields. Little by little the sky was darkened by the mixing dust, and the wind felt over the earth, loosened the dust and carried it away.'

J. Steinbeck, *The Grapes of Wrath*, 1939

What is a desert?

'The deserts of the world, which occur in every continent including Antarctica, are areas where there is a great deficit of moisture, predominantly because rainfall levels are low. In some deserts this situation is in part the result of high temperatures, which mean that evaporation rates are high. It is the shortage of moisture which determines many of the characteristics of the soils, the vegetation, the landforms, the animals, and the activities of humans' (Goudie and Watson, 1990).

A desert environment has conventionally been described in terms of its deficiencies – water, soils, vegetation and population. Deserts include those parts of the world that produce the smallest amount of organic matter and have the lowest net primary production (NPP, page 306). In reality, many desert areas have potentially fertile soils, evidenced by successful irrigation schemes; all have some plant and animal life, even if special adaptations are necessary for their survival; and some are populated by humans, occasionally only seasonally by nomads but elsewhere permanently, e.g. in large cities like Cairo and Karachi.

The traditional definition of a desert is an area receiving less than 250 mm of rain per year. While very few areas receive no rain at all (Places 24, page 180), amounts of precipitation are usually small and occurrences are both infrequent and unreliable. Climatologists have sometimes tried to differentiate between cold deserts where for at least one month a year the mean temperature is below 6°C, and hot deserts. Several geomorphologists have used this to distinguish the landforms found in the hot sub-tropical deserts – our usual mental image of a desert – from those found in colder latitudes, e.g. the Gobi Desert and the tundra.

Modern attempts to define deserts are more scientific and are specifically linked to the water balance (page 60). This approach is based on the relationship between the input of water as precipitation (P), the output of moisture resulting from evapotranspiration (E), and changes in water held in storage in the ground. In parts of the world where there is little precipitation annually or where there is a seasonal drought, the **actual evapotranspiration** (AE) is compared with **potential evapotranspiration** (PE) – the amount of water loss that would occur if sufficient moisture was always available to the vegetation cover. C.W. Thornthwaite in 1931 was the first to define an **aridity index** using this relationship (Figure 7.1).

Figure 7.1

The index of aridity

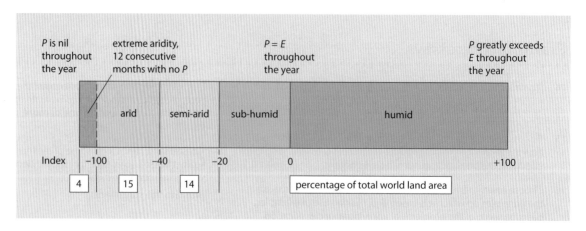

Location and causes of deserts

On the basis of climatic characteristics, including Thornthwaite's aridity index, one-third of the world's land surface can be classified as desert, i.e. arid and semi-arid. Alarmingly, this figure, and therefore the extent of deserts, may be increasing (Case Study 7).

As shown in Figure 7.2, the majority of deserts lie in the centre or on the west coast of continents between 15° and 30° north and south of the Equator. This is the zone of sub-tropical high pressure where air is subsiding (the descending limb of the Hadley cell, Figure 9.34). On page 226 there is an explanation of how warm, tropical air is forced to rise at the Equator, producing convectional rain, and how later that air, once cooled and stripped of its moisture, descends at approximately 30° north and south of the Equator. As this air descends it is compressed, warmed and produces an area of permanent high pressure. As the air warms, it can hold an increasing amount of water vapour which causes the lower atmosphere to become very dry. The low relative humidity, combined with the fact that there is little surface water for evaporation, gives clear skies.

A second cause of deserts is the rainshadow effect produced by high mountain ranges. As the prevailing winds in the sub-tropics are the trade winds, blowing from the north-east in the northern hemisphere and the south-east in the southern hemisphere, then any barrier, such as the Andes, prevents moisture from reaching the western slopes. Where plate movements have pushed up mountain ranges in the east of a continent, the **rainshadow effect** creates a much larger extent of desert (e.g. 82 per cent of the land area of Australia) than when the mountains are to the west, as in South America.

Aridity is increased as the trade winds blow towards the Equator, becoming warmer and therefore drier. Where the trade winds blow from the sea, any moisture which they might have held will be precipitated on eastern coasts leaving little moisture for mid-continental areas. The three major deserts in the northern hemisphere which lie beyond the sub-tropical high pressure zone (the Gobi and Turkestan in Asia and the Great Basins of the USA) are mid-continental regions far removed from any rain-bearing winds, and surrounded by protective mountains.

A third combination of circumstances giving rise to deserts is also shown in Figure 7.2. Several deserts lie along western coasts where the ocean water is cold. In each case, the prevailing winds blow parallel to the coastline and, due to the Earth's rotation, they tend to push surface water seaward at right-angles to the wind direction. The Coriolis force (page 224) pushes air and water coming from the south towards the left in the southern hemisphere and water from the north to the right in the northern hemisphere. Consequently, very cold water is drawn upwards to the ocean surface, a process called **upwelling**, to replace that driven out to sea. Any air which then crosses this cold water is cooled and its capacity to hold moisture is diminished. Where these cooled winds from the sea blow onto a warm land surface, advection fogs form (page 222 and Places 24).

Figure 7.2

Arid lands of the world

- extreme aridity
- arid
- semi-arid

H high pressure
R rainshadow
M mid-continent
U upwelling of cold water

1 Australia (e.g. Simpson Great Sandy) **H R M**
2 Gobi **M**
3 Thar **H**
4 Iran **H M**
5 Turkestan **M**
6 Arabia **H**
7 Somalia **H**
8 Kalahari **H M**
9 Namib **H U**
10 Sahara **H** (**U** in west, **M** in centre)
11 Patagonia **R**
12 Monte **H**
13 Atacama **H R U**
14 Sonora **H**
15 Mojave **H R** (**U** on coasts)
16 Great Basins **R**

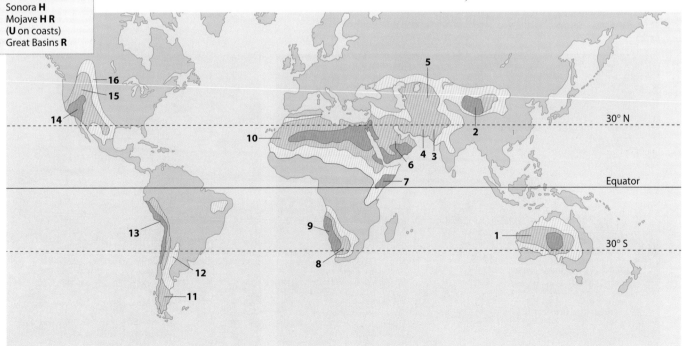

Figure 7.3

The Atacama Desert

The prevailing winds in the Atacama, which lies in the sub-tropical high pressure belt, blow northwards along the South American coast. These winds, and the northward-flowing Humboldt (Peruvian) current over which they blow, are pushed westwards (to the left) and out to sea by the Coriolis force as they approach the Equator. This allows the upwelling of cold water from the deep Peru–Chile sea trench (Figure 1.12) that provides the rich nutrients to nourish the plankton which form the basis of Peru's fishing industry. The upwelling also cools the air above which then drifts inland and over the warmer desert. The meeting of warm and cold air produces advection fogs (page 222) which provide sufficient moisture for a limited vegetation cover. Inland, parts of the Atacama are alleged to be the only truly rainless desert in the world, but even here the occasional rainfall event does occur.

Desert landscapes: what does a desert look like?

Deserts provide a classic example of how easy it is to portray or to accept an inaccurate mental picture of different places (or people) in the world. What is your image of a desert? Is it a land-scape of sand dunes similar to those shown in Figures 7.15–7.18, perhaps with a camel or palm tree somewhere in the background? Large areas of dunes, known as **erg**, do exist – but they cover only about one-quarter of the world's deserts. Most deserts consist either of bare rock, known as **hammada** (Figure 7.4), or stone-covered plains, called **reg** (Figure 7.5). Deserts contain a great diversity of landscapes. This diversity is due to geological factors (tectonics and rock type) as well as to climate (temperature, rainfall and wind) and resultant weathering processes.

Figure 7.4

A rocky (hammada) desert, Wadi Rum, Jordan

Figure 7.5

A stony (reg) desert, Sahara

Arid processes and landforms

In their attempts to understand the development of arid landforms, geographers have come up against three main difficulties:

1 How should the nature of the weathering processes be assessed? Desert weathering was initially assumed to be largely mechanical and to result from extreme diurnal ranges in temperature. More recently, the realisation that water is present in all deserts in some form or other has led to the view that chemical weathering is far more significant than had previously been thought. Latest opinions seem to suggest that the major processes, e.g. exfoliation and salt weathering, may involve a combination of both mechanical and chemical weathering.

2 What is the relative importance of wind and water as agents of erosion, transportation and deposition in deserts?

3 How important have been the effects of climatic change on desert landforms? During some phases of the Quaternary, and previously when continental plates were in different latitudes, the climate of present arid areas was much wetter than it is today. How many of the landforms that we see now are, therefore, relict and how many are still in the process of being formed?

Mechanical weathering

Traditionally, weathering in deserts was attributed to mechanical processes resulting from extremes of temperature. Deserts, especially those away from the coast, are usually cloudless and are characterised by daily extremes of temperature. The lack of cloud cover can allow day temperatures to exceed 40°C for much of the year; while at night, rapid radiation often causes temperatures to fall to zero. Although in some colder, more mountainous deserts, frost shattering is a common process, it was believed that the major process in most deserts was **insolation weathering**. Insolation weathering occurs when, during the day, the direct rays of the sun heat up the surface layers of the rock. These surface layers, lacking any protective vegetation cover, may reach 80°C. The different types and colours of minerals in most rocks, especially igneous rocks, heat up and cool down at different rates, causing internal stresses and fracturing. This process was thought to cause the surface layers of exposed rock to peel off – **exfoliation** – or individual grains to break away – **granular disintegration** (page 41). Where surface layers do peel away, newly exposed surfaces experience pressure release (page 41). This is believed to be a contributory process in the formation of rounded exfoliation domes such as Uluru (Figure 7.6) and Sugarloaf Mountain (Figure 2.3).

Doubts about insolation weathering began when it was noted that the 4500-year-old ancient monuments in Egypt showed little evidence of exfoliation, and that monuments in Upper Egypt, where the climate is extremely arid, showed markedly fewer signs of decay than those located in Lower Egypt, where there is a limited rainfall. D.T. Griggs (1936) conducted a series of laboratory experiments in which he subjected granite blocks to extremes of temperature in excess of 100°C. After the equivalent of almost 250 years of diurnal temperature change, he found no discernible difference in the rock. Later, he subjected the granite to the same temperature extremes while at the same time spraying it with water. Within the equivalent of two and a half years of diurnal temperature change, he found the rock beginning to crack. His conclusions, and those of later geomorphologists, suggest that some of the weathering previously attributed to insolation can now be ascribed to chemical changes caused by moisture. Although rainfall in deserts may be limited, the rapid loss of temperature at night frequently produces dew (175 nights a year in Israel's Negev) and the mingling of warm and cold air on coasts (e.g. of the Atacama) causes advection fog (page 222). There is sufficient moisture, therefore, to combine with certain minerals to cause the rock to swell (hydration) and the outer layers to peel off (exfoliation). At present, it would appear that the case for insolation weathering is neither proven nor disproven and that it may be a consequence of either mechanical weathering, or chemical weathering, or both.

Figure 7.6

An exfoliation dome: Uluru (formerly called Ayers Rock), Uluru-Kata Tjuta National Park, Australia (compare with Figure 2.3)

The second mechanical process in desert environments, **salt weathering**, is more readily accepted although the action of salt can cause chemical, as well as physical, changes in the rock (page 40). Salts in rainwater, or salts brought to the surface by capillary action, form crystals as the moisture is readily evaporated in the high temperatures and low relative humidities. Further evaporation causes the salt crystals to expand and mechanically to break off pieces of the rock upon which they have formed (page 40). Subsequent rainfall, dew or fog may be absorbed by salt minerals causing them to swell (hydration) or chemically to change their crystal structure (page 42). Where salts accumulate near or on the surface, particles may become cemented together to form **duricrusts**. These hard crusts are classified according to the nature of their chemical composition. (Students with a special interest in geology or chemistry may wish to research the meaning of the terms **calcretes**, **silcretes** and **gypcretes**.) Another form of crust, **desert varnish**, is a hard, dark glazed surface found on exposed rocks which have been coated by a film composed largely of oxides of iron and manganese (Figure 7.7) and, possibly, bacterial action. It is hoped that the dating of desert varnish may help to establish a chronology of climatic changes in arid and semi-arid environments.

Figure 7.7

Carvings in desert varnish, Wadi Rum, Jordan

The importance of wind and water

Geomorphologists working in Africa at the end of the last century believed the wind to be responsible for most desert landforms. Later fieldwork, carried out mainly in the higher and wetter semi-arid regions of North America, recognised and emphasised the importance of running water and, in doing so, de-emphasised the role of wind. Today, it is more widely accepted that both wind and water play a significant, but locally varying, part in the development of the different types of desert landscape.

Aeolian (wind) processes
Transport

The movement of particles is determined by several factors. Aeolian movement is greatest where winds are strong (usually over 20 km/hr), turbulent, come from a constant direction and blow steadily for a lengthy period of time. Of considerable importance, too, is the nature of the regolith. It is more likely to be moved if there is no vegetation to bind it together or to absorb some of the wind's energy; if it is dry and unconsolidated; if particles are small enough to be transported; and if material has been loosened by farming practices. While such conditions do occur locally in temperate latitudes, e.g. coastal dunes, summits of mountains and during dry summers in arable areas, the optimum conditions for transport by wind are in arid and semi-arid environments.

Wind can move material by three processes: suspension, saltation and surface creep. The effectiveness of each method is related to particle size (Figure 7.8).

Suspension Where material is very fine, i.e. less than 0.15 mm in diameter, it can be picked up by the wind, raised to considerable heights and carried great distances. There have been occasions, though perhaps recorded only once a decade, when red dust from the Sahara has been carried northwards and deposited as 'red rain' over parts of Britain. Visibility in deserts is sometimes reduced to less than 1000 m and this is called a **dust storm** (Figure 7.9). The number of recorded dust storms on the margins of the Sahara has increased rapidly in the last 25 years as the drought of that region has intensified. In Mauritania during the early part of the 1960s, there was an average of only 5 days/yr with dust storms compared with an average of 80 days/yr over a similar period in the early 2000s.

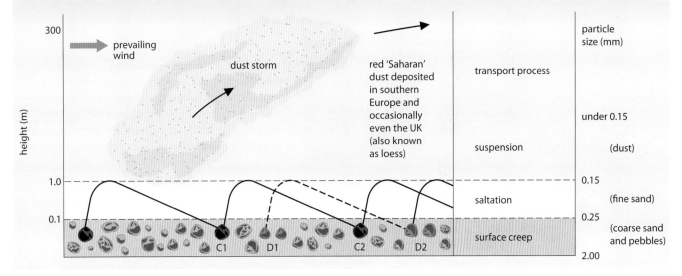

height (m)
300
1.0
0.1

prevailing wind

dust storm

red 'Saharan' dust deposited in southern Europe and occasionally even the UK (also known as loess)

C1 D1 C2 D2

particle size (mm)

transport process

under 0.15 — suspension — (dust)

0.15 — saltation — (fine sand)

0.25 — surface creep — (coarse sand and pebbles)

2.00

Saltation When wind speeds exceed the threshold velocity (the speed required to initiate grain movement), fine and coarse-grained sand particles are lifted. They may rise almost vertically for several centimetres before returning to the ground in a relatively flat trajectory of less than 12° (Figure 7.8). As the wind continues to blow, the sand particles bounce along, leap-frogging over one another. Even in the worst storms, sand grains are rarely lifted higher than 2 m above the ground.

Surface creep Every time a sand particle, transported by saltation, lands, it may dislodge and push forward larger particles (more than 0.25 mm in diameter) which are too heavy to be uplifted. This constant bombardment gradually moves small stones and pebbles over the desert surface.

Figure 7.10

A desert pavement with ventifacts in Jordan, created by deflation

4x4 vehicles are being accused of damaging the ecology of the Sahara Desert and contributing to the world's growing dust storm problem. Since the 1990s, 4x4 Land Cruisers have replaced the camel as the vehicle of choice (a process referred to as 'Toyotarisation'). These vehicles, according to Professor Goudie, are gradually destroying the thin layer of lichen and gravel that keeps the desert surface stable in high winds. In the worst-affected regions, estimates suggest that 1270 million tonnes of dust are thrown up each a year – ten times more than half a century ago. The dust, which may contain harmful microbes and pesticides, is transported high into the atmosphere during storms and deposited (known as blood rain in certain places) as far afield as the Alps (seen as a red layer on top of the snow), the Caribbean (where fungal pores carried with it have been blamed for destroying coral reefs) and on cars and property in southern England.

Figure 7.9

Dust storms created by human activity

Erosion

There are two main processes of wind erosion: deflation and abrasion.

Deflation is the progressive removal of fine material by the wind leaving pebble-strewn desert pavements or reg (Figures 7.10 and 7.11). Over much of the Sahara, and especially in Sinai in Egypt, vast areas of monotonous, flat and colourless pavement are the product of an earlier, wetter climate. Pebbles were transported by water from the surrounding highlands and deposited with sand, clay and silt on the lowland plains. Later, the lighter particles were removed by the wind, causing the remaining pebbles to settle and to interlock like cobblestones.

Elsewhere in the desert, dew may collect in hollows and material may be loosened by chemical weathering and then removed by wind to leave **closed depressions** or **deflation hollows**. Closed depressions are numerous and vary in size from a few metres across to the extensive Qattara

Figure 7.11

The process of deflation

silt and sand removed by wind, leaving stones

land surface is lowered

leaving desert pavement: a coarse mosaic of stones resembling a cobbled street, which protects against further erosion

Deserts **183**

Figure 7.12

Landshore yardangs, Western Desert, Egypt

Depression in Egypt which reaches a depth of 134 m below sea-level. Closed depressions may also have a tectonic origin (the south-west of the USA) or a solution origin (limestone areas in Morocco). The Dust Bowl, formed in the American Mid-West in the 1930s, was a consequence of deflation following a severe drought in a region where inappropriate farming techniques had been introduced. Vast quantities of valuable topsoil were blown away, some of which was deposited as far away as Washington, DC.

Abrasion is a sandblasting action effected by materials as they are moved by saltation. This process smooths, pits, polishes and wears away rock close to the ground. Since sand particles cannot be lifted very high, the zone of maximum erosion tends to be within 1 m of the Earth's surface. Abrasion produces a number of distinctive landforms which include ventifacts, yardangs and zeugen.

Ventifacts are individual rocks with sharp edges and, due to abrasion, smooth sides. The white rock in the foreground of Figure 7.10 has a long axis of 25 cm.

Yardangs are extensive ridges of rock, separated by grooves (troughs), with an alignment similar to that of the prevailing winds

(Figure 7.12). In parts of the Sahara, Arabian and Atacama Deserts, they are large enough to be visible on air photographs and satellite imagery.

Zeugen are tabular masses of resistant rock separated by trenches where the wind has cut vertically through the cap into underlying softer rock.

Deposition

Dunes develop when sand grains, moved by saltation and surface creep, are deposited. Although large areas of dunes, known as ergs, cover about 25 per cent of arid regions, they are mainly confined to the Sahara and Arabian Deserts, and are virtually absent in North America. Much of the early field-work on dunes was carried out by R.A. Bagnold in North Africa in the 1920s. He noted that some, but by no means all, dunes formed around an obstacle – a rock, a bush, a small hill or even a dead camel; and most dunes were located on surfaces that were even and sandy and not on those which were irregular and rocky. He concentrated on two types of dune: the barchan and the seif. The **barchan** is a small, crescent-shaped dune, about 30 m high, which is moved by the wind (Figures 7.13 and 7.15). The **seif**, named after an Arab curved sword, is much larger (100 km in length and 200 m in height) and more common (Figure 7.17), although the process of its formation is more complex than initially thought by Bagnold. Textbooks often over-emphasise these two dunes, especially the barchan which is a relatively uncommon feature.

While Bagnold had to travel the desert in specially converted cars, modern geographers derive their picture of desert landforms from aerial photographs and Landsat images. These new techniques have helped to identify several types of dune, and the modern classification, still based on morphology, contains several additional types (Figure 7.14). Dune morphology depends upon the supply of sand, wind direction, availability of vegetation and the nature of the ground surface.

Figure 7.13

The movement of a crescent-shaped barchan

a in plan

prevailing wind

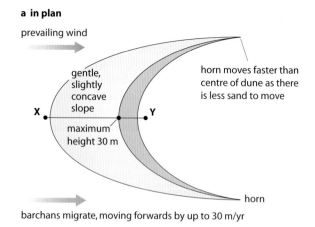

gentle, slightly concave slope

horn moves faster than centre of dune as there is less sand to move

X   Y

maximum height 30 m

horn

barchans migrate, moving forwards by up to 30 m/yr

b in profile

prevailing wind

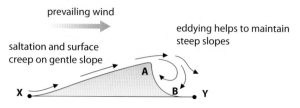

eddying helps to maintain steep slopes

saltation and surface creep on gentle slope

A

X   B   Y

A steep, upper slip slope of coarse grains and with continual sand avalanches due to unconsolidated material (unlike a river, coarse grains are at the top)

B gentle, basal apron with sand ripples: the finer grains, as on a beach, give a gentler gradient than coarser grains

Type of dune	Description	Supply of sand	Wind direction and speed	Vegetation cover	Speed of dune movement
barchan 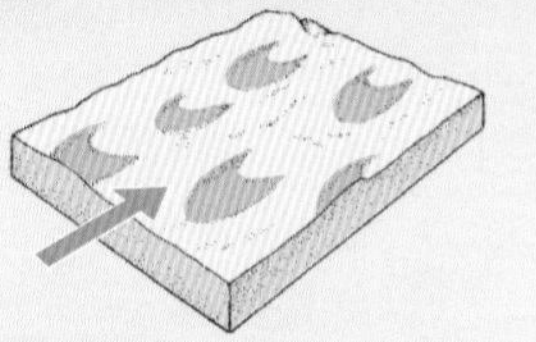	individual dunes, crescent shape with horns pointing downwind (Figures 7.13 and 7.15)	limited	constant direction, at right-angles to dune	none	highly mobile
barchanoid ridges 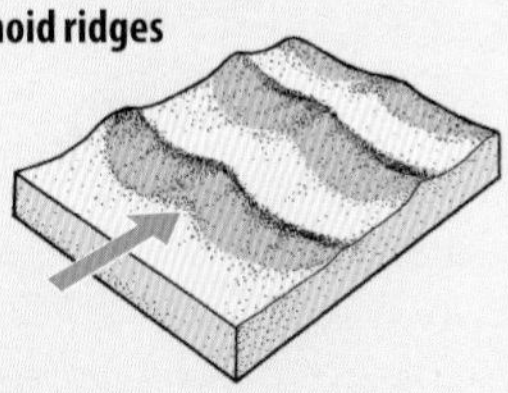	asymmetrical, oriented at right-angles to wind, rows of barchans forming parallel ridges	limited	constant direction, at right-angles to dune	none	mobile
transverse 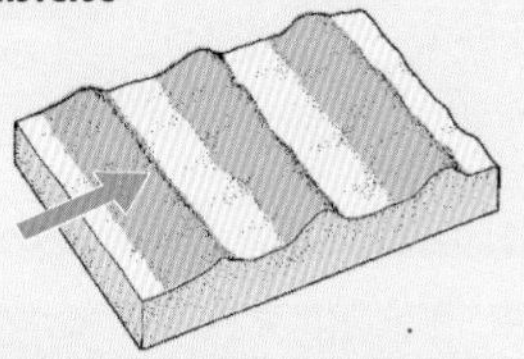	oriented at right-angles to wind but lacking barchanoid structure, resemble ocean waves (Figure 7.16)	abundant (thick) sand cover	steady winds (trades), constant direction but with reducing speeds, at right-angles to dune	vegetation stabilises sand	sand checked by barriers, limited mobility
dome 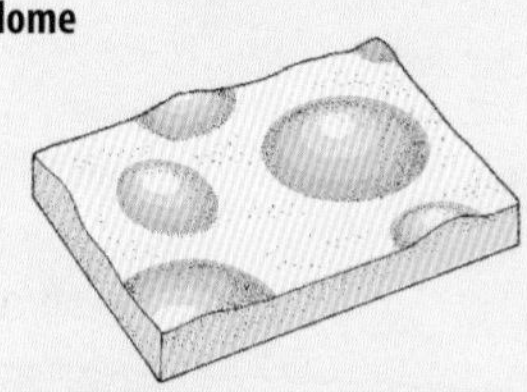	dome-shaped (height restricted by wind)	appreciable amounts of coarse sand	strong winds limit height of dune	none	virtually no movement
seif (linear) 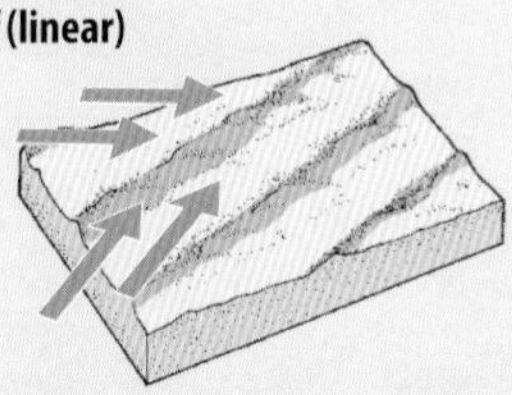	longitudinal, parallel dunes with slip faces on either side, can extend for many km (Figure 7.17)	large	persistent, steady winds (trades), with slight seasonal or diurnal changes in direction	none	regular (even) surface, virtually no movement
parabolic 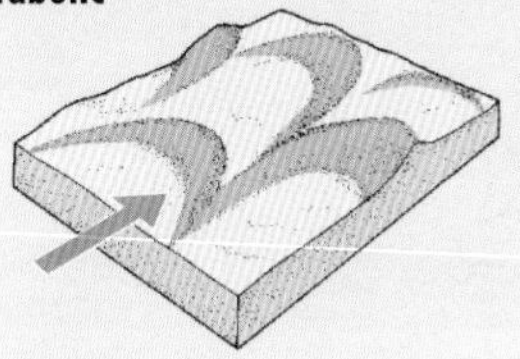	hairpin-shaped with noses pointing downwind, a type of blowout (eroded) dune where middle section has moved forward, may occur in clusters	limited	constant direction	where present, can anchor sand	highly mobile (by blowouts in nose of dune)
star 	complex dune with a star (starfish) shape (compare arêtes radiating from central peak) (Figure 7.18)	limited	effective winds blow from several directions	none	virtually no movement
reversing 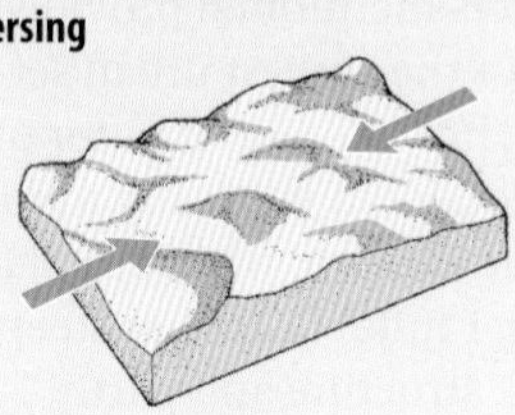	undulating, haphazard shape	limited	winds of equal strength and duration from opposite directions	none	virtually no movement

Figure 7.14

Classification of sand dunes (*after* Goudie)

Figure 7.15

Barchan dunes near Lüderitz, Namibia

Figure 7.16

Transverse dunes Djanet, Algeria

Figure 7.17

Seif (linear) dunes, Sossusvlei, Namibia

Figure 7.18

Star dunes, Sossus Namibia

The effects of water

It has already been noted that, in arid areas, moisture must be present for processes of chemical weathering to operate. We have also seen that often rainfall is low, irregular and infrequent, with long-term fluctuations. Although most desert rainfall occurs in low-intensity storms, the occasional sudden, more isolated, heavy downpour, does occur. There are records of several extreme desert rainfall events, each equivalent to the three-monthly mean rainfall of London. The impact of water is, therefore, very significant in shaping desert landscapes.

Rivers in arid environments fall into three main categories.

Exogenous Exogenous rivers are those like the Colorado, Nile, Indus, Tigris and Euphrates, which rise in mountains beyond the desert margins. These rivers continue to flow throughout the year even if their discharge is reduced by evaporation when they cross the arid land. (The last four rivers mentioned provided the location for some of the earliest urban settlements – page 388.) The Colorado has, for over 300 km of its course, cut down vertically to form the Grand Canyon. The canyon, which in places is almost 2000 m (over 1 mile) deep, has steep sides partly due to rock structure and partly due to insufficient rainfall to degrade them (Figure 7.19).

Studies in Kenya, Israel and Arizona suggest that surface runoff is likely to occur within 10 minutes of the start of a downpour (Figure 7.20). This may initially be in the form of a **sheet flood** where the water flows evenly over the land and is not confined to channels. Much of the sand, gravel and pebbles covering the desert floor is thought to have been deposited by this process; yet, as the event has rarely been witnessed, it is assumed that deposition by sheet floods occurred mainly during earlier wetter periods called **pluvials**.

Very soon, the collective runoff becomes concentrated into deep, steep-sided ravines known as **wadis** (Figure 7.22) or **arroyos**. Normally dry, wadis may be subjected to irregular flash floods (Figure 7.20 and Places 25). The average occurrence of these floods is once a year in the semi-arid margins of the Sahara, and once a decade in the extremely arid interior. This infrequency of floods compared with the great number and size of wadis, suggests that they were created when storms were more frequent and severe – i.e. they are a relict feature.

Figure 7.20

A flash flood

Pediments and playas

Stretching from the foot of the highlands, there is often a gently sloping area either of bare rock or of rock covered in a thin veil of debris (Figures 7.21 and 7.24). This is known as a **pediment**. There is often an abrupt break of slope at the junction of the highland area and the pediment. Two main theories suggest the origin of the pediment, one involving water. This theory proposes that weathered material from cliff faces, or debris from alluvial fans, was carried during pluvials by sheet floods. The sediment planed the lowlands before being deposited, leaving a gently concave slope of less than 7° (Figure 7.24). The alternative theory involves the parallel retreat of slopes resulting from weathering (King's hypothesis, Figure 2.24c).

Endoreic Endoreic drainage occurs where rivers terminate in inland lakes. Examples are the River Jordan into the Dead Sea and the Bear into the Great Salt Lake.

Ephemeral Ephemeral streams, which are more typical of desert areas, flow intermittently, or seasonally, after rainstorms. Although often shortlived, these streams can generate high levels of discharge due to several local characteristics. First, the torrential nature of the rain exceeds the infiltration capacity of the ground and so most of the water drains away as surface runoff (overland flow, page 59). Second, the high temperatures and the frequent presence of duricrust combine to give a hard, impermeable surface which inhibits infiltration. Third, the lack of vegetation means that no moisture is lost or delayed through interception and the rain is able to hit the ground with maximum force. Fourth, fine particles are displaced by rainsplash action and, by infilling surface pore spaces, further reduce the infiltration capacity of the soil. It is as a result of these minimal infiltration rates that slopes of less than 2° can, even under quite modest storm conditions, experience extensive overland flow.

Playas are often found at the lowest point of the pediment. They are shallow, ephemeral, saline lakes formed after rainstorms. As the rain water rapidly evaporates, flat layers of either clay, silt or salt are left. Where the dried-out surface consists of clay, large **desiccation cracks**, up to 5 m deep, are formed. When the surface is salt-covered, it produces the 'flattest landform on land'. Rogers Lake, in the Mojave Desert, California, has been used for spacecraft landings, while the Bonneville saltflats in Utah have been the location for land-speed record attempts.

Figure 7.21

Pediment at foot of highlands, Wadi Rum, Jordan

Places 25 Wadis: flash floods

Figure 7.22

A wadi near Qumran, Israel

Camping in a wadi is something that experienced desert travellers avoid. It is possible to be swept away by a flash flood which occurs virtually without warning – there may have been no rain at your location, and perhaps nothing more ominous than a distant rumble of thunder. Indeed, the first warning may be the roar of an approaching wall of water. One minute the bed of the wadi is dry, baked hard under the sun and littered with weathered debris from the previous flood or from the steep valley sides (Figure 7.22), and the next minute it is a raging torrent.

The energy of the flood enables large boulders to be moved by traction, and enormous amounts of coarse material to be taken into suspension – some witnesses have claimed it is more like a mudflow. Friction from the roughness of the bed, the large amounts of sediment and the high rates of evaporation soon cause a reduction in the stream's velocity. Deposition then occurs, choking the channel, followed by braiding as the water seeks new outlets. Within hours, the floor of the wadi is dry again (Figure 7.23).

The rapid runoff does not replenish groundwater supplies, and without the groundwater contribution to base flow, characteristic of humid climates, rivers cease to flow. At the mouth of the wadi, where the water can spread out and energy is dissipated, material is deposited to form an **alluvial fan** or **cone** (Figure 7.24). If several wadis cut through a highland close to each other, their semi-circular fans may merge to form a **bahada (bajada)**, which is an almost continuous deposit of sand and gravel.

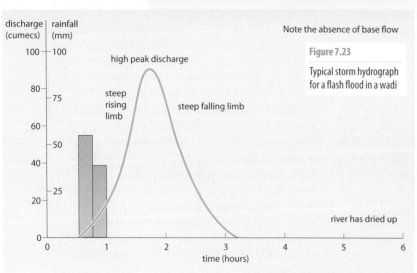

Figure 7.23

Typical storm hydrograph for a flash flood in a wadi

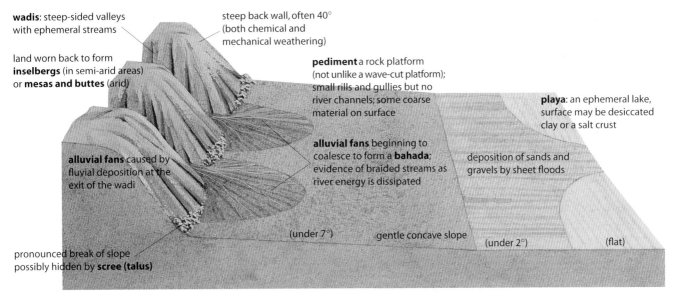

wadis: steep-sided valleys with ephemeral streams

steep back wall, often 40° (both chemical and mechanical weathering)

land worn back to form **inselbergs** (in semi-arid areas) or **mesas and buttes** (arid)

pediment a rock platform (not unlike a wave-cut platform); small rills and gullies but no river channels; some coarse material on surface

playa: an ephemeral lake, surface may be desiccated clay or a salt crust

alluvial fans caused by fluvial deposition at the exit of the wadi

alluvial fans beginning to coalesce to form a **bahada**; evidence of braided streams as river energy is dissipated

deposition of sands and gravels by sheet floods

pronounced break of slope possibly hidden by **scree (talus)**

(under 7°)

gentle concave slope

(under 2°)

(flat)

Figure 7.24

Pediments and playas

Occasionally, isolated, flat-topped remnants of former highlands, known as **mesas**, rise sheer from the pediment. Some mesas, in Arizona, have summits large enough to have been used as village sites by the Hopi Indians. **Buttes** are smaller versions of mesas. The most spectacular mesas and buttes lie in Monument Valley Navajo Tribal Park in Arizona (Figure 7.25).

Relationship between wind and water

Some desert areas are dominated by wind, others by water. Areas where wind appears to be the dominant geomorphological agent are known as **aeolian domains**. The effectiveness of the wind increases where, and when, amounts of rainfall decrease. As rainfall decreases, so too does any vegetation cover. This allows the wind to transport material unhindered, and rates of erosion (abrasion and deflation) and deposition (dunes)

to increase. **Fluvial domains** are those where water processes are dominant or, as evidence increasingly suggests, have been dominant in the past. Vegetation, which stabilises material, increases as rainfall increases or where coastal fog and dew are a regular occurrence.

Evidence also suggests that wind and water can interact in arid environments and that landforms produced by each do co-exist within the same locality. However, the balance between their relative importance has often altered, mainly due to climatic change either over lengthy periods of time (e.g. the 18 000 years since the time of maximum glaciation) or during shorter fluctuations (e.g. since the mid-1960s in the Sahel). At present, and especially in Africa, the decrease in rainfall in the semi-arid desert fringes means that the role of water is probably declining, while that of the wind is increasing.

Figure 7.25

Mesas and buttes, Monument National Park, Arizona, USA

Climatic change

There have already been references to pluvials within the Sahara Desert (page 181). Prior to the Quaternary era, these may have occurred when the African Plate lay further to the south and the Sahara was in a latitude equivalent to that of the present-day savannas. In the Quaternary era, the advance of the ice sheets resulted in a shift in windbelts which caused changes in precipitation patterns, temperatures and evaporation rates. At the time of maximum glaciation (18 000 years ago), desert conditions appear to have been more extensive than they are today (Figure 7.26). Since then, as suggested by radiocarbon dating (page 248), there have been frequent, relatively short-lived pluvials, the last occurring about 9000 years ago. Evidence for a once-wetter Sahara is given in Figure 7.27.

Herodotus, a historian living in Ancient Greece, described the Garamantes civilisation which flourished in the Ahaggar Mountains 3000 years ago. This people, who recorded their exploits in cave paintings at Tassili des Ajjers, hunted elephants, giraffes, rhinos and antelope. Twenty centuries ago, North Africa was the 'granary of the Roman Empire'. Wadis are too large and deep and alluvial cones too widespread to have been formed by today's occasional storms, while sheet floods are too infrequent to have moved so much material over pediments. Radiating from the Ahaggar and Tibesti Mountains, aerial photographs and satellite imagery have revealed many dry valleys which once must have held permanent rivers (compare Figure 6.44). Lakes were also once much larger and deeper. Around Lake Chad, shorelines 50 m above the present level are visible, and research suggests that lake levels might once have been over 100 m higher. (Lake Bonneville in the USA is only one-tenth of its former maximum size and, like Lake Chad, is drying up rapidly.) Small crocodiles found in the Tibesti must have been trapped in the slightly wetter uplands as the desert advanced. Also, pollen analysis has shown that oak and cedar forests abounded in the same region 10 000 years ago. Groundwater in the Nubian sandstone has been dated, by radio-isotope methods, to be over 25 000 years old, and may have accumulated at about the same time as fossil laterite soils (page 321).

Figure 7.26

Extent of sand dunes in Africa

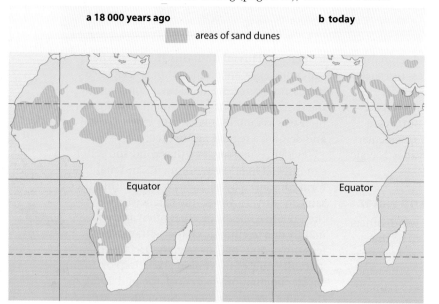

a 18 000 years ago **b today**

areas of sand dunes

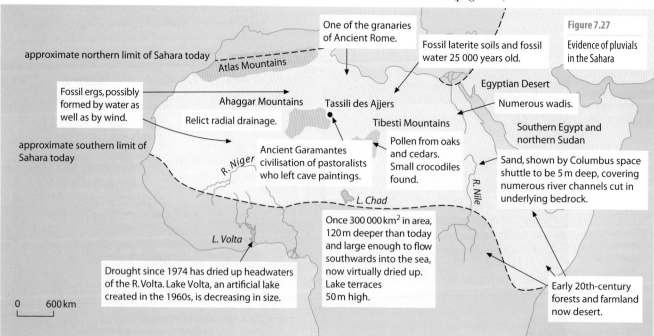

Figure 7.27

Evidence of pluvials in the Sahara

One of the granaries of Ancient Rome.

approximate northern limit of Sahara today

Atlas Mountains

Fossil laterite soils and fossil water 25 000 years old.

Egyptian Desert

Fossil ergs, possibly formed by water as well as by wind.

Ahaggar Mountains

Tassili des Ajjers

Numerous wadis.

Relict radial drainage.

Tibesti Mountains

Southern Egypt and northern Sudan

approximate southern limit of Sahara today

R. Niger

Ancient Garamantes civilisation of pastoralists who left cave paintings.

Pollen from oaks and cedars. Small crocodiles found.

R. Nile

Sand, shown by Columbus space shuttle to be 5 m deep, covering numerous river channels cut in underlying bedrock.

L. Chad

L. Volta

Once 300 000 km² in area, 120 m deeper than today and large enough to flow southwards into the sea, now virtually dried up. Lake terraces 50 m high.

Drought since 1974 has dried up headwaters of the R. Volta. Lake Volta, an artificial lake created in the 1960s, is decreasing in size.

Early 20th-century forests and farmland now desert.

0   600 km

Desertification: fact or fiction?

In the mid-1970s, desertification, not global warming, was perceived as the world's major environmental issue. Since then the nature, extent, causes and effects of desertification have become shrouded in controversy. Taken literally, desertification means 'the making of a desert'. More helpfully, it has been defined as 'the turning of the land, often through physical processes and human mismanagement, into desert'. Even so, although the term has been in use for over half a century, few can agree on exactly what it means. The diversity of definitions – there are over 100 – is due largely to uncertainty over its causes.

Goudie says that 'the question has been asked whether this process is caused by temporary drought periods of high magnitude, is due to longer-term climatic change towards aridity, is caused by man-induced climatic change, or is the result of human action through man's degradation of the biological environments in arid zones. Most people now believe that it is produced by a combination of increasing human and animal populations, which cause the effects of drought years to become progressively more severe so that the vegetation is placed under increasing stress.'

Those places perceived to be at greatest risk from desertification are shown in Figure 7.28. In 2005 the UN claimed that desertification directly affected over 250 million people and threatened another 1 billion living in at-risk countries. It is generally agreed that the desert is encroaching into semi-arid, desert margins, especially in the Sahel – a broad belt of land on the southern side of the Sahara (2–4 in Figure 7.28).

Some of the main interrelationships between the believed causes of desertification are shown in Figure 7.29.

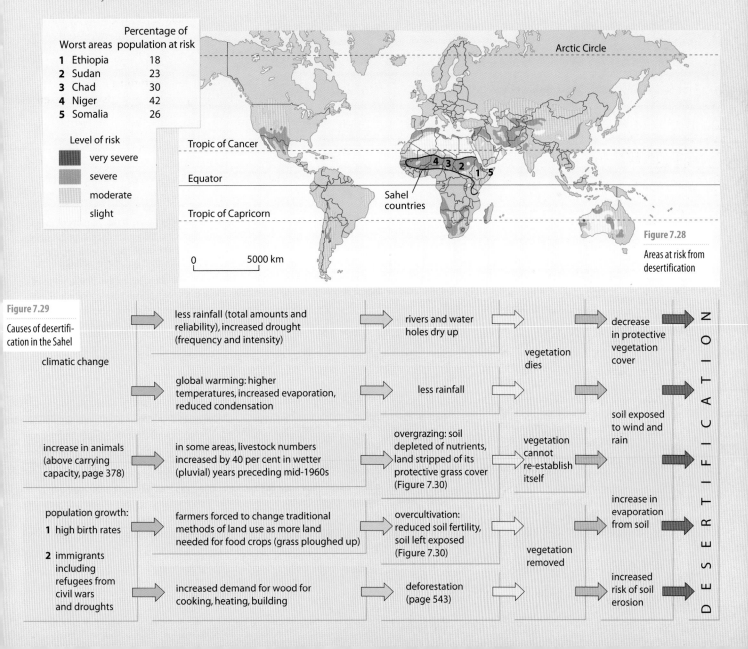

Worst areas	Percentage of population at risk
1 Ethiopia	18
2 Sudan	23
3 Chad	30
4 Niger	42
5 Somalia	26

Level of risk
- very severe
- severe
- moderate
- slight

Arctic Circle
Tropic of Cancer
Equator
Tropic of Capricorn

Sahel countries

0   5000 km

Figure 7.28

Areas at risk from desertification

Figure 7.29

Causes of desertification in the Sahel

climatic change → less rainfall (total amounts and reliability), increased drought (frequency and intensity) → rivers and water holes dry up → vegetation dies → decrease in protective vegetation cover

climatic change → global warming: higher temperatures, increased evaporation, reduced condensation → less rainfall → vegetation dies

increase in animals (above carrying capacity, page 378) → in some areas, livestock numbers increased by 40 per cent in wetter (pluvial) years preceding mid-1960s → overgrazing: soil depleted of nutrients, land stripped of its protective grass cover (Figure 7.30) → vegetation cannot re-establish itself → soil exposed to wind and rain

population growth:
1 high birth rates
2 immigrants including refugees from civil wars and droughts
→ farmers forced to change traditional methods of land use as more land needed for food crops (grass ploughed up) → overcultivation: reduced soil fertility, soil left exposed (Figure 7.30) → vegetation removed → increase in evaporation from soil

→ increased demand for wood for cooking, heating, building → deforestation (page 543) → increased risk of soil erosion

DESERTIFICATION

Figure 7.30

Desertification and
a overgrazing
b overcultivation

In 1975, Hugh Lamprey, a bush pilot and environmentalist, claimed that, since his previous study 17 years earlier, the desert in the Sudan had advanced southwards by 90–100 km. In 1982 and at the height of one of Africa's worst-ever recorded droughts, UNEP (United Nations Environmental Programme) claimed that the Sahara was advancing southwards by 6–10 km a year and that, globally, 21 million hectares of once-productive soil were being reduced each year to zero productivity, that 850 million people were being affected, and 35 per cent of the world's surface was at risk (figures quoted by UNEP at the 1992 Rio Earth Summit). Since then scientific studies using satellite imagery and more detailed fieldwork (Figure 7.31) have thrown considerable doubt on the causes, effects and extent of desertification. Today, certain early statistics regarding its advance have proven to be unreliable. It is believed that overgrazing is no longer considered so important, fuelwood has not become exhausted as previously predicted, while famine and drought are more likely to result from poverty, poor farming techniques, civil unrest and war than from natural causes (page 503). In contrast, the extent and effects of salinisation (page 273 and Figure 16.53) appear to have increased.

The semi-arid lands are a fragile environment whose boundaries change due to variations in rainfall and land use. It is often difficult to separate natural causes from human ones and short-term fluctuations from long-term trends (Figure 7.32). The effects of global warming are as yet an unknown factor, although computer models suggest that the climate will get even drier.

Figure 7.31

Scientific evaluation
in the mid-1990s

Researchers at the University of Lund, in Sweden, carried out field surveys and examined satellite pictures of Sudan in an attempt to confirm Lamprey's findings. In a report published in the mid-1990s they stated 'no major shifts in the northern cultivation limit, no major sand dune transformation, no major changes in vegetation cover beyond the dramatic but short-term effects of variable rainfall'. A belt of sand dunes that Lamprey said formed the advancing front of the Sahara had shown no sign of movement since 1962, nor was there any evidence of patches of desert growing around wells, waterholes or villages – a phenomenon frequently claimed to be the result of overgrazing by herds of cattle [Places 65]. The report ended by stressing the need for recordings of a high scientific standard.

Figure 7.32

a Desert retreat or
b desert advance?

a

The southern Sahara Desert is in retreat, making farming again viable in parts of the Sahel. Satellite images taken this summer show that sand dunes are retreating the whole 6000 km across the Sahel region between Mauritania to Eritrea. Nor does it appear to be a short-term trend – analysts claim it has been happening unnoticed since the mid-1980s. In parts of Burkina Faso, devastated by the droughts of the 1980s, some of the landscape is now showing green, with more trees for firewood and more grassland for livestock. Farmers also claim their yields of sorghum and millet have nearly doubled, though this may partly be due to improved farming methods [Figure 10.40].

Adapted from *New Scientist*, 2002

b

Our 21st-century civilisation is being squeezed between advancing desert and rising seas, leaving less land to support a growing human population. This is illustrated by the heavy losses of land to advancing deserts in Nigeria and China, the most populous countries in Africa and Asia respectively. Nigeria is losing 3500 km² a year, whereas China, which lost on average 1500 km² a year between 1950 and 1975, has been losing 3600 km² a year since 2000. Satellite images have shown two deserts in Inner Mongolia and Gansu provinces expanding and merging, as are two larger ones to the west in Xinjiang province. To the east the Gobi Desert has advanced to within 250 km of Beijing. Chinese scientists report that some 24 000 villages in the north and west of the country have been abandoned or partly depopulated as they were overrun by drifting sand.

Adapted from *Earth Policy Institute*, 2006

Further reference

Goudie, A.S. (2001) *The Nature of the Environment*, WileyBlackwell.

Goudie, A.S. (2007) 'Dust storms' in *Geography Review* Vol 21 No 1 (September).

Goudie, A.S. and Watson, A. (1990) *Desert Geomorphology*, Macmillan.

Cooke, R.U., Warren, A. and Goudie, A.S. (1993) *Desert Geomorphology*, Routledge.

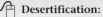 **Desertification:**
www.fao.org/desertification/default.asp?lang=en

Desert processes and landforms:
www.ux1.eiu.edu/~cfjps/1300/desert.html

http://geoweb.tamu.edu/courses/geo1101/grossman/Deserts.html

Unitarian Service Committee of Canada:
www.usc-canada.org/

UN Convention to Combat Desertification:
www.un.org/ecosocdev/geninfo/sustdev/desert.htm

UN Environment Programme Global Deserts Outlook:
www.unep.org/Geo/gdoutlook/

Questions & Activities

Activities

1 a Describe the characteristics that define a hot desert climate. *(4 marks)*

b Study Figure 7.2 (page 179) and describe the location of the world's deserts. *(4 marks)*

c Explain two causes of a desert climate. *(4 marks)*

d Write a paragraph to explain to someone why the typical view of a desert as a 'sea of sand' is often not true. *(4 marks)*

e What is 'exfoliation' weathering? *(4 marks)*

f Explain one other denudation process that operates in hot desert areas. *(5 marks)*

2 a Describe and name an example of a wadi. *(4 marks)*

b i Sometimes a 'flash flood' rushes through a wadi. Explain what a flash flood is. *(3 marks)*

ii Why is there little or no warning that a flash flood is about to happen? *(3 marks)*

iii Why do rivers stop flowing very soon after a flood in a desert area? *(3 marks)*

c In the area where a wadi opens onto lowland there is often an alluvial fan. Describe an alluvial fan and explain how it is formed. *(6 marks)*

d Describe a playa and explain how playas are formed. *(6 marks)*

Exam practice: basic structured questions

3 a Describe how wind transports material in a desert environment. *(6 marks)*

b Why is wind transportation a more important method of movement in deserts than in wet environments? *(3 marks)*

c Choose **one** type of sand dune.

i Draw an annotated diagram to show its main features.

ii Explain how the dune has been formed. *(8 marks)*

d Choose **one** desert landform created by wind erosion.

i Describe its shape and size.

ii Explain the processes that have formed it. *(8 marks)*

4 a On a sketch or copy of Figure 7.25 page 189, add labels to show: caprock; free face; bare rock; rectilinear slope; loose scree; gently sloping plain. *(6 marks)*

b Explain why the loose material you can see in the photograph has not been moved away. *(5 marks)*

c i In the Sahara Desert in North Africa there is evidence that the climate has not always been like this. Choose **one** piece of evidence to show that the climate has changed, state it and explain how it shows climate change. *(7 marks)*

ii Choose **one** piece of evidence to suggest that the climate of North Africa is changing now. State it and explain how it shows climate is changing. *(7 marks)*

Exam practice: structured question and essays

5 a Why do arid conditions occur in continental areas in the tropics? *(10 marks)*

b Making good use of examples, describe **two** ways in which plants adapt to drought conditions in desert areas. *(8 marks)*

c Explain the term 'water balance' used to identify the extent of tropical desert climates. *(7 marks)*

6 'Semi-arid lands are fragile environments.'

Discuss this statement with reference to semi-arid areas that you have studied. *(25 marks)*

7 Using Figures 7.3, 7.4 and 7.5 (page 180), describe and account for the range of surface conditions found in desert areas. *(25 marks)*

8

Rock types and landforms

· ·

'At first sight it may appear that rock type is the dominant influence on most landscapes … As geomorphologists, we are more concerned with the ways in which the characteristics of rocks respond to the processes of erosion and weathering than with the detailed study of rocks themselves.'

R. Collard, *The Physical Geography of Landscape*, 1988

Previous chapters have demonstrated how landscapes at both local and global scales have developed from a combination of processes. Plate tectonics, weathering and the action of moving water, ice and wind both create and destroy landforms. Yet these processes, however important they are at present or have been in the past, are insufficient to explain the many different and dramatic changes of scenery which can occur within short distances, especially in the British Isles.

Lithology refers to the physical characteristics of a rock. As each individual rock type has different characteristics, so it is capable of producing its own characteristic scenery. Landforms are greatly influenced by a rock type's vulnerability to weathering, its permeability and its structure.

To show how these three factors affect different rocks and to explain their resultant landforms and potential economic use, five rock types have been selected as exemplars. Carboniferous limestone, chalk and sandstone (sedimentary rocks), and granite and basalt (both igneous) have been chosen because, arguably, these produce some of the most distinctive types of landform and scenery.

Lithology and geomorphology
Vulnerability to weathering

Mechanical weathering in Britain occurs more readily in rocks that are jointed. Water can penetrate either down the **joints** or along the **bedding planes** (Figure 8.1) of Carboniferous limestone, or into cracks resulting from pressure release or contraction on cooling within granite and basalt (page 41 and Figure 1.31). Subsequent freezing and thawing along these lines of weakness causes frost shattering (page 40).

Chemical weathering is a major influence in limestone and granite landforms. Limestone, composed mostly of calcium carbonate, is slowly dissolved by the carbonic acid in rainwater, i.e. the process of carbonation (page 43). Granite consists of quartz, feldspar and mica. It is susceptible to hydration, where water is incorporated into the rock structure causing it to swell and crumble (page 42), and to hydrolysis, when the feldspar is chemically changed into clay (pages 42–43). Quartz, in comparison with other minerals, is one of the least prone to chemical weathering.

Mottershead has emphasised that 'the mechanical resistance of rocks depends on the strength of the individual component minerals and the bonds between them, and that chemical resistance depends on the individual chemical resistances of the component minerals. Mechanical strength decreases if just one of these component minerals becomes chemically altered.'

Figure 8.1

Bedding planes with joints and angle of dip

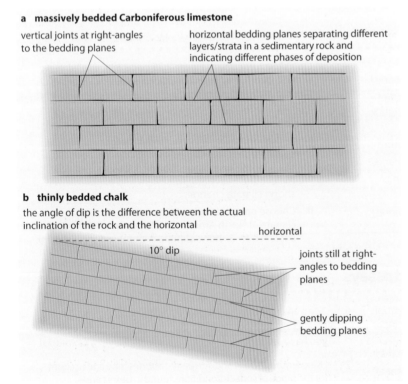

a   massively bedded Carboniferous limestone

vertical joints at right-angles to the bedding planes

horizontal bedding planes separating different layers/strata in a sedimentary rock and indicating different phases of deposition

b   thinly bedded chalk

the angle of dip is the difference between the actual inclination of the rock and the horizontal

horizontal

10° dip

joints still at right-angles to bedding planes

gently dipping bedding planes

Permeability

Permeability is the rate at which water may be stored within a rock or is able to pass through it. Permeability can be divided into two types.

1 **Primary permeability or porosity** This depends on the texture of the rock and the size, shape and arrangement of its mineral particles. The areas between the particles are called **pore spaces** and their size and alignment determine how much water can be absorbed by the rock. Porosity is usually greatest in rocks that are coarse-grained, such as gravels, sands, sandstone and oolitic limestone, and usually lowest in those that are fine-grained, such as clay and granite. (It is possible to have fine-grained sandstone and coarse-grained granite.) **Infiltration capacity** is the maximum rate at which water percolates into the ground. The infiltration capacity of sands is estimated to average 200 mm/hr, whereas in clay it is only 5 mm/hr. Pore spaces are larger where the grains are rounded rather than angular and compacted (Figure 8.2). Porosity can be given as an index value based upon the percentage of the total volume of the rock which is taken up by pore space, e.g. clay 20 per cent, gravel 50 per cent. When all the pore spaces are filled with water, the rock is said to be **saturated**. The water table marks the upper limit of saturation (Figure 8.9). Permeable rocks which store water are called **aquifers**.

2 **Secondary permeability** or **perviousness** This occurs in rocks that have joints and fissures along which water can flow. The most pervious rocks are those where the joints have been widened by solution, e.g. Carboniferous limestone, or by cooling, e.g. basalt. A rock may be pervious because of its structure, though water may not be able to pass through the rock mass itself. Where rocks are porous or pervious, water rapidly passes downwards to become groundwater, leaving the surface dry and without evident drainage – chalk and limestone regions have few surface streams. **Impermeable rocks**, e.g. granite, neither absorb water nor allow it to pass through them. These rocks therefore have a higher drainage density (page 67).

Structure

Resistance to erosion depends on whether the rock is massive and stratified, folded or faulted. Usually the more massive the rock and the fewer its joints and bedding planes, the more resistant it is to weathering and erosion. Conversely, the softer, more jointed and less compact the rock, the more vulnerable it is to denudation processes. Usually, more resistant rocks remain as upland areas (granite), while those that are less resistant form lowlands (clay).

However, there are exceptions. Chalk, which is relatively soft and may be well-jointed, forms rolling hills because it allows water to pass through it and so fluvial activity is limited. Carboniferous or Mountain limestone, having joints and bedding planes, produces jagged karst scenery because although it is pervious it has a very low porosity.

Figure 8.2

Pore spaces and infiltration capacity

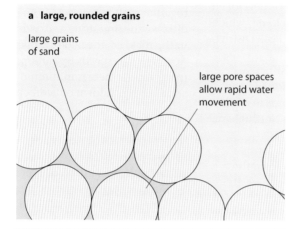

a large, rounded grains

large grains of sand

large pore spaces allow rapid water movement

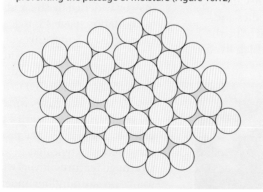

b small, rounded grains

although there are more pore spaces, they are much smaller: water clings to grains (surface tension) preventing the passage of moisture (Figure 10.12)

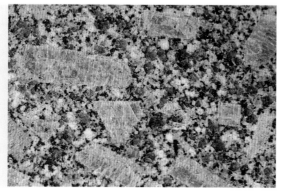

c crystals in granite

these fit together more closely than rounded grains, limiting the amount of water held and inhibiting the movement of moisture

Limestone

Limestone is a rock consisting of at least 80 per cent calcium carbonate. In Britain, most limestone was formed during four geological periods, each of which experienced different conditions. The following list begins with the oldest rocks. Use an atlas to find their location.

Carboniferous limestone This is hard, grey, crystalline and well-jointed. It contains many fossils, including corals, crinoids and brachiopods. These indicate that the rock was formed on the bed of a warm, clear sea and adds to the evidence that the British Isles once lay in warmer latitudes. Carboniferous limestone has developed its own unique landscape, known as **karst**, which in Britain is seen most clearly in the Peak District and Yorkshire Dales National Parks.

Magnesian limestone This is distinctive because it contains a higher proportion of magnesium carbonate. In Britain, it extends in a belt from the mouth of the River Tyne to Nottingham. In the Alps, it is known as **dolomite**.

Jurassic (oolitic) limestone This forms a narrow band extending southwards from the North Yorkshire Moors to the Dorset coast. Its scenery is similar to that typical of chalk.

Cretaceous chalk This is a pure, soft, well-jointed limestone. Stretching from Flamborough Head in Yorkshire (Figure 6.19), it forms the escarpment of the Lincoln Wolds, the East Anglian Heights and the North and South Downs, before ending up as the 'White Cliffs' at Dover and at Beachy Head, the Needles and Swanage. Cretaceous chalk is assumed to be the remains of small marine organisms which lived in clear, shallow seas.

The most distinctive of the limestone landforms are found in Carboniferous limestone and chalk.

Carboniferous limestone

This rock develops its own particular type of scenery primarily because of three characteristics. First, it is found in thick beds separated by almost horizontal bedding planes and with joints at right-angles (Figure 8.1). Second, it is pervious but not porous, meaning that water can pass along the bedding planes and down joints but not through the rock itself. Third, calcium carbonate is soluble. Carbonic acid in rainwater together with humic acid from moorland plants, dissolve the limestone and widen any weaknesses in the rock, i.e. the bedding planes and joints. Acid rain also speeds up carbonation and solution (page 43). As there is minimum surface drainage and little breakdown of bedrock to form soil, the vegetation cover tends to be thin or absent. In winter, this allows frost shattering to produce scree at the foot of steep cliffs.

It is possible to classify Carboniferous limestone landforms into four types:

1 **Surface features caused by solution**
   Limestone pavements are flat areas of exposed rock. They are flat because they represent the base of a dissolved bedding plane, and exposed because the surface soil may have been removed by glacial activity and never replaced. Where joints reach the surface, they may be widened by the acid rainwater (carbonation, page 43) to leave deep gashes called **grikes**. Some grikes at Malham in north-west Yorkshire are 0.5 m wide and up to 2 m deep. Between the grikes are flat-topped yet dissected blocks referred to as **clints** (Figure 2.8). In time, the grikes widen and the clints are weathered down until a lower bedding plane is exposed and the process of solution–carbonation is repeated.

2 **Drainage features** Rivers which have their source on surrounding impermeable rocks, such as the shales and grits of northern England, may disappear down **swallow holes** or **sinks** as soon as they reach the limestone (Figure 8.3). The streams flow underground finding a pathway down enlarged joints, forming **potholes**, and along bedding planes. Where solution is more active, underground **caves** may form. While most caves develop above the water table (**vadose caves**, Figure 8.8), some may form beneath it (**phreatic caves**).

Figure 8.3

A stream disappearing down a swallow hole near Hunt Pot, Pen-y-Ghent, Yorkshire Dales National Park

Figure 8.5

The Watlowes dry valley
above Malham Cove

Figure 8.4

The resurgence at the foot
of Malham Cove, Yorkshire
Dales

Corrosion often widens the caverns until parts of the roof collapse, providing the river with angular material ideal for corrasion. Heavy rainfall very quickly infiltrates downwards, so caverns and linking passages may become water-filled within minutes. The resultant turbulent flow can transport large stones and the floodwater may prove fatal to cavers and potholers. Rivers make their way downwards, often leaving caverns abandoned as the water finds a lower level, until they reach underlying impermeable rock. A **resurgence** occurs where the river reappears on the surface, often at the junction of permeable and impermeable rocks (Figure 8.4).

3   **Surface features resulting from underground drainage** Steep-sided valleys are likely to have been formed as rivers flowed over the surface of the limestone, probably during periglacial times when permafrost acted as an impermeable layer. When the rivers were able to revert to their subterranean passages, the surface valleys were left dry (Figure 8.5). Many dry valley sides are steep and gorge-like, e.g. Cheddar Gorge. If the area above an individual cave collapses, a small surface depression called a **doline** is formed. **Shakeholes** are smaller doline-like features found in the northern Pennines where glacial material has subsided into underground cavities (Figure 8.8). In the former Yugoslavia, where the term 'karst' originated, huge depressions called **poljes** may have formed in a similar way. Poljes may be up to 400 km^2 in area. In the tropics, the landscape may be composed of either cone-shaped hills and polygonal depressions known as 'cockpit country' (e.g. Jamaica) or tall isolated 'towers' rising from wide plains (e.g. near Guilin, China – Places 26).

4   **Underground depositional features** Groundwater may become saturated with calcium bicarbonate, which is formed by the chemical reaction between carbonic acid in rainwater and calcium carbonate in the rock. However, when this 'hard' water reaches a cave, much of the carbon dioxide bubbles out of solution back into the air – i.e. the process of carbonation in reverse. Aided by the loss of some moisture by evaporation, calcium carbonate (calcite) crystals are subsequently precipitated. Water dripping from the ceiling of the cave initially forms pendant soda straws which, over a very long period of time, may grow into icicle-shaped **stalactites** (Figure 8.6). Experiments in Yorkshire caves suggest that stalactites grow at about 7.5 mm per year. As water drips onto the floor, further deposits of calcium carbonate form the more rounded, cone-shaped **stalagmites** which may, in time, join the stalactites to give **pillars**.

Figure 8.6

Stalactites, stalagmites and pillars, Carlsbad Caverns, New Mexico, USA

Figure 8.7

The karst towers of Guilin, south China

The limestones that outcrop near Guilin have formed a unique karst landscape. The massively bedded, crystalline rock, which in places is 300 m thick, has been slowly pushed upwards from its seabed origin by the same tectonic movements that formed the Himalayas and the Tibetan Plateau far to the west. The heavy summer monsoon rain, sometimes exceeding 2000 mm, has led to rapid fluvial erosion by such rivers as the Li Jiang (Li River). The availability of water together with the high sub-tropical temperatures (Guilin is at 25°N) encourage highly active chemical weathering (solution–carbonation, page 43).

The result has been the formation of a landscape which for centuries has inspired Chinese artists and, recently, has attracted growing numbers of tourists. To either side of the river are natural domes and towers, some of which rise almost vertically 150 m from surrounding paddy fields (Figure 8.7), giving the valley its gorge-like profile. Caves, visible on the sides of the towers, were formed by underground tributaries to the Li Jiang when the main river was flowing at levels considerably higher than those of today.

Limestone covers some 300 000 km² of China – an area larger than that of the UK. Its scenery is seen at its most spectacular in the Three Gorges section of the Yangtze River and where it forms the karst towers in the Guilin region of Guangxi Province.

Figure 8.8

Characteristic features of Carboniferous limestone (karst) scenery

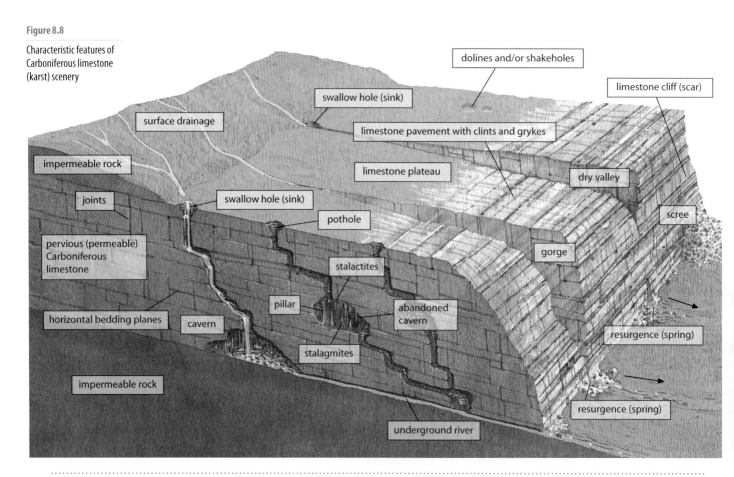

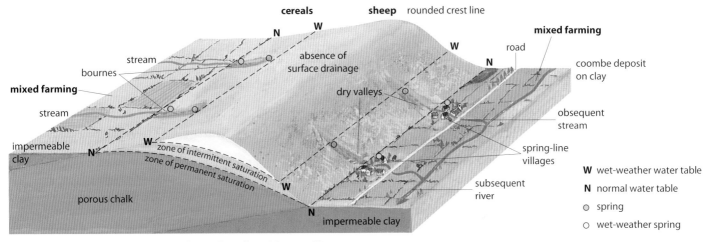

escarpment or cuesta

dip slope · scarp slope · clay vale

cereals · sheep · rounded crest line

mixed farming

N · W

absence of surface drainage

W · N · road · coombe deposit on clay

stream

bournes

dry valleys · W

mixed farming

stream

impermeable clay

zone of intermittent saturation
zone of permanent saturation

porous chalk

W

obsequent stream

spring-line villages

subsequent river

impermeable clay

W wet-weather water table
N normal water table
○ spring
○ wet-weather spring

Figure 8.9

Scarp and vale scenery: an idealised section through a chalk escarpment in south-east England

Economic value of Carboniferous limestone

Human settlement on this type of rock is usually limited and dispersed (page 397) due to limited natural resources, especially the lack of water and good soil. Villages such as Castleton (Derbyshire) and Malham (Yorkshire) have grown up near to a resurgence.

Limestone is often quarried as a raw material for the cement and steel industries or as ornamental stone, but the resultant scars have led to considerable controversy (Case Study 8). The conflict is between the economic advantages of extracting a valuable raw material and providing local jobs, versus the visual eyesore, noise, dust and extra traffic resulting from the operations, e.g. the Hope valley, Derbyshire.

Farming is hindered by the dry, thin, poorly developed soils for, although most upland limestone areas of Britain receive high rainfall totals, water soon flows underground. The rock does not readily weather into soil-forming particles, such as clay or sand, but is dissolved and the residue is then leached (page 261). On hard limestones, rendzina soils may develop (page 274). These soils are unsuitable for ploughing and their covering of short, coarse, springy grasses favours only sheep grazing. In the absence of hedges and trees, drystone walls were commonly built

as field boundaries. The scenery attracts walkers and school parties, while underground features lure cavers, potholers and **speleologists** (scientists who study caves).

Chalk

Chalk, in contrast to Carboniferous limestone scenery, consists of gently rolling hills with rounded crest lines. Typically, chalk has steep, rather than gorge-like, dry valleys and is rarely exposed on the surface (Figure 8.9).

The most distinctive feature of chalk is probably the **escarpment**, or **cuesta**, e.g. the North Downs and South Downs (Figures 8.10 and 14.4). Here the chalk, a pure form of limestone, was gently tilted by the earth movements associated with the collision of the African and Eurasian Plates. Subsequent erosion has left a steep scarp slope and a gentle dip slope. In south-east England, clay vales are found at the foot of the escarpment (Figure 3.51b).

Although chalk – like Carboniferous limestone – has little surface drainage, apart from rivers like the Test and Itchen, its surface is covered in numerous dry valleys (Figure 8.11). Given that chalk can absorb and allow rainwater to percolate through it, how could these valleys have formed?

Figure 8.10

South Downs chalk escarpment, Poynings, Sussex

Figure 8.11

A dry valley in chalk: Devil's Dyke, South Downs, Sussex

Goudie lists 16 different hypotheses that have been put forward regarding the origins of dry valleys. These he has grouped into three categories:

1 **Uniformitarian** These hypotheses assume that there have been no major changes in climate or sea-level and that 'normal' – i.e. fluvial – processes of erosion have operated without interruption. A typical scenario would be that the drainage system developed on impermeable rock overlying the chalk, and subsequently became superimposed upon it (page 85).

2 **Marine** These hypotheses are related to relative changes in sea-level or base level (page 81). One, which has a measure of support, suggests that when sea-levels rose eustatically at the end of the last ice age (page 123), water tables and springs would also have risen. Later, when the base level fell, so too did the water table and spring line, causing valleys to become dry.

3 **Palaeoclimatic** This group of hypotheses, based on climatic changes during and since the ice age, is the most widely accepted. One hypothesis claims that under periglacial conditions any water in the pore spaces would have been frozen, causing the chalk to behave as an impermeable rock (page 135). As temperatures were low, most precipitation would fall as snow. Any meltwater would have to flow over the surface, forming valleys that are now relict landforms (Figure 8.11).

An alternative hypothesis stems from occasions when places receive excessive amounts of rainfall and streams temporarily reappear in dry valleys. Climatologists have shown that there have been times since the ice age when rainfall was considerably greater than it is today. Figure 8.9 shows the normal water table with its associated spring line. If there is a wetter than average winter, or longer period, when moisture loss through evaporation is at its minimum, then the level of permanent saturation will rise. Notice that the wet-weather water table causes a rise in the spring line and so seasonal rivers, or **bournes**, will flow in the normally dry valleys. Remember also that there will be a considerable lag time (Figure 3.5 and page 61) between the peak rainfall and the time when the bournes will begin to flow (throughflow rather than surface runoff on chalk). The springs are the source of obsequent streams (page 84).

The presence of coombe deposits, resulting from solifluction (pages 47 and 135), also links chalk landforms with periglacial conditions.

Economic value of chalk

The main commercial use of chalk is in the production of cement, but there are objections on environmental grounds to both quarries and the processing works. Settlement tends to be in the form of nucleated villages strung out in lines along the foot of an escarpment, originally to take advantage of the assured water supply from the springs (Figures 8.9, 8.10 and 14.4). Water-storing chalk aquifers have long been used as a natural, underground reservoir by inhabitants of London. Despite recent increases in demand for this artesian water, the water table under London has actually risen in recent decades.

Chalk weathers into a thin, dry, calcareous soil with a high pH. Until this century, the springy turf of the Downs was mainly used to graze sheep and to train race horses. Horse racing is still important locally, as at Epsom and Newmarket, but much of the land has been ploughed and converted to the growing of wheat and barley. In places, the chalk is covered by a residual deposit of **clay-with-flints** which may have been an insoluble component of the chalk or may have been left from a former overlying rock. This soil is less porous and more acidic than the calcareous soil and several such areas are covered by beech trees – or were, before the violent storm of October 1987 (Places 29, page 232). Flint has been used as a building material and was the major source for Stone Age tools and weapons.

Figure 8.12

Bedding planes in Old Red Sandstone, Old Man of Hoy, Orkney

Sandstone

Sandstone is the most common rock in Britain. It is a sedimentary rock composed mainly of grains of quartz, and occasionally feldspar and even mica, which have been compacted by pressure and cemented by minerals such as calcite and silica. This makes it a more coherent and resistant, but less porous, rock than sands. The sands, before compaction, may have been deposited in either **a** shallow seas, **b** estuaries and deltas, or **c** hot deserts. The presence of bedding planes (Figure 8.12) indicates the laying down of successive layers of sediment. Sandstone can vary in colour from dark brown or red through to yellow, grey and white (Figure 6.52), depending on the degree of oxidation or hydration (page 42). Like limestone (page 196), sandstone has formed in several geological periods (Figure 8.13), of which perhaps the most significant have been the following:

- The **Devonian**, or **Old Red Sandstone** (Figure 1.1), when sand was deposited in a shallow sea which covered present-day south-west England, South Wales and Herefordshire. These deposits, which were often massively bedded, were contorted and uplifted by subsequent earth movements. Landforms, indicative of an often resistant rock, vary from spectacular coastal cliffs to the plateau-like Exmoor, the north-facing scarp slope of the Brecon Beacons and the flatter lowlands of Herefordshire.
- The **Carboniferous** period, during part of which **Millstone Grit** was formed under river delta conditions. This is a darker, coarser and more resistant rock interbedded with shales. In the southern Pennines it can form either a plateau (Kinder Scout) or steep escarpments (Stanage Edge).
- The **Permian**, or **New Red Sandstone**, when sand was deposited under hot desert conditions, often in shallow water (i.e. when Britain lay in the latitude of the present-day Sahara). The rock is red, due to oxidation, and, being less resistant than the Old Red Sandstone, tends to form valleys (Exe and Eden) or low-lying hills (English Midlands).

Economic value of sandstone

Sandstone is the most common building material in Britain. In the past it was often used as stone for castles and cathedrals and, later, converted into brick for housing. Much of the New Red Sandstone has weathered into a warm, red, light and easily worked soil of high agricultural value, in contrast to the Old Red Sandstone which, being

Figure 8.13

Geological periods of various British sandstones

Geological period/ epoch	Type of sandstone	Examples: location in the UK
Post-Eocene	See Figure 1.1	
Eocene		London and Hampshire basins
Cretaceous	Greensand	The Weald (south-east England)
Jurassic		
Triassic	Bunter and Keuper sandstone	English Midlands, Cheshire
Permian	New Red Sandstone	Exe and Eden valleys, south Arran
Carboniferous	Millstone Grit	Southern Pennines
Devonian	Old Red Sandstone	South-west England, South Wales, Herefordshire, central and north-east Scotland
Silurian, Ordovician and Cambrian		
Pre-Cambrian	Torridon	Wester Ross, Scotland

more resistant, weathers to form uplands that have largely been left as moorland. Millstone Grit areas provided grindstones for Sheffield's cutlery industry in the past, and today these areas are popular for walking, rock-climbing, grouse moors and reservoirs.

Granite

Granite was formed when magma was intruded into the Earth's crust. Initially, as on Dartmoor and in northern Arran, the magma created deep-seated, dome-shaped batholiths (page 29). Since then the rock has been exposed by various processes of weathering and erosion. Having been formed at a depth and under pressure, the rate of cooling was slow and this enabled large crystals of quartz, mica and feldspar to form. As the granite continued to cool, it contracted and a series of cracks were created vertically and horizontally, at irregular intervals. These cracks may have been further enlarged, millions of years later, by pressure release as overlying rocks were removed (Figure 8.14).

The coarse-grained crystals render the rock non-porous but, although many texts quote granite as an example of an impermeable rock, water can find its way along the many cracks making some areas permeable. Despite this, most granite areas usually have a high drainage density and, as they occur in upland parts of Britain which have a high rainfall, they are often covered by marshy terrain.

Although a hard rock, granite is susceptible to both physical and chemical weathering. The joints, which can hold water, are widened by frost shattering (page 40), while the different rates of expansion and cooling of the various minerals within the rock cause granular disintegration (page 41). The feldspar and, to a lesser extent, mica can be changed chemically by hydrolysis (page 42). This means that calcium, potassium, sodium, magnesium and, if the pH is less than 5.0, iron and aluminium, are released from the chemical structure. Where the feldspar is changed near to the surface it forms a whitish clay called **kaolinite**. Where the change occurs at a greater depth (perhaps due to hydrothermal action), it produces **kaolin**. Quartz, which is not affected by chemical weathering, remains as loose crystals (Figure 2.7).

The most distinctive granite landform in temperate countries is the **tor** (Figure 8.14) and, in tropical regions, the **inselberg** (Figures 2.3 and 7.6). There are two major theories concerning their formation, based on physical and chemical weathering respectively. Both, however, suggest the removal of material by solifluction and hence lead to the opinion that tors and inselbergs are relict features.

The first hypothesis suggests that blocks of exposed granite were broken up, subaerially, by frost shattering during periglacial times. The weathered material was then moved downhill by solifluction to leave the more resistant rock upstanding on hill summits and valley sides.

The second, proposed by D.L. Linton, suggests that joints in the granite were widened by sub-surface chemical weathering (Figure 8.15). He suggested that deep weathering occurred during the warm Pliocene period (Figure 1.1) when rainwater penetrated the still-unexposed granite. As the joints widened, roughly rectangular blocks or core-stones were formed. The weathered rock is believed to have been removed by solifluction during periglacial times to leave outcrops of granite tors, separated by shallow depressions. The spacing of the joints is believed to be critical in tor formation: large, resistant core-stones have been left where joints were spaced far apart; where they were closely packed and weathering was more active, clay-filled depressions have developed. The rounded nature of the core-stones (Figure 8.15), especially in tropical regions, is caused by **spheroidal weathering**, a form of exfoliation (page 41).

Figure 8.14

Hound Tor, Dartmoor

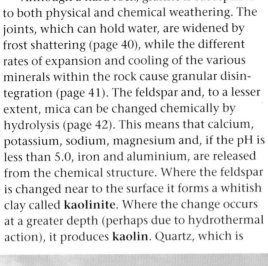

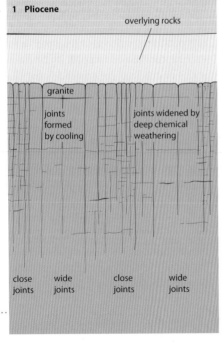
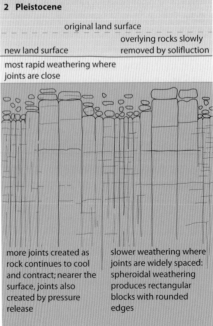
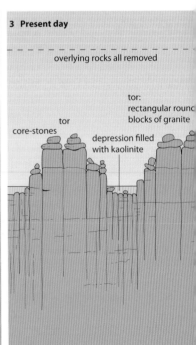

Figure 8.15

The formation of tors (*after* D.L. Linton)

1 Pliocene

overlying rocks

granite

joints formed by cooling

joints widened by deep chemical weathering

close joints | wide joints | close joints | wide joints

2 Pleistocene

original land surface

overlying rocks slowly removed by solifluction

new land surface

most rapid weathering where joints are close

more joints created as rock continues to cool and contract; nearer the surface, joints also created by pressure release

slower weathering where joints are widely spaced: spheroidal weathering produces rectangular blocks with rounded edges

3 Present day

overlying rocks all removed

tor: rectangular round blocks of granite

tor

core-stones

depression filled with kaolinite

Economic value of granite

As a raw material, granite can be used for building purposes; Aberdeen, for example, is known as 'the granite city'. Kaolin, or china clay, is used in the manufacture of pottery. Peat, which overlies large areas of granite bedrock, is an acidic soil which is often severely gleyed (page 275) and saturated with water, forming blanket bogs. The resultant heather-covered moorland is often unsuitable for farming but provides ideal terrain for grouse, and for army training. With so much surface water and heavy rainfall, granite areas provide ideal sites for reservoirs. Tors, such as Hound Tor on Dartmoor (Figure 8.14), may become tourist attractions, but granite environments tend to be inhospitable for settlement.

Basalt

Unlike granite, basalt formed on the Earth's surface, usually at constructive plate margins. The basic lava, on exposure to the air, cooled and solidified very rapidly. The rapid cooling produced small, fine-grained crystals and large cooling cracks which, at places like the Giant's Causeway in Northern Ireland (Figure 1.27) and Fingal's Cave on the Isle of Staffa, are characterised by perfectly shaped hexagonal, columnar jointing. Basalt can be extruded from either fissures or a central vent (page 25). When extruded from fissures, the lava often covers large areas of land – hence the term flood basalts – to produce flat plateaus such as the Deccan Plateau in India and the Drakensbergs in South Africa. Successive eruptions often build upwards to give, sometimes aided by later erosion, stepped hillsides beneath flat, tabular summits (e.g. the Drakensbergs, Lanzarote and Antrim). When extruded from a central vent, the viscous lava produces gently sloping shield volcanoes (Figure 1.22b). Shield volcanoes can reach considerable heights – Mauna Loa (Hawaii) rises over 9000 m from the Pacific seabed making it, from base to summit, the highest mountain on Earth.

Economic value of basalt

Basaltic landforms can sometimes be monotonous, such as places covered in flood basalts, and sometimes scenic and spectacular, as the Giant's Causeway, the Hawaiian volcanoes and the Iguaçu Falls in Brazil (Places 11, page 76). Basaltic lava can weather relatively quickly into a deep, fertile soil as on the Deccan in India and in the coffee-growing region of south-east Brazil. It can also be used for road foundations.

Quarrying in northern India

The 1960s

Dehra Dun, the main town in the Dun Valley with a population exceeding 400 000, is situated in the foothills of the Himalayas some 200 km north of Delhi (Figure 8.16). Until the 1960s, the rich soil of the valley allowed farmers to produce high-quality basmati rice, and the lush green forest surrounding the town had been used sustainably by local people for centuries. That changed in the 1960s when several large quarries were allowed to open up in the valley without any regard for either the inhabitants of the area or the environment.

As India's economy grew, there was increasing conflict between development and the environment. The extraction of rocks and minerals was necessary to provide the new manufacturing industries with raw materials and to provide people with jobs, but mining and quarrying can be very damaging to the environment and to fragile ecosystems. The limestone that was quarried in the Dun Valley was either crushed and used in India's steel industry or used for road building, concrete and whitewash.

Figure 8.16

Location of Dehra Dun

Figure 8.17

Air pollution resulting from quarrying in India

Figure 8.18

Conservation methods include working on flat terraces to stop boulders and waste material sliding downhill, and replanting areas where quarrying has finished

The effects

- As new quarries developed, many of the trees growing on the hillsides were removed. Steep hillsides and deforestation in an area with a monsoon climate (page 239) meant that when the heavy summer rains fell, the soil was seriously eroded. Surface runoff led to the fertile soils being covered in debris and caused landslides, especially where unstable quarry waste had been dumped. Deforestation also meant there was less fuelwood for people living in nearby villages.
- Material carried downhill often ended up in rivers, where it not only polluted water supplies but also blocked the river with boulders and waste. Before quarrying began, one bridge had an arch nearly 20 m above the river, but after quarrying it was reduced to less than 5 m.
- Before quarrying, settlements in the area had an all-year supply of clean water obtained from springs and resurgences formed when underground rivers in the limestone

hills re-appeared at the surface (page 197). The increase in surface runoff due to quarrying and deforestation caused the water table to fall by 5 m in seven years. This meant that Dehra Dun often received water for only a few hours a day. Without enough water to irrigate their fields, local farmers were unable to provide enough food for their families.

- The blasting of rock created noise and air pollution and caused nearby buildings to vibrate.
- The trucks and lorries – many old and badly maintained – that transported the limestone down the steep, narrow roads caused the road surface to break up, released poisonous fumes and created more dust (Figure 8.17).
- The kilns that processed the limestone also added to the air pollution.

Local protests

In the 1980s, many local people grouped together to form the 'Friends of Dun'. The group, led mainly by wealthy and influential business and retired people, submitted a

petition to the Supreme Court which led, in 1988, to all the quarries (with one exception) being closed down. By the end of the 20th century, trees planted by school children and local people had begun to mature into forest, although farmers still found much of their soil unusable.

Should the one quarry remain open?

The Supreme Court allowed one quarry to operate until its lease ran out. This was partly because the quarry provided hundreds of jobs for local people, although they were poorly paid, and partly because the quarry owners attempted to implement conservation techniques, such as working on flat terraces to stop boulders and waste material sliding downhill (Figure 8.18) and replanting areas where quarrying had finished. The argument now appears to be between the wealthy conservation group who want to protect and restore the Dun Valley and the poorer workers who, without the quarry and with few alternative jobs available, would have no income if it closed.

Further reference

Goudie, A.S. (2001) *The Nature of the Environment*, WileyBlackwell.

 Michigan Karst Conservancy Group:
www.caves.org/conservancy/mkc/
michigan_karst_conservancy.htm

Pretoria Portland Cement Co. Ltd:
www.ppc.co.za

Activities

1. **a** Describe the characteristics of each of the following rock types in terms of chemical composition, rock structure and origin: Carboniferous limestone; chalk; granite; basalt. *(12 marks)*

   b Choose **one** of the rock types in **a** and draw an annotated diagram to identify the characteristic landscape features associated with it. *(9 marks)*

   c For **each** of the rock types identified in **a**, suggest one reason why it may be of value as a resource for human use. *(4 marks)*

2. Study the OS map extract of the area around Malham in Figure 8.19.

   a i How high above sea-level is the minor road at GR 907649? *(1 mark)*

   ii What is the feature at GR 906655? *(1 mark)*

   b Identify and give grid references for **two** pieces of evidence that large parts of this area have limestone rock outcropping at the surface.

   Justify **each** of your choices. *(6 marks)*

   c Explain, using one or more diagrams, why there are large areas of bare flat rock in the area shown on the map extract. *(6 marks)*

   d A stream flows from the edge of the map at 893661 to 894657. South of this point is a dry valley. Suggest why this dry valley is here. *(6 marks)*

   e Farming in this area has been described as 'marginal; it could not exist without subsidies'.

   Suggest why the physical geography makes farming so difficult. *(5 marks)*

3. **a** Making good use of annotated diagrams, describe the surface features of a chalk cuesta. *(6 marks)*

   b Describe and explain the location of the water table within an area of chalk hills. *(6 marks)*

   c Describe and suggest reasons for the location of settlements close to the foot of a chalk cuesta. *(4 marks)*

   d Suggest two reasons why some chalk downs have prehistoric carved figures on them. *(4 marks)*

   e Chalk escarpments may have 'hangers' (areas of beech woodland on the brow of the scarp). Suggest why these woodlands are found here. *(5 marks)*

Exam practice: structured questions

4. Study the OS map extract of the area around Malham in Figure 8.19.

   a Identify and locate two pieces of evidence to suggest that this area is limestone rock. For **each** explain how the evidence shows it to be limestone. *(10 marks)*

   b Why is there so much settlement and other ancient remains visible in an area such as this? *(7 marks)*

   c This area is both a tourist area and a working farming area. Identify **one** way these two land uses are in conflict and explain the reasons for this conflict. *(8 marks)*

5. **a i** Explain how areas of granite rock, such as Dartmoor, were formed. *(6 marks)*

   ii Describe the processes of weathering in granite. *(6 marks)*

   b 'Granite tors form as a result of the nature and structure of the rock and the nature of the weathering processes.'

   Describe a typical granite tor and explain its formation. *(13 marks)*

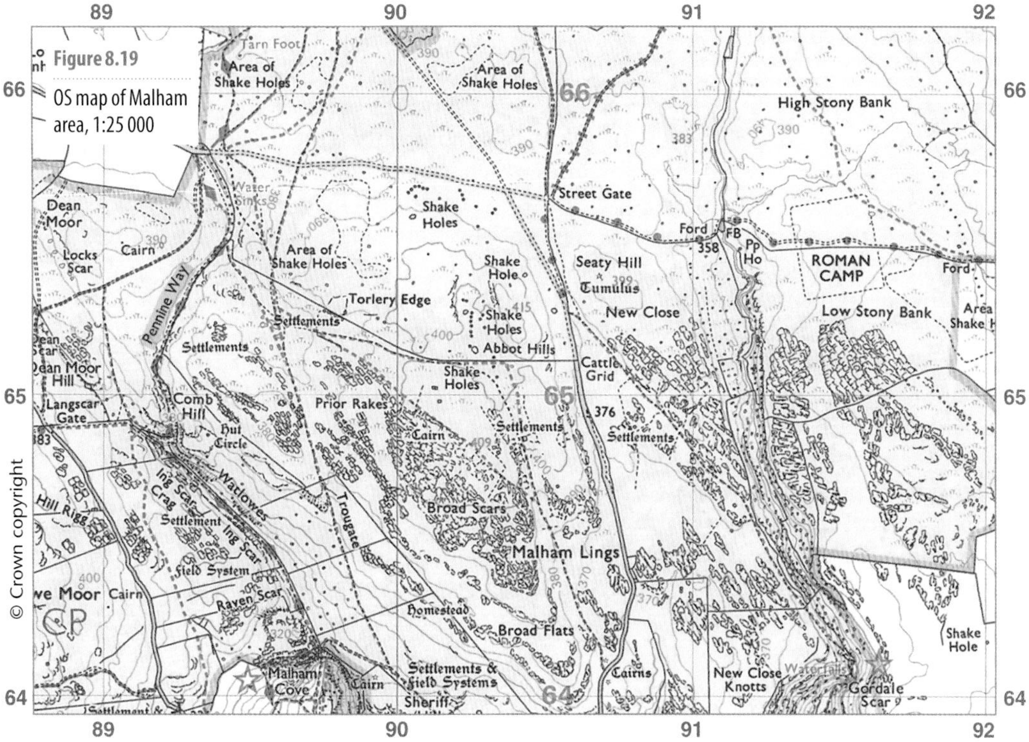

Exam practice: essay

6. Should the quarry in the Dun valley be allowed to remain open?

   Present the arguments for **two** groups that think the quarry should continue and for **two** groups that think it should be closed.

   Then present a conclusion taking the arguments of both sides into consideration. *(25 marks)*

Figure 8.19 OS map of Malham area, 1:25 000

© Crown copyright

9
Weather and climate

• •

'There is really no such thing as bad weather, only different types of good weather.'

John Ruskin, Quote from Lord Avebury

'When two Englishmen meet, their first talk is of the weather.'

Samuel Johnson, *The Idler*

The science of meteorology is the study of atmospheric phenomena; it includes the study of both weather and climate. The distinction between climate and weather is one of scale. **Weather** refers to the state of the atmosphere at a local level, usually on a short timescale of minutes to months. It emphasises aspects of the atmosphere that affect human activity, such as sunshine, cloud, wind, rainfall, humidity and temperature. **Climate** is concerned with the long-term behaviour of the atmosphere in a specific area. Climatic characteristics are represented by data on temperature, pressure, wind, precipitation, humidity, etc. which are used to calculate daily, monthly and yearly averages (Framework 8, page 246) and to build up global patterns (Chapter 12).

Structure and composition of the atmosphere

The atmosphere is an envelope of transparent, odourless gases held to the Earth by gravitational attraction. While the furthest limit of the atmosphere is said by international convention to be at 1000 km, most of the atmosphere, and therefore our climate and weather, is concentrated within 16 km of the Earth's surface at the Equator and 8 km at the poles. Fifty per cent of atmospheric mass is within 5.6 km of sea-level and 99 per cent is within 40 km. Atmospheric pressure decreases rapidly with height but, as recordings made by radiosondes, weather balloons and more recently weather satellites have shown, temperature changes are more complex. Changes in temperature mean that the atmosphere can be conveniently divided into four distinctive layers

(Figure 9.1); moving outwards from the Earth's surface:

1 **Troposphere** Temperatures in the troposphere decrease by 6.4°C with every 1000 m increase in altitude (environmental lapse rate, page 216). This is because the Earth's surface is warmed by incoming solar radiation which in turn heats the air next to it by conduction, convection and radiation. Pressure falls as the effect of gravity decreases, although wind speeds usually increase with height. The layer is unstable and contains most of the atmosphere's water vapour, cloud, dust and pollution. The tropopause, which forms the upper limit to the Earth's climate and weather, is marked by an isothermal layer where temperatures remain constant despite any increase in height.

2 **Stratosphere** The stratosphere is characterised by a steady increase in temperature (temperature inversion, page 217) caused by a concentration of **ozone** (O_3) (Places 27, page 209). This gas absorbs incoming **ultra-violet (UV) radiation** from the sun. Winds are light in the lower parts, but increase with height; pressure continues to fall and the air is dry. The stratosphere, like the two layers above it, acts as a protective shield against meteorites which usually burn out as they enter the Earth's gravitational field. The **stratopause** is another isothermal layer where temperatures do not change with increasing height.

3 **Mesosphere** Temperatures fall rapidly as there is no water vapour, cloud, dust or ozone to absorb incoming radiation. This layer experiences the atmosphere's lowest temperatures (–90°C) and strongest winds (nearly 3000 km/hr). The **mesopause**, like the tropopause and stratopause, shows no change in temperature.

4 **Thermosphere** Temperatures rise rapidly with height, perhaps to reach 1500°C. This is due to an increasing proportion of atomic oxygen in the atmosphere which, like ozone, absorbs incoming ultra-violet radiation.

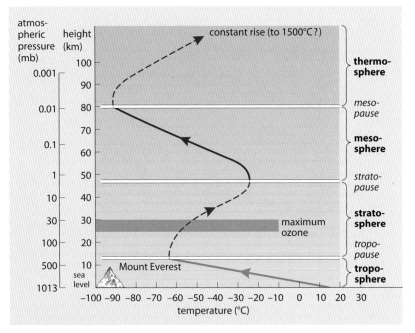

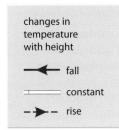

changes in
temperature
with height

→ fall

— constant

--▶-- rise

Figure 9.1

The vertical structure of
the atmosphere

Atmospheric gases

The various gases which combine to form the
atmosphere are listed in Figure 9.2. Of these,
nitrogen and oxygen together make up 99 per cent
by volume. Of the others, water vapour (lower
atmosphere), ozone (O_3) (upper atmosphere) and
carbon dioxide (CO_2) have an importance far
beyond their seemingly small amounts. It is the
depletion of O_3 (Places 27) and the increase in
CO_2 (Case Study 9B) which are causing concern
to scientists.

Energy in the atmosphere

The sun is the Earth's prime source of energy. The
Earth receives energy as incoming **short-wave**
solar radiation (also referred to as **insolation**). It
is this energy that controls our planet's climate
and weather and which, when converted by photosynthesis in green plants, supports all forms of
life. The amount of incoming radiation received
by the Earth is determined by four astronomical
factors (Figure 9.3): the solar constant, the distance from the sun, the altitude of the sun in
the sky, and the length of night and day. Figure
9.3 is theoretical in that it assumes there is no
atmosphere around the Earth. In reality, much
insolation is absorbed, reflected and scattered as it
passes through the atmosphere (Figure 9.4).

   Absorption of incoming radiation is mainly by
ozone, water vapour, carbon dioxide and particles
of ice and dust. It occurs in, and is limited to, the
infra-red part of the spectrum. Clouds and, to a
lesser extent, the Earth's surface **reflect** considerable amounts of radiation back into space. The
ratio between incoming radiation and the amount
reflected, expressed as a percentage, is known as
the **albedo**. The albedo varies with cloud type from
30–40 per cent in thin clouds, to 50–70 per cent in
thicker stratus and 90 per cent in cumulo-nimbus
(when only 10 per cent reaches the atmosphere
below cloud level). Albedos also vary over different land surfaces, from less than 10 per cent over

Figure 9.2

The composition of the atmosphere

Gas		Percentage by volume	Importance for weather and climate	Other functions/source
Permanent gases:	nitrogen	78.09	Mainly passive	Needed for plant growth.
	oxygen	20.95		Produced by photosynthesis; reduced by deforestation.
Variable gases:	water vapour	0.20–4.0	Source of cloud formation and precipitation, reflects/absorbs incoming long-wave radiation. Keeps global temperatures constant. Provides majority of natural 'greenhouse effect'.	Essential for life on Earth. Can be stored as ice/snow.
	carbon dioxide	0.03	Absorbs long-wave radiation from Earth and so contributes to 'greenhouse effect'. Its increase due to human activity is a major cause of global warming.	Used by plants for photosynthesis; increased by burning fossil fuels and by deforestation.
	ozone	0.00006	Absorbs incoming short-wave ultra-violet radiation.	Reduced/destroyed by chlorofluorocarbons (CFCs).
	pollutants	trace	Sulphur dioxide, nitrogen oxide, methane. Absorb long-wave radiation, cause acid rain and contribute to the greenhouse effect.	From industry, power stations and car exhausts.
Inert gases:	argon	0.93		
	helium, neon, krypton	trace		
Non-gaseous:	dust	trace	Absorbs/reflects incoming radiation. Forms condensation nuclei necessary for cloud formation.	Volcanic dust, meteoritic dust, soil erosion by wind.

Note: the figures refer to dry air and so the variable amount of water vapour is not usually taken into consideration.

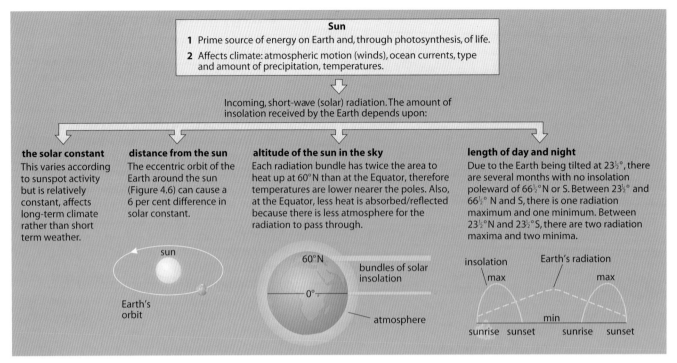

Sun

1 Prime source of energy on Earth and, through photosynthesis, of life.

2 Affects climate: atmospheric motion (winds), ocean currents, type and amount of precipitation, temperatures.

Incoming, short-wave (solar) radiation. The amount of insolation received by the Earth depends upon:

the solar constant
This varies according to sunspot activity but is relatively constant, affects long-term climate rather than short term weather.

distance from the sun
The eccentric orbit of the Earth around the sun (Figure 4.6) can cause a 6 per cent difference in solar constant.

altitude of the sun in the sky
Each radiation bundle has twice the area to heat up at 60°N than at the Equator, therefore temperatures are lower nearer the poles. Also, at the Equator, less heat is absorbed/reflected because there is less atmosphere for the radiation to pass through.

length of day and night
Due to the Earth being tilted at 23½°, there are several months with no insolation poleward of 66½°N or S. Between 23½° and 66½° N and S, there is one radiation maximum and one minimum. Between 23½°N and 23½°S, there are two radiation maxima and two minima.

Figure 9.3

Incoming radiation received by the Earth (assuming that there is no atmosphere)

oceans and dark soil, to 15 per cent over coniferous forest and urban areas, 25 per cent over grasslands and deciduous forest, 40 per cent over light-coloured deserts and 85 per cent over reflecting fresh snow. Where deforestation and overgrazing occur, the albedo increases. This reduces the possibility of cloud formation and precipitation and increases the risk of **desertification** (Case Study 7). **Scattering** occurs when incoming radiation is diverted by particles of dust, as from volcanoes and deserts, or by molecules of gas. It takes place in all directions and some of the radiation will reach the Earth's surface as **diffuse** radiation.

As a result of absorption, reflection and scattering, only about 24 per cent of incoming radiation reaches the Earth's surface directly, with a further 21 per cent arriving at ground-level as diffuse radiation (Figure 9.4). Incoming radiation is converted into heat energy when it reaches the Earth's surface. As the ground warms, it radiates energy back into the atmosphere where 94 per cent is absorbed (only 6 per cent is lost to space), mainly by water vapour and carbon dioxide – the greenhouse effect (Case Study 9B). Without the natural greenhouse effect, which traps so much of the outgoing radiation, world temperatures would be 33°C lower than they are at present and life on Earth would be impossible. (During the ice age, it was only 4°C cooler.) This outgoing (terrestrial) radiation is **long-wave** or **infra-red** radiation.

Figure 9.4

The solar energy cascade

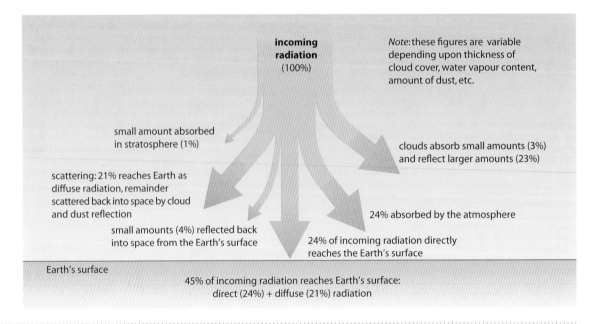

incoming radiation (100%)

Note: these figures are variable depending upon thickness of cloud cover, water vapour content, amount of dust, etc.

small amount absorbed in stratosphere (1%)

clouds absorb small amounts (3%) and reflect larger amounts (23%)

scattering: 21% reaches Earth as diffuse radiation, remainder scattered back into space by cloud and dust reflection

small amounts (4%) reflected back into space from the Earth's surface

24% absorbed by the atmosphere

24% of incoming radiation directly reaches the Earth's surface

Earth's surface

45% of incoming radiation reaches Earth's surface: direct (24%) + diffuse (21%) radiation

The major concentration of ozone is in the stratosphere, 25–30 km above sea-level (Figure 9.1). Ozone acts as a shield protecting the Earth from the damaging effects of ultra-violet (UV) radiation from the sun. An increase in UV radiation means an increase in sunburn and skin cancer (fair skin is at greater risk than dark skin), snow-blindness, cataracts and eye damage, ageing and skin wrinkling in humans, as well as having a major impact on Antarctic organisms.

A depletion in ozone above the Antarctic was first observed, by chance, by the British Antarctic Survey in 1977, and the first 'hole' was described in a scientific paper published in 1985. The term 'hole' is misleading as it means a depletion in ozone of over 50 per cent (not a 100 per cent loss). Each Antarctic spring (September to November) the temperature falls so low that it causes ozone to be destroyed in a chemical reaction with chlorine. At the time there were two main sources of chlorine:

- the release of chlorofluorocarbons (CFCs) from aerosols such as hairsprays, deodorants, refrigerator coolants and manufacturing processes that produced foam packaging (a long-term effect)
- from major volcanic eruptions, e.g. Mount Pinatubo (Case Study 1– a short-term effect).

The 1985 paper was followed by a spate of experiments aimed at trying to establish the causes and probable effects of ozone depletion. Within two years – a remarkably short time for international action –

the Montreal Protocol was signed by which the more industrialised countries agreed to set much lower limits for CFC production, and subsequently to reduce this to zero. The agreement came so quickly, and CFC production dropped so rapidly, that the Montreal Protocol has been held up as a 'model' international environmental agreement.

Initially, ozone depletion continued. The first Arctic 'hole' was observed in 1989 following the coldest-ever recorded January in that region. The 'hole' over Antarctica continued to grow each year until 2003, by which time it had reached its maximum extent and was affecting populated parts of Chile and New Zealand. Since then, mainly due to most of the harmful CFCs having been replaced by gases less toxic to ozone (though still greenhouse gases), there have been encouraging signs of ozone replacement and hopes are high that ozone concentrations will return to normal by the middle or latter part of this century – a rare success story for international environment management.

In contrast, vehicle exhaust systems generate dangerous quantities of ozone close to the Earth's surface, especially during calm summer anticyclonic conditions (page 234). Under extreme conditions, nitrogen oxide from exhausts reacts with VOCs (volatile organic compounds) in sunlight to create a petrochemical smog. This can cause serious damage to the health of people (especially those with asthma) and animals.

The heat budget

Since the Earth is neither warming up nor cooling down, there must be a balance between incoming insolation and outgoing terrestrial radiation. Figure 9.5 shows that:

- there is a net gain in radiation everywhere on the Earth's surface (curve A) except in polar latitudes which have high albedo surfaces
- there is a net loss in radiation throughout the atmosphere (curve B)
- after balancing the incoming and outgoing radiation, there is a net surplus between 35°S and 40°N (the difference in latitude is due to the larger land masses of the northern hemisphere) and a net deficit to the poleward sides of those latitudes (curve C).

This means that there is a **positive heat balance** within the tropics and a **negative heat balance** both at high latitudes (polar regions) and high altitudes. Two major **transfers** of heat, therefore, take place to prevent tropical areas from over-heating (Figure 9.6).

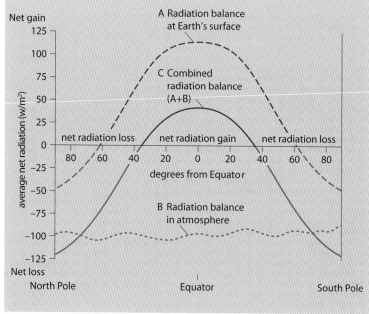

Figure 9.5

The heat budget

1 **Horizontal heat transfers** Heat is transferred away from the tropics, thus preventing the Equator from becoming increasingly hotter and the poles increasingly colder. Winds (air movements including jet streams, page 227; hurricanes, page 235; and depressions, page 230) are responsible for 80 per cent of this heat transfer, and ocean currents for 20 per cent (page 211).

2 **Vertical heat transfers** Heat is also transferred vertically, thus preventing the Earth's surface from getting hotter and the atmosphere colder. This is achieved through **radiation, conduction, convection** and the transfer of **latent heat**. Latent heat is the amount of heat energy needed to change the state of a substance without affecting its temperature. When ice changes into water or water into vapour, heat is taken up to help with the processes of melting and evaporation. This absorption of heat results in the cooling of the atmosphere. When the process is reversed – i.e. vapour condenses into water or water freezes into ice – heat energy is released and the atmosphere is warmed.

Variations in the radiation balance occur at a number of spatial and temporal scales. Regional differences may be due to the uneven distribution of land and sea, altitude, and the direction of prevailing winds. Local variations may result from **aspect** and amounts of cloud cover. Seasonal and diurnal variations are related to the altitude of the sun and the length of night and day.

Global factors affecting insolation

Factors that influence the amount of insolation received at any point, and therefore its radiation balance and heat budget, vary considerably over time and space.

Long-term factors

These are relatively constant at a given point.

■ **Height above sea-level** The atmosphere is not warmed directly by the sun, but by heat radiated from the Earth's surface and distributed by conduction and convection. As the height of mountains increases, they present a decreasing area of land surface from which to heat the surrounding air. In addition, as the density or pressure of the air decreases, so too does its ability to hold heat (Figure 9.1). This is because the molecules in the air which receive and retain heat become fewer and more widely spaced as height increases.

■ **Altitude of the sun** As the angle of the sun in the sky decreases, the land area heated by a given ray and the depth of atmosphere through which that ray has to pass both increase. Consequently, the amount of insolation lost through absorption, scattering and reflection also increases. Places in lower latitudes therefore have higher temperatures than those in higher latitudes.

■ **Land and sea** Land and sea differ in their ability to absorb, transfer and radiate heat energy. The sea is more transparent than the land, and is capable of absorbing heat down to a depth of 10 metres. It can then transfer this heat to greater depths through the movements of waves and currents. The sea also has a greater **specific heat capacity** than that of land. Specific heat capacity is the amount of energy required to raise the temperature of 1 kg of a substance by 1°C, expressed in kilojoules per kg per °C. Expressed in kilocalories, the specific heat capacity of water is 1.0, that of land is 0.5 and that of sand 0.2.

Figure 9.6

Heat transfers in the atmosphere

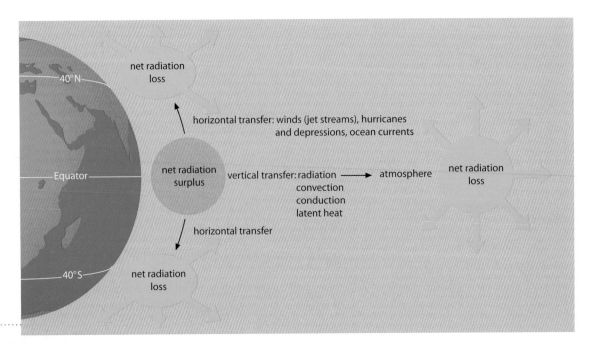

Figure 9.7

Mean annual ranges in global temperature (°C)

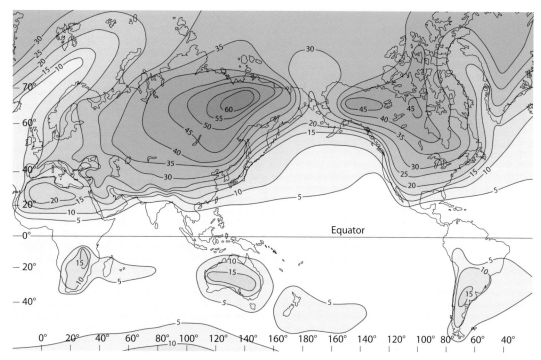

This means that water requires twice as much energy as soil and five times more than sand to raise an equivalent mass to the same temperature. During summer, therefore, the sea heats up more slowly than the land. In winter, the reverse is the case and land surfaces lose heat energy more rapidly than water. The oceans act as efficient 'thermal reservoirs'. This explains why coastal environments have a smaller annual range of temperature than locations at the centres of continents (Figure 9.7).

■ **Prevailing winds** The temperature of the wind is determined by its area of origin and by the characteristics of the surface over which it subsequently blows (Figure 9.8). A wind blowing from the sea tends to be warmer in winter and cooler in summer than a corresponding wind coming from the land.

■ **Ocean currents** These are a major component in the process of horizontal transfer of heat energy. Warm currents carry water polewards and raise the air temperature of the maritime environments where they flow. Cold currents carry water towards the Equator and so lower the temperatures of coastal areas (Figure 9.9).

The main ocean currents follow circular routes – clockwise in the northern hemisphere, anti-clockwise in the southern hemisphere.

Figure 9.10 shows the difference between the mean January temperature of a place and the mean January temperatures of other places with the same latitude; this difference is known as a **temperature anomaly**. (The term 'temperature anomaly' is used specifically to describe temperature differences from a mean. It should not be confused with the more general definition of 'anomaly' which refers to something that does not fit into a general pattern.) For example, Stornoway (Figure 9.10) has a mean January temperature of 4°C, which is 20°C higher than the average for other locations lying at 58°N. Such anomalies result primarily from the uneven heating and cooling rates of land and sea and are intensified by the horizontal transfer of energy by ocean currents and prevailing winds. Remember that the sun appears overhead in the southern hemisphere at this time of year (January) and isotherms have been reduced to sea-level – i.e. temperatures are adjusted to eliminate some of the effects of relief, thus emphasising the influence of prevailing winds, ocean currents and continentality.

Season	SEA	West coast	LAND	East coast	SEA	Season
Winter	*Warm*	warm wind →	**COLD**	cool wind →	*Warm*	Winter
Summer	*Cool*	cool wind →	**WARM**	warm wind →	*Cool*	Summer

Figure 9.9

Major ocean currents

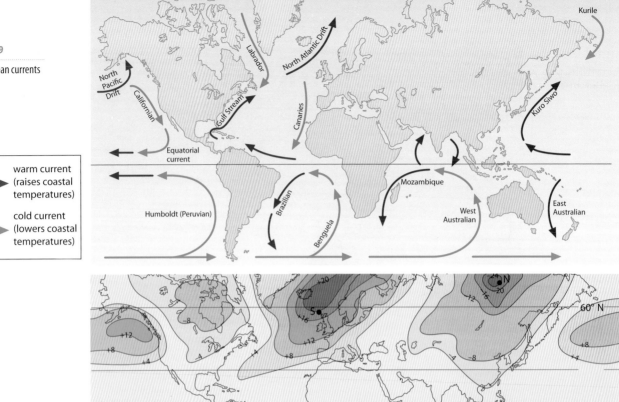

Figure 9.10

Temperature anomalies for January (*after* D.C. Money)

Short-term factors

- **Seasonal changes** At the spring and autumn equinoxes (21 March and 22 September) when the sun is directly over the Equator, insolation is distributed equally between both hemispheres. At the summer and winter solstices (21 June and 22 December) when, due to the Earth's tilt, the sun is overhead at the tropics, the hemisphere experiencing 'summer' will receive maximum insolation.

- **Length of day and night** Insolation is only received during daylight hours and reaches its peak at noon. There are no seasonal variations at the Equator, where day and night are of equal length throughout the year. In extreme contrast, polar areas receive no insolation during part of the winter when there is continuous darkness, but may receive up to 24 hours of insolation during part of the summer when the sun never sinks below the horizon ('the lands of the midnight sun').

Local influences on insolation

- **Aspect** Hillsides alter the angle at which the sun's rays hit the ground (Places 28).

In the northern hemisphere, north-facing slopes, being in shadow for most or all of the year, are cooler than those facing south. The steeper the south-facing slope, the higher the angle of the sun's rays to it and therefore the higher will be the temperature. North- and south-facing slopes are referred to, respectively, as the **adret** and **ubac**.

- **Cloud cover** The presence of cloud reduces both incoming and outgoing radiation. The thicker the cloud, the greater the amount of absorption, reflection and scattering of insolation, and of terrestrial radiation. Clouds may reduce daytime temperatures, but they also act as an insulating blanket to retain heat at night. This means that tropical deserts, where skies are clear, are warmer during the day and cooler at night than humid equatorial regions with a greater cloud cover. The world's greatest diurnal ranges of temperature are therefore found in tropical deserts.

- **Urbanisation** This alters the albedo (page 207) and creates urban 'heat islands' (page 242).

Atmospheric moisture

Water is a liquid compound which is converted by heat into vapour (gas) and by cold into a solid (ice). The presence of water serves three essential purposes:

1 It maintains life on Earth: flora, in the form of natural vegetation (biomes) and crops; and fauna, i.e. all living creatures, including humans.

2 Water in the atmosphere, mainly as a gas, absorbs, reflects and scatters insolation to keep our planet at a habitable temperature (Figure 9.4).

3 Atmospheric moisture is of vital significance as a means of transferring surplus energy from tropical areas either horizontally to polar latitudes or vertically into the atmosphere to balance the heat budget (Figure 9.5). Despite this need for water, its existence in a form readily available to plants, animals and humans is limited. It has been estimated that 97.2 per cent of the world's water is in the oceans and seas; in this form, it is only useful to plants tolerant of saline conditions (**halophytes**, page 291) and to the populations of a few wealthy countries that can afford desalinisation plants (the Gulf oil states).

Approximately 2.1 per cent of water in the hydrosphere is held in storage as polar ice and snow. Only 0.7 per cent is fresh water found either in lakes and rivers (0.1 per cent), as soil moisture and groundwater (0.6 per cent), or in the atmosphere (0.001 per cent).

Places 28 An alpine valley: aspect

Many alpine valleys in Switzerland and Austria have an east–west orientation which means that their valley sides face either north or south. South-facing adret slopes are much warmer and drier than those facing north (Figure 9.11). The south-facing slopes have more plant species, a higher tree-line, and a greater land use with alpine pastures at higher altitudes and fruit and hay lower down; also, they usually provide the best sites for settlement. In contrast, north-facing ubac slopes are snow-covered for a much longer period, they are less suited to farming, the tree-line is lower, and they tend to be left forested. However, on the valley floors, as severe frosts are likely to occur during times of temperature inversion (page 217), sensitive plants and crops do not flourish.

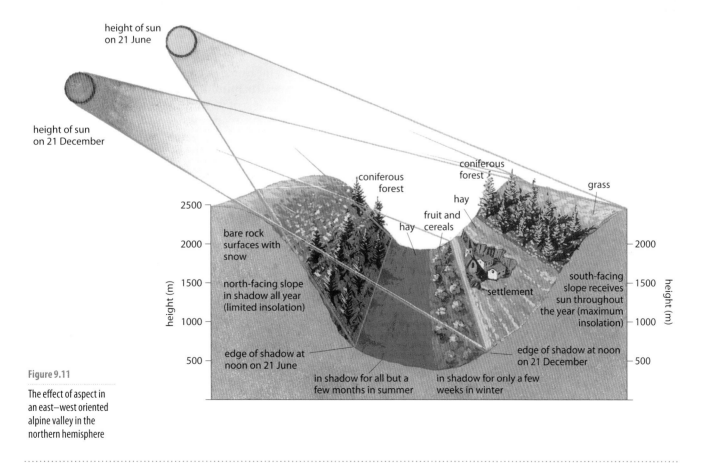

Figure 9.11

The effect of aspect in an east–west oriented alpine valley in the northern hemisphere

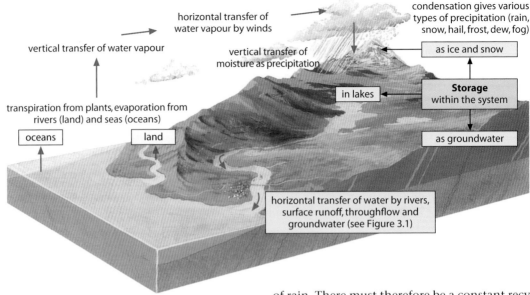

Figure 9.12

The hydrological cycle
(compare with Figure 3.1)

horizontal transfer of
water vapour by winds

vertical transfer of water vapour

condensation gives various
types of precipitation (rain,
snow, hail, frost, dew, fog)

vertical transfer of
moisture as precipitation

as ice and snow

Storage
within the system

in lakes

as groundwater

transpiration from plants, evaporation from
rivers (land) and seas (oceans)

oceans

land

horizontal transfer of water by rivers,
surface runoff, throughflow and
groundwater (see Figure 3.1)

At any given time, the atmosphere only holds, on average, sufficient moisture to give every place on the Earth 2.5 cm (about 10 days' supply) of rain. There must therefore be a constant recycling of water between the oceans, atmosphere and land (Figure 9.13). This recycling is achieved through the **hydrological cycle** (Figure 9.12).

Figure 9.13

The world's water
balance

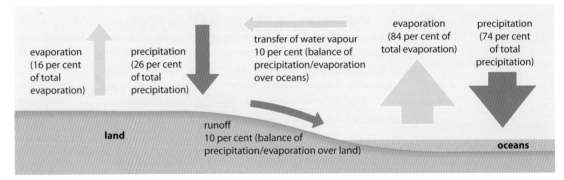

evaporation
(16 per cent
of total
evaporation)

precipitation
(26 per cent
of total
precipitation)

transfer of water vapour
10 per cent (balance of
precipitation/evaporation
over oceans)

evaporation
(84 per cent of
total evaporation)

precipitation
(74 per cent
of total
precipitation)

land

runoff
10 per cent (balance of
precipitation/evaporation over land)

oceans

Humidity

Humidity is a measure of the water vapour content in the atmosphere. **Absolute humidity** is the mass of water vapour in a given volume of air measured in grams per cubic metre (g/m^3). **Specific humidity** is similar but is expressed in grams of water per kilogram of air (g/kg). Humidity depends upon the temperature of the air. At any given temperature, there is a limit to the amount of moisture that the air can hold. When this limit is reached, the air is said to be **saturated**. Cold air can hold only relatively small quantities of vapour before becoming saturated but this amount increases rapidly as temperatures rise (Figure 9.14). This means that the amount of precipitation obtained from warm air is generally greater than that from cold air. **Relative humidity** (RH) is the amount of water vapour in the air at a given temperature expressed as a percentage of the maximum amount of vapour that the air could hold at that temperature. If the RH is 100 per cent, the air is saturated. If it lies between 80 and 99 per cent, the air is said to be 'moist' and the weather is humid or clammy. When the RH drops to 50 per cent, the air is 'dry'– figures as low as 10 per cent have been recorded over hot deserts.

Figure 9.14

Air temperatures and
absolute humidity for
saturated air

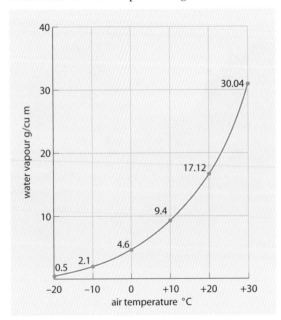

If unsaturated air is cooled and atmospheric pressure remains constant, a critical temperature will be reached when the air becomes saturated (i.e. RH = 100 per cent). This is known as the **dew point**. Any further cooling will result in the condensation of excess vapour, either into water droplets where condensation nuclei are present, or into ice crystals if the air temperature is below 0°C. This is shown in the following worked example.

1 The early morning air temperature was 10°C. Although the air could have held 100 units of water at that temperature, at the time of the reading it held only 90. This meant that the RH was 90 per cent.

2 During the day, the air temperature rose to 12°C. As the air warmed it became capable of holding more water vapour, up to 120 units. Owing to evaporation, the reading reached a maximum of 108 units which meant that the RH remained at 90 per cent – i.e. $(108 \div 120) \times 100$.

3 In the early evening, the temperature fell to 10°C at which point, as stated above, it could hold only 100 units. However, the air at that time contained 108 units so, as the temperature fell, dew point was reached and the 8 excess units of water were lost through condensation.

Condensation

This is the process by which water vapour in the atmosphere is changed into a liquid or, if the temperature is below 0°C, a solid. It usually results from air being cooled until it is saturated. Cooling may be achieved by:

1 **Radiation** (contact) **cooling** This typically occurs on calm, clear evenings. The ground loses heat rapidly through terrestrial radiation and the air in contact with it is then cooled by conduction. If the air is moist, some vapour will condense to form radiation fog, dew, or – if the temperature is below freezing point – hoar frost (page 221).

2 **Advection cooling** This results from warm, moist air moving over a cooler land or sea surface. Advection fogs in California and the Atacama Desert (Places 24, page 180 and page 122) are formed when warm air from the land drifts over cold offshore ocean currents (Figure 9.9).

As both radiation and advection involve horizontal rather than vertical movements of air, the amount of condensation created is limited.

3 **Orographic** and **frontal uplift** Warm, moist air is forced to rise either as it crosses a mountain barrier (orographic ascent, page 220) or when it meets a colder, denser mass of air at a front (page 229).

4 **Convective** or **adiabatic cooling** This is when air is warmed during the daytime and rises in pockets as **thermals** (Figure 9.15). As the air expands, it uses energy and so loses heat and the temperature drops. Because air is cooled by the reduction of pressure with height rather than by a loss of heat to the surrounding air, it is said to be adiabatically cooled (see lapse rates, page 216).

As both orographic and adiabatic cooling involve vertical movements of air, they are more effective mechanisms of condensation.

Condensation does not occur readily in clean air. Indeed, if air is absolutely pure, it can be cooled below its dew point to become **supersaturated** with an RH in excess of 100 per cent. Laboratory tests have shown that clean, saturated air can be cooled to –40°C before condensation or, in this case, **sublimation**. Sublimation is when vapour condenses directly into ice crystals without passing through the liquid state. However, air is rarely pure and usually contains large numbers of condensation nuclei. These microscopic particles, referred to as **hygroscopic nuclei** because they attract water, include volcanic dust (heavy rain always accompanies volcanic eruptions); dust from windblown soil; smoke and sulphuric acid originating from urban and industrial areas; and salt from sea spray. Hygroscopic nuclei are most numerous over cities, where there may be up to 1 million per cm^3, and least common over oceans (only 10 per cm^3). Where large concentrations are found, condensation can occur with an RH as low as 75 per cent – as in the smogs of Los Angeles (Figure 9.25 and Case Study 15A).

Figure 9.15

Convective cooling

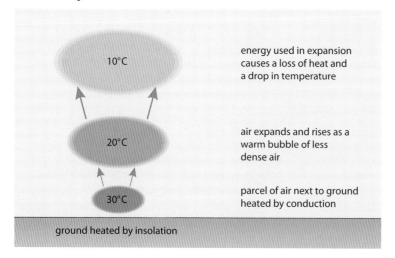

energy used in expansion causes a loss of heat and a drop in temperature

air expands and rises as a warm bubble of less dense air

parcel of air next to ground heated by conduction

10°C

20°C

30°C

ground heated by insolation

Figure 9.16

Examples of lapse rates shown in temperature –height diagrams (tephigrams)

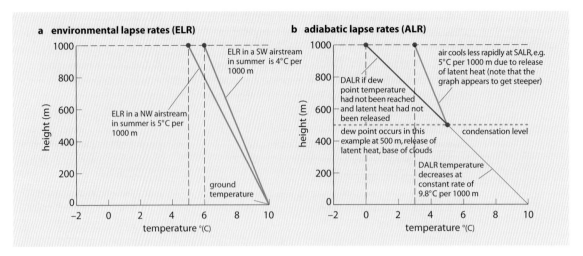

a environmental lapse rates (ELR)

ELR in a SW airstream in summer is 4°C per 1000 m

ELR in a NW airstream in summer is 5°C per 1000 m

ground temperature

b adiabatic lapse rates (ALR)

air cools less rapidly at SALR, e.g. 5°C per 1000 m due to release of latent heat (note that the graph appears to get steeper)

DALR if dew point temperature had not been reached and latent heat had not been released

dew point occurs in this example at 500 m, release of latent heat, base of clouds

condensation level

DALR temperature decreases at constant rate of 9.8°C per 1000 m

Lapse rates

The **environmental lapse rate** (ELR) is the decrease in temperature usually expected with an increase in height through the troposphere (Figure 9.1). The ELR is approximately 6.5°C per 1000 m, but varies according to local air conditions. It may vary due to several factors: **height** – ELR is lower nearer ground-level; **time** – it is lower in winter or during a rainy season; over different **surfaces** – it is lower over continental areas; and between different **air masses** (Figure 9.16a).

The **adiabatic lapse rate** (ALR) describes what happens when a parcel of air rises and the decrease in pressure is accompanied by an associated increase in volume and a decrease in temperature (Figure 9.15). Conversely, descending air will be subject to an increase in pressure causing a rise in temperature. In either case, there is negligible mixing with the surrounding air. There are two adiabatic lapse rates:

1 If the upward movement of air does not lead to condensation, the energy used by expansion will cause the temperature of the parcel of air to fall at the **dry adiabatic lapse rate** (DALR on Figure 9.16b). The DALR, which is the rate at which an unsaturated parcel of air cools as it rises or warms as it descends, remains constant at 9.8°C per 1000 m (i.e. approximately 1°C per 100 m).

2 When the upward movement is sufficiently prolonged to enable the air to cool to its dew point temperature, condensation occurs and the loss in temperature with height is then partly compensated by the release of latent heat (Figure 9.16b and page 210). Saturated air, which therefore cools at a slower rate than unsaturated air, loses heat at the **saturated adiabatic lapse rate** (SALR). The SALR can vary because the warmer the air the more moisture it can hold, and so the greater the amount of latent heat released following

condensation. The SALR may be as low as 4°C per 1000 m and as high as 9°C per 1000 m. It averages about 5.4°C per 1000 m (i.e. approximately 0.5°C per 100 m). Should temperatures fall below 0°C, then the air will cool at the **freezing adiabatic lapse rate** (FALR). This is the same as the DALR as very little moisture is present at low temperatures.

Air stability and instability

Parcels of warm air which rise through the lower atmosphere cool adiabatically. The rate and maintenance of any vertical uplift depend on the temperature–density balance between the rising parcel and the surrounding air. In a simplified form, this balance is the relationship between the environmental lapse rate and the dry and saturated adiabatic lapse rates.

Stability

The state of **stability** is when a rising parcel of unsaturated air cools more rapidly than the air surrounding it. This is shown diagrammatically when the ELR lies to the right of the DALR, as in Figure 9.17. In this example the ELR is 6°C per 1000 m and the DALR is 9.8°C per 1000 m. By the time the rising air has reached 1000 m, it has cooled to 10.2°C which leaves it colder and denser than the surrounding air which has only cooled to 14°C. If there is nothing to force the parcel of air to rise, e.g. mountains or fronts, it will sink back to its starting point. The air is described as stable because dew point may not have been reached and the only clouds which might have developed would be shallow, flat-topped cumulus which do not produce precipitation (Figure 9.20). Stability is often linked with anticyclones (page 234), when any convection currents are suppressed by sinking air to give dry, sunny conditions.

Figure 9.18

Instability and cloud
development

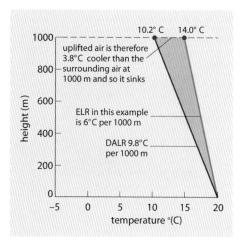

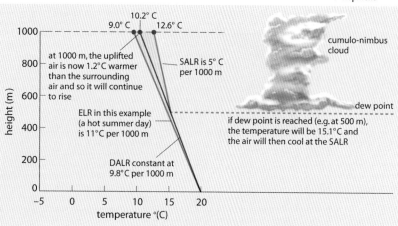

Instability

Conditions of **instability** arise in Britain on hot days. Localised heating of the ground warms the adjacent air by conduction, creating a higher lapse rate. The resultant parcel of rising unsaturated air cools less rapidly than the surrounding air. In this case, as shown in Figure 9.18, the ELR lies to the left of the DALR. The rising air remains warmer and lighter than the surrounding air. Should it be sufficiently moist and if dew point is reached, then the upward movement may be accelerated to produce towering cumulus or cumulo-nimbus type cloud (Figure 9.20). Thunderstorms are likely (Figure 9.21) and the saturated air, following the release of latent heat, will cool at the SALR.

Conditional instability

This type of instability occurs when the ELR is lower than the DALR but higher than the SALR. In Britain, it is the most common of the three conditions. The rising air is stable in its lower layers and, being cooler than the surrounding air, would normally sink back again. However, if the mechanism which initially triggered the uplift remains, then the air will be cooled to its dew point. Beyond this point, cooling takes place at the slower SALR and the parcel may become warmer than the surrounding air (Figure 9.19).

It will now continue to rise freely, even if the uplifting mechanism is removed, as it is now in an unstable state. Instability is conditional upon the air being forced to rise in the first place, and later becoming saturated so that condensation occurs. The associated weather is usually fine in areas at altitudes below condensation level, but cloudy and showery in those above.

Temperature inversions

As the lapse rate exercises have shown, the temperature of the air usually decreases with altitude, but there are certain conditions when the reverse occurs. **Temperature inversions**, where warmer air overlies colder air, may occur at three levels in the atmosphere. Figure 9.1 showed that temperatures increase with altitude in both the stratosphere and the thermosphere. Inversions can also occur near ground-level and high in the troposphere. High-level inversions are found in depressions where warm air overrides cold air at the warm front or is undercut by colder air at the cold front (page 229). Low-level, or ground, inversions usually occur under anticyclonic conditions (page 234 and Figure 9.24) when there is a rapid loss of heat from the ground due to radiation at night, or when warm air is advected over a cold surface. Under these conditions, fog and frost (page 221 and Figure 9.23) may form in valleys and hollows.

Figure 9.19

Conditional instability

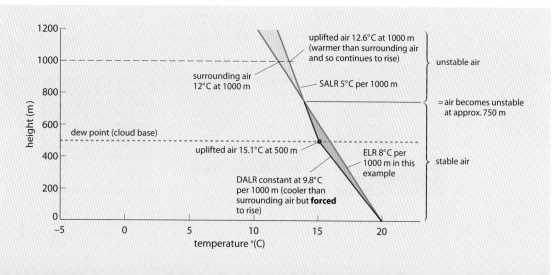

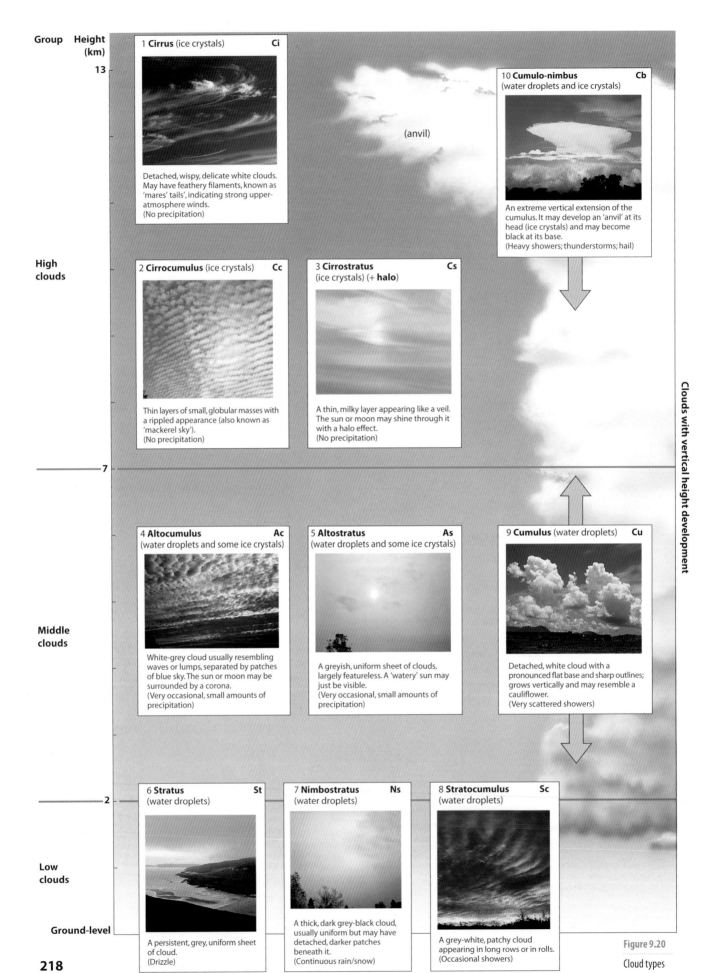

Figure 9.20

Cloud types

Clouds

Clouds form when air cools to dew point and vapour condenses into water droplets and/or ice crystals. There are many different types of cloud, but they are often difficult to distinguish as their form constantly changes. The general classification of clouds was proposed by Luke Howard in 1803. His was a descriptive classification, based on cloud shape and height (Figure 9.20). He used four Latin words: **cirrus** (a lock of curly hair); **cumulus** (a heap or pile); **stratus** (a layer); and **nimbus** (rain-bearing). He also compiled composite names using these four terms, such as cumulo-nimbus, cirrostratus; and added the prefix 'alto-' for middle-level clouds.

Precipitation

Condensation produces minute water droplets, less than 0.05 mm in diameter, or, if the dew point temperature is below freezing, ice crystals. The droplets are so tiny and weigh so little that they are kept buoyant by the rising air currents which created them. So although condensation forms clouds, clouds do not necessarily produce precipitation. As rising air currents are often strong, there has to be a process within the clouds which enables the small water droplets and/or ice crystals to become sufficiently large to overcome the uplifting mechanism and fall to the ground.

There are currently two main theories that attempt to explain the rapid growth of water droplets:

1   The **ice crystal mechanism** is often referred to as the Bergeron–Findeisen mechanism. It appears that when the temperature of air is between –5°C and –25°C, supercooled water droplets and ice crystals exist together. Super-cooling takes place when water remains in the atmosphere after temperatures have fallen below 0°C – usually due to a lack of condensation nuclei. Ice crystals are in a minority because the **freezing nuclei** necessary for their formation are less abundant than condensation nuclei. The relative humidity of air is ten times greater above an ice surface than over water. This means that the water droplets evaporate and the resultant vapour condenses (subli-mates) back onto the ice crystals which then grow into hexagonal-shaped snowflakes. The flakes grow in size – either as a result of further condensation or by fusion as their numerous edges interlock on collision with other flakes. They also increase in number as ice splinters break off and form new nuclei. If the air temperature rises above freezing point as the snow falls to the ground, flakes melt into raindrops. Experiments to produce rainfall artificially by cloud-seeding are based upon this process. The Bergeron–Findeisen theory is supported by evidence from temperate latitudes where rainclouds usually extend vertically above the freezing level. Radar and high-flying aircraft have reported snow at high altitudes when it is raining at sea-level. However, as clouds rarely reach freezing point in the tropics, the forma-tion of ice crystals is unlikely in those latitudes.

2   The **collision and coalescence** process was suggested by Longmuir. 'Warm' clouds (i.e. those containing no ice crystals), as found in the tropics, contain numerous water droplets of differing sizes. Different-sized droplets are swept upwards at different velocities and, in doing so, collide with other droplets. It is thought that the larger the droplet, the greater the chance of collision and subsequent coalescence with smaller droplets. When coa-lescing droplets reach a radius of 3 mm, their motion causes them to disintegrate to form a fresh supply of droplets. The thicker the cloud (cumulo-nimbus), the greater the time the droplets have in which to grow and the faster they will fall, usually as thundery showers.

Latest opinions suggest that these two theories may complement each other, but that a major process of raindrop enlargement has yet to be understood.

Types of precipitation

Although the definition of precipitation includes sleet, hail, dew, hoar frost, fog and rime, only rain and snow provide significant totals in the hydrological cycle.

Rainfall

There are three main types of rainfall, distin-guished by the mechanisms which cause the initial uplift of the air. Each mechanism rarely operates in isolation.

1   **Convergent** and **cyclonic** (**frontal**) rainfall results from the meeting of two air streams in areas of low pressure. Within the tropics, the trade winds, blowing towards the Equator, meet at the inter-tropical convergence zone or ITCZ (page 226). The air is forced to rise and, in con-junction with convection currents, produces the heavy afternoon thunderstorms associ-ated with the equatorial climate (page 316). In temperate latitudes, depressions form at the boundary of two air masses. At the associated fronts, warm, moist, less dense air is forced to rise over colder, denser air, giving periods of prolonged and sometimes intense rainfall. This is often augmented by orographic precipitation.

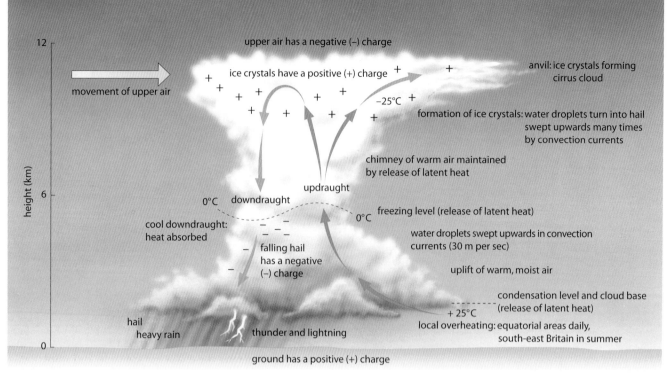

height (km)
12
6
0

movement of upper air

upper air has a negative (–) charge

ice crystals have a positive (+) charge

anvil: ice crystals forming cirrus cloud

–25°C

formation of ice crystals: water droplets turn into hail swept upwards many times by convection currents

chimney of warm air maintained by release of latent heat

downdraught   updraught

0°C

0°C   freezing level (release of latent heat)

cool downdraught: heat absorbed

water droplets swept upwards in convection currents (30 m per sec)

falling hail has a negative (–) charge

uplift of warm, moist air

condensation level and cloud base (release of latent heat)

+ 25°C

local overheating: equatorial areas daily, south-east Britain in summer

hail
heavy rain   thunder and lightning

ground has a positive (+) charge

Figure 9.21

Convectional rainfall: the development of a thunderstorm

2 **Orographic** or **relief** rainfall results when near-saturated, warm maritime air is forced to rise where confronted by a coastal mountain barrier. Mountains reduce the water-holding capacity of rising air by enforced cooling and can increase the amounts of cyclonic rainfall by retarding the speed of depression movement. Mountains also tend to cause air streams to converge and funnel through valleys. Rainfall totals increase where mountains are parallel to the coast, as is the Canadian Coast Range, and where winds have crossed warm offshore ocean currents, as they do before reaching the British Isles. As air descends on the leeward side of a mountain range, it becomes compressed and warmed and condensation ceases, creating a **rainshadow** effect where little rain falls.

3 **Convectional** rainfall occurs when the ground surface is locally overheated and the adjacent air, heated by conduction, expands and rises. During its ascent, the air mass remains warmer than the surrounding environmental air and it is likely to become unstable (page 217) with towering cumulonimbus clouds forming. These unstable conditions, possibly augmented by frontal or orographic uplift, force the air to rise in a 'chimney' (Figure 9.21). The updraught is maintained by energy released as latent heat at both condensation and freezing levels. The cloud summit is characterised by ice crystals in an anvil shape, the top of the cloud being flattened by upper-air movements. When the ice crystals and frozen water droplets,

i.e. hail, become large enough, they fall in a downdraught. The air through which they fall remains cool as heat is absorbed by evaporation. The downdraught reduces the warm air supply to the 'chimney' and therefore limits the lifespan of the storm. Such storms are usually accompanied by thunder and lightning. How storms develop immense amounts of electric charge is still not fully understood. One theory suggests that as raindrops are carried upwards into colder regions, they freeze on the outside. This ice-shell compresses the water inside it until the shell bursts and the water freezes into positively-charged ice crystals while the heavier shell fragments, which are negatively charged, fall towards the cloud base inducing a positive charge on the Earth's surface (Figure 9.21). **Lightning** is the visible discharge of electricity between clouds or between clouds and the ground. **Thunder** is the sound of the pressure wave created by the heating of air along a lightning flash. Convection is one process by which surplus heat and energy from the Earth's surface are transferred vertically to the atmosphere in order to maintain the heat balance (Figure 9.6).

Thunderstorms associated with the so-called **Spanish plume** can affect southern England several times during a hot, sultry summer. They occur when very hot air over the Sierra Nevada mountains (southern Spain) moves northwards over the Bay of Biscay where it draws in cooler, moist air. Should the resultant storm reach Britain, it can cause flash flooding, landslips and electricity blackouts.

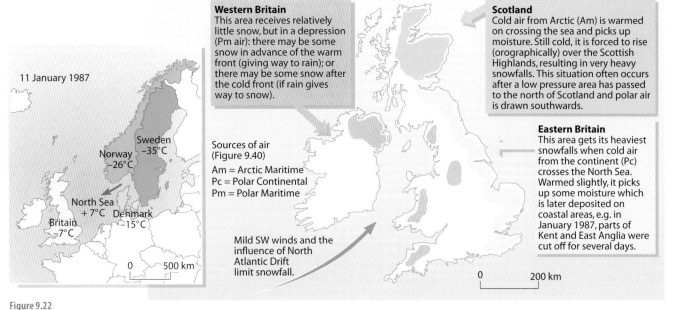

Western Britain
This area receives relatively little snow, but in a depression (Pm air): there may be some snow in advance of the warm front (giving way to rain); or there may be some snow after the cold front (if rain gives way to snow).

Scotland
Cold air from Arctic (Am) is warmed on crossing the sea and picks up moisture. Still cold, it is forced to rise (orographically) over the Scottish Highlands, resulting in very heavy snowfalls. This situation often occurs after a low pressure area has passed to the north of Scotland and polar air is drawn southwards.

Eastern Britain
This area gets its heaviest snowfalls when cold air from the continent (Pc) crosses the North Sea. Warmed slightly, it picks up some moisture which is later deposited on coastal areas, e.g. in January 1987, parts of Kent and East Anglia were cut off for several days.

11 January 1987

Sweden −35°C
Norway −26°C
North Sea +7°C
Denmark −15°C
Britain −7°C

0   500 km

Sources of air (Figure 9.40)

Am = Arctic Maritime
Pc = Polar Continental
Pm = Polar Maritime

Mild SW winds and the influence of North Atlantic Drift limit snowfall.

0   200 km

Figure 9.22

Causes of uneven snowfall patterns across Britain

Snow, sleet, glazed frost and hail

Snow forms under similar conditions to rain (Bergeron–Findeisen process) except that as dew point temperatures are under 0°C, then the vapour condenses directly into a solid (sublimation, page 215). Ice crystals will form if hygroscopic or freezing nuclei are present and these may aggregate to give snowflakes. As warm air holds more moisture than cold, snowfalls are heaviest when the air temperature is just below freezing. As temperatures drop, it becomes 'too cold for snow'. Figure 9.22 shows the typical conditions under which snow might fall in Britain.

Sleet is a mixture of ice and snow formed when the upper air temperature is below freezing, allowing snowflakes to form, and the lower air temperature is around 2 to 4°C, which allows their partial melting.

Glazed frost is the reverse of sleet and occurs when water droplets form in the upper air but turn to ice on contact with a freezing surface. When glazed frost forms on roads, it is known as 'black ice'.

Hail is made up of frozen raindrops which exceed 5 mm in diameter. It usually forms in cumulo-nimbus clouds, resulting from the uplift of air by convection currents, or at a cold front. It is more common in areas with warm summers where there is sufficient heat to trigger the uplift of air, and less common in colder climates. Hail frequently proves a serious climatic hazard in cereal-growing areas such as the American Prairies.

Dew, hoar frost, fog and rime

Dew, hoar frost and radiation fog all form under calm, clear, anticyclonic conditions when there is rapid terrestrial radiation at night. Dew point is reached as the air cools by conduction and moisture in the air, or transpired from plants, condenses. If dew point is above freezing, **dew** will form; if it is below freezing, **hoar frost** develops. Frost may also be frozen dew. Dew and hoar frost usually occur within 1 m of ground-level.

If the lower air is relatively warm, moist and contains hygroscopic nuclei, and if the ground cools rapidly, **radiation fog** may form. Where visibility is more than 1 km it is mist, if less than 1 km, fog. In order for radiation fog to develop, a gentle wind is needed to stir the cold air adjacent to the ground so that cooling affects a greater

Figure 9.23

Temperature inversion: radiation fog in a valley, Iceland

Figure 9.24

A low-level temperature inversion

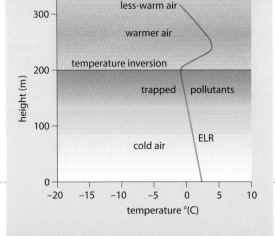

less-warm air

warmer air

temperature inversion

trapped   pollutants

cold air

ELR

height (m)

300

200

100

0

−20   −15   −10   −5   0   5   10
temperature °(C)

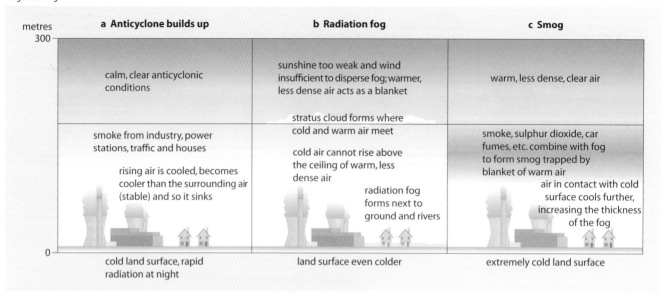

Figure 9.25

Formation of radiation
fog and smog

metres 300	**a Anticyclone builds up**	**b Radiation fog**	**c Smog**
	calm, clear anticyclonic conditions	sunshine too weak and wind insufficient to disperse fog; warmer, less dense air acts as a blanket	warm, less dense, clear air
	smoke from industry, power stations, traffic and houses	stratus cloud forms where cold and warm air meet	smoke, sulphur dioxide, car fumes, etc. combine with fog to form smog trapped by blanket of warm air
	rising air is cooled, becomes cooler than the surrounding air (stable) and so it sinks	cold air cannot rise above the ceiling of warm, less dense air	air in contact with cold surface cools further, increasing the thickness of the fog
		radiation fog forms next to ground and rivers	
0	cold land surface, rapid radiation at night	land surface even colder	extremely cold land surface

Figure 9.26

Rime frost, North
Carolina, USA

thickness of air. Radiation fogs usually occur in valleys, are densest around sunrise, and consist of droplets which are sufficiently small to remain buoyant in the air. Fog is likely to thicken if temperature inversion takes place (Figures 9.23 and 9.24), i.e. when cold surface air is trapped by overlying warmer, less dense air. It is under such conditions, in urban and industrial areas, that smoke and other pollutants released into the air are retained as smog (Figures 9.25 and 15.55).

Advection fog forms when warm air passes over or meets with cold air to give rapid cooling. In the coastal Atacama Desert (Places 24, page 180), sufficient droplets fall to the ground as 'fog-drip' to enable some vegetation growth.

Rime (Figure 9.26) occurs when supercooled droplets of water, often in the form of fog, come into contact with, and freeze on, solid objects such as telegraph poles and trees.

Acid rain

This is an umbrella term for the presence in rainfall of a series of pollutants which are produced mainly by the burning of fossil fuels. Coal-fired power stations, heavy industry and vehicle exhausts emit sulphur dioxide and nitrogen oxides. These are carried by prevailing winds across seas and national frontiers to be deposited either directly onto the Earth's surface as dry deposition or to be converted into acids (sulphuric and nitric acid) which fall to the ground in rain as wet deposition. Clean rainwater has a pH value of 5.6, which is slightly acidic due to the natural presence of carbonic acid (dissolved carbon dioxide). Today, rainfall over most of north-west Europe has a pH of about 5, the lowest ever recorded being 2.2 (the same as lemon juice).

The effects of acid rain include the increase in water acidity which caused the deaths of fish and plant life, mainly in Scandinavian rivers and lakes, and the pollution of fresh water supplies. Forests can be destroyed as important soil nutrients (calcium and potassium) are washed away and replaced by manganese and aluminium, both of which are harmful to root growth. In time trees shed their needles (coniferous) and leaves (deciduous) and become less resistant to drought, frost and disease.

However, between 1980 and 2000 emissions of sulphur dioxide were reduced by nearly 60 per cent in Western Europe and by about 30 per cent in North America (although in China and Southeast Asia they nearly doubled, albeit from a low base). Although the problem of acid rain still exists, it is becoming less prominent, especially in Western Europe where rivers and lakes are beginning to recover.

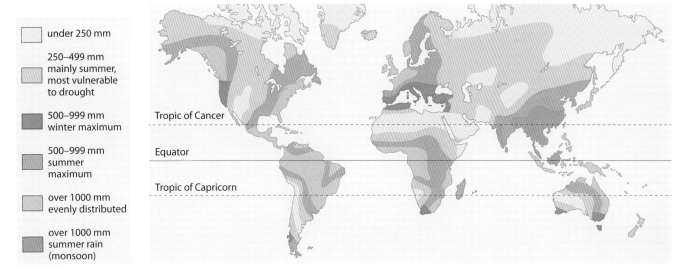

under 250 mm

250–499 mm mainly summer, most vulnerable to drought

500–999 mm winter maximum

500–999 mm summer maximum

over 1000 mm evenly distributed

over 1000 mm summer rain (monsoon)

Tropic of Cancer

Equator

Tropic of Capricorn

Figure 9.27

World precipitation: mean annual totals and seasonal distribution

World precipitation: distribution and reliability

Geographers are interested in describing distributions and in identifying and accounting for any resultant patterns. Where precipitation is concerned, geographers have, in the past, concentrated on long-term distributions which show either mean annual amounts or seasonal variations. Long-term fluctuations vary considerably across the globe but, nevertheless, a map showing world precipitation does show identifiable patterns (Figure 9.27).

Equatorial areas have high annual rainfall totals due to the continuous uplift of air resulting from the convergence of the trade winds and strong convectional currents (page 226). The presence of the ITCZ ensures that rain falls throughout the year. Further away from the Equator, rainfall totals decrease and the length of the dry season increases. These tropical areas, especially those inland, experience convectional rainfall in summer, when the sun is overhead, followed by a dry winter. Latitudes adjacent to the tropics receive minimal amounts as they correspond to areas of high pressure caused by subsiding, and therefore warming, air (Figure 7.2).

To the poleward side of this arid zone, rainfall quantities increase again and the length of the dry season decreases. These temperate latitudes receive large amounts of rainfall, spread evenly throughout the year, due to cyclonic conditions and local orographic effects. Towards the polar areas, where cold air descends to give stable conditions, precipitation totals decrease and rain gives way to snow. Between 30° and 40° north and south (in the west of continents) the Mediterranean climate is characterised by winter rain and summer drought. This general latitudinal zoning of rainfall is interrupted locally by the apparent movement of the overhead sun, the presence of mountain ranges or ocean currents, the monsoon, and continentality (distance from the sea).

More recently, geographers have become increasingly concerned with shorter-term variations. In many parts of the world, economic development and lifestyles are more closely linked to the duration, intensity and reliability of rainfall than to annual amounts. Precipitation is more valuable when it falls during the growing season (Canadian Prairies) and less effective if it occurs when evapotranspiration rates are at their highest (Sahel countries). In the same way, lengthy episodes of steady rainfall as experienced in Britain provide a more beneficial water supply than storms of a short and intensive duration which occur in tropical semi-arid climates. This is because moisture penetrates the soil more gradually and the risks of soil erosion, flooding and water shortages are reduced.

Of utmost importance is the reliability of rainfall. There appears to be a strong positive correlation (Framework 19, page 612) between rainfall totals and rainfall reliability – i.e. as rainfall totals increase, so too does rainfall reliability. In Britain and the Amazon Basin, rainfall is reliable with relatively little variation in annual totals from year to year (Figure 9.28).

Elsewhere, especially in monsoon or tropical continental climates, there is a pronounced wet and dry season. Consequently, if the rains fail one year, the result can be disastrous for crops, and possibly also for animals and people. The most vulnerable areas, such as north-east Brazil and the Sahel countries, lie near to desert margins (Figure 9.28). Here, where even a small variation of 10 per cent below the mean can be critical, many places often experience a variation in excess of 30 per cent.

Figure 9.28

World rainfall reliability

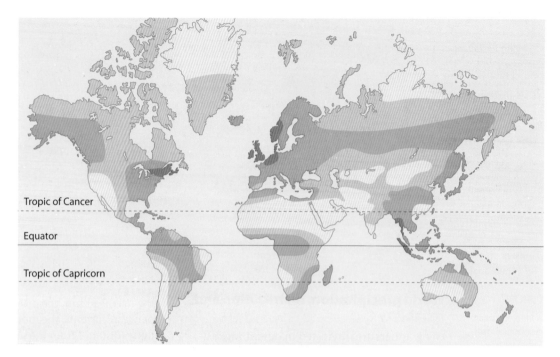

Percentage departure from the mean

- over 30
- 21–30
- 11–20
- 10 and under

Tropic of Cancer

Equator

Tropic of Capricorn

Atmospheric motion

The movement of air in the atmospheric system may be vertical (i.e. rising or subsiding) or horizontal; in the latter case it is commonly known as **wind**. Winds result from differences in air pressure which in turn may be caused by differences in temperature and the force exerted by gravity, as pressure decreases rapidly with height (Figure 9.1). An increase in temperature causes air to heat, expand, become less dense and rise, creating an area of low pressure below. Conversely, a drop in temperature produces an area of high pressure. Differences in pressure are shown on maps by **isobars**, which are lines joining places of equal pressure. To draw isobars, pressure readings are normally reduced to represent pressure at sea-level. Pressure is measured in **millibars** (mb) and it is usual for isobars to be drawn at 4 mb intervals.

Average pressure at sea-level is 1013 mb. However, the isobar pattern is usually more important in terms of explaining the weather than the actual figures. The closer together the isobars, the greater the difference in pressure – **the pressure gradient** – and the stronger the wind. Wind is nature's way of balancing out differences in pressure as well as temperature and humidity.

Figure 9.29 shows the two basic pressure systems which affect the British Isles. In addition to the differences in pressure, wind speed and wind direction, the diagrams also show that winds blow neither directly at right-angles to the isobars along the pressure gradient, nor parallel to them. This is due to the effects of the Coriolis force and of friction, which are additional to the forces exerted by the pressure gradient and gravity.

The Coriolis force

If the Earth did not rotate and was composed entirely of either land or water, there would be one large convection cell in each hemisphere (Figure 9.30). Surface winds would be parallel to pressure gradients and would blow directly from high to low pressure areas. In reality, the Earth does rotate and the distribution of land and sea is uneven. Consequently, more than one cell is created (Figures 9.34 and 9.35) as rising air, warmed at the Equator, loses heat to space – there is less cloud cover to retain it – and as it travels further from its source of heat. A further consequence is that moving air appears to be deflected to the right in the northern hemisphere and to the left in the southern hemisphere. This is a result of the Coriolis force.

Figure 9.29

The two basic pressure systems affecting Britain

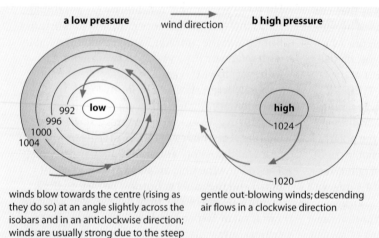

a low pressure wind direction **b high pressure**

992 **low**
996
1000
1004

high
1024

1020

winds blow towards the centre (rising as they do so) at an angle slightly across the isobars and in an anticlockwise direction; winds are usually strong due to the steep pressure gradient

gentle out-blowing winds; descending air flows in a clockwise direction

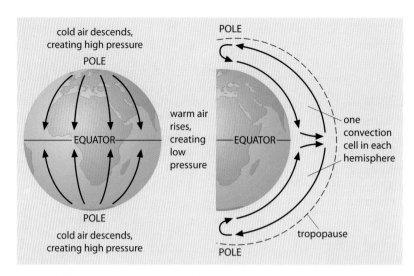

Figure 9.30

Air movement on a rotation-free Earth

Figure 9.31

The Coriolis force in the northern hemisphere

an astronaut in a space shuttle, the path would look straight). This helps to explain why the prevailing winds blowing from the tropical high pressure zone approach Britain from the south-west rather than from the south. In theory, if the Coriolis force acted alone, the resultant wind would blow in a circle.

Winds in the upper troposphere, unaffected by friction with the Earth's surface, show that there is a balance between the forces exerted by the pressure gradient and the Coriolis deflection. The result is the **geostrophic wind** which blows parallel to isobars (Figure 9.32). The existence of the geostrophic wind was recognised in 1857 by a Dutchman, Buys Ballot, whose law states that 'if you stand, in the northern hemisphere, with your back to the wind, low pressure is always to your left and high pressure to your right'.

Friction, caused by the Earth's surface, upsets the balance between the pressure gradient and the Coriolis force by reducing the effect of the latter. As the pressure gradient becomes relatively more important when friction is reduced with altitude, the wind blows across isobars towards the low pressure (Figure 9.29). Deviation from the geostrophic wind is less pronounced over water because its surface is smoother than that of land.

Imagine that Person A stands in the centre of a large rotating disc and throws a ball to Person B, standing on the edge of that moving disc. As Person A watches, the ball appears to take a curved path away from Person B – due to the fact that, while the ball is in transit, Person B has been moved to a new position by the rotation of the disc (Figure 9.31). Similarly, the Earth's rotation through 360° every 24 hours means that a wind blowing in a northerly direction in the northern hemisphere appears to have been diverted to the right on a curved trajectory by 15° of longitude for every hour (though to

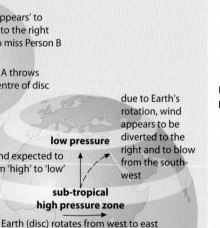

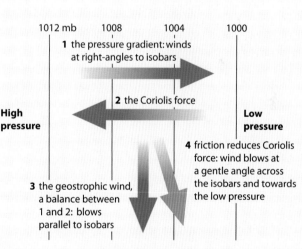

Figure 9.32

The geostrophic wind and the effect of friction (in the northern hemisphere)

A hierarchy of atmospheric motion

An appreciation of the movement of air is fundamental to an understanding of the workings of the atmosphere and its effects on our weather and climate. The extent to which atmospheric motion influences local weather and climate depends on winds at a variety of scales and their interaction in a hierarchy of patterns. One such hierarchy, which is useful in studying the influence of atmospheric motion, was suggested by B.W. Atkinson in 1988.

Although defining four levels, he stressed that there were important interrelationships between each (Figure 9.33).

Figure 9.33

A hierarchy of atmosphere motion systems (*after* Atkinson, 1988)

Scale	Characteristic horizontal size (km)	Systems
1 Planetary	5000–10 000	Rossby waves, ITCZ
2 Synoptic (macro)	1000–5000	Monsoons, hurricanes, depressions, anticyclones
3 Meso-scale	10–1000	Land and sea breezes, mountain and valley winds, föhn, thunderstorms
4 Small (micro)	0.1–10	Smoke plumes, urban turbulence

Planetary scale: atmospheric circulation

It has already been shown that there is a surplus of energy at the Equator and a deficit in the outer atmosphere and nearer to the poles (Figure 9.6). Therefore, theoretically, surplus energy should be transferred to areas with a deficiency by means of a single convective cell (Figure 9.30). This would be the case for a non-rotating Earth, a concept first advanced by Halley (1686) and expanded by Hadley (1735). The discovery of three cells was made by Ferrel (1856) and refined by Rossby (1941). Despite many modern advances using **radiosonde** readings, satellite imagery and computer modelling, this tricellular model still forms the basis of our understanding of the general circulation of the atmosphere.

The tricellular model

The meeting of the trade winds in the equatorial region forms the inter-tropical convergence zone, or **ITCZ**. The trade winds, which pick up latent heat as they cross warm, tropical oceans, are forced to rise by violent convection currents. The unstable, warm, moist air is rapidly cooled adiabatically to produce the towering cumulo-nimbus clouds, frequent afternoon thunderstorms and low pressure characteristic of the equatorial climate (page 316). It is these strong upward currents that form the 'power-house of the general global circulation' and which turn latent heat first into sensible heat and later into potential energy. At ground-level, the ITCZ experiences only very gentle, variable winds known as the **doldrums**.

As rising air cools to the temperature of the surrounding environmental air, uplift ceases and it begins to move away from the Equator. Further cooling, increasing density, and diversion by the Coriolis force cause the air to slow down and to subside, forming the descending limb of the **Hadley cell** (Figures 9.34 and 9.35). In looking at the northern hemisphere (the southern is its mirror image), it can be seen that the air subsides at about 30°N of the Equator to create the sub-tropical high pressure belt with its clear skies and dry, stable conditions (Figure 9.36). On reaching the Earth's surface, the cell is completed as some of the air is returned to the Equator as the north-east trade winds.

Figure 9.34

Tricellular model showing atmospheric circulation in the northern hemisphere

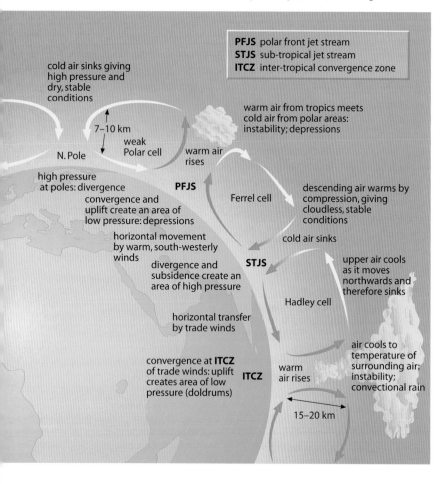

PFJS polar front jet stream
STJS sub-tropical jet stream
ITCZ inter-tropical convergence zone

cold air sinks giving high pressure and dry, stable conditions

7–10 km

weak Polar cell

N. Pole

warm air from tropics meets cold air from polar areas: instability; depressions

warm air rises

high pressure at poles: divergence

PFJS

convergence and uplift create an area of low pressure: depressions

Ferrel cell

descending air warms by compression, giving cloudless, stable conditions

horizontal movement by warm, south-westerly winds

cold air sinks

divergence and subsidence create an area of high pressure

STJS

upper air cools as it moves northwards and therefore sinks

Hadley cell

horizontal transfer by trade winds

convergence at **ITCZ** of trade winds: uplift creates area of low pressure (doldrums)

ITCZ

warm air rises

air cools to temperature of surrounding air: instability; convectional rain

15–20 km

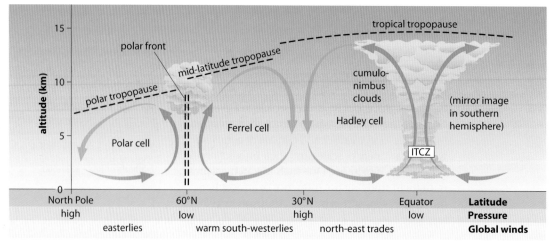

Figure 9.35

Tricellular model to show atmospheric circulation in the northern hemisphere and within the tropopause

North Pole high	60°N low	30°N high	Equator low	**Latitude** **Pressure**
easterlies	warm south-westerlies	north-east trades		**Global winds**

The remaining air is diverted polewards, forming the warm south-westerlies which collect moisture when they cross sea areas. These warm winds meet cold Arctic air at the polar front (about 60°N) and are uplifted to form an area of low pressure and the rising limb of the **Ferrel** and **Polar cells** (Figures 9.34 and 9.35). The resultant unstable conditions produce the heavy cyclonic rainfall associated with mid-latitude **depressions**. Depressions are another mechanism by which surplus heat is transferred. While some of this rising air eventually returns to the tropics, some travels towards the poles where, having lost its

heat, it descends to form another stable area of high pressure. Air returning to the polar front does so as the cold easterlies.

This overall pattern is affected by the apparent movement of the overhead sun to the north and south of the Equator. This movement causes the seasonal shift of the heat Equator, the ITCZ, the equatorial low pressure zone and global wind and rainfall belts. Any variation in the characteristics of the ITCZ – i.e. its location or width – can have drastic consequences for the surrounding climates, as seen in the Sahel droughts of the early 1970s and most of the 1980s (Case Study 7).

Figure 9.36

Image taken by the Meteosat geosynchronous satellite. Notice the clouds resulting from uplift at the ITCZ (not a continuous belt), the clear skies over the Sahara, the polar front over the north Atlantic, and a depression over Britain

Rossby waves and jet streams

Evidence of strong winds in the upper troposphere first came when First World War Zeppelins were blown off-course, and several inter-war balloons were observed travelling at speeds in excess of 200 km/hr. Pilots in the Second World War, flying at heights above 8 km, found eastward flights much faster and their return westward journeys much slower than expected, while north–south flights tended to be blown off-course. The explanation was found to be the **Rossby waves**, which often follow a meandering path (Figure 9.37a), distorting the upper-air westerlies. The number of meanders, or waves, varies seasonally, with usually

four to six in summer and three in winter. These waves form a complete pattern around the globe (Figure 9.37b and c).

Further investigation has shown that the velocity of these upper westerlies is not internally uniform. Within them are narrow bands of extremely fast-moving air known as **jet streams**. Jet streams, which help in the rapid transfer of energy, can exceed speeds of 230 km/hr, which is sufficient to carry a balloon, or ash from a volcano, around the Earth within a week or two (Figure 9.39 and Case Study 1). Of five recognisable jet streams, two are particularly significant, with a third having seasonal importance.

The **polar front jet stream** (PFJS, Figure 9.34) varies between latitudes 40° and 60° in both hemispheres and forms the division between the Ferrel and Polar cells, i.e. the boundary between warm tropical and cold polar air. The PFJS varies in extent, location and intensity and is mainly responsible for giving fine or wet weather on the Earth's surface. Where, in the northern hemisphere, the jet stream moves south (Figure 9.38a), it brings with it cold air which descends in a clockwise direction to give dry, stable conditions associated with areas of high pressure (**anticyclones**, page 234). When the now-warmed jet stream backs northwards, it takes with it warm air which rises in an anticlockwise direction to give the strong winds and heavy rainfall associated with areas of low pressure (**depressions**, page 230). As the usual path of the PFJS over Britain is oblique – i.e. towards the north-east – this accounts for our frequent wet and windy weather. Occasionally, this path may be temporarily altered by a stationary or **blocking anticyclone** (Figures 9.38b and 9.48) which may produce extremes of climate such as the hot, dry summers of 1976 and 1989 or the cold January of 1987.

The **subtropical jet stream**, or STJS, occurs about 25° to 30° from the Equator and forms the boundary between the Hadley and Ferrel cells (Figures 9.34 and 9.35). This meanders less than the PFJS, has lower wind velocities, but follows a similar west–east path.

The **easterly equatorial jet stream** is more seasonal, being associated with the summer monsoon of the Indian subcontinent (page 239).

Figure 9.37

Rossby waves and jet streams (northern hemisphere)

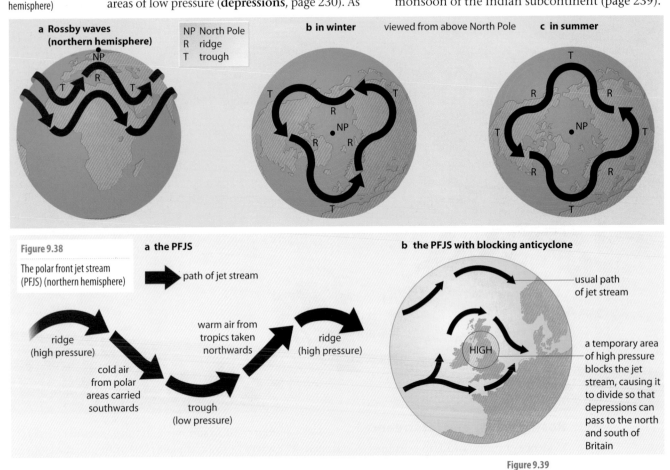

Figure 9.38

The polar front jet stream (PFJS) (northern hemisphere)

Figure 9.39

Orbiter 3's first balloon flight around the world

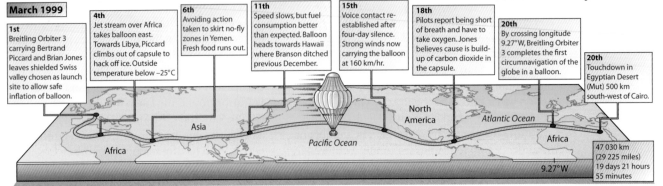

Macro-scale: synoptic systems

The concept of air masses is important because air masses help to categorise world climate types (Chapter 12). In regions where one air mass is dominant all year, there is little seasonal variation in weather, for example at the tropics and at the poles. Areas such as the British Isles, where air masses constantly interchange, experience much greater seasonal and diurnal variation in their weather.

Air masses and fronts: how they affect the British Isles

If air remains stationary in an area for several days, it tends to assume the temperature and humidity properties of that area. Stationary air is mainly found in the high pressure belts of the subtropics (the Azores and the Sahara) and in high latitudes (Siberia and northern Canada). The areas in which homogeneous air masses develop are called **source regions**. Air masses can be classified according to:

- the **latitude** in which they originate, which determines their temperature – Arctic (*A*), Polar (*P*), or Tropical (*T*)
- the nature of the **surface** over which they develop, which affects their moisture content – maritime (*m*), or continental (*c*).

The five major air masses which affect the British Isles at various times of the year (*Am, Pm, Pc, Tm* and *Tc*) are derived by combining these characteristics of latitude and humidity (Figure 9.40). When air masses move from their source region they are modified by the surface over which they pass and this alters their temperature, humidity and stability. For example, tropical air moving northwards is cooled and becomes more stable, while polar air moving south becomes warmer and increasingly unstable. Each air mass therefore brings its own characteristic weather conditions to the British Isles. The general conditions expected with each air mass are given in Figure 9.41. However, it should be remembered that each air mass is unique and dependent on: the climatic conditions in the source region at the time of its development; the path which it subsequently follows; the season in which it occurs; and, since it has a three-dimensional form, the vertical characteristics of the atmosphere at the time.

When two air masses meet, they do not mix readily, due to differences in temperature and density. The point at which they meet is called a **front**. A **warm front** is found where warm air is advancing and being forced to override cold air. A **cold front** occurs when advancing cold air undercuts a body of warm air. In both cases, the rising air cools and usually produces clouds, easily seen on satellite weather photographs (Figures 9.67 and 9.68); these clouds often generate precipitation. Fronts may be several hundred kilometres wide and they extend at relatively gentle gradients up into the atmosphere. The most notable type of front, the **polar front**, occurs when warm, moist, *Tm* air meets colder, drier, *Pm* air. It is at the polar front that **depressions** form. Depressions are areas of low pressure. They form most readily over the oceans in mid-latitudes, and track eastwards bringing cloud and rain to western margins of continents.

Figure 9.40

Air masses that affect the British Isles

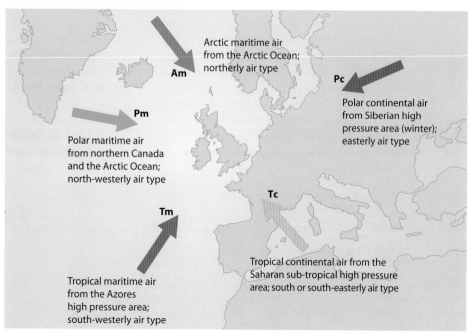

Arctic maritime air from the Arctic Ocean; northerly air type

Am

Pc

Polar continental air from Siberian high pressure area (winter); easterly air type

Pm

Polar maritime air from northern Canada and the Arctic Ocean; north-westerly air type

Tc

Tm

Tropical continental air from the Saharan sub-tropical high pressure area; south or south-easterly air type

Tropical maritime air from the Azores high pressure area; south-westerly air type

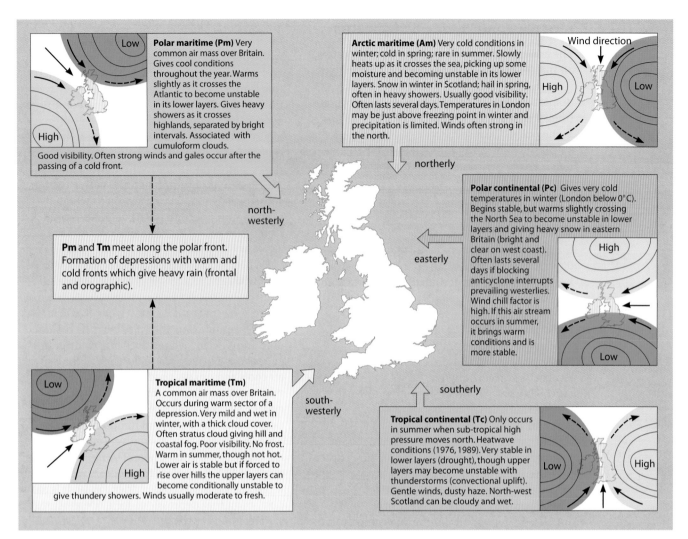

Figure 9.41

Air masses and the British weather

Depressions

The polar front theory was put forward by a group of Norwegian meteorologists in the early 1920s. Although some aspects have been refined since the innovation of radiosonde readings and satellite imagery, the basic model for the formation of frontal depressions remains valid. The following account describes a 'typical' or 'model' depression (Framework 12, page 352). It should be remembered, however, that individual depressions may vary widely from this model.

Depressions follow a life-cycle in which three main stages can be identified: embryo, maturity and decay (Figures 9.42, 9.43 and 9.44).

1 The **embryo depression** begins as a small wave on the polar front. It is here that warm, moist, tropical (*Tm*) air meets colder, drier, polar (*Pm*) air (Figure 9.42). Recent studies have shown the boundary between the two air masses to be a zone rather than the simple linear division claimed in early models. The convergence of the two air masses results in the warmer, less dense air being forced to rise in a spiral movement. This upward movement results in 'less' air at the Earth's surface, creating an area of below-average or low pressure. The developing depression, with its warm front (the leading edge of the tropical

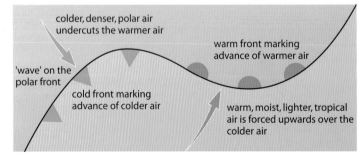

colder, denser, polar air undercuts the warmer air

warm front marking advance of warmer air

'wave' on the polar front

cold front marking advance of colder air

warm, moist, lighter, tropical air is forced upwards over the colder air

Figure 9.42

Life-cycle of a depression: Stage 1 – embryo depression

air) and cold front (the leading edge of the polar air), usually moves in a north-easterly direction under the influence of the upper westerlies, i.e. the polar front jet stream.

2 A **mature depression** is recognised by the increasing amplitude of the initial wave (Figure 9.43). Pressure continues to fall as more warm air, in the warm sector, is forced to rise. As pressure falls and the pressure gradient steepens, the inward-blowing winds increase in strength. Due to the Coriolis force (page 224), these anticlockwise-blowing winds come from the south-west. As the relatively warm air of the warm sector continues to rise along the warm front, it eventually cools to dew point. Some of its vapour will condense to release large amounts of latent heat, and clouds will develop. Continued uplift and cooling will cause precipitation as the clouds become both thicker and lower.

Satellite photographs have shown that there is likely to be a band of 'meso-scale precipitation' extending several hundred kilometres in length and up to 150 km in width along, and just in front of, a warm front. As temperatures rise and the uplift of air decreases within the warm sector, there is less chance of precipitation and the low cloud may break to give some sunshine. The cold front moves faster and has a steeper gradient than the warm front (Figure 9.45).

Progressive undercutting by cold air at the rear of the warm sector gives a second episode of precipitation – although with a greater intensity and a shorter duration than at the warm front. This band of meso-scale precipitation may be only 10–50 km in width. Although the air behind the cold front is colder than that in advance of the warm front (having originated in and travelled through more northerly latitudes), it becomes unstable, forming cumulo-nimbus clouds and heavy showers. Winds often reach their maximum strength at the cold front and change to a more north-westerly direction after its passage (Figure 9.45).

3 The depression begins to **decay** when the cold front catches up the warm front to form an **occlusion** or **occluded front** (Figure 9.44). By this stage, the *Tm* air will have been squeezed upwards leaving no warm sector at ground-level. As the uplift of air is reduced, so too are (or will be) the amount of condensation, the release of latent heat and the amount and pattern of precipitation – there may be only one episode of rain. Cloud cover begins to decrease, pressure rises and wind speeds decrease as the colder air replaces the uplifted air and 'infills' the depression.

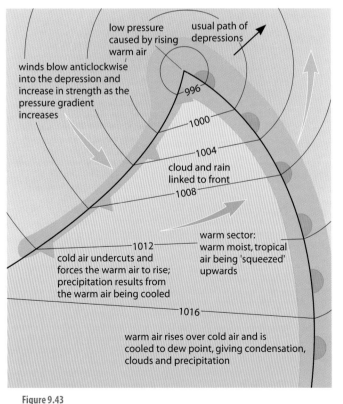

Figure 9.43

Life-cycle of a depression: Stage 2 – maturity

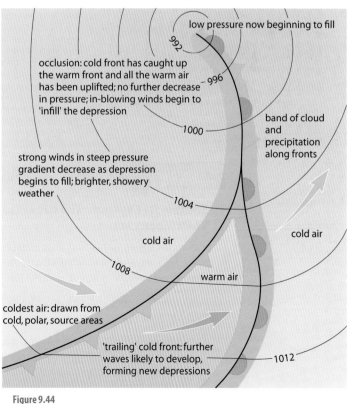

Figure 9.44

Life-cycle of a depression: Stage 3 – decay

Figure 9.45

Weather associated with the passing of a typical mid-latitude depression

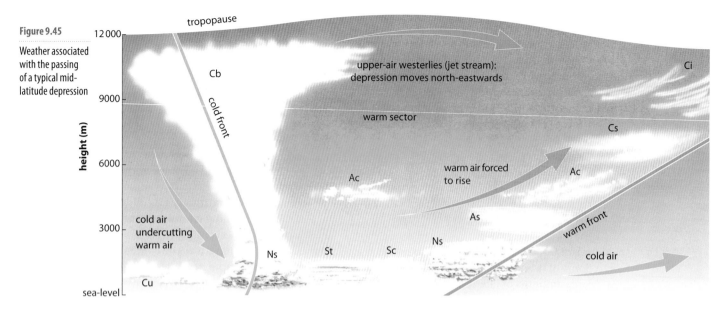

read from right to left (i.e. from 1 to 5)

	5. Behind the cold front	4. Passing of the cold front	3. Warm sector	2. Passing of the warm front	1. Approach of depression
Pressure	rise continues more slowly	sudden rise	steady	fall ceases	steady fall
Wind direction	NW	veers from SW to NW	SW	veers from SSE to SW	SSE
Wind speed	squally; speed slowly decreases (e.g. force 3–6)	very strong to gale force (e.g. force 6–8)	decreases (e.g. force 2–4)	strong (e.g. force 5–6)	slowly increases (e.g. force 1–3)
Temperature (e.g. winter)	cold (e.g. 3°C)	sudden decrease	warm/mild (e.g. 10°C)	sudden rise	cool (e.g. 6°C)
Relative humidity	rapid fall	high during precipitation	steady and high	high during precipitation	slow rise
Cloud (Figure 9.20)	decreasing; in succession, Cb and Cu	very thick and towering Cb	low or may clear; St, Sc, Ac	low and thick Ns	high and thin; in succession, Ci, Cs, Ac, As
Precipitation	heavy showers	short period of heavy rain or hail	drizzle or stops raining	continuous rainfall, steady and quite heavy	none
Visibility	very good; poor in showers	poor	often poor	decreases rapidly	good but beginning to decrease

Places 29 Storms in southern England

South-east England: 'The Great Storm', 16 October 1987

This storm, the worst to affect south-east England since 1703, developed so rapidly that its severity was not predicted in advance weather forecasts.

11 October: High winds and heavy rain forecast for the end of the week.

15 October **1200 hrs:** Depression expected to move along the English Channel with fresh to strong winds.

2130 hrs: TV weather forecast: strong winds gusting to 50 km/hr.

16 October **0030 hrs:** Radio weather forecast: warning of severe gales.

0130 hrs: Police and fire services alerted about extreme winds.

0500 hrs: Winds reached 94 km/hr at Heathrow and 100 km/hr on parts of the south coast.

0800 hrs: Centre of depression reached the North Sea. Winds over southern England dropped to 50–70 km/hr.

1200 hrs: 'The Great Storm' was over.

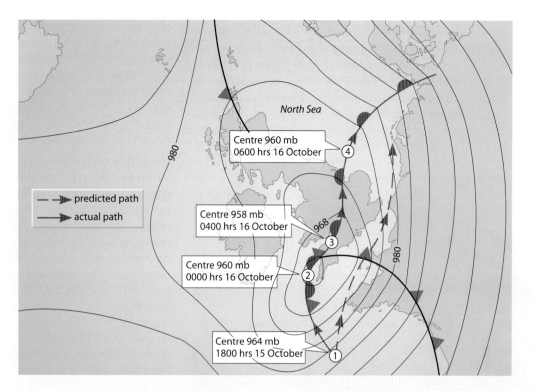

Figure 9.46

'The Great Storm', 16 October 1987

Centre 960 mb
0600 hrs 16 October

North Sea

Centre 958 mb
0400 hrs 16 October

Centre 960 mb
0000 hrs 16 October

Centre 964 mb
1800 hrs 15 October

predicted path
actual path

The storm began on 15 October as a small wave on a cold front in the Bay of Biscay, where the few weather ships give only limited information. It was caused by contact between very warm air from Africa and cold air from the North Atlantic. It appeared to be a 'typical' depression until, at about 1800 hrs on 15 October, it unexpectedly deepened giving a central pressure reading of 964 mb and creating an exceptionally steep pressure gradient. The exact cause of this is unknown but it was believed to result from a combination of an exceptionally strong jet stream (initiated on 13 October by air spiralling upwards along the east coast of North America in Hurricane Floyd) and extreme warming over the Bay of Biscay (see hurricanes, page 235). Together, these could have caused an excessive release of latent heat energy which North American meteorologists compare with the effect of detonating a bomb. It was this unpredicted deepening, combined with the change of direction from the English Channel towards the Midlands, which caught experts by surprise.

The depression moved rapidly across southern England, clearing the country in six hours (Figure 9.46). Winds remained light in and around the centre (Birmingham 13 km/hr), but the strong pressure gradient on its southern flank resulted in severe winds from Portland Bill (102 km/hr, gusting to 141 km/hr) to Dover (115 km/hr, gusting to 167 km/hr).

Although the storm passed within a few hours, and luckily during the night when most people were asleep, it left a trail of death and destruction. There were 16 deaths; several houses collapsed and many others lost walls, windows and roofs; an estimated 15 million trees were blown over, blocking railways and roads; one-third of the trees in Kew Gardens were destroyed; power lines were cut and, in some remote areas, not restored for several days; few commuters managed to reach London the next day; a ferry was blown ashore at Folkestone; and insurance claims set an all-time record.

Once every 50 years, winds exceeding 100 km/hr with gusts of over 165 km/hr can be expected north of a line from Cornwall to Durham, and even stronger winds, gusting to 185 km/hr, once in 20 years in western and northern Scotland. The winds associated with the Great Storm were remarkable not so much for their strength as for their occurrence over south-east England. Here, the predicted return period can be measured in centuries rather than decades.

10 March 2008

Southern Britain experienced the worst storm for over 20 years with winds of 150 km/hr recorded on the Isle of Wight and torrential rain falling over Wales and southern England. Flights to and from Heathrow were either cancelled or diverted and there were delays at other London airports. Cross-Channel ferries to France and Ireland were also cancelled and over 10 000 homes in south-west England lost their electricity.

Anticyclones

An anticyclone is a large mass of subsiding air which produces an area of high pressure on the Earth's surface (Figure 9.47). The source of the air is the upper atmosphere, where amounts of water vapour are limited. On its descent, the air warms at the DALR (page 216), so dry conditions result. Pressure gradients are gentle, resulting in weak winds or calms (Figure 9.29b). The winds blow outwards and clockwise in the northern hemisphere. Anticyclones may be 3000 km in diameter – much larger than depressions – and, once established, can give several days or, under extreme conditions, several weeks, of settled weather. There are also differences, again unlike in a depression, between the expected weather conditions in a summer and a winter anticyclone.

Weather conditions over Britain

Summer Due to the absence of cloud, there is intense insolation which gives hot, sunny days (up to 30°C in southern England) and an absence of rain. Rapid radiation at night, under clear skies, can lead to temperature inversions and the formation of dew and mist, although these rapidly clear the following morning. Coastal areas may experience advection fogs and land and sea breezes, while highlands have mountain and valley winds (pages 240–241). If the air has its source over North Africa – that is, if it is a *Tc* air mass (Figure 9.40) – then heatwave conditions tend to result. Often, after several days of increasing thermals, there is an increased risk of thunderstorms and the so-called Spanish plume (page 220).

Winter Although the sinking air again gives cloudless skies, there is little incoming radiation during the day due to the low angle of the sun. At night, the absence of clouds means low temperatures and the development of fog and frost. These may take a long time to disperse the next day in the weak sunshine. Polar continental (*Pc*) air (Figure 9.40), with its source in central Asia and a slow movement over the cold European land mass, is cold, dry and stable until it reaches the North Sea where its lower layers acquire some warmth and moisture. This can cause heavy snowfalls on the east coast (Figure 9.22).

Blocking anticyclones

These occur when cells of high pressure detach themselves from the major high pressure areas of the subtropics or poles (Figure 9.38b). Once created, they last for several days and 'block' eastward-moving depressions (Figure 9.48) to create anomalous conditions such as extremes of temperature, rainfall and sunshine – as in Britain in the summer of 1995 and the winter of 1987.

Figure 9.47

Anticyclone over the British Isles

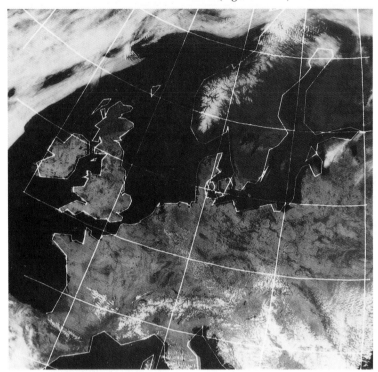

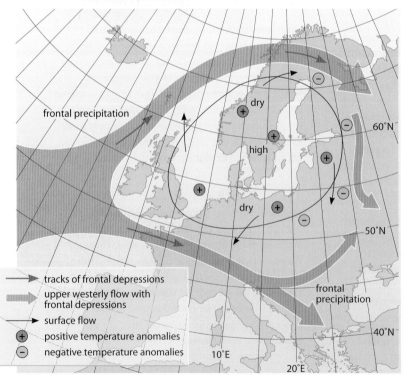

legend:
- → tracks of frontal depressions
- ⟶ upper westerly flow with frontal depressions
- → surface flow
- ⊕ positive temperature anomalies
- ⊖ negative temperature anomalies

Figure 9.48

A blocking anticyclone over Scandinavia: the upper westerlies divide upwind of the block and flow around it with their associated rainfall; there are positive temperature anomalies within the southerly flow to the west of the block and negative anomalies to the east

Tropical cyclones

Tropical cyclones are systems of intense low pressure known locally as **hurricanes**, **typhoons** and **cyclones** (Figure 9.49). They are characterised by winds of extreme velocity and are accompanied by torrential rainfall – two factors that can cause widespread damage and loss of life (Places 31, page 238). As yet, there is still insufficient conclusive evidence as to the process of their formation, although knowledge has been considerably improved recently due to airflights through and over individual systems, and the use of weather satellites. Tropical cyclones tend to develop:

- over warm tropical oceans, where sea temperatures exceed 26°C and where there is a considerable depth of warm water
- in autumn, when sea temperatures are at their highest
- in the trade wind belt, where the surface winds warm as they blow towards the Equator
- between latitudes 5° and 20° north or south of the Equator (nearer to the Equator the Coriolis force is insufficient to enable the feature to 'spin' – page 225).

Once formed, they move westwards – often on erratic, unpredictable courses – swinging poleward on reaching land, where their energy is rapidly dissipated (Figure 9.49). They are another mechanism by which surplus energy is transferred away from the tropics (Figure 9.6).

Hurricanes

Hurricanes are the tropical cyclones of the Atlantic. They form after the ITCZ has moved to its most northerly extent enabling air to converge at low levels, and can have a diameter of up to 650 km. Unlike depressions, hurricanes occur when temperatures, pressure and humidity are uniform over a wide area in the lower troposphere for a lengthy period, and anticyclonic conditions exist in the upper troposphere. These conditions are essential for the development, near the Earth's surface, of intense low pressure and strong winds. To enable the hurricane to move, there must be a continuous source of heat to maintain the rising air currents. There must also be a large supply of moisture to provide the latent heat, released by condensation, to drive the storm and to provide the heavy rainfall. It is estimated that in a single day a hurricane can release an amount of energy equivalent to that released by 500 000 atomic bombs the size of the one dropped on Hiroshima in the Second World War. Only when the storm has reached maturity does the central **eye** develop. This is an area of subsiding air, some 30–50 km in diameter, with light winds, clear skies and anomalous high temperatures (Figure 9.50). The descending air increases instability by warming and exaggerates the storm's intensity.

The hurricane rapidly declines once the source of heat is removed, i.e. when it moves over colder water or a land surface; these increase friction and cannot supply sufficient moisture. The average lifespan of a tropical cyclone is 7 to 14 days. The characteristic weather conditions associated with the passage of a typical hurricane are shown diagrammatically in Figure 9.50, and from space in Figure 9.51.

Figure 9.49

Global location and mean frequency of tropical cyclones

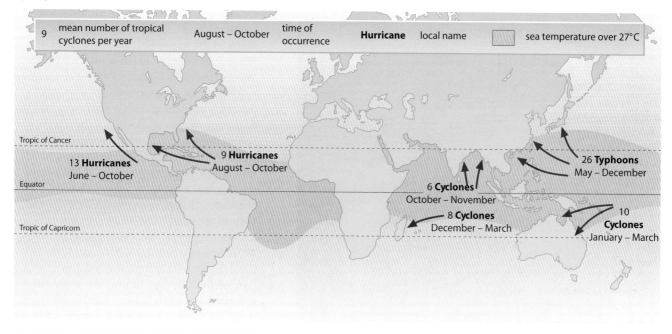

9 mean number of tropical cyclones per year | August – October time of occurrence | **Hurricane** local name | sea temperature over 27°C

Tropic of Cancer

13 **Hurricanes** June – October

9 **Hurricanes** August – October

26 **Typhoons** May – December

Equator

6 **Cyclones** October – November

8 **Cyclones** December – March

10 **Cyclones** January – March

Tropic of Capricorn

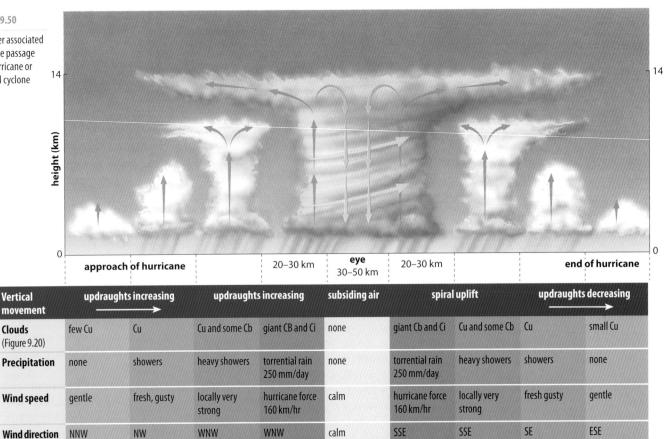

Vertical movement	updraughts increasing →				subsiding air	spiral uplift			updraughts decreasing →	
Clouds (Figure 9.20)	few Cu	Cu	Cu and some Cb	giant CB and Ci	none	giant Cb and Ci	Cu and some Cb	Cu	small Cu	
Precipitation	none	showers	heavy showers	torrential rain 250 mm/day	none	torrential rain 250 mm/day	heavy showers	showers	none	
Wind speed	gentle	fresh, gusty	locally very strong	hurricane force 160 km/hr	calm	hurricane force 160 km/hr	locally very strong	fresh gusty	gentle	
Wind direction	NNW	NW	WNW	WNW	calm	SSE	SSE	SE	ESE	
Temperatures (plus examples)	high (30°C)	still high (30°C)	falling (26°C)	low (24°C)	high (32°C)	low (24°C)	rising (26°C)	high (28°C)	high (30°C)	
Pressure	average, 1012 mb	steady, 1010 mb	slowly falling, 1006 mb	rapid fall	low, 960 mb	rapid rise	slowly rising, 1004 mb	steady, 1010 mb	average, 1012 mb	

Column headers along top axis: approach of hurricane — 20–30 km — eye 30–50 km — 20–30 km — end of hurricane

Figure 9.51

Satellite image of Hurricane Mitch, October 1998. The 'eye' is very noticeable

Tropical cyclones are a major natural hazard which often cause considerable loss of life and damage to property and crops (Places 31). There are four main causes of damage.

1 **High winds**, which often exceed 160 km/hr and, in extreme cases, 300 km/hr. Whole villages may be destroyed in economically less developed countries (of which there are many in the tropical cyclone belt), while even reinforced buildings in the south-east USA may be damaged. Countries whose economies rely largely on the production of a single crop (bananas in Nicaragua) may suffer serious economic problems. Electricity and communications can also be severed.

2 **Ocean storm (tidal) surges**, resulting from the high winds and low pressure, may inundate coastal areas, many of which are densely populated (Bangladesh, Places 19, page 148).

3 **Flooding** can be caused either by a storm (tidal) surge or by the torrential rainfall. In 1974, 800 000 people died in Honduras as their flimsy homes were washed away.

4 **Landslides** can result from heavy rainfall where buildings have been erected on steep, unstable slopes (Hong Kong, Figure 2.33).

'The Number 8 signal may be raised today as Typhoon Leo moves closer to Hong Kong. Its approach forced the Hong Kong observatory to hoist the strong wind signal Number 3 yesterday afternoon [Figure 9.52] – the first time it had ever been raised in April [Figure 9.55]. Leo intensified into a typhoon yesterday, with central wind speeds of up to 130 km/hr. At midnight, it was 310 km south-south-east of Hong Kong, and was moving at about 8 km/hr [Figure 9.53]. The typhoon is expected to be closest to Hong Kong early tomorrow morning, by which time weather will deteriorate further and average rainfall could exceed 500 mm [Figure 9.54].

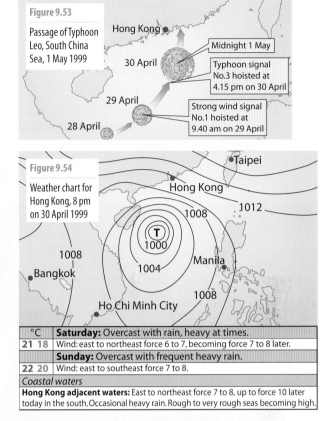

Figure 9.53

Passage of Typhoon Leo, South China Sea, 1 May 1999

Hong Kong

30 April

Midnight 1 May

Typhoon signal No.3 hoisted at 4.15 pm on 30 April

29 April

Strong wind signal No.1 hoisted at 9.40 am on 29 April

28 April

Figure 9.54

Weather chart for Hong Kong, 8 pm on 30 April 1999

Taipei

Hong Kong

1012

1008

T 1000

1008

Manila

1004

Bangkok

1008

Ho Chi Minh City

Figure 9.52

Typhoon warning system, Hong Kong

Typhoon No. 1 is hoisted.
一號風球現正懸掛

The Education Department has ordered kindergartens, schools for the mentally and physically handicapped, and nursing schools to remain closed. The Home Affairs Department's temporary shelters will open if Signal 8 goes up. People in need of shelter can make enquiries by calling the hotline.'

Source: South China Morning Post

Figure 9.55

Typhoon warning signals, Hong Kong

°C		
	Saturday: Overcast with rain, heavy at times.	
21 18	Wind: east to northeast force 6 to 7, becoming force 7 to 8 later.	
	Sunday: Overcast with frequent heavy rain.	
22 20	Wind: east to southeast force 7 to 8.	

Coastal waters

Hong Kong adjacent waters: East to northeast force 7 to 8, up to force 10 later today in the south. Occasional heavy rain. Rough to very rough seas becoming high.

Signal (four categories of tropical cyclone based on wind speed)		Meaning of the signal	What you should do — Specific advice is contained in weather broadcasts, but the following general precautions can be taken
Stand by	1	A tropical cyclone is centred within about 800 km of Hong Kong. Hong Kong is placed on a state of alert because the tropical cyclone is a potential threat and may cause destructive winds later.	**Listen to weather broadcasts.** Some preliminary precautions are desirable and you should take the existence of the tropical cyclone into account in planning your activities.
Strong wind – A Tropical depression	3	Strong wind expected or blowing, with a sustained speed of 41–62 km/hr and gusts that may exceed 110 km/hr. The timing of the hoisting of the signal is aimed to give about 12 hours' advance warning of a strong wind in Victoria Harbour but the warning period may be shorter for more exposed waters.	**Take all necessary precautions.** Secure all loose objects, particularly on balconies and rooftops. Secure hoardings, scaffolding and temporary structures. Clear gutters and drains. Take full precautions for the safety of boats. Ships in port normally leave for typhoon anchorages or buoys. Ferry services may soon be affected by wind or waves. Even at this stage heavy rain accompanied by violet squalls may occur.
Gale or storm – B Tropical storm	4–8	Gale or storm expected or blowing, with a sustained wind speed of 63–117 km/hr from the quarter indicated and gusts that may exceed 180 km/hr. The timing of the replacement of the Strong Wind Signal No.3 by the appropriate one of these four signals, is aimed to give about 12 hours' advance warning of a gale in Victoria Harbour, but the sustained wind speed may reach 63 km/hr within a shorter period over more exposed waters. Expected changes in the direction of the wind will be indicated by corresponding changes of these signals.	**Complete all precautions** as soon as possible. It is extremely dangerous to delay precautions until the hoisting of No.9 or No.10 signals as these are signals of great urgency. Windows and doors should be bolted and shuttered. **Stay indoors** when the winds increase to avoid flying debris, but if you must go out, keep well clear of overhead wires and hoardings. All schools and law courts close and ferries will probably stop running at short notice. The sea-level will probably be higher than normal, particularly in narrow inlets. If this happens near the time of normal high tide then low-lying areas may have to be evacuated very quickly. Heavy rain may cause flooding, rockfalls and mudslips.
Increasing gale or storm – C Severe tropical storm	9	Gale or storm expected to increase significantly in strength. This signal will be hoisted when the sustained wind speed is expected to increase and come within the range 88–117 km/hr during the next few hours.	**Stay where you are** if reasonably protected and **away from exposed windows and doors.** These signals imply that the centre of a severe tropical storm or a typhoon will come close to Hong Kong. If the eye passes over there will be a lull lasting from a few minutes to some hours, but be prepared for a sudden resumption of destructive winds from a different direction.
Hurricane – D Typhoon	10	Hurricane-force winds expected or blowing, with a sustained wind speed reaching upwards from 118 km/hr and with gusts that may exceed 220 km/hr.	

West Indies, September 2004

The year 2004 experienced the 'mother of hurricanes season'. Following hurricanes Charlie, which killed 16 people and caused damage in Florida only once previously exceeded, and Frances, Hurricane Ivan began its destructive course.

Hurricane Ivan, deservedly nicknamed 'the Terrible', began its trail of destruction on Grenada on 5 September – the first time the island had been affected by a major hurricane since 1955. Reports put the death toll at 34; water, electricity and air transport were disrupted for several days, and two-thirds of the island's 100 000 residents were made homeless (Figure 9.56).

After several days of warning, Ivan hit Jamaica on 11 September. The laid-back approach of many Jamaicans contrasted strongly with the well-practised response of people in Florida. Many of those Jamaicans who lived in shanty settlements refused to leave their flimsy, often makeshift homes, and only a few thousand of the half million ordered to evacuate heeded the government's warning, many preferring to protect what might be left of their possessions from post-hurricane looting. The resultant death toll was put at 20. By the time Ivan ravaged the Cayman Islands a day later, it had become a category 5 event – one of only a handful of that intensity in the last 100 years. Winds reached 260 km/hr while torrential rain and 6 m waves caused extensive flooding but, fortunately, no deaths were reported. In Cuba, next in Ivan's path, 2 million people were evacuated in advance of what was considered the most violent hurricane for over 50 years but at almost the last minute it veered sufficiently for the eye to pass just to the west of the island. Ivan, by now slightly reduced in strength, made landfall in the USA between Mobile (Alabama) and Pensacola (Florida) on 16 September, with wind speeds of 210 km/hr and a tidal surge of 4 m. Although

2 million people had been evacuated along a 675 km stretch of the Gulf coast, 12 deaths were reported. This might have been worse had Ivan veered westwards where parts of the Louisiana coast lie 3 m or more below sea-level and are protected by huge levées.

Myanmar, May 2008

Bangladesh frequently experiences tropical cyclones which move northwards, accompanied by winds with speeds exceeding 200 km/hr, up the narrowing, shallowing Bay of Bengal. These cyclones can create storm surges of over 8 m that affect the flat delta region of the Ganges–Brahmaputra (Places 19, page 148). Improvements in coastal defences and early warning systems have reduced considerably the amount of damage and the number of deaths from 200 000 after the 1970 storm to 140 000 in 1990, 135 000 in 1991, 40 000 in 1994 and 10 000 in 1999. However, in 2008 tropical cyclone Nargis hit the still unprotected Irrawaddy delta lying to the south in Myanmar.

Little warning was given before Nargis, with wind speeds of 200 km/hr, swept over the flat Irrawaddy delta before affecting the former capital city of Rangoon. Unlike other recent catastrophes such as the Indian Ocean tsunami (Places 4) and the China earthquake (Places 2) where the world was immediately aware of the event, here, due to a lack of contact with the military regime, it was two days before news began to leak out of Myanmar and then only to admit to 350 deaths.

Later it became known that a tidal surge that followed the cyclone created devastation of tsunami proportions. Crops had been totally destroyed in the country's so-called rice bowl, as had coastal shrimp farms and fishing boats. Huge areas were left without fresh water, electricity or transport. Although the military junta made a rare appeal for help, outside aid workers were not to be allowed into the country and a week later many isolated areas had received no internal relief of any kind. By this time it was announced that the death toll was 22 000 with a further 40 000 missing in a declared disaster zone of 24 million people. Reports talked of flood waters receding to leave rotting, bloated bodies, both human and animal, reminiscent of the 2004 post-tsunami scenes. Indeed two weeks after Nargis hit the country and with overseas aid still being rejected, the UN suggested that up to 200 000 Burmese had either died at the time, afterwards through a failure to provide relief, or were unaccounted for – a figure close to that of the 2004 tsunami.

Figure 9.56

The path of Hurricane Ivan, September 2004

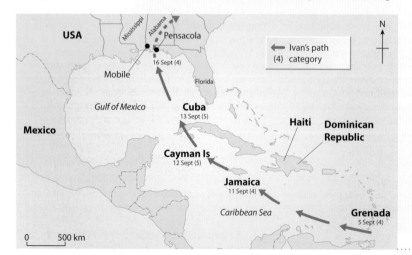

Figure 9.57

The monsoon in the Indian subcontinent

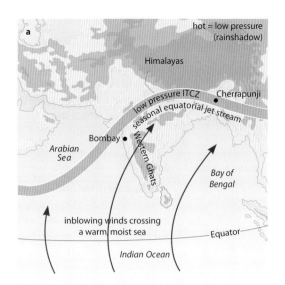

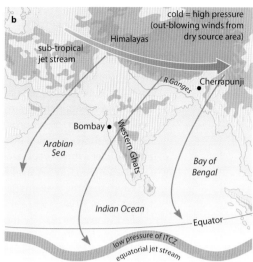

The monsoon

The word **monsoon** is derived from the Arabic word for 'a season', but the term is more commonly used in meteorology to denote a seasonal reversal of wind direction.

The major monsoon occurs in south-east Asia and results from three factors:

1   The extreme heating and cooling of large land masses in relation to the smaller heat changes over adjacent sea areas (page 210). This in turn affects pressure and winds.

2   The northward movement of the ITCZ (page 226) during the northern hemisphere summer.

3   The uplift of the Himalayas which, some 6 million years ago, became sufficiently high to interfere with the general circulation of the atmosphere (Places 5, page 20).

The south-west or summer monsoon

As the overhead sun appears to move northwards to the Tropic of Cancer in June, it draws with it the convergence zone associated with the ITCZ (Figure 9.57a). The increase in insolation over northern India, Pakistan and central Asia means that heated air rises, creating a large area of low pressure. Consequently, warm moist *Em* (equatorial maritime) and *Tm* air, from over the Indian Ocean, is drawn first northwards and then, because of the Coriolis force, is diverted north-eastwards (page 224). The air is humid, unstable and conducive to rainfall. Amounts of precipitation are most substantial on India's west coast, where the air rises over the Western Ghats, and on the windward slope of the Himalayas: Bombay has 2000 mm and Cherrapunji 13 000 mm in four summer months. The advent of monsoon storms allows the planting of rice (Places 67, page 481). Rainfall totals are accentuated as the air rises by both orographic and convectional uplift and the 'wet' monsoon is maintained by the release of substantial amounts of latent heat. The average arrival date is 10 May in Sri Lanka and 5 July at the Pakistan border – a time-lapse of seven weeks (Places 32).

The north-east or winter monsoon

During the northern winter, the overhead sun, the ITCZ and the subtropical jet stream all move southwards (Figure 9.57b). At the same time, central Asia experiences intense cooling which allows a large high pressure system to develop. Airstreams that move outwards from this high pressure area are dry because their source area is semi-desert. They become even drier as they cross the Himalayas and adiabatically warmer as they descend to the Indo-Gangetic plain. Bombay receives less than 100 mm of rain during these eight months. The south-west monsoon usually begins its retreat from the extreme north-west of India on 1 September and takes until 15 November, i.e. 11 weeks, to clear the southern tip.

The monsoon, which in reality is much more complex than the model described above, affects the lives of one-quarter of the world's population. Unfortunately, monsoon rainfall, especially in the Indian subcontinent, is unreliable (Figure 9.28). If the rains fail, then drought and famine ensue: 1987 was the ninth year in a decade when the monsoon failed in north-west India. If, on the other hand, there is excessive rainfall then large areas of land experience extreme flooding (Bangladesh in 1987, 1988 and 1998).

June

'Rain brought welcome relief to the Indian capital yesterday, a day after 18 people collapsed and died on the streets in the blistering heat, pushing the summer death toll in northern India to nearly 350. Heavy showers cooled the furnace-like city, reeling under a three-week heatwave that has kept daytime temperatures at an almost constant 45°C and which had, the previous day, experienced its hottest day in 50 years when the mercury soared to 42.6°C. It was the first pre-monsoon rain of the season to lash Delhi, and children celebrated by soaking themselves in the rain, with many elderly citizens joining them in the belief that monsoon rains help cure blisters and skin diseases caused by extreme heat. More thunderstorms are expected by the weekend, which should mark the onset of the summer monsoon.'

July

'The July death toll from relentless monsoon rains across India and Pakistan rose to more than 590 as several waves of severe storms passed across the subcontinent. Many streets in Delhi are still under water.'

Meso-scale: local winds

Of the three meso-scale circulations described here, two – **land and sea breezes** and **mountain and valley winds** – are caused by local temperature differences; the third – the **föhn** – results from pressure differences on either side of a mountain range.

The land and sea breeze

This is an example, on a diurnal timescale, of a circulation system resulting from differential heating and cooling between land surfaces and adjacent sea areas. The resultant pressure differences, although small and localised, produce gentle breezes which affect coastal areas during calm, clear anticyclonic conditions. When the land heats up rapidly each morning, lower pressure forms and a gentle breeze begins to blow from the sea to the land (Figure 9.58a). By early afternoon, this breeze has strengthened sufficiently to bring a freshness which, in the tropics particularly, is much appreciated by tourists at the beach resorts. Yet by sunset, the air and sea are both calm again.

Although the circulation cell rarely rises above 500 m in height or reaches more than 20 km inland in Britain, the sea breeze is capable of lowering coastal temperatures by 15°C and can produce advection fogs such as the 'sea-fret' or 'haar' of eastern Britain.

At night, when the sea retains heat longer than the land, there is a reversal of the pressure gradient and therefore of wind direction (Figure 9.58b). The land breeze, the gentler of the two, begins just after sunset and dies away by sunrise.

The mountain and valley wind

This wind is likely to blow in mountainous areas during times of calm, clear, settled weather. During the morning, valley sides are heated by the sun, especially if they are steep, south-facing (in the northern hemisphere) and lacking in vegetation cover. The air in contact with these slopes will heat, expand and rise (Figure 9.59a), creating a pressure gradient. By 1400 hours, the time of maximum heating, a strong uphill or **anabatic wind** blows up the valley and the valley sides – ideal conditions for hang-gliding! The air becomes conditionally unstable (Figure 9.19), often producing cumulus cloud and, under very warm conditions, cumulo-nimbus with the possibility of thunderstorms on the mountain ridges. A compensatory sinking of air leaves the centre of the valley cloud-free.

Figure 9.58

Land and sea breezes in Britain

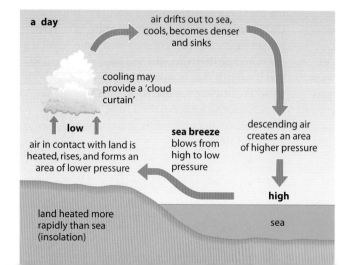

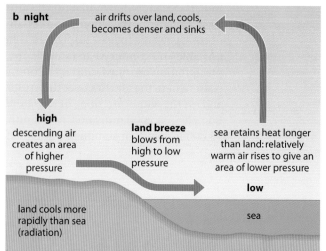

Figure 9.59

Mountain and
valley winds

a day (anabatic flow)

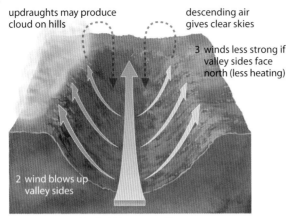

updraughts may produce
cloud on hills

descending air
gives clear skies

3 winds less strong if
valley sides face
north (less heating)

2 wind blows up
valley sides

1 wind blows up-valley

b night (katabatic flow)

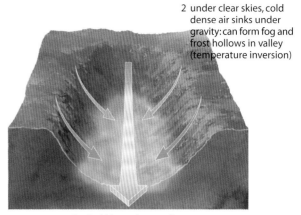

2 under clear skies, cold
dense air sinks under
gravity: can form fog and
frost hollows in valley
(temperature inversion)

1 wind blows down-valley

During the clear evening, the valley loses
heat through radiation. The surrounding air now
cools and becomes denser. It begins to drain,
under gravity, down the valley sides and along
the valley floor as a mountain wind or **katabatic
wind** (Figure 9.59b). This gives rise to a temperature inversion (Figure 9.24) and, if the air is moist
enough, in winter may create fog (Figure 9.23) or
a **frost hollow**. Maximum wind speeds are generated just before dawn, normally the coldest time
of the day. Katabatic winds are usually gentle in
Britain, but are much stronger if they blow over
glaciers or permanently snow-covered slopes. In
Antarctica, they may reach hurricane force.

The föhn

The föhn is a strong, warm and dry wind which
blows periodically to the lee of a mountain
range. It occurs in the Alps when a depression
passes to the north of the mountains and draws
in warm, moist air from the Mediterranean. As
the air rises (Figure 9.60), it cools at the DALR of
1°C per 100 m (page 216). If, as in Figure 9.60,

condensation occurs at 1000 m, there will be a
release of latent heat and the rising air will cool
more slowly at the SALR of 0.5°C per 100 m. This
means that when the air reaches 3000 m it will
have a temperature of 0°C instead of the –10°C
had latent heat not been released. Having crossed
the Alps, the descending air is compressed
and warmed at the DALR so that, if the land
drops sufficiently, the air will reach sea-level at
30°C. This is 10°C warmer than when it left the
Mediterranean. Temperatures may rise by 20°C
within an hour and relative humidity can fall to
10 per cent.

This wind, also known as the **chinook** on the
American Prairies, has considerable effects on
human activity. In spring, when it is most likely
to blow, it lives up to its Native American name
of 'snow-eater' by melting snow and enabling
wheat to be sown; and in Switzerland it clears the
alpine pastures of snow. Conversely, its warmth
can cause avalanches, forest fires and the prema-
ture budding of trees (Case Study 4a).

Figure 9.60

The föhn

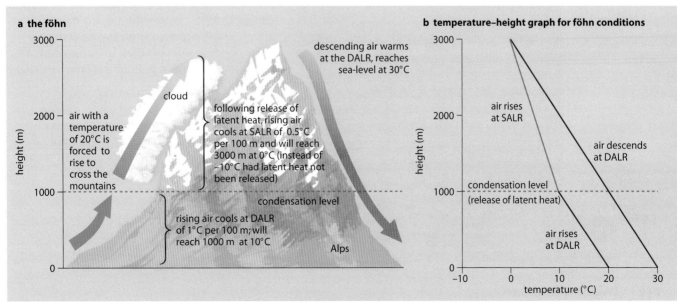

a the föhn

air with a
temperature
of 20°C is
forced to
rise to
cross the
mountains

cloud

following release of
latent heat, rising air
cools at SALR of 0.5°C
per 100 m and will reach
3000 m at 0°C (instead of
–10°C had latent heat not
been released)

descending air warms
at the DALR, reaches
sea-level at 30°C

condensation level

rising air cools at DALR
of 1°C per 100 m; will
reach 1000 m at 10°C

Alps

b temperature–height graph for föhn conditions

air rises
at SALR

air descends
at DALR

condensation level
(release of latent heat)

air rises
at DALR

temperature (°C)

Microclimates

Microclimatology is the study of climate over a small area. It includes changes resulting from the construction of large urban centres as well as those existing naturally between different types of land surface, e.g. forests and lakes.

Urban climates

Large cities and conurbations experience climatic conditions that differ from those of the surrounding countryside. They generate more dust and condensation nuclei than natural environments; they create heat; they alter the chemical composition and the moisture content of the air above them; and they affect both the albedo and the flow of air. Urban areas therefore have distinctive climates.

Temperature

Although tower blocks cast more shadow, normal building materials tend to be non-reflective and so absorb heat during the daytime. Dark-coloured roofs, concrete or brick walls and tarmac roads all have a high thermal capacity which means that they are capable of storing heat during the day and releasing it slowly during the night. Further heat is obtained from car fumes, factories, power stations, central heating and people themselves. The term **urban heat island** acknowledges that, under calm conditions, temperatures are highest in the more built-up city centre and decrease towards the suburbs and open countryside (Figure 9.61). In urban areas:

- daytime temperatures are, on average, 0.6°C higher
- night-time temperatures may be 3° or 4°C higher as dust and cloud act like a blanket to reduce radiation and buildings give out heat like storage radiators

- the mean winter temperature is 1° to 2°C higher (rural areas are even colder when snow-covered as this increases their albedo)
- the mean summer temperature may be 5°C higher
- the mean annual temperature is higher by between 0.6°C in Chicago and 1.3°C in London compared with that of the surrounding area.

Note how, in Figure 9.61, temperatures not only decrease towards London's boundary but also beside the Thames and Lea rivers. The urban heat island explains why large cities have less snow, fewer frosts, earlier budding and flowering of plants and a greater need, in summer, for air-conditioning than neighbouring rural areas.

Sunlight

Despite having higher mean temperatures, cities receive less sunshine and more cloud than their rural counterparts. Dust and other particles may absorb and reflect as much as 50 per cent of insolation in winter, when the sun is low in the sky and has to pass through more atmosphere, and 5 per cent in summer. High-rise buildings also block out light (Figure 9.62).

Wind

Wind velocity is reduced by buildings which create friction and act as windbreaks. Urban mean annual velocities may be up to 30 per cent lower than in rural areas and periods of calm may be 10–20 per cent more frequent. In contrast, high-rise buildings, such as the skyscrapers of New York and Hong Kong (Figure 9.62), form 'canyons' through which wind may be channelled. These winds may be strong enough to cause tall buildings to sway and pedestrians to be blown over and troubled by dust and litter. The heat island effect may cause local thermals and reduce the wind chill factor. It also tends to generate considerable small-scale turbulence and eddies. In 19th-century Britain, the most sought-after houses were usually on the western and south-western sides of cities, to be up-wind of industrial smoke and pollution (Mann's model, pages 422–423).

Relative humidity

Relative humidity is up to 6 per cent lower in urban areas where the warmer air can hold more moisture and where the lack of vegetation and water surface limits evapotranspiration.

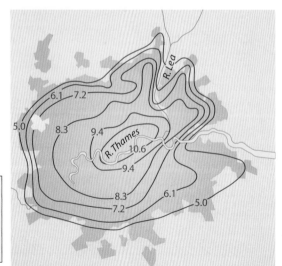

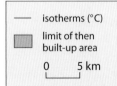

Figure 9.61

An urban heat island: minimum temperatures over London, 14 May 1959 (*after* Chandler)

Cloud

Urban areas appear to receive thicker and up to 10 per cent more frequent cloud cover than rural areas. This may result from convection currents generated by the higher temperatures and the presence of a larger number of condensation nuclei.

Precipitation

The mean annual precipitation total and the number of days with less than 5 mm of rainfall are both between 5 and 15 per cent greater in major urban areas. Reasons for this are the same as for cloud formation. Strong thermals increase the likelihood of thunder by 25 per cent and the occurrence of hail by up to 400 per cent. The higher urban temperatures may turn the snow of rural areas into sleet and limit, by up to 15 per cent, the number of days with snow lying on the ground. On the other hand, the frequency, length and intensity of fog, especially under anticyclonic conditions, is much greater – there may up to 100 per cent more in winter and 25 per cent more in summer, caused by the concentration of condensation nuclei (Figure 9.63).

Figure 9.62

Narrow streets with high-rise buildings are more likely to develop micro-climates than those that are wider and have lower buildings; New York City

Thick fog is continuing to cause travel chaos among those looking forward to spending Christmas abroad. Over the last few days, thousands of passengers have experienced severe delays or cancellations of flights at numerous UK airports.

Yesterday 350 flights, 40 per cent of the total, were cancelled from Heathrow alone and, with fog set to remain today, British Airways has already decided to cancel all domestic flights to and from that airport. The problem with fog is that it means, for safety reasons, the distance between aircraft on approach to runways has to be doubled, thus reducing the number of landings.

22 December 2006

Figure 9.63

Fog causes Christmas chaos

Atmospheric composition

There may be three to seven times more dust particles over a city than in rural areas. Large quantities of gaseous and solid impurities are emitted into urban skies by the burning of fossil fuels, by industrial processes and from car exhausts. Urban areas may have up to 200 times more sulphur dioxide and 10 times more nitrogen oxide (the major components of acid rain) than rural areas, as well as 10 times more hydrocarbons and twice as much carbon dioxide. These pollutants tend to increase cloud cover and precipitation, cause smog (Figure 9.25), give higher temperatures and reduce sunlight.

Forest and lake microclimates

Different land surfaces produce distinctive local climates. Figure 9.64 summarises and compares some of the characteristics of microclimates found in forests and around lakes. As with urban climates, research and further information are still needed to confirm some of the statements.

Figure 9.64

Microclimates of forests and water surfaces

Microclimate feature	Forest (coniferous and deciduous)	Water surface (lake, river)
Incoming radiation and albedo	Much incoming radiation is absorbed and trapped. Albedo for coniferous forest is 15%; deciduous 25% in summer and 35% in winter; and desert scrub 40%.	Less insolation absorbed and trapped. Albedo may be over 60%, i.e. higher than over seas/oceans (page 207). Higher on calm days.
Temperature	Small diurnal range due to blanket effect of canopy. Forest floor is protected from direct sunlight. Some heat lost by evapotranspiration.	Small diurnal range because water has a higher specific heat capacity. Cooler summers and milder winters. Lakesides have a longer growing season.
Relative humidity	Higher during daytime and in summer, especially in deciduous forest. Amount of evapotranspiration depends on length of day, leaf surface area, wind speed, etc.	Very high, especially in summer when evaporation rates are also high.
Precipitation	Heavy rain can be caused by high evapotranspiration rates, e.g. in tropical rainforests. On average, 30–35% of rain is intercepted: more in deciduous woodland in summer than in winter.	Air is humid. If forced to rise, air can be unstable and produce cloud and rain. Amounts may not be great due to fewer condensation nuclei. Fogs form in calm weather.
Wind speed and direction	Trees reduce wind speeds, especially at ground level. (They are often planted as windbreaks.) Trees can produce eddies.	Wind may be strong due to reduced friction. Large lakes (e.g. L. Victoria) can create land and sea breezes (page 240).

Weather maps and forecasting in Britain

A weather map or **synoptic chart** shows the weather for a particular area at one specific time (Figures 9.67 and 9.68). It is the result of the collection and collation of a considerable amount of data at numerous weather stations, i.e. from a number of sample points (Framework 6, page 159). These data are then refined, usually as quickly as possible and now using computers, and are plotted using internationally accepted weather symbols. A selection of these symbols is shown in Figure 9.65. Weather maps are produced for different purposes and at various scales.

1 The daily weather map, as seen on television or in a national newspaper, aims to give a clear, but highly simplified, impression of the weather.

2 At a higher level, a synoptic map shows selected meteorological characteristics for specific **weather stations**. The **station model** in Figure 9.66 shows six elements: temperature, pressure, cloud cover, present weather (e.g. type of precipitation), wind direction and wind speed.

3 At the highest level, the Meteorological Office produces maps showing finite detail, e.g. amounts of various types of cloud at low, medium and high levels, dew point temperatures, barometric tendency (i.e. trends of pressure change), etc.

The role of the weather forecaster is to try to determine the speed and direction of movement of various air masses and any associated fronts, and to try to predict the type of weather these movements will bring. Forecasters now make considerable use of **satellite images** (Figures 9.67 and 9.68). Satellite images are photos taken by weather satellites as they continually orbit the Earth. These photos, which are relayed back to Earth, are invaluable in the prediction of short-term weather trends. Although forecasting is increasingly assisted by information from satellites, radar and computers, which show upper air as well as surface air conditions in a three-dimensional model, the complexity and unpredictability of the atmosphere can still catch the forecaster by surprise (Places 29, page 232). Part of this problem is related to the fact that meteorological information is a sample (Framework 6, page 159) rather than a total picture of the atmosphere, and so there is always a risk of the anomaly becoming the reality.

Figure 9.65

Weather symbols for cloud, precipitation, wind speed, temperature, pressure and wind direction

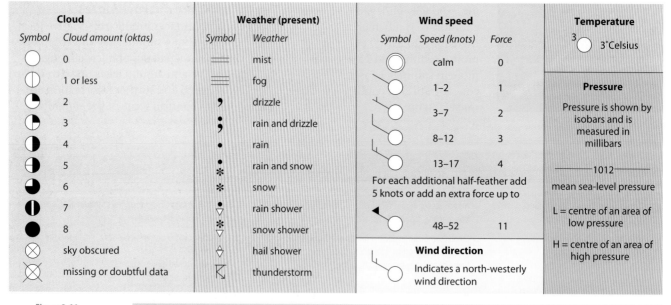

Cloud		Weather (present)		Wind speed			Temperature
Symbol	Cloud amount (oktas)	Symbol	Weather	Symbol	Speed (knots)	Force	3°Celsius
	0	≡	mist		calm	0	
	1 or less	≡	fog		1–2	1	**Pressure**
	2	𝅘	drizzle		3–7	2	Pressure is shown by isobars and is measured in millibars
	3	𝅘	rain and drizzle		8–12	3	
	4	•	rain		13–17	4	——1012——
	5	• / ✳	rain and snow				mean sea-level pressure
	6	✳	snow	For each additional half-feather add 5 knots or add an extra force up to			L = centre of an area of low pressure
	7	▽•	rain shower				
	8	▽✳	snow shower		48–52	11	H = centre of an area of high pressure
⊗	sky obscured	▽△	hail shower	**Wind direction**			
⊗	missing or doubtful data	⎏	thunderstorm	Indicates a north-westerly wind direction			

Figure 9.66

A weather station model and an example

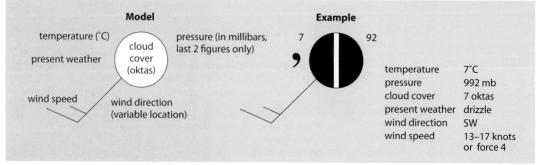

temperature	7°C
pressure	992 mb
cloud cover	7 oktas
present weather	drizzle
wind direction	SW
wind speed	13–17 knots or force 4

Figure 9.67

Synoptic chart and
satellite image,
17 September 1983

Figure 9.67 shows the synoptic chart (weather map) and satellite (infra-red) image of a depression approaching the British Isles (compare Figure 9.43). Figure 9.68 shows the same depression 24 hours later, by which time it had passed over the British Isles (compare Figure 9.44).

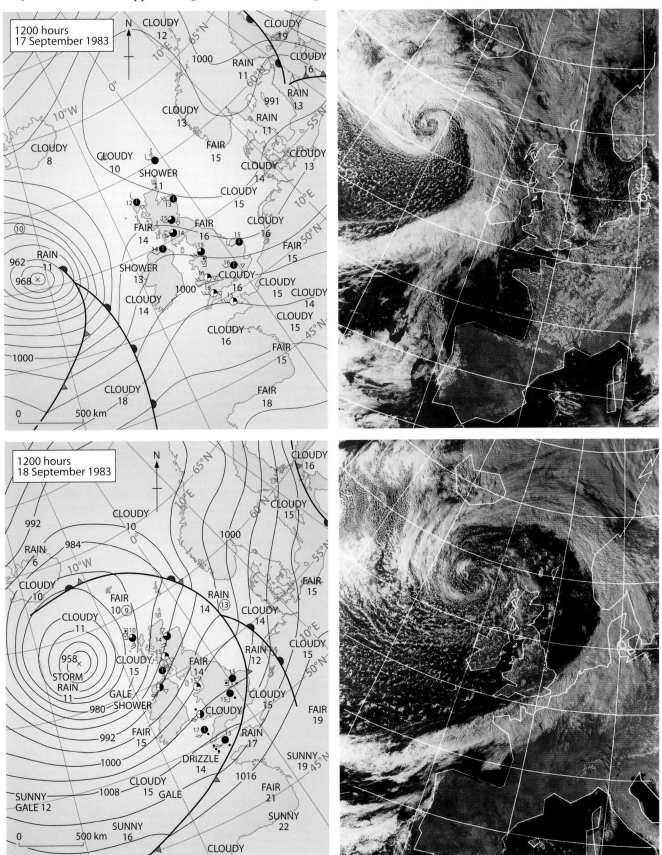

Figure 9.68

Synoptic chart and satellite
image, 18 September 1983

Throughout this chapter on weather and climate, mean climatic figures have been quoted. To build up these pictures of global, regional and local climate patterns, statistics have been obtained by averaging readings, usually for temperature and precipitation, over a 30-year timescale. However, these averages themselves are often not as significant as the range or the degree to which they vary from, or are dispersed about, the mean.

For example, two tropical weather stations may have equal annual rainfall totals when measured over 30 years. Station A may lie on the Equator and experience reliable rainfall with little variation from one year to the next. Station B may experience a monsoon climate where in some years the rains may fail entirely while in others they cause flooding.

The measure of dispersion from the mean can be obtained by using any one of three statistical techniques:

- the range
- the interquartile range, or
- the standard deviation.

These techniques are included here because meteorological data both require and benefit from their use, but they may be applied to most branches of geography where there is a danger that the mean, taken alone, may be misleading (the problems of overgeneralisation are discussed in Framework 11, page 347). Again, it must be stressed that use of a quantitative technique does not guarantee objective interpretation of data: great care must be taken to ensure that an appropriate method of manipulating the data is chosen.

It has already been seen how it is possible, given a data set, to calculate the mean and the median (Framework 5, page 112). However, neither statistic gives any idea of the spread, or range, of that data. As the example above of two tropical weather stations shows, mean values on their own give only part of the full picture. The spread of the data around the mean should also be considered.

Range

This very simple method involves calculating the difference between the highest and lowest values of the sample population, e.g. the annual range in temperature for London is 14°C (July 18°C, January 4°C). The range emphasises the extreme values and ignores the distribution of the remainder.

Interquartile range

The interquartile range consists of the middle 50 per cent of the values in a distribution, 25 per cent each side of the median (middle value). This calculation is useful because it shows how closely the values are grouped around the median (Figure 9.69). It is easy to calculate; it is unaffected by extreme values; and it is a useful way of comparing sets of similar data.

The example in Figure 9.69 gives temperatures for 19 weather stations in the British Isles at 0600 on 14 January 1979. These temperatures have been ranked in the table.

Figure 9.69

The interquartile range

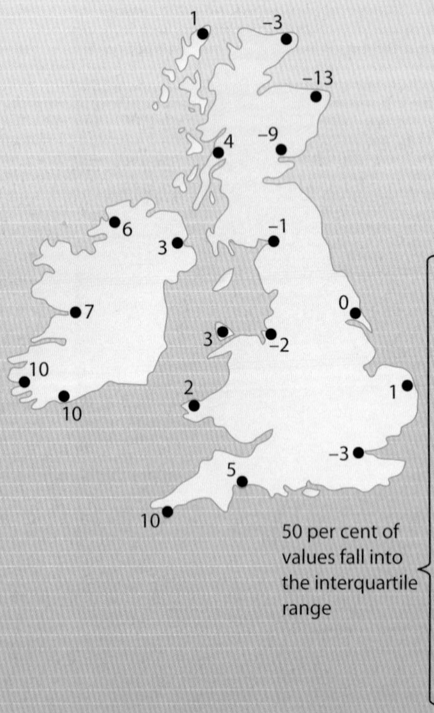

Rank	Temperatures 0°C (ranked)	
	10	
	10	
	10	
	7	
5	6	upper quartile
	5	
	4	
	3	
	3	
10	2	median (middle quartile)
	1	
	1	
	0	
	−1	
15	−2	lower quartile
	−3	
	−3	
	−9	
	−13	

50 per cent of values fall into the interquartile range

The upper quartile (UQ) is obtained by using the formula:

$$UQ = \left(\frac{n+1}{4}\right) \qquad \text{i.e.} \quad \left(\frac{19+1}{4}\right) = 5$$

This means that the UQ is the fifth figure from the top of the ranking order, i.e. 6°C. The lower quartile (LQ) is found by using a slightly different formula:

$$LQ = \left(\frac{n+1}{4}\right) \times 3 \qquad \text{i.e.} \quad \left(\frac{19+1}{4}\right) \times 3 = 15$$

This shows the LQ to be the 15th figure in the ranking order, i.e. –2°C. You will notice that the middle quartile is the same as the median. The interquartile range is the difference between the upper and lower quartiles, i.e. 6°C − −2°C = 8°C.

Another measure of dispersion, the quartile deviation, is obtained by dividing the interquartile range by two, i.e. 8°C ÷ 2 = 4°C

The smaller the interquartile range, or quartile deviation, the greater the grouping around the median and the smaller the dispersion or spread.

Standard deviation

This is the most commonly used method of measuring dispersion and although it may involve lengthy calculations it can be used with the arithmetic mean and it removes extreme values. The formula for the standard deviation is:

$$\sigma = \sqrt{\frac{\Sigma (x - \bar{x})^2}{n}}$$

where: σ = standard deviation

x = each value in the data set

$\bar{x}$ = mean of all values in the data set, and

n = number of values in the data set.

Let us suppose that the minimum temperatures for 10 weather stations in Britain on a winter's day were, in °C, 5, 8, 3, 2, 7, 9, 8, 2, 2 and 4. The standard deviation of this data set is worked out in Figure 9.70, proceeding as follows:

1 Find the mean ($\bar{x}$).

2 Subtract the mean from each value in the set:

$x - \bar{x}$.

3 Calculate the square of each value in 2, to remove any minus signs: $(x - \bar{x})^2$.

4 Add together all the values obtained in 3:

$\Sigma (x - \bar{x})^2$.

5 Divide the sum of the values in 4 by n:

$$\frac{\Sigma (x - \bar{x})^2}{n}$$

6 Take the square root of the value obtained in 5 to obtain the standard deviation:

$$\sqrt{\frac{\Sigma (x - \bar{x})^2}{n}}$$

The resulting standard deviation of σ = 2.65 is a low value, indicating that the data are closely grouped around the mean.

Figure 9.70

Finding the standard deviation

Minimum temperatures for 10 weather stations in Britain on a winter's day

The mean of 5, 8, 3, 2, 7, 9, 8, 2, 2, 4:

$$\bar{x} = \frac{50}{10} = 5$$

Weather station	Temperature at each station (x)	$x - \bar{x}$	$(x - \bar{x})^2$
1	5	5 − 5 = 0	0
2	8	8 − 5 = 3	9
3	3	3 − 5 = −2	4
4	2	2 − 5 = −3	9
5	7	7 − 5 = 2	4
6	9	9 − 5 = 4	16
7	8	8 − 5 = 3	9
8	2	2 − 5 = −3	9
9	2	2 − 5 = −3	9
10	4	4 − 5 = −1	1
			$\Sigma (x - \bar{x})^2 = 70$

$$\sigma = \sqrt{\frac{70}{10}}$$

$$\sigma = \sqrt{7} \qquad \therefore \text{ standard deviation} = 2.65$$

Climatic change

Climates have changed and still are constantly changing at all scales, from local to global, and over varying timespans, both long-term and short-term (Case Studies 9A and 9B). However, there have been surges of change over time which meteorologists and earth scientists are continually trying to clarify and explain.

Evidence of past climatic changes

- **Rocks** are found today which were formed under climatic conditions and in environments that no longer exist (Figure 1.1). In Britain, for example, coal was formed under hot, wet tropical conditions; sandstones were laid down during arid times; various limestones accumulated on the floors of warm seas; and glacial deposits were left behind by retreating ice sheets.
- **Fossil landscapes** exist, produced by certain geomorphological processes which no longer operate. Examples include glacially eroded highlands in north and west Britain (Chapter 4), granite tors on Dartmoor (page 202) and wadis formed during wetter periods (pluvials) in deserts (Places 25, page 188).
- Evidence exists of **changes in sea-level** (both isostatic as on Arran – Places 23, page 166) and eustatic (as at present in the Maldives – page 169) and changes in **lake levels** (Sahara, Figure 7.27).
- **Vegetation belts** have shifted through some 10° of latitude, e.g. changes in the Sahara Desert (Figure 7.27).
- **Pollen analysis** shows which plants were dominant at a given time. Each plant species has a distinctively-shaped pollen grain. If these grains land in an oxygen-free environment, such as a peat bog, they resist decay. Although pollen can be transported considerable distances by the wind and by wildlife, it is assumed that grains trapped in peat form a representative sample of the vegetation that was growing in the surrounding area at a given time; also, that this vegetation was a response to the climatic conditions prevailing at that time. Vertical sections made through peat show changes in pollen (i.e. vegetation), and these changes can be used as evidence of climatic change (the vegetation–climatic timescale in Figures 11.18 and 11.19).
- **Dendrochronology**, or tree-ring dating, is the technique of obtaining a core from a tree-trunk and using it to determine the age of the tree. Tree growth is rapid in spring, slower by the autumn and, in temperate latitudes, stops in winter. Each year's growth is shown by a single ring. However, when the year is warm and wet, the ring will be larger because the tree grows more quickly than when the year is cold and dry. Tree-rings therefore reflect climatic changes. Recent work in Europe has shown that tree growth is greatest under intense cyclonic activity and is more a response to moisture than to temperature. Tree-ring timescales are being established by using the remains of oak trees, some nearly 10 000 years old, found in river terraces in south-central Europe. Bristlecone pines, still alive after 5000 years, give a very accurate measure in California (page 294).
- **Chemical methods** include the study of oxygen and carbon isotopes. An isotope is one of two or more forms of an element which differ from each other in atomic weight (i.e. they have the same number of protons in the nucleus, but a different number of neutrons). For example, two isotopes in oxygen are O-16 and O-18. The O-16 isotope, which is slightly lighter, vaporises more readily; whereas O-18, being heavier, condenses more easily. During warm, dry periods, the evaporation of O-16 will leave water enriched with O-18 which, if it freezes into polar ice, will be preserved as a later record (Places 14, page 104). Colder, wetter periods will be indicated by ice with a higher level of O-16. The most accurate form of dating is based on C-14, a radioactive isotope of carbon. Carbon is taken in by plants during the carbon cycle (Figure 11.25). Carbon-14 decays radioactively at a known rate and can be compared with C-12, which does not decay. Using C-12 and C-14 from a dead plant, scientists can determine the date of death to a standard error of ± 5 per cent. This method can accurately date organic matter up to 50 000 years old.
- **Historical records** of climatic change include:
  - cave paintings of elephants in central Sahara (Figure 7.27) and giraffes in Jordan (Figure 7.7)
  - vines growing successfully in southern England between AD 1000 and 1300
  - graves for human burial in Greenland which were dug to a depth of 2 m in the 13th century, but only 1 m in the 14th century, and could not be dug at all in the 15th century due to the extension of permafrost – in contrast to its retreat in the 2000s (Case Study 5)
  - fairs held on the frozen River Thames in Tudor times
  - the measurement of recent advances and retreats of alpine glaciers and polar sea-ice.

Causes of climatic change

Several suggestions have been advanced to try to explain climatic change over different timescales (Figure 4.2) and epochs (Figure 1.1). Most climatologists now accept that each of the causes of climatic change described below has a role to play in explaining change in the past, whether over long or short periods of time.

1 **Variations in solar energy** Although it was initially believed that solar energy output did not vary over time (hence the term 'solar constant' in Figure 9.3), increasing evidence suggests that sunspot activity, which occurs in cycles, may significantly affect our climate – times of high annual temperatures on Earth appear to correspond to periods of maximum sunspot activity.

2 **Astronomical relationships between the sun and the Earth** There is increasing evidence supporting Milankovitch's cycles of change in the Earth's orbit, tilt and wobble (Figure 4.6), which would account for changes in the amounts of solar radiation reaching the Earth's surface. This evidence is mainly from cores that have been drilled through undisturbed ocean-floor sediment which has accumulated over thousands of years (compare Places 14, page 104).

3 **Changes in oceanic circulation** Changes in oceanic circulation affect the exchange of heat between the oceans and the atmosphere. This can have both long-term effects on world climate (where currents at the onset of the Quaternary ice age flowed in opposite directions to those at the end of the ice age) and short-term effects (El Niño, Case Study 9A). The latest theory compares the North Atlantic Drift with a conveyor belt that brings water to northwest Europe. Should this conveyor belt be closed down, possibly by a huge influx of fresh water into the sea, then the climate will become dramatically colder.

4 **Meteorites** A major extinction event, which included the dinosaurs, took place about 65 million years ago. This event was believed to have been caused by one or more meteors colliding with the Earth. This seems to have caused a reduction in incoming radiation, a depletion of the ozone layer and a lowering of global temperatures.

5 **Volcanic activity** It has been accepted for some time that volcanic activity has influenced climate in the past, and continues to do so. World temperatures are lowered after any large single eruption, e.g. Mount Pinatubo (Case Study 1) and Krakatoa (Figure 1.29 and Places 35, page 289) or after a series of volcanic eruptions. This is due to the increase in dust particles in the lower atmosphere which will absorb and scatter more of the incoming radiation (Figure 9.4). Evidence suggests that these major eruptions may temporarily offset the greenhouse effect. Precipitation also increases due to the greater number of hygroscopic nuclei (dust particles) in the atmosphere (page 215).

6 **Plate tectonics** Plate movements have led to redistributions of land masses and to long-term effects on climate. These effects may result from a land mass 'drifting' into different latitudes (British Isles, page 22); or from the seabed being pushed upwards to form high fold mountains (page 19). The presence of fold mountains can lead to a colder climate (a suggested cause of the Quaternary ice age, page 103) and can act as a barrier to atmospheric circulation – the Asian monsoon was established by the creation of the Tibetan Plateau (page 239).

7 **Composition of the atmosphere** Gases in the atmosphere can be increased and altered following volcanic eruptions. At present there is increasing concern at the build-up of CO_2 and other greenhouse gases in the atmosphere (Case Study 9B), together with the use of aerosols and the release of CFCs (Places 27, page 209), which are blamed for the depletion of ozone in the upper atmosphere.

Climatic change in Britain

Britain's climate has undergone changes in the longest term (page 22 and Figure 1.1); during and since the onset of the Quaternary (Figure 4.2); and in the more recent short term (Figure 11.18). Following the 'little ice age' (which lasted from about AD 1540 to 1700), temperatures generally increased to reach a peak in about 1940. After that time, there was a tendency for summers to become cooler and wetter, springs to be later, autumns milder and winters more unpredictable. However, since the onset of the 1980s there appears to have been a considerable warming, with eight of the ten warmest years on record being in the last decade. This, together with the apparent increase in variations from the norm for Britain's expected autumn, winter, spring and even, since 2005, summer weather, tends to add further evidence to the concept of global warming (Case Study 9B).

A Short-term change: El Niño and La Niña

The oceans, as we have seen, have a considerable heat storage capacity which makes them a major influence on world climates. If ocean temperatures change, this will have a considerable effect upon weather patterns in adjacent land masses. Interactions between the ocean and the atmosphere have become, recently, a major scientific study.

The most important and interesting example of the ocean–atmosphere interrelationship is provided by the El Niño and La Niña events which occur periodically in the Pacific Ocean. Under normal atmospheric conditions, pressure rises over the eastern Pacific Ocean (off the coast of South America) and falls over the western Pacific Ocean (towards Indonesia and the Philippines). The descending air over the eastern Pacific gives the clear, dry conditions that create the Atacama Desert in Peru (Figure 7.2 and Places 24, page 180), while the warm, moist ascending air over the western Pacific gives that region its heavy convectional rainfall (page 226). This movement of air creates a circulation cell, named after Walker who first described it, in which the upper air moves from west to east, and the surface air from east to west as the trade winds (Figure 9.71). The trade winds:

- push surface water westwards so that sea-level in the Philippines is normally 60 cm higher than in Panama and Colombia
- allow water, flowing westward as the equatorial current, to remain near to the ocean surface where it can gradually heat. This gives the western Pacific the world's highest ocean temperature, usually above 28°C. In contrast, as warm water is pushed away from South America, it is replaced by an upwelling of colder, nutrient-rich water. This colder water lowers temperatures, sometimes to below 20°C, but does provide a plentiful supply of plankton which forms the basis of Peru's fishing industry.

Figure 9.71

The Walker circulation cell

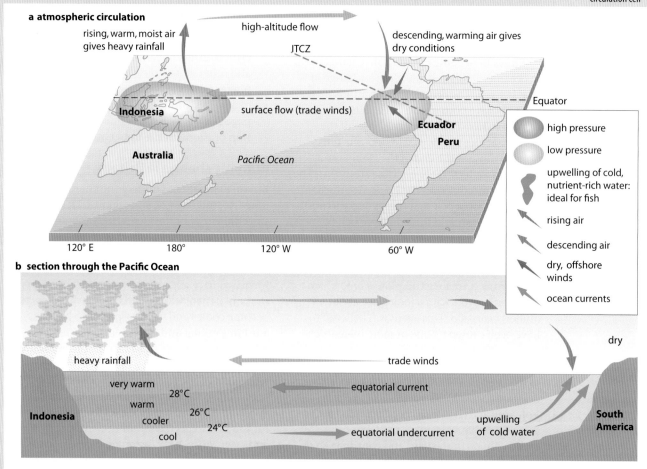

a atmospheric circulation

rising, warm, moist air gives heavy rainfall

high-altitude flow

ITCZ

descending, warming air gives dry conditions

Indonesia

surface flow (trade winds)

Ecuador
Peru

Equator

Australia

Pacific Ocean

high pressure

low pressure

upwelling of cold, nutrient-rich water: ideal for fish

rising air

descending air

dry, offshore winds

ocean currents

120° E   180°   120° W   60° W

b section through the Pacific Ocean

heavy rainfall

trade winds

dry

very warm

28°C

warm

26°C

cooler

24°C

Indonesia

equatorial current

equatorial undercurrent

upwelling of cold water

South America

cool

El Niño

An El Niño event, scientifically referred to as an El Niño Southern Oscillation (ENSO), occurs periodically – on average every three to four years. It is called 'El Niño', which means 'little child' in Spanish, because, in those years that it does occur, it appears just after Christmas. An El Niño event usually lasts for 12–18 months.

In contrast to normal conditions (Figure 9.71) there is a reversal, in the equatorial Pacific region, in pressure, precipitation and, often, winds and ocean currents (Figure 9.72). Pressure rises over the western Pacific and falls over the eastern Pacific. This allows the ITCZ (Figure 9.34) to migrate southwards and causes the trade winds to weaken in strength, or, sometimes, even to be reversed in their direction. The descending air, now over South-east Asia, gives that region much drier conditions than it usually experiences and, on extreme occasions, even causing drought. In contrast the air over the eastern Pacific is now rising, giving much wetter conditions in places, like Peru, that normally experience desert conditions. The change in the direction of the trade winds means that:

- surface water tends to be pushed eastwards so that sea-level in South-east Asia falls, while it rises in tropical South America

- surface water temperatures in excess of 28°C extend much further eastwards and the upwelling of cold water off South America is reduced, allowing sea temperatures to rise by up to 6°C. The warmer water in the eastern Pacific lacks oxygen, nutrients and, therefore, plankton and so has an adverse effect on Peru's fishing industry.

NASA-Mir astronauts were able, during the record-breaking 1997–98 El Niño, to observe, photograph and document the global impacts of the event. These, together with ground observations and recordings, are summarised in Figure 9.73.

Figure 9.72

An El Niño event

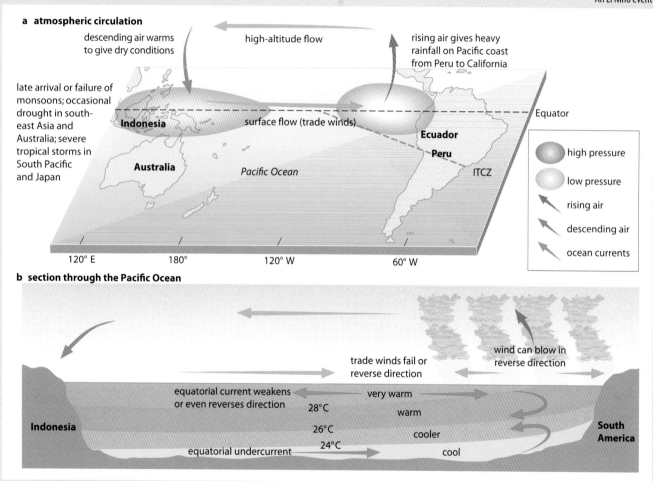

a atmospheric circulation

descending air warms to give dry conditions

high-altitude flow

rising air gives heavy rainfall on Pacific coast from Peru to California

late arrival or failure of monsoons; occasional drought in south-east Asia and Australia; severe tropical storms in South Pacific and Japan

Indonesia

surface flow (trade winds)

Equator

Ecuador

Peru

Australia

Pacific Ocean

ITCZ

120° E   180°   120° W   60° W

high pressure
low pressure
rising air
descending air
ocean currents

b section through the Pacific Ocean

trade winds fail or reverse direction

wind can blow in reverse direction

equatorial current weakens or even reverses direction

very warm

28°C

warm

Indonesia

26°C

cooler

24°C

cool

South America

equatorial undercurrent

Evidence collected during the El Niño events of 1982–83 (at the time the biggest ever recorded), 1986 and 1992–93, increasingly suggested that the ENSO had a major effect on places far beyond the Pacific margins as well as on those bordering the ocean itself in its low latitudes. Apart from the drier conditions in South-east Asia and the wetter conditions in South America:

- severe droughts were experienced in the Sahel (Case Study 7) and southern Africa as well as across the Indian subcontinent

- there were extremely cold winters in central North America, and stormy conditions with floods in California

- exceptionally wet, mild and windy winters were experienced in Britain and north-west Europe.

The 1997–98 event: the biggest yet experienced

Early 1997	Evidence of a rapid rise in sea temperatures in the eastern Pacific.
July	El Niño conditions intense.
September	Over 24 million km^2 of warm water (size of North and Central America) extended from the International Dateline to South America.
1998 April	Evidence of El Niño weakening.
June	NASA satellite surveillance showed a significant drop in sea temperatures in the eastern Pacific.
Autumn	Signs of a La Niña event (page 253).

Figure 9.73

The effects of the
1997–98 El Niño event

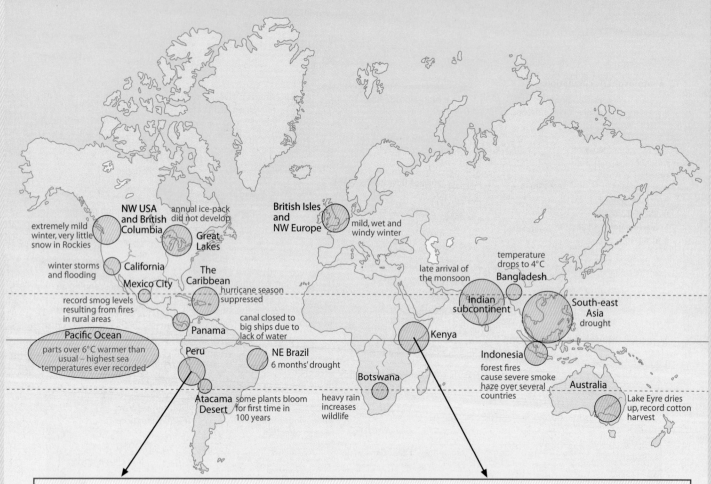

PERU For *each* of 12 days in early March, Peru received the equivalent of six months of normal rain. Over several months, flash flooding caused 292 deaths, injured more than 16 000 people, left 400 missing, destroyed 13 200 houses, wrecked 250 000 km of roads, swept away bridges, damaged crops and schools and disrupted the lives of up to half a million Peruvians.

KENYA Parts of Kenya received over 1000 mm of rainfall during six months (up to 50 times more than the average) at a time normally considered to be the 'dry season'. Roads and the mainline railway were swept away, the latter causing the derailment of the Nairobi–Mombasa train. Later, more than 500 people died of malaria as the receding floodwaters created ideal mosquito-spawning pools.

A mild El Niño episode: 2006–07

In September 2006, NASA's Jason altimetric satellite detected a rise in the sea-level of the Pacific Ocean which indicated the return of El Niño. However, the rise was slight, suggesting that the event might be short-lived and, being far less intense than the 1997–98 El Niño episode, unlikely to have a great effect on global weather patterns. It declined within six months without ending the drought in the south-west of the USA.

La Niña

Just as El Niño was ending in June 1998, forecasters were predicting – based on an 8°C fall in sea temperatures in the eastern Pacific in May – the arrival that winter of a La Niña event. La Niña, or 'little girl', has climatic conditions that are the reverse of those of El Niño. However, although when La Niña does appear it is just before or just after El Niño, its occurrence has been less frequent (the last was between June 1988 and February 1989) and, consequently, it is less easy to predict its possible effects because there is less evidence.

In a La Niña event, in contrast to normal conditions in the Pacific Ocean (Figure 9.71), the low pressure over the western Pacific becomes even lower and the high pressure over the eastern Pacific even higher (Figure 9.74). This means that rainfall increases over South-east Asia (was the La Niña event of 1988 responsible for the severe flooding at that time in Bangladesh?), there are drought conditions in South America and, due to the increased difference in pressure between the two places, the trade winds strengthen. The stronger trade winds:

- push large amounts of water westwards, giving a higher than normal sea-level in Indonesia and the Philippines
- increase the equatorial under-current and significantly enhance the upwelling of cold water off the Peruvian coast.

Scientists suggest that La Niña can be linked with increased hurricane activity in the Caribbean (Places 31) and that it can interrupt the jet stream over Britain to give stormier (Places 29), wetter (Case Study 3C) and cooler conditions.

Figure 9.74

A La Niña event

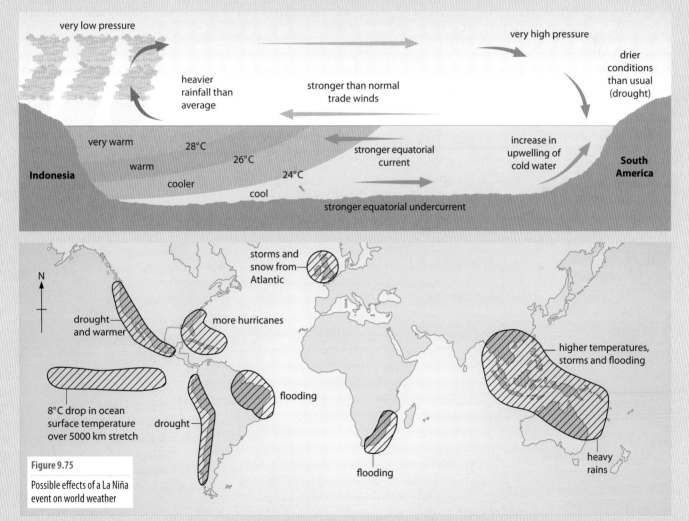

Figure 9.75

Possible effects of a La Niña event on world weather

A La Niña episode: 2007–08

The Jason altimetric satellite noted, in February 2007, a transition from the warm El Niño to the cool La Niña, a change not welcomed by the parched south-west of the USA. This La Niña episode, the strongest for several years, lasted for over 12 months until it began to weaken in April 2008. By then, it had caused torrential rain in Australia, breaking a long crop-ruining drought, and had given central China an exceptionally cold, snow-covered winter.

B Long-term change: global warming – an update

2005 and 2007: the warmest two years on record

Scientists claimed it was clear that temperatures around the world were continuing their upward climb. The global average for these years was 14.76°C in 2005 and 14.73°C in 2007 – the two warmest since reliable instrumental records began 126 years earlier and, according to palaeo-climatologists using evidence from ancient tree-rings (page 248), probably the highest in over 1200 years. Records collected by NASA GISS also showed that eight of the ten warmest years have been in the last decade and that 2007 was the 31st consecutive year when the global mean surface temperature exceeded the long-term average (Figure 9.76). More alarmingly, whereas the global mean rose by only 0.23°C in the 100 years between 1880 and 1979, in the 27 years since then it has increased by 0.62°C. Although the main reason for the rise in global temperature (Figure 9.76) is the longer-term effect of the continued release of greenhouse gases into the atmosphere (Figures 9.77 and 9.78), there is increasing evidence suggesting that temperatures increase more rapidly during an El Niño rather than in a La Niña episode (Case Study 9A).

Figure 9.76

Average global temperatures, 1880–2007

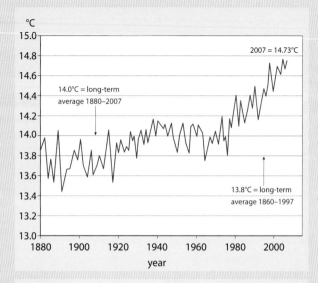

Figure 9.77

Atmospheric concentration of carbon dioxide, 1000–2007

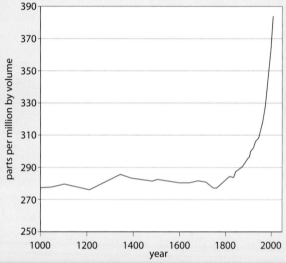

Figure 9.78

The major greenhouse gases

Gas	Sources (natural and human)
water vapour	evaporation from the ocean, evapotranspiration from land
carbon dioxide	burning of fossil fuels (power houses, industry, transport), burning rainforests, respiration
methane	decaying vegetation (peat and in swamps), farming (fermenting animal dung and rice-growing), sewage disposal and landfill sites
nitrous oxide	vehicle exhausts, fertiliser, nylon manufacture, power stations
CFCs	refrigerators, aerosol sprays, solvents and foams

a the radiation balance

incoming short-wave radiation (ultra-violet) passes directly through the natural greenhouse gases

most outgoing long-wave radiation (infrared) is radiated back into space

natural greenhouse gases

some outgoing radiation is absorbed by, or trapped beneath, the greenhouse gases

previously a balance:
CO₂ given off by humans and animals = CO₂ taken in by trees
O₂ given out by trees = O₂ used by humans and animals

b the greenhouse effect

less heat escapes into space

Figure 9.79

The radiation balance and the greenhouse effect

increase in greenhouse gases due to human activity

as more heat is trapped and retained, so the Earth's atmosphere becomes warmer (global warming)

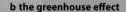

short-wave radiation is transformed into long-wave radiation (heat) on contact with the Earth's surface

The Earth is warmed during the day by incoming, short-wave radiation (insolation) from the sun and cooled at night by out-going, longer-wave, infra-red radiation (page 207). As, over a lengthy period of time, the Earth is neither warming up nor cooling down, there must be a balance between incoming and outgoing radiation (page 209). While incoming radiation is able to pass through the atmosphere (which is 99 per cent nitrogen and oxygen, Figure 9.2), some of the outgoing radiation is trapped by a blanket of trace gases. Because they trap heat as in a greenhouse, these are referred to as **greenhouse gases** (Figure 9.79). Without these natural greenhouse gases, the Earth's average temperature would be 33°C lower than it is today – far too cold for life in any form. (During the last ice age, temperatures were only 4°C lower.) Water vapour provides the majority of the natural greenhouse effect, with lesser contributions from carbon dioxide, methane, nitrous oxide and ozone.

During the last 150 years there has been, with the exception of water vapour which remains a constant in the system, a rise in greenhouse gas concentrations (Figure 9.78). This has been due largely to the increase in world population and a corresponding growth in human activity, especially agricultural and industrial activities.

By adding these gases to the atmosphere, we are increasing its ability to trap heat (Figure 9.79). Most scientists now accept that the greenhouse effect is causing global warming. World temperatures have risen by 0.9°C in the last 100 years. Latest predictions suggest that they are likely to increase by between 1°C and 6°C by the year 2100. Some of the predicted global effects of this climate change are shown in Figure 9.81.

Britain's weather forecast for the 2080s

The latest government report predicts, in general, an increasingly grim forecast for the next 70 years. Heavy winter rains, up to 30 per cent in excess of today, will lead to more frequent flooding, as was seen in the English Midlands in 2007 (Case Study 3C) and destructive gales will be more frequent and severe. With a predicted rise in sea-level of between 2 and 10 cm, storm surges and higher tides will threaten coastal areas (Case Study 6). However, the chances of extremely cold winters, and the risk of fog and heavy snowfalls, will decrease. Days with more than 25 mm of rain, at present an extreme event, could occur three or four times a year. Summers will be drier with a decrease in rain of up to 30 per cent in the south-east where drought will become more common. With a

predicted increase in summer temperatures of over 3°C, heat waves will become a more regular occurrence and there will be many more days when thermometers exceed 25°C. Changes in the weather will be greater in the south-east than in the north-west.

However, some computer predictions are suggesting that Britain's climate could, over a long period of time, get colder. This could happen if the release of fresh water from Greenland's melting ice-cap pushed the North Atlantic Drift further south so that it no longer affected all, or certainly parts, of Britain.

Effects of climate change in the UK

DEFRA's claims, based on the predicted forecast of milder, wetter, stormier winters and warmer, drier summers, are summarised in Figure 9.80. Its two main concerns are:

- the potential effects of changing rainfall patterns on hydrology and ecosystems
- rising sea-levels and more frequent storms in coastal areas where there is a large proportion of Britain's population, its manufacturing industry, energy production, mineral extraction, valued natural environments and recreational amenities.

Soils	Higher temperatures could reduce water-holding capacities and increase soil moisture deficits, affecting the types of crops and trees. Less organic matter due to drier summers (less produced) and wetter winters (more lost).
Flora/fauna	Higher temperatures and increased water deficit could mean loss of several native species. Warmer climate would allow plants to grow further north and at higher altitudes. Earlier flowering plants and arrival of migrant birds.
Agriculture	Grasses helped by longer growing season (extra 15 days) but cereals hit by drier summers. Increase in number of pests. Maize and vines in the south. Need for irrigation in summer.
Forestry	Certain trees able to grow at higher altitudes. New species could be introduced from warmer climates. Threats from fires, diseases and pests.
Coastal regions	Rise in sea-level plus increase in frequency/number of gales and frequency/height of storm surges would mean more flooding, especially around estuaries, and increased erosion. Major impact on housing, industry, farming, energy, transport and wildlife, including marine eco-systems.
Water resources	Water resources would benefit from wetter winters, but hotter, drier summers would increase demands/pressures. Need for irrigation in summer in south-east. More frequent river flooding.
Energy	Space heating demand would fall in winter but need for air-conditioning would rise in summer. Probable overall fall in demand. Many power stations are in threatened coastal areas.
Manufacturing/construction	Problem for coastal industries. Fewer days lost in construction due to less snow/frost.
Transport	Many types of transport are sensitive to extreme weather conditions. Benefit of less snow, ice and perhaps fog. Loss due to more frequent and severe storms and flooding, including flash floods.
Recreation/tourism	Tourism would benefit from longer, warmer, drier summers, but insufficient snow for skiing in Scotland.

Figure 9.80

Specific effects of climate change in the UK

Source: DEFRA

Figure 9.81

Some predicted effects of global warming

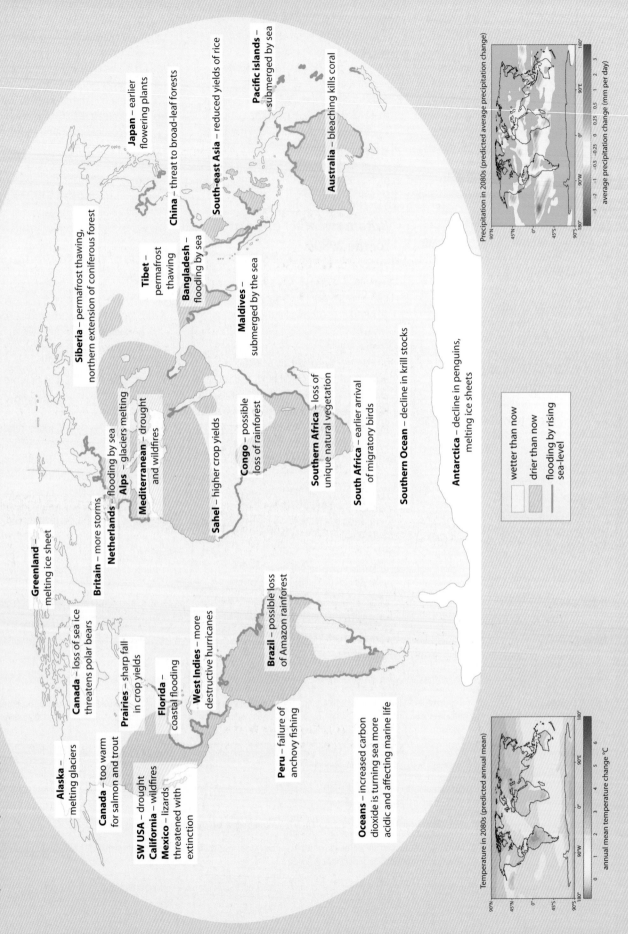

Global increase in droughts, floods and storms

Greenland – melting ice sheet

Alaska – melting glaciers

Canada – loss of sea ice threatens polar bears

Canada – too warm for salmon and trout

Prairies – sharp fall in crop yields

SW USA – drought
California – wildfires
Mexico – lizards threatened with extinction

Florida – coastal flooding

West Indies – more destructive hurricanes

Brazil – possible loss of Amazon rainforest

Peru – failure of anchovy fishing

Oceans – increased carbon dioxide is turning sea more acidic and affecting marine life

Siberia – permafrost thawing, northern extension of coniferous forest

Japan – earlier flowering plants

China – threat to broad-leaf forests

Tibet – permafrost thawing

Bangladesh – flooding by sea

South-east Asia – reduced yields of rice

Pacific islands – submerged by sea

Australia – bleaching kills coral

Maldives – submerged by the sea

Britain – more storms

Netherlands – flooding by sea

Alps – glaciers melting

Mediterranean – drought and wildfires

Sahel – higher crop yields

Congo – possible loss of rainforest

Southern Africa – loss of unique natural vegetation

South Africa – earlier arrival of migratory birds

Southern Ocean – decline in krill stocks

Antarctica – decline in penguins, melting ice sheets

☐ wetter than now
▨ drier than now
— flooding by rising sea-level

Precipitation in 2080s (predicted average precipitation change)

average precipitation change (mm per day)
-3 -2 -1 -0.5 -0.25 0 0.25 0.5 1 2 3

90°N 45°N 0° 45°S 90°S
180° 90°W 0° 90°E 180°

Temperature in 2080s (predicted annual mean)

annual mean temperature change °C
0 1 2 3 4 5 6

90°N 45°N 0° 45°S 90°S
180° 90°W 0° 90°E 180°

Further reference

Barry, R.G. and Chorley, R.J. (2003) *Atmosphere, Weather and Climate*, Routledge.

Bentley, M. (2005) 'Antarctic ozone hole', *Geography Review* Vol 18 No 3 (January).

Burt, T. (2005) 'Rain in the hills', *Geography Review* Vol 18 No 4 (March).

Digby, B. (2005) 'El Niño Part 1', *Geography Review* Vol 19 No 2 (November).

Digby, B. (2006) 'El Niño Part 2' *Geography Review* Vol 19 No 3 (January).

Goudie, A.S. (2001) *The Nature of the Environment*, WileyBlackwell.

Holden, J. (2008) *Introduction to Physical Geography and the Environment*, Prentice Hall.

Middleton, N. 'Acid shock', *Geography Review* Vol 18 No 4 (March).

O'Hara, G., Sweeney, J. and O'Hare, G. (1986) *The Atmospheric System*, Oliver & Boyd.

Smithson, P., Addison, K. and Atkinson, K. (2008) *Fundamentals of Physical Geography*, Routledge.

Center for Ocean-Atmospheric Prediction Studies, resources:
http://coaps.fsu.edu/lib/elninolinks/themes

Earth Space Research Group, Indian monsoons:
www.icess.ucsb.edu/esrg/IOM2/Start2_IOM.html

Jet Propulsion Laboratory, NASA, El Niño:
www.jpl.nasa.gov/earth/ocean_motion/el_nino_index.cfm

UK Climate Impacts Programme:
www.ukcip.org.uk/index.php

UK Meterological Office (Met Office):
www.metoffice.gov.uk/

UK Met Office weather charts:
www.meto.gov.uk/education/data/charts.html

Union of Concerned Scientists (UCS), global warming:
http://ucsusa.org/warming/index.html

Union of Concerned Scientists (UCS), ozone depletion:
www.ucsusa.org/global_warming/science_and_impacts/science/faq-about-ozone-depletion-and.html

US Environmental Protection Agency, global warming:
www.epa.gov/climatechange/index.html

US Environmental Protection Agency, glossary of climate change terms:
http://yosemite.epa.gov/oar/globalwarming.nsf/content/glossary.html

US Environmental Protection Agency, ozone science:
www.epa.gov/ozone/strathome.html

US National Oceanographic and Atmospheric Administration (NOAA) Climate Prediction Center:
www.cpc.noaa.gov/

US NOAA, El Niño / La Niña:
www.cpc.noaa.gov/products/analysis_monitoring/ensostuff/
www.elnino.noaa.gov/

US NOAA, hurricanes:
http://hurricanes.noaa.gov/

US NOAA, research:
www.cdc.noaa.gov/ENSO

Questions & Activities

Activities

1 **a** What is the 'atmosphere' of the Earth? *(3 marks)*

   b What is the *difference* between 'weather' and 'climate'? *(4 marks)*

   c Describe the 'solar cascade of energy' to the Earth. *(4 marks)*

   d What is the importance of **i** carbon dioxide and **ii** clouds in the energy balance of the Earth? *(4 marks)*

   e Ozone in the troposphere is a danger to health. Why is there concern that ozone in the stratosphere is being depleted? *(5 marks)*

   f What measures can be taken to restrict the potential damage due to ozone depletion? *(5 marks)*

2 **a** How does a meteorologist get information to forecast the weather? *(4 marks)*

   b Use Places 29 (page 232) to answer the following questions:

   i What was the weather forecast on 11–15 October 1987? *(3 marks)*

   ii Describe the meteorological conditions over the Western Approaches and Bay of Biscay at 6.00 pm on 15 October. *(3 marks)*

   iii Describe the track of the storm over the next 12 hours. *(4 marks)*

   iv What happened to the weather over southern England during this 12-hour period? *(4 marks)*

   v Describe three effects of the storm on people. *(3 marks)*

   c Explain two reasons why meteorologists failed to forecast the very strong winds of 15 October. *(4 marks)*

Exam practice: basic structured questions

3 **a** Explain how each of the following factors affects the winds that cross them:

     i a large body of water (e.g. a sea) *(4 marks)*

     ii a mountain range. *(6 marks)*

   b On a field course in Switzerland a geography student noted: 'On the north-facing side of the valley the forests came close to the valley floor while the settlement huddled at the foot of the south-facing slope and here there were ploughed fields. There were forests but they started higher up the slope.'

     Suggest the cause of these differences in land use. *(6 marks)*

   c A January weather forecast for the UK stated: 'Although it will be cool today, temperatures will stay above freezing tonight because of the cloud cover'.

     Explain the effect of cloud on temperature. *(4 marks)*

   d Why is it warmer in summer than in winter? *(5 marks)*

4 **a** **i** What is 'stratus' cloud? *(2 marks)*

     ii What is 'cumulo-nimbus' cloud? *(2 marks)*

   b Making good use of diagrams, explain why rain falls when an onshore wind blows over an upland area. *(7 marks)*

   c Why does fog often form over a coastal area in the autumn? *(6 marks)*

   d Explain the formation of smog over an urban area. *(8 marks)*

5 **a** Describe the causes of the ITCZ. *(5 marks)*

   b What weather conditions are associated with the ITCZ? *(10 marks)*

   c Why does the ITCZ move with the seasons? *(10 marks)*

6 Study Figure 9.82 and answer the following questions.

   a What is the name of the pressure system shown? *(2 marks)*

   b What is the weather like at place A (Doncaster)? *(4 marks)*

   c What is the red line with half circles on it? *(5 marks)*

   d Locate the warmest and the coolest place in the British Isles. *(2 marks)*

   e **i** Over the next 12 hours the pressure system moves so that it is in the North Sea.

     Give a weather forecast for place A (Doncaster) over this period. *(6 marks)*

     ii Why would you expect this to happen? *(6 marks)*

Exam practice: structured questions

7 **a** Study Figures 9.82 and 9.83. Describe the changes in the weather being experienced at Limerick (place C) over this 24-hour period. *(8 marks)*

   b Explain what has happened to the frontal system over this period of time. *(8 marks)*

   c Describe, and explain the causes of, the types and distribution of the precipitation shown in Figure 9.83. *(9 marks)*

8 **a** Describe **three** mechanisms that are likely to trigger upward movement of a parcel of air from sea level. *(6 marks)*

   b Study Figure 9.84.

     i What is meant by the term 'ELR'? *(4 marks)*

     ii Identify the height of the base of clouds. *(1 mark)*

     iii Explain why this height is the cloud base. *(4 marks)*

     iv Identify the air stream(s) (A, B, C) that would have cloud cover. State why this is so. *(2 marks)*

     v At what height would condensation in a cloud be in the form of ice? *(2 marks)*

   c Choose **either** stability **or** instability. Describe and explain the weather conditions normally associated with that atmospheric condition. *(6 marks)*

9 **a** **i** Using an annotated diagram only, illustrate the variation of temperature and pressure with altitude in the atmosphere. *(6 marks)*

     ii Explain the variations in temperature with altitude in the atmosphere. *(6 marks)*

   b **i** Study Figure 9.5 (page 209). Making good use of the data, explain why there is a general trend of movement of heat energy from the Equator to the poles. *(6 marks)*

     ii Describe how heat is transferred from the tropics towards the poles. *(7 marks)*

10 **a** Describe and explain what happens to *incoming solar radiation* (**insolation**) once it reaches the edge of the Earth's atmosphere. *(10 marks)*

   b Explain the importance of **each** of the following in relation to heat energy in the atmosphere:

     i latitude

     ii altitude

     iii land and sea. *(10 marks)*

   c The greatest amount of insolation is experienced close to the Equator. Why does this area not become increasingly hot? *(5 marks)*

11 **a** Suggest **one** way you could test the hypothesis that the temperatures in an urban area are different from those in the surrounding countryside. Describe the method you would use to collect and record the data to carry out the proposed test. *(7 marks)*

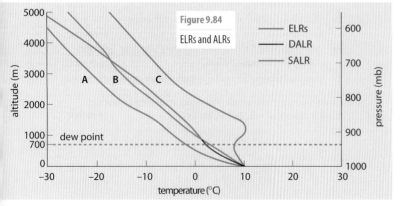

Figure 9.84
ELRs and ALRs
— ELRs
— DALR
— SALR

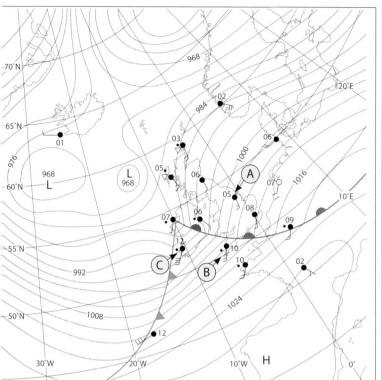

Figure 9.82

Weather map for 1200 hrs,
12 January 1984

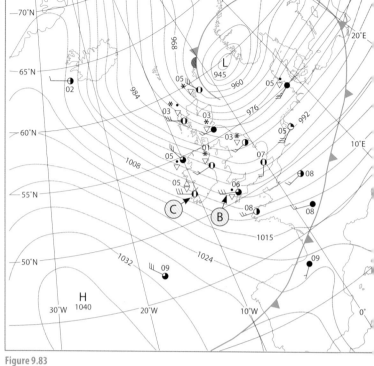

Figure 9.83

Weather map for 1200 hrs,
13 January 1984

b Explain **two** reasons why temperatures in urban areas may be higher than those in surrounding rural areas.
(10 marks)

c Suggest **two** ways in which planning policies can reduce the problems caused by microclimatic features of urban areas. *(8 marks)*

12 a Explain the **difference** between absolute humidity and relative humidity. *(8 marks)*

b Making good use of diagrams, show how condensation occurs as air rises through the atmosphere. *(10 marks)*

c Explain the cause of low-level clouds (mist) as shown in Figure 9.23 (page 221). *(7 marks)*

13 The following are meteorological conditions that develop a range of weather conditions over the British Isles:

a an anticyclone centred over the English Midlands in winter

b a mature depression with its centre over the Central Valley of Scotland in summer

c a depression centred over Paris and an anticyclone to the north of Scotland in January.

Choose two of the situations a–c and, in both cases, describe how weather conditions would vary in two contrasting locations in the British Isles.

Explain these variations. *(12 + 13 marks)*

14 a Study Figure 9.49 (page 235). Describe the major distribution of tropical storms as shown on the map.
(6 marks)

b Choose any **one** type of tropical storm. Describe and explain the sequence of weather associated with the passage of the storm. *(10 marks)*

c Explain how people respond to the hazard posed by tropical storms. In your answer refer to countries at different stages of economic development. *(9 marks)*

Exam practice: essays

15 'The polar front jet stream is one of the most important influences on the climate of the British Isles.'
Discuss this statement. *(25 marks)*

16 The passage of a depression over the British Isles leads to predictable changes in the weather over a period of time. Describe and explain the sequence of weather experienced in Liverpool over a 12-hour period as a mature depression passes from west to east. *(25 marks)*

17 'There is now overwhelming scientific evidence that human activity is causing major changes to the global climate.'
Is this statement true? Justify your answer. *(25 marks)*

10 Soils

··

'To many people who do not live on the land, soil appears to be an inert, uniform, dark-brown coloured, uninteresting material in which plants happen to grow. In fact little could be further from the truth.'

Brian Knapp, *Soil Processes*, 1979

Soil forms the thin surface layer of the Earth's crust. It can be defined as the unconsolidated mineral and organic material on the Earth's surface, often characterised by horizons or layers (Figure 10.5), that serves as a natural medium for the growth of plants and therefore the support of animal life on land. It has been subjected to, and shows the effects of, genetic and environmental factors of: climate (including water and temperature), macro- and micro-organisms, relief and the underlying parent rock (Figure 10.1). It develops over a period of time through the interaction of several physical, chemical, biological and morphological properties and characteristics.

The study of soil, its origins and characteristics (**pedology**) is a science in itself.

Soil formation

The first stage in the formation of soil is the accumulation of a layer of loose, broken, unconsolidated parent material known as **regolith**. Regolith may be derived from either the *in situ* weathering of bedrock (i.e. the parent or underlying rock) or from material that has been transported from elsewhere and deposited, e.g. as alluvium, glacial drift, loess or volcanic ash. The second stage, the formation of **true soil** or **topsoil**, results from the addition of water, gases (air), living organisms (biota) and decayed organic matter (humus).

Pedologists have identified five main factors involved in soil formation (Figure 10.1). As all of these are closely interconnected and interdependent, their relationship may be summarised as follows:

soil = *f* (parent material + climate + topography + organisms + time)
where: *f* = function of.

Parent material

When a soil develops from an underlying rock, its supply of minerals is largely dependent on that rock. The minerals are susceptible to different rates and processes of weathering – see the example of granite, Figure 10.2. Parent material contributes to control of the depth, texture, drainage (permeability) and quality (nutrient content) of a soil and also influences its colour. In most of Britain, parent material is the major factor in determining the soil type, e.g. limestone, granite or, most commonly, drift.

Figure 10.1

Factors affecting the formation of soil

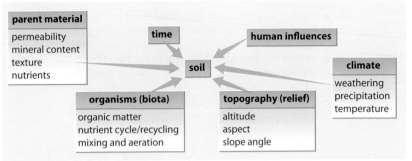

Figure 10.2

The influence of a parent rock – granite – on soil formation

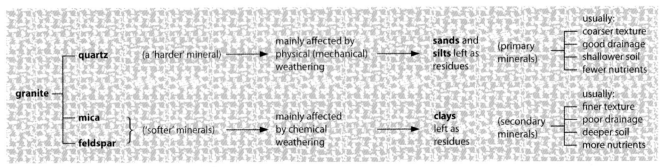

Climate

Climate determines the type of soil at a global scale. The distribution of world soil types corresponds closely to patterns of climate and vegetation. Climate affects the rate of weathering of the parent rock, with the most rapid breakdown being in hot, humid environments. Climate also affects the amount of humus (organic material) in the soil. The amount is a balance between the input and output, the input and output being a function of the effects of temperature and moisture on biological activity. One might expect tropical rainforest soils to have more humus than tundra soils because of the greater mass of vegetation. However, it is possible for some tundra soils to have more humus accumulation due to a lower output, and some tropical rainforest soils to have less because of greater humus breakdown.

Rainfall totals and intensity are also important. Where rainfall is heavy, the downward movement of water through the soil transports mineral salts (i.e. soluble minerals) with it, a process known as **leaching**. Where rainfall is light or where evapotranspiration exceeds precipitation, water and mineral salts may be drawn upwards towards the surface by the process of **capillary action**.

Temperatures determine the length of the growing season and affect the supply of humus. The speed of vegetation decay is fastest in hot, wet climates as temperatures also influence (i) the activity and number of soil organisms and (ii) the rate of evaporation, i.e. whether leaching or capillary action is dominant.

Topography (relief)

As the height of the land increases, so too do amounts of precipitation, cloud cover and wind, while temperatures and the length of the growing season both decrease. Aspect is an important local factor in mid-latitudes (page 212), with south-facing slopes in the northern hemisphere being warmer and drier than those facing north. The angle of slope affects drainage and soil depth. Greater moisture flows and the increased effect of gravity on steeper slopes can accelerate mass movement and the risk of soil erosion. Soils on steep slopes are likely to be thin, poorly developed and relatively dry. The more gentle the slope, the slower the rate of movement of water through the soil and the greater the likelihood of waterlogging and the formation of peat on plateau-like surfaces at the top of the slope (Figure 10.3). There is little risk of soil erosion but the increased rate of weathering, due to the extra water, and the receipt of material moved downslope, tend to produce deep soils at the foot of the slope. A **catena** is where soils are related to the topography of a hillside and is a sequence of soil types down a slope. The catena (Figure 10.3) is described in more detail on page 276.

Organisms (biota)

Plants, micro-organisms such as bacteria and fungi, and animals all interact in the **nutrient cycle** (page 300). Plants take up mineral nutrients from the soil and return them to it after they die. This recycling of plant nutrients (Figure 12.7) is achieved by the activity of micro-organisms, which assist in nitrogen fixation (page 268) and the decomposition and decay of dead vegetation. At the same time, macro-organisms, which include worms and termites, mix and aerate the soil. Human activity is increasingly affecting soil development through the addition of fertiliser, the breaking up of horizons by ploughing, draining or irrigating land, and by unwittingly accelerating or deliberately controlling soil erosion.

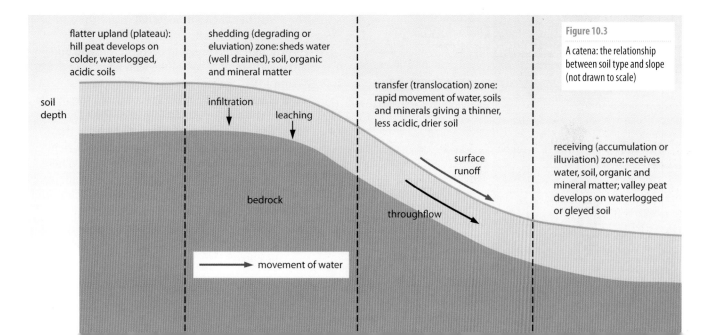

flatter upland (plateau): hill peat develops on colder, waterlogged, acidic soils

shedding (degrading or eluviation) zone: sheds water (well drained), soil, organic and mineral matter

transfer (translocation) zone: rapid movement of water, soils and minerals giving a thinner, less acidic, drier soil

Figure 10.3

A catena: the relationship between soil type and slope (not drawn to scale)

soil depth

infiltration

leaching

surface runoff

receiving (accumulation or illuviation) zone: receives water, soil, organic and mineral matter; valley peat develops on waterlogged or gleyed soil

bedrock

throughflow

movement of water

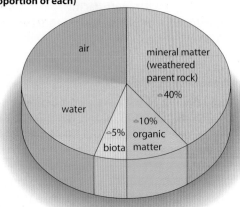

pore space containing **air** and/or water = **45%** (can be **45% water**, or **45% air**, but is more usually a proportion of each)

mineral matter + organic matter + biota = **55%**

air

mineral matter (weathered parent rock) ≈40%

water

≈10% organic matter

≈5% biota

Figure 10.4

Relative proportions, by volume, of components in a 'normal' soil (*after* Courtney and Trudgill)

Time

Soils usually take a long time to form, perhaps up to 400 years for 10 mm and, under extreme conditions, 1000 years for 1 mm. It can take 3000 to 12 000 years to produce a sufficient depth of mature soil for farming, although agriculture can be successful on newly deposited alluvium and volcanic ash. Newly forming soils tend to retain many characteristics of the parent material from which they are derived. With time, they acquire new characteristics resulting from the addition of organic matter, the activity of organisms, and from leaching. **Horizons**, or layers (Figure 10.5), reflect the balance between soil processes and the time that has been available for their development. In northern Britain, upland soils must be less than 10 000 years old, as that was the time of the last glaciation, when any existing soil cover was removed by ice. The time taken for a mature soil to develop depends primarily on parent material and climate. Soils develop more rapidly where parent material derived from

in situ weathering consists of sands rather than clays, and in hot, wet climates rather than in colder and/or drier environments.

A mature, fully-developed soil consists of four components: mineral matter, organic matter including biota (page 268), water and air. The relative proportions of these components in a 'normal' soil, by volume, is given in Figure 10.4.

The soil profile

The **soil profile** is a vertical section through the soil showing its different horizons (Figure 10.5). It is a product of the balance between soil system inputs and outputs (Figure 10.6) and the redistribution of, and chemical changes in, the various soil constituents. Different soil profiles are described in Chapter 12, but an idealised profile is given here to aid familiarisation with several new terms.

The three major soil horizons, which may be subdivided, are referred to by specific letters to indicate their genetic origin.

- The upper layer, or **A horizon**, is where biological activity and humus content are at their maximum. It is also the zone that is most affected by the leaching of soluble materials and by the downward movement, or **eluviation**, of clay particles. Eluviation is the washing out of material, i.e. the removal of organic and mineral matter from the *A* horizon (Figure 10.5).
- Beneath this, the **B horizon** is the zone of accumulation, or **illuviation**, where clays and other materials removed from the *A* horizon are redeposited. Illuviation is the process of inwashing, i.e. the redeposition of organic and mineral matter in the *B* horizon. The *A* and *B* horizons together make up the true soil.
- The **C horizon** consists mainly of recently weathered parent material (regolith) resting on the bedrock.

Although this threefold division is useful and convenient, it is, as will be seen later, oversimplified. Several examples show this:

- Humus may be mixed throughout the depth of the soil, or it may form a distinct layer. Where humus is incorporated within the soil to give a crumbly, black, nutrient-rich layer it is known as **mull** (page 266). Where humus is slow to decompose, as in cold, wet upland areas, it produces a fibrous, acidic and nutrient-deficient surface horizon known as **mor** (page 266) (peat moorlands).
- The junctions of horizons may not always be clear.

Figure 10.5

An idealised soil profile in Britain

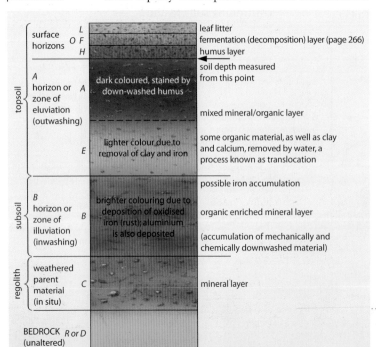

surface horizons	L O F H	leaf litter / fermentation (decomposition) layer (page 266) / humus layer
		soil depth measured from this point
topsoil — A horizon or zone of eluviation (outwashing)	A	dark coloured, stained by down-washed humus
		mixed mineral/organic layer
	E	lighter colour due to removal of clay and iron
		some organic material, as well as clay and calcium, removed by water, a process known as translocation
subsoil — B horizon or zone of illuviation (inwashing)	B	brighter colouring due to deposition of oxidised iron (rust); aluminium is also deposited
		possible iron accumulation
		organic enriched mineral layer
		(accumulation of mechanically and chemically downwashed material)
regolith — weathered parent material (in situ)	C	mineral layer
BEDROCK (unaltered)	R or D	

- All horizons need not always be present.
- The depth of soil and of each horizon vary at different sites. Local conditions produce soils with characteristic horizons differing from the basic *A*, *B*, *C* pattern: for example, a waterlogged soil, having a shortage of oxygen, develops a gleyed (*G*) horizon (page 275).

The soil system

Figure 10.6 is a model showing the soil as an open system where materials and energy are gained and lost at its boundaries. The system comprises inputs, stores, outputs and recycling or feedback loops (Framework 3, page 45). Inputs include:
- water from the atmosphere or throughflow from higher up the slope
- gases from the atmosphere and the respiration of soil animals and plants
- mineral nutrients from weathered parent material, which are needed as plant food
- organic matter and nutrients from decaying plants and animals, and
- solar energy and heat.

Outputs include:
- water lost to the atmosphere through evapotranspiration
- nutrients lost through leaching and throughflow, and
- loss of soil particles through soil creep and erosion.

Recycling

Plants, in order to live, take up nutrients from the soil (page 268). Some of the nutrients may be stored until:
- either the vegetation sheds its leaves (during the autumn in Britain), or
- the plants die and, over time, decompose due to the activity of micro-organisms (biota, page 268).

These two processes release the stored nutrients, allowing them to be returned to the soil ready for future use – the so-called **nutrient** (or humus) **cycle**.

Soil properties

The four major components of soil – water, air, mineral and organic matter (Figure 10.4) – are all closely interlinked. The resultant interrelationships produce a series of 'properties', ten of which are listed and described below.
1. mineral (inorganic) matter
2. texture
3. structure
4. organic matter (including humus)
5. moisture
6. air
7. organisms (biota)
8. nutrients
9. acidity (pH value)
10. temperature.

It is necessary to understand the workings of these properties to appreciate how a particular soil can best be managed.

1 Mineral (inorganic) matter

As shown in Figure 10.2, soil minerals are obtained mainly by the weathering of parent rock. Weathering is the major process by which nutrients, essential for plant growth, are released. **Primary minerals** are minerals that were present in the original parent material and which remain unaltered from their original state. They are present throughout the soil-forming process, mainly because they are insoluble, e.g. quartz. **Secondary minerals** are produced by weathering reactions and are therefore produced within the soil. They include oxides and hydroxides of primary minerals (e.g. iron) which result from the exposure to air and water (page 40).

Figure 10.6

The 'open' soil system

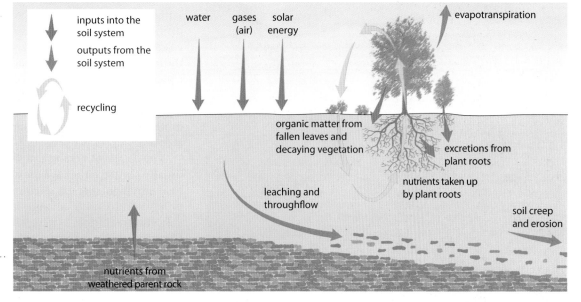

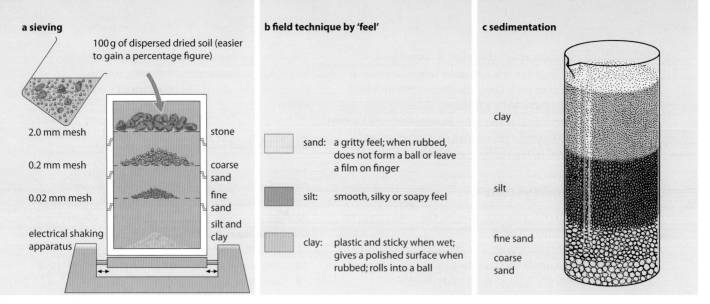

a sieving

100 g of dispersed dried soil (easier to gain a percentage figure)

2.0 mm mesh — stone

0.2 mm mesh — coarse sand

0.02 mm mesh — fine sand

— silt and clay

electrical shaking apparatus

b field technique by 'feel'

sand: a gritty feel; when rubbed, does not form a ball or leave a film on finger

silt: smooth, silky or soapy feel

clay: plastic and sticky when wet; gives a polished surface when rubbed; rolls into a ball

c sedimentation

clay

silt

fine sand

coarse sand

Figure 10.7

Measuring soil texture (*after* Courtney and Trudgill)

2 Soil texture

The term 'texture' refers to the degree of coarseness or fineness of the mineral matter in the soil. It is determined by the proportion of **sand**, **silt** and **clay** particles. Particles larger than sand are grouped together and described as stones. In the field, it is possible to decide whether a soil sample is mainly sand, silt or clay by its 'feel'. As shown in Figure 10.7b, a sandy soil feels gritty and lacks cohesion; a silty soil has a smoother, soaplike feel as well as having some cohesion; and a clay soil is sticky and plastic when wet and, being very cohesive, may be rolled into various shapes.

This method gives a quick guide to the texture, but it lacks the precision needed to determine the proportion of particles in a given soil with any accuracy. This precision may be obtained from either of two laboratory measurements, both of which are dependent upon particle size. The Soil Survey of England and Wales uses the British Standards classification, which gives the following diameter sizes:

Heading	Description from case study
coarse sand	between 2.0 and 0.6 mm
medium sand	between 0.6 and 0.2 mm
fine sand	between 0.2 and 0.06 mm
silt	between 0.06 and 0.002 mm
clay	less than 0.002 mm

One method of measuring texture involves the use of sieves with different meshes (Figure 10.7a). The sample must be dry and needs to be well-shaken. A mesh of 0.2 mm, for example, allows fine sand, silt and clay particles to pass through it, while trapping the coarse sand. The weight of particles remaining in each sieve is expressed as a percentage of the total sample.

In the second method, sedimentation (Figure 10.7c), a weighed sample is placed in a beaker of water, thoroughly shaken and then allowed to settle. According to **Stoke's Law**, 'the settling rate of a particle is proportional to the diameter of that particle'. Consequently, the larger, coarser, sand grains settle quickly at the bottom of the beaker and the finer, clay particles settle last, closer to the surface (compare Figure 3.22). The Soil Survey and Land Research Centre tends to use both methods because sieving is less accurate in measuring the finer material and sedimentation is less accurate with coarser particles.

The results of sieving and sedimentation are usually plotted either as a pie chart (Figure 10.8) or as a triangular graph (Figure 10.9). As the proportions of sand, silt and clay vary considerably, it is traditional to have 12 texture categories (Figure 10.9).

Figure 10.8

The texture of different soil types

silt loam
15%
15%
70%

sandy loam
15%
20%
65%

clay
15%
15%
70%

loam
20%
40%
40%

silt   sand   clay

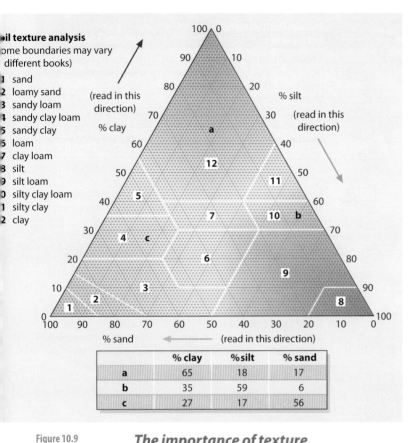

1 sand
2 loamy sand
3 sandy loam
4 sandy clay loam
5 sandy clay
6 loam
7 clay loam
8 silt
9 silt loam
10 silty clay loam
11 silty clay
12 clay

	% clay	% silt	% sand
a	65	18	17
b	35	59	6
c	27	17	56

Figure 10.9

Soil texture analysis:
the use of a
triangular graph

The importance of texture

As texture controls the size and spacing of soil pores, it directly affects the soil water content, water flow and extent of aeration. Clay soils tend to hold more water and are less well drained and aerated than sandy soils (page 267).

Texture also controls the availability and retention of nutrients within the soil. Nutrients stick to – i.e. are adsorbed onto – clay particles and are less easily leached by infiltration or throughflow than in sandy soils (page 268).

Plant roots can penetrate coarser soils more easily than finer soils, and 'lighter' sandy soils are easier to plough for arable farming than 'heavier' clays.

Texture greatly influences soil structure.

How does texture affect farming?

The following comments are generalised as it must be remembered that soils vary enormously.

Sandy soils, being well drained and aerated, are easy to cultivate and permit crop roots (e.g. carrots) to penetrate. However, they are vulnerable to drought, mainly because, due to their relatively large particle size (Figure 8.2a), they lack the micropores that would retain moisture (page 267) and partly because they usually contain limited amounts of organic matter. They also need considerable amounts of fertiliser because nutrients and organic matter are often leached out and not replaced.

Silty soils also tend to lack mineral and organic nutrients. The smaller pore size means that more moisture is retained than in sands but heavy rain tends to 'seal' or cement the surface, increasing the risk of sheetwash and erosion.

Clay soils tend to contain high levels of nutrient and organic matter but they are difficult to plough and, after heavy rain and due to their small particle size (Figure 8.2b) which helps to retain water (page 267), are prone to waterlogging and may become gleyed (pages 272 and 275). Plant roots find difficulty in penetration. Clays expand when wet, shrink when dry and take the longest time to warm up.

The ideal soil for agriculture is a **loam** (Figures 10.8 and 10.9). This has sufficient clay (20 per cent) to hold moisture and retain nutrients; sufficient sand (40 per cent) to prevent waterlogging, to be well aerated and to be light enough to work; and sufficient silt (40 per cent) to act as an adhesive, holding the sand and clay together. A loam is likely to be least susceptible to erosion.

3 Soil structure

It is the aggregation of individual particles that gives the soil its structure. In undisturbed soils, these aggregates form different shapes known as **peds**. It is the shape and alignment of the peds which, combined with particle size/texture, determine the size and number of the pore spaces through which water, air, roots and soil organisms can pass. The size, shape, location and suggested agricultural value of each of the six ped types are given in Figure 10.10. It should be noted, however, that some soils may be structureless (e.g. sands), some may have more than one ped structure (Figure 10.11), and most are likely to have a distinctive ped in each horizon. It is accepted that soils with a good crumb structure give the highest agricultural yield, are more resistant to erosion and develop best under grasses – which is why fallow should be included in a farming crop rotation. Sandy soils have the weakest structures as they lack the clays, organic content and secretions of organisms needed to cause the individual particles to aggregate. A crumb structure is ideal as it provides the optimum balance between air, water and nutrients.

Type of structure (ped)	Size of structure (mm)	Description of peds	Shape of peds	Location (horizon: texture) and formation	Agricultural value
crumb	1–5	small individual particles similar to breadcrumbs; porous	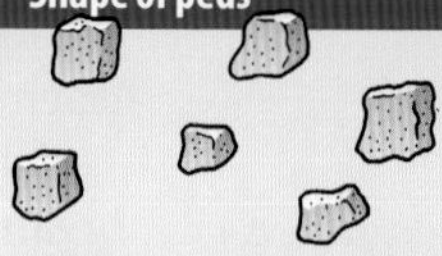	A horizon: loam soil; formed by action of soil fauna (e.g. earthworms, mites and termites), high content of fibrous roots (grasses) and excretion of micro-organisms	the most productive; well aerated and drained – good for roots
granular	1–5	small individual particles; usually non-porous		A horizon: clay soil; formation as for crumb structure	fairly productive; problems with drainage and aeration
platy	1–10	vertical axis much shorter than horizontal, like overlapping plates; restrict flow of water	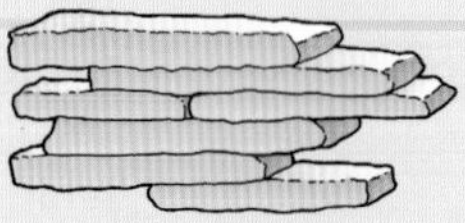	B horizon: silts and clays; formed by contraction by tree roots, especially when trees (e.g. Scots pine) sway in wind. Also due to ice lens, and compaction due to farm machinery	the least productive; hinders water and air movement; restricts roots
blocky	10–75	irregular shape with horizontal and vertical axes about equal; may be rounded or angular but closely fitting	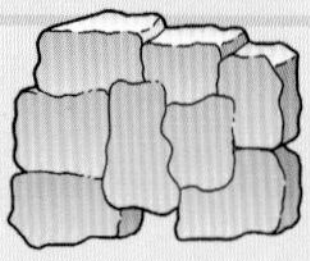	B horizon: clay-loam soils; formation associated with wetting–drying and freeze–thaw processes	productive: usually well drained and aerated
prismatic	20–100	vertical axis much larger than horizontal; angular caps and sides to columns	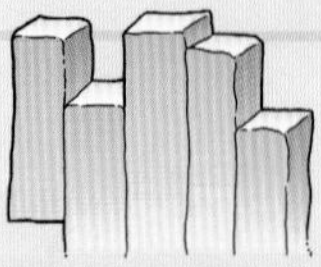	B and C horizons: often limestones or clays; formation associated with wetting–drying and freeze–thaw processes	usually quite productive: formed by wetting and drying; adequate water movement and root development
columnar	20–100	vertical axis much larger than horizontal; rounded caps and sides to columns	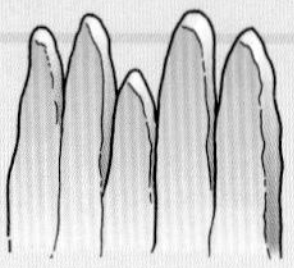	B and C horizons; alkaline soils; formation associated with accumulation of sodium	quite productive (if water available)

Figure 10.10

Different soil structures

Figure 10.11

Differences in peds (*after* Courtney and Trudgill)

4 Organic matter

Organic matter, which includes humus, is derived mainly from decaying plants and animals, or from the secretions of living organisms. Fallen leaves and decaying grasses and roots are the main source of organic matter. Soil organisms, such as bacteria and fungi, break down the organic matter and, depending on the nature of the soil-forming processes (Figure 10.17), help develop up to three distinct organic layers at the surface of the soil profile (Figure 10.5):

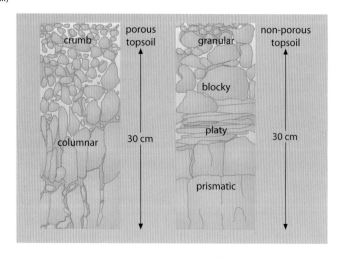

1 L or **leaf litter** layer: plant remains are still visible.
2 F or **fermentation (decomposition)** layer: decay, which biochemically involves yeast, is most rapid, although some plant remains are still visible.
3 H or **humus** layer: primarily organic in nature where, following decomposition, all recognisable plant and animal remains have been broken down into a black, slimy, amorphous organic material.

Wherever soil biological activity is low (due to one or a combination of acidity, low temperatures, wetness or the difficulty in decomposing organic matter), soil organism activity is greatly reduced or absent. As the litter layer cannot be mixed into the soil, then organic horizons build up to give the distinct L, F and H layers of a **mor**.

Where soil organisms are active, they will readily mix the litter into the soil, dispersing it throughout the A horizon where it decomposes into an A horizon rich in humus – the **mull** layer. Where organic material and mineral matter do mix, mainly due to earthworm activity, the result is the **clay–humus complex** (page 268). The clay–humus complex is essential for a fertile soil as it provides it with a high water- and nutrient-holding capacity and, by binding particles together, helps reduce the risk of erosion.

Humus gives the soil a black or dark-brown colour. The highest amounts are found in the **chernozems**, or black earths (page 327), of the North American Prairies, Russian Steppes and Argentinean Pampas. In tropical rainforests, heavy rainfall and high biological activity cause the rapid decomposition of organic matter which releases nutrients ready for their uptake and storage by plants (Figures 10.6 and 11.29c) or, if the forest is cleared, for leaching out of the system. In drier climates there may be insufficient vegetation to give an adequate supply.

5 Soil moisture

Soil moisture is important because it affects the upward and downward movement of water and nutrients. It helps in the development of horizons; it supplies water for living plants and organisms; it provides a solvent for plant nutrients; it influences soil temperature; and it determines the incidence of erosion. The amount of water in a soil at a given time can be expressed as:

$$W \propto R - (E + T + D)$$
(input) – (outputs)

where: W = water in the soil
$\propto$ = proportional to
R = rainfall/precipitation
T = transpiration
E = evaporation
D = drainage.

Drainage depends on the balance between the **water retention capacity** (water storage in a soil) and the infiltration rate. This is controlled by porosity and permeability which in turn is controlled by the soil's texture and structure. It has already been shown how texture and structure affect the size and distribution of pore spaces. Clays have numerous small pores (**micropores**) which can retain water for long periods, giving it a high water retention capacity, but which also restrict

infiltration rates (page 59). Sands have fewer but much larger **macropores** which permit water to pass through more quickly (a rapid infiltration rate), but have a low water retention capacity. A loam provides a more balanced supply of water, in the micropores, and air, in the macropores.

The presence of moisture in the soil does not necessarily mean that it is available for plant use. Plants growing in clays may still suffer from water stress even though clay has a high water-holding capacity. Soil water can be classified according to the tension at which it is held. Following a heavy storm or a lengthy episode of rain or snowmelt, all the pore spaces may be filled, with the result that the soil becomes saturated. When infiltration ceases, water with a low surface tension drains away rapidly under gravity. This is called **gravitational** or **free** water which is available to plants when the soil is wet, but unavailable when water has drained away. Once this excess water has drained away, the remaining moisture that the soil can hold is said to be its **field capacity** (Figures 3.3 and 10.12).

Moisture at field capacity is held either as **hygroscopic water** or as **capillary water**. Hygroscopic water is always present, unless the soil becomes completely dry, but is unavailable for plant use. It is found as a thin film around the soil particles to which it sticks due to the strength of its surface tension. Capillary water is attracted to, and forms a film around, the hygroscopic water, but has a lower cohesive strength. It is capillary water that is freely available to plant roots. However, this water can be lost to the soil by evapotranspiration. When a plant loses more water through transpiration than it can take up through its roots it is said to suffer **water stress** and it begins to wilt. At **wilting point**, photosynthesis (page 295) is reduced but, provided water can be obtained relatively soon or if the plant is adapted to drought conditions, this need not be fatal. Figure 10.12 shows the different water-holding characteristics of soil.

Figure 10.12

Availability of soil moisture for plant use

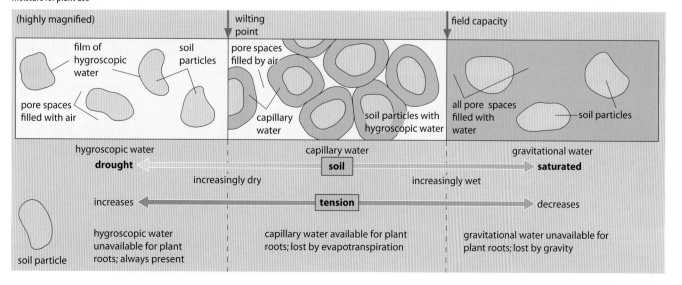

6 Air

Air fills the pore spaces left unoccupied by soil moisture. It is oxygen in the air that is essential for plant growth and living organisms. Compared with atmospheric air, air in the soil contains more carbon dioxide, released by plants and soil biota, and more water vapour; but less oxygen, as this is consumed by bacteria. Biota need oxygen and give off carbon dioxide by respiration and through the oxidation of organic matter. These gases are exchanged through the process of diffusion.

7 Soil organisms (biota)

Soil organisms include bacteria, fungi and earthworms. They are more active and plentiful in warmer, well-drained and aerated soils than they are in colder, more acidic and less well-drained and aerated soils.

Figure 10.13

Nutrients needed by plants

Macro-nutrients	Needed in large quantities	Carbon	C	Needed for basic cell construction. Obtained from air and water.
Macro-nutrients	Needed in large quantities	Hydrogen	H	Needed for basic cell construction. Obtained from air and water.
Macro-nutrients	Needed in large quantities	Oxygen	O	Needed for basic cell construction. Obtained from air and water.
Macro-nutrients	Needed in large quantities	Nitrogen	N	Basis of plant proteins. Promotes rapid growth. Improves quality and quantity of leaf growth.
Macro-nutrients	Needed in large quantities	Phosphorus	P	Encourages rapid seedling growth and early root formation. Helps in flowering and with seed formation.
Macro-nutrients	Needed in smaller quantities	Sulphur	S	Especially important for root crops.
Macro-nutrients	Needed in smaller quantities	Potassium	K	Helps with production of proteins and in overcoming disease. Strengthens stems and stalks.
Macro-nutrients	Needed in smaller quantities	Calcium	Ca	Reduces acidity. Helps with growth of roots and new shoots.
Macro-nutrients	Needed in smaller quantities	Magnesium	Mg	Used in photosynthesis, being a basic constituent of chlorophyll. Important for arable crops.
Micro-nutrients (trace elements)	Needed in very small quantities	Sodium	Na	Helps to increase yields.
Micro-nutrients (trace elements)	Needed in very small quantities	Manganese	Mn	Used in respiration, protein synthesis and enzyme reactions.
Micro-nutrients (trace elements)	Needed in very small quantities	Copper	Cu	Reduces toxicity of other elements in soil. Helps enzyme reactions.
Micro-nutrients (trace elements)	Needed in very small quantities	Zinc	Zn	Helps in fruit production.
Micro-nutrients (trace elements)	Needed in very small quantities	Molybdenum	Mo	Needed in nitrogen fixation by activating enzymes.
Micro-nutrients (trace elements)	Needed in very small quantities	Silicon	Si	Important constituent of grasses.
Micro-nutrients (trace elements)	Needed in very small quantities	Boron	B	Helps growth.
Micro-nutrients (trace elements)	Needed in very small quantities	Chlorine	Cl	Can increase yields of some crops.
Micro-nutrients (trace elements)	Needed in very small quantities	Cobalt	Co	Helps fruit trees and bushes.

Organisms are responsible for three important soil processes:

- **Decomposition:** detritivores, such as earthworms, ants, termites, mites, woodlice and slugs, begin this process by burying leaf litter (detritus), which hastens its decay, and eating some of it. Their faeces (wormcasts, etc.) increase the surface area of detritus upon which fungi and bacteria can act. Fungi and bacteria secrete enzymes which break down the organic compounds in the detritus. This releases nutrient ions essential for plant growth (soil nutrients, Figure 10.13), into the soil while some organic compounds remain as humus.
- **Fixation:** by this process, bacteria can transform nitrogen in the air into nitrate, which is an essential nutrient for plant growth.
- **Development of structure:** fungi help to bind individual soil particles together to give a crumb structure, while burrowing animals create passageways that help the circulation of air and water and facilitate root penetration.

8 Soil nutrients

Nutrient is the term given to chemical elements found in the soil which are essential for plant growth and the maintenance of the fertility of a soil (Figure 10.13). The two main sources of nutrients are:

1 the weathering of minerals in the soil, and
2 the release of nutrients on the decomposition of organic matter and humus by soil organisms.

Nutrients can also be obtained through:

3 rainwater, and
4 the artificial application of fertiliser.

Nutrients occur in the soil solution as positively charged (+) ions called **cations** and negatively charged (-) ions known as **anions**. It is largely in the ionic form that plants can utilise nutrients in the soil. Both clay and humus, which have negative charges, attract the positively charged minerals in the soil solution, notably Ca^{2+}, Mg^{2+}, K^+ and Na^+. This results in the cations being adsorbed (i.e. they become attached) to the clay and humus particles. The process of **cation exchange** allows cations to be moved between:

- soil particles of clay and/or humus and the soil solution
- plant roots and either the surface of the soil particles or from the soil solution (Figure 10.14).

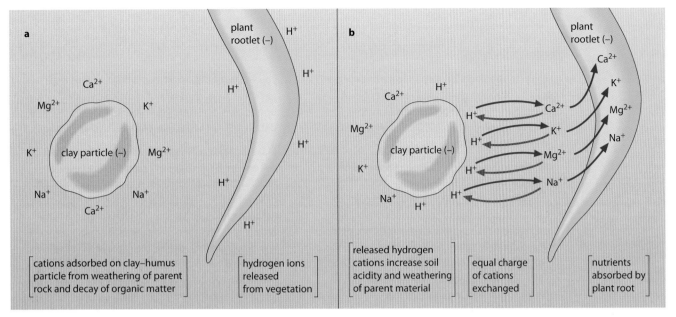

Figure 10.14

The process of cation exchange (*after* Courtney and Trudgill)

a
plant rootlet (–)
Ca²⁺
Mg²⁺
K⁺
K⁺   clay particle (–)   Mg²⁺
Na⁺   Na⁺
Ca²⁺

H⁺
H⁺
H⁺
H⁺
H⁺
H⁺

[cations adsorbed on clay–humus particle from weathering of parent rock and decay of organic matter]   [hydrogen ions released from vegetation]

b
plant rootlet (–)
Ca²⁺
Ca²⁺   H⁺
Mg²⁺
K⁺   clay particle (–)
Na⁺   H⁺

Ca²⁺   K⁺
K⁺   Mg²⁺
Mg²⁺   Na⁺
Na⁺

[released hydrogen cations increase soil acidity and weathering of parent material]   [equal charge of cations exchanged]   [nutrients absorbed by plant root]

As well as providing nutrients for plant roots, the cation exchange releases hydrogen which in turn increases acidity in the soil (see below). Acidity accelerates weathering of parent rock, releasing more minerals to replace those used by plants or lost through leaching. The **cation exchange capacity** (CEC) is a measure of the ability of a soil to retain cations for plant use. Soils with a low CEC, such as sands, are less able to keep essential plant nutrients than those with a high CEC, like clays and humus; consequently they are less fertile.

9 Acidity (pH)

As mentioned in the previous section, soil contains positively charged hydrogen cations. **Acidity** or **alkalinity** is a measure of the degree of concentration of these cations. It is measured on the pH scale (Figure 10.15), which is logarithmic (compare the Richter scale, Figure 1.3). This means that a reading of 6 is 10 times more acidic than a reading of 7 (which is neutral), and 100 times more acidic than one of 8 (which is alkaline). Most British soils are slightly acidic,

although in upland Britain acidity increases as the heavier rainfall leaches out elements such as calcium faster than they can be replaced by weathering. Acid soils therefore tend to need constant liming if they are to be farmed successfully.

A slightly acid soil is the optimum for farming in Britain as this helps to release secondary minerals. However, if a soil becomes too acidic it releases iron and aluminium which, in excess, may become toxic and poisonous to plants and organisms. Increased acidity makes organic matter more soluble and therefore vulnerable to leaching; and it discourages living organisms, thus reducing the rate of breakdown of plant litter and so is a factor in the formation of peat.

In areas where there is a balance between precipitation and evapotranspiration, soils are often neutral, as in the American Prairies (page 327); while in areas with a water deficiency, as in deserts (page 323), soils are more alkaline.

10 Soil temperature

Incoming radiation can be absorbed, reflected or scattered by the Earth's surface (Figure 9.4).

The topsoil, especially if vegetation cover is limited, heats up more rapidly than the subsoil during the daytime and loses heat more rapidly at night. A 'warm', moist soil will have greater biota activity, giving a more rapid breakdown of organic matter; it will be more likely to contain nutrients because the chemical weathering of the parent material will be faster; and seeds will germinate more readily in it than in a 'cold', dry soil.

Figure 10.15

The pH scale showing soil acidity and alkalinity

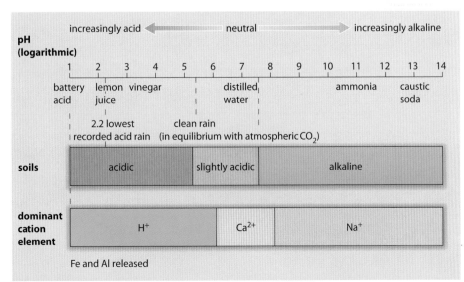

increasingly acid ← neutral → increasingly alkaline

pH (logarithmic)

1   2   3   4   5   6   7   8   9   10   11   12   13   14

battery acid   lemon juice   vinegar   distilled water   ammonia   caustic soda

2.2 lowest recorded acid rain   clean rain (in equilibrium with atmospheric CO_2)

soils   acidic   slightly acidic   alkaline

dominant cation element H^+ Ca^{2+} Na^+

Fe and Al released

Begin by reading a book that describes in detail how to dig a soil pit and how to describe and explain the resultant profile (e.g. Courtney and Trudgill, 1984, or O'Hare, 1988; see References at end of chapter).

First, make sure you obtain permission to dig a pit. The site must be carefully chosen. You will need to find an undisturbed soil – so avoid digging near to hedges, trees, footpaths or on recently ploughed land. Ideally, make the surface of the pit approximately 0.7 m², and the depth 1 m (unless you hit bedrock first). Carefully lay the turf and soil on plastic sheets. Clear one face of the pit, preferably one facing south as this will get the maximum light, to get a 'clean' profile so that you can complete your recording sheet. (The one in Figure 10.16 is a very detailed example.) Sometimes you will not be able to take all the readings due to problems such as lack of clarity between boundaries, time and equipment; sometimes some details will not be relevant to a particular enquiry.

Make a detailed fieldsketch before replacing the soil and turf. You may have to complete several tasks in the laboratory before writing up your description. You can gather information from a soil without needing to know how it formed or what type it is. Remember, it is unlikely that your answer will exactly fit a model profile. It may show the characteristics of a podsol (Figure 12.40) if you live in a cooler, wetter and/or higher part of Britain; or of a brown earth (Figure 12.34) if you live in a warmer, drier and/or lower part of the country – but you must not *force* your profile to fit a model.

Figure 10.16

Soil recording sheets

a soil site

Recorded by		Date		Locality			Six-figure grid reference	
Parent rock (geological map)		Altitude (estimated from Ordnance Survey map)		Angle of slope (Abney level)		Aspect (bearing or compass point)		Relief (uniform, concave or convex slope, terrace)
Exposure (exposed, sheltered)		Drainage (shedding or receiving site, floodplain, terrace, boggy)		Natural vegetation or type of farming (tree species, ground vegetation, crops, animals)		Previous few days' weather (warm, cold, wet, dry)		Other local details (remember your labelled fieldsketch)

b soil profile

Horizon	Depth of horizon (cm)	Lower boundary of horizon	Colour	Texture	Stoniness	Structure (peds)	Consistency	pH	Moisture content	Porosity	Organic matter	Roots	Carbonates	Soil biota and/or animals
How to read, estimate and measure	measure from top of soil surface	sharp, abrupt, clear, indistinct, gradual, irregular, smooth, broken	use Munsell colour chart	percentage clay, silt or sand; 'feel'; sieves; sedimentation	size of stones, number of stones, shape of stones	structureless crumb, etc.	loose, friable, firm, hard, plastic, sticky, soft	pH paper or soil-testing kit	weigh sample, evaporate water, reweigh sample, or use a moisture meter	time taken for a beakerful of water to infiltrate	type, estimate percentage, measure depth	weigh, burn sample (and roots), reweigh sample, calculate percentage	add dilute (10%) hydrochloric acid; if it effervesces, sample is over 1% carbonate	number, types
A														
B														
C														

Figure 10.17

Soil-forming processes

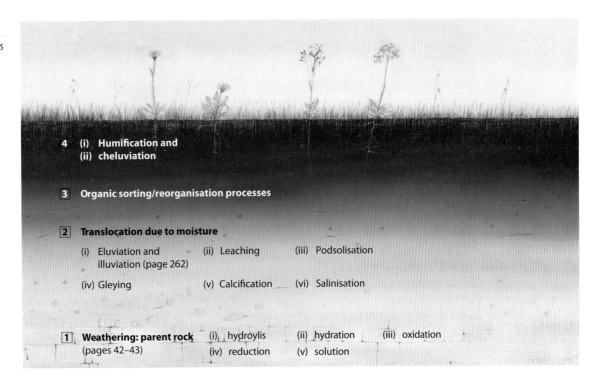

4 (i) **Humification and**
(ii) **cheluviation**

3 **Organic sorting/reorganisation processes**

2 **Translocation due to moisture**

(i) Eluviation and — (ii) Leaching (iii) Podsolisation
illuviation (page 262)

(iv) Gleying (v) Calcification (vi) Salinisation

1 **Weathering: parent rock** (i) hydroylis (ii) hydration (iii) oxidation
(pages 42–43) (iv) reduction (v) solution

Processes of soil formation

Numerous processes are involved in the formation of soil and the creation of the profiles, structures and other features described above. Soil-forming processes depend on all the five factors described on pages 260–262. Some of the more important processes are shown in Figure 10.17.

1 Weathering

As described on page 263 and in Figure 10.2, weathering leaves primary minerals as residues and produces secondary minerals as well as determining the rates of release of nutrients and the soil depth, texture and drainage. In systems terms, this means that minerals are released as inputs into the soil system from the bedrock store and transferred into the soil store (Figure 10.6).

2 Humification and cheluviation

Humification is the process by which organic matter is decomposed to form humus (page 266) – a task performed by soil organisms. Humification is most active either in the *H* horizon of the soil profile (Figure 10.5) where it can result in mull (pH 5.5 to 6.5), or in the upper *A* horizon where it can produce mor (pH 3.5 to 4.5) (page 266). Moder (pH 4.5 to 5.5) is transitional between the mor and mull (page 262).

As organic matter decomposes, it releases nutrients and organic acids. These acids, known as **chelating agents**, attack clays and other minerals, mainly in the *A* horizon, releasing iron and aluminium. The chelating agents then combine with the cations of the iron and aluminium to form organic-metal compounds known as **chelates**. Chelates are soluble and are readily transported downwards through the soil profile – the process of **cheluviation**. The iron and aluminium may be deposited in the lower profile as they become less soluble in the slightly higher pH levels found there (Figure 10.5).

3 Organic sorting

Several processes operate within the soil to re-organise mineral and organic matter into horizons, and to contribute to the aggregation of particles and the formation of peds.

4 Translocation of soil materials

Translocation is the movement of soil components in any form (solution, suspension, or by animals) or direction (downward, upward). It usually takes place in association with soil moisture.

In Britain, there is:
- usually a soil moisture budget surplus due to an annual excess of precipitation over evapotranspiration (water balance – Figure 3.3)
- locally, an increase in soil moisture due to poor drainage.

The increase in soil moisture, resulting from these two factors, can lead to:
- either the translocation processes of leaching and podsolisation, or
- gleying associated with areas of poor drainage.

(i) Eluviation and illuviation

See page 262.

(ii) Leaching

Leaching is the removal of soluble material in solution. Where precipitation exceeds evapotranspiration and soil drainage is good, rainwater – containing oxygen, carbonic acid and organic acids, collected as it passes through the surface vegetation – causes chemical weathering, the breakdown of clays and the dissolving of soluble salts (bases). Ca and Mg are eluviated from the *A* horizon, making it increasingly acid as they are replaced by hydrogen ions, and are subsequently illuviated to the underlying *B* horizon, or are leached out of the system (Figure 10.18).

(iii) Podsolisation

Podsolisation is more common in cool climates where precipitation is greatly in excess of evapotranspiration and where soils are well drained or sandy. Podsolisation is also defined as the removal of iron and aluminium oxides, together with humus. As the surface vegetation is often coniferous forest, heathland or moors, rain percolating through it becomes progressively more acidic and may reach a pH of 5.0 or less (Figure 10.15). This in turn dissolves an increasing amount and number of bases (Ca, Mg, Na and K), silica and, ultimately, the sesquioxides of iron and aluminium (Figure 10.19). The resultant **podsol soil** (Figure 12.40) therefore has two distinct horizons: the bleached *A* horizon, drained of coloured minerals by leaching; and the reddish-brown *B* horizon where the sesquioxides have been illuviated. Often the iron deposits form an **iron pan** which is a characteristic of a podsol.

(iv) Gleying

This occurs when the output of water from the soil system is restricted, giving **anaerobic** or **waterlogged** conditions (page 275). This is most likely to occur on gentle slopes, in depressions where the underlying rock is impermeable, where the water table is high enough to enter the soil profile (e.g. along river floodplains) or in areas with very heavy rainfall and poor drainage. Under such conditions the pore spaces fill with stagnant water which becomes de-oxygenised. The reddish-coloured oxidised iron, iron III (Fe^{3+} or ferric iron), is chemically reduced to form iron II (Fe^{2+} or ferrous iron) which is grey-blue in colour. Occasionally, pockets of air re-oxygenise the iron II to give scatterings of red mottles (Figure 10.26). Although many British soils show some evidence of gleying, the conditions develop most extensively on moorland plateaus.

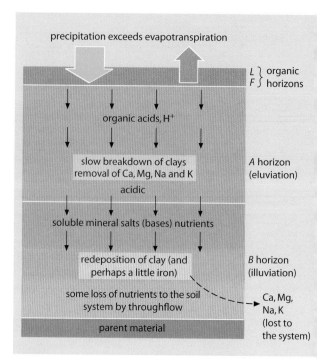

Figure 10.18

The processes of leaching

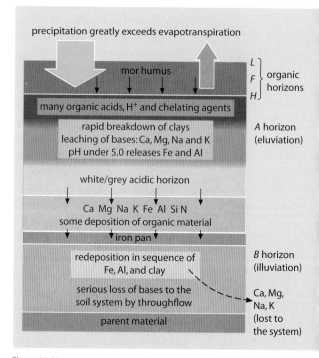

Figure 10.19

The process of podsolisation

Courtney and Trudgill (Figure 10.20) have summarised the relationship between leaching, podsolisation and gleying, and precipitation and drainage.

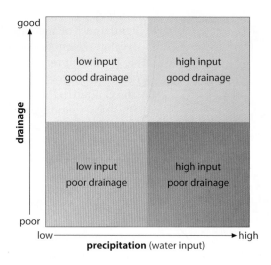

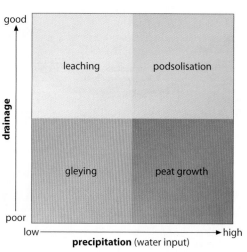

(v) Calcification

Calcification is a process typical of low-rainfall areas where precipitation is either equal to, or slightly higher than, evapotranspiration. Although there may be some leaching, it is insufficient to remove all the calcium which then accumulates, in relatively small amounts, in the *B* horizon (Figure 10.21; and chernozems, page 327).

(vi) Salinisation

This occurs when potential evapotranspiration is greater than precipitation in places where the water table is near to the surface. It is therefore found locally in dry climates and is not a characteristic of desert soils. As moisture is evaporated from the surface, salts are drawn upwards in solution by capillary action. Further evaporation results in the deposition of salt as a hard crust (Figure 10.22). Salinisation has become a critical problem in many irrigated areas, such as California (Figure 16.53).

Zonal, azonal and intrazonal soils

Zonal soils

Zonal soils are mature soils. They result from the maximum effects of climate and living matter (vegetation) upon parent rock in areas where there are no extremes of weathering, relief or drainage and where the landscape and climate have been stable for a long time. Consequently, zonal soils have had time to develop distinctive profiles and, usually, clear horizons. However, it is misleading to imply that all zonal soils have distinct horizons; brown earths (page 329), chernozems (page 327) and prairie soils (page 328) have indistinct horizons which merge into each other. A description of the major zonal soils, and how their formation can be linked to climate and vegetation, is given in Chapter 12 and Figure 12.2. It should be stressed that this linkage is regarded by soil scientists as greatly outdated and a grossly simplified model – but it is still the one used in all the latest AS, A-level and Scottish Higher syllabuses that examine soils!

Azonal soils

Azonal soils, in contrast to zonal soils, have a more recent origin and occur where soil-forming processes have had insufficient time to operate fully. As a consequence, these soils usually show the characteristics of their origin (i.e. parent material, which may have resulted from *in situ* weathering of parent rock or have been transported from elsewhere and deposited), do not have well defined horizons, and are not associated with specific climatic–vegetational zones. Azonal soils, in Britain, include **scree** (weathering), **alluvium** (fluvial), **till** (glacial), **sands** and **gravels** (glaciofluvial), **sand dunes** (aeolian and marine), **saltmarsh** (marine), and **volcanic** (tectonic) **soils**.

Figure 10.21

e process of calcification

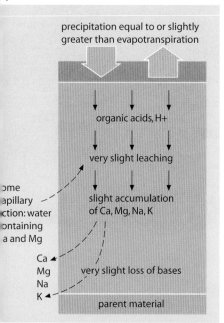

Figure 10.22

The process of salinisation

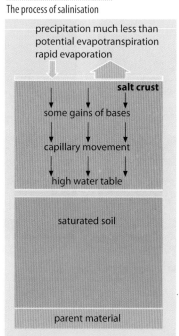

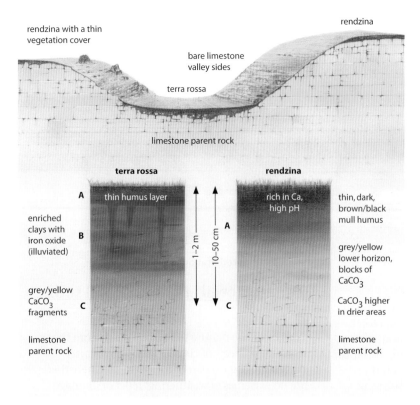

Figure 10.23

Calcimorphic soils: terra rossa and rendzina

Intrazonal soils

Intrazonal soils reflect the dominance of a single local factor, such as parent rock or extremes of drainage. As they are not related to general climatic controls, they are not found in zones. They can be divided into three types:

- **Calcimorphic** or **calcareous** soils develop on a limestone parent rock (rendzina and terra rossa, Figure 10.23).
- **Hydromorphic** soils are those having a constantly high water content (gleyed soils and peat – Figures 10.26 and 10.27).
- **Halomorphic** soils have high levels of soluble salts which render them saline.

Figure 10.24

A rendzina, Kent

Calcimorphic

1 **Rendzina** The rendzina (Figure 10.24) develops where softer limestones or chalk are the parent material and where grasses (the English Downs) and beech woodland (the Chilterns) form the surface vegetation. The grasses produce a leaf litter that is rich in bases. This encourages considerable activity by organisms which help with the rapid recycling of nutrients. The *A* horizon therefore consists of a black/dark-brown mull humus. Due to the continual release of calcium from the parent rock and a lack of hydrogen cations, the soil is alkaline with a pH of between 7.0 and 8.0. The calcium-saturated clays, with a crumb or blocky structure, tend to limit the movement of water and so there is relatively little leaching. Consequently there is no *B* horizon. The underlying limestones, affected by chemical weathering, leave very little insoluble residue and this, together with the permeable nature of the bedrock, results in a thin soil with limited moisture reserves.

2 **Terra rossa** As its name suggests, terra rossa (Figure 10.25) is a red-coloured soil (it has been called a 'red rendzina'). It is found in areas of heavy, even if seasonal, rainfall where the calcium carbonate parent rock is chemically weathered (carbonation) and silicates are leached out of the soil to leave a residual deposit rich in iron hydroxides. It usually occurs in depressions within the limestone and in Mediterranean areas where the vegetation is garrigue (Figure 12.24).

Figure 10.25

Terra rossa, Cuba

Hydromorphic

1 **Gley soils** Gleying occurs in saturated soils when the pore spaces become filled with water to the exclusion of air. The lack of oxygen leads to anaerobic conditions (page 272) and the reduction (chemical weathering) of iron compounds from a ferric (Fe^{3+}) to a ferrous (Fe^{2+}) form. The resultant soil has a grey-blue colour with scatterings of red mottles (Figure 10.26). Because gleying is a result of poor drainage and is almost independent of climate, it can occur in any of the zonal soils. Pedologists often differentiate between **surface gleys**, caused by slow infiltration rates through the topsoil, and **groundwater gleys**, resulting from a seasonal rise in the water table or the presence of an impermeable parent rock.

2 **Peat** Where a soil is waterlogged and the climate is too cold and/or wet for organisms to break down vegetation completely, layers of peat accumulate (Figure 10.27). These conditions mean that litter input (supply) is greater than the rate of decomposition by organisms whose activity rates are slowed down by the low temperatures and the anaerobic conditions. Peat is regarded as a soil in its own right when the layer of poorly decomposed material exceeds 40 cm in depth. Peat can be divided according to its location and acidity. **Blanket peat** is very acidic; it covers large areas of wet upland plateaus in Britain (Kinder Scout in the Peak District); and it is believed to have formed 5000 to 8000 years ago during the Atlantic climatic phase (Figure 11.18). **Raised bogs**, also composed of acidic peat, occur in lowlands with a heavy rainfall. Here the peat accumulates until it builds up above the surrounding countryside. **Valley**, or **basin**, **peat** may be almost neutral or only slightly acidic if water has drained off surrounding calcareous uplands (the Somerset Levels and the Fens); otherwise, it too will be acid (Rannoch Moor in Scotland). Fen peat is a high-quality agricultural soil.

Halomorphic

Halomorphic soils contain high levels of soluble salts and have developed through the process of salinisation (page 273 and Figure 16.53). They are most likely to occur in hot, dry climates where, in the absence of leaching, mineral salts are brought to the surface by capillary action and where the parent rock or groundwater contains high levels of carbonates, bicarbonates and sulphates, especially as salts of calcium and magnesium and some sodium chloride (common salt). The water, on reaching the surface, evaporates to leave a thick crust (e.g. Bonneville saltflats in Utah,

page 188) in which only salt-resistant plants (halophytes, page 291) can grow.

Figure 10.26

Gleying: a hydromorphic soil

Figure 10.27

Peat in the Flow Country, Sutherland, Scotland

The soil catena

A **catena** (Latin for 'chain') is a sequence of soil types down a slope where each **soil type**, or **facet** is different from, but linked to, its adjacent facets (Figure 10.3). Catenas therefore illustrate the way in which soils can change down a slope where there are no marked changes in climate or parent material. Each catena is an example of a small-scale, open system involving inputs, processes and outputs. The slope itself is in a delicate state of dynamic equilibrium (Figure 2.12) with the soils and landforms being in a state of flux and where the ratio of erosion and deposition varies between the different slope facets. Soils on lower slopes tend to be deeper and wetter than those on upper slopes, as well as being more enriched by a range of leached materials. The thinnest and driest soils are likely to be found on central parts of the slope. It takes a considerable period of time for catenary relationships to become established and therefore the best catenas can be found in places with a stable environment, such as in parts of Africa, where there have been relatively few recent changes in either the landscape or the climate.

Places 34 Arran: a soil catena

Figure 10.28 shows a catena based on fieldwork conducted on the Isle of Arran. The transect was taken from a relatively flat, peat-covered upland area above the glaciated Glen Rosa valley, down a steep valley side to the Rosa Water (parallel to, and south of, the Garbh Allt tributary located on Figure 4.37).

Notice, with reference to Figure 10.3, the location on the transect of the shedding (eluviation or input), transfer (translocation) and receiving (illuviation or output) zones, and the relationships between the angles of slope and (i) soil depth, (ii) pH and (iii) soil moisture.

Figure 10.28

Readings taken along a catena in Glen Rosa

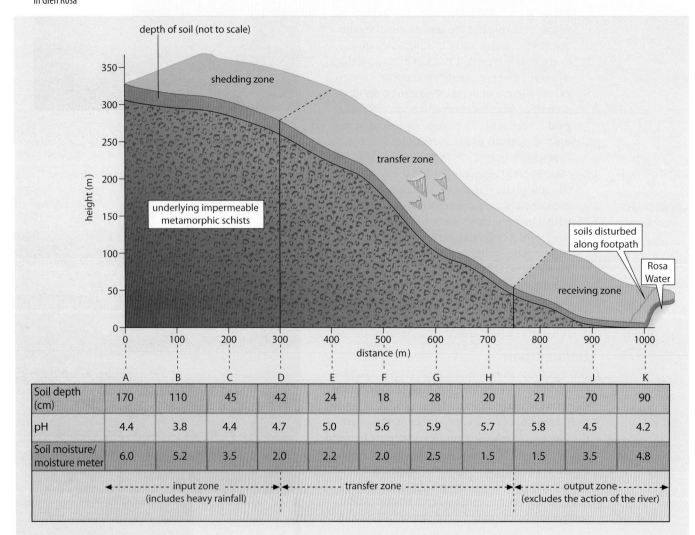

	A	B	C	D	E	F	G	H	I	J	K
Soil depth (cm)	170	110	45	42	24	18	28	20	21	70	90
pH	4.4	3.8	4.4	4.7	5.0	5.6	5.9	5.7	5.8	4.5	4.2
Soil moisture/ moisture meter	6.0	5.2	3.5	2.0	2.2	2.0	2.5	1.5	1.5	3.5	4.8

input zone (includes heavy rainfall) — transfer zone — output zone (excludes the action of the river)

Framework 9 Geographic Information Systems (GIS)

For centuries cartographers and geographers have been drawing and analysing maps by hand but, with recent technological developments, this work is increasingly being carried out by computers. Advances in geomatics – the science of handling geographic information – mean that huge amounts of data can be combined with digital maps and computer graphics in Geographic Information Systems (GIS).

Figure 10.29

Google Earth image of London, overlain with geographic information

It is estimated that around 80 per cent of all digitally stored information has a spatial element or is tied to a certain place. Powerful GIS software packages enable geographers to view, analyse, interpret, question and display this data in order to reveal relationships, patterns and trends that may otherwise be hidden.

Increasing numbers of businesses now use GIS to make decisions about a wide range of subjects. Examples include:

- where to site gas and electricity services
- the optimal place to build a wind farm
- the most efficient way to route emergency vehicles
- how to protect and conserve sensitive wetland areas.

In the home, through basic internet-based packages such as Google Earth (Figure 10.29), many people use GIS to learn about the world and to plan their leisure time and holidays.

A computer-based GIS needs three main components:

- a computerised map – used as a backdrop on which to place all the other information; this can be a conventional map, an aerial photograph or a satellite image

- a GIS software package – this will contain the tools for manipulating the map and the information
- the information itself – contained in a database, as photographs, text or any other kind of digital data.

The base map can be made up of a number of layers showing geographical components such as height, soils, settlement patterns or vegetation. These maps come from many different sources including remote sensing companies or mapping organisations such as Britain's Ordnance Survey.

The GIS software is the link that enables data to be positioned on the base map (Figure 10.30) and contains tools to manipulate the base map, add information layers and display the results. The data added can be tailored to fit the end users of the GIS. Public utilities such as electricity, gas and water companies, for instance, can add information layers showing the locations of their cables and manholes. Data is not just limited to the surface but can include features such as underground pipelines, and computers can display the information as a three-dimensional representation of reality. Technicians can enter this virtual environment on their screens, walking underneath the streets of our cities to analyse the problems that occur within such complex networks (Figure 10.31).

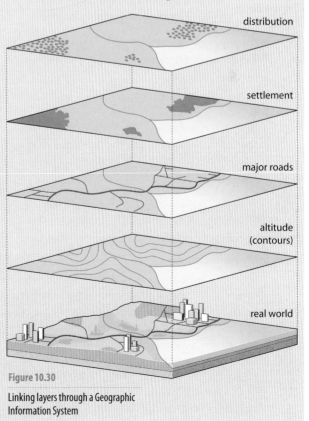

Figure 10.30

Linking layers through a Geographic Information System

GIS in the real world

'The application of GIS is limited only by the imagination of those who use it.'

Dr Jack Dangermond, President of pioneering GIS company ESRI

Across the globe, governments, local councils, the military, private companies and individuals use GIS daily to provide the services we take for granted. Problems such as finding the best position for a new power station or where to build a new cinema or housing estate are all analysed using GIS. At an individual level, self-employed businessmen can use home PC-based systems to improve their productivity. Farmers, for instance, log on to analyse information on weather patterns, soil type and economic trends in order to determine the best time to plant crops.

GIS in the future

'Imagine looking down a street but instead of simply seeing houses, shops and offices, your view has added extras like travel news, tours and even games.'

Ordnance Survey website

As the capability of computers increases, software developers are looking at ways in which GIS can provide information in the future. Businesses are constantly on the lookout for more digital data, especially if it is available in real time, and many are looking to run increasingly complex simulations

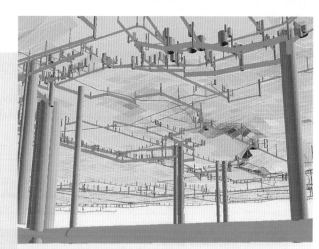

Figure 10.31

Underground water pipe network in Ballerup, Denmark shown in ArcGIS

through GIS before spending huge sums of money on big building projects. Governments and international organisations, meanwhile, are using systems to model the effects of climate change, sea-level rise, pollution incidents and other environmental disasters.

For personal users, the Ordnance Survey is developing The Magic Window, a handheld device that will superimpose geographic data on real-world images using a virtual 1:1 scale map of Great Britain (Figure 10.33). Developments in the sharing of data through the internet will also influence the availability of free GIS packages, bringing the easy-to-understand analysis of geographic information to millions of homes around the world.

Figure 10.32

How the emergency services and the police use GIS

Emergency services

Avoiding delays when sending an ambulance on an emergency call can be a matter of life or death. When operators take a phone call they ask for the location where help is needed and input the information into a GIS. The system quickly identifies the nearest available ambulance (sometimes by receiving data from satellites), builds a picture of expected traffic patterns based on the time of day and analyses the data to determine the quickest route the ambulance crew should take.

Mapping crime

Criminal analysts working for the police use GIS to locate, track and analyse incidents and help the police predict where and when crimes are likely to take place. Car thefts, for instance, often happen at night but are not reported until the morning when the owners wake up. By looking at patterns on their databases of abandoned cars, overlain with information on known offenders, the police are able to target resources and have had notable successes in catching criminals.

Figure 10.33

The Magic Window

Further reference

DeMers, M. (2009) *GIS for Dummies*, John Wiley and Sons.

Sommer, S. and Wade, T. (2006) *A to Z GIS: An Illustrated Dictionary of Geographic Information Systems*, ESRI Press.

 http://mapzone.ordnancesurvey.co.uk/mapzone/giszone.html

www.gis.com/whatisgis/index.html

www.gis.rgs.org/whatisgis.html

As we have seen (page 262), soil can takes thousands of years to become sufficiently deep and developed for economic use (exceptions include alluvium deposited by rivers and ash ejected from volcanoes). During that time, there is always some natural loss through leaching, mass movement and erosion by either water or wind. Normally there is an equilibrium, however fragile, between the rate at which soil forms and that at which it is eroded or degraded. That natural balance is being disturbed by human mismanagement with increasing frequency and with serious consequences.

Recent estimates suggest that 7 per cent of the world's topsoil is lost each year. The World Resources Institute claims that Burkina Faso loses 35 tonnes of soil per hectare per year. Other comparable figures are Ethiopia 42, Nepal 70, and the loess plateau of North China 251 (Figure 10.35). Soil removed during a single rain storm or dust storm may never be replaced. The Soil Survey of England and Wales claims that 44 per cent of arable soils in the UK, an area once considered not to be under threat, are now at risk (Figure 10.34).

Soil degradation

Degradation is the result of human failures to understand and manage the soil. The major cause of soil erosion is the removal of the natural vegetation cover, leaving the ground exposed to the elements. The most serious of such removals is deforestation. In countries such as Ethiopia (Places 76, page 520), the loss of trees, resulting from population growth and the extra need for farmland and fuelwood, means that the heavy rains, when they do occur, are no longer intercepted by the vegetation. Rainsplash (the direct impact of raindrops, Figure 2.12) loosens the topsoil and prepares it for removal by sheetwash (overland flow). Water flowing over the surface has little time to infiltrate into the soil or recharge the soil moisture store (pages 59–60). More topsoil tends to be carried away where there is little vegetation because there are neither plant roots nor organic matter to bind it together. Small

UK soil degradation

Soil degradation involves both the physical loss (erosion) and the reduction in quality of topsoil. Currently, 2.2 million tonnes of topsoil is eroded annually in the UK and over 17 per cent of arable land shows signs of erosion. Degradation can result from one or more of several factors:

- Physical degradation is when soil erosion results from the action of the wind or water. It is a natural process, accelerated by human activity. Erosion by wind is less widespread and less frequent than erosion by water but when it does occur it is often more severe. Estimates suggest that 44 per cent of arable land is at risk of being eroded by physical processes.
- Chemicals carried by water can cause diffuse pollution, while biological degradation is when organic matter, in the form of plant remains or organic manure, is washed out of the soil.

- Climate change suggests that Britain will experience more seasonal extremes with wetter, stormier winters and warmer, drier summers [page 255]. Wetter winters may mean waterlogged soils and an increase in water erosion, while drier soils are more likely to be susceptible to wind erosion.
- Land use can affect the soil, for example when grass is removed to expose the soil and, without roots to bind it together, the soil becomes unstable.

Figure 10.34

Soil erosion in Britain

channels or rills may be formed which, in time, may develop into large gulleys, making the land useless for agriculture (Figure 10.35).

Even where the soil is not actually washed away, heavy rain may accelerate leaching and remove nutrients and organic matter at a rate faster than that at which they can be replaced by the weathering of bedrock and parent material and the decomposition of vegetation (e.g. the Amazon Basin, Figure12.7 and Places 66, page 480). The loss of trees also reduces the rate of transpiration and therefore the amount of moisture in the air. There are fears that large-scale deforestation will turn areas at present under rainforest into deserts.

Although the North American Prairies and the African savannas were grassland when the European settlers first arrived, it is now believed that these areas too were

once forested and were cleared by fire – mainly natural due to lightning, but partly by the local people (Case Study 12B). The burning of vegetation initially provides nutrients for the soil, but once these have been leached by the rain or utilised by crops there is little replacement of nutrients. Where the grasslands have been ploughed up for cereal cropping, the breakdown of soil structure (peds) has often led to their drying out and becoming easy prey to wind erosion (Figure 10.34). Large quantities of topsoil were blown away to create the American Dust Bowl in the 1930s, while a similar fate has more recently been experienced by many of the Sahel countries. In Britain, the removal of hedges to create larger fields – easier for modern machinery – has led to accelerated soil erosion by wind (page 495).

Loess plateau of North China

Figure 10.35

Loess in China

This region, more than 2.5 times the size of the UK, experiences the most rapid soil loss in the world. During and following the ice age, Arctic winds transported large amounts of loess and deposited this fine, yellow material to a depth of 200 m in the Huang He basin. Following the removal of the subsequent vegetation cover of trees and grasses to allow cereal farming (especially under the directions of Chairman Mao), the unconsolidated material has been washed away by the heavy summer monsoon rains, or blown by yellow dust storms, at the rate of 1 cm per year. It is estimated that 1.6 bn tonnes of soil reach the Huang He River during each annual summer flood. This material, the most carried by any river in the world, has given the Huang He its name – i.e. the 'Yellow River'. A further problem is that 6 cm of silt settles annually on the river's bed so that it now flows 10 m above its floodplain. Should the large flood banks be breached, the river can drown thousands of people (over 1 million in the 1939 flood) and ruin all crops.

Ploughing can have adverse effects on soils. Deep ploughing destroys the soil structure by breaking up peds (page 265) and burying organic material too deep for plant use. It also loosens the topsoil for future wind and water erosion. The weight of farm machinery can compact the soil surface or produce platy peds, both of which reduce infiltration capacity and inhibit aeration of the soil. Ploughing up- and down-hill creates furrows which increase the rate of surface runoff and the process of gullying.

Overgrazing, especially on the African savannas, also accelerates soil erosion. Many African tribes have long measured their wealth in terms of the numbers, rather than the quality, of their animal herds. As the human populations of these areas continue to expand rapidly, so too do the numbers of herbivorous animals needed to support them. This almost inevitably leads to overgrazing and the reduction of grass cover (Case Study 7). When new shoots appear after the rains, they are eaten immediately by cattle, sheep, goats and camels. The arrival of the rains causes erosion; the failure of the rains results in animal deaths.

Where there is a rapid population growth, land that was previously allowed a fallow resting period now has to be culti-vated each year (Figure 10.36) – as are other areas that were previously considered to be too marginal for crops. Monoculture – the cultivation of the same crop each year on the same piece of land – repeatedly uses up the same soil nutrients.

Burkina Faso

As the size of cattle and goat herds has grown, the already scant dry scrub savanna vegetation on the southern fringes of the Sahara has been totally removed over increasingly large areas. As the Sahara 'advances', the herders are forced to move southwards into moister environments where they compete for land with sedentary farmers who are already struggling to produce sufficient food for their own increasing numbers. This disruption of equilibrium further reduces the land carrying capacity [page 378] – i.e. the number of people that the soil and climate of an area can permanently support when the land is planted with staple crops. These farmers have long been aware that three years' cropping had to be followed by at least eight fallow years in order for grass and trees to re-establish themselves and organic matter to be replenished. The arrival of the herders has brought a land shortage resulting in crops being grown on the same plots every year, and the nutrient-deficient soil, typical of most of tropical Africa, is rapidly becoming even less productive. This overcropping, a problem in many of the world's subsistence areas, uses up organic matter and other nutrients, weakens soil structures and leaves the surface exposed and thus susceptible to accelerated erosion.

Figure 10.36

Overgrazing: Burkina Faso

The Soil Protection Review is carried out by Britain's farmers as part of cross-compliance. It involves identifying soil issues, deciding on measures to manage and protect soils, and then reviewing the results. The 2006 review concluded with the following recommended options to protect the soil from physical decline and erosion:

- reducing mechanical operations on wet ground
- planting crops early in autumn to protect the soil during the winter from water erosion
- ploughing across slopes where it is safe to do so (compare Figure 10.38)
- using low ground-pressure set-ups on machinery
- shepherding livestock and rotating forage areas
- planting and/or maintaining hedges or shelter belts to reduce wind erosion

and measures to protect the soil's organic matter:

- leaving straw and other crop residues on the land after a crop has been harvested
- including grass in crop rotations
- applying animal manure, compost and sewage sludge
- using reduced or shallow cultivation to maintain or increase near-surface organic matter.

Many farmers suggest that these options are often already adopted but need better co-ordination together with continued targeted advice, information and monitoring.

Figure 10.37

Mitigation strategies for soil degradation

In many parts of the world where livestock are kept and firewood is at a premium, dung has to be used as a fuel instead of being applied to the land. In parts of Ethiopia, the sale of dung – mixed with straw and dried into 'cakes' – is often the only source of income for rural dwellers. If this dung were to be applied to the fields, rather than sold to the towns, harvests could be increased by over 20 per cent.

Water is essential for a productive soil. The early civilisations, which grew up in river valleys (Figure 14.1), relied on irrigation, as do many areas of the modern world. Unfortunately, irrigation in a hot, dry climate tends to lead to salinisation, with dissolved salts being brought, by capillary action, into the root zone of agricultural trees and crops (Figure 16.53). Wells, sunk in dry climates, use up reserves of groundwater which may have taken many centuries to accumulate and which cannot be replaced quickly (fossil water stores, page 190). The resultant lowering of the water table makes it harder for plant roots to obtain moisture. The sinking of wells in sub-Saharan Africa, following the drought of the early 1980s, has unintentionally created difficulties. The presence of an assured water supply has attracted numerous migrants and their animals and this has accelerated the destruction of the remaining trees and exacerbated the problems of overgrazing (Places 65, page 479). Even well-intentioned aid projects may therefore be environmentally damaging.

Fertiliser and pesticides are not always beneficial if applied repeatedly over long periods. Chemical fertiliser does not add organic material and so fails to improve or maintain soil structure. There is considerable concern over the leaching of nitrate fertiliser into streams and underground water supplies. Where nitrates reach rivers they enrich the water and encourage the rapid growth of algae and other aquatic plants which use up oxygen, through the process of eutrophication, to leave insufficient for plant life (Figure 16.50). The use of pesticides (including insecticides and fungicides) can increase yields by up to 100 per cent by killing off insect pests. However, their excessive and random use also kills vital soil organisms, which means organic matter decomposes more slowly and the release of nutrients is retarded. Chemical pesticides are blamed for the decline in Britain's bee population.

Soil management

Fertility refers to the ability of a soil to provide for the unconstrained or optimum growth of plants. The capacity to produce high or low yields depends upon the nutrient content, structure, texture, drainage, acidity and organic content of a particular soil as well as the relief, climate and farming techniques. For ideal growth, plants must have access to nine macro-nutrients and nine micro-nutrients (Figure 10.13). Under normal recycling (Figure 10.6), these nutrients will

be returned to the soil as the vegetation dies and decomposes. When a crop is harvested there is less organic material left to be recycled. As nutrients are taken out of the soil system and not replaced, there will be an increasing shortage of macro-nutrients, particularly nitrogen, calcium, phosphorus and potassium. Where this occurs, and when other nutrients are dissolved and leached from the soil, fertiliser is essential if yields are to be maintained. Soils need to be managed carefully if they are to produce maximum agricultural yields and cause least environmental damage (Figure 10.37).

If the most serious cause of erosion is the removal of vegetation cover, the best way to protect the soil is likely to be by the addition of vegetation. Afforestation provides a long-term solution because, once the trees have grown, their leaves intercept rainfall while their roots help to bind the soil together and reduce surface runoff. The growing of ground-cover crops reduces rainsplash and surface runoff, and can protect newly ploughed land from exposure to climatic extremes. Marram grass anchors sand, while gulleys can be seeded and planted with brushwood. Certain crops and plants, especially leguminous species such as peas, beans, clover and gorse, are capable of fixing atmospheric nitrogen in the soil, thus improving its quality. Trees can also be planted to act as windbreaks and shelterbelts. This reduces the risk of wind erosion as well as providing habitats for wildlife.

Soil can also be managed by improving farming methods. Most arable areas benefit from a rotation of crops, including grasses, which improve soil structures and reduce the likelihood of soil-borne diseases which may develop under monoculture. Many tropical soils need a recovery period of 5–15 years under shrub or forest for each 3–6 years under crops. In areas where slopes reach up to 12°, ploughing should follow the contours to prevent excessive erosion. On even steeper slopes (Figures 10.41 and 16.29), terracing helps to slow down runoff, giving water more time to infiltrate and thus reducing its erosive ability.

Strip cropping can involve either the planting of crops in strips along the contours or the intercropping of different crops in the same field. Both methods are illustrated in Figure 10.38. The crops may differ in height, time of harvest and use of nutrients.

Where evapotranspiration exceeds precipitation, dry farming can be adopted. This entails covering the soil with a mulch of straw and/or weeds to reduce moisture loss and limit erosion. In the Sahel countries, the drastic depopulation of cattle following the droughts of the 1980s has given herders a chance to restock with smaller (reducing overgrazing), better-quality (giving more meat and milk) herds so that incomes do not fall and the soils are given time to recover.

The addition of organic material helps to bind loose soil and so reduces its vulnerability to erosion (Figures 10.38 and 10.39). Soil structure and texture may be improved, theoretically, by adding lime to acid soils, which reduces their acidity and helps to make them warmer; by adding humus, clay or peat to sands, to give body and to

Organic milk has more healthy benefits

A study of organic milk, conducted by Professor Carlo Leifert of Newcastle University, has shown that drinking organic milk has greater health benefits than drinking normal milk. The study showed that organic milk contained 67 per cent more antioxidants and vitamins than ordinary milk and 60 per cent more of a healthy fatty acid called conjugated linoleic acid (CLA9) which tests have shown can shrink tumours. Similar levels of vaccenic acid, which has been shown to cut the risk of heart disease, diabetes and obesity, were also found as was an extra 39 per cent of the fatty acid Omega-3 which has also been shown to cut the risk of heart disease.

Gillian Butler, the livestock project manager, pointed out the health benefits even if consumers did not switch completely to organic milk. She pointed out that organic milk is more expensive to produce, as you get less milk per unit of land, and to buy, but because it is higher in all these beneficial compounds you do not need to buy as much to get health benefits.

Figure 10.39

Organic farming in Washington State, USA

Figure 10.38

Strip farming along contours in the southern USA

improve their water-holding capacity; and by adding sand to heavy clays, so improving drainage and aeration and making them lighter to work. In practice, such methods are rarely used due to the expense involved.

Chemical (inorganic) fertilisers help to replenish deficient nutrients, especially nitrogen, potassium and phosphorus. However, their use is expensive, especially to farmers in economically less developed countries, and can cause environmental damage. Many farmers in poorer countries cannot afford such fertilisers and have to rely upon organic fertiliser. Animal dung and straw left after the cereal harvest are mixed together and spread over the ground. This improves soil structure and, as it decays, returns nutrients to the soil. Where crop rotations are practised, grasses add organic matter, and legumes provide nitrogen.

Stone lines in Burkina Faso

Figure 10.40

Stone lines in Burkina Faso

This project, begun by Oxfam in 1979, aimed to introduce water-harvesting techniques for tree planting. It met with resistance from local people who were reluctant to divert land and labour from food production, or to risk dry-season water needed for drinking.

Attention was therefore diverted to improving food production by using the traditional local technique of placing lines of stones across slopes to reduce runoff [Figures 10.40 and 16.64]. When aligned with the contours, these lines dammed rainfall, giving it time to infiltrate. Unfortunately, most slopes were so gentle, under 2°, that local farmers could not determine the contours. A device costing less than £3 solved the problem. A calibrated transparent hose, 15 m long, is fixed at each end to the tops of stakes of equal lengths and

filled with water. When the water level is equal at both ends of the hose, the bottom of the stakes must be on the same contour. The lines can be made during the dry season when labour is not needed for farming. Although they take up only 1 or 2 per cent of cropland, they can increase yields by over 50 per cent. They also help to replenish falling water tables and can regenerate the barren, crusted earth because soil, organic matter and seeds collect on the upslope side of the stone lines and plants begin to grow again.

Since 2000, Practical Action has been financing the construction of crescent-shaped terraces which, built of earth along the contours of the land, last longer and hold on to vital rainwater more efficiently than traditional square dams. Crops grown here thrive in soil, rich in nutrients, that was previously washed away.

In Britain and North America, a growing number of farmers are turning to organic farming for environmental reasons (Figure 10.39 and page 497 and Case Study 16B).

Many soils suffer from either a shortage or a surfeit of water. In irrigated areas, water must be continually flushed through the system to prevent salinisation. In areas of heavy and/or seasonal rainfall, dams may be built to control flooding and to store surplus water. The drainage of waterlogged soils can be improved by adding field drains.

In several Sahelian countries, people use stones to build small dams which trap water for long enough for some to infiltrate into the ground; they also collect the soil carried away by surface runoff (Figures 10.40 and 16.64).

Soil conservation in northern Shaanxi (China)

According to historic records, the northern province of Shaanxi was once a region with plenty of water, fertile loess soil, lush grass and livestock. Since then, overcultivation and deforestation have led to severe soil erosion [Figure 10.35]. This has in turn caused serious desertification [Case Study 7], creating drifting sand dunes which have buried farmland and villages, while frequent droughts, floods and dust-storms have hindered the development of the local farming economy. Agriculture fell into a vicious circle: people, because of their poverty, reclaimed land but the more land they reclaimed, the poorer they became because this land was also subject to erosion.

Since the early 1980s, however, the central government has encouraged and supported a comprehensive programme for erosion control on the loess plateau. The two main aims have been to control and stabilise drifting sand in northern Shaanxi and to transform the soil throughout the province. This has involved the development of irrigation projects, the terracing of hillsides [Figure 10.41], the planting of trees as a shelter-forest network against the shifting sand [Figure 10.42] and the construction of check-dams [Figure 10.43].

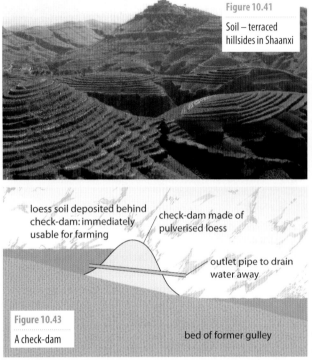

Figure 10.41

Soil – terraced hillsides in Shaanxi

loess soil deposited behind check-dam: immediately usable for farming

check-dam made of pulverised loess

outlet pipe to drain water away

Figure 10.43

A check-dam

bed of former gulley

Figure 10.42

The shelter-forest network

During the 1990s, the Loess Plateau Rehabilitation Project was set up with the twin aims of increasing the income of local farmers and reducing sediment flow into the Huang He. The project has involved engaging the local farmers and government officials in planting more trees, bushes and grasses that were previously native to the region; creating terraces for agriculture; planting orchards and vineyards; and constructing more sediment control dams and irrigation networks. At first the farmers and officials were sceptical about restoring so much land and leaving it for nature, but the desperate poverty of the plateau region led them to co-operate.

Each year more than 4 per cent of the eroded area is targeted for soil and water conservation projects.

The success of the Loess Project can be seen in the huge tracts of land that are now nurturing young forests; the crops grown in newly created fields along valley floors; the reduction in the amount of soil washed into the Huang He or blown towards Beijing in dust-storms; the restoration of an ecosystem; and, within a decade, the quadrupling of the income of local people. The project has helped promote sustainable and productive agriculture and improved the standard of living and quality of life of local people.

Further reference

Bridges, E.M. (1997) *World Soils*, Cambridge University Press.

Courtney, F.M. and Trudgill, S.T. (1984) *The Soil: An Introduction to Soil Study*, Hodder Arnold.

Ellis, S. and Mellor, A. (1995) *Soils and Environment*, Routledge.

Knapp, B. (1979) *Soil Processes*, Allen & Unwin.

O'Hare, G. (1988) *Soils, Vegetation and the Ecosystem*, Oliver & Boyd.

Trudgill, S. (1988) *Soil and Vegetation Systems*, Clarendon Press.

Department of Environment (Malaysia), controlling soil erosion:
www.jas.sains.my/doe/new/index.html

Nature journal:
www.nature.com/nature/

Soil salinity and erosion control in Alberta, Canada:
www.agric.gov.ab.ca/app21/rtw/index.jsp – use search option

UN Convention to Combat Desertification:
www.unced.int/main.php

UN Food and Agriculture Organisation, desertification:
www.fao.org/desertification/default.asp?lang=eng

Activities

1 a i   What are the two main components of a soil?   *(2 marks)*
      ii Study Figure 10.1 (page 260) and describe how **two** of these factors affect the formation of a soil.   *(4 marks)*
      iii Why does the water content of a soil vary from the top of a slope to the bottom?   *(4 marks)*
   b What is a 'soil horizon'?   *(4 marks)*
   c Choose **one** soil that you have studied.
      i   Name the soil.
      ii Draw an annotated soil profile to show the main characteristics of the soil.   *(6 marks)*
   d Why do farmers plough their arable land?   *(5 marks)*

2 a What can happen to water when it lands on the surface of a soil?   *(4 marks)*
   b i   What does it mean when 'precipitation exceeds evapotranspiration'?   *(4 marks)*
      ii What happens to the soil when leaching occurs? *(5 marks)*
   c Name and describe a soil that results from the process of leaching.   *(4 marks)*
   d i   Why would a farmer want to change soil acidity? *(2 marks)*
      ii What can a farmer do to change the pH of a soil? *(2 marks)*
      iii How does the activity you have described in **ii** change the pH?   *(4 marks)*

3 a What is a 'soil horizon'?   *(3 marks)*
   b Draw an **annotated** diagram to show the main features of a brown earth soil.   *(5 marks)*
   c What natural vegetation type and climatic type is associated with formation of a brown earth soil?   *(3 marks)*

   d Explain the processes by which a brown earth is formed.   *(6 marks)*
   e In what type of area would you expect to find a brown earth within the British Isles?   *(3 marks)*
   f What effect is a farmer trying to achieve when ploughing a brown earth?   *(5 marks)*

4 Choose **one** example of soil you have studied in the field.
   a i   Identify the aims and objectives of the study.   *(3 marks)*
      ii Describe the main features of the area where the fieldwork was carried out.   *(3 marks)*
      iii Explain how the fieldwork was planned before the trip took place.   *(3 marks)*
   b Describe the methods used to collect the data (your response should include 'what', 'why', 'where', 'how' and 'how it was recorded').   *(8 marks)*
   c i   For **one** piece of analysis you have carried out, explain how the data were sorted to prepare them for analysis.   *(4 marks)*
      ii How were results prepared for presentation after the fieldwork trip?   *(4 marks)*

5 a Identify and explain the **five** main factors affecting the formation of a soil.   *(10 marks)*
   b What is:
      i   soil texture
      ii soil structure?   *(8 marks)*
   c For **either** soil structure **or** soil texture, describe how you would identify it in a soil. In your answer you should identify equipment used and explain how to interpret the results.   *(7 marks)*

Exam practice: basic structured questions

6 a Study Figure 10.9 on page 265.
      i   Identify the constituents of soils a, b and c, and suggest a name for each soil.   *(3 marks)*
      ii Plot the soil textures from Figure 10.44 onto a triangular graph.   *(5 marks)*
   b Explain how soil texture and soil structure can influence farming.   *(9 marks)*
   c Identify **two** ways in which a farmer can improve the fertility of the soil. In your answer you should explain the effect of the activity on the farmer's output.   *(8 marks)*

7 Study Figures 10.45 and 10.46 which show four soils and their locations.
   a i   Describe how the depths of soil vary across this area.   *(4 marks)*
      ii Account for the differences that you observed in **a i**.   *(6 marks)*

   b Study soil profile B.
      i   Describe the humus layer in soil B and explain how it has been formed.   *(5 marks)*
      ii Describe the texture of the *A* horizon in soil B and explain how the texture affects farming.   *(5 marks)*
   c Explain why a farm on the Charnwood Forest would be different from one on the Lincoln Edge.   *(5 marks)*

Sample	Clay (%)	Silt (%)	Sand (%)
d	61	26	13
e	33	7	60
f	8	79	13
g	5	5	90
h	34	36	30

Figure 10.44

Five soil samples

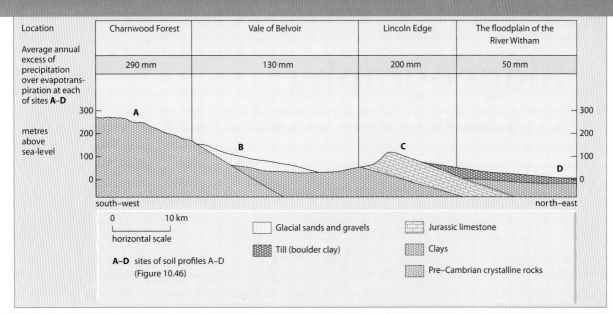

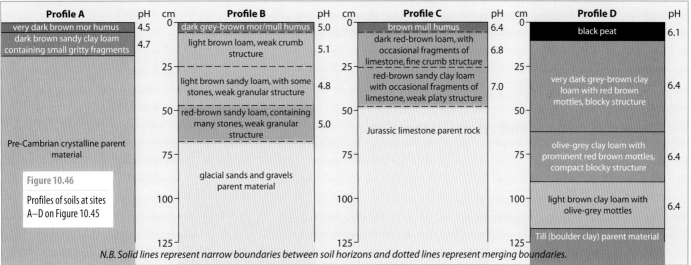

Figure 10.46

Profiles of soils at sites A–D on Figure 10.45

N.B. Solid lines represent narrow boundaries between soil horizons and dotted lines represent merging boundaries.

Exam practice: structured questions and essay

8 a Using Figures 10.45 and 10.46, identify which of the soils are zonal, azonal and intrazonal. *(4 marks)*

  b Select either soil B or soil C.

  i Describe the main characteristics of your chosen soil. *(4 marks)*

  ii Account for the nature of the A horizon in your chosen soil. *(6 marks)*

  c Soil D is a peaty gley. Explain **two** aspects of this soil that make it difficult for a farmer to cultivate. *(6 marks)*

  d Suggest **two** reasons why soil A is a very shallow soil. *(5 marks)*

9 a Why does soil move downhill? *(5 marks)*

  b Describe **two** unintended effects of human activity on soils. *(10 marks)*

  c Explain **two** ways in which farmers can combat accelerated soil erosion. *(10 marks)*

10 a i Choose **one** azonal soil you have studied and draw an annotated diagram to show the characteristics of the soil. *(7 marks)*

  ii Explain why it is classified as azonal. *(3 marks)*

  b Why do geographers and others classify soils? *(5 marks)*

  c Identify **one** scientific soil classification system you have studied. Making use of example soils, explain the basis on which the classification is made. *(10 marks)*

11 a What is a 'soil catena'? *(3 marks)*

  b Explain how and why soil depth varies down the slope of a catena. *(7 marks)*

  c Peat can develop in both the upland and the lowland areas of a soil catena.

  Explain how this happens, making reference to the differences in the nature of the peat in the two areas. *(15 marks)*

12 With reference to countries at different stages of development, explain why farmers need to manage their soils more carefully if farming is to be sustainable. *(25 marks)*

11

Biogeography

'The Earth's green cover is a prerequisite for the rest of life. Plants alone, through the alchemy of photosynthesis, can use sunlight energy, and convert it to the chemical energy animals need for survival.'

James Lovelock, *The Gaia Atlas of Planet Management*, 1985

Biogeography may be defined as the study of the distribution of plants and animals over the Earth's surface. The biogeographer is interested in describing and explaining meaningful patterns of plant and animal distributions in a given area, either at a particular time or through a time-period.

Seres and climax vegetation

A **sere** is a stage in a sequence of events by which the vegetation of an area develops over a period of time. The first plants to colonise an area and develop in it are called the **pioneer community** (or **species**). A **prisere** is the complete chain of successive seres beginning with a pioneer community and ending with a **climax vegetation** (Figure 11.1a). F.E. Clements suggested, in 1916, that for each climatic zone only one type of climax vegetation could evolve. He referred to this as the **climatic climax vegetation**, now known as the **monoclimax concept**. The climatic climax occurs when the vegetation is in harmony or equilibrium with the local environment, i.e. when the natural vegetation has reached a delicate but stable balance with the climate and soils of an area (Chapter 12). Each successive seral community usually shows an increase in the number of species and the height of the plants, an increase in carbon storage and enhanced biogeochemical cycling and soil formation.

Each individual sere is referred to by one or more of the larger species within that community – the so-called **dominant species**. The dominant species may be the **largest** plant or tree in the community which exerts the maximum influence on the local environment or habitat, or the most **numerous** species in the community. In parts of the world where the climatic climax is forest – i.e. areas with higher rainfall – the plant community tends to be structured in layers (Figures 11.2 and 12.4). It can take several thousand years to reach a climatic climax. Communities are, however, relatively ephemeral on timescales of millennia. When climatic change does occur, temperature and/or precipitation alterations often only affect individual species rather than changing the community as a whole. This concept, the 'individualistic concept of plant association', was originated by H.A. Gleason in 1928. In recent years this has become widely accepted as a result of the analysis of pollen taken from lake sediments and peats (page 294).

Figure 11.1

A seral progression, with possible interruptions

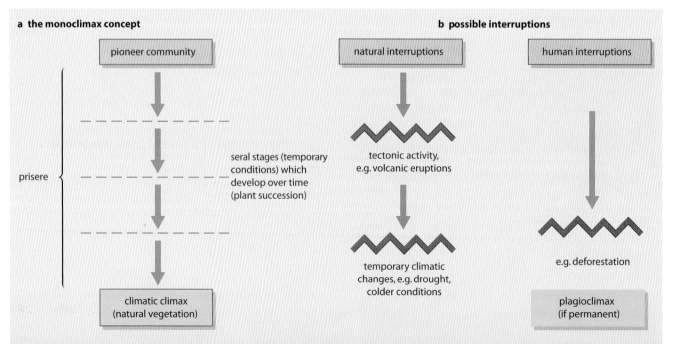

a the monoclimax concept

pioneer community

prisere

seral stages (temporary conditions) which develop over time (plant succession)

climatic climax (natural vegetation)

b possible interruptions

natural interruptions

tectonic activity, e.g. volcanic eruptions

temporary climatic changes, e.g. drought, colder conditions

human interruptions

e.g. deforestation

plagioclimax (if permanent)

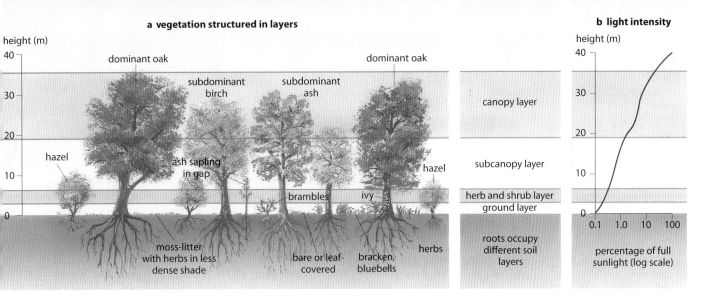

a vegetation structured in layers

height (m)

dominant oak
dominant oak
subdominant birch
subdominant ash
hazel
ash sapling in gap
brambles
ivy
hazel
moss-litter with herbs in less dense shade
bare or leaf-covered
bracken, bluebells
herbs

b light intensity

height (m)

canopy layer
subcanopy layer
herb and shrub layer
ground layer
roots occupy different soil layers

percentage of full sunlight (log scale)

Figure 11.2

Vegetation structure and light intensity typical of a temperate deciduous woodland (*after* O'Hare)

There are, however, very few parts of today's world with a climatic climax. This is partly because few physical environments remain stable sufficiently long for the climax to be reached: most are affected by tectonic or temporary climatic changes (an area becomes warmer, colder, wetter or drier). More recently, however, instability has resulted from such human activities as deforestation, the ploughing of grassland, and acid rain. Where human activity has permanently arrested and altered the natural succession and the ensuing vegetation is maintained through management, the resultant community is said to be a **plagioclimax** (Figure 11.1b) – examples of which include heather moorlands in Britain, and the temperate grasslands (page 326).

While it is still accepted that climate exerts a major influence upon vegetation, the linear monoclimax concept has been replaced by the **polyclimax theory**. This theory acknowledges the importance not only of climate, but of several (poly) local factors including drainage, parent rock,

Figure 11.4

Primary successions

new inorganic (non-vegetated) surface

plagioclimax subclimax (after deforestation, ploughing, burning, volcanic eruptions)

primary succession

climax vegetation

secondary succession

biotic subclimax

edaphic subclimax

1 primary succession
2 **natural variations** due to local conditions
3 **retrogressive succession** due to disturbance (natural or human)
4 secondary succession

Figure 11.3

The polyclimax theory

relief, microclimate and source of plants. The polyclimax theory, therefore, relates the climax vegetation to a variety of factors. Figure 11.3 shows how the climax vegetation may result from a **primary** or a **secondary succession**. A primary succession occurs on a new or previously unvegetated land surface, or in water. Figure 11.4 shows how the four more commonly accepted non-vegetated environments in Britain develop until they all reach the same climax vegetation: the oak woodland. A secondary succession is more likely to occur on land on which the previous management has been discontinued, e.g. abandoned farmland due to shifting cultivation in the tropical rainforest (Places 66, page 480). A **subclimax** occurs when the vegetation is prevented from reaching its climax due to interruptions by local factors such as soils and human interference.

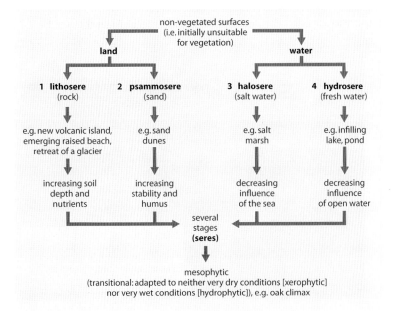

non-vegetated surfaces (i.e. initially unsuitable for vegetation)

land

water

1 **lithosere** (rock)
2 **psammosere** (sand)
3 **halosere** (salt water)
4 **hydrosere** (fresh water)

e.g. new volcanic island, emerging raised beach, retreat of a glacier

e.g. sand dunes

e.g. salt marsh

e.g. infilling lake, pond

increasing soil depth and nutrients

increasing stability and humus

decreasing influence of the sea

decreasing influence of open water

several stages (**seres**)

mesophytic (transitional: adapted to neither very dry conditions [xerophytic] nor very wet conditions [hydrophytic]), e.g. oak climax

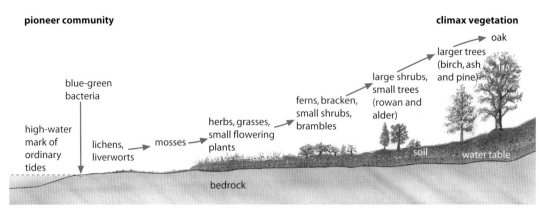

pioneer community

climax vegetation

oak

larger trees (birch, ash and pine)

large shrubs, small trees (rowan and alder)

ferns, bracken, small shrubs, brambles

herbs, grasses, small flowering plants

blue-green bacteria

mosses

lichens, liverworts

high-water mark of ordinary tides

bedrock

soil

water table

Four basic seres forming a primary succession

1 Lithoseres

Areas of bare rock will initially be colonised by blue-green bacteria and single-celled photosynthesisers that have no root system and can survive where there are few mineral nutrients. Blue-green bacteria are autotrophs (page 296), photosynthesising and producing their own food source. Lichens and mosses also make up the pioneer community (Figure 11.5). These plants are capable of living in areas lacking soil, devoid of a permanent supply of water and experiencing extremes of temperature. Lichen and various forms of weathering help to break up the rock to form a veneer of soil in which more advanced plant life can then grow. As these plants die, they are converted by bacteria into humus which helps in the development of increasingly richer soils and aids water retention. Seeds, mainly of grasses and herbs, then colonise the area. As these plants are taller than the pioneer species, they will replace the lichen and mosses as the dominants, although the lichens and mosses will still continue to grow in the community. As the plant succession evolves over a period of time, the grasses will give way as

dominants to fast-growing shrubs, which in turn will be replaced by relatively fast-growing trees (rowan). These will eventually face competition from slower-growing trees (ash) and, finally, the oak which forms the climax vegetation. It should be noted that although each stage of the succession is marked by a new dominant, many of the earlier species continue to grow there, although some are shaded out.

Figure 11.5 shows an idealised primary succession across a newly emerging rocky coastline. It excludes the increasing number of species found at each stage of the seral succession. The species are determined by local differences in rainfall, temperature and sunlight, bedrock and soil type, aspect and relief. Lithoseres can develop on bare rock exposed by a retreating glacier (page 294), on ash or lava following a volcanic eruption on land (Krakatoa, Places 35) or forming a new island (Surtsey, Places 3, page 16), or, as in Figure 11.5, on land emerging from the sea as a result of isostatic uplift following the melting of an icecap (page 163).

Over time, the area shown to have the pioneer community passes through several stages until the climatic climax is reached – assuming that the land continues to emerge from the sea, that

Figure 11.7

Primary succession on the same lithosere in Arran: bracken and deciduous woodland behind the rocky coastline

there is no significant change in the local climate, and that there is no human interference. Figures 11.6 and 11.7 are photos showing two stages in the succession, taken on a raised beach on the east coast of Arran. Figure 11.6 shows lichen, favouring a south-facing aspect on gently dipping rocks, and mosses, growing in darker north-facing hollows. Beyond, where soil has begun to form and where the water table is high, grasses and bog myrtle have entered the succession. Figure 11.7 was taken where the soil depth and amount of humus have increased and the water table is lower, as indicated by the presence of bracken. To the right, but not clearly visible on the photo, reeds are growing in a hollow where the water table is nearer to the surface. In the middle distance are small deciduous trees with, behind them, taller oaks indicating a climax vegetation.

Places 35 Krakatoa: a lithosere

In August 1883, a series of volcanic eruptions reduced the island of Krakatoa to one-third of its previous size and left a layer of ash over 50 m deep. No vegetation or animal life was left on the island or in the surrounding sea. Yet within three years (Figure 11.8), 26 species had reappeared and, in 1933, 271 plant and 720 insect species, together with several reptiles, were recorded. The first recolonisers arrived in three ways. Most were seeds blown from surrounding islands by the wind, while others drifted in from the sea or were carried by birds. However, on Krakatoa the plant succession, as defined by F.E. Clements in 1916 (page 286), was influenced by another variable: chance. For example, a piece of driftwood with a particular seed type just happened to be washed ashore onto the new ash, whereas it could just as easily have missed the island altogether.

Figure 11.8

Primary succession, Krakatoa: vegetation distribution according to height above sea-level, 1983

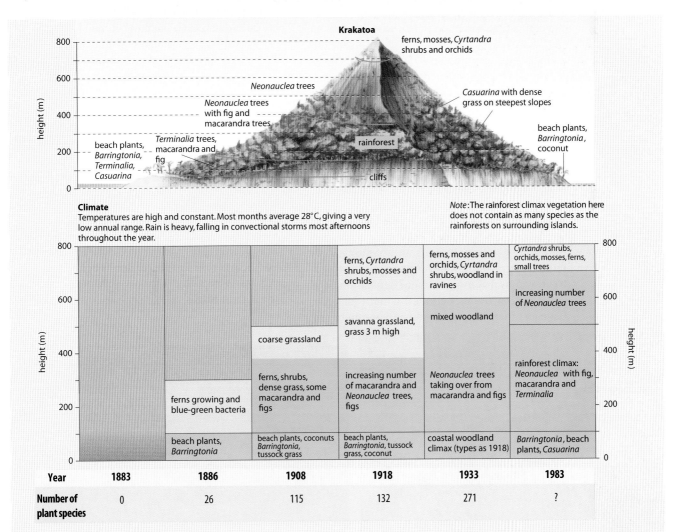

Climate
Temperatures are high and constant. Most months average 28°C, giving a very low annual range. Rain is heavy, falling in convectional storms most afternoons throughout the year.

Note: The rainforest climax vegetation here does not contain as many species as the rainforests on surrounding islands.

Year	1883	1886	1908	1918	1933	1983
Number of plant species	0	26	115	132	271	?

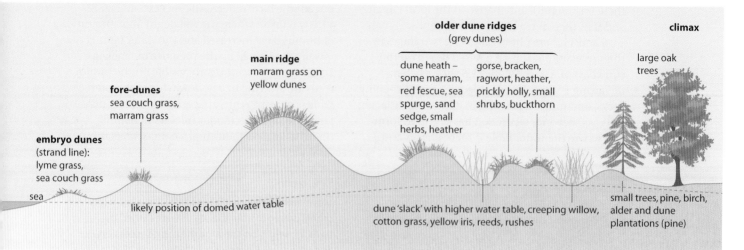

older dune ridges
(grey dunes)

climax

large oak
trees

main ridge
marram grass on
yellow dunes

dune heath –
some marram,
red fescue, sea
spurge, sand
sedge, small
herbs, heather

gorse, bracken,
ragwort, heather,
prickly holly, small
shrubs, buckthorn

fore-dunes
sea couch grass,
marram grass

embryo dunes
(strand line):
lyme grass,
sea couch grass

sea

likely position of domed water table

dune 'slack' with higher water table, creeping willow,
cotton grass, yellow iris, reeds, rushes

small trees, pine, birch,
alder and dune
plantations (pine)

Figure 11.9

Transect across sand
dunes to show a
psammosere, Morfa
Harlech, north Wales

2 Psammoseres

A psammosere succession develops on sand and is
best illustrated by taking a transect across coastal
dunes (Figure 11.9). The first plants to colonise,
indeed to initiate dune formation, are usually
lyme grass, sea couch grass and marram grass.
Sea couch grass grows on berms around the tidal
high-water mark and is often responsible for the
formation of embryo dunes (Figure 6.31). On the
yellow fore-dunes, which are arid, being above the
highest of tides and experiencing rapid percolation by rainwater, marram grass becomes equally
important.

The main dune ridge, which is extremely
arid and exposed to wind, is likely to be vegetated exclusively by marram grass. Marram has
adapted to these harsh conditions by having
leaves that can fold to reduce surface area,
which are shiny and which can be aligned to the
wind direction: three factors capable of limiting
evapotranspiration. Marram also has long roots
to tap underground water supplies and is able
to grow upwards as fast as sand deposition can
cover it. Grey dunes, behind the main ridge,
have lost their supply of sand and are sheltered

from the prevailing wind. Their greater humus
content, from the decomposition of earlier
marram grass, enables the soil to hold more
moisture. Although marram is still present, it
faces increasing competition from small flowering plants and herbs such as sea spurge (with
succulent leaves to store water) and heather.

The older ridges, further from the water, have
both more and taller species. Dune slacks may
form in hollows between the ridges if the water
table reaches the surface. Plants such as creeping
willow, yellow iris, reeds and rushes and shrubs
are indicators of a deeper and wetter soil. On the
landward side of the dunes, perhaps 400 m from
the beach, are small deciduous trees including
ash and hawthorn and, as the soil is sandy, pine
plantations. Furthest inland comes the oak
climax. Figure 11.9 shows a psammosere based
on sand dunes at Morfa Harlech, north Wales.
Figures 11.10 and 6.32 show marram and lyme
grass forming the yellow fore-dunes, with gorse
and heather on the greyer dunes behind.
Figures 11.11 and 6.33 show vegetation on
the inland ridges.

Figure 11.10

Primary succession on a
psammosere: colonisation
of fore-dunes, Winterton,
Norfolk (compare Figures
6.32 and 6.33)

Figure 11.11

Primary succession on a psammosere:
vegetation on a grey dune ridge and on
a dune slack, Braunton Burrows, Devon

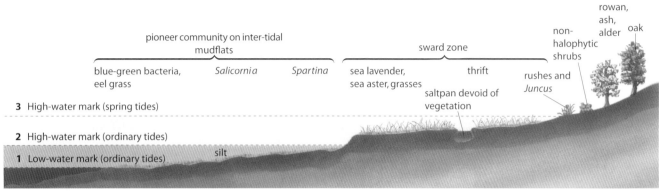

pioneer community on inter-tidal mudflats

blue-green bacteria, eel grass   *Salicornia*   *Spartina*

sward zone

sea lavender, sea aster, grasses   thrift

saltpan devoid of vegetation

rushes and *Juncus*

non-halophytic shrubs

rowan, ash, alder   oak

3 High-water mark (spring tides)

2 High-water mark (ordinary tides)

1 Low-water mark (ordinary tides)   silt

Figure 11.12

Transect showing a primary succession in a halosere, Llanrhidian Marsh, Gower Peninsula, south Wales

3 Haloseres

In river estuaries, large amounts of silt are deposited by the ebbing tide and inflowing rivers. The earliest plant colonisers are green algae and eel grass which can tolerate submergence by the tide for most of the 12-hour cycle and which trap mud, causing it to accumulate. Two other colonisers are *Salicornia* and *Spartina* which are **halophytes** – i.e. plants that can tolerate saline conditions. They grow on the inter-tidal mudflats (Figure 6.34), with a maximum of 4 hours' exposure to the air in every 12 hours. *Spartina* has long roots enabling it to trap more mud than the initial colonisers of algae and *Salicornia*, and so, in most places, it has become the dominant vegetation. The inter-tidal flats receive new sediment daily, are waterlogged to the exclusion of oxygen, and have a high pH value.

The sward zone (page 158), in contrast, is inhabited by plants that can only tolerate a maximum of 4 hours' submergence in every 12 hours. Here the dominant species are sea lavender, sea aster and grasses, including the 'bowling green turf' of the Solway Firth. However, although the vegetation here tends to form a thick mat, it is not continuous. Hollows may remain where the seawater becomes trapped leaving, after evaporation, saltpans in which the salinity is too great for plants (Figure 11.13). As the tide ebbs, water draining off the land may be concentrated into creeks (Figure 6.35). The upper sward zone is only covered by spring tides and here *Juncus* and other rushes grow. Further inland, non-halophytic grasses and shrubs enter the succession, to be followed by small trees and ultimately by the climax oak vegetation. Figure 11.12 is a transect based on the saltmarshes on the north coast of the Gower Peninsula in south Wales. Figure 11.14 shows several stages in the halosere succession.

Figure 11.13

Primary succession in a halosere: a saltpan on the Suffolk coast, covered only by the highest of tides

Figure 11.14

Primary succession in a halosere: Blakeney Point, Norfolk

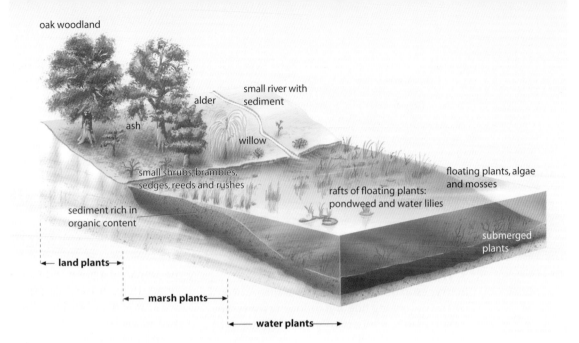

Figure 11.15

Idealised primary succession in a hydrosere

oak woodland

ash

alder

small river with sediment

willow

small shrubs, brambles, sedges, reeds and rushes

rafts of floating plants: pondweed and water lilies

floating plants, algae and mosses

submerged plants

sediment rich in organic content

← **land plants** →

← **marsh plants** →

← **water plants** →

Figure 11.16

Primary succession in a hydrosere at the head of a reservoir in Cumbria

4 Hydroseres

Lakes and ponds originate as clear water which contains few plant nutrients. Any sediment carried into the lake will enrich its water with nutrients and begin to infill it. The earliest colonisers will probably be algae and mosses whose spores have been blown onto the water surface by the wind. These grow to form vegetation rafts which provide a habitat for bacteria and insects. Next will be water-loving plants which may either grow on the surface, e.g. water lilies and pondweed, or be totally submerged (Figure 11.15). Bacteria recycle the nutrients from the pioneer community, and marsh plants such as bulrushes, sedges and reeds begin to encroach into the lake. As these marsh plants grow outwards into the lake and further sediment builds upwards at the expense of the water, small shrubs and trees will take root forming a marshy thicket. In time, the lake is likely to contract in size, to become deoxygenised by the decaying vegetation and eventually to disappear and be replaced by the oak climax vegetation. This primary succession is shown in Figure 11.15. Figure 11.16 shows land plants encroaching at the head of a reservoir, while Figure 11.17 illustrates the water, marsh and land plant succession in and around a small lake.

Incidentally, it is not necessary to be an expert botanist to recognise the plants named in these primary successions; you just need a good plant recognition book!

Figure 11.17

Primary succession in a small lake, Sussex

Secondary succession

A climatic climax occurs when there is stability in transfers of material and energy in the eco-system (page 295) between the plant cover and the physical environment. However, there are several factors that can arrest the plant succession before it has achieved this dynamic equilibrium, or which may alter the climax after it has been reached. These include:

- a mudflow or landslide (Places 36)
- deforestation or afforestation
- overgrazing by animals or the ploughing-up of grasslands
- burning grasslands, moorlands, forests and heaths
- draining wetlands
- disease (e.g. Dutch elm), and
- changes in climate (page 286).

Places 36 Arran: secondary plant succession

The mudflow shown in Figure 2.16 occurred in October 1981 and completely covered all the existing vegetation. Twelve months later it was estimated that 20 per cent of the flow had been recolonised, a figure that had grown to 40 per cent in 1984 and 70 per cent in 1988. Had this been a primary succession, lichens and mosses would have formed the pioneer community and they would probably have covered only a small area. The pioneer plants would probably also have been randomly distributed and, even after seven years, the species would have been few in number and small in height.

Instead, by 1988, much of the flow had already been recolonised. It could be seen that most of the plants were found near the edges of the flow and were not randomly distributed, and there were already several species including grasses, heather, bog myrtle and mosses, some of which exceeded 50 cm in height.

These observations suggest a secondary succession with plants from the surrounding climax community having invaded the flow, mainly due to the dispersal of their seeds by the wind.

The effect of fire

The severity of a fire and its effect on the eco-system depend largely upon the climatic conditions at the time. The fire is likely to be hottest in dry weather and, in the northern hemisphere, on sunny south-facing slopes where the vegetation is driest. The spread of a fire is fastest when the wind is strong and blowing uphill and where there is a build-up of combustible material. The extent of disruption also depends upon the type and the state of the vegetation. The following is a list of examples, in rank order of severity.

1 Areas with a Mediterranean climate, where the chaparral of California and the maquis/ garrigue of southern Europe are densest and tinder-dry in late summer after the seasonal drought. Since 2005, major bush fires, which are occurring more often, have threatened Sydney in Australia, Olympia (site of the first Olympics) in Greece, parts of the south of France and, in California, Los Angeles (Case Study 15A). In early 2009, over 200 people lost their lives in bushfires, in the Australian state of Victoria.
2 Coniferous forests where the leaf litter burns readily.
3 Ungrazed grasslands and, especially, the savannas, which have a low biomass but a thick litter layer (Figure 11.28). **Biomass** is the total mass of living organisms present in a community at any given time, expressed in terms of oven-dry weight (mass) per unit area.

4 Intensively grazed grasslands in semi-arid areas which have a lower biomass and a limited litter layer.
5 Deciduous woodlands which, despite the presence of a thick litter layer, are often slow to burn.

Following a fire, the blackened soil has a lower albedo and absorbs heat more readily and, without its protective vegetation cover, the soil is more vulnerable to erosion. Ash initially increases considerably the quantity of inorganic nutrients in the soil and bacterial activity is accelerated. Any seedlings left in the soil will grow rapidly as there is now plenty of light, no smothering layer of leaf litter, plenty of nutrients, a warmer soil and, at first, less competition from other species. Heaths and moors that have been fired are conspicuous by their greener, more vigorous growth. A fire climax community, known as **pyrophytic** vegetation, contains plants with seeds which have a thick protective coat and which may germinate because of the heat of the fire. The community may have a high proportion of species that can sprout quickly after the fire – plants that are protected by thick, insulating bark (cork oak in the chaparral (page 324) and baobab in the savannas (Figure 12.14)), or which have underground tubers or rhizomes insulated by the soil. It has been suggested that the grasslands of the American Prairies and the African savannas are not climatic climax vegetation, but are the result of firing by indigenous Indian and African tribes (Case Study 12).

Vegetation changes in the Holocene

The Holocene is the most recent of the geological periods (Figures 1.1 and 11.18). The last glacial advance in Britain ended about 18 000 years ago. Although the extreme south of England remained covered with hardy tundra plants, most of northern Britain was left as bare rock or glacial till. Had the climate gradually and constantly ameliorated, a primary succession would have taken place, from south to north, as previously described for a lithosere. It has been established that there have been several major fluctuations in climate during those 18 000 years which have resulted in significant changes in the climax vegetation (Figure 11.18).

There are several techniques that help to determine vegetation change: pollen analysis, dendrochronology, radio-carbon dating, and historical evidence (page 248). Families of plants have the same pollen grain in terms of its shape and pattern. Where pollen is blown by the wind onto peat bogs, such as at Tregaron in west Wales, the grains are trapped by the peat. As more peat accumulates over the years, the pollen of successively later times indicates which were the dominant and subdominant plants of the period (Figures 11.18 and 11.19). As each plant grows best within certain defined temperature and precipitation limits, it is possible to determine when the climate either improved (ameliorated) or deteriorated. Dendrochronology – dating by means of the annual growth-rings of trees – has shown that the bristlecone pine of California can be dated back some 5000 years, while European dendrochronology, based on bog oaks in Ireland and Germany, extends back some 10 000 years. Radio-carbon dating is based on changing amounts of radioactivity in the atmosphere and in plants. Notice in Figure 11.18, which links climatic and vegetation changes, how forests increase as the climate ameliorates, and how heathland and peat moors take over when the climate deteriorates.

Figure 11.19

Changes in the surface of lowland England, Wales and Scotland over the last 12 000 years (*after* Wilkinson)

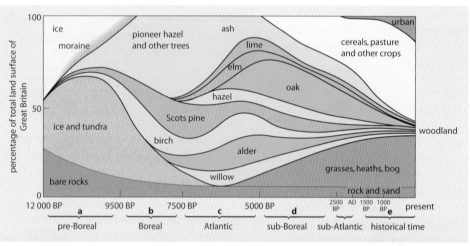

Figure 11.18

Climatic and vegetation change in Britain since the Holocene (BP = Before Present)

Date BP	Phase/period	Climate	Vegetation	Cultures
pre-17 000	final glaciation	glacial	none in northern Britain; tundra in southern England	none
17 000–14 000	periglacial	cold, 6°C in summer	tundra	Palaeolithic
14 000–12 000	Allerød	warming slowly to 12°C in summer	tundra with hardy trees, e.g. willow and birch	Palaeolithic
12 000–10 000	pre-Boreal	glacial advance: colder, 4°C in summer	Arctic/Alpine plants, tundra	Mesolithic
10 000–8000	Boreal	continental: winters colder and drier, summers warmer than today	forests: juniper first then pine and birch and finally oak, elm and lime	Mesolithic
8000–5000	Atlantic	maritime: warm summers, 20°C; mild winters, 5°C; wet	our 'optimum' climate and vegetation: oak, alder, hazel, elm and lime (too cold for lime today); peat on moors	beginning of Neolithic; first deforestation about 3500 BC
5 000–2500	sub-Boreal	continental: warmer and drier	elm and lime declined; birch flourished; peat bogs dried out	Neolithic period, settled agriculture; beginning of Bronze Age
2500–2000	sub-Atlantic	maritime: cooler, stormy and wet	peat bogs re-formed; decline in forests due to climate and farming	settled agriculture
2000–1000	historical times	improvement: warmer and drier	clearances for farming	Roman occupation during early part
1000–450		decline: much cooler and wetter	further clearances: little climax vegetation left; medieval farming	
450–300		'little ice age': colder than today		
post-300–present		gradual improvement	recently some afforestation: coniferous trees	Agrarian and Industrial Revolutions

Ecology and ecosystems

The term **ecology**, which comes from the Greek word *oikos* meaning 'home', refers to the study of the interrelationships between organisms and their habitats. An organism's home or **habitat** lies in the biosphere, i.e. the surface zone of the Earth and its adjacent atmosphere in which all organic life exists. The scale of each home varies from small **micro-habitats**, such as under a stone or a leaf, to **biomes**, which include tropical rainforests and deserts (Figure 11.20). Fundamental to the four ecological units listed in Figure 11.20 is the concept of the **environment**. The environment is a collective term to include all the conditions in which an organism lives. It can be divided into:

1 the physical, non-living or **abiotic environment**, which includes temperature, water, light, humidity, wind, carbon dioxide, oxygen, pH, rocks and nutrients in the soil, and
2 the living or **biotic environment**, which comprises all organisms: plants, animals, humans, bacteria and fungi.

The ecosystem

An ecosystem is a natural unit in which the life-cycles of plants, animals and other organisms are linked to each other and to the non-living constituents of the environment to form a natural system (Framework 3, page 45). The **community** consists of all the different species within a habitat or ecosystem. The **population** comprises all the individuals of a particular species in a habitat. An ecosystem depends on two basic processes: **energy flows** and **material cycling**. As the flow of energy is only in one direction and because it crosses the system boundaries, this aspect of the ecosystem behaves as an **open** system. Nutrients, which are constantly recycled for future use, are circulated in a series of **closed** systems.

1 Energy flows

The sun is the primary source of energy for all living things on Earth. As energy is retained only briefly in the biosphere before being returned to space, ecosystems have to rely upon a continual supply. The sun provides *heat energy* which cannot be captured by plants or animals but which warms up the communities and their non-living surroundings. The sun is also a source of *light energy* which can be captured by green plants and transformed into chemical energy through the process of **photosynthesis**. Without photosynthesis, there would be no life on Earth. Light, chlorophyll, warmth, water and carbon dioxide are required for this process to operate. Carbon dioxide, which is absorbed through stomata in the leaves of higher plants, reacts indirectly with water taken up by the roots when temperatures are suitably high, to form carbohydrate. The energy needed for this comes from sunlight which is 'trapped' by chlorophyll. Oxygen is a by-product of the process. The carbohydrate is then available as food for the plant.

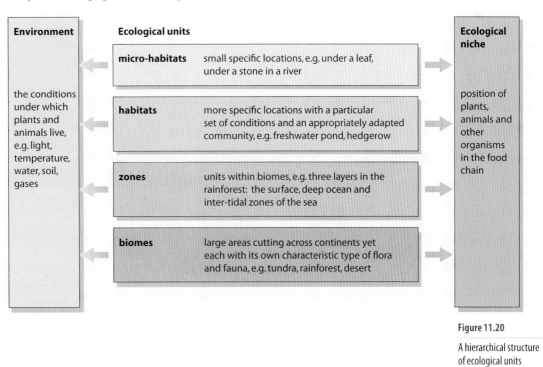

Figure 11.20

A hierarchical structure of ecological units

Food chains and trophic levels

A food chain arises when energy, trapped in the carbon compounds initially produced by plants through photosynthesis, is transferred through an ecosystem. Each link in the chain feeds on and obtains energy from the one preceding it, and in turn is consumed by and provides energy for the following link (Figure 11.21).

Figure 11.21

Three examples of food chains through four trophic levels

Level 1	Level 2	Level 3	Level 4
grass	worm	blackbird	hawk
leaf	caterpillar	shrew	badger
phytoplankton	zooplankton	fish	human

There are usually, but not always, four links in the chain. Each link or stage is known as a **trophic** or **energy level** (Figure 11.22). In order for the first link in the chain to develop, the non-living environment has to receive both energy from the sun and the other factors (water, CO_2, etc.) needed for photosynthesis.

The **first trophic level** is occupied by the **producers** or **autotrophs** ('self-feeders') which include green plants capable of producing their own food by photosynthesis. All other levels are occupied by **consumers** or **heterotrophs** ('other-feeders'). These include animals that obtain their energy either by eating green plants directly or by eating animals that have previously eaten green plants. The **second trophic level** is where **herbivores**, the primary consumers, eat the producers. The **third trophic level** is where smaller **carnivores** (meat-eaters) act as secondary consumers feeding upon the herbivores. The **fourth trophic level** is occupied by the larger carnivores, the tertiary consumers. Also known as **omnivores** (or diversivores), this group – which includes humans – eat both plants and animals and so have two sources of food. Figure 11.22 shows the main trophic or feeding levels in a food chain. **Detritivores**, such as bacteria and fungi, are consumers that operate at all trophic levels.

Figure 11.22

Trophic levels

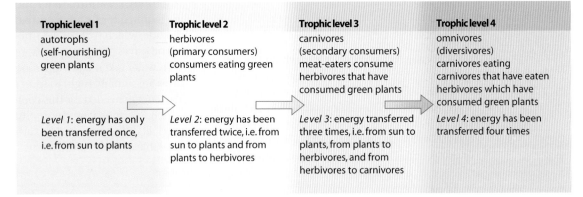

Trophic level 1	Trophic level 2	Trophic level 3	Trophic level 4
autotrophs (self-nourishing) green plants	herbivores (primary consumers) consumers eating green plants	carnivores (secondary consumers) meat-eaters consume herbivores that have consumed green plants	omnivores (diversivores) carnivores eating carnivores that have eaten herbivores which have consumed green plants
Level 1: energy has only been transferred once, i.e. from sun to plants	*Level 2*: energy has been transferred twice, i.e. from sun to plants and from plants to herbivores	*Level 3*: energy transferred three times, i.e. from sun to plants, from plants to herbivores, and from herbivores to carnivores	*Level 4*: energy has been transferred four times

Figure 11.23

Energy flows in the ecosystem

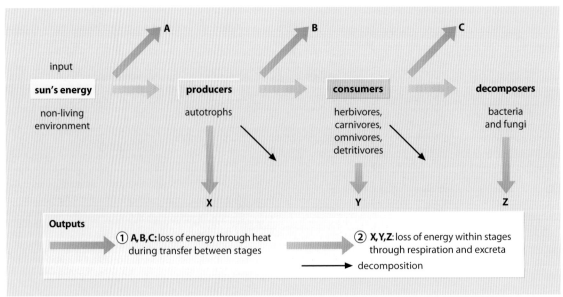

However, no transfer of energy is 100 per cent efficient and, as Figure 11.23 shows, energy is lost through respiration, by the decay of dead organisms and in excreta within each unit of the food chain, and also as heat given off when energy is passed from one trophic level to the next. Consequently, at each higher level, fewer organisms can be supported than at the previous level, even though their individual size generally increases. Simple food chains are rare; there is usually a variety of plants and animals at each level forming a more complicated **food web**. This range of species is necessary since a sole species occupying a particular trophic level in a simple food chain could be 'consumed' and this would adversely affect the organisms in the succeeding stages.

The progressive loss of energy through the food chain imposes a natural limit on the total mass of living matter (the **biomass**) and on the number of organisms that can exist at each level. It is convenient to show these changes in the form of a pyramid (Figure 11.24). A pyramid of organism numbers is of limited value for comparing ecosystems for two reasons. First, it is difficult to count the numbers of grasses or algae per unit area. Secondly, it does not take into account the relative sizes of organisms – a bacterium would count the same as a whale! A pyramid of biomass takes into account the difference in size between organisms, but cannot be used to compare masses at different trophic levels in the same ecosystem or at similar trophic levels in different ecosystems. This is because biomass will have accumulated over different periods of time.

Humans are found at the end of a food chain and human population is dependent upon the length of the chain (and therefore the amount of energy lost). In other words, in a shorter food chain, less energy will have been lost by the time it reaches humans and so the land can support a higher density of population. In a longer food chain, more energy will have been lost by the time the food is consumed by humans, which means that the carrying capacity (page 378) is lower and fewer people can be supported by a given area of land – as in western Europe, where most of the population are accustomed to animal products as well as crops.

Figure 11.24

The trophic pyramid

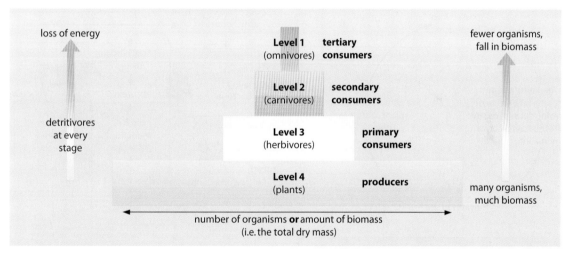

2 Nutrient cycling

Chemicals needed to produce organic material are circulated around the ecosystem and are continually recycled. Various chemicals can be absorbed by plants either as gases from the atmosphere or as soluble salts from the soil. Each cycle consists, at its simplest, of plants taking up chemical nutrients which, once they have been used, are passed on to the herbivores and then the carnivores that feed upon them. As organisms at each of these trophic levels die, they decompose and nutrients are returned to the system. Two of these cycles, the carbon and nitrogen cycles, are illustrated in Figures 11.25 and 11.26. In each case, the most basic cycle is given (diagram **a**); followed by a more detailed example, although still not in its entire complexity (diagram **b**).

Figure 11.25

The carbon cycle (*after* M.B.V. Roberts)

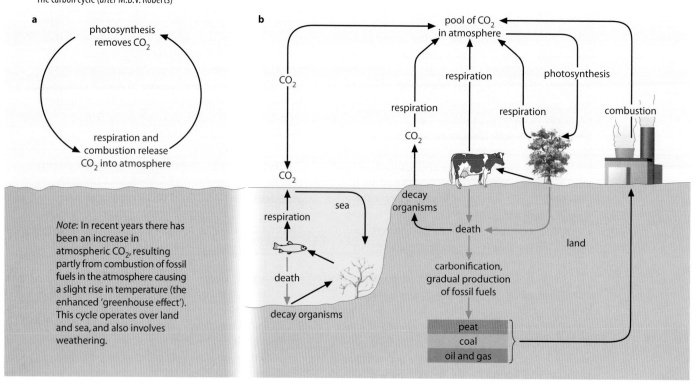

Note: In recent years there has been an increase in atmospheric CO_2, resulting partly from combustion of fossil fuels in the atmosphere causing a slight rise in temperature (the enhanced 'greenhouse effect'). This cycle operates over land and sea, and also involves weathering.

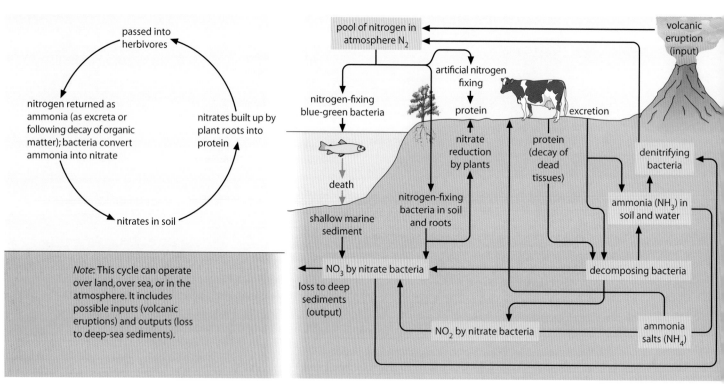

Note: This cycle can operate over land, over sea, or in the atmosphere. It includes possible inputs (volcanic eruptions) and outputs (loss to deep-sea sediments).

Figure 11.26

The nitrogen cycle (*after* M.B.V. Roberts)

Recent investigations, mainly in New Zealand and the Andes, have shown that nitrogen from seawater, or released by plants and animals as they die on the seabed, can be channelled upwards, together with magma, at subduction (destructive) plate margins. The nitrogen can later be released back into the atmosphere, either as water or as a gas, through volcanic eruptions. Once in the atmosphere, the nitrogen can return to Earth and the sea in rainwater – so completing another nitrogen cycle.

Framework 10 Scientific enquiry: hypothesis

Since the 1960s, geographers have felt an increasing need to adopt a more scientific approach to their studies. This stemmed from a number of changes that were taking place in attitudes to the study of geography and to science in a broader sense:

- The increasing scale and complexity of the subject's material and the data available.
- The rapid development of theory, often using computer modelling, from which predictions could be made.
- The realisation that, despite great care, all human observers have their own, subjective, opinions which influence an assessment or conclusion (i.e. scientific objectivity could not be guaranteed).

The scientific approach to geography involves a series of logical steps, already practised in the physical sciences, which enabled conclusions to be drawn from precise and unbiased data (Framework 8, page 246). This approach is summarised in the flow diagram (Figure 11.27).

Figure 11.27

Hypothesis testing

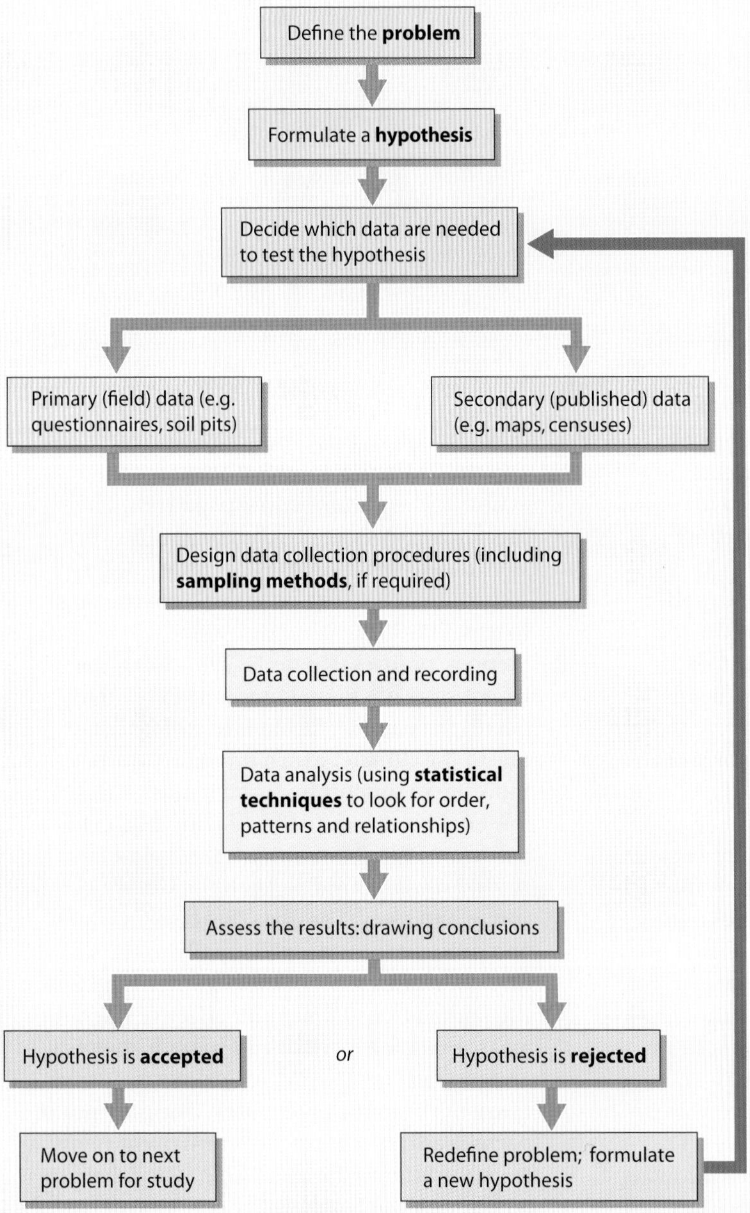

During a sixth-form field week on the Isle of Arran, one day was set aside for hypothesis testing. This involved seeking possible relationships between several variables on Goatfell (Figure 11.37). The hypotheses included:

- Vegetation density decreases as altitude increases.
- Soil acidity increases as altitude increases.
- Soil acidity increases as the angle of slope increases.
- Soil moisture increases as the angle of slope increases.
- Depth of soil increases as altitude decreases.
- Height of vegetation increases as altitude decreases.
- Number of species increases as altitude increases.
- Soil temperature increases as altitude decreases.

Data collection required the taking of readings at a minimum of 15 sites from sea-level to the top of Goatfell. It is important that the selection of sites is made without introducing bias (see Framework 6, page 159).

Data analysis may include drawing a scattergraph to investigate the possibility of any correlation between the two variables; calculating the strength of the relationship between the variables by using the Spearman's rank correlation coefficient (Framework 19, page 613); and then testing the result to see how likely it is that the correlation occurred by chance (page 614).

It should then be possible to determine whether the original hypothesis is acceptable as an explanation of the data, or not. If it is rejected, then a new hypothesis should be formulated.

Figure 11.28

A model of the mineral nutrient cycle (*after* Gersmehl)

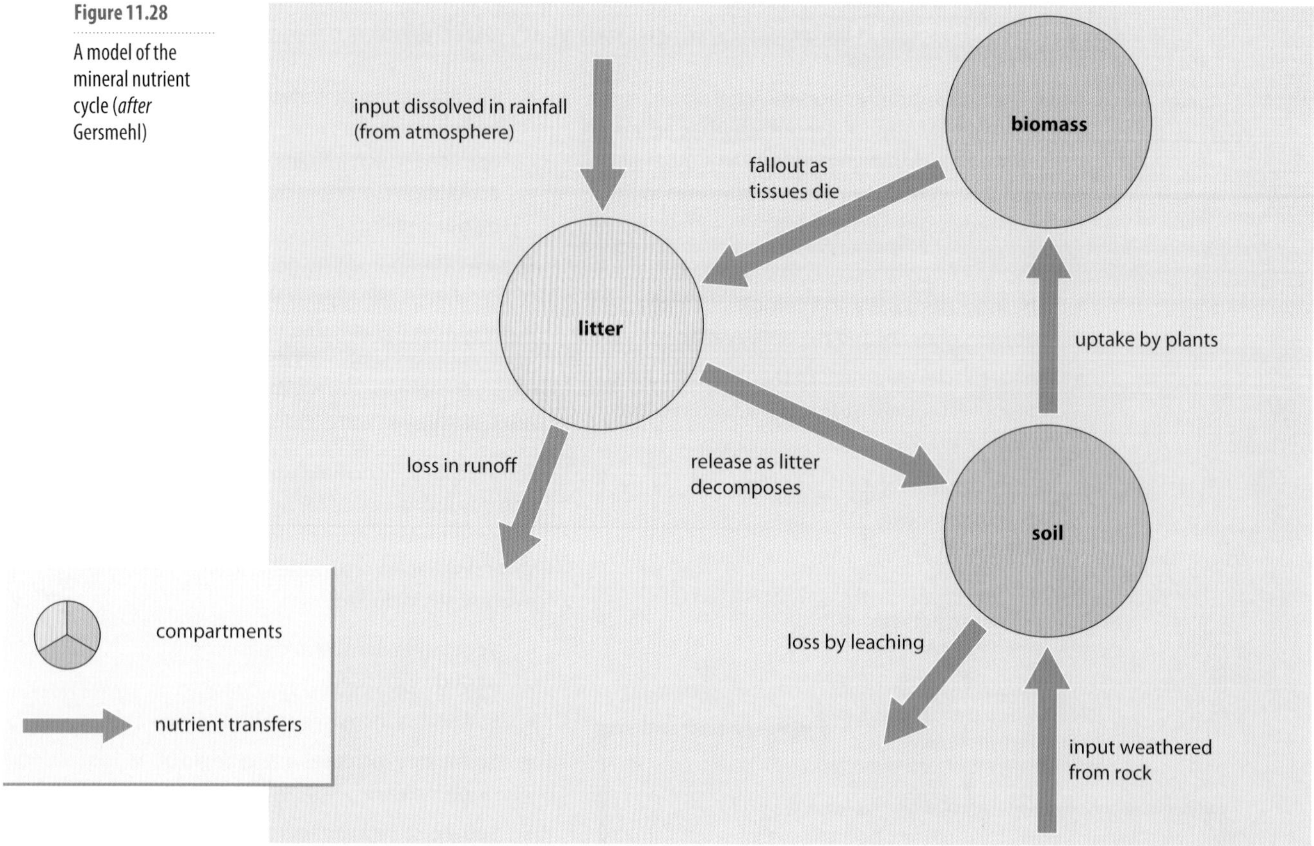

Model of the mineral nutrient cycle

This model, developed by P.F. Gersmehl in 1976, attempts to show the differences between ecosystems in terms of nutrients stored in, and transferred between, three compartments (Figure 11.28):

1 **Litter** – the total amount of organic matter, including humus and leaf litter, in the soil (it is, therefore, more than just the *L* or litter layer as shown in the soil profile in Figure 10.5).
2 **Biomass** – the total mass of living organisms, mainly plant tissue, per unit area.
3 **Soil**.

Figure 11.29 shows the mineral nutrient cycles for three selected biomes: the coniferous forest (taiga), the temperate grassland (prairies and steppes), and the tropical rainforest (selvas).

1 **Taiga** (Figure 11.29a) Litter is the largest store of mineral nutrients in the taiga. Although forest, the biomass is relatively low because the coniferous trees form only one layer, have little undergrowth, contain a limited variety of species, and have needle-like leaves. The soil contains few nutrients because, following their loss through leaching and as surface runoff (after snowmelt when the ground is still frozen), replacement is slow: the low temperatures restrict the rate of chemical weathering of parent rock. The layer of needles is often thick, but their thick cuticles and the low temperatures discourage the action of the decomposers (page 268). The breakdown of litter into humus is thus very slow. These factors account for the low fertility potential of the podsol soils of the taiga (pages 331–332).

2 **Steppes/prairies** (Figure 11.29b) Soil is the largest store of mineral nutrients in the temperate grasslands. The biomass store is small due to the climate, which provides insufficient moisture to support trees and temperatures low enough to reduce the growing season to approximately six months. Indeed, much of the biomass is found beneath the surface as rhizomes and roots. The grass dies back in winter and nutrients are returned rapidly to the soil. The soil retains most of these nutrients because the rainfall is insufficient for effective leaching and the climate is conducive to both chemical and physical weathering which release further nutrients from the parent rock. The presence of bacteria also speeds up the return of nutrients from the litter to the soil. These factors help to account for the high fertility potential of the black chernozem soils associated with temperate grasslands (pages 327 and 340).

a taiga (northern coniferous forest)

B biomass   **L** litter   **S** soil

compartments (circle size proportional to amount stored)

nutrient transfers (arrows are proportional to amount of flow)

b steppe and prairie (mid-latitude continental grassland)

c selvas (tropical rainforest)

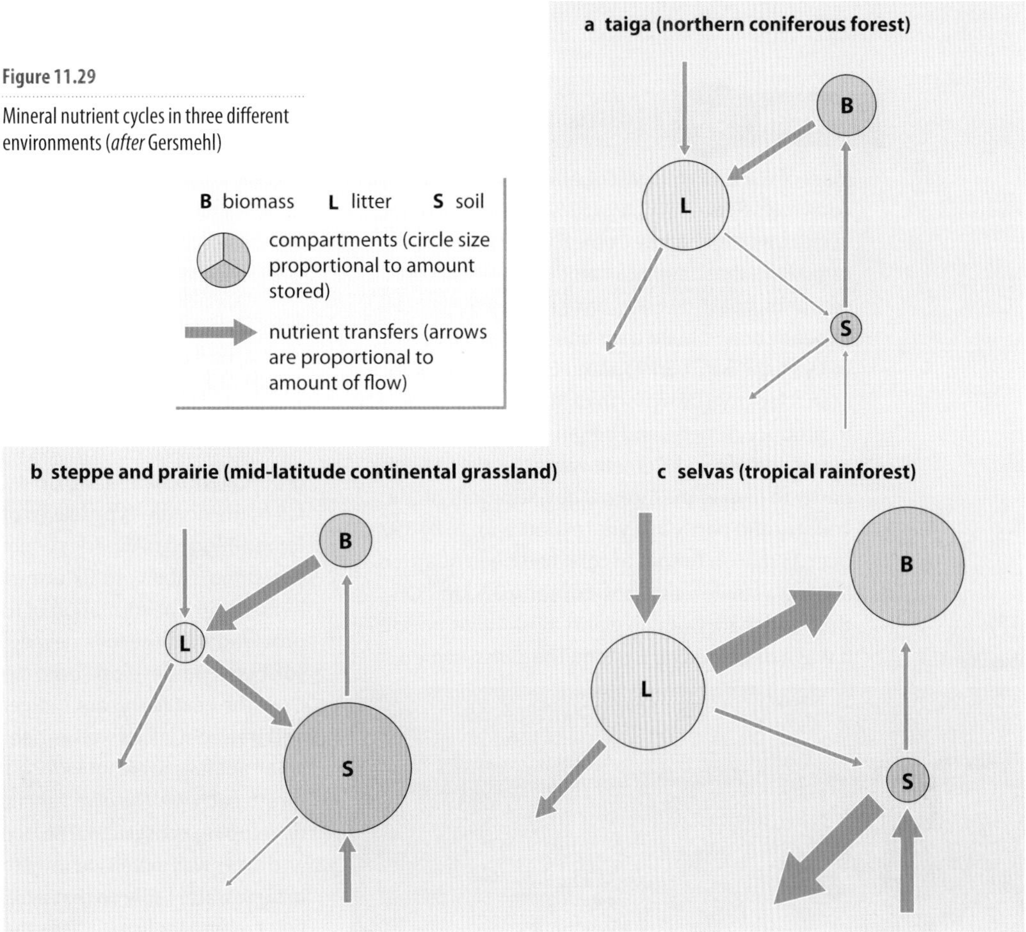

3 **Selvas** (Figure 11.29c) The tropical rainforests have, of all the major environments, the highest rates of transfer – an annual rate ten times greater than that of the taiga. The biomass is the largest store of mineral nutrients in the tropical rainforests. High annual temperatures, the heavy, evenly distributed rainfall and the year-long growing season all contribute to the tall, dense and rapid growth of vegetation. The biomass is composed of several layers of plants and countless different species. The many plant roots take up vast amounts of nutrients. In comparison, the litter store is limited, despite the continuous fall of leaves, because the hot, wet climate provides the ideal environment for bacterial action (both in numbers and type) and the decomposition of dead vegetation. In areas where the forest is cleared, the heavy rain soon removes the nutrients from the soil by leaching or surface runoff. The leaf litter content rapidly decomposes due to the high temperatures and heavy rainfall. The rainforests are characterised by 'tight' biogeochemical cycling between the litter and the top layers of the soil in which most tropical species are rooted, and the biomass. This means that the soil component, and by proxy the bedrock that is usually found at some considerable depth (Figure 12.10), is only a small component in the nutrient cycle. Initially nutrients such as phosphorus may increase if the forest is burnt, but deforestation usually leads to a rapid decline in soil fertility (pages 317–318).

Figure 11.30 compares the storage and transfer of nutrients in four major biomes (i.e. ecosystems on a large scale). Remember that these figures refer to natural cycles which, in reality, have often been interrupted or modified by human activity.

Figure 11.30

Storage and transfer of nutrients within selected biomes

		Ecosystem type	Nutrient storage			Annual nutrient transfer		
			Stored in biomass	Stored in litter	Stored in soil	Soil to biomass	Biomass to litter	Litter to soil
Forest cycles	**A**	Equatorial rainforest	11 081	178	352	2 028	1 540	4 480
		Coniferous forest	3 350	2 100	142	178	145	86
Grassland cycles	**B**	Tropical savanna	978	300	502	319	312	266
		Temperate prairie/steppe	540	370	5 000	422	426	290

All measurements in kg/ha

Most of the eastern coast of Africa is protected by coral reefs (Places 80, page 526). Coral, which live in clear, warm, shallow tropical waters, are small organisms that have a calcareous skeleton. For centuries, coast-dwellers have hacked out blocks of dead coral to build their houses and mosques. In 1954, the Bamburi Portland Cement Company built a factory 10 km north of Mombasa, Kenya, to produce cement, and began the open-cast extraction of coral. Cement was essential to Kenya, partly to help in the internal development of the country and partly as a vital export earner. By 1971, over 25 million tonnes of coral had been quarried, leaving a sterile wasteland covering 3.5 km². On that land there were no plants, no wildlife, no soil: it was a degraded ecosystem. The Swiss-owned transnational cement company then appointed Dr Rene Haller to restore the environment from what he himself described as 'a lunar landscape filled with saline pools' (Figure 11.31).

After trying 26 different types of tree, Dr Haller found the key to be the *Casuarina* tree (Figure 11.32). This pioneer tree grew by 3 m a year, flourished in the coral rubble, and was able to withstand both the high salinity and the high ground temperatures (up to 40°C). The constant fall of the needle-type leaves provided a habitat for red-legged millipedes which, together with the *Casuarina*'s ability to 'fix' atmospheric nitrogen, helped with the formation of the first soil and provided the base for a new ecosystem. As the soil began to develop, more trees were planted. Over the next few years, indigenous herbs, grasses and tree species, as well as beetles, spiders and small animals, were introduced into the young forest, each with its own function (niche) in the developing ecosystem. The depth of the ponds and lakes was increased until they reached the groundwater table so that a freshwater habitat was created for fish (initially the local tilapia which are tolerant of saline water), crocodiles and hippopotami. Hippopotami excrement stimulated the growth of algae which oxygenated the water, preventing eutrophication. After only 20 years, the soil depth had reached 20 cm and the rainforest, with over 220 tree species, had become sufficiently restored to be home for over 180 recorded species of bird. The ecosystem was completed with the introduction of grazing animals (herbivores) such as the buffalo, oryx, antelope and giraffe. The re-creation of the rainforest (Figure 11.33) had been completed without the use of artificial fertiliser and insecticides, as Dr Haller considered these to be incompatible with his concept of a complex, balanced ecosystem.

The project has not only been an environmental success, it has also become a sustainable commercial venture with income derived from, for example, the sale of timber, bananas, vegetables, crocodiles and honey. The main source of the economy is the integrated aquaculture system (Figure 11.34) with, at its centre, the tilapia fish farm. The nutrients in the effluent water are used as fertiliser in the adjacent fruit plantation and for biogas to operate the pumps. From here, the water is led through a rice field into settlement ponds, where 'Nile cabbage' is grown for use in clearing the fish ponds. A crocodile farm is attached to the

Figure 11.31

The Bamburi Quarry

Figure 11.32

Casuarina trees planted in coral rubble, Bamburi Quarry

Figure 11.33

The re-created rain-forest ecosystem, Haller Park

water system, as crocodile waste, which is rich in phosphate and nitrogen, is a valuable fertiliser. The crocodiles are part of a planned food chain. Not only are they fed on surplus tilapia, but their eggs are eaten by monitor lizards that help to control the snake population which in turn controls the rodent population. Tourism has become a recent source of income. Haller Park, the name of the restored area, is open to school parties each morning and to other visitors in the afternoon. In 1992 it received over 100 000 visitors, making it one of the largest attractions in the Mombasa area. In brief, the once-barren quarry is now an ecologically and economically sound enterprise (Figure 11.35).

Dr Haller also believes that his intensive aquaculture and agroforestry techniques, geared to maximum yield of food, fuel and income from minimum land area and inputs, offer significant hope for small-scale African farmers who may be short of fertile land in a continent with an explosive population growth and which is ravaged by environmental and human-created disasters. He suggests that these methods could easily be adapted by Africans since their genesis lies in tribal techniques taught to him by local farmers.

Figure 11.34

The Haller Park integrated aquaculture system

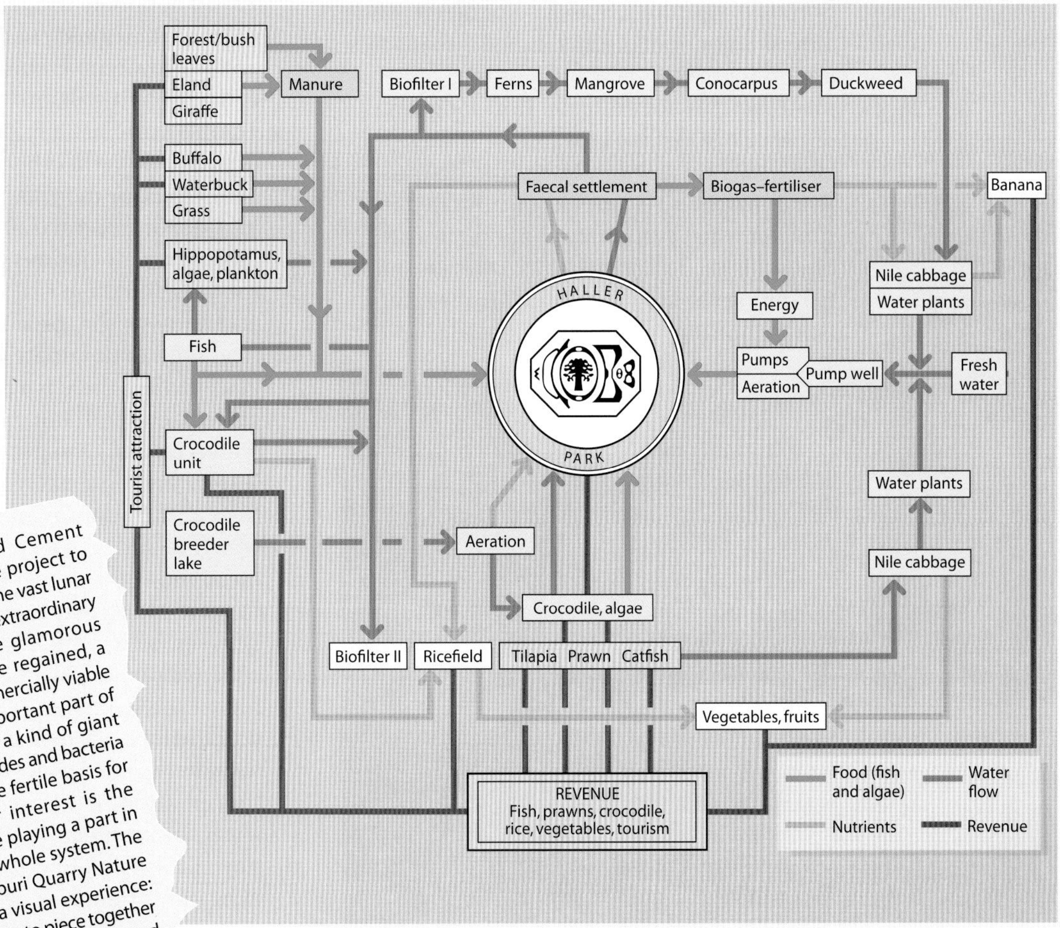

Figure 11.35

From the Bamburi Quarry Nature Trail leaflet

In 1971 the Bamburi Portland Cement Company embarked on a unique project to re-create a living environment in the vast lunar landscape of its quarry. Today an extraordinary change is evident. Behind the glamorous façade of the tropical paradise regained, a completely balanced and commercially viable aquaculture complex is an important part of the established ecosystem. In a kind of giant jig-saw, casuarina trees, millipedes and bacteria work together, to provide the fertile basis for reclamation. Of particular interest is the surprising variety of wildlife playing a part in the natural balance of the whole system. The living creation of the Bamburi Quarry Nature Trail holds more than just a visual experience: a walk around will enable you to piece together the giant jig-saw puzzle of wasteland rehabilitation.

Biomes

A biome is a large global ecosystem. Each biome gets its name from the dominant type of vegetation found within it (temperate grassland, coniferous forest, etc.). Each contains climax communities of plants and animals and can be closely linked to zonal soil types and animal communities. Climate has usually been the major controlling factor in the location and distribution of biomes, but economic development has transformed many of these natural systems. A biome can extend across a large part of a continent while its characteristics may be found in several continents (deserts and tropical rainforests). Although some authorities suggest that it is 'old-fashioned' to link together climate, vegetation and soils in a 'natural region', the concept is still useful and convenient as a framework of study and as a valid hypothesis for investigation. Four main factors – **climatic**, **topographic**, **edaphic** and **biotic** – interrelate to produce and control each biome.

1 Climatic factors

■ **Precipitation** largely determines the vegetation type, e.g. forest, grassland or desert. The annual amount of precipitation is usually less important than its effectiveness for plant growth – for example: How long is any dry season? Does the area receive steady, beneficial rain or short, heavy and destructive downpours? Is rainfall concentrated in summer when evapotranspiration rates are higher? Is the rainfall reliable? Does most rain fall during the growing season? Is there sufficient moisture for photosynthesis? Heavy

Figure 11.36

Wind-distorted tree, Mauritius

rainfall throughout the year enables forests to grow. These may be tropical rainforests, where the plants need a constant and heavy supply of water, or coniferous forests, where trees can grow due to the lower rates of evapotranspiration. Many other parts of the world receive seasonal rainfall. Rainfall is more effective, as in places with a Mediterranean climate, when it falls in winter rather than in summer, as this coincides with the time of year when evapotranspiration rates are at their lowest. However, as Mediterranean areas receive little summer rainfall, trees and shrubs growing there have to be **xerophytic** (drought-resistant) in order to survive. Rain is less effective when it falls in the summer because much of the moisture is lost through surface runoff and evapotranspiration. Effective precipitation is insufficient for trees, and so savanna grasses grow in tropical latitudes and prairie grasses in more temperate areas. Places where rainfall is limited throughout the year have either a desert biome, where **ephemerals** (plants with very short life-cycles, Figure 12.19) dominate the vegetation, or a tundra biome, where precipitation falling as snow and the low temperatures combine to discourage plant growth.

■ **Temperature** has a major influence on the flora – i.e. whether the forest is tropical or coniferous, or the grassland is temperate (prairie) or tropical (savanna). Where mean monthly temperatures remain above 21°C for the year and there is a continuous growing and rainy season, broad-leaved evergreen trees tend to dominate (tropical rainforests). Places where there is a resting period in tree growth, either in hot climates with a dry season or cool climates with a short growing season, are more likely to have coniferous trees as their dominant vegetation. Grasses, which include most cereals, need a minimum mean monthly temperature of 6°C in order to grow. Many plants prefer temperatures between 10°C, which is the minimum for effective photosynthesis, and 35°C. The higher the temperature, the sooner wilting point will be reached and the greater the need for water to combat losses through evapotranspiration. The lower the temperature, the fewer the number of soil organisms and the slower the breakdown of humus and recycling of nutrients needed for plant growth (Figure 12.7).

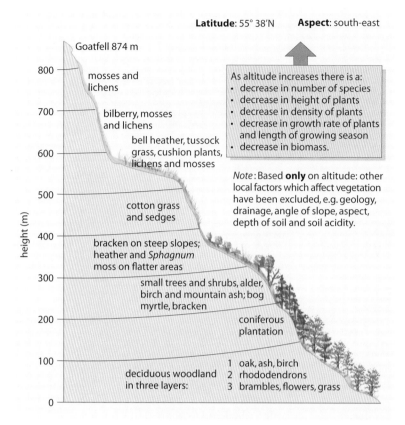

Latitude: 55° 38′N   **Aspect:** south-east

Goatfell 874 m

As altitude increases there is a:
- decrease in number of species
- decrease in height of plants
- decrease in density of plants
- decrease in growth rate of plants and length of growing season
- decrease in biomass.

Note: Based **only** on altitude: other local factors which affect vegetation have been excluded, e.g. geology, drainage, angle of slope, aspect, depth of soil and soil acidity.

mosses and lichens

bilberry, mosses and lichens

bell heather, tussock grass, cushion plants, lichens and mosses

cotton grass and sedges

bracken on steep slopes; heather and *Sphagnum* moss on flatter areas

small trees and shrubs, alder, birch and mountain ash; bog myrtle, bracken

coniferous plantation

deciduous woodland in three layers:
1 oak, ash, birch
2 rhododendrons
3 brambles, flowers, grass

Figure 11.37

The effect of altitude on vegetation, Goatfell, Arran

■ **Light intensity** affects the process of photosynthesis. Tropical ecosystems, receiving most incoming radiation, have higher energy inputs than do ecosystems nearer to the poles. Where the amount of light decreases, as on the floor of the tropical rainforests or with increasing depth in the oceans, plant life decreases. Quality of light also affects plant growth, e.g. the increase in ultra-violet light on mountains reduces the number of species found there.

■ **Winds** increase the rate of evapotranspiration and the wind-chill factor. Trees are liable to 'bend' if exposed to strong, prevailing winds (Figure 11.36).

2 Topographic factors

■ As **altitude** increases, there are fewer species; they grow less tall; and they provide a less dense cover (Figures 11.37 and 16.4b). Relief may provide protection against heavy rain (rainshadow) and wind.

■ **Slope angle** influences soil depth, acidity (pH) and drainage. Steeper slopes usually have thinner soils, are less waterlogged and less acidic than gentler slopes (soil catena, page 276).

■ **Aspect** (the direction in which a slope faces) affects sunlight, temperatures and moisture. South-facing slopes in the northern hemisphere are more favourable to plant growth than those facing north because they are brighter, warmer and drier (Places 28, page 213).

3 Edaphic (soil) factors

In Britain, there is considerable local variation in vegetation due to differences in soil and underlying parent rock, e.g. grass on chalk, conifers on sand, and deciduous trees on clay. Plant growth is affected by soil texture, structure, acidity, organic content, depth, water and oxygen content, and nutrients (Chapter 10).

4 Biotic factors

Biotic factors include the element of competition: between plants for light, root space and water, and between animals. Competition increases with density of vegetation. Natural selection is an important biotic factor. The composition of seral communities and the degree of reliance upon other plants and animals either for food (parasites) or energy (heterotrophs feeding on autotrophs) are also biotic factors. Today, there are very few areas of climax vegetation or biomes left in the world, as most have either been altered by human activity or even entirely replaced by human-created environments. The landscape has been altered by subsidence from mining, urbanisation, the construction of reservoirs and roads, exhaustion of soils, deforestation and afforestation, fires, the clearing of land for farming, and the effects of tourism. The ecological balance has been upset by the use of fertiliser and pesticides, the grazing of domestic animals, and acid rain.

The spatial pattern of world biomes

Figure 11.38 shows the distribution of the world's major biomes. When looking at maps of biomes in an atlas (they usually come under the heading 'Vegetation'), remember that all vegetation maps are very generalised (Framework 11, page 347). Vegetation maps do not show local variations, transition zones or, except in extreme cases, the influence of relief. Nor is there any universal consensus among geographers and biogeographers as to the precise number of biomes. Bradshaw has suggested 16 land biomes and 5 marine; Simmons describes 13 (11 land biomes plus islands and seas); O'Hare accepts 11; while Goudie (in common with most examination syllabuses) restricts his list, as does this text, to 8 land biomes.

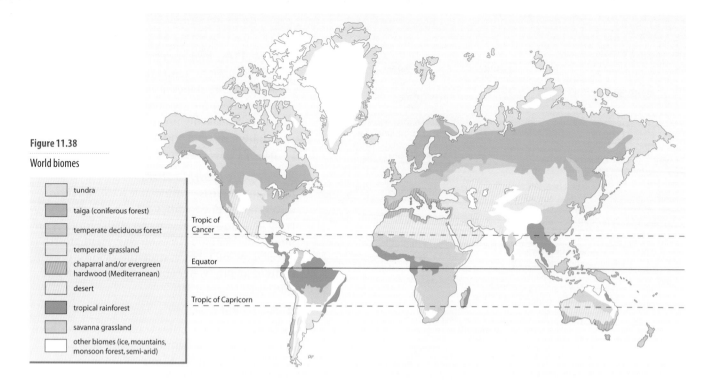

Figure 11.38

World biomes

- tundra
- taiga (coniferous forest)
- temperate deciduous forest
- temperate grassland
- chaparral and/or evergreen hardwood (Mediterranean)
- desert
- tropical rainforest
- savanna grassland
- other biomes (ice, mountains, monsoon forest, semi-arid)

Tropic of Cancer

Equator

Tropic of Capricorn

The eight major biomes, as shown in Figure 11.38, can be determined using a variety of criteria; two examples are discussed below and summarised in Figure 11.39.

1 **The traditional method** This links the type and global distribution of vegetation with that of the major world climatic types and zonal soils. This method was based on the understanding that it is climate that exerts the major influence and control over both vegetation and soils. The interactions between climate, soils and vegetation are described and explained in Chapter 12.

2 **The modern method** This is based upon differentiating between relative rates of producing organic matter – i.e. the speed at which vegetation grows. The rate at which organic matter is produced is known as the **net primary production** or NPP, expressed in grams of dry organic matter per square metre per year ($g/m^2/yr$). As shown in Figure 11.40, it is the tropical rainforests, with their large biomass resulting from constant high temperatures, heavy rainfall and year-round growing season, that produce on average the greatest amount of organic matter annually. The tundra (too cold) and the deserts (too dry) produce the least. It may be noted that the average NPP for arable land is 650, lakes and rivers 400 and oceans 125.

Figure 11.40

Net primary production (NPP) of eight major biomes

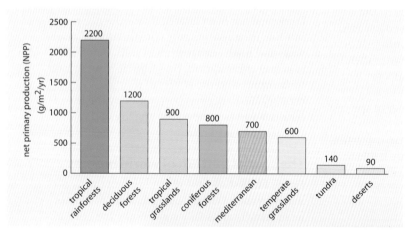

	1 Traditional method (vegetation, climate and soils subjectively linked)		
Tropical	1	Rainforests
	2	Tropical grasslands
	3	Deserts
Warm temperate	4	Mediterranean
Cool temperate	5	Deciduous forest
	6	Temperate grasslands
Cold	7	Coniferous forest
	8	Tundra

	2 Modern method (scientifically based on net primary production)		
High energy	1	Rainforests
	2	Deciduous forest
Average energy	3	Tropical grasslands
	4	Coniferous forest
	5	Mediterranean
	6	Temperate grasslands
Low energy	7	Tundra
	8	Deserts

Figure 11.39

Two methods of clarifying the major biomes (*after* I. Simmonds)

The situation before 2000

Western Australia is ten times the size of the UK, and about 2 per cent of the state was forested before white settlement began in 1829. The forested area stretches from Gingin, 75 km north of Perth, to Walpole, 400 km to the south (Figure 11.41). The Darling and Stirling ranges form the edge of the Darling Plateau and consist mainly of ancient igneous and metamorphic rocks. A number of river valleys cut into the plateau edge. These have broad, flat valley floors.

East of the plateau the old river valleys (now largely dry) are very broad and flat. At the western edge of the scarp, the drainage has been rejuvenated and recaptured by newer fast-flowing streams.

The Blackwood River is an exception. It has maintained enough flow to continue erosion of its bed as the plateau was uplifted. Therefore it has an old meandering course within which is a new cross-sectional V-shaped profile.

Figure 11.42
Jarrah forest

Figure 11.41

South-western Australia

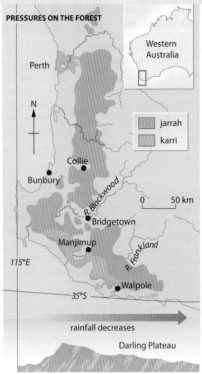

PRESSURES ON THE FOREST

Perth

N

Collie

Bunbury

R. Blackwood

Bridgetown

Manjimup

R. Frankland

115°E

35°S

Walpole

rainfall decreases

Darling Plateau

Western Australia

jarrah
karri

0   50 km

Agricultural clearing
Up to 500 m to allow sheep rearing; wheat grown on well-drained soils to east of forest area; forest now half extent of 165 years ago.

Settlement
Small towns expanding; most densely populated area of state outside Perth; infrastructure damages forest.

Commercial logging
280 000 m³ p.a. sawn timber for building; timber for woodchips; originally used waste offcuts and damaged timber; 150 000 tonnes jarrah sent to Kemerton for charcoal in silicon manufacture; clear felling now extensive; greatest pressures in the south, but jarrah forest ecosystem under threat.

Deforestation
Leading to soil erosion, higher water table and salinisation.

Quarrying
Limestone, sand, gravel.

Mining
Bauxite, gold, tin and tantalite; 800 ha forest lost each year; little rehabilitation.

Dieback
Soil-borne fungal disease *Phytophthora cinnamomi* affects 14% of forests spreading because of winter logging and other human disturbance.

Prescribed burning
Frequent burns in spring reduce flora species and damage food supply for breeding birds; jarrah forest not adapted to short intervals between burns.

Pest infestations
Jarrah leaf miner, gumleaf skeletoniser, affects 62 000 ha; thinning forest canopy (logging and spring burning) stimulates young foliage, attracts pests.

Loss of habitat
Affects flora and fauna; 26 species of plants and animals in jarrah forests lost or in need of protection; 5 fauna species extinct in karri forests.

The climate of this region is Mediterranean in type, with most rainfall in winter from May to October (700 mm); rainfall is highest (1100 mm) on the western edge of the plateau and decreases rapidly to the east. Temperatures are high in the summer (18–27°C) and lower in the winter (7–15°C). Snow has been known to fall in the Stirling Range!

These conditions allowed the development of high forests, unique to Western Australia, of hardwood trees: varieties of eucalyptus known as karri, jarrah and marri. Jarrah forest is the only tall forest in the world to grow in a truly Mediterranean type.

The great karri trees, which grow to over 80 m in height, are found in the south-west where the soils have a higher moisture content and are more fertile (Figure 11.17). The quality of the forest deteriorates to the east, with a variety of eucalypts reflecting lower rainfall. The jarrah forest is more extensive and has a very high species diversity (Figure 11.42). The forests provide important wild-life habitats for birds and animals – over 50 species live in the hollows of the trees.

Since the coming of the white settlers in 1829, half the tall forest cover has been removed (nearly 2 million ha). Much of the early clearance was for agriculture, with pastures of clover and grasses for sheep and cattle replacing the 500-year-old trees. Although the timber provided a valuable secondary source of income for the farmers, they were never able to sell it for themselves at a commercial rate. Instead, the state sold it for 'royalty' to timber industry firms as the commercial value of the tall forests was realised.

The situation in 2000

In 2000, the Western Australian government controlled 1260 ha of native trees in so-called 'State Forests'. The Department of Environment and Conservation (DEC) claimed that there was 139 000 ha of 'old growth' forest left (unlogged virgin forest) and 1 120 000 ha of 'regrowth' forest (areas that had been logged in the last 100 years). Despite opposition from conservation groups, including the Western Australian

HERITAGE FORESTS FACE THE AXE

IRREPLACABLE FORESTS TO BE CLEAR FELLED FOR WOODCHIPS

The Australian Heritage Commission (AHC) officially includes 40 areas of WA's world-unique native forests on the interim list of the Registers of the National Estate, the highest national recognition of the ecological, aesthetic, scientific or cultural value of an area. Once an area has been interim-listed, it is considered to be on the Register and entitled to protection. The Federal Minister for the environment is legally bound to prevent logging in these areas until an examination has shown that there are no 'prudent and feasible alternative log sources'.

In spite of this protection, the Department of Environment and Conservation (DEC) plans to clear-fell many interim-listed forests, mainly to produce export woodchips. Some of the listed areas are already being clear-felled, roaded and burnt with the full knowledge of the AHC and the Federal government.

In addition, there is supposed to be a moratorium on logging in all high-conservation value forests. Now that at least some of the best of WA's remaining native forests have been given official recognition, each of these agencies must back up their self-congratulations with action.

The only action they can reasonably take is to stop all roading and logging in WA's heritage forests immediately.

Figure 11.43

By the Western Australian Forest Alliance, Perth (Adapted)

of forests, with their unique wild species of small mammals, birds and flora, raised the question of sustainability. Fears were raised that, at the then present rate of deforestation, all the 'old growth' forest would have disappeared by 2030.

The DEC now has total responsibility for the logging and regeneration of felled areas within State Forests (Figure 11.44). It invites tenders for cutting and hauling and then selling the logs to sawmillers and woodchippers. The chief market for Western Australian timber is Japan. Since 1976 over 15 million tonnes of karri have been exported as woodchip through the port of Bunbury.

The main method of removal is by clear felling (Figure 11.45). An area of land is divided into sections referred to as coupes. Coupes vary in size from 60 ha in karri forests down to 10 ha in the jarrah. In clear felling, every tree in the coupe is felled and the logged area is then burnt. Most coupes are in the 'old growth' native forests, areas not previously touched, where the trees have reached their greatest height. Each of the felled giant karri needs a double trailer to take it to Bunbury, and often 12 of these can be seen on the main road to the port every hour. The DEC regeneration programme involves the hand-planting of karri seeds as they grow more quickly than jarrah. This is leading to a growing concern over what appears to be a deliberate phasing out of the jarrah, especially as in drought years, which are increasing in frequency, the karri is the less likely to survive.

Forest Alliance and the Global Warming Forest Group, the annual cut had increased to over 1 500 000 m³ with the large timber companies using the timber to produce woodchips, saw-logs and poles. The residue was sent in large quantities to be used as charcoal in a silicon smelter. The timber mills provided work for 2000 people, an important source of employment in a sheep-rearing region adversely affected by the low world price for wool. At the same time, the state was encouraging agroforestry, a form of plantation agriculture (page 482).

Meanwhile, conservationists were trying to stop the rapid increase in logging in the virgin forests (Figure 11.43). The rate of loss

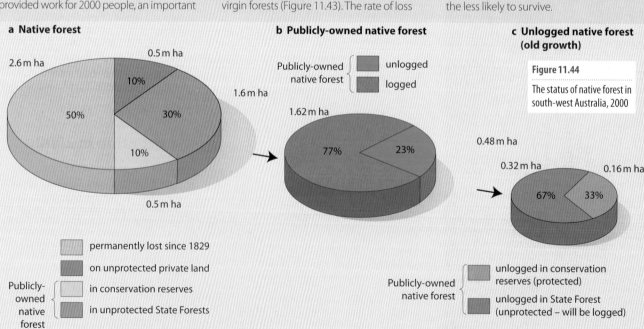

a Native forest

2.6 m ha — 50%
0.5 m ha — 10%
30%
10%
0.5 m ha

b Publicly-owned native forest

Publicly-owned native forest { unlogged / logged }

1.6 m ha
1.62 m ha — 77% / 23%

c Unlogged native forest (old growth)

Figure 11.44

The status of native forest in south-west Australia, 2000

0.48 m ha
0.32 m ha — 67% / 33% — 0.16 m ha

Publicly-owned native forest {
permanently lost since 1829
on unprotected private land
in conservation reserves
in unprotected State Forests
}

Publicly-owned native forest {
unlogged in conservation reserves (protected)
unlogged in State Forest (unprotected – will be logged)
}

Jarrah timber is commercially valuable for its dark-red colour, hardness and durability. However, it grows far more slowly, and is less in demand, than karri – hence the difficulty in maintaining sustainable production – even though the state government has restricted extraction to 500 000 m³ per year. Marri, the third type of eucalyptus growing in Western Australia, tends to be found within the jarrah forest and, like the karri, its main use is for woodchip.

Effects of deforestation

Visual and physical degradation of the landscape

This is especially bad in clear-felled areas. Where the land is steep, tree removal means there is no canopy to intercept heavier rainfall, nor roots to hold the soil together. This results in an increase in surface runoff and consequent problems of soil erosion, the sedimentation of rivers and a greater risk of flooding (page 63). Any nutrients in the soil, including those released by burning the cleared forest, will be lost due to leaching.

Loss of native flora and fauna

The south-west of Western Australia is noted for its wildflowers, typical of other regions with a Mediterranean-type climate (page 324). These are threatened, as are birds and small animals that at present rely on the groundcover of the forest. In total, 27 native species are listed as rare, including the chuditch, which is a marsupial, and the Western ring-tailed possum.

Salinisation of streams

This, resulting from the loss of the forest canopy, has become a serious problem in the region (page 496). Salts, previously trapped by the laterite soils (page 321), can be transported relatively easily by the increase in groundwater which itself becomes more saline. In time this water finds its way into streams and, eventually, the main watercourses.

Eutrophication

As forest land is cleared for agriculture, the nitrates used in fertilisers are also transferred by groundwater to rivers (page 281 and Figure 16.50). The nitrates enrich plant life which uses up more oxygen. This leaves less for fish and other water-inhabiting organisms.

The situation since 2000

In early 2001, the state government ended logging in all the 'old growth' forests in the care of the Conservation Commission of Western Australia and began, under the DEC, a process of creating the conservation parks and the 12 National Parks proposed in its 'Protecting our old growth forests' policy. A major capital works programme was established to upgrade visitor facilities, and to encourage tourism and leisure along with nature conservation.

The Forest Management Plan 2004–13 came into effect in 2004. This provided for increased protection of forest values, improved forest management and, coming into being later that year, 29 National Parks and other conservation reserves and forest areas. At the same time, landowners were encouraged to practise agroforestry by planting fast-growing trees on agricultural land in belts separated by grass pasture usable for sheep grazing. This was to use up surplus fertiliser in the soil and to reduce nitrates flowing into streams.

Although deforestation in Western Australia may not be on the scale of that in the Amazon rainforest or Indonesia, to the people living in the south-west corner of the state it is just as damaging. To some people deforestation means the destruction of a non-replaceable ecosystem and a loss for future generations. To others logging means employment in an area with relatively few job opportunities. It is easy to become emotive on a topic such as this, especially if the question is oversimplified to 'Which is the more important – jobs provided by the production of paper or the protection of trees and wildlife?' It revives a question long asked in Geography of which is the more important: economic gain or environmental loss? At present the answer appears to lie in the prospect of 'sustainable development' (Framework 16, page 499).

Figure 11.46 describes the viewpoints given in 2008 by, on one hand, the state government and representatives of the timber workers and, on the other, conservation groups.

Figure 11.45

Clear felling of karri, near Bridgetown

The **Global Warming Forest Group** claimed that the logging of a 62 m tall, 500-year-old karri tree near Pemberton showed that the old growth protection policy was a sham and that they, and other environmental groups, had been double-crossed on definitions as, according to present government policy, a single stump in a hectare of virgin forest disqualifies it as 'old growth'. To them, forests containing huge centuries-old trees have a high conservation value and it is absurd that these old trees should be logged before they die and fall naturally. Such trees are more valuable as wildlife habitats than as woodchip or sawdust, which is the end product of most harvested timber.

The **Forest Industries Federation** stated that it had ensured that 1.2 million ha of 'old growth' forest was now totally protected by state law in the south-west corner of Western Australia. However, it also said that there was never a commitment to protect individual trees, but rather to conserve areas as a whole. Admittedly, there were still old karri trees that had not been logged in 'regrowth forests' which might in time might have to be felled, but these were outside 'old growth protection areas'. The federation also said that over a dozen karri trees, both bigger in diameter and taller in height than the felled Pemberton tree, were under protection, including one growing near Manjimup (Figure 11.41) which was 61 m tall and had a diameter of 291 cm – 26 cm greater than that of the Pemberton tree.

Austwest, the biggest karri miller in the state, said it was rare to receive timber from trees the size of the one near Pemberton. When it did, it was put to the most valuable use which was usually for flooring or staircases (Figure 11.47).

Figure 11.47

A single large karri tree

Figure 11.46

Adapted from material on the official Serengeti website (www.serengeti.org)

Further reference

Bradbury, I.K. (1998) *The Biosphere*, WileyBlackwell.

Brown, J.H., Riddle, B.R. and Lomolino, M.V. (2005) *Biogeography*, Sinauer Associates Inc.

Huggett, R.J. (2004) *Fundamentals of Biogeography*, Routledge.

MacDonald, G. (2003) *Biogeography: Introduction to Space, Time and Life*, Wiley.

O'Hare, G. (1988) *Soils, Vegetation and the Ecosystem*, Oliver & Boyd.

Biosphere basics:
www.geography4kids.com

Bridgetown-Greenbushes Friends of the Forest:
http://members.westnet.com.au/bgff/

Cary Institute of Ecosystem Studies:
www.ecostudies.org/

International Biogeography Society:
www.biogeography.org/

National Association of Forest Industries (Australia):
www.nafi.com.au

Radford University Virtual Geography Department's 'Biome':
www.runet.edu/~swoodwar/CLASSES/GEOG235/biomes/main.html#tabcont

Union of Concerned Scientists (UCS), 'Understanding biodiversity':
www.ucsusa.org/ – use search option

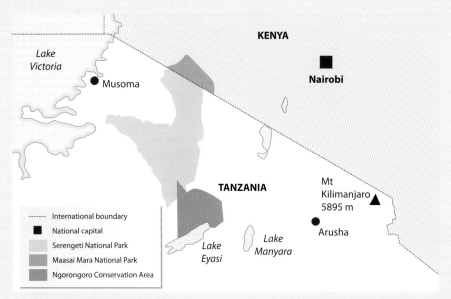

Figure 11.48

Location of the
Serengeti

Before starting this exercise, read pages 319–321, Tropical grasslands, and pages 335–338, Tropical grasslands in Kenya.

Serengeti National Park's website is at: www.serengeti.org

The Serengeti Shall Not Die area is useful for this exercise.

The Serengeti grasslands

The Serengeti grasslands lie just south of the Tanzanian/Kenyan border, between 2° and 4° South (Figure 11.48). Mean maximum temperatures are 24° to 27°C, and mean minimum temperatures 15° to 21°C. Mean annual rainfall varies from 1050 mm in the north-west to 550 mm in the south-east. Rainfall peaks in March to May, and November to December (compare Figure 12.49).

The soils are formed from volcanic ash. The eco-region consists of slightly undulating grassy plains, interrupted by scattered rocky areas (kopjes) which are parts of the Precambrian basement rocks protruding through the ash layers.

Biodiversity features

The Serengeti grasslands are vital to the cyclical movement of millions of large mammals. Populations fluctuate, but about 1.3 million blue wildebeest, 200 000 plains zebra, and 400 000 Thomson's gazelle migrate between Serengeti and southern

Kenya each year. Many associated predators are also involved in these movements. By the onset of the dry season (late May), the grasses on the plains have either dried out or been eaten down to stubble, and water is scarce. This triggers the massive migration from the plains northwards. Then, at the start of the wet season, the animals complete the cycle, and return to the plains.

Fires, usually set by humans, are an important disturbance in this eco-region. The burning helps provide accessible pasture for the herds of cattle that are kept here but other species, including wildebeest, also favour grazing on the green flush that emerges after burning.

Current status

Much of the eco-region occurs within protected areas, most of which are joined into a continuous block. The protected area includes Serengeti National Park (SNP) and Ngorongoro Conservation Area (NCA), both of which are World Heritage Sites (page 596). This area is probably large enough to ensure the survival of the habitat and its biodiversity. There has been little loss of habitat within the protected areas, except for small areas used for tourist hotels. Outside the protected areas, however, there has been a rapid expansion of human settlement and agriculture in recent years.

Types and severity of threats

While Maasai pastoralists occupy the NCA, there are no people living within the SNP. However, the western frontier of this park has a dense population, growing at 4 per cent a year. Livestock numbers are increasing, and much of the area is being converted into cropland. Agriculture is the main source of income, but many people have been attracted to the area by the wildlife resources and tourism opportunities that the park presents.

Many animals within the SNP are killed by poachers, who may be local people hunting 'bush meat' for subsistence, organised commercial hunters taking meat for sale in the cities, or Big Game hunters taking part in organised illegal safaris.

However, it is hoped that schemes to give local communities legal rights to manage the wildlife around their villages will reduce the worst excesses of the hunting. There are also plans to channel more money earned from tourist activities within the park back into the community as, so far, the contribution from tourism to the local economy has been relatively low.

Are the Serengeti grasslands natural?

The Serengeti changed from a grassland state to woodland twice in the last century. The few old, large trees dotting the landscape started life about 1900, followed by a slow decline in numbers due to elephants, fire, disease, and natural thinning, leaving the few that we see today. The second group of smaller trees established themselves between 1976 and 1983, and these trees are still growing in abundance. Both groups were able to grow because for two periods there were neither elephants nor fires.

Rinderpest, a cattle disease, came to East Africa in about 1896. Most of the Serengeti wildebeest died in a few years, as did the cattle herds. There was famine, followed by

Figure 11.49

Information from the Serengeti website

In recent years human population has increased, putting pressure on park resources. Conflicts arise as wild animals damage property and even threaten life. Illegal poaching activities create conflict. In some sections cultivation is right on the park border and this fuels conflict as animals destroy the crops on one side or are illegally hunted on the other.

The Serengeti is a prime example of how many natural ecosystems are being eroded by human population effects, irrespective of legal boundaries. The original Serengeti-Ngorongoro 'undisturbed' ecosystem (which included indigenous hunters with traditional weapons), set aside in the 1950s, has declined steadily. Some 40% of the natural ecosystem has been lost to farming and herding. Today, there are signs that this loss may be accelerating.

The Serengeti is also losing species. Thus, rhinoceros, once abundant, have been effectively exterminated from the ecosystem, and elephants were reduced by 80%, both by poaching. Wild dogs went extinct in the early 90s, due to contact with domestic dogs and infection with diseases like distemper and rabies. Unregulated hunting of large predators in areas around Serengeti has had dramatic impacts. Over-hunting of male lions alters the local adult sex ratio, draws males out from the park, and thus disrupts populations within in it.

The 1989 worldwide ivory ban almost completely stopped the poaching of elephants and their numbers are recovering. However, meat poaching continues. In an average year, local people living around the park illegally kill about 40,000 animals, mainly wildebeest and zebra, but also giraffe, buffalo and impala. The populations of these animals seem to be able to survive this poaching without any long-term decline but the killing is a manifestation of growing antagonism between the impoverished villagers and the authorities of the SNP. This conflict did not exist two decades ago; there was land enough for everyone and every animal. What we must all face – poachers, tourists, farmers, conservationists and pastoralists – is the fact that the land does not go on forever.

In an effort to harmonize the pastoralists with the wildlife in the Serengeti, locally administered reserves – Wildlife Management Areas – are now created on the borders of the park, where villagers are given a far greater degree of control over the land and its resources. In situations where protection of biodiversity is not seen to be of clear economic benefit to the community, outside assistance must attempt to bring change by:

- increasing community pride in their natural environment
- increasing the economic benefits of conservation, e.g. by fostering ecotourism, hiring community members as resource stewards, rangers, etc.
- rehabilitating depleted resource systems
- increasing the community's ability to control the use of the resource by outside interests.

Figure 11.50

Scenes in the Serengeti

the grasslands did not dry and burn during the 'dry season'. During this time there was an enormous upswing in the illegal ivory trade. With fire and elephants removed, the trees again established themselves in a burst. These trees are now about 30 years old and range from 2 to 5 m tall, often forming dense thickets.

There has been a large increase of impala inside the park. They seem to be much more successful in the woodlands than in the grasslands, and have increased as the woodlands have increased. In the past, elephants and fire have controlled the establishment of new trees. Today, both elephants and fire are monitored closely. The Park Ecology Department burns fire-breaks to stop the spread of large fires, and conducts 'cool' early burns in fire-prone areas. It is also monitoring the ecosystem carefully to see how all aspects interact.

emigration. With no people there was no one to light fires and the Serengeti went un-burnt. At the same time, the trade in ivory was at its peak. With no fires and no elephants, young trees were able to grow and flourish in the first big establishment of the century. Then, gradually, the wildebeest and cattle recovered and by the 1930s elephants started to return, and growth of new trees ceased.

Between 1976 and 1984 the weather patterns in and around Serengeti changed. The seasonal rains became more spread out, so

> The Serengeti National Park Authorities have two main aims:
> 1 to conserve the natural environment of the SNP
> 2 to support the traditional way of life of the people who live around the SNP.
> Draw up a list of management objectives for the Park, justifying each of your objectives and explaining how individual objectives combine to form a coherent management plan for the area.

Activities

1 **a** What are: **i** herbs **ii** shrubs **iii** trees? *(3 marks)*

   b What is plant succession? *(3 marks)*

   c How do herbs and shrubs help to prepare the ground so that trees can grow? *(6 marks)*

   d How would you carry out a field survey to discover the distribution of plants in the area of a playing field? *(5 marks)*

   e What kinds of plants would you expect to find on an abandoned urban railway track? Suggest reasons for your answer. *(4 marks)*

   f Flowers that grow in deciduous woodland are early spring flowers such as bluebell and primrose. Why do these plants flower so early in the year? *(4 marks)*

2 **a** Study Figure 11.25 (page 298).

   i Explain the roles played by plants in the carbon cycle. *(4 marks)*

   ii Human activity (combustion) releases CO_2 into the air. What is the source of this carbon? *(3 marks)*

   b **i** Study Figure 11.26 (page 298). Why is nitrogen important for plant life? *(2 marks)*

   ii What is the main source of new nitrogen into the nitrogen system? *(2 marks)*

   iii What is the main cause of loss of nitrogen from the system? *(2 marks)*

   c What is the meaning of the term 'biomass'? *(2 marks)*

   d What is the role of humans in the food chain? *(2 marks)*

   e As CO_2 builds up in the atmosphere, plant growth is increased. Suggest **two** effects of this on the material cycles. *(4 marks)*

   f Explain the 'greenhouse effect'. *(4 marks)*

3 Study Case Study 11 (pages 307–310).

   a **i** What is the extent of deforestation in south-west Australia since white settlement started? *(2 marks)*

   ii Identify the proportion of:
   (i) conserved native forest (ii) public ownership of the forest (iii) forest in danger of being logged. *(3 marks)*

   iii Identify and explain **three** reasons for deforestation in south-west Australia. *(6 marks)*

   b Explain **two** impacts of deforestation on areas such as south-west Australia. *(6 marks)*

   c Describe **two** advantages of the native forest to south-west Australia and its people. *(4 marks)*

   d Explain **one** way of protecting the forest lands of south-west Australia. *(4 marks)*

Exam practice: basic structured question

4 **a** What is meant by:

   i seral change

   ii climatic climax vegetation cover? *(6 marks)*

   b Why is vegetation cover within an urban area different from the climatic climax vegetation in a similar rural area? *(7 marks)*

   c Assume that there has been a landslide on an area of non-calcareous rock in lowland Britain. Describe and explain the sequence of vegetation that would occur so that the area eventually achieved a climatic climax vegetation cover. *(12 marks)*

Exam practice: structured questions

5 **a** Explain the meaning of:

   i seral progression *(2 marks)*

   ii dominant species. *(2 marks)*

   b Choose **one** of a *psammosere*, a *halosere*, or a *hydrosere*.

   i Draw an annotated diagram **only** to show the variation in vegetation cover across the environment. *(6 marks)*

   ii Explain the variation in vegetation cover shown on your diagram. *(15 marks)*

6 **a** Study Figure 11.28 (page 300).

   i Explain the meaning of the term 'litter'. *(2 marks)*

   ii Explain what the arrows show. *(2 marks)*

   b Figure 11.29 (page 301) shows the nutrient cycles in three different environments.

   i Why are the transfers in the taiga so small? *(6 marks)*

   ii Explain the differences between the tropical forests and the mid-latitude grasslands in terms of their nutrient stores and flows. *(15 marks)*

Exam practice: essay

7 Explain why the 'polyclimax' theory of vegetation progression is now generally considered to be better than the climatic climax theory of F.E. Clements as a way of explaining the distribution of vegetation types. *(25 marks)*

12

World climate, soils and vegetation

•••••••••••••••••••••••••••••••••••••••

'There was … an instant in the distant past when the living things, the rocks, the air and the oceans merged to form the new entity, Gaia.'

James Lovelock, *The Ages of Gaia*, 1989

Although it is possible to study climatic phenomena in isolation (Chapter 9), an understanding of the development of soils (Chapter 10) and vegetation (Chapter 11) necessitates an appreciation of the interrelationships between all three (Figure 12.1a). This chapter attempts to show how the integration and interaction of climate, soils and vegetation give the world its major ecosystems, or biomes, and how these have often been modified, in part or almost totally, by human activity (Figure 12.1b).

Soils can be grouped, at the simplest of levels, under zonal, azonal and intrazonal (page 273) with each group, in turn, being subdivided (zonal Figure 12.2, azonal page 273, and intrazonal page 274). Likewise, the major vegetation and fauna groupings (biomes) were listed on page 306 and their generalised global locations and distributions shown in Figure 11.38. In a similar way, geographers seek – despite the difficulties and limitations – to classify different world climates (Framework 7, page 167).

Classification of climates

By studying the weather – the atmospheric conditions prevailing at a given time or times in a specific place or area – it is possible to make generalisations about the climate of that place or area, i.e. the average, or 'normal' conditions over a period of time (usually 35 years). Any area may experience short-term departures from its 'normal' climate, especially if the 35-year mean coincided with an unusually wet/dry or hot/cold period, but, at the same time, it may have long-term similarities with regions in other parts of the world.

In seeking a sense of order, the geographer tries to group together those parts of the world that have similar measurable climatic characteristics (temperature, rainfall distribution, winds, etc.) and to identify and to explain similarities and differences in spatial and temporal distributions and patterns. Areas may then be compared on a global scale – bearing in mind the problems resulting from short-term and long-term climatic change – to help to identify and to explain distributions of soil, vegetation and crops.

Bases for classification

The early Greeks divided the world into three zones based upon a simple temperature description: torrid (tropical), temperate, and frigid (polar); they ignored precipitation.

In 1918, **Köppen** advanced the first modern classification of climate. To support his claim that natural vegetation boundaries were determined by climate, he selected as his basis what he considered were appropriate temperature and seasonal precipitation values. His resultant classification is still used today, although a modification by **Trewartha**, with 23 climatic regions, has become more widely accepted.

Figure 12.1

Relationship between climate, vegetation and soils

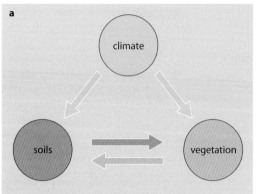

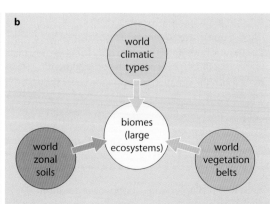

Thornthwaite, in the 1930s and 1940s, suggested and later modified a classification with a more quantitative basis. He introduced the term 'effectiveness of precipitation' (his P/E index – page 178) which he obtained by dividing the mean monthly precipitation of a place by its mean monthly evapotranspiration, and taking the sum of the 12 months. The difficulty was, and still is, in obtaining accurate evapotranspiration figures. (How can you measure transpiration loss from a forest?) This classification resulted in 32 climatic regions.

In Britain, in the 1930s, **Miller** proposed a relatively simple classification based on five latitudinal temperature zones which he determined by using just three temperature figures: 21°C (the limit for growth of coconut palms); 10°C (the minimum for tree growth); and 6°C (the minimum for grasses and cereals). He then subdivided these zones longitudinally according to seasonal distributions of precipitation. The advantages of this classification include its ease of use and convenience; and its close relationship to vegetation zones and also, as these are a response on a global scale to climate and vegetation, to zonal soils.

All classifications have weaknesses: none is perfect.

■ They do not show transition zones between climates, and often the division lines are purely arbitrary.

■ They do not allow for mesoscale variation (the Lake District and London do *not* have exactly the same climate) or microscale (local) variation.

■ They can be criticised for being either too simplistic (Miller) or too complex (Thornthwaite).

■ They ignore human influence and climatic change, both in the long term and the short term.

■ Most tend to be based upon temperature and precipitation figures, and neglect recent studies in heat and water budgets, air-mass movement and the transfer of energy.

■ All suffer from the fact that some areas still lack the necessary climatic data to enable them to be categorised accurately.

However, climatic classifications such as those named above are rarely used today. Instead, as we saw in Chapter 11, the relationship between climate, vegetation and soils can best be described and understood at this level through the study of ecosystems, especially the largest of the ecosystems: the **biomes** (Figure 12.1b). Figure 12.2 lists eight of the more important biomes and shows, simplistically, the links between climate, vegetation and soils. These links are described in more detail and explained in the remainder of this chapter, using knowledge and understanding gained from Chapters 9, 10 and 11.

Climate type		Text reference number	Climatic characteristics	Biome (based on NPP)	Soil (zonal type)
arctic		8	very cold all year	tundra	tundra
cold		7	cold all year	coniferous forest (taiga)	podsols
cool temperate	western margin	6	rain all year, winter maximum	temperate deciduous forest	brown earths
	continental	5	summer rainfall maximum	temperate grassland	chernozems prairie chestnut
warm temperate	western margins: Mediterranean	4	winter rain	Mediterranean	Mediterranean
	eastern margins: monsoon	4A	some rain all year, summer maximum	tropical deciduous forest	
tropical	desert	3	little rain	desert (xerophytes)	desert
	continental	2	summer rain	tropical grassland (savanna)	ferruginous
	monsoon	1B		jungle	
	tropical eastern margins	1A	rain all year	rainforest	ferralitic
	equatorial	1			

Figure 12.2

World biomes: the relationship between climate, vegetation and soils at the global scale

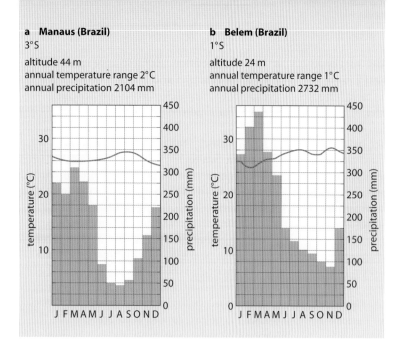

a Manaus (Brazil)
3°S

altitude 44 m
annual temperature range 2°C
annual precipitation 2104 mm

b Belem (Brazil)
1°S

altitude 24 m
annual temperature range 1°C
annual precipitation 2732 mm

Figure 12.3

Climate graphs for the equatorial biome

Figure 12.4

Emergents rising above the rainforest canopy, Borneo

1 Tropical rainforests

The rainforest biome is located in the tropics and principally within the equatorial climate belt, 5° either side of the Equator. It includes the Amazon and Congo basins and the coastal lands of Ecuador, West Africa, and extreme south-east Asia (Figure 11.38).

Equatorial climate

Temperatures are high and constant throughout the year because the sun is always high in the sky. The annual temperature range is under 3°C inland (Manaus, Figure 12.3a) and 1°C on the coast (Belem, Figure 12.3b). Mean monthly temperatures, ranging from 26°C to 28°C, reflect the lack of seasonal change. Slightly higher temperatures may occur during any 'drier' season. Insolation is evenly distributed throughout the year, with each day having approximately 12 hours of daylight and 12 hours of darkness. The diurnal temperature range is also small, about 10°C. Evening temperatures rarely fall below 22°C while, due to the presence of afternoon cloud, daytime temperatures rarely rise above 32°C. It is the high humidity, with its sticky, unhealthy heat, that is least appreciated by Europeans.

Annual rainfall totals usually exceed 2000 mm (Belem, 2732 mm) and most afternoons have a heavy shower (Belem has 243 rainy days per year). This is due to the convergence of the trade winds at the ITCZ and the subsequent enforced ascent of warm, moist, unstable air in strong convection currents (Figure 9.34). Evapo-transpiration is rapid from the many rivers, swamps and trees. Most storms are violent, with the heavy rain, accompanied by thunder and lightning, falling from cumulo-nimbus clouds. Some areas may have a drier season when the ITCZ moves a few degrees away from the Equator at the winter and summer solstices (Belem), and others have double maxima when the sun is directly overhead at the spring and autumn equinoxes. The high daytime humidity needs only a little night-time radiation to give condensation in the form of dew. The winds at ground-level at the ITCZ are light and variable (doldrums) allowing land and sea breezes to develop in coastal areas (page 240).

Rainforest vegetation

It is estimated that the rainforests provide 40 per cent of the net primary production of terrestrial energy (NPP, page 306). This is a result of high solar radiation, an all-year growing season, heavy rainfall, a constant moisture budget surplus, the rapid decay of leaf litter and the recycling of nutrients.

Figure 12.5

Buttress roots, Queensland, Australia

In just one hectare of rainforest in Amazonian Ecuador, researchers recorded 473 species of tree, including rosewood, mahogany, ebony, greenheart, palm and rubber, which is more than twice the total number found in all of North America.

The trees, which are mainly hardwoods, have an evergreen appearance for, although deciduous, they can shed their leaves at any time during the continuous growing season. The tallest trees, **emergents**, may reach up to 50 m in height and form the habitat for numerous birds and insects. Below the emergents are three layers, all competing for sunlight (Figure 12.4).

The top layer, or **canopy**, forms an almost continuous cover which absorbs over 70 per cent of the light and intercepts 80 per cent of the rainfall. The crowns of these trees merge some 30 m above ground-level. They shade the underlying species, protect the soil from erosion, and provide a habitat for most of the birds, animals and insects of the rainforest.

The second layer, or **undercanopy**, consists of trees growing up to 20 m (similar in height to deciduous trees in Britain). The lowest, or **shrub layer**, consists of shrubs and small trees which

are adapted to living in the shade of their taller neighbours.

The climate is at the optimum for photosynthesis. The trees grow tall to try to reach the sunlight, and the tallest have buttress roots which emerge over 3 m above ground-level to give support (Figure 12.5). The trunks are usually slender and branchless. Some, like the cacao, have flowers growing on them, and their bark is thin as there is no need for protection against adverse climatic conditions. Tree trunks also provide support for lianas, vine-like plants, which can grow to 200 m in length. Lianas climb up the trunk and along branches before plunging back down to the forest floor. Leaves are dark green, smooth and often have **drip tips** to shed excess water.

Epiphytes – plants that do not have their roots in the soil – grow on trunks, branches and even on the leaves of trees and shrubs. Epiphytes simply 'hang on' to the tree: they derive no nourishment from the host and are *not* parasites. Less than 5 per cent of insolation reaches the forest floor, with the result that undergrowth is thin except in areas where trees may have been felled by shifting cultivators or where a giant emergent has fallen, dragging with it several of the top canopy trees. Vegetation is also dense along the many river banks, again because sunlight can penetrate the canopy here. Alongside the Amazon, many trees spend several months of the year growing in water as the river and its tributaries rise over 15 m in the rainy season. Huge water lilies with leaves exceeding 2 m in width are found in flooded areas adjacent to rivers (Figure 12.6). Mangrove swamps occur in coastal areas.

Figure 12.7

The rainforest nutrient cycle

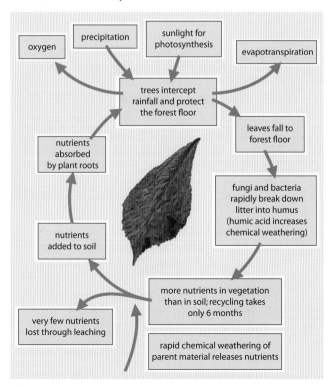

Figure 12.8

The interrupted nutrient cycle

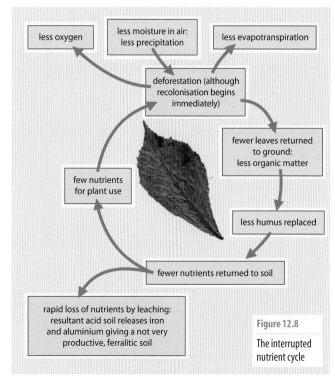

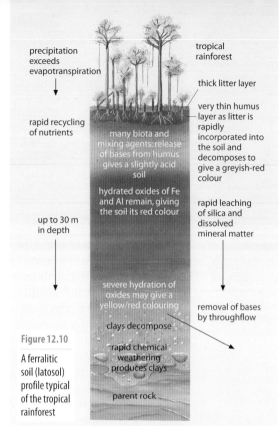

precipitation exceeds evapotranspiration

tropical rainforest

thick litter layer

rapid recycling of nutrients

very thin humus layer as litter is rapidly incorporated into the soil and decomposes to give a greyish-red colour

many biota and mixing agents: release of bases from humus gives a slightly acid soil

hydrated oxides of Fe and Al remain, giving the soil its red colour

rapid leaching of silica and dissolved mineral matter

up to 30 m in depth

severe hydration of oxides may give a yellow/red colouring

removal of bases by throughflow

clays decompose

rapid chemical weathering produces clays

parent rock

Figure 12.10

A ferralitic soil (latosol) profile typical of the tropical rainforest

Although ground animals are relatively few in number, the rainforests of Brazil alone are said to be the habitat for 2000 species of birds, 600 species of insects and mosquitoes, and 1500 species of fish.

The productivity of this biome, upon which the world depends to replace much of its used oxygen, is due largely to the rapid and unbroken recycling of nutrients. Figure 12.7 shows the natural nutrient cycle and Figure 12.8 the consequences of breaking the system, e.g. by felling the forest. In areas where the forest has been cleared, the secondary succession differs from that of the original climax vegetation. The new dominants are less tall; the trees are less stratified; there are fewer species and many are intolerant of shade – even though there is more light at ground-level which encourages a dense undergrowth.

Ferralitic soils (latosols)

These soils result from the high annual temperature and rainfall which cause rapid chemical weathering of bedrock and create the optimum conditions for breaking down the luxuriant vegetation. Continuous leaf fall within the forest gives a thick litter layer, but the underlying humus is thin due to the rapid decomposition and mixing of organic matter by intensive biota activity, e.g. ants and termites. A key feature of these soils is a dense root mat in the top 20–30 cm of the A horizon. According to research, this intercepts and can take up as much as 99.9 per cent of the nutrients released by the decomposition of organic matter. The root map helps the rapid recycling of nutrients in the humus cycle (Figure 12.7). Even so, many soils have a low nutrient status (94 per cent of soils in the Amazon Basin have a nutrient deficiency) and fertility is only maintained by the rapid and continuous replacement from the lush vegetation. Where the tree canopy is absent, or is removed, the heavy rainfall causes the release of iron (giving the soil its characteristic red colour – Figure 12.9) and aluminium (most ferralitic soils

suffer from aluminium toxicity) from the parent material. Leaching results in the removal of silica.

The continual leaching and abundance of mixing agents inhibit the formation of horizons (Figure 12.10). The lower parts of the profile may have a more yellowish-red tint due to the extreme hydration of aluminium and iron oxides. The clay-rich soils are also very deep, often up to 20 m, due to the rapid breakdown of parent material. Ferralitic soils have a loose structure and, if exposed to heavy rainfall, are easily gullied and eroded. Despite their depth, the soils of the rainforest are not agriculturally productive. Once the source of nutrients (the trees) has been removed, the soil rapidly loses its fertility and local farmers, often shifting cultivators, have to move to clear new plots (Places 66, page 480).

1A Tropical eastern margins

Located within the tropics, the eastern coasts of central America, Brazil, Madagascar and Queensland (Australia) receive rain throughout the year. The rain is brought by the trade winds which blow across warm, offshore ocean currents (Figure 9.9) before being forced to rise by coastal mountains. Temperatures are generally very high, although there is a slightly cooler season when the overhead sun appears to have migrated into the opposite hemisphere. The resultant vegetation and soil types are, therefore, similar to those found in the equatorial belt, i.e. rainforest and ferralitic.

2 Tropical grasslands

These are mainly located between latitudes 5° and 15° north and south of the Equator and within central parts of continents, i.e. the Llanos (Venezuela), the Campos (Brazilian Highlands), most of central Africa surrounding the Congo Basin, and parts of Mexico and northern Australia (Figure 11.38).

Tropical continental climate

Although temperatures are high throughout the year, there is a short, slightly cooler season (in comparison with the equatorial) when the sun is overhead at the tropic in the opposite hemisphere (Figure 12.11). The annual range is also slightly greater (Kano 8°C) due to the sun's slightly reduced angle in the sky for part of the year, the greater distance from the sea, and the less complete cloud and vegetation cover. Temperatures may drop slightly at the onset of the rainy season. For most of the year, cloud amount is limited, allowing diurnal temperatures to exceed 25°C.

The main characteristic of this climate is the alternating wet and dry seasons. The wet season occurs when the sun moves overhead bringing with it the heat equator, the ITCZ, and the equatorial low pressure belt (Figure 12.12). Heavy convectional storms can give 80 per cent of the annual rainfall total in four or five months. The dry season corre-

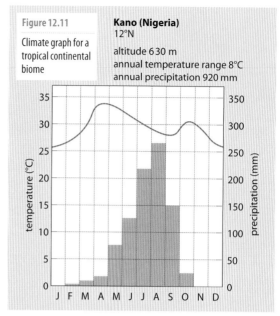

Kano (Nigeria)
12°N

altitude 630 m
annual temperature range 8°C
annual precipitation 920 mm

sponds with the moving away of the ITCZ, leaving the area with the strong, steady trade winds. The trades are dry because they are warming as they blow towards the Equator and they will have shed any moisture on distant eastern coasts. Places nearer to the desert margins tend to experience dry, stable conditions (the subtropical high pressure) caused by the migration of the descending limb of the Hadley cell (page 226). Humidity is also low during this season.

Tropical or savanna grassland vegetation

The tropical grasslands are estimated to have a mean NPP of 900 g/m²/yr (page 306). This is considerably less than the rainforest, partly because of the smaller number of trees, species and layers and partly because, although grasslands have the potential to return organic matter back to the soil, the rate of decomposition is reduced during the winter drought leaving considerable amounts left stored in the litter.

As shown in Figure 12.13, the savanna includes a series of transitions between the rainforest and the desert. At one extreme, the 'closed' savanna is mainly trees with areas of grasses; at the other, the 'open' savanna is vegetated only by scattered tufts of grass. The trees are deciduous and, like those in Britain, lose their leaves to reduce transpiration, but, unlike in Britain, this is due to the winter drought rather than to cold. Trees are xerophytic, or drought-resistant. Even when leaves do appear, they are small, waxy and sometimes thorn-like. Roots are long and extend to tap any underground water. Trunks are gnarled and the bark is usually thick to reduce moisture loss.

Figure 12.12

Causes of seasonal rainfall in places with a tropical continental (savanna) climate

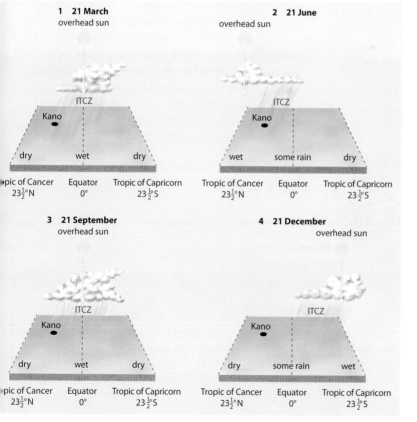

Figure 12.13

Transect across the
savanna grasslands

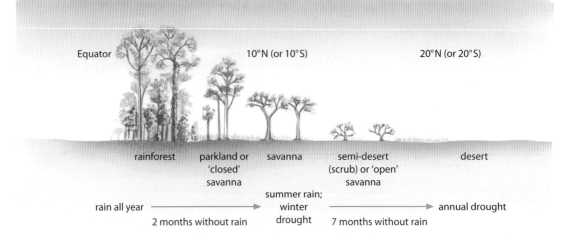

Equator   10°N (or 10°S)   20°N (or 20°S)

rainforest   parkland or   savanna   semi-desert   desert
      'closed'   (scrub) or 'open'
      savanna   savanna

summer rain;
rain all year   →   winter   →   annual drought
    2 months without rain   drought   7 months without rain

The baobab tree (also known as the 'upside-down tree') has a trunk of up to 10 m in diameter in which it stores water. Its root-like branches hold only a minimum number of tiny leaves in order to restrict transpiration (Figure 12.14). Some baobabs are estimated to be several thousand years old and, like other savanna trees, are pyrophytic, i.e. their trunks are resistant to the many local fires. Acacias, with their crowns flattened by the trade winds (Figure 12.15), provide welcome though limited shade – as do the eucalyptus in Australia. Savanna trees reach 6–12 m in height. Many have Y-shaped, branching trunks – ideal for the leopard to rest in after its meal! The number of trees increases near to rivers and waterholes. Grasses grow in tufts and tend to have inward-curving blades and silvery spikes. After the onset of the summer rains, they grow very quickly to over 3 m in height: elephant grass reaches 5 m (Figure 12.15). As the sun dries up the vegetation, it becomes yellow in colour (Figure 12.46). By early winter, the straw-like grass has died down, leaving seeds dormant on the surface until the following season's rain. By the end of winter, only the roots remain and the surface is exposed to wind and rain.

Over 40 different species of large herbivore graze on the grasslands, including wildebeest, zebra and antelope, and it is the home of several carnivores – both predators, such as lions, and scavengers, such as hyenas. Termites and microbes are the major decomposers. As previously mentioned (page 293), fire is possibly the major determinant of the savanna biome – either caused deliberately by farmers or resulting from lightning associated with summer electrical storms.

It is the fringes of the savannas, those bordering the deserts, which are at greatest risk of desertification (Case Study 7). As more trees are removed for fuel and overgrazing reduces the productivity of grasslands, the heavy rain forms gulleys and wind blows away the surface soil. Where the savanna is not farmed, there are usually more trees, suggesting that grass may not be the natural climatic climax vegetation.

Figure 12.14

A baobab tree,
Malawi

Figure 12.15

Savanna grassland during the wet season in the Maasai Mara, Kenya

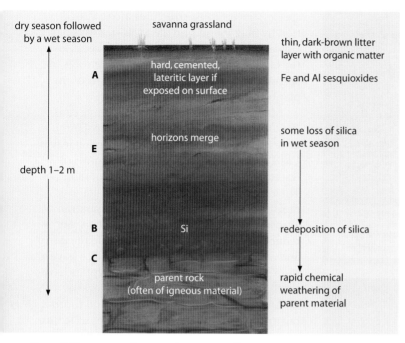

savanna grassland

dry season followed by a wet season

A — thin, dark-brown litter layer with organic matter

hard, cemented, lateritic layer if exposed on surface — Fe and Al sesquioxides

depth 1–2 m

E — horizons merge — some loss of silica in wet season

B — Si — redeposition of silica

C

parent rock (often of igneous material) — rapid chemical weathering of parent material

Figure 12.16

A ferruginous soil profile

Ferruginous soils

As savanna grasses die back during the dry season, they provide organic matter which is readily broken down to give a thin, dark-brown layer of humus (Figure 12.16). During the wet season, rapid leaching removes silica from the upper profile, leaving behind the red-coloured oxides of iron and aluminium. As these soils contain few nutrients, they tend to be acidic and lacking in bases. Although the process of capillary action might be expected to operate during the dry season, in practice it rarely does as the water table invariably falls too low at this time of year.

Ferruginous soils tend to be soft unless exposed at the surface where, being subject to wet and dry seasons, they can harden to form a cemented crust known as **laterite**. The term laterite is derived from the Latin for 'brick'. Indeed this deposit is used as a building material because, being initially soft, it can easily be dug from the soil, shaped into bricks and left to harden by exposure to cycles of wetting and drying. It is only when the laterite crust forms that drainage and plant root penetration is impeded.

As these soils hold few nutrients and tend to dry out during the dry season, they are not particularly suited to agriculture; together with the grassland they support, they are better suited to animal rearing than to arable farming. Where a lateritic crust forms on the surface, or when deep ploughing removes the surface vegetation, the upper soil tends to dry out during the dry season, becoming highly vulnerable to erosion by wind and, when the rains return, by water.

3 Hot deserts

The hot deserts of the Atacama and Kalahari-Namib and those in Mexico and Australia, are all located in the trade wind belt, between 15° and 30° north or south of the Equator, and on the west coasts of continents where there are cold, off-shore, ocean currents (Figures 7.2, 9.9 and 11.38). The exception is the extensive Sahara-Arabian-Thar desert which owes its existence to the size of the Afro-Asian landmass.

Climate

Desert temperatures are characterised by their extremes. The annual range is often 20–30°C and the diurnal range over 50°C (Figure 12.17). During the daytime, especially in summer, there are high levels of insolation from the overhead sun, intensified by the lack of cloud cover and the bare rock or sand ground surface. In contrast, nights may be extremely cold with temperatures likely to fall below 0°C. Coastal areas, however, have much lower monthly temperatures (Arica in the Atacama has a warmest month of only 22°C) due to the presence of offshore, cold, ocean currents (Figure 9.9).

Although all deserts suffer an acute water shortage, none is truly dry. Aridity and extreme aridity have been defined by using Thornthwaite's P/E index (Figure 7.1), and four of the main causes of deserts are described on page 179. Amounts of moisture are usually small and precipitation is extremely unreliable. Death Valley, California, averages 40 mm a year, yet rain may fall only once every two or three years. Whereas mean annual totals vary by less than 20 per cent a year in north-west Europe, the equivalent figure for the Sahel is 80–150 per cent (Figure 9.28). Rain,

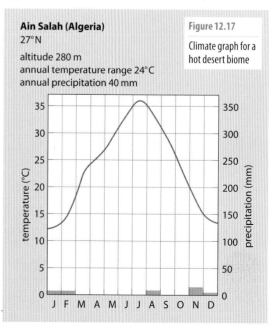

Ain Salah (Algeria)
27°N

altitude 280 m
annual temperature range 24°C
annual precipitation 40 mm

Figure 12.17

Climate graph for a hot desert biome

Figure 12.18

Saguaros cacti in the Arizona desert, USA

Figure 12.19

Ephemerals in flower following a desert rainstorm

species, have simple structures, no stratification by height and provide a low-density cover. However, plants must be xerophytic because the lack of water hinders the ability of roots to absorb nutrients and of any green parts of the plants to photosynthesise.

Many plants are **succulents**, i.e. they can store water in their tissues. Many succulents have fleshy stems and some have swollen leaves. Cacti (Figure 12.18) absorb large amounts of water during the infrequent periods of rain. Their stems swell up, only to contract later as moisture is slowly lost through transpiration. Transpiration takes place from the stems, but is reduced by the stomata closing during the day and opening nocturnally. The stems also have a thick, waxy cuticle. Australian eucalyptids have thick, protective bark for the same purpose.

Most plants, for example cactus and thorn-bush, have small, spiky or waxy leaves to reduce transpiration and to deter animals. Roots are either very long to tap groundwater supplies – those of the acacia exceed 15 m – or they spread out over wide areas near to the surface to take the maximum advantage of any rain or dew, like those of the creosote bush. Bushes are, therefore, widely spaced to avoid competition for water. Some plants have bulbous roots for storing water. Seeds, which usually have a thick case protecting a pulpy centre, can lie dormant for months or several years until the next rainfall.

Following a storm, the desert blooms (Figure 12.19). Many plants are **ephemerals** and can complete their life-cycles in two or three weeks. Others, like the saltbush, are **halophytic** and can survive in salty depressions; yet others, like the date-palm, survive where the water table is near enough to the surface to form oases. Due to the lack of grass and the limited number of green plants, there are very few food chains: desert biomes have a low capacity to sustain life. There is insufficient plant food to support an abundance of animal life. Food chains (page 296) are simple, often just a single linear sequence (in contrast to the interlocking webs characteristic of, for example, forests). This is why the desert ecosystem is 'fragile': organisms do not have the alternative sources of food which are available in more complex ecosystems. Many animals are small and nocturnal (the camel is an exception) and burrow into the sand during the heat of the day. Reptiles are more adaptable, but bird life is limited. The desert fringes form a delicately balanced ecosystem which is being disturbed by human activity and population growth which are, together, increasing the risk of desertification (Case Study 7).

when it does fall, produces rapid surface runoff which, together with low infiltration and high evaporation rates, minimises its effectiveness for vegetation. The Atacama, an almost rainless desert, has some vegetation as moisture is available in the form of advection fog (Places 24, page 180). The subsiding air, forming the descending limb of the Hadley cell, creates high pressure and produces the trade winds which are strong, persistent and likely to cause localised dust storms (Figures 7.9 and 9.34).

Desert vegetation

Deserts have the lowest organic productivity levels of any biome (Figure 11.40). The average NPP is 90 g/m²/yr, most of which occurs underground away from the direct heat of the sun. Vegetation has to have a high tolerance to the moisture budget deficit, intense heat and, often, salinity. Few areas are totally devoid of vegetation, although desert plants are few in

Desert soils

In desert areas, the climate is too dry and the vegetation too sparse for any significant chemical weathering of bedrock or the accumulation of organic material. In the relatively few places where the water table is near to the surface, soil moisture is likely to be drawn upwards by capillary action. This process causes salts and bases, such as magnesium, sodium and calcium, to be deposited in the upper profile to give a slightly alkaline soil. Many desert soils are grey in colour as the lack of moisture often restricts hydrolysis and, therefore, the release of red-coloured iron (page 42). Soils, which tend to lack both structure and horizons (Figure 12.20), are often thin, although their depth can vary depending upon the origin of the parent material, i.e. *in situ* weathering or the deposition of material by wind or water (Chapter 7). A characteristic of many desert soils is the presence of either a thin crust, 2 to 3 mm thick, caused by the impact of high-intensity rainfall, and/or a 'desert pavement' (Figure 7.10) which consists of small stones, often ventifacted and covered in desert varnish (page 182), which help stabilise the surface.

Desert soils are unproductive mainly because of the lack of moisture and humus, but potentially they are not particularly infertile. Areas under irrigation are capable of producing high-quality crops, although this farming technique is being threatened by salinisation (Figures 10.22 and 16.53).

4 Mediterranean (warm temperate, western margins)

This type of biome is found on the west coasts of continents between 30° and 40° north and south of the Equator, i.e. in Mediterranean Europe (which is the only area where the climate penetrates far inland), California, central Chile, Cape Province (South Africa) and parts of southern Australia (Figure 11.38).

Climate

The climate is noted for its hot, dry summers and warm, wet winters (Figure 12.21). Summers in southern Europe are hot. The sun is high in the sky, though never directly overhead, and there is little cloud. Winters are mild, partly because the sun's angle is still quite high but mainly due to the moderating influence of the sea. Other 'Mediterranean' areas are less warm in summer and have a smaller annual range due to cold, offshore currents (compare San Francisco, 8°C in January and 15°C in July, with Malta). Diurnal temperature ranges are often high due to the fact that many days, even in winter, are cloudless.

As the ITCZ moves northwards in the northern summer, the subtropical high pressure areas migrate with it to affect these latitudes. The trade winds bring arid conditions, with the length of the dry season increasing towards the desert margins. In winter, the ITCZ, and subsequently the subtropical jet stream (page 228), move southwards allowing the westerlies, which blow from the sea, to bring moisture. Most areas are backed by coastal mountains and so the combined effects of orographic and frontal precipitation give high seasonal totals. Areas with adjacent, cold, offshore currents experience advection fogs (California). The Mediterranean Sea region is noted for its local winds (Figure 12.22). The **sirocco** and **khamsin** are two of the hot, dry winds that blow from the

Figure 12.20

A desert soil profile

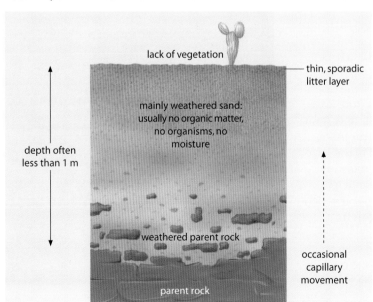

Figure 12.21

Climate graph for a Mediterranean biome

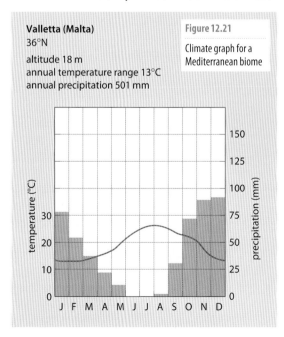

Valletta (Malta)
36°N
altitude 18 m
annual temperature range 13°C
annual precipitation 501 mm

Figure 12.22

Mediterranean winds

Sahara and can raise temperatures to over 40°C. The **mistral** is a cold wind which originates over the Alps and is funnelled at considerable speed down the Rhône valley.

Vegetation

The NPP of Mediterranean ecosystems is about 700 g/m²/yr (Figure 11.40). It is limited by the summer drought and has probably been reduced considerably over the centuries by human activity. Indeed, human activity, together with frequent fires, has left very little of any original climatic climax vegetation. The climax vegetation was believed to have been, in Europe at least, open woodland comprising a mixture of broad-leaved, evergreen trees (e.g. cork oak and holm oak) and conifers (e.g. aleppo pines, cypresses and cedars). The sequoia, or giant redwood, is native in California.

The present vegetation, which is mainly **xerophytic** (drought-resistant), is described as 'woodland and sclerophyllous scrub'. **Sclerophyllous** means 'hard-leaved' and is used to describe those evergreen trees or shrubs that have small, hard, leathery, waxy or even thorn-like leaves and which are efficient at reducing transpiration during the dry summer season. Many of the trees are evergreen, maximising the potential for photosynthesis. Trees such as the cork oak have thick and often gnarled bark to help reduce transpiration. Others, such as the olive and eucalyptus, have long tap roots to reach groundwater supplies and, in some cases, may have bulbous roots in which to store water. High temperatures during the dry summer limit the amount and quality of grass. Citrus fruits, although not indigenous, are suited to the climate as their thick skins preserve moisture. Most trees only grow from 3 to 5 m in height. They provide little shade, as they grow at widely spaced intervals, and they are **pyrophytic** (fire-resistant, page 293).

Where the natural woodland has been replaced, and in areas too dry for tree growth, a scrub vegetation has developed. The scrub is known as *chaparral* in California, *maquis* or *garrigue* in Europe, *fynbus* in South Africa and *mallee* in Australia. In Mediterranean Europe, the type of scrub depends on the underlying parent rock. **Maquis** (Figure 12.23), which is taller, denser and more tangled, grows in areas of impermeable rock (granite). It consists of shrubs, such as heathers and broom, which reach a height of 3 m. **Garrigue** (Figure 12.24) grows on drier and more permeable rocks (limestone). It is less tall and less dense than maquis. Apart from gorse, with its prickles, the more common plants include aromatic shrubs such as thyme, lavender and rosemary.

The limited leaf litter tends to decompose slowly during the dry summer, even though temperatures are high enough for year-round bacterial activity. Wildlife and climax vegetation have retreated as human activity has advanced. Arguably, the Mediterranean regions of Europe and California (together with the temperate deciduous forests) form the biome most altered by human activity.

Figure 12.23

Maquis vegetation

Figure 12.24

Garrigue vegetation

Soils

Mediterranean soils are transitional between brown earths on the wetter margins and desert soils at the drier fringes. Initially formed under broad-leaved and coniferous woodland, the soil is partly a relict feature from a previously forested landscape.

There are often sufficient roots and decaying plant material to provide a significant humus layer. Winter rains cause some leaching of bases, sesquioxides of iron and aluminium and the translocation of clays. The *B* horizon is therefore clay-enriched and may be coloured a bright red

by the redeposition of iron and aluminium. The soils, which are often thin, are less acid than the brown earths as there is less leaching in the dry season and calcium is often released, especially in limestone areas (Figure 12.25).

In many Mediterranean areas, parent rock is locally a more important factor in soil formation than climate. This leads to the development of intrazonal soils such as rendzina and terra rossa (Figures 10.23 and 10.24).

4A Eastern margin climates in Asia (monsoon)

South-east and eastern Asia are dominated by the monsoon (page 239). Temperature figures and rainfall distributions are similar to those of places having a tropical continental climate with a very warm and dry season from November to May and a hot and very wet season from June to October (Places 32, page 240). The major difference between the two climates is that monsoon areas receive appreciably higher annual amounts of rain. The natural vegetation is jungle (tropical deciduous forest) and the dominant soil type is ferralitic. Both vegetation and soils, therefore, share many similarities with the tropical rainforest.

5 Temperate grasslands

The temperate grassland biome lies in the centre of continents approximately between latitudes 40° and 60° north of the Equator. The two main areas are the North American Prairies and the Russian Steppes (Figure 11.38).

Cool temperate continental climate

The annual range of temperature is high as there is no moderating influence from the sea (38°C at Saskatoon, Figure 12.26). The land warms up rapidly in summer to give maximum mean monthly readings of around 20°C. However, the rapid radiation of heat from mid-continental areas in winter means there are several months when the temperature remains below freezing point. The clear skies also result in a large diurnal temperature range.

In Russia, precipitation decreases rapidly towards the east as distance from the sea – and therefore from the rain-bearing winds – increases; in North America, however, totals are lowest to the west which is directly in the rainshadow of the Rockies. Annual amounts in both areas only average 500 mm and there is a threat of drought, as experienced in North America in 1988. Although, fortuitously, 75 per cent of

Figure 12.25

A Mediterranean soil profile

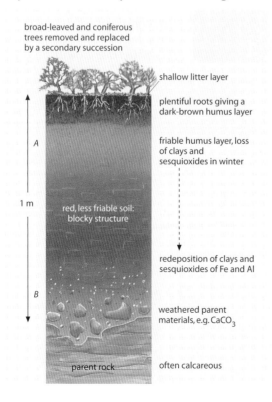

broad-leaved and coniferous trees removed and replaced by a secondary succession

shallow litter layer

plentiful roots giving a dark-brown humus layer

A

friable humus layer, loss of clays and sesquioxides in winter

1 m

red, less friable soil: blocky structure

redeposition of clays and sesquioxides of Fe and Al

B

weathered parent materials, e.g. CaCO₃

parent rock

often calcareous

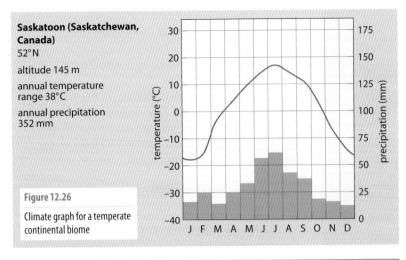

Saskatoon (Saskatchewan, Canada)

52°N

altitude 145 m

annual temperature range 38°C

annual precipitation 352 mm

Figure 12.26

Climate graph for a temperate continental biome

Figure 12.27

Tufted grasses on the North American Prairies, USA

precipitation falls during the summer growing season, it can occur in the form of harmful thunderstorms and hailshowers. The ground can be snow-covered for several months between October and April. Overall, there is a close balance between precipitation and evapotranspiration. In winter, both areas are open to cold blasts of arctic air, although the chinook may bring temporary warmer spells to the Prairies (page 241).

Temperate grassland vegetation

This type of vegetation lies to the south of the coniferous forest belt in the dry interiors of North America and Russia. Temperate grasslands are, however, also found sporadically in parts of the southern hemisphere, where they usually lie between 30° and 40°S. The Pampas (South America) and the Canterbury Plains (New Zealand) are towards the eastern coast, while the Murray–Darling basin (Australia) and the Veld (South Africa) are further inland. The NPP of 600 g/m²/yr is considerably less than that of the tropical grasslands because the vegetation grows neither as rapidly nor as tall (Figure 11.40). Whatever the original climax vegetation of the biome may have been, the ecosystem has been significantly altered by fire and human exploitation to leave, today, grama and buffalo grass as the dominants. There are two main types of grass. Feather grasses grow to 50 cm and form a relatively even coverage, whereas tufted (tussock) grasses, reaching up to 2 m, are found in more compact clumps (Figure 12.27). The grass forms a tightly knit sod which may have restricted tree growth, and certainly made early

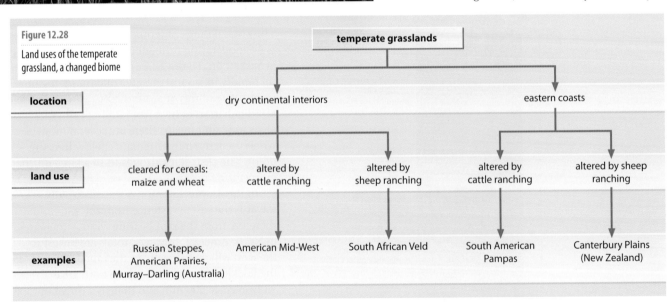

Figure 12.28

Land uses of the temperate grassland, a changed biome

		temperate grasslands			
location	dry continental interiors			eastern coasts	
land use	cleared for cereals: maize and wheat	altered by cattle ranching	altered by sheep ranching	altered by cattle ranching	altered by sheep ranching
examples	Russian Steppes, American Prairies, Murray–Darling (Australia)	American Mid-West	South African Veld	South American Pampas	Canterbury Plains (New Zealand)

Figure 12.29

A chernozem (black earth) soil profile

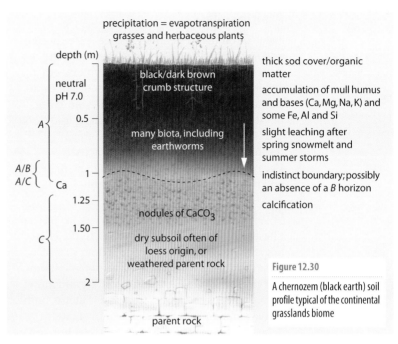

precipitation = evapotranspiration
grasses and herbaceous plants

depth (m)

neutral
pH 7.0

A

0.5

A/B
A/C

Ca

1

1.25

C

1.50

2

black/dark brown
crumb structure

many biota, including
earthworms

nodules of CaCO₃

dry subsoil often of
loess origin, or
weathered parent rock

parent rock

thick sod cover/organic
matter

accumulation of mull humus
and bases (Ca, Mg, Na, K) and
some Fe, Al and Si

slight leaching after
spring snowmelt and
summer storms

indistinct boundary; possibly
an absence of a B horizon

calcification

Figure 12.30

A chernozem (black earth) soil
profile typical of the continental
grasslands biome

ploughing difficult. The deep roots, which often extend to a depth of 2 m in order to reach the water table, help to bind the soil together and so reduce erosion. Most of the organic material is in the grass roots and it is the roots and rhizomes that provide the largest store of nutrients (Case Study 12B).

During autumn, the grasses die down to form a turf mat in which seeds lie dormant until the snowmelt, rains and higher temperatures of the following spring. Growth in early summer is rapid and the grasses produce narrow, inward-curving blades to limit transpiration. By the end of summer, their blue-green colour may have turned more parched. Herbaceous plants and some trees (willow) grow along water courses. In response to the windy climate, many prairie and steppe farms are protected by trees planted as windbreaks. The decay of grasses in summer causes a rapid accumulation of humus in the soil, making the area ideal for cereals or, in drier areas, for cattle ranching (Figures 12.27 and 12.28).

The temperate grasslands are a resilient eco-system. The grasses provide food for burrowing animals such as rabbits and gophers, and for large herbivores such as antelopes, bison and kangaroos. These, in turn, may be consumed by carnivores (wolves and coyotes) or by predatory birds (hawks and eagles).

Chernozems or black earths

The thick grass cover and the importance of roots as a source of organic matter together provide a plentiful supply of mull humus which forms a black, crumbly topsoil (Figure 12.29). While the abundance of biota, especially earthworms, causes the rapid decay and mixing of organic matter during the warm summer, decomposition is arrested during drier spells and in the long, cold winter. Due to rapid mixing, humus is spread throughout the A horizon, and as a result of rapid decomposition there is effective recycling as the grasses take up and return nutrients to the soil. The late spring snowmelt and early summer storms cause some leaching (Figure 12.30), and bases such as potassium and magnesium may be slowly moved downwards. In late summer, and in places where the water table is near to the surface, capillary water may bring bases nearer to the surface to maintain a neutral or slightly alkaline soil (pH 7 to 7.5). The grasses have an extensive root system which gives a deep (up to 1 m) dark-brown to black A horizon.

The alternating dry and wet seasons immob-ilise iron and aluminium sesquioxides and clay within aggregates (peds) in the upper horizon and this, together with the large number of mixing agents, limits the formation of a recog-nisable B horizon. The subsoil, often of loess origin (page 136), is usually porous and this, together with the capillary moisture movement in summer, means that it remains dry. This upward movement of moisture causes calcium carbonate to be deposited, often in the form of nodules, in the upper C horizon.

Chernozems are regarded as the optimum soil for agriculture as they are deep, rich in organic matter, retain moisture, and have an ideal crumb structure with well-formed peds. After intensive ploughing, chernozems may require the addition of potassium and nitrates.

Prairie soils

These lie on the wetter margins of the chernozems and form a transition between them and the brown forest earths. As precipitation exceeds evapotranspiration, there is an absence of capillary action and the soil lacks the accumulation of calcium carbonate associated with chernozems. The *A/B* horizons tend to merge, as there is limited leaching and strong biota activity. Decaying grasses provide much organic material and the soils are ideal for cereal crops.

Chestnut soils

These are found in juxtaposition with the chernozems, but where the climate is drier so that evapotranspiration slightly exceeds precipitation and the resultant vegetation is sparser and more xerophytic. As the root system is less dense, both the amount and the depth of organic matter decrease, as does the thickness of the *A* horizon, and the colour becomes a lighter brown than in chernozems. Chestnut soils are more alkaline, due to increased capillary action, and suffer from more frequent summer droughts. Deposits of calcium carbonate are found near to the surface and the soil is generally shallower than a chernozem. Chestnut soils are agriculturally productive if aided by irrigation, but mismanagement can quickly lead to their exhaustion and erosion.

6 Temperate deciduous forests

Temperate deciduous forests are located on the west coasts of continents between approximately latitudes 40° and 60° north and south of the Equator. Apart from north-west Europe (which includes the British Isles), other areas covered by this biome include the north-west of the USA, British Columbia, southern Chile, Tasmania and South Island, New Zealand (Figure 11.38).

Cool temperate western margins climate

Summers are cool (Figure 12.31) with the warmest month between 15°C and 17°C. This is a result of the relatively low angle of the sun in the sky, combined with frequent cloud cover and the cooling influence of the sea. Winters, in comparison, are mild. Mean monthly temperatures remain a few degrees above freezing due to the warming effect of the sea, the presence of warm, offshore ocean currents and the insulating cloud cover. Diurnal temperature ranges are low; autumns are warmer than springs; and seasonal temperature variations depend on prevailing air masses (Figure 9.41).

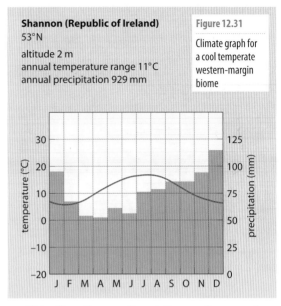

Shannon (Republic of Ireland)
53°N

altitude 2 m
annual temperature range 11°C
annual precipitation 929 mm

Figure 12.31

Climate graph for a cool temperate western-margin biome

This climatic zone lies at the confluence of the Ferrel and Polar cells (Figures 9.34 and 9.35), where tropical and polar air converge at the Polar Front. Warmer tropical air is forced to rise, creating an area of low pressure and forming depressions with their associated fronts. The prevailing south-westerlies, laden with vapour after crossing warm, offshore currents, give heavy orographic and frontal rain. Precipitation, often exceeding 2000 mm annually, falls throughout the year but with a winter maximum when depressions are more frequent and intense. Although snow is common in the mountains, it rarely lies for long at sea-level. Fog, most common in the autumn, forms under anticyclonic conditions (page 234).

Deciduous forests

Although having the second-highest NPP of all biomes (1200 g/m²/yr), the temperate deciduous forest falls well short of the figure for tropical rainforests, mainly because of the dormant winter season when the deciduous trees in temperate latitudes shed their leaves (Figure 11.40). Leaf fall has the effect of reducing transpiration when colder weather reduces the effectiveness of photosynthesis and when roots find it harder to take up water and nutrients.

In Britain, oaks, which can reach heights of 30 to 40 m, became the dominant species as the climax vegetation developed through a series of several primary successions (Figure 11.4). Other trees, such as the elm (common before its population was diminished by Dutch elm disease), beech, sycamore, ash and chestnut, grow a little less tall. They all develop large crowns and have broad but thin leaves (Figure 12.32). Unlike

Figure 12.32

Broad-leaved deciduous oak woodland in Surrey, England

Figure 12.33

A brown earth soil profile

the rainforests, the temperate deciduous forests contain relatively few species. The maximum number of species per km² in southern Britain is eight, and some woodlands, such as beech, may only have a single dominant. The trees have a growing season of 6 to 8 months in which to bud, leaf, flower and fruit, and may only grow by about 50 cm a year.

Most woodlands show some stratification (Figure 11.2). Beneath the canopy is a lower shrub layer varying between 5 m (holly, hazel and hawthorn) and 20 m (ash and birch). This layer can be quite dense because the open mosaic of branches of the taller trees allows more light to penetrate than in the rainforests. The forest floor, if the shrub layer is not too dense, is often covered in a thick undergrowth of brambles, grass, bracken and ferns. Many flowering plants (bluebells) bloom early in the year before the taller trees have developed their full foliage.

Epiphytes, which include mosses, lichens and algae, often grow on tree trunks.

The forest floor has a reasonably thick leaf litter which is readily broken down by the numerous mixing agents living in the relatively warm soil. There is a rapid recycling of nutrients, although some are lost through leaching. The leaching of humus and nutrients and the mixing by biota produce a brown-coloured soil. Soil type contributes to determine the dominant tree: oaks and elms prefer loams; beech the more acid gravels and the drier chalk; ash the lime-rich soils; and willows and alder wetter soils. There is a well-developed food chain in these forests, with many autotrophs, herbivores (rabbits, deer and mice) and carnivores (foxes).

Most of Britain's natural primary deciduous woodland has been cleared for farming, for use as fuel and in building, and for urban development. Deciduous trees give way to conifers towards polar latitudes and where there is an increase in either altitude or steepness of slope.

Brown earths

The considerable leaf litter, which accumulates in autumn, decomposes relatively quickly due to the activity of soil biota. Organic matter is incorporated as mull into the A horizon by the action of earthworms, giving it a dark-brown colour (Figure 12.33). Precipitation exceeds evapotranspiration sufficiently to allow leaching. Bases, especially calcium and magnesium, are absent in the upper horizons and, in some instances, there may be a loss of clay and sesquioxides (Figure 12.34). Because there is greater biota activity, the horizons merge more gradually than in a podsol (Figure 12.39), while the colour may become increasingly reddish-brown with depth if iron and aluminium are redeposited.

Figure 12.34

A brown earth soil profile typical of a temperate deciduous woodland biome

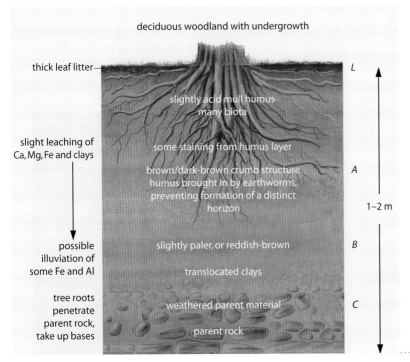

deciduous woodland with undergrowth

thick leaf litter — L

slightly acid mull humus many biota

some staining from humus layer

slight leaching of Ca, Mg, Fe and clays

brown/dark-brown crumb structure humus brought in by earthworms, preventing formation of a distinct horizon — A

1–2 m

possible illuviation of some Fe and Al

slightly paler, or reddish-brown — B

translocated clays

tree roots penetrate parent rock, take up bases

weathered parent material — C

parent rock

Brown earths tend to be free-draining as they do not have a hard pan. There is considerable recycling as the deciduous trees take up large amounts of nutrients from the soil in summer, only to return them through leaf-fall the following autumn. Brown earths are usually deeper than podsols, partly because tree roots can penetrate and break up the bedrock (Figure 2.5) and are more fertile, mainly because of the higher content of organic matter and clay (although they often benefit from liming).

7 Coniferous forests

The coniferous forest, or taiga, biome occurs in cold climates to the poleward side of 60°N in Eurasia and North America as well as at high altitudes in more temperate latitudes and in southern Chile (Figure 11.38).

Cold climates

Winters are long and cold. Minimum mean monthly temperatures may be as low as –25°C (–24°C at Fairbanks, Figure 12.35) – there is little moderating influence from the sea and no insolation as, at this time of year, the sun never rises in places north of the Arctic Circle. Strong winds mean there is a high wind-chill factor (frostbite is a hazard to humans); any moisture is rapidly evaporated (or frozen); and snow is frequently blown about in blizzards. Summers are short, but the long hours of daylight and clear skies mean that they are relatively warm. Precipitation is light throughout the year because the air can hold only limited amounts of moisture, and

Figure 12.36

Coniferous forest, British Columbia

most places are a long way from the sea. The slight summer maximum is caused by isolated convectional rainstorms.

Coniferous forest or taiga

The coniferous forest has an average NPP of 800 g/m^2/yr (Figure 11.40). The coniferous trees have developed distinctive adaptations which enable them to tolerate long, cold winters; cool summers with a short growing season; limited precipitation; and podsolic soils. The size of the dominant trees and the fact that they are evergreen – giving them the potential for year-round photosynthesis – result in their relatively high NPP. The trees, which are softwoods, rarely number more than two or three species per km^2. Often there may be extensive stands of a single species, such as spruce, fir or pine. In colder areas, like Siberia, the larch tends to dominate. Although larches are cone-bearing, the European larch is deciduous and sheds its leaves in winter. All trees in the taiga, some of which attain a height of 40 m, are adapted to living in a harsh environment (Figure 12.36).

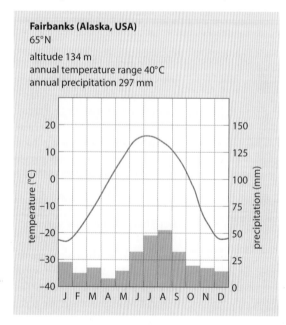

Fairbanks (Alaska, USA)
65°N

altitude 134 m
annual temperature range 40°C
annual precipitation 297 mm

Figure 12.35

Climate graph for a cold climate biome

Figure 12.37

Forest floor in a coniferous forest, Cumbria

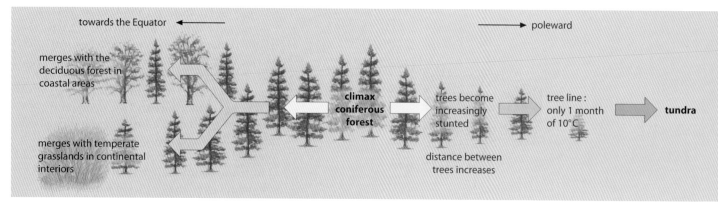

towards the Equator ←   → poleward

merges with the deciduous forest in coastal areas

merges with temperate grasslands in continental interiors

climax coniferous forest

trees become increasingly stunted

tree line : only 1 month of 10°C

tundra

distance between trees increases

Figure 12.38

The coniferous forest and its transition zones

Conditions for photosynthesis become favourable in spring as incoming radiation increases and water becomes available through snowmelt (days in winter are long and dark and soil moisture is frozen). The needle-like leaves are small and the thick cuticles help to reduce transpiration during times of strong winds and during the winter when moisture is in a form unavailable for absorption by tree roots. Cones shield the seeds and thick, resinous bark protects the trunk from the extreme cold of winter and the threat of summer forest fires. The conical shape of the tree and its downward-sloping, springy branches allow the winter snows to slide off without breaking the branches. The conical shape also gives some stability against strong winds as the tree roots are usually shallow. There is usually only one layer of vegetation in the coniferous forest. The amount of ground cover is limited, due partly to the lack of sunlight reaching the forest floor and partly to the deep, acidic layer of non-decomposed needles (Figure 12.37). Plants that can survive on the forest floor include mosses, lichens and wood sorrel. The cold climate and acid soil discourage earthworms and bacteria. Needles decompose very slowly to give an acid mor humus (page 262) with most of the nutrients held within the litter (Figure 11.29a). Evapotranspiration rates are very low and, as they are usually less than precipitation totals, leaching occurs and the few nutrients that are returned to the podsol soil are soon lost. Conifers require few nutrients, taking only 225 kg of plant nutrient annually from each hectare compared with the 430 kg taken by deciduous trees. The limited food supply means that animal life is not abundant. The dark woods are not favoured by bird life, although deer, wolves, brown bears, moose, elk and beavers are found in certain areas.

In North America and Eurasia, the coniferous forest merges into the tundra on its northern fringes (Figure 12.38). The tree line, the point above which trees are unable to grow, is often clearly marked in mountainous areas (Figure 12.36). South of the taiga lie either the deciduous forest or the temperate grassland biomes (Figure 11.38), depending upon whether the location is coastal or inland.

Podsols

Podsols develop in areas where precipitation exceeds evapotranspiration; under coniferous forest, heathland and other vegetation tolerant of low-nutrient-status soils; and where parent materials produce coarse-textured soils. Although podsols usually occur in places with a cool climate, they can be found virtually anywhere between the Equator and the Arctic, providing the required conditions are present.

BSSS WYE 1973

PROFILE   B2

HOTHFIELD   SERIES

HUMO-FERRIC   PODZOL

Figure 12.39

Soil profile of a podsol

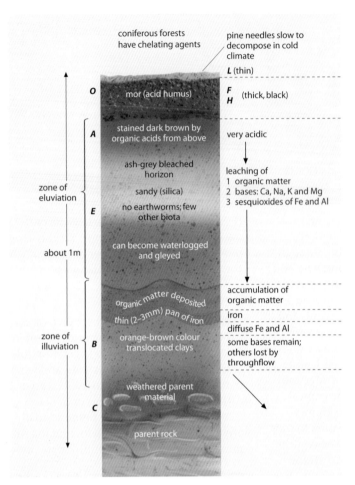

coniferous forests have chelating agents

pine needles slow to decompose in cold climate

L (thin)

O — mor (acid humus)

F **H** (thick, black)

A — stained dark brown by organic acids from above — very acidic

zone of eluviation

ash-grey bleached horizon

sandy (silica)

leaching of
1 organic matter
2 bases: Ca, Na, K and Mg
3 sesquioxides of Fe and Al

E — no earthworms; few other biota

about 1m — can become waterlogged and gleyed

organic matter deposited

accumulation of organic matter

thin (2–3mm) pan of iron

iron

diffuse Fe and Al

zone of illuviation

B — orange-brown colour translocated clays

some bases remain; others lost by throughflow

weathered parent material

C — parent rock

Figure 12.40

Soil profile of a podsol, typical of coniferous forests

Pine needles, with their thick cuticles, provide only a thin leaf litter and inhibit the formation of humus. Any humus formed is very acid (mor) and provides chelating agents and fulvic acid which help to make the iron and aluminium minerals more soluble. The cold climate discourages organisms and the soil is too acidic for earthworms. Consequently, well-defined horizons develop due to the slow decomposition of leaf litter and the lack of mixing agents. The downward percolation of water through the soil, especially following snowmelt, causes the leaching of bases, the translocation of organic matter, and the eluviation of the sesquioxides of iron and aluminium. This leaves an ash-grey, bleached *A* horizon (podsol is Russian for 'ash-like') composed mainly of quartz sand and silica (Figures 12.39 and 12.40).

Pedologists accept that different processes (physical, chemical and biological) can be employed in the translocation of materials, e.g. humus and clay in suspension, bases in solution, sesquioxides by biochemical agents in solution and, perhaps most significantly, movement caused by soil fauna mixing the soil.

The dark-coloured humus is redeposited at the top of the *B* horizon. Beneath this the sesquioxides of iron and aluminium are often – though not always – deposited as a thin, rust-coloured, hard pan. Where it is developed, this pan is rarely more than 2 or 3 mm in depth and often has a convoluted shape. It acts as an impermeable layer restricting the downward movement of moisture and the penetration of plant roots. This can cause some waterlogging in the *E* horizon to give a gleyed podsol. The lower *B* horizon, an area of diffuse accumulation of iron and aluminium, has an orange-brown colour and overlies weathered parent material. Any throughflow from this horizon is likely to contain bases in solution. Although these soils are not naturally fertile, they can be improved by the addition of lime and fertiliser, or by ripping the iron pan with a deep, single-line plough.

8 The tundra

The tundra, which lies to the north of the taiga, includes the extreme northern parts of Alaska, Canada and Russia, together with all of Greenland (Figure 11.38). The ground, apart from the top few centimetres in summer when temperatures are high enough for some plant growth, remains permanently frozen (the permafrost, Chapter 5).

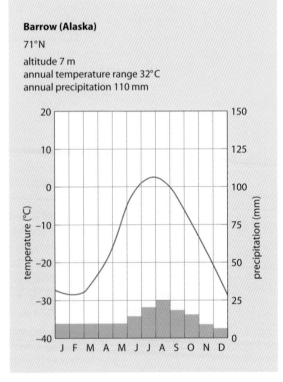

Barrow (Alaska)

71°N

altitude 7 m
annual temperature range 32°C
annual precipitation 110 mm

Figure 12.41

Climate graph for a tundra biome

Arctic climate

Summers may have lengthy periods of continuous daylight but, with the angle of the sun so low in the sky, temperatures struggle to rise above freezing-point (Barrow 3°C, Figure 12.41) and the growing season is exceptionally short. Nearer the poles, the climate is one of perpetual frost. Although winters are long, dark and severe, and the sea freezes, the water has a moderating effect on temperatures, keeping them slightly higher than inland places further south (Siberia). Precipitation, which falls as snow, is light – indeed, Barrow with 110 mm would be classified as a desert if temperatures were high enough for plant growth.

Tundra vegetation

The tundra ecosystem is one with very low organic productivity. The NPP of only 140 g/m²/yr is the second-lowest of the major land biomes (Figure 11.40). In Finnish, *tundra* means a 'barren or treeless land', which accurately describes its winter appearance, and in Russian a 'marshy plain', which is what large areas are in summer. Any vegetation must have a high degree of tolerance of extreme cold and of moisture-deficient conditions – the latter because water is unavailable for most of the year when it is stored as ice or snow. There are fewer species of plants in the tundra than in any other biome. Most are very slow- and low-growing, compact and rounded to gain protection against the wind (plants as well as people are affected by wind-chill), and most have to complete their life-cycles within 50 to 60 days. There is no stratification of vegetation by height.

The five main dominants, each with its specialised local habitat, are lichens, mosses, grasses, cushion plants, and low shrubs (Figure 12.42). Most have small leaves to limit

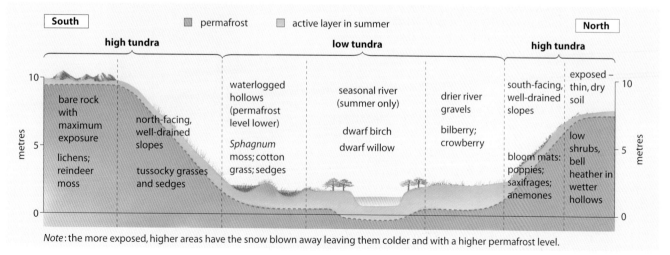

transpiration and short roots to avoid the permafrost. Lichens are pioneer plants in areas where the ice is retreating, and they can help date the chronology of an area following deglaciation (page 288). Much of the tundra is waterlogged in summer (Figures 5.18 and 12.43) due to the impermeable permafrost preventing infiltration. Where relief is gentle and evaporation rates are low, mosses, cotton grass and sedges thrive. On south-facing slopes and in better-drained soils, cushion plants provide a mass of bright colour in summer (Figure 12.44). These 'bloom mats' include arctic poppies, anemones, orchids, pink saxifrages and gentians. Where decaying vegetation accumulates (there is little bacterial action to decompose dead plants), the resultant peat is likely to be covered in heather, whereas

Figure 12.42

Relationship between vegetation and site factors in the tundra

| South | | permafrost | active layer in summer | | North |

high tundra — **low tundra** — **high tundra**

bare rock with maximum exposure — lichens; reindeer moss

north-facing, well-drained slopes — tussocky grasses and sedges

waterlogged hollows (permafrost level lower) — *Sphagnum* moss; cotton grass; sedges

seasonal river (summer only) — dwarf birch — dwarf willow

drier river gravels — bilberry; crowberry

south-facing, well-drained slopes — bloom mats: poppies; saxifrages; anemones

exposed – thin, dry soil — low shrubs, bell heather in wetter hollows

metres: 10 — 5 — 0 (both sides)

Note: the more exposed, higher areas have the snow blown away leaving them colder and with a higher permafrost level.

Figure 12.44

'Bloom mats' at Prudhoe Bay, Alaska

Tundra soils

The limited plant growth of this biome only produces a small amount of litter and, as there are few soil biota in the cold soil, organic matter decomposes only very slowly to give a thin peaty layer of humus or mor. There are many sites where there is free drainage. Where this occurs, water is able to percolate downwards, usually as meltwater in late spring, giving limited leaching and, due to the fulvic acid within it (the pH can be under 4.5), allowing the release of iron. Underlying the soil, at a very variable depth but usually under 50 cm, is the permafrost. This, acting as an impermeable layer, severely restricts moisture percolation and causes extreme water-logging and gleying (Figure 12.45). Few mixing agents can survive in the cold, wet, tundra soils, which are thin and have no developed horizons (an exception is the arctic brown soil which develops on better-drained sites). Where bedrock is near to the surface, the parent material is physically weathered by freeze–thaw action. The shattered angular fragments are raised to the surface by frost-heave, preventing the formation of horizons and creating a range of periglacial landforms (Figure 5.21).

on drier gravels, berried plants (e.g. bilberry and crowberry) are the dominants. Adjacent to the seasonal snowmelt rivers, dwarf willows, horizontal junipers and stunted birch grow, but only to a maximum of about 30 cm; even so their crowns are often distorted and misshapen by the wind. In winter, the whole biome is covered in snow, which acts as insulation for the plants.

The lack of nitrogen-fixing plants, other than in the pioneer community (page 286), limits fertility, and the cold, wet conditions inhibit the breakdown of plant material. Photosynthesis is hindered by the lack of sunlight and water for most of the year, though the presence of autotrophs, such as lichens and mosses, does provide the basis for a food chain longer than might be expected. Herbivores such as reindeer, caribou and musk-ox survive because plants like reindeer moss have a high sugar content. However, these animals have to migrate in winter to find pasture that is not covered by snow. The major carnivores are wolves and arctic fox; owls are also found here.

The tundra is an extremely fragile ecosystem in a delicate balance. Once it is disturbed by human activity, such as tourism or oil exploration and extraction, it may take many years before it becomes re-established.

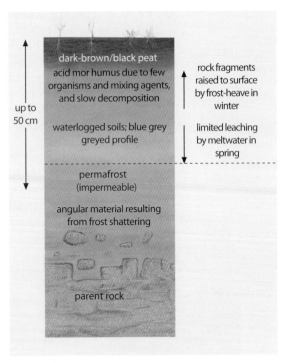

Figure 12.45

Soil profile of a typical tundra soil

Figure 12.46

Tropical savanna grassland on the Loita Plain, Kenya

A Tropical grasslands in Kenya

The main expanses of tropical grassland in Kenya lie within the Rift Valley and on the adjacent plains of the Mara (an extension of the Serengeti) and Loita (Figure 12.48). Their appearance is one of open savanna (Figures 12.13 and 12.46) with small acacia and evergreen trees (Figure 12.15). There is evidence, however, that the original climax vegetation was forest, but that this has been altered by fires, started both naturally and by humans (page 293), by overgrazing (Figure 12.47) and by climatic change.

The climate is very warm and dry for most of the year with, usually, a short season (three months) of fairly reliable and abundant rainfall and an even shorter period known as the 'little rains' (Figure 12.49). Both periods of rainfall follow soon after the ITCZ and the associated overhead sun have passed over the

Figure 12.47

Scattered trees and over-grazed land in the central Rift Valley, Kenya

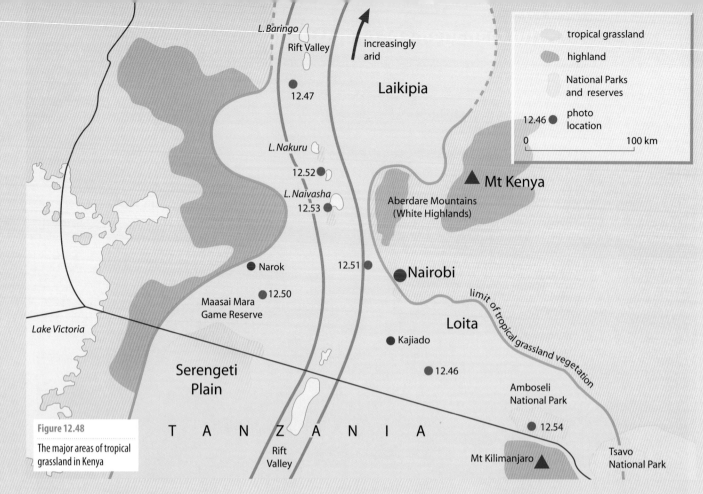

tropical grassland

highland

National Parks and reserves

12.46 ● photo location

0   100 km

L. Baringo

Rift Valley

increasingly arid

12.47

Laikipia

L. Nakuru

12.52

L. Naivasha

12.53

▲ Mt Kenya

Aberdare Mountains (White Highlands)

12.51

Narok

Nairobi

12.50

Maasai Mara Game Reserve

Loita

Lake Victoria

Kajiado

12.46

limit of tropical grassland vegetation

Serengeti Plain

Amboseli National Park

T A N Z A N I A

12.54

Rift Valley

Mt Kilimanjaro ▲

Tsavo National Park

Figure 12.48

The major areas of tropical grassland in Kenya

Nairobi (Kenya)

1°S

altitude 1820 m
annual temperature range 3°C
annual precipitation 958 mm

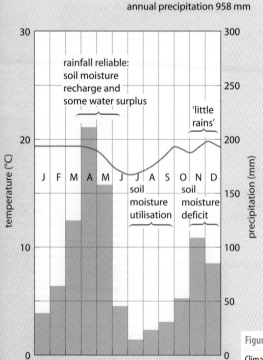

rainfall reliable: soil moisture recharge and some water surplus

'little rains'

soil moisture utilisation

soil moisture deficit

Figure 12.49

Climate graph and water balance for Nairobi (note that, due to its higher altitude, Nairobi is cooler and wetter than the surrounding grasslands)

Equator (Figure 12.12). The annual water balance shows a deficit (Figure 3.3) so that, although there is some leaching during the rainy season, for most of the year capillary action occurs. This has resulted in the development of ferruginous soils with, in places, a lateritic crust (page 321). Water supply is therefore a major management problem in this part of Kenya.

Water is obtained from springs at the foot of Mount Kilimanjaro – the mountain itself is in Tanzania – which are fed by melting snow; from several of the Rift Valley lakes (not all, as some are highly saline); from rivers (many of which are seasonal); and from waterholes. Even so, there have been, in the last 100 years alone, several major droughts when the carrying capacity of the region was exceeded. The **carrying capacity** (page 378) is the maximum number of a population (people, animals, plants, etc.) that can be supported by the resources of the environment in which they live, e.g. the greatest number of cattle that can be fed adequately on the available amount of grassland.

Human pressure on the natural resources

Maasai pastoralists

Maasai are defined as 'people who speak the Maa language'. Their ancestors were Nilotic, coming from southern Sudan during the first millennium AD. They kept cattle and grew sorghum and millet. The present Maasai may be descendants of the last of several migration waves. Latest evidence suggests that they may have only been in Kenya for 300 years. Over time, they specialised more in cattle and came to see themselves, and to be seen by others, historically and ethnically, as 'people of cattle'. Figure 12.50 is a stereotype photo of the Maasai, dressed in their red cloaks and with their humped zebu cattle. While all Maasai are Maa speakers, not all Maa speakers are Maasai – nor, today, are all Maasai pastoralists! The Maasai became semi-nomadic, moving seasonally with their cattle in search of water and pasture (two wet seasons and two dry seasons meant four moves a year; Figure 12.49). Herds had to be large enough to provide sufficient milk and meat for their owners and to reproduce themselves over time, including the ability to recover from drought and disease.

Kikuyu (Bantu) farmers

The Kikuyu were one of several Bantu tribes who arrived in Kenya, from the south, some 2000 years ago. They became subsistence farmers growing crops on the higher land which bounded the eastern side of the Rift Valley. The Kikuyu and Maasai often lived a complementary life-style. For the Maasai, Kikuyu in the highlands were a secure source of foodstuffs and a place of refuge during times of drought and cattle disease. For the Kikuyu, Maasai provided a constant supply of cattle products and wives. The division between them only appeared in early colonial times when the Maasai were forcibly moved from places like Laikipia (Figure 12.48) southwards onto the newly created Maasai reservation (the districts of Narok and Kajiado). The vacuum left was filled by newly arrived European settlers, and by Kikuyu (their rising numbers were causing a land shortage in the highlands). Figure 12.51 shows numerous, small, Kikuyu shambas (smallholdings) on the eastern edge of the Rift Valley to the north-west of Nairobi.

Colonial (European) settlers

While many Europeans settled in the so-called 'White Highlands', others developed huge estates within the Rift Valley. The most famous was Lord Delamere from Cheshire. He introduced, in turn, Australian sheep (they died, as the local grass was mineral-deficient); British sheep and clover (the sheep died, as African bees did not pollinate British clover); British cattle (wiped out by local diseases); wheat (which was more successful unless trampled by wild animals); and, finally and successfully, drought-resistant beef cattle. The present

Figure 12.50

Maasai herdsmen

Figure 12.51

Kikuyu shambas near Nairobi

Delamere estate (Figure 12.52) covers 22 600 hectares (divided into 180-hectare paddocks); it has 10 900 long-horned Boran cattle (the carrying capacity is 12 000) crossed with 300 Friesian bulls; and 280 permanent workers. Although the estate

is managed by 'whites', the stockmen are Maasai. More recently, transnational firms have set up large flower farms (Figure 12.53) and vegetable (especially peas and beans) farms in and near the Rift Valley. The closeness to Nairobi airport means that these perishable products can be transported to and sold in European markets, out of season, the day after they are picked.

Figure 12.52

Commercial cattle ranching, Delamere Estate, Kenya

Figure 12.53

Flower-growing estate near Lake Naivasha, Kenya

Figure 12.54

Amboseli National Park, Kenya, watered by melting snow from Mt Kilimanjaro (Tanzania)

Population growth and urbanisation

Kenya, an economically less developed country, has one of the world's fastest-growing population rates. This means increased pressure on the land, especially the grasslands, to grow more subsistence crops to feed the growing domestic market; more cash crops to earn needed money from increased exports; and more land lost to urban growth.

Maasai in the late 2000s

The traditional Maasai way of life and their grassland habitat are under constant threat. Figure 12.55 summarises, but does insufficient justice to, some of the present-day problems. Change, as in many societies, is being forced upon the Maasai. While many values and traditions are still known and held, the basis of their economy – the concept of land as territory – has been so transformed that the survival of the herding system is in jeopardy. For some years, many Maasai have tried either to buy individual ranches (IRs) or to amalgamate to create group ranches (GRs), a practice which seems to fail at times of severe drought. The Maasai are also having to come to terms with a sedentary rather than a semi-nomadic lifestyle. Practical Action (PA), a British development group (Places 90, page 577), has been working with Maasai people to improve the standard of housing. In response to the main complaint of Maasai women, PA has helped to design a watertight cement skin which can be laid over an old mud roof (it was the women's job to apply more dung and mud onto a leaking roof during a wet night), and have improved ventilation within the house (where all the cooking is done). The government have laid a pipeline from Kilimanjaro to Kajaido, to ensure a more reliable water supply. The quality of Maasai herds has improved, with some cattle being sold for meat in Nairobi. The improvement to herds has been aided by PA which has helped train local villagers to become 'vets' (*wasaidizi*, Figure 21.5) capable of vaccinating animals and dealing with common diseases. Some Maasai have begun to grow crops, while others have begun to benefit from tourism. In Amboseli National Park, the Maasai are allowed to sell artefacts from their own shop. They can retain all the income which had, previously, gone to the government.

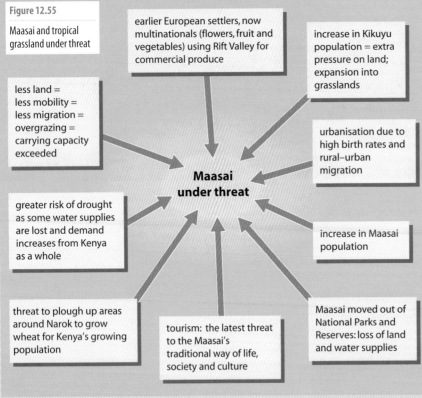

Figure 12.55

Maasai and tropical grassland under threat

- earlier European settlers, now multinationals (flowers, fruit and vegetables) using Rift Valley for commercial produce
- increase in Kikuyu population = extra pressure on land; expansion into grasslands
- less land = less mobility = less migration = overgrazing = carrying capacity exceeded
- urbanisation due to high birth rates and rural–urban migration
- greater risk of drought as some water supplies are lost and demand increases from Kenya as a whole
- increase in Maasai population
- threat to plough up areas around Narok to grow wheat for Kenya's growing population
- tourism: the latest threat to the Maasai's traditional way of life, society and culture
- Maasai moved out of National Parks and Reserves: loss of land and water supplies

Maasai under threat

National parks and reserves

The passing of the National Parks Ordinance in 1945 meant that specific areas were set aside either exclusively for wildlife (no permanent settlement in National Parks other than at tourist lodges) or where other types of land use were permitted only at the discretion of local councils. While wildlife has become a major source of income for Kenya, it has meant less land being available for crops and, to the Maasai, denial of access to important resources of dry-season water and pasture (Amboseli; Figure 12.54). Maasai herds were heavily depleted during the droughts of 1952 and 1972–76.

B The temperate grasslands in North America: the Prairies

Early travellers such as the Spaniard Coronado in the 16th century, who rode into Kansas from Mexico, and later French trappers and explorers in Canada, reported vast extents of waist-high, green grasses sometimes so tall that men on horseback stood in their stirrups to see where they were going. The plains seemed so vast that no limit could be found. Nineteenth-century settlers moving westwards across the Mississippi–Missouri in their wagons or drawing their handcarts, must have wondered if they would ever see woods, forests and mountains again. Today, the extent of the interior grasslands of North America is well known (Figure 12.56).

The Native Americans, who used the ecosystem, did little to alter the grassland, which remained in its original state of natural balance until the late 19th century. One visitor described it thus:

'It is a wild garden. Each week from April through September, about a dozen new kinds of flowers come into bloom. Once the

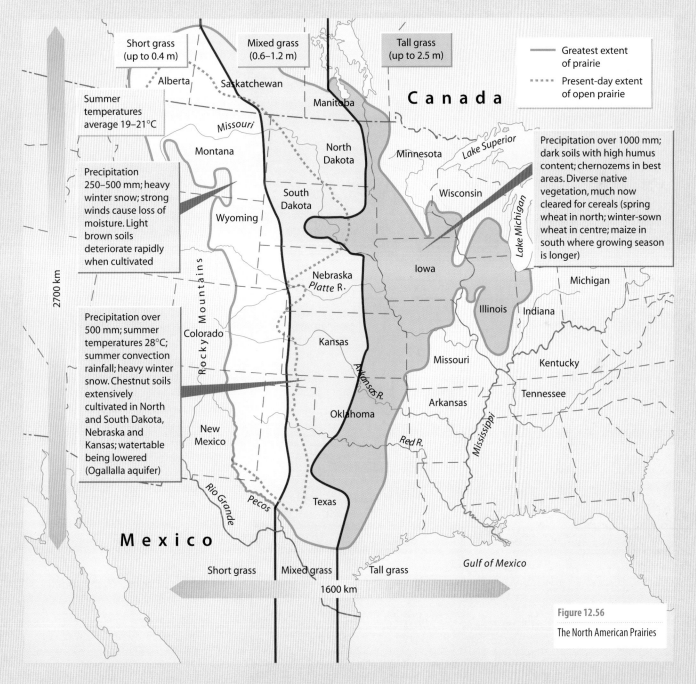

Figure 12.56

The North American Prairies

Figure 12.57

Short-grass prairie with deciduous trees (cottonwoods and aspen) and some conifers: west central Wyoming

Figure 12.58

Long-grass prairie (buffalo grass): Saskatchewan

layer of dead grass gets too thick, though, it starts to choke off the smaller grasses and wild flowers. Meantime, woody plants – they like shade and moisture – can gain a foothold in the sod and spread. If you go long enough without fire, much of this countryside will be covered with trees'.

Grasses such as blue stem and buffalo grass have a network of roots which can extend to considerable depth to absorb water and obtain nutrients. Root systems may make up over 80 per cent of the vegetative biomass in the prairie. These, together with the smaller herbs, have helped to develop a thick sod close to the surface. Plants can survive from year to year because they die back to the ground and lie dormant during the cold winters (page 327).

Soils grade in colour and fertility from brown in the western short-grass prairie (Figure 12.57) through chestnut in the mixed-grass zone to the fertile black chernozems (millisols) of the tall-grass eastern zone (Figure 12.58). The chernozems have a high humus content (page 327).

Decaying humus releases minerals slowly for the grasses. The soils are kept light and aerated which helps to prevent compaction under heavy rain (summer convection storms) and the weight of heavy animals (bison and humans). The presence of humus also helps to conserve moisture. Due to frequent droughts, the vegetation has developed protective mechanisms, such as leaves that curl up to prevent evaporation loss, and well-developed root fibres which can obtain moisture from deep in the soil.

The rapid spring growth and early maturity of grass allows it to produce seeds early. It then becomes semi-dormant until autumn and can survive heat and drought. Late-growing species may not be able to compete and this has led to an extension of short grass into the mixed-grass zone during a succession of long dry periods.

In the 17th century, there were estimated to be 60–70 million bison roaming the grasslands with 50 million antelope, plus grizzly bears, wolves and prairie dogs, together with many species of birds – hawks, larks, buntings, etc. – and insects and reptiles such as snakes and lizards. Today there are few of the larger mammals left except in wilderness refuges such as National Parks.

Figure 12.59

Cattle farming in the Prairies

Figure 12.60

Cattle feed lots in Denver

Nutrient cycling within the prairie ecosystem

In Figure 11.29b, the small litter store reflects the relatively small amount of vegetative matter and low leaf fall. Litter decomposes into humus and nutrients are released to the soil, giving it good crumb structure. Moderate rainfall reduces loss from runoff. The large soil storage is a result of the weathering of rock and the presence of deep, rich chernozems (in the eastern and central prairies) which have accumulated a high proportion of humus (organic matter) in the temperate continental conditions. There is little or no leaching because the rainfall is exceeded by evaporation in the summer months.

How and why may this ecosystem change?

There is little tall-grass prairie left, and estimates by government agencies indicate that there is less than 34 per cent true mixed-grass prairie and less than 23 per cent true short-grass prairie in existence. This is mainly due to conversion to crop production, damming of major rivers together with flood control and irrigation systems, and in favourable areas the draining of wetlands for crops.

1 Natural conditions

- **Unpredictable rainfall** and **drought** have been a major factor in change

in the North American Prairies. Low rainfall in the 1930s allied to bad farming practices led to the creation of the American Dust Bowl; reduced the natural fodder for animals; and permitted an eastward extension of the short-grass prairies into the eastern tall grass.

- **Lightning** is a frequent cause of fire in the grasslands in summer. This destroys the surface vegetation; kills small animals; and damages the food supply. In the years following serious fires, lower bird numbers have been recorded, as many nest on the ground.

- **Bison herds** have been effective in change. They are heavy grazers and reduce the coarser medium grasses, leaving short grasses. In the spring and early summer when mosquitoes hatch they plague the bison, causing them to roll on the ground to reduce the itching! This forms depressions in the prairie surface. These bare soils may be re-colonised later by seeds carried by birds.

2 Human activity

- **Hunting** The earliest inhabitants were the Native Americans who hunted animals for food, using fire and traps to kill unselectively. With the coming of the Europeans and the introduction of the horse and the rifle, they were able

to kill large numbers of bison almost to extinction. It was only in the late 20th century that the number of bison began to increase, as a result of careful management in the National Parks, such as the Houck Ranch in South Dakota.

- **Cereal farming** Initially the prairie sod proved too difficult to remove using only wooden or iron ploughs, but after the 1840s this became possible following the development of the steel plough. Cereal crops were successfully introduced and were soon to be exported in huge quantities to Western Europe (Places 70, page 486). Overcultivation by the 1930s, when there was also a severe drought, led to extensive soil erosion and, especially in the southern Prairie states of Kansas and Oklahoma, the creation of the 'dust bowl'. Despite soil conservation methods on a large scale, which reduced some of the damaged areas, further droughts during the late 20th century caused an estimated loss of up to 1 m of topsoil in some ploughed areas.

- **Cattle ranching** became the main farming activity on the western Prairies (Figure 12.59) on land previously grazed by bison. Serious problems of overgrazing occurred in areas with lower rainfall Although irrigation is used to grow fodder cops in states such as Alberta, Montana and the western

Dakotas, the extraction of water has led to a lowering of the water table. Intensive ranching now takes place on huge feed lots close to railheads such as Denver (Figure 12.60). Young cattle are fattened in stockyards on grain transported from the eastern Prairies before being moved further east to the slaughter yards.

- **Mineral extraction** has increased since the 1970s, with extensive strip mining for coal and the construction of over 50 wells for the extraction of oil and natural gas. This, together with the roads, railways and pipelines needed to transport both workers and the minerals across the Prairies, has had an adverse effect on parts of the grasslands.
- **Other land uses** include areas set aside for military training and recreational activities including camping and bird watching.

Can the grassland ecosystem be saved?

Many species of grass found only in temperate grassland ecosystems have declined substantially in recent years. This has been blamed on a combination of factors including poor grazing management, the effects of fire, the spread of invasive, non-native plants and, in places, urban development. Estimates suggest that the prairie grasslands decreased by almost 30 million hectares between 1986 and 2002. The Grassland Reserve Programme (GRP) came into effect in 2002 as a result of the unlikely co-operation of the Nature Conservancy and the National Cattlemen's Beef Association (Figure 12.61). It was designed to be a buffer against the continuing loss of grassland areas and to ensure that the native prairie remained a viable, productive and important ecosystem for present, and future, generations. This was necessary if indigenous plant and animal species were to survive. The GRP programme imposes no regulations on grazing and allows private entities, such as ranching land trusts, to have rights of way over other properties.

Under the GRP, ranchers and other private grassland owners who enrolled had to agree to placing 10, 15, 20 or 30-year contracts on their land, prohibiting changes in land use, such as the growing of crops, and other activities incompatible with conserving the grassland ecosystem. In return, the landowners receive annual payments for short-term contracts or a one-time payment for which they agree to rights of way over their property. The GRP also provided additional resources to assist landowners wishing to restore former grassland areas.

All went well until 2007, when America's open plains and prairies were threatened by soaring global grain prices that increased the land's value as cropland. Grain prices were driven up partly by an increase in world demand for food, especially in the emerging economies of China and India, and partly by the American Congress voting to double the production of corn-based ethanol, a cleaner-burning fuel that can reduce greenhouse gas emissions. Both of these factors appear to work contrary to the government's programmes designed to conserve the grasslands. Landowners in many parts of the Prairies, especially the Dakotas, began converting to cropland some marginally productive areas that for decades – in some case centuries – had remained uncultivated as it would have been unprofitable to turn them into arable.

In 2008 a growing number of farmers chose not to re-enroll when their GRP contracts expired, potentially enabling over 2 million hectares of grassland, 15 per cent of the GRP total, to become available for cropland by 2010. With wheat that earned $4 or $5 a bushel in 2006 getting $12 a bushel in 2008, it is a case of short-term incentives overtaking long-term benefits. Conservationists warn that the hard work of the last few years could easily be undone and can only hope that grain prices will drop again in the near future – another example of global uncertainty.

Cattle ranchers believe that ...	Conservationists believe that ...
controlled burning to renew pasture should be allowed (this was a Native American custom)	there is too much burning; the prairie does not recover; burns are too frequent
overgrazing can be avoided by careful pasture management	new information gained from research will help both graziers and conservation
soil and water conservation are already practised	the prairie needs restoration to maintain its ecosystem
tourists, picnic sites and more roads will damage the environment	the prairie has already been damaged by cattle grazing

Figure 12.61

Conflict before the introduction of GRP in 2002

Further reference

Goudie, A.S. (2001) *The Nature of the Environment*, WileyBlackwell.

Money, D.C. (1978) *Climate, Soils and Vegetation*, Harper Collins.

 Oxfam's Cool Planet tropical rainforest: www.oxfam.org.uk/coolplanet/ontheline/explore/nature/trfindex.htm

West Tisbury School, Massachusetts biome:

www.blueplanetbiomes.org/climate.htm

World Wildlife Fund, ecoregions: www.worldwildlife.org/ecoregions/

Activities

1 **a** What is the 'climate' of a place? *(3 marks)*

   b What is the reason for wanting to classify climates? *(3 marks)*

   c Why do many geographers use the natural vegetation of a place as an indication of the climate? *(3 marks)*

   d For any **one** world climatic zone:

   i Name the climatic zone and identify **two** places which experience this climate.

   ii Draw and annotate a graph to show the pattern of temperature and precipitation which is typical of the climatic zone.

   iii Name the typical natural vegetation cover of the climatic zone and a typical zonal soil type. *(10 marks)*

   e Explain the causes of **one** of the climatic characteristics (temperature or precipitation) you have identified in **d ii**. *(6 marks)*

2 **a** Describe the climate of the areas which have natural temperate deciduous forests. *(3 marks)*

   b Draw a diagram to show the characteristic structure and composition of the vegetation of temperate deciduous forests. *(6 marks)*

   c Explain **one** way in which the vegetation of the temperate deciduous forests is adapted to the climate of the area. *(4 marks)*

   d Describe **one** zonal soil type of the temperate deciduous forests. *(4 marks)*

   e Why is there litter on the forest floor in the temperate deciduous forests? *(3 marks)*

   f Explain what has happened to most of the world's temperate deciduous forests since the settlement of these areas by people. *(5 marks)*

Exam practice: basic structured questions

3 Choose **one** of the world biomes that you have studied.

   a **i** Describe the main characteristics of the climate. *(5 marks)*

   ii Describe and explain the nutrient cycle in your chosen biome. You should include a diagram of the mineral nutrient cycle in your answer. *(6 marks)*

   b Describe the zonal soil of your chosen biome. *(6 marks)*

   c How is the natural vegetation of the biome adapted to the climatic conditions there? *(8 marks)*

4 **a** Describe the climate of the tropical rainforest. *(5 marks)*

   b Draw a diagram to show the composition and structure of the characteristic vegetation of the tropical rainforest. *(6 marks)*

   c Explain how the vegetation of the tropical rainforest is adapted to the climate of the area. *(8 marks)*

   d Describe **one** zonal soil type of tropical rainforest areas and explain how it developed. *(6 marks)*

Exam practice: structured questions

5 **a** Describe and account for the climatic pattern experienced in areas with a Mediterranean climate. *(8 marks)*

   b Describe the vegetation and explain **two** ways in which it is adapted to the climatic conditions of the Mediterranean. *(9 marks)*

   c How has the long-term presence of people affected the relationship between climate, soils and vegetation? *(8 marks)*

6 Choose **one** biome and answer the following questions about it.

   a Describe and explain the relationships within the nutrient cycle of the biome. *(10 marks)*

   b Describe **one** way the natural vegetation of the area is used by people and the effect of this use on the structure and composition of the vegetation. *(10 marks)*

   c How can damage due to past human uses of the biome be reduced? *(5 marks)*

Exam practice: essays

7 Choose **one** grassland biome and discuss the comparative importance of climate and human activity in influencing the nature of the vegetation cover. *(25 marks)*

8 Outline the basic features of **one** system of climate classification that you have studied and assess the importance of climate classifications in the study of geography. *(25 marks)*

13

Population

'There is a real danger that in the year 2000 a large part of the world's population will still be living in poverty. The world may become overpopulated and will certainly be overcrowded.'

Willy Brandt, *North–South: A Programme for Survival*, 1980

'In 1999, 600 million children in the world lived in poverty – 50 million more than in 1990.'

United Nations

In demography – the study of human population – it is important to remember that the situation is dynamic, not static. Population numbers, distributions, structures and movements constantly change in time, in space and at different levels (the micro-, meso- and macro-scales).

Distribution and density

Population distribution describes the way in which people are spread out across the Earth's surface. The distribution is uneven and there are often considerable changes over periods of time.

Population distributions can be shown by means of a dot map, where each dot represents a given number of people. For example, in Figure 13.1 this method effectively shows the concentration of people in the Nile valley in Egypt, where 99 per cent of the country's population live on 4 per cent of the total land area. However, Figure 13.1 is also misleading because it suggests, incorrectly, that areas away from the Nile are totally uninhabited. In fact, parts are populated, but have insufficient numbers to warrant a symbol. When drawing a dot map, therefore, it is important to select the best possible dot value and to bear in mind its limitations.

Figure 13.1

World distribution of population, 2008

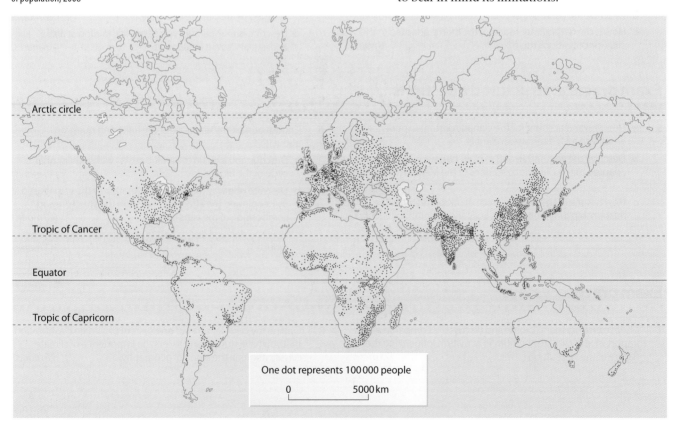

Arctic circle

Tropic of Cancer

Equator

Tropic of Capricorn

One dot represents 100 000 people

0   5000 km

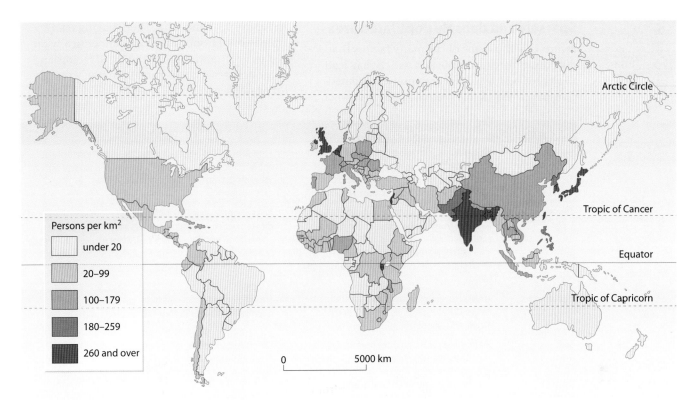

Figure 13.2

World density of
population, 2008

Persons per km²

under 20

20–99

100–179

180–259

260 and over

0   5000 km

Arctic Circle

Tropic of Cancer

Equator

Tropic of Capricorn

Population density describes the number of people living in a given area, usually a square kilometre (km²). Population densities are often shown by means of a choropleth map, of which Figures 13.2 and 13.5 are examples. Densities are obtained by dividing the total population of a country (or administrative area) by the total area of that country (or region). The densities are then grouped into classes, each of which is coloured lighter or darker to reflect lesser or greater density. Although these maps are easy to read, they hide concentrations of population within each unit area. Figure 13.2, for example, gives the impression that the population of Egypt is equally distributed across the country;

it also suggests that there is an abrupt change in population density at the national boundary. A poorly designed system of colouring or shading can make quite small spatial differences seem large – or make huge differences look smaller.

Figures 13.1 and 13.2 both show that there are parts of the world which are sparsely populated and others which are densely populated. One useful generalisation that may be made – remembering the pitfalls of generalisation (Framework 11, page 347) – is that, at the global scale, this distribution is affected mainly by physical opportunities and constraints; whereas, at regional and local scales, it is more likely to be influenced by economic, political and social factors.

Land accounts for about 30 per cent of the Earth's surface (70.9 per cent is water). Of the land area, only about 11 per cent presents no serious limitations to settlement and agriculture (Figure 13.3). Much of the remainder is desert, snow and ice, high or steep-sided mountains, and forest. Usually there are several reasons why an area is sparsely or densely populated.

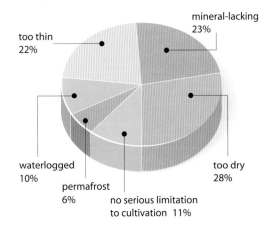

mineral-lacking
23%

too thin
22%

waterlogged
10%

permafrost
6%

no serious limitation
to cultivation 11%

too dry
28%

Figure 13.3

The uninhabitable Earth: how valuable are the world's soils for food production?

Sparsely and densely populated areas

Figure 13.4 lists some of the many factors that operate at the global scale and which may lead to an area being sparsely or densely populated. Compare these factors with the patterns shown in Figures 13.1 and 13.2. Then, having read Framework 11 opposite, comment critically on the accuracy and value of the listed factors and suggest, for each factor, an alternative example (or examples).

Figure 13.4

Major factors affecting population density

Factors	Sparsely populated areas	Densely populated areas
Physical	Rugged mountains where temperature and pressure decrease with height; active volcanoes (the Andes); high plateau (Tibet) and worn-down shield lands (the Canadian Shield, Places 48, page 377).	Flat, lowland plains are attractive to settlement (the Netherlands and Bangladesh, Places 48, page 377) as are areas surrounding some volcanoes (Mt Pinatubo, Case Study 1 and Mt Etna).
Climate	Areas receiving very low annual rainfall (the Sahara Desert, page 178); areas having a long seasonal drought or unreliable, irregular rainfall (the Sahel countries, page 223); areas suffering high humidity (the Amazon Basin, page 316); very cold areas, with a short growing season (northern Canada, page 333).	Areas where the rainfall is reliable and evenly distributed throughout the year; with no temperature extremes and a lengthy growing season (north-west Europe, page 223); where sunshine (the Costa del Sol) or snow (the Alps) is sufficient to attract tourists (Chapter 20); and areas with a monsoon climate (South-east Asia, page 239).
Vegetation	Areas such as the coniferous forests of northern Eurasia and northern Canada (page 330), and the rainforests of the tropics (page 317).	Areas of grassland tend to have higher population densities than places with dense forest or desert.
Soils	The frozen soils of the Arctic (the permafrost in Siberia, Case Study 5); the thin soils of mountains (Nepal); the leached soils of the tropical rainforest (the Amazon Basin, Places 66, page 480); also, increasingly large areas are experiencing severe soil erosion resulting from deforestation and over-grazing (the Sahel, Case Study 7).	Deep, humus-filled soils (the Paris Basin) and, especially, river-deposited silt (the Ganges delta, Places 67, page 481, and Nile delta, Places 73, page 490 – both favour farming).
Water supplies	Many areas lack a permanent supply of clean fresh water: mainly due either to insufficient, irregular rainfall or to a lack of money and technology to build reservoirs and wells or lay pipelines (Malawi, Places 97, page 611). Contamination by sewage, nitrates and salt.	Population is more likely to increase in areas with a reliable water supply. This may result from either a reliable, evenly distributed rain-fall (northern England) or where there is the wealth and technology to build reservoirs and to provide clean water (California). Places with heavy seasonal rainfall (the monsoon lands of South-east Asia, page 239) also support many people.
Diseases and pests	These may limit the areas in which people can live or may seriously curtail the lives of those who do populate such areas (malaria in central Africa; HIV/AIDS in southern Africa, Places 100, page 623).	Some areas were initially relatively disease- and pest-free; others had the capital and medical expertise to eradicate those which were a problem (the formerly malarial Pontine Marshes, near Rome).
Resources	Areas devoid of minerals and easily obtainable sources of energy rarely attract people or industry (Tibet).	Areas having, or formerly having, large mineral deposits and/or energy supplies (the Ruhr) often have major concentrations of population; these resources often led to the development of large-scale industry (South Wales, Places 87, page 570).
Communications	Areas where it is difficult to construct and maintain transport systems tend to be sparsely populated, e.g. mountains (Bolivia), deserts (the Sahara) and forests (the Amazon Basin and northern Canada).	Areas where it is easier to construct canals, railways, roads and air-ports have attracted settlements (the North European Plain), as have large natural ports which have been developed for trade (Singapore, Places 104, page 636).
Economic	Areas with less developed, subsistence economies usually need large areas of land to support relatively few people (although this is not applicable to South-east Asia). Such areas tend to fall into three belts: tundra (the Lapps), desert fringes (the Rendille,Places 65, page 479) and tropical rain-forests (shifting cultivators, Places 66, page 480).	Regions with intensive farming or industry can support large numbers of people on a small area of land (as in the Netherlands, Places 71, page 487).
Political	Areas where the state fails to invest sufficient money or to encourage development – either economically or socially (in parts of the interior of Brazil, Places 38).	Decisions may affect population distribution, e.g. by creating new cities, such as Brasilia; or by opening up 'pioneer' lands for development, as in Israel (page 391).

Over 50 per cent of the world's population live in six countries: China, India, the USA, Indonesia, Brazil and Pakistan.

The study of an environment, whether natural or altered by human activity, involves the study of numerous different and interacting processes. The relative importance of each process may vary according to the scale of the study, i.e. global or **macro-scale**; intermediate or **meso-scale**; and local or **micro-scale**. It may also vary according to the timescale chosen, i.e. whether processes are studied through **geological time**, **historical time**, or **recent time**.

In the study of soils (Chapters 10 and 12), it is clearly climate that tends to impose the greatest influence upon the formation and distribution of the major global (zonal) types (the podsol and chernozem). At the regional level, rock type may be the major influencing factor (Mediterranean areas with their terra rossas and rendzinas). Within a small area, such as a river valley with homogeneous climate and rock type, relief may be dominant (the catena, pages 261 and 276).

In the study of erosion, time is a major variable: a stretch of coastline may be eroded by the sea during a period of several decades or centuries; footpath erosion may occur during a single summer.

A common problem with spatial and time scales, as with models (Framework 12, page 352), is that a chosen level of detail may become inappropriate to all or part of the problem under study: it may become either too large and generalised, or too small and complex. For example, population distributions and densities may be studied at a variety of spatial and time scales. At the world scale (Figures 13.1 and 13.2), the pattern shown is so general and deterministic that it may lead the student into an over-simplified understanding of the processes that produced the apparent distribution and/or density. Such generalised patterns usually break down into something more complex when examined at a more local level or over a period of time.

Although it may often be easier to identify and account for distributions, densities, anomalies and changes at the national level, it is more difficult in the case of a country the size of Brazil (Figure 13.5) than it is for a smaller country such as Uruguay. It is often only when looking at a smaller region (Figure 13.6) or an urban area (Figure 13.7), perhaps over a relatively short time period, that the complexities of the various processes can be readily understood.

Places **38** Brazil: population densities at the national level

Even a quick look at the population density map of Brazil (Figure 13.5) shows a relatively simple, generalised pattern. Over 90 per cent of Brazilians live in a discontinuous strip about 500 km wide, adjacent to the east coast. This strip accounts for less than 25 per cent of the country's total area. The density declines very rapidly towards the north-west, where several remote areas are almost entirely lacking in permanent settlement.

The area marked **1A** on Figure 13.5 is the dry north-east (the Sertao). Here the long and frequent water balance deficit (drought), high temperatures and poor soils combine to make the area unsuitable for growing high-yield crops or rearing good-quality animals. The Sertao also lacks known mineral or energy reserves; communications are poor; and the basic services of health, education, clean water and electricity are lacking. Although birth rates are exceptionally high (many mothers have more than ten babies), there is a rapid outward migration to the urban areas (page 366), a high infant mortality rate (page 354), and a short life expectancy (page 353).

Area **1B** is the tropical rainforest, drained by the River Amazon and its tributaries. Here the climate is hot, wet and humid; rivers flood annually; and there is a high incidence of disease. In the past, the forest has proved difficult to clear, but once the protective trees have gone, soils are rapidly leached and become infertile. Land communications are difficult to build and maintain. The area has suffered, as has **1A**, from a lack of federal investment until recently when parts have been developed commercially for ranching, logging and growing soya.

Figure 13.5

Population density in Brazil: the national scale

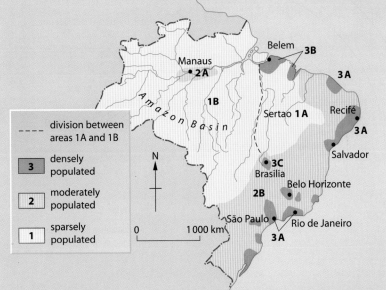

There are, however, two anomalies in Amazonia. The first is a zone along the River Amazon centred on Manaus (**2A** on Figure 13.5). Originally a Portuguese trading post, Manaus has had two growth periods. The first was associated with the rubber boom at the turn of the 19th/20th centuries, while the second began in the 1980s with the development of tourism and the granting of its free port status (Places 104, page 636). The second anomaly has followed the recent exploitation of several minerals (iron ore at Carajas and bauxite at Trombetas) and energy resources (hydro-electricity at Tucuri).

The more easterly parts of the Brazilian Plateau are moderately populated (area **2B**). The climate is cooler and it is considerably healthier than on the coast and in the rainforest. The soil, in parts, is a rich terra rossa (page 274) which here is a weathered volcanic soil ideal for the growing of coffee. However, rainfall is irregular with a long winter drought; communications are still limited; and federal investment has been insufficient to stimulate much population growth.

Except where the highland reaches the sea, the eastern parts of the plateau around São Paulo and Belo Horizonte and the east coast have the highest population densities (area **3A**). Although the coastal area is often hot and humid, the water supply is good. Several natural harbours proved

ideal for ports and this encouraged trade and the growth of industry. Salvador, the first capital, was the centre of the slave trade. Rio de Janeiro became the second capital, developing as an economic, cultural and administrative centre. More recently, it has received increasing numbers of tourists from overseas and migrants from the north of Brazil.

One of the world's fastest-growing cities is São Paulo. The cooler climate and terra rossa soils initially led to the growth of commercial farming based on coffee. Access to minerals such as iron ore and to energy supplies later made it a major industrial centre. The São Paulo region has had high levels of federal investment, leading to the development of a good communications network and the provision of modern services.

Area **3B** is another focus of recent growth based on the discovery and exploitation of vast deposits of iron ore and bauxite, the construction of hydro-electric power stations and the advantages of access along the coastal strip and Amazon corridor. **3C** is the new federal capital, Brasilia, built in the early 1960s to try to redress the imbalance in population density and wealth between the south-east of the country and the interior. Figures 13.6 and 13.7 show population densities at different levels of scale from that of Brazil.

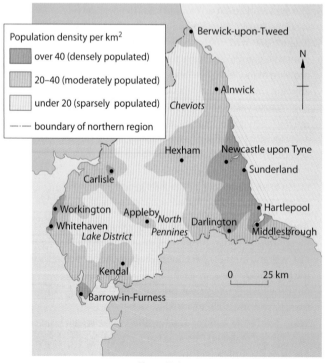

Figure 13.6

Population density in the 'North' economic planning region of England: the regional scale

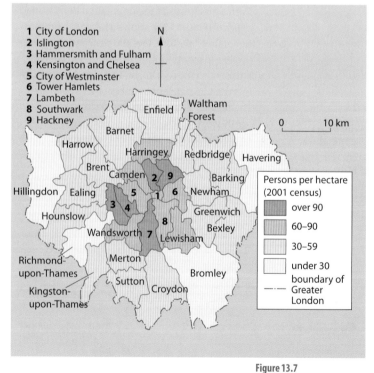

Figure 13.7

Population density in Greater London: the urban scale

Lorenz curves

Figure 13.8

A Lorenz curve: the distribution of world population in mid-2008

Lorenz curves are used to show inequalities in distributions. Population, industry and land use are three topics of interest to the geographer which show unequal distributions over a given area. Figure 13.8 illustrates the unevenness of population distribution over the world. The diagonal line represents a perfectly even distribution, while the concave curve (it may be convex in other examples) illustrates the degree of concentration of population within the various continents. The greater the concavity of the slope, the greater the inequality of population distribution (or industry, land use, etc.).

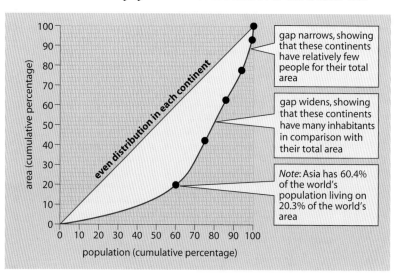

Continents ranked in descending order of population (1998)	Population (%)	Population (cumulative %)	Area (%)	Area (cumulative %)
Asia	60.4	60.4	20.3	20.3
Africa	14.4	74.8	22.3	42.6
Europe/ Russian Federation	11.0	85.8	20.1	62.7
Latin America	8.6	94.4	15.2	77.9
North America	5.1	99.5	15.8	93.7
Oceania	0.5	100.0	6.3	100.0

Population changes in time

It has already been stated (page 344) that populations are dynamic, i.e. their numbers, distributions, structure and movement (migration) constantly change over time and space. Population change is another example of an open system (Framework 3, page 45) with inputs, processes and outputs (Figure 13.9).

Birth rates, death rates and natural increase

The total population of an area is the balance between two forces of change: **natural increase** and **migration** (Figure 13.9). The natural increase is the difference between birth rates and death rates. The **crude birth rate** is the number of live births per 1000 people per year and the **crude death rate** is the number of deaths per 1000 people per year. Throughout history, until the last few years in a small number of the economically most developed countries, birth rates have nearly always exceeded death rates. Exceptions have followed major outbreaks of disease (the bubonic plague and AIDS, page 622) or wars (as in Rwanda). Any natural change in the population, either an increase or a decrease, is usually expressed as a percentage and referred to as the **annual growth rate**. Population change is also affected by migration. Although migration does not affect world population totals, it does affect the way people are distributed across the world. Migration leads to *either* an increase in the population – when the number of immigrants exceeds the number of emigrants (as in Spain and Congo) – *or* a decrease in population – when the number of emigrants exceeds the number of immigrants (as in Iraq and Rwanda).

Figure 13.9

Simple model showing population change as an open system

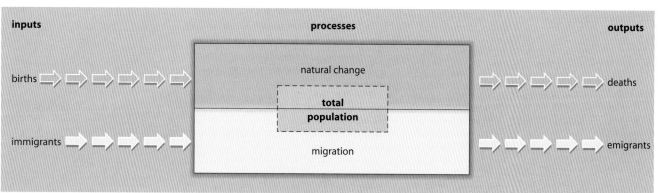

The demographic transition model

The demographic transition model describes a sequence of changes over a period of time in the relationship between birth and death rates and overall population change. The model, based on population changes in several industrialised countries in western Europe and North America, suggests that *all* countries pass through similar demographic transition stages or **population cycles** – or will do, given time. Figure 13.10 illustrates the model and gives reasons for the changes at each transition stage. It also gives examples of countries that appear to 'fit' the descriptions of each stage.

Figure 13.10

The demographic transition model

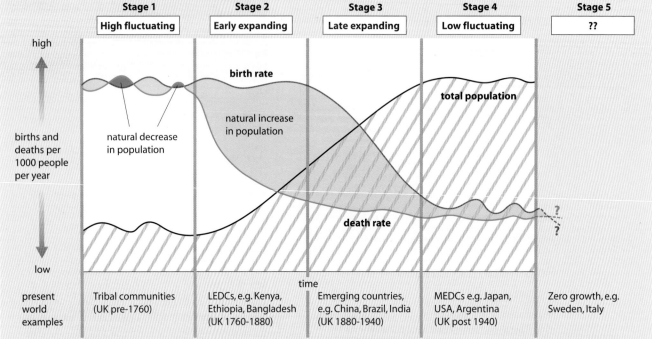

Stage 1: Here both birth rates and death rates fluctuate at a high level (about 35 per 1000) giving a small population growth.

Birth rates are high because:

- no birth control or family planning
- so many children die in infancy that parents tend to produce more in the hope that several will survive
- many children are needed to work on the land
- children are regarded as a sign of virility
- Some religious beliefs (Roman Catholics, Muslims and Hindus) encourage large families.

High death rates, especially among children, are due to:

- disease and plague (bubonic, cholera, kwashiorkor)
- famine, uncertain food supplies, poor diet
- poor hygiene: no piped, clean water and no sewage disposal
- little medical science: few doctors, hospitals, drugs.

Stage 2: Birth rates remain high, but death rates fall rapidly to about 20 per 1000 people giving a rapid population growth.

The fall in death rates results from:

- improved medical care: vaccinations, hospitals, doctors, new drugs and scientific inventions
- improved sanitation and water supply
- improvements in food production, both quality and quantity
- improved transport to move food, doctors, etc
- a decrease in child mortality.

Stage 3: Birth rates now fall rapidly, to perhaps 20 per 1000 people, while death rates continue to fall slightly (15 per 1000 people) to give a slowly increasing population.

The fall in birth rates may be due to:

- family planning: contraceptives, sterilisation, abortion and government incentives
- a lower infant mortality rate leading to less pressure to have so many children
- increased industrialisation and mechanisation meaning fewer labourers are needed
- increased desire for material possessions (cars, holidays, bigger homes) and less desire for large families
- an increased incentive for smaller families
- emancipation of women, enabling them to follow their own careers rather than being solely child-bearers.

Stage 4: Both birth rates (16 per 1000) and death rates (12 per 1000) remain low, fluctuating slightly to give a steady population.

(Will there be a **Stage 5** where birth rates fall below death rates to give a declining population? Evidence suggests that this is occuring in several Western European countries although growth rates here are augmented by immigration.)

Like all models, the demographic transition model has its limitations (Framework 12, page 352). It failed to consider, or to predict, several factors and events:

1 Birth rates in several of the most economically developed countries have, since the model was put forward, fallen below death rates (Germany, Sweden). This has caused, for the first time, a population decline which suggests that perhaps the model should have a fifth stage added to it.

2 The model, being more or less Eurocentric, assumed that in time all countries would pass through the same four stages. It now seems unlikely, however, that many of the economically less developed countries, especially in Africa, will ever become industrialised.

3 The model assumed that the fall in the death rate in Stage 2 was the consequence of industrialisation. Initially, the death rate in many British cities rose, due to the insanitary conditions which resulted from rapid urban growth, and it only began to fall after advances were made in medicine. The delayed fall in the death rate in many developing countries has been due mainly to their inability to afford medical facilities. In many countries, the fall in the birth rate in Stage 3 has been *less* rapid than the model suggests

due to religious and/or political opposition to birth control (Brazil), whereas the fall was much *more* rapid, and came earlier, in China following the government-introduced 'one-child' policy (Case Study 13).

4 The timescale of the model, especially in several South-east Asian countries such as Hong Kong and Malaysia, is being squashed as they develop at a much faster rate than did the early industrialised countries.

The model can be used:

■ to show how the population growth of a country changes over a period of time (the UK, in Figure 13.11)

■ to compare rates of growth between different countries at a given point in time (Figure 13.12).

Figure 13.11 shows certain changes in Britain:

■ 1700–1760: high birth and death rates giving a slow natural increase.

■ 1760–1880: a rapidly falling death rate and a high birth rate giving a fast natural increase.

■ 1880–1940: rapidly declining birth and death rates giving a slower natural increase.

■ 1940–2000: low, fluctuating birth and death rates giving a small natural increase.

■ Since 2000: a rising birth rate amongst new and first generation immigrants giving a faster natural increase.

Figure 13.11

The demographic transition cycle: changes in Britain's population, 1700–2008

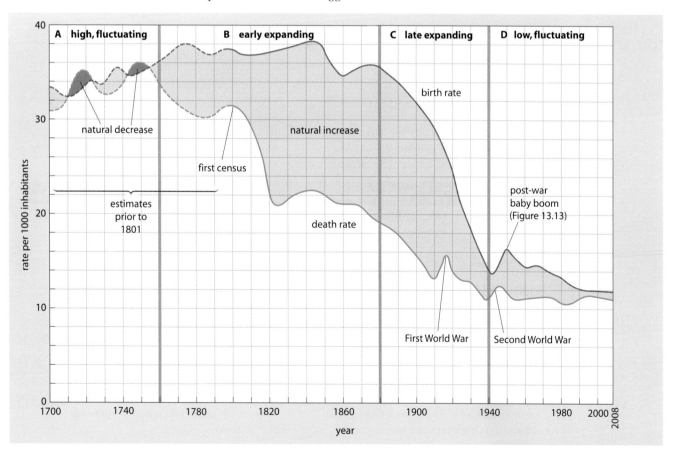

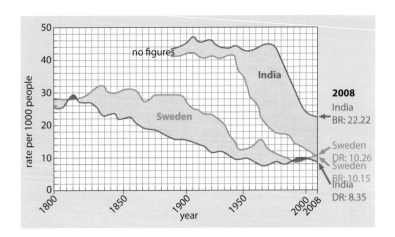

Figure 13.12 shows how Sweden has long since reached Stage 4 of the demographic transition model – a characteristic of most economically more developed countries – whereas India is still at Stage 3 – a characteristic of many economically less developed countries (remember that some of the least economically developed countries are still at Stage 2).

Population structure

The rate of natural increase or decrease, resulting from the difference between the birth and death rates of a country, represents only one aspect of the

Figure 13.12

A comparison between the demographic cycles of Sweden and India, 1800–2008

Framework **12** Models

Models form an integral and accepted part of present-day geographical thinking and teaching. Nature is highly complex and, in an attempt to understand this complexity, geographers try to develop simplified models of it.

Chorley and Haggett described a model as:

> a simplified structuring of reality which presents supposedly significant features or relationships in a generalised form ... as such they are valuable in obscuring incidental detail and in allowing fundamental aspects of reality to appear.

They stated that a model:

> can be a theory or a law, an hypothesis or structured idea, a role, a relationship, or equation, a synthesis of data, a word, a graph, or some other type of hardware arranged for experimental purposes.

A good model will stand up to being tested in the real world and should fall between two extremes:

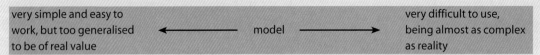

| very simple and easy to work, but too generalised to be of real value | ← model → | very difficult to use, being almost as complex as reality |

To achieve this balance, several – though sometimes only one – critical criteria or variables are selected as a basis for the model. For example, J.H. von Thünen (page 471) chose distance from a market as his critical variable and then tried to show the relationship between this variable and the intensity of land use. If necessary, other variables may be added which, as in the case of von Thünen's navigable river and a rival market, may add both greater reality and greater complexity. Models can be used in all fields of geography. Some applications are shown in the following table.

Physical (landforms)	Page	Climate, soils and vegetation	Page	Human and economic	Page
beach profile	143	atmospheric circulation	226	cities (Burgess)	420
slope development	51	heat budget	209	land use (von Thünen)	471
corrie development	110	seres	286	industrial location (Weber)	557
sand dune development	157	food chains and tropic levels	296	settlement size/distribution (Christaller)	407
glacier system	106	depressions	230	gravity models	410
drainage basins	58	soil profiles (e.g. podsols)	332	demographic transition	350
limestone scenery	198	soil catena	261	economic growth (Rostow)	615

Throughout this book, models and theories are presented; their advantages and limitations are examined; and their applications to real-world situations are demonstrated, together with their usefulness and accuracy in explaining that situation.

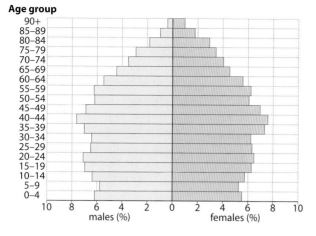

Figure 13.13

Constructing the population pyramid for the UK, mid-2007

Age group	Males Number (000s)	Males Percentage	Females Number (000s)	Females Percentage
0–4	1781	6.13	1696	5.65
5–9	1691	5.80	1618	5.36
10–14	1835	6.31	1746	5.79
15–19	2003	6.78	1885	6.25
20–24	2057	7.08	1952	6.47
25–29	1934	6.55	1916	6.35
30–34	1888	6.50	1891	6.27
35–39	2183	7.51	2223	7.37
40–44	2268	7.81	2315	7.68
45–49	2040	7.02	2090	6.93
50–54	1790	6.16	1835	6.08
55–59	1798	6.19	1854	6.15
60–64	1657	5.70	1736	5.76
65–69	1263	4.35	1364	4.52
70–74	1074	3.69	1226	4.06
75–79	841	2.89	1082	3.59
80–84	557	1.91	859	2.85
85–89	286	0.98	565	1.87
90+	106	0.36	311	1.03
	Total 29 054		**Total 30 162**	

study of population structure. A second important aspect is population. This is important because the make-up of the population by its age and gender, together with its life expectancy, has implications for the future growth, economic development and social policy of a country. **Life expectancy** is the number of years that the average person born in a given area may expect to live. Differences in language, race, religion, family size, etc. can all affect a country's socio-economic welfare.

Population pyramids

The population structure of a country is best illustrated by a **population** or **age–gender pyramid**. The technique normally divides the population into 5-year age groups (e.g. 0–4, 5–9, 10–14) on the vertical scale, and into males and females on the horizontal scale (Figure 13.13). The number in each age group is given as a percentage of the total population and is shown by horizontal bars, with males located to the left and females to the right of the central axis. As well as showing past changes, the pyramid can predict both short-term and long-term future changes in population.

Whereas the demographic transition model shows only the natural increase or decrease resulting from the balance between births and deaths, the population pyramid shows the effects of migration, the age and gender of migrants (Figure 13.45) and the effects of large-scale wars and major epidemics of disease. Figure 13.13 is the population pyramid for the United Kingdom in mid-2007.

Figure 13.14

Population pyramids characteristic of each stage of the demographic transition model

Notice the following:
- a narrow pyramid showing approximately equal numbers in each age group
- a low birth rate (meaning fewer school places will be needed) and a low death rate (suggesting a need for more elderly people's homes) which together indicate a steady, almost static, population growth
- the greater number of boys in the younger age groups (a higher birth rate) but more females than males in the older age groups (women having the longer life expectancy)
- a relatively large proportion of the population in the pre- and post-reproductive age groups, and a relatively small number in the 15–64 age groups which produce most of the national wealth (see dependency ratios, page 354).

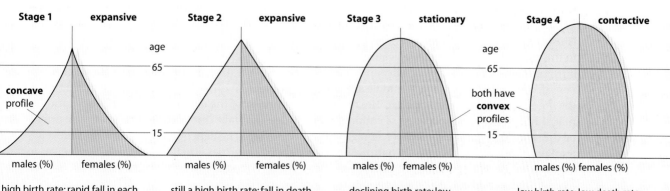

Stage 1 expansive — **Stage 2** expansive — **Stage 3** stationary — **Stage 4** contractive

high birth rate; rapid fall in each upward age group due to high death rates; short life expectancy

still a high birth rate; fall in death rate as more living in middle age; slightly longer life expectancy

declining birth rate; low death rate; more people living to an older age

low birth rate; low death rate; higher dependency ratio; longer life expectancy

economically least developed countries ⟶ economically more developed countries

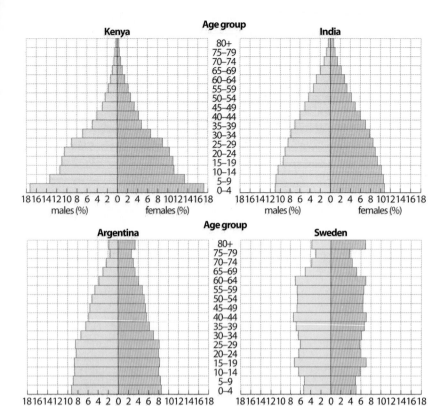

Figure 13.15

Population pyramids for four selected countries, mid-2007

A model has also been produced to try to show the characteristics of four basic types of pyramid (Figure 13.14). As with most models, many countries show a transitional shape which does not fit precisely into any pattern. Figure 13.15 shows the pyramids for *selected* countries – chosen because they *do* conform closely to the model! Stage 1 only occurs in isolated tribal communities.

Stage 2 Kenya's pyramid has a concave shape, showing that the birth rate is very high. Almost half the inhabitants (42 per cent) are under 15 years old (the corresponding figure for 1990 was 51 per cent); there is a rapid fall upwards in each age group showing a high death rate (including infant mortality) and a low life expectancy, with less than 3 per cent who can expect to live beyond 65. The **infant mortality rate** is the average number of children out of every 1000 born alive who die under the age of one year.

Stage 3 India appears to have reached Stage 3 (shown by the more uniform sides). All pyramids in this stage have a broad base indicating a high birth rate but, as the infant mortality and death rates decline, more people reach middle age and the life expectancy is slightly longer. The result is that although the actual numbers of children may be the same, they form a smaller percentage of the total population (as shown by the narrower base). The large youthful population will soon enter the reproductive period and become economically

active. India has 32 per cent under 15; and 5 per cent over 65 (the corresponding figures for 1990 were 39 and 3 per cent respectively).

Stage 4 Argentina has probably just reached this stage as its birth rate is declining – as shown by the almost equal numbers in the lower age groups. As the death rate is much lower, more people are able to live to a greater age, and the actual growth rate becomes stable. Argentina has 25 per cent under 15; and 11 per cent over 65 (the 1990 figures were 26 and 6).

Stage 5 Sweden has a smaller proportion of its population in the pre-reproductive age groups (16 per cent under 15) and a larger proportion in the post-reproductive groups (18 per cent over 65), indicating low birth, infant mortality and death rates and a long life expectancy (the equivalent figures for 1990 were 22 and 16). As the numbers entering the reproductive age groups decline there will be, in time, a fall in the total population.

Dependency ratios

The population of a country can be divided into two categories according to their contribution to economic productivity. Those aged 15–65 years are known as the **economically active** or **working population**; those under 15 (the youth dependency ratio) and over 65 (the old age dependency ratio) are known as the **non-economically active population**. (Perhaps in Britain the division should be made at 16, the school-leaving age; in developing countries, however, the cut-off point is much lower as many children have to earn money from a very young age.)

The dependency ratio can be expressed as:

$$\frac{\text{children (0–16) and}}{\text{those of working age}} \times 100$$

e.g. UK 1971 (figures in millions):

$$\frac{13\ 387\ +\ 7\ 307}{31\ 616} \times 100 = 65.45$$

So for every 100 people of working age there were 65.45 people dependent on them.

By 2007 the dependency ratio had changed to:

$$\frac{11\ 537\ +\ 11\ 344}{37\ 707} \times 100 = 60.68$$

So the drop in the number of children was more than offset by the larger increase in the number of the elderly (the dependency ratio does not take into account those who are unemployed). The dependency ratio for most developed countries is between 50 and 70, whereas for less economically developed countries it is often over 100.

Trends in population growth

1 Global trends

In Mother Earth's 46 years (Places 1, page 9), it 'was only in the last hour that man began to live in settlements' and 'the human population slowly started to increase'. In the absence of any census, the population has been crudely estimated to have been about 500 million in 1650. It was only after the Industrial Revolution in Western Europe, one minute ago in the Earth's history, that numbers began to 'multiply prodigiously' in what is now the developed world while the so-called 'population explosion' only extended to developing continents after the middle of last century (Figure 13.18).

The United Nations Fund for Population Activities (UNFPA) designated October 1999 as the date when the world's population reached six billion (6,000 million). This 'celebration' – and many people would disagree with that – was fictitious as no one knew the exact figure, due to either inaccurate or non-existent census figures and the often non-recorded migration of people. Bearing in mind the approximation of population figures (Framework 15, page 448), the world clock suggests that numbers are increasing by:

	World	MEDCs	LEDCS
Year	82 million	2 million	80 million
Month	570 000	7000	563 000
Day	18 773	233	18 540
Minute	156	4	152

During 2008, the UN claimed that 139 million babies would be born and that 57 million people would die giving the natural increase of 82 million shown in the table above.

Fertility rates in many economically developing countries are slowly beginning to decline although they remain considerably higher than those in developed countries (Figure 13.16). The **fertility rate** is the number of children born to women of child-bearing age. The UN claim that the annual growth rate of the world's population, which had been 2.1 per cent between 1964 and 1970, had fallen to 1.2 per cent by 2008 – a fall mainly credited to China's 'one child per family policy' (Case Study 13). The consequence of this slowing-down has led to the present revised prediction that the world's population will now only pass the 7 billion mark in 2012, rather than in that same year reaching the 7.6 billion predicted in 1992, or the 8.4 billion had the 1950–80 growth rate not declined. The UNFPA now predicts that the annual growth rate will fall to 1.0 per cent in 2020 and to 0.5 per cent in 2050. By that later date, the world's expected population is predicted to be between 7.41 billion (lowest) to 10.63 billion (highest), with a medium variant of 8.92 billion (Figure 13.17).

2 Regional trends

What these figures fail to show is the marked variations between different parts of the world, especially between the economically developed and the economically developing continents, bearing in mind that it is likely there will also be considerable variations within the continents themselves. At present, the average population growth rate for all countries referred to as developing is 1.92 per cent per year compared with only 0.52 per cent for those described as developed (Figures 13.16, 13.17 and 13.18). In 2008 (the 1998 comparable figures are shown in brackets) the population of Asia was increasing by over 48 million a year (50); Africa by 23 million (17) and Latin America by nearly 9 million (nearly 8). Africa had the highest growth rate at 2.4 per cent per year compared with Europe which actually showed a decline of 0.01 per cent.

Figure 13.16

Global and regional trends in population growth, 2008

In global terms the major trend has been a decline in the rate of population growth from a peak of 2.1 per cent between 1965 and 1970 to approximately 1.2 per cent in 2008, although there are still 82 million more people alive at the end of each year. While the distinction between the low/no population growth of most of the 'developed' countries and the high population growth of the developing countries continues, a major feature of the last five decades has been the widening trajectories of the least developed countries. Making broad generalisations to a complex world pattern, low birth rates in Europe have led to the very real prospect of a population decrease, despite continuing net immigration. The USA and Canada are anomalies for MEDCs as they have a robust population growth, mainly due to their high immigration rates. Of the developing countries, those in Latin America and the Middle East have the lowest (though still moderately high) growth rates while the least developed countries, mostly located in sub-Saharan Africa, have by far the greatest growth. Asia has extremes from low growth rates in the newly industrialised countries (NICs, page 578) and China (one-child policy) to the continued rapid growth in parts of India.

Figure 13.17

Predictions on world population growth, after 2008

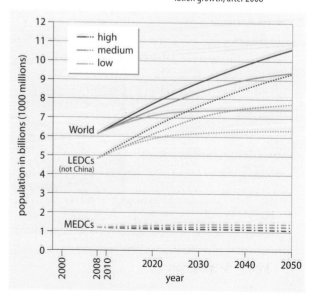

Figure 13.18

World population growth (2008 data)

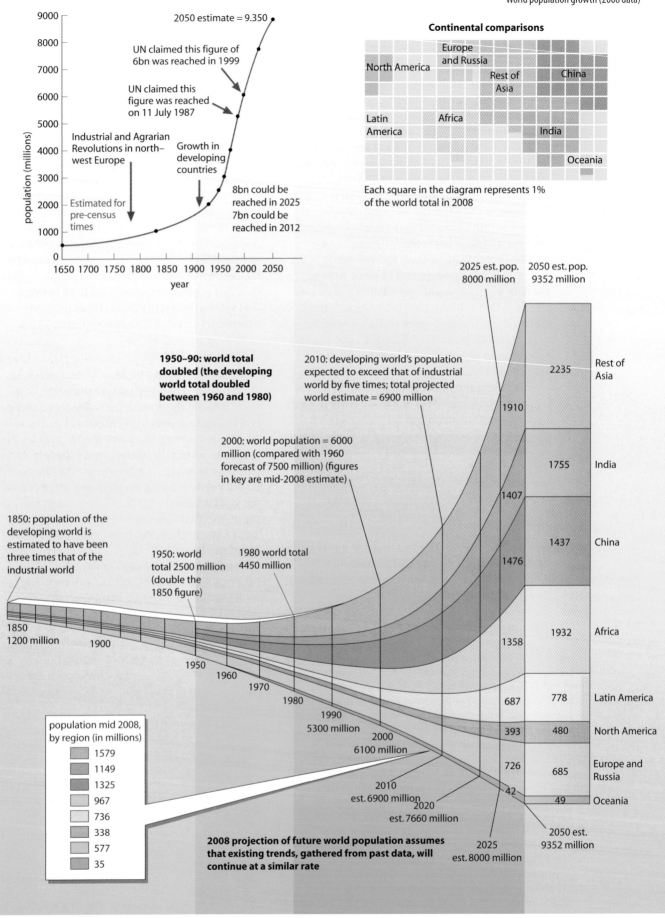

Continental comparisons

Each square in the diagram represents 1% of the world total in 2008

2050 estimate = 9.350

UN claimed this figure of 6bn was reached in 1999

UN claimed this figure was reached on 11 July 1987

Industrial and Agrarian Revolutions in north–west Europe

Growth in developing countries

Estimated for pre-census times

8bn could be reached in 2025
7bn could be reached in 2012

population (millions)

year

2025 est. pop. 8000 million

2050 est. pop. 9352 million

1950–90: world total doubled (the developing world total doubled between 1960 and 1980)

2010: developing world's population expected to exceed that of industrial world by five times; total projected world estimate = 6900 million

2000: world population = 6000 million (compared with 1960 forecast of 7500 million) (figures in key are mid-2008 estimate)

1850: population of the developing world is estimated to have been three times that of the industrial world

1950: world total 2500 million (double the 1850 figure)

1980 world total 4450 million

1850
1200 million

1900

1950

1960

1970

1980

1990
5300 million

2000
6100 million

2010
est. 6900 million

2020
est. 7660 million

2025
est. 8000 million

2050 est.
9352 million

2235   Rest of Asia

1910

1755   India

1407

1437   China

1476

1358   Africa

1932

687   778   Latin America

393   480   North America

726   685   Europe and Russia

42   49   Oceania

population mid 2008, by region (in millions)

1579
1149
1325
967
736
338
577
35

2008 projection of future world population assumes that existing trends, gathered from past data, will continue at a similar rate

births per woman
fewer than 1.5
1.5–2.1
2.2–2.9
3.0–4.9
5.0 or more

Figure 13.19

Total fertility
rate, 2008

3 Birth rates, total fertility rates (TFR) and replacement rates

The world's crude birth rate in 2008 was 21 per 1000. Germany had the lowest with 8 per 1000 followed by several other Western European countries together with Taiwan and Japan, which had 9 per 1000. In contrast, of 23 countries with a birth rate exceeding 40 per 1000, 21 were located in Africa.

The **total fertility rate (TFR)** is the average number of children a woman is likely to have if she lives to the end of her child-bearing age, based on current birth rates. The present world average is 2.6, varying between 1.6 in developed countries and China to 3.2 in developing countries and 4.7 in those that are the least developed (Figure 13.19). The TFR is one of the best indicators of future population growth. In most economically developed countries the TFR is low and still declining and, while it is still much higher in economically less developed countries, it appears that changing attitudes there will eventually lead to lower TFR in the future. High birth and fertility rates have been considered characteristic of 'underdevelopment'. Indeed although there does seem to be a close correlation between a country's birth rate and its GNP (Framework 19, page 612), the UN have claimed that 'a high birth rate is a consequence, not a cause, of poverty'. Typically, the lower the use of contraceptives, the higher the TFR, and the higher the level of female education, the lower the TFR (Figure 13.20). Government policies can also have an enormous impact on the number of children women are likely to have (Places 39).

It is now recognised that the three key factors influencing fertility decline are improvements in family planning programmes, in health care, and in women's education and status (arguably in that order, although they are all interrelated). The major world movement is now towards 'children by choice rather than chance', a goal that can only be achieved by giving women reproductive options.

The causes of this unmet need for contraception include lack of knowledge of contraception methods and/or sources of supply; limited access to and low quality of family planning services; lack of education, especially among women; cost of contraception commodities; disapproval of husbands and family members; and opposition by religious groups. Improvements in health care include safer abortions and a reduction in infant mortality – the latter meaning that fewer children need to be born as more of them survive. Improved education raises the status of women and postpones the age of marriage. Several governments, especially in South-east Asia, have attempted in recent years to encourage couples to have fewer children unless, as in the case of Singapore, prospective parents belonged to selective groups (Places 39). Other governments, notably that of China (Case Study 13), have attempted to reduce birth rates through coercion, a method that has not gained international approval.

Figure 13.20

Family planning as
a human right

Reaping the rewards of family planning

The freedom to choose how many children to have, and when, is a fundamental human right. Better access to safe and affordable contraceptive methods is key to achieving the MDGs [page 609]. Family planning has proven benefits in terms of gender equality, maternal health, child survival and preventing HIV [page 622]. It can also reduce poverty and promote economic growth by improving family well-being, raising female productivity and lowering fertility. It is one of the most cost-effective investments a country can make towards a better quality of life. Limited access to contraception, in contrast, constrains women's opportunities to pull themselves and their families out of poverty [page 609]. Reproductive health, including voluntary family planning, should be at the centre of initiatives to promote the human rights of women and should replace earlier efforts that focused more on curbing rapid population growth, in some cases at the expense of women's rights. Freedom to make reproductive decisions is essential for achieving gender equality and sustainable development.

Source: UNFPA, 2007

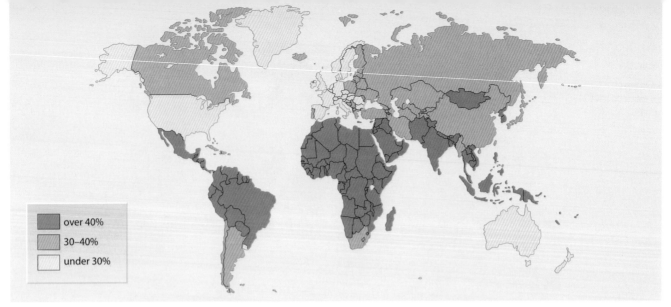

Figure 13.21

Percentage of total population under 15 years, 2008

High birth and fertility rates result in a high proportion of the total population being aged 15 or under (see Kenya's population structure in Figure 13.15). Countries where this occurs – many of them in Africa, Latin America and southern Asia (Figure 13.21) – are likely, in the future, to:

■ need greater health care and education – two services many can ill afford

■ have more women reaching child-bearing age. In contrast, many of the more economically developed countries have such low birth and fertility rates that there is a growing problem of 'too few' rather than 'too many' children: is this the possible Stage 5 in the demographic transition model (Figure 13.10)?

The UN stated that, in 2008, there were 71 countries with a TFR below 2.1, the figure needed by a country to replace its population

(the **replacement rate**, said to be 2.1 children per woman, is when there are just sufficient children born to balance the number of people who die). Throughout history, except during times of plague or war, the replacement rate has always been exceeded – hence the growth in world population. At present, many African countries have a TFR of over 5.0 (in Liberia, Somalia and Uganda it is 6.8) whereas in most European and eastern Asian countries it is below 1.8 (Taiwan 1.1, Japan and Italy 1.3 and Singapore and Spain 1.4). Where the replacement rate is not being met, there are fears that, in time, there will be too few consumers and skilled workers to maintain national economies and to support an ageing population; a reduction in any competitive advantage in science and technology; and schools and colleges closed for a lack of students.

Places 39 Singapore: family planning

When, in 1965, Singapore had a birth rate of 29.5 and a TFR of 4.6, the government introduced a massive family planning scheme in which the main objectives were:

- to establish family planning clinics and to provide contraceptives at minimal charge

- to advertise through the media the need for, and the advantages of, smaller families – a voluntary 'stop at two' policy

- to legislate so that under certain circumstances both abortions and sterilisation could be allowed

- to introduce social and economic incentives such as paid maternity leave, income tax relief, housing priority, cheaper health care and free education, all of which would cease as the size of a family grew.

By 1995 the policy had been so successful that the birth rate had fallen to 15.2 and the TFR to 1.7

so that already there was an insufficient supply of labour to fill the job vacancies and fewer people to support an increasingly ageing population – hence the changed slogan of 'stop at three if you can afford three'. The government became concerned that it was the middle-class elite that was having fewest children, partly because women were pursuing their own careers and either staying single, or marrying and having children at a later age. As a result, female graduates were encouraged to have three or more children through financial benefits and larger tax exemptions, while low-income non-graduates only received housing benefits if they stopped at two children. This seems to have done little to reverse the trend as, in 2008, the birth rate had fallen to 5.8 and the TFR to 1.4, and the government sent out Valentine Day messages encouraging people to 'make love not money'.

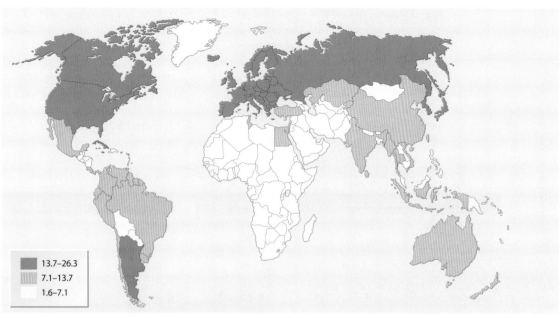

Figure 13.23

Percentage of total population aged 60 and over in 2005

Legend:
- 13.7–26.3
- 7.1–13.7
- 1.6–7.1

	1970		1998		2025 (estimate)	
	Male	Female	Male	Female	Male	Female
Japan (highest)	71	76	79	85	84	88
Italy	69	75	78	83	82	86
UK	69	75	76	81	78	83
USA	68	75	75	80	78	83
China	63	64	70	74	76	80
India	51	49	62	63	68	70
Bangladesh	46	44	61	62	66	68
Kenya	49	53	49	47	52	53
Zambia (lowest)	39	42	38	37	41	41

Figure 13.22

Life expectancies in selected countries

Figure 13.24

Growth in the percentage of population aged 65 and over in selected countries, 1950–2020

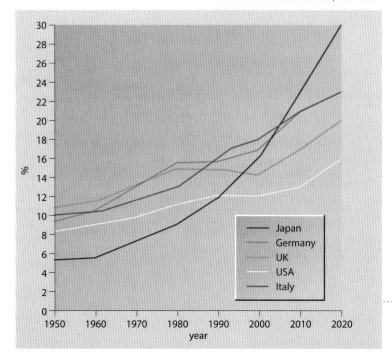

4 Death rates and life expectancy

Death rates, whether it be infant mortality or among children and adults, have, traditionally, declined as a country develops (i.e. Stage 2 onwards in the demographic transition model – Figure 13.10). Due to improvements in medical facilities, hygiene and the increased use of vaccines, the decline in the death rate has led to a sharp increase in life expectancy, initially in the economically more developed countries but, more recently, also in many of the economically less developed countries (Figure 13.22). Already, several of the more developed countries have over 20 per cent of their population aged over 65 (Figures 13.23 and 13.24) and several others have, for the first time in history, more people aged over 65 than they have children aged under 15 (Places 40). In Europe – the major area to be affected by ageing until overtaken recently by eastern Asia, notably Japan – the proportion of children is projected to decline from 16 per cent in 2008 to 14 per cent by 2050, while the proportion of people aged over 65 is expected to rise from 20 per cent to 35 per cent in the same period (the most rapid increases being in Spain and Italy). By 2025 the UN predict that as the number of old people in the world increases, this will mean:

- a greater demand for services (e.g. pensions, medical care and residential homes) which will have to be provided (i.e. paid for) by a smaller percentage of people of working age (i.e. in the economically active age group) in the more developed countries, and
- a rapid increase in population size with an associated strain on the often already overstretched resources of the less developed countries.

The UN have, since 1998, provided population data for what they call the 'oldest-old' (Figure 13.25), an age-group that was then divided into octogenarian (aged 80–89), nonagenarian (aged 90–99) and centenarian (aged over 100). Unlike Figure 13.13, Figure 13.25 does not show the high female proportion of the over-90s group, which has a female–male ratio of 5:1. An exception to increased life expectancy has occurred in those countries where the AIDS epidemic has had its greatest impact (page 622).

Of the ten countries with the world's lowest life expectancy, all are in sub-Saharan Africa (Places 100, page 623). From the onset of the pandemic, life expectancy in these countries fell to an average of 42.5 years in 2007 – a decrease of 10 years – whereas it might have been expected to have reached 60 years had AIDS not occurred (Figure 13.26). Despite the decline in life expectancy, the total population of countries in the region has not decreased as the number of deaths has been offset by the high TFR.

Figure 13.25

Age composition of the world's 'oldest old'

Age group	Millions		%	
	2005	2050	2005	2050
Oldest-old: 80+	79.4	379.0	100	100
Octogenarian: 80–89	71.1	314.4	88.2	83.0
Nonagenarian: 90–99	8.1	61.4	11.5	16.2
Centenarian: 100+	0.2	3.2	0.3	0.8

UN Population Division

	1970	1998	2008
Botswana	51	45	34
Malawi	41	36	45
South Africa	54	50	47
Zimbabwe	52	41	37

Figure 13.26

The effect of AIDS on life expectancy in selected countries

Places 40 Japan: an ageing population

Japan has developed an ominously top-heavy demographic profile (Figure 13.27) which by 2040 is predicted to be the inverse of that of a developing country at stage 2 of economic development (Figure 13.15). As Japanese women are both marrying and having children at a later age, if at all, so the country's TRF has fallen to 1.3 – one of the lowest in the world – compared with over 5 in 1928 and 1.7 in 1988. In contrast, the Japanese, who on average can expect to live to 83 years of age, have the world's greatest longevity. By 2050 Japan is projected to have the world's highest proportion of centenarians – 960 000 or 0.8 per cent of the total population. Of these, 91 per cent will be women. With a birth rate of 8.6 and a death rate of 8.8, Japan has a negative natural increase, the figure being –0.1 in 2007. With fewer births and people living longer, this means an increasingly greater proportion of the population is aged over 65, having risen from 5 per cent in 1980 (Figure 13.27) to 21.5 per cent in 2007, and to a predicted 30 per cent by 2020 and 39.6 per cent in 2040.

The potential to the Japanese economy in terms of the demand for extra resources to look after the elderly, and the reduced revenue from taxes as the proportion of people in the working-age group decreases (Figure 13.27) has led the government to implement major reforms in its elderly care programme and to offer inducements to encourage women to have more children.

Figure 13.27

Changes in the population structure of Japan, 1950–2040

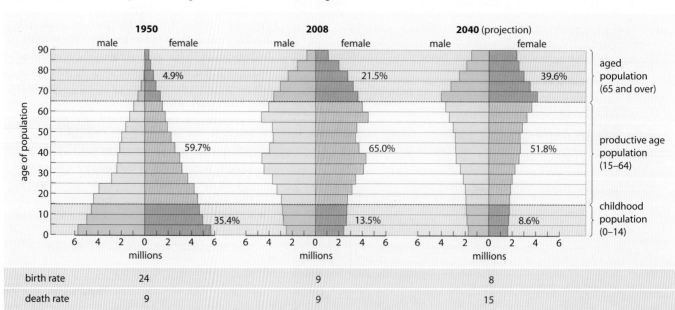

Migration: change in space and time

Migration is a movement and in human terms usually refers to a permanent change of home. It can also, however, be applied more widely to include temporary changes involving seasonal and daily movements. It includes movements both between countries and within a country. Migration affects the distribution of people over a given area as well as the total population of a region and the population structure of a country or city. The various types of migration are not easy to classify, but one method is given in Figure 13.28.

Internal and external (international) migration

Internal migration refers to population movement within a country, whereas external migration involves a movement across national boundaries and between countries. External migration, unlike internal movement, affects the total population of a country. The **migration balance** is the difference between the number of **emigrants** (people who leave the country) and **immigrants** (newcomers arriving in the country). Countries with a **net migration loss** lose more through emigration than they gain by immigration and, depending upon the balance between birth and death rates, may have a declining population. Countries with a **net migration gain** receive more by immigration than they lose through emigration and so are likely to have an overall population increase (assuming birth and death rates are evenly balanced).

Voluntary and forced migration

Voluntary migration occurs when migrants move from choice, e.g. because they are looking for an improved quality of life or personal freedom. Such movements are usually influenced by '**push and pull' factors** (page 366). **Push** factors are those that cause people to leave because of pressures which make them dissatisfied with their present home, while **pull** factors are those perceived qualities that attract people to a new settlement. When people have virtually no choice but to move from an area due to natural disasters or because of economic, religious or social impositions (Figure 13.29), migration is said to be **forced**.

Times and frequency

Migration patterns include people who may move only once in a lifetime, people who move annually or seasonally, and people who move daily to work or school. Figure 13.28 shows the considerable variations in timescale over which migration processes can operate.

Distance

People may move locally within a city or a country or they may move between countries and continents: migration takes place at a range of spatial scales.

Migration laws and a migration model

In 1885, E.G. Ravenstein put forward seven 'laws of migration' based on his studies of migration within the UK. These laws stated that:
1. Most migrants travel short distances and their numbers decrease as distance increases (distance decay, page 410).
2. Migration occurs in waves and the vacuum left as one group of people moves out will later be filled by a counter-current of people moving in.
3. The process of dispersion (emigration) is the inverse of absorption (immigration).
4. Most migrations show a two-way movement as people move in and out: net migration flows are the balance between the two movements.
5. The longer the journey, the more likely it is that the migrant will end up in a major centre of industry or commerce.
6. Urban dwellers are less likely to move than their rural counterparts.
7. Females migrate more than males within their country of birth, but males are more likely to move further afield.

Permanent	External (international):	between countries
	1 voluntary	West Indians to Britain
	2 forced (refugees)	African slaves to America, Kurds, Rwandans
	Internal:	within a country
	1 rural depopulation	most developing countries
	2 urban depopulation	British conurbations
	3 regional	from north-west to south-east of Britain
Semi-permanent	for several years	migrant workers in France and (former West) Germany
Seasonal	for several months or several weeks	Mexican harvesters in California, holidaymakers, university students
Daily	commuters	south-east England

Figure 13.28

Types of migration

Forced migration	Prevention of voluntary movement
Religious: Jews; Pilgrim Fathers to New England	Government restrictions: immigration quotas, Berlin Wall, work permits
Wars: Muslims and Hindus in India and Pakistan; Rwanda, Chechnya	Lack of money: unable to afford transport to and housing in new areas
Political persecution: Ugandan Asians, Kosovar Albanians	Lack of skills and education
Slaves or forced labour: Africans to south-east USA	Lack of awareness of opportunities
Lack of food and famine: Ethiopians into the Sudan	Illness
Natural disasters: floods, earthquakes, volcanic eruptions (Mt Pinatubo, Case Study 1)	Threat of family division and heavy family responsibilities
Overpopulation: Chinese in South-east Asia	**Reasons for return**
Redevelopment: British inner-city slum clearance	Racial tension in new area
Resettlement: Native Americans (USA) and Amerindians (Brazil) into reservations	Earned sufficient money to return
Environmental: Chernobyl (Ukraine), Bhopal (India)	To be reunited with family
Dam construction: Three Gorges (China)	Foreign culture proved unacceptable
Voluntary migration	Causes of initial migration removed (political or religious persecution)
Jobs: Bantus into South Africa, Polish workers to the UK (Places 44), Mexicans into California	Retirement
Higher salaries: British doctors to the USA	**Barriers to return**
Tax avoidance: British pop/rock and film stars to the USA	Insufficient money to afford transport
Opening up of new areas: American Prairies; Israelis into Negev Desert; Brasilia	Standard of living lower in original area
Territorial expansion: Roman and Ottoman Empires, Russians into Eastern Europe	Racial, religious or political problems in original area
Trade and economic expansion: former British colonies	Loss of family ties
Retirement to a warmer climate: Americans to Florida	
Social amenities and services: better schools, hospitals, entertainment	

Figure 13.29

Causes of migration, with examples

More recent global migration studies have largely accepted Ravenstein's 'laws', but have demonstrated some additional trends:

8 Most migrants follow a step movement which entails several small movements from the village level to a major city, rather than one traumatic jump.

9 People are leaving rural areas in ever-increasing numbers, especially in China.

10 People move mainly for economic reasons, e.g. jobs and the opportunity to earn more money. Growing numbers of short-term migrant workers send remittances home – a major factor of globalisation.

11 Most migrants fall into the 20–34 age range.

12 With the exception of short journeys in developed countries, males are the more mobile. (In many societies, females are still expected to remain at home.)

13 There are increasing numbers of migrants who are unable to find accommodation in the place to which they move; this forces them to live on the streets, in shanties (Places 57, page 443) and in refugee camps (page 367).

14 There are increasing numbers of refugees, displaced persons and economic and illegal immigrants (page 367).

The examples in Figure 13.29 help to explain the migration model shown in Figure 13.30.

Migration within developed countries

Certain patterns of internal migration are more characteristic of developed countries than economically less developed countries. Three examples have been chosen to illustrate this: rural–urban movement; regional movement; and movements within and out of large urban areas.

Rural–urban movement

Although rural depopulation is now a worldwide phenomenon, it has been taking place for much longer in the more developed, industrialised countries. Figure 13.31 and Places 41 describe and explain the changing balance between rural and urban dwellers in China since 1980.

Figure 13.30

A migration model (*after* Hornby and Jones)

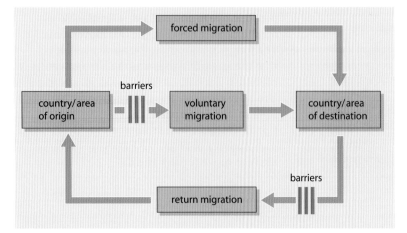

Since about 1980, rural–urban migration has become not only a major socio-economic phenomenon in China, but probably the world's greatest ever internal movement of people.

The 1940s and 1950s were a time when rural labourers were encouraged to participate in urban development. Although 40 million were recruited, only just over 10 per cent of the country's total population then lived in urban areas. Between 1958 and 1983, under the system of *hukou*, rural labourers were forbidden to leave their home villages to seek jobs or to run businesses without official permission. Rural poverty increased.

In 1984, an official document was issued which allowed rural workers to enter cities to seek work. This complete change in policy was closely related to other socio-economic and institutional changes, such as the replacement of the commune system and the creation of the Responsibility Scheme (Places 63, page 468); the reform of the system of purchase and sale of food products; the beginnings of mechanisation of agriculture; and the setting-up of 5 Special Economic Zones (SEZs) and 15 Open Cities in coastal areas (Case Study 19). Although policies at central level now encouraged a regulated movement of people to urban areas, some coastal areas and larger cities tried, at a local level, to impose restrictions limiting a totally free movement of workers. Even so, the urban population had increased to 36 per cent by 2000, and with between another 13 to 16 million workers now moving each

year to cities, to 44 per cent by 2008. To keep pace with China's industrial growth, the government now hopes that between 300 and 500 million people will leave their rural homes and settle in coastal provinces and cities by 2020, by which time the urban population could be almost 60 per cent of the total.

Seventy per cent of migrant labourers are between 16 and 35 years of age; most have received up to nine years of education and about one-third are female (Case Study 21). They contribute significantly to the development of China's industries, especially those producing cheap goods intended for global markets.

Figure 13.31

Migration within China

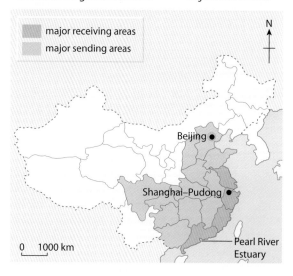

Regional movement in Britain

For over a century there has been a drift of people from the north and west of Britain to the south-east of England. The early 19th century was the period of the Industrial Revolution when large numbers of people moved into large urban settlements on the coalfields of northern England, central Scotland and South Wales, and to work in the textile, steel, heavy engineering and shipbuilding industries. However, since the 1920s there has been more than a steady drift of population away from the north of Britain to the south (Figures 13.32, 13.33 and 13.34). Some of the major reasons for this movement are listed here.

■ A decline in the farming workforce and rural population, for reasons similar to those quoted in Places 41 on rural–urban movements.
■ The exhaustion of supplies of raw materials (coal and iron ore).

■ The decline of the basic heavy industries such as steel, textiles and shipbuilding. Many industrial towns had relied not only on one form of industry but, in some cases, on one individual firm. With no alternative employment, those wishing to work had to move south.
■ Higher birth rates in the industrial cities meant more potential job-seekers.
■ New post-war industries, which included car manufacturing, electrical engineering, food processing and, since the 1980s, micro-electronics and high-tech industries, have tended to be market-oriented. They are said to be footloose, in the sense that they have a free choice of location – unlike the older industries which had to locate near to sources of raw materials and/or energy supplies.
■ The ageing population is attracted to the south coast for retirement.

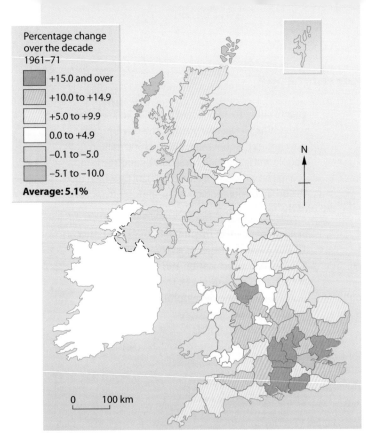

Percentage change over the decade 1961–71

- +15.0 and over
- +10.0 to +14.9
- +5.0 to +9.9
- 0.0 to +4.9
- −0.1 to −5.0
- −5.1 to −10.0

Average: 5.1%

N

0   100 km

Figure 13.32

Population changes in the UK, 1961–71 (boundaries adjusted to the 1974 changes)

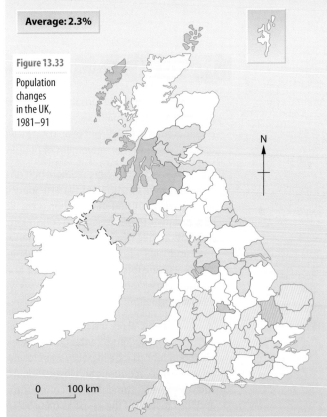

Average: 2.3%

Figure 13.33

Population changes in the UK, 1981–91

N

0   100 km

- The growth of service industries has been mainly in the south-east. This has resulted from the many office firms wanting a prestigious London address, the growth of government offices, the demand for hospitals, schools and shops in a region where one in five British people live, and tourism taking advantage of Britain's warmer south coast.
- Joining the EU meant increased job opportunities in the south and east, while traditional industrial areas and ports such as Glasgow, Liverpool and Bristol, which had links with the Americas, have declined.
- Salaries were higher in the south.
- With so much older housing, derelict land and waste tips, the quality of life is often perceived as being lower in the north, despite the beauty of its natural scenery and slower pace of life.
- There are more social, sporting and cultural amenities in the south.
- Communications were easier to construct in the flatter south. Motorways, railways, international airports, cross-Channel ports and the Channel Tunnel were built and/or improved as this region had the greatest wealth and population size.

Movements within urban areas

Since the 1930s there has been, in Britain, a movement away from the inner cities to the suburbs – a movement accelerated first by improved public transport provision and then by the increase in private car ownership. The former included more bus services, the extension of the London Underground and the construction of the Tyne and Wear Metro. Some of the many stereotyped reasons for this outward movement are summarised in Figure 13.35. The result, in human terms, has been a polarisation of groups of people within society and the accentuation of inequalities (wealth and skills) between them. (Beware, however, of the dangers of stereotyping when discussing these inequalities; Framework 14, page 437.)

- The inner cities tended to be left with a higher proportion of low-income families, handicapped people, the elderly, single-parent families, people with few skills and limited qualifications, first-time home-buyers, the unemployed, recent immigrants and ethnic minorities.
- The suburbs tended to attract people moving towards middle age, married with a growing family, possessing higher skills and qualifications, earning higher salaries, in secure jobs and capable of buying their own homes and car. Recently there has been, in part at least, a reversal of this movement and parts of some inner cities have become regenerated and 'fashionable'. This re-urbanisation is partly due to energy conservation, partly to changes in housing markets, partly to planning initiatives such as refurbished waterfronts (London, Places 56, page 440; Baltimore) and partly to new employment growth (leisure, financial services).

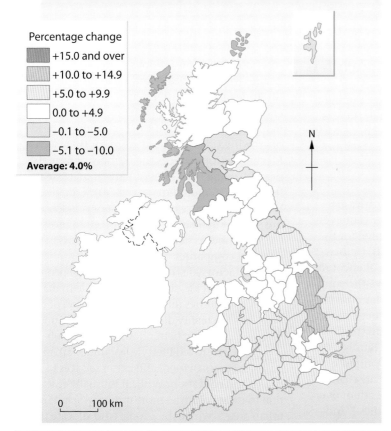

Percentage change

▓	+15.0 and over
▒	+10.0 to +14.9
░	+5.0 to +9.9
□	0.0 to +4.9
░	−0.1 to −5.0
▒	−5.1 to −10.0

Average: 4.0%

N

0   100 km

Movements away from conurbations

After the mid-1950s, large numbers of people moved out of London altogether (Figure 13.36). Initially these were people who were virtually forced to move as large areas of 19th-century inner-city housing were demolished. Many of these people were rehoused in one of the several planned new towns that were created around London. More recently, even the outer suburbs have lost population (Figure 13.35) as people moved, often voluntarily, to smaller towns, or into commuter and suburbanised villages, with a more rural environment. This process of counter-urbanisation became characteristic of all Britain's conurbations (Figure 13.37) until a reverse movement began, initially in the 1990s mainly due to the regeneration of inner-city areas (especially those with a quayside location) and in the 2000s as an increasing number of migrants moved in.

Figure 13.34

Population changes in the UK, 2001–2006

Figure 13.35

Some causes of migration from the inner cities to the suburbs

	Inner city	Suburbs
Housing	Poor quality; lacking basic amenities; high density; overcrowding	Modern; high quality; with amenities; low density
Traffic	Congestion; noise and air pollution; narrow, unplanned streets; parking problems	Less congestion and pollution; wider, well-planned road system; close to motorways and ring roads
Industry	Decline in older secondary industries; cramped sites with poor access on expensive land	Growth of modern industrial estates; footloose and service industries; hypermarkets and regional shopping centres; new office blocks and hotels on spacious sites
Jobs	High unemployment; lesser-skilled jobs in traditional industries	Lower unemployment; cleaner environment; often more skilled jobs in newer high-tech industries
Open space	Limited parks and gardens	Individual gardens; more, larger parks; nearer countryside
Environment	Noise and air pollution from traffic and industry; derelict land and buildings; higher crime rate; vandalism	Cleaner; less noise and air pollution; lower crime rate; less vandalism
Social factors	Fewer, older services, e.g. schools and hospitals; ethnic and racial problems	Newer and more services; fewer ethnic and racial problems
Planning and investment	Often wholesale redevelopment/clearance; limited planning and investment	Planned, controlled development; public and private investment
Family status/wealth	Low incomes; often elderly and young; large family or none	Improved wealth and family/professional status

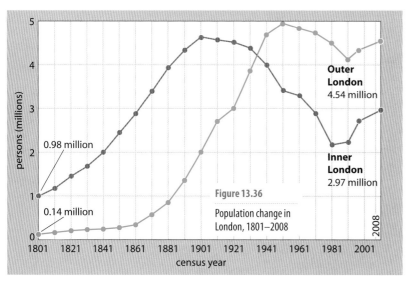

Figure 13.36

Population change in London, 1801–2008

Conurbation	1961–71	1971–81	1981–91	1991–96	2001–06
Greater London	−6.8	−11.3	−4.9	+5.9	+2.6
Inner London	−13.2	−20.0	−6.6	+8.1	+4.0
Outer London	−1.8	−5.0	−4.1	+7.7	+1.7
Greater Manchester	+0.3	−5.6	−5.5	+3.0	+0.1
Merseyside	−3.6	−9.3	−9.1	+1.2	−3.0
South Yorkshire	+1.5	−2.3	−4.1	+3.3	+0.1
Tyne and Wear	−2.6	−6.3	−5.4	+2.9	−2.6
West Midlands	+17.8	−5.9	−5.5	+3.6	+2.2
West Yorkshire	+3.1	−2.2	−2.7	+4.8	+2.2
Glasgow City	−13.8	−23.1	−15.5	+2.6	−5.5

Figure 13.37

Population change in UK conurbations (percentage change per decade)

Internal migration in economically less developed countries

There is usually a much greater degree of migration within developing countries than there is in more developed countries. Two examples have been chosen to illustrate this: rural–urban movement and political resettling.

Rural–urban movement

Many large cities in developing countries are growing at a rate of more than 20 per cent every decade. This growth is partly accounted for by rural 'push' and partly by urban 'pull' factors.

Push factors are those that force or encourage people to move – in this case, to leave the countryside. Many families do not own their own land or, where they do, it may have been repeatedly divided by inheritance laws until the plots have become too small to support a family. Food shortages develop if the agricultural output is too low to support the population of an area, or if crops fail. Crop failure may be the result of overcropping and overgrazing (Case Studies 7 and 10), or natural disasters such as drought (the Sahel countries), floods (Bangladesh), hurricanes (the Caribbean) and earth movements (in Andean countries). Elsewhere, farmers are encouraged to produce cash crops for export to help their country's national economy instead of growing sufficient food crops for themselves. Mechanisation reduces the number of farmers needed, while high rates of natural increase may lead to overpopulation (page 376). Some people may move because of a lack of services (schools, hospitals, water supply) or be forced to move by governments or the activities of transnational companies.

Pull factors are those that encourage people to move – in this case, to the cities. People in many rural communities may have a perception of the city which in reality does not exist. People migrate to cities hoping for better housing, better job prospects, improved lifestyle (aspirational), more reliable sources of food, and better services in health and education. While it is usually true that in most countries more money is spent on the urban areas – where the people who allocate the money themselves live – the present rate of urban growth far exceeds the amount of money available to provide accommodation for all the new arrivals. Recent studies seem to confirm that many migrants make a stepped movement from their rural village first to small towns, then to larger cities and finally to a major city.

Places 42 — Tunisia: migration patterns

Figure 13.38 shows migration patterns in Tunisia. There are several points to notice:

- There is a greater movement of rural than of urban dwellers.
- Most migrants move to Tunis, the capital city.
- Most migrants tend to travel short distances: relatively few make long journeys (distance decay factor, page 410).
- Most move from rural, inland, desert areas to urban, coastal areas.
- A few move from Tunis to coastal towns, such as the holiday resorts of Sousse and Monastir.
- Very few migrants return to rural districts.
- There is evidence of a twofold movement into and out of Tunis and Sfax.

Figure 13.38

Migration patterns

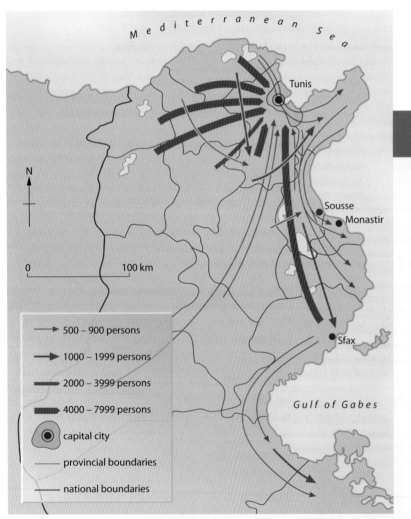

Legend:
- 500 – 900 persons
- 1000 – 1999 persons
- 2000 – 3999 persons
- 4000 – 7999 persons
- capital city
- provincial boundaries
- national boundaries

0   100 km

Mediterranean Sea
Tunis
Sousse
Monastir
Sfax
Gulf of Gabes

Political resettling

National governments may, for political reasons, direct, control or enforce movement as a result of decisions which they believe to be in the country's (or their own) best interests. Some governments have actively encouraged the development of new community settlements, especially in areas which were, at the time, sparsely populated, e.g. the creation of *kibbutzim* in Israel and of *ujamaa* in Tanzania. Other governments have founded new capital cities in an attempt to develop new growth regions, e.g. Brasilia, Dodoma (Tanzania) and Abuja (Nigeria); while others have built settlements to try to strengthen their claims to an area (e.g. Israeli settlements in the West Bank) or to rehouse people moved for flood control and the production of energy (Three Gorges Project in China – Places 82, page 544).

In Brazil and the USA, minority groups of indigenous people – the Amerindians and Native Americans respectively – have been forced off their tribal lands and onto reservations. In South Africa, under *apartheid*, the black population was forced to live either in shanty settlements in urban townships or on homelands in rural areas, which lacked resources (Places 45, page 372). In the last few years, an increasing number of people have been forced to move due to so-called 'ethnic cleansing' policies enforced by several governments, as in the former Yugoslavia.

External migration

- **Refugees** The United Nations High Commission for Refugees (UNHCR) defines a refugee as 'a person who cannot return to his or her own country because of a well-founded fear of persecution for reasons of race, religion, nationality, political association or political opinion'. The term is often expanded to include people forced to leave their home country due to internal strife (civil wars) or environmental disasters (e.g. earthquakes, famine) in order to seek security or help.
- **Asylum seekers** are people who have left their country of origin, have applied for recognition as a refugee, and are awaiting a decision on their application. International law recognises the right of individuals to seek asylum but does not force states to provide it.
- **Economic migrants** make a conscious choice to leave their country of origin knowing that they will be able to return to it without any problems at a future date. This group includes **migrant workers** (page 369).
- **Illegal immigrants** enter a country without meeting the legal requirements for entry or residence. They often arrive with only the barest necessities and without personal documents or passports. Many become part of the 'hidden economy', having to rely on people for shelter and work which leaves them vulnerable to exploitation.
- **Internally Displaced Persons (IDPs)** are included here for although they have not left their country of origin, they may have been forced to flee their home for similar reasons to those of refugees. Many IDPs exist in the same conditions and face the same problems as do refugees. Globally, IDPs outnumber refugees.

Before the Second World War, the majority of refugees tended to become assimilated in their new host country but since then, and beginning with the setting-up of Palestinian Arab camps following the creation of the state of Israel in 1948, the number of permanent refugees has risen rapidly. According to UNHCR, the number of global refugees reached a peak of 17.6 million in 1992 before falling to a trough of 13.2 million in 1999 before rising once again. The UNHCR claim that as most refugees are illegal immigrants, accurate figures are impossible to give but they believed that in 2008 there were 16 million refugees and 51 million IDPs, of whom 26 million were conflict-generated IDPs and 25 million were natural disaster IDPs (Figure 13.40). The refugee and IDP problem has intensified due to conflicts in countries such as Iraq and Afghanistan, and food shortages and political unrest in much of sub-Saharan Africa including Darfur, Somalia and Zimbabwe (Figure 13.39).

Figure 13.39

Part of a displacement camp at Nyaconga, Rwanda

The UNHCR have, in previous years, pointed out that half of the world's refugees are children of school age; most adult refugees are female; and four-fifths of the total are in developing countries which have the fewest resources to deal with the problem. Refugees usually live in extreme poverty and lack food, shelter, clothing, education and medical care – the basic Millennium Development Goals (page 609). They rarely have citizenship and few (if any) civil, legal or human rights. There is little prospect of their returning home and the long periods spent in camps means that they often lose their sense of identity and purpose.

Figure 13.40

World refugees at the end of 2007
a World total
b Major refugee hosting countries

a

Region	Total refugees at end 2007
UNHCR regions	
Central Africa and Great Lakes	1 100 100
East and Horn of Africa	815 200
Southern Africa	181 200
West Africa	174 700
Total Africa (excluding North Africa)	*2 271 200*
Americas	987 500
Asia and Pacific	3 825 000
Europe	1 585 300
Middle East and North Africa	2 721 600
Total UNHCR regions	**11 390 600**
UNRWA regions (Palestinians)	**4 622 000**
Total all refugees	*16 012 600*

b

Pakistan — 2 033 100
Syria — 1 503 800
Iran — 963 500
Germany — 578 900
Jordan — 500 300
Tanzania — 435 600
China — 301 100
UK — 299 700
Chad — 294 000
USA — 281 200

A 2008 analysis of refugee data by UNHCR revealed two major patterns:

1 The vast majority of refugees are hosted by neighbouring countries, with over 80 per cent remaining in their region of origin i.e. within Africa or the Middle East. This conflicts with the perception that many seek protection in North America or Western Europe.

2 The number of refugees living in urban areas continues to grow and now exceeds 50 per cent of the total.

Apart from 4.6 million Palestinian refugees who come under UNRWA which is a different UN department from UNHCR, Afghanistan continues to be the leading country of origin. At the end of 2007 (Figure 13.41), there were almost 3.1 million Afghan refugees, of whom 96 per cent were to be found in neighbouring Pakistan and Iran, and 2.3 million Iraqis, most of whom have sought refuge in Syria or Jordan. Afghans and Iraqis together account for almost half of the world's total refugees under UNHCR responsibility, followed by half a million Colombians. Other main source countries were the Sudan, Somalia, Burundi and the Democratic Republic of the Congo. At that same time, about 60 per cent of all refugees were residing in Asia, particularly Pakistan, Syria and Iran (Figure 13.40b). Of the remainder, Africa, Europe and North America respectively hosted 20, 14 and 9 per cent.

Figure 13.41

Major source countries of refugees, 2007

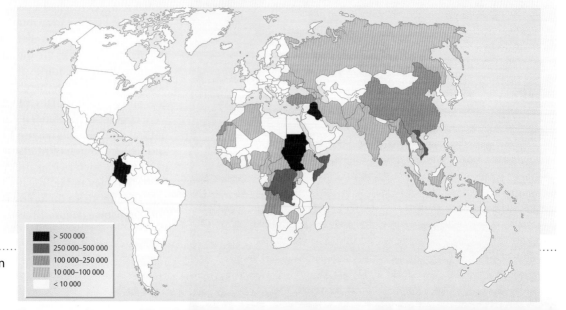

> 500 000
250 000–500 000
100 000–250 000
10 000–100 000
< 10 000

Migrant workers

As economic development has taken place at different rates in different countries, supplies of and demand for labour are uneven, and due to improvements in transport there has been an increase in the number of people who move from one country to another in search of work. Such cross-border movements in search of work can operate at different timescales. For example:

■ **Permanent** For a century and a half, the UK has received Irish workers and, since the 1950s, West Indians and citizens from the Indian subcontinent. Most of these migrants have made Britain their permanent home.

■ **Semi-permanent** After the Second World War, several European countries experienced a severe labour shortage. In order to help rebuild their economies, countries such as France accepted cheap semi-skilled labour mainly from North Africa while the then West Germany did the same for workers from Turkey, Yugoslavia and the Middle East. Recently, similar types of workers from Eastern Europe, particularly Poland, have been attracted to the UK (Places 44).

■ **Short-term and seasonal** The South African economy depends largely on migrant black labour from adjacent nation states. In North America, large numbers of Mexicans find seasonal employment picking fruit and vegetables in California (Case Study 15A).

■ **Daily** With the introduction of free movement for EU nationals within the EU, an increasing number of workers commute daily into adjacent countries.

Places 44 UK: Polish migrant workers

In 2004, Poland and seven other former Eastern European countries gained entry into the EU. Of the existing members, only the UK and the Republic of Ireland allowed unlimited immigration from the new members. This led, in the UK, to the largest influx for centuries with, by early 2008, the arrival of over 800 000 migrant workers – an average in excess of 200 000 a year. Of these an estimated 500 000 had come from Poland (Figure 13.42).

Migrants from Poland were largely welcomed as they came with a wide range of skills, many of which were currently lacking in the British workforce. At one end were people who gained senior jobs in administration, business and management such as computing, IT support, teaching and the National Health Service. At the other were those prepared to work for long hours either as health care workers, as shop assistants, or as manual labour in either factories or on farms. Somewhere in the middle were people such as plumbers, electricians, bricklayers and decorators – four other professions in which Britain had a severe skills shortage.

Whereas earlier migrants into Britain tended to concentrate in certain urban areas, and then within specific districts in those areas, it was claimed that by May 2008 Polish workers were living in every local authority area of Britain. As a group, they set up their own radio stations, printed their own newspapers, celebrated church Mass in their own language, produced Polish bread and processed other food products which they then sold in their own shops (Figure 13.43).

Why did they come to the UK?

Most came to find better-paid jobs as, when they arrived, the average wage in the UK was several times higher than that in Poland. To some the idea was to work hard, earn as much money as they could and then return home, hopefully with the finance needed to set up their own business. To others it was a case of earning sufficient money to live on themselves and to send the remainder back to Poland to help their families there improve their standard of living. The majority of immigrants were men, of whom over 80 per cent were aged between 18 and 34 years.

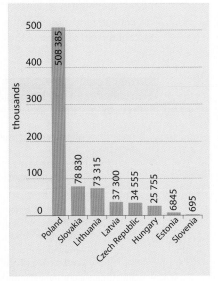

Figure 13.42

Nationality of foreign workers, May 2004 – February 2008

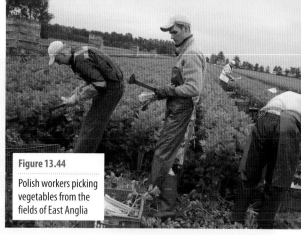

Figure 13.44

Polish workers picking vegetables from the fields of East Anglia

Figure 13.43

One of the many Polish shops to be found in the UK

Peterborough is one of several British towns where Poles now comprise over 10 per cent of the total population. Many have concentrated in the Lincoln Road area where houses were less expensive. Whereas in 2004 the local primary school had had to make provision for children of Pakistani origin, now it has to provide for Polish speakers who, in the four years after 2004, grew from 0 per cent of the school population to over 30 per cent. Although many migrant workers in Peterborough find jobs on building sites or in factories and superstores, the majority seek work on farms in the nearby Fens picking and packing fruit and vegetables (Figure 13.44). Farm labourers are likely to work from 7 am to 5 pm, six days a week. Even with overtime they may only earn between £300 and £500 a week.

The Poles are generally well accepted by local communities in Britain. Perhaps this is because they are European, or is an acknowledgement of just how hard they work and how valuable they are at filling vacancies in the British skills market. However, that is not to say that their presence does not create problems. In large numbers they can 'swamp' schools, hospitals and other services; by buying property at the lower price range they compete with local first-time buyers; those with fewer skills compete with local job-seekers, because they are prepared to accept lower wages and longer hours; and money they earn is sent out of the country and so is lost to the British economy.

In contrast, while families in Poland benefit from remittances sent back to them, the country as a whole may lose its most skilled and educated workers; has to train women to fill job vacancies; and sees families divided with so many males working abroad.

Why are they returning home?

By 2008, the migration pattern began to alter. Since the first arrivals in 2004, prices in the UK have increased far more than they have in Poland. Also, as the pound has become weaker in comparison with the Polish zloty, the UK is less attractive as a place to live and work. Meanwhile the Eastern European economy has grown and both investment and wages within Poland have increased. The result is that many Poles are now beginning to return home to build their own houses, set up their own businesses and to start families. They are also needed to provide facilities, including stadiums for the 2012 UEFA football tournament which is to be held in Poland. Some predictions suggest that half the Polish workers will have returned home by 2010 leaving Britain, once again, with a shortage of skills.

Figure 13.45 lists some of the advantages and disadvantages to both the home and the host country with respect to migrant workers. The same can be applied to migrant workers from North Africa into France, Turkey into Germany, and Mexico into California, as from Poland into the UK.

Figure 13.45

Advantages and disadvantages of migrant workers

		Advantages	Disadvantages
Home country		• Reduces pressure on jobs and local resources • Birth rate may be lowered as people of child-bearing age leave • Money may be received as remittances from abroad • Migrants may develop new skills which they can bring back home	• People of working age migrate • Those with skills and education are most likely to leave • It is mainly males who migrate and this divides families • An elderly population is left with fewer people to look after them • Can create a dependency on money being sent back as remittances
Host country		• May receive highly skilled migrants to fill specialised vacancies in the job market • Labour shortage overcome, especially in dirty, poorly paid, unskilled jobs • Provides cheaper labour who work for longer hours • Cultural advantages of discovering new foods, music, pastimes, etc.	• Migrants can put a strain on local services and resources • Resentment towards migrants if they take the best jobs • Some migrant groups do not mix and try to retain their own culture • Mainly young males which can create social problems • Migrants may feel discriminated against which can cause racial tension

Multicultural societies

This is often a sensitive and emotive issue. Attempts here to explain terms are not intended to cause insult or resentment.

The latest scientific research suggests that humans evolved in central Africa about 200 000 years ago and began, 100 000 years later, to migrate to other parts of the world. This common origin, identified by the study of genes, shows that humans are genetically homogeneous to a degree unparalleled in the animal kingdom.

Previous scientific opinion suggested that the many peoples of the modern world had descended from three main races. These were the Negroid, Mongoloid and Caucasoid. The dictionary definition of *race* is 'a group of people having their own inherited characteristics distinguishing them from people of other races', e.g. colour of skin and physical features. In reality, often because of intermarriage, the distinction between races is now so blurred that the word 'race' has little significant scientific value. Today, while colour still remains the most obvious visible characteristic, groups of people differ from one another in religion, language, nationality and culture. These differences have led to the identification of many ethnic groups.

What criteria do members of various ethnic groups prefer to use when identifying themselves?

- **Colour of skin** Whereas people of 'European' stock have long accepted being called 'white', it is only in more recent years that, in Britain, people from Africa and the Caribbean have preferred to be known collectively as 'black'. The 1971 UK census divided immigrants born in Commonwealth countries into the Old (white) and New (black) Commonwealth. (It made no allowance for children born in the UK of New Commonwealth parentage.)

- **Place of birth (nationality)** *The Annual Abstract of Statistics* for the UK lists immigrants under the heading 'country of last residence' – thus avoiding a reference to colour. Most groups of people, in the USA for example, have been identified by their place of birth, or that of their ancestors, and are known as Chinese, Puerto Rican, etc. There is currently a major movement in the USA (and to a lesser extent in the UK) by blacks, also wishing to be identified by place of origin, to be referred to as African-Americans. Will black people in the UK eventually prefer to be known as African-Caribbean, African-British, or another term not yet invented?

- **Language** At present, the largest group of migrants moving into the USA is the Hispanics, i.e. Spanish speakers. These migrants, mainly from Mexico, Central America and the West Indies, have been identified and grouped together by their common language and higher fertility.

- **Religion** Other ethnic groups prefer to be linked with, and are easily recognised by, their religion, e.g. Jews, Sikhs, Hindus and Muslims.

The 1991 UK census asked respondents, for the first time, to identify themselves by ethnic group. Figure 13.46 lists these groups, and gives the results of this question, which was repeated in the 2001 census. The increase in Asian or Asian British was due to their high birth rates, not to new immigrants.

The migrations of different ethnic groups have led to the creation of multicultural societies in many parts of the world. In most countries there is at least one minority group. While such a group may be able to live in peace and harmony with the majority group, unfortunately it is more likely that there will be prejudice and discrimination leading to tensions and conflict. Four multicultural countries with differing levels of integration and ethnic tension are: South Africa (Places 45), the USA and Brazil (Places 46), and Singapore (Places 47). Remember, though, that when we look at these countries from a distance we can rarely appreciate the feelings generated by, or the successes/failures of, different state or government policies.

Figure 13.46

Ethnic groups in Britain at the 1991 and 2001 censuses

Ethnic groups	Percentage in each group	
	1991	2001
White	94.5	92.1
Mixed*	–	1.1
Asian or Asian British		
Indian	1.5	1.8
Pakistani	0.9	1.3
Bangladeshi	0.3	0.5
Other Asian	0.4	0.4
Black or black British		
Black-Caribbean	0.9	1.0
Black-African	0.4	0.8
Black Other	0.3	0.2
Chinese	0.3	0.4
Other	0.5	0.4
Total non-white	5.5	7.9

* New category for 2001 census for people considering themselves to belong to more than one group

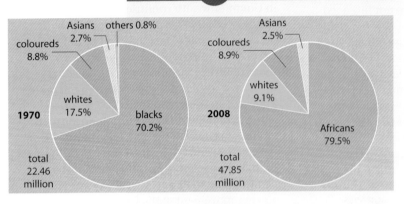

Figure 13.47

Ethnic groups in South Africa, 1970 and 2008

As shown in Figure 13.47, the population of South Africa doubled between 1970 and 2008 and the proportion of Africans (referred to as blacks in 1970) had increased considerably.

The first inhabitants in this region were the San (Bushmen) and Khoi-Khoin. Today's African majority originated as Bantu speakers who migrated into the area many centuries ago, while the white population is descended from Dutch, German, French and British immigrants who arrived after the 1650s. Asians, mainly from India, Malaysia and Indonesia, began arriving after 1860. The coloured ethnic group result from mixed relations between European settlers, Asian migrants and indigenous peoples.

A policy of segregation between black and white originated in the first Dutch settlement, the Cape, in 1652. This practice became customary, and was established legally as apartheid by the first National Party government in 1948 when some members of the Party united to protect their language, culture and heritage from a perceived threat by the black majority and to assert their economic and political independence from British colonial domination.

Statutory apartheid regulated the lives of all groups, but particularly of blacks, coloureds and Indians. The Population Registration Act categorised the nation into White, Black, Indian, Malay and Coloured citizens. Further Acts made mixed marriages illegal, and prescribed segregation in restaurants, transport, schools, places of entertainment and political parties. The Group Areas Act stipulated where and with whom people could live; and the Black Authorities Act established black homelands.

The outcome of all this legislation was the unequal division of rights and resources. This included the disproportionate division of land; the unequal distribution of funding for education; and the general denial of constitutional rights for the majority of South Africans.

Legalised racial discrimination was abolished in the early 1990s and the first free all-party elections, held in 1994, established a multi-party Government of National Unity. This ended the existence of the homelands and set out to improve standards and to reduce inequality in human rights, housing, health care, education and land ownership. It was expected that the legacy of apartheid, some aspects of which are described below, would take many years to eradicate.

Housing

The Group Areas Act (1950) ensured that white, coloured and Asian communities lived in different parts of the city (Figure 13.48) with the whites having the best residential areas (Figure 13.49). Buffer zones at least 100 m wide, often along main roads or railway lines, were created to try to prevent contact between the three groups. Blacks were

Figure 13.48

Segregated residential areas in two South African cities

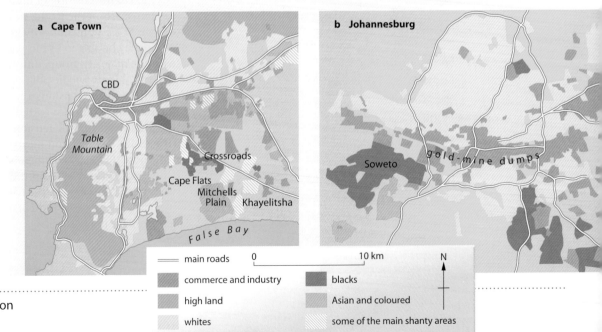

Figure 13.51

Shanty settlement, Khayelitsha, Cape Town

Figure 13.49

A white residential area in Cape Town

Figure 13.50 (centre)

Housing in the Soweto township, Johannesburg

treated differently. Those who had lived in the city since birth, or had worked for the same employer for 10 years, were moved to newly created townships on the urban fringes. The remainder were forced away from the cities to live on one of ten designated reserves or homelands, where the environmental advantages were minimal (drought, poor soils and a lack of raw materials). The homelands took up 13 per cent of South Africa's land; held 72 per cent of its total population; and produced 3 per cent of the country's wealth. Most blacks living in the homelands were employed on one-year contracts, to prevent them gaining urban residential rights.

Life in the townships was no less difficult. These were built far away from white residential areas, which meant that those blacks who found jobs in the cities had long and expensive journeys to work. Many of the original shanty towns have been bulldozed and replaced by rows of identical, single-storey houses (Figure 13.50). These have four rooms and a backyard toilet, but only 20 per cent have electricity. Corrugated-iron roofs make the buildings hot in summer and cold in winter. The settlements lack infrastructure and services and, due to rapid population growth (high birth rates and in-migration), are surrounded by vast shanty settlements (Figure 13.51). Two of the better-known townships are Soweto in Johannesburg (an estimated 4 million inhabitants) and Cape Flats in Cape Town.

Although the African National Congress (ANC) had managed to build 700 000 new houses by 2000, thousands of Africans were still living either in the squalid poverty-stricken squatter camps which had developed during the apartheid era, or in new, but mainly one-roomed, low-cost housing which, their owners claimed, were often poorly constructed and too small for their large families.

Guguletu is part of Cape Flats (Figure 13.48). In the late 1990s, a typical shack was small, 3 m square, and built from discarded wood and corrugated iron (compare Figure 13.51). Doors and windows were held together by nails or string while bricks and rope held down the flat roof. Up to six people might live in a shack which may have contained, as furniture, only a bed, some seating and a table. Although electricity was often available, most shacks lacked running water and sewerage and up to six families were obliged to share

an outdoor toilet and had to queue each morning for their daily water supply. Roads were rarely maintained. However, since then several self-help schemes, most run by women, have developed skills, created jobs and improved the quality of some of the housing.

In 2004 people in Soweto celebrated the centenary of the township. They were also celebrating its transformation from a hopeless ghetto to both a tourist attraction and a desirable suburb. Most of the residents lived in new homes, although they were still small. The relatively few remaining old shacks housed the newest arrivals who tended to be migrants who had fled the poverty of rural South Africa. Local people have, in the last decade or so, developed a sense of optimism for the future despite the fact that unemployment in Soweto is about 40 per cent and violent crime and AIDS are still major problems.

Employment

Under apartheid, blacks were severely restricted in mobility and type of job. Male workers had to return to their homeland in order to apply for a job. If successful, they were given contracts to work in 'white' South Africa for 11 months, after which they had to return to their homeland – a policy that prevented migrant workers becoming permanent city residents. Throughout the 1990s unemployment remained the core cause of poverty and social division. In 2007, unemployment was still high although since 1998 it had fallen for Africans from 38 to 27 per cent and for Asians from 11 to 9 per cent (it had remained the same for coloureds and whites at 11 per cent and 4 per cent respectively). It was much higher for women than it was for men.

Education

Under apartheid, schooling was free and compulsory for whites and Asians, but not for coloureds or blacks – the 1996 census showed over one-quarter of black children did not receive any formal education. Despite attempts by the ANC to improve school buildings and to encourage school attendance, in 2007 most whites attended private schools, coloured children went to schools in the suburbs and Africans to those in the townships. White schools still have a better teacher : pupil ratio and a higher proportion of qualified teachers.

The USA

According to the US Census Bureau, the proportion of 'racial and ethnic minority groups' increased from 24 per cent in 1996 to 33 per cent in 2006 (Figure 13.52). Since 45 per cent of under-5-year-olds in the USA belong to this group, as these children reach child-bearing age, together with the half to one million immigrants per year from Mexico, it is predicted that by 2050 over half of the country's population will be from racial and ethnic groups. Already in more than 10 per cent of America's 3140 counties this sector of the population exceeds 50 per cent of the population, especially with blacks in the south-eastern states and Hispanics in the south-west.

Although Americans have long prided themselves that their country is a 'melting pot' in which people of all ethnic groups can be assimilated into one nation, problems have, and do, exist. The indigenous Native American population has been granted reservations where they can maintain their culture, but as these are usually in areas lacking resources, many have drifted to urban areas. Likewise many black African-Americans, released from slavery after the Civil War, could not find jobs on the land and so moved to large urban areas where they congregated in inner-city 'ghettos' (Chicago Places 52, page 421; and Los Angeles Case Study 15B). Hispanics are the largest growing group, most arriving from Mexico and other Spanish-speaking countries in Latin America.

Despite the US claim that it has an 'open-door' policy, strong restrictive laws have frequently been imposed as a barrier to immigration (Figure 13.30), e.g. against Chinese in the 1920s, the Japanese during the Second World War, Mexicans since the 1980s and, currently, illegal immigrants. Meanwhile many immigrant groups still identify themselves with their 'home country' and its culture, living and marrying within their own ethnic or national group (Puerto Ricans in New York) or congregating to form ethnic areas (Chinatown, Japantown, Koreatown and Filipinotown in Los Angeles).

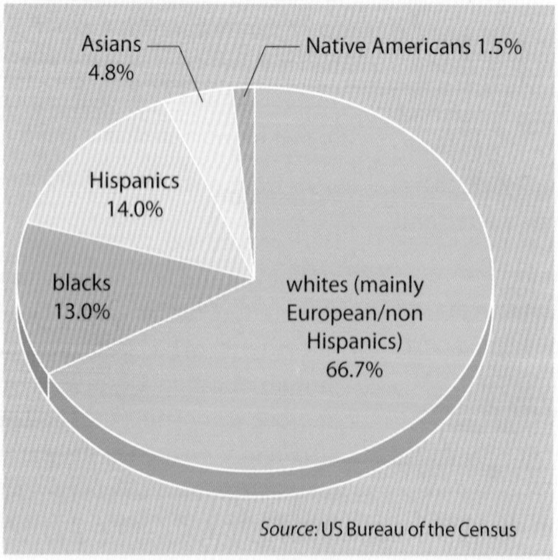

Brazil

Most of the inhabitants of Brazil, having almost every colour of skin conceivable, regard themselves as Brazilians, and the country rightly claims that it has little racial discrimination or prejudice. The Census Department does, however, recognise the following divisions based on colour:

1   Whites (*Branco*): anyone with a lighter-coloured skin. This group includes many of the European migrants who came from Portugal (the original colonists), Italy, Germany and Spain.

2   Mulatto (*Pardo*): darker skins but with a discernible trace of European ancestry. They are the result of mixed marriages or 'liaisons' between the early Portuguese male settlers and either female Indians or African slaves. There is pride rather than prejudice in coming from two racial backgrounds.

3   Blacks (*Preto*): those of pure African descent.

4   Orientals (*Amarelo*): recent emigrants from south and east Asia.

5   Amerindians: a continually declining, yet still distinctive, indigenous group.

All these groups mix freely, especially at football matches, in carnivals and on the beach. Yet despite the lack of racial tension there tends to be a correlation between colour and social status and employment. Walking into a hotel on arrival in Rio, it is apparent that the baggage-carriers are black, hotel porters a slightly lighter colour and the receptionists and cashiers white. In the army, officers are usually white and the ranks black or mulatto. Similarly, the lighter the colour of skin, the more likely it is that a person will become a doctor, bank manager, solicitor or airline pilot.

Figure 13.52

Ethnic groups in the USA, 2006

Asians 4.8%
Native Americans 1.5%
Hispanics 14.0%
blacks 13.0%
whites (mainly European/non Hispanics) 66.7%

Source: US Bureau of the Census

The three main races of Singapore have separate religions, yet each is completely tolerant of the others, with most people even celebrating all three 'New Years' (Figure 13.53), Although by 1994 there was still a Chinatown (restricted to ten streets – Figure 15.48), Arab Street (four streets) and Little India (six streets), the government had pulled down most of the old houses in those areas. Ethnic concentrations had been broken up and now almost 90 per cent of Singaporeans live in modern high-rise flats either within the city itself or in surrounding new towns (Places 60, page 450). Posters promote racial harmony (Figure 13.54) and all races, religions and income groups live together in what appears to be a most successful attempt to create a national unity – a unity best seen on National Day.

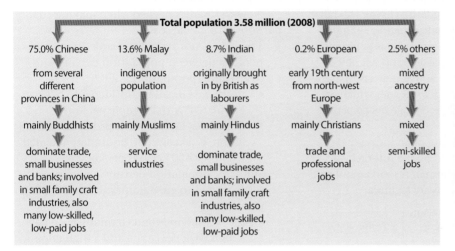

Total population 3.58 million (2008)

75.0% Chinese	13.6% Malay	8.7% Indian	0.2% European	2.5% others
from several different provinces in China	indigenous population	originally brought in by British as labourers	early 19th century from north-west Europe	mixed ancestry
mainly Buddhists	mainly Muslims	mainly Hindus	mainly Christians	mixed
dominate trade, small businesses and banks; involved in small family craft industries, also many low-skilled, low-paid jobs	service industries	dominate trade, small businesses and banks; involved in small family craft industries, also many low-skilled, low-paid jobs	trade and professional jobs	semi-skilled jobs

Figure 13.53

Ethnic and religious groups in Singapore

Figure 13.54

Racial harmony poster, Singapore

Daily migration: commuting

A commuter is a person who lives in one community and works in another. There are two types of commuting:

1 **Rural–urban**, where the commuter lives in a small town or village and travels to work in a larger town or city. There is rarely much movement in the reverse direction. The **commuter village** is sometimes also referred to as a **dormitory village** or a **suburbanised village** (page 398).

2 **Intra-urban**, where people who live in the suburbs travel into the city centre for work. This category now includes inhabitants of inner-city areas who have to make the reverse journey to edge-of-city industrial estates and regional shopping centres.

A **commuter hinterland**, or **urban field**, is the area surrounding a large town or city where the work-force lives. Patterns of commuting are likely to develop where:

■ hinterlands are large, communications are fast and reliable (the London Underground), public transport is highly developed and private car ownership is high (south-east England)

■ modern housing is a long way from either the older inner-city industrial areas or from the CBD (as in the New Towns in central Scotland)

■ there is a nearby city or conurbation with plenty of jobs, especially in service industries (London)

■ there is no rival urban centre within easy reach (Plymouth)

■ salaries are high so that commuters can afford travelling costs

■ people feel that their need to live in a cleaner environment outweighs the disadvantages of time and cost of travel to work (people living in the Peak District and working in Sheffield or Manchester)

■ housing costs are high so that younger people are forced to look for cheaper housing further away from their work (as in south-east England)

■ flexible working hours allow people to travel during non-rush-hour times

■ the more elderly members of the workforce buy homes in the country or near to the coast and commute until they retire (the Sussex coast)

■ there have been severe job losses which force people to look for work in other areas/towns (some of the inhabitants of Cleveland work in south-east England).

Optimum, over- and under-population

Optimum population

The **optimum population** of an area is a theoretical state in which the number of people, when working with all the available resources, will produce the highest per capita economic return, i.e. the highest standard of living and quality of life. If the size of the population increases or decreases from the optimum, the output per capita and standard of living will fall. This concept is of a dynamic situation changing with time as technology improves, as population totals and structure change (age and sex ratios), trade opportunities alter, and as new raw materials are discovered to replace old ones which are exhausted or whose values change over a period of time.

The **standard of living** of an individual or population is determined by the interaction between physical and human resources and can be expressed in the following formula:

$$\text{Standard of living} = \frac{\text{Natural resources} \left\{ \begin{array}{l} \text{minerals,} \\ \text{energy,} \\ \text{soils, etc.} \end{array} \right\} \times \text{Technology}}{\text{Population}}$$

Overpopulation

Overpopulation occurs when there are too many people relative to the resources and technology locally available to maintain an 'adequate' standard of living. Bangladesh, Ethiopia and parts of China, Brazil and India are often said to be overpopulated as they have insufficient food, minerals and energy resources to sustain their populations. They suffer from localised natural disasters such as drought and famine; and are characterised by low incomes, poverty, poor living conditions and often a high level of emigration. In the case of Bangladesh (Places 48), where the population density increased from 282 people per km^2 in 1950, to 704 in 1985, and to 1062 in 2008, it is easier to appreciate the problem of 'too many people' than in the case of the north-east of Brazil where the density is less than 2 persons per km^2 (Places 38, page 347).

Underpopulation

Underpopulation occurs when there are far more resources in an area, e.g. of food, energy and minerals, than can be used by the number of people living there. Canada, with a total population of 33 million in 2008, could theoretically double its population and still maintain its standard of living (Places 48). Countries like Canada and Australia can export their surplus food, energy and mineral resources, have high incomes, good living conditions, and high levels of technology and immigration. It is probable that standards of living would rise, through increased production and exploitation of resources, if population were to increase.

However, care is needed when making comparisons on a global scale.

1. There does not seem to be any direct correlation between population density and over-/underpopulation:
    - north-east Brazil is considered to be 'overpopulated' with 2 people per km^2
    - California, despite water problems and pollution, is perceived to be 'underpopulated' with over 600 persons per km^2.
2. Similarly, population density is not necessarily related to gross domestic product (GDP) per capita:
    - the Netherlands and Germany both have a high GDP per capita and a high population density
    - Canada and Australia have a high GDP per capita and a low population density
    - Bangladesh and Puerto Rico have a low GDP per capita and a high population density
    - Sudan and Bolivia have a low GDP per capita and low population density.

The balance of population and resources within a country may also be uneven. For example:

- a country may have a population that is too great for one resource such as energy, yet too small to use fully a second, such as food supply, e.g. Saudi Arabia
- some parts of a country may be well off, e.g. south-east Brazil, while others may be relatively poor, e.g. north-east Brazil.

The relationships between population and resources are highly complex and the terms 'overpopulation' and 'underpopulation' must therefore be used with extreme care.

The latest term to be introduced to try to illustrate the relationship between the increase in the world's population and its effect on the Earth's resources is the **ecological footprint**. This is explained on page 379.

Is Bangladesh overpopulated?

Bangladesh, with 153.5 million inhabitants (2008), has one of the world's highest population densities with 1062 person per km² (Figure 13.55). It has a high, but falling, birth rate (49 per 1000 in 1970, 29 per 1000 in 2008) and fertility rate (7 per woman in 1970, 3 in 2008) together with a falling death rate (28 per 1000 in 1970, 8 per 1000 in 2008). This led to a high and accelerating natural increase from 1.6 per cent in 1950 to 2.7 per cent in 1990 but this fell back to 2.0 per cent in 2008 (page 349). Infant mortality is also falling, but is still very high (140 per 1000 in 1970, 57 in 2008), and life expectancy is increasing (45 years in 1970, 63 in 2008). In 2008, 37 per cent of the population was under 15 years of age but only 3.5 per cent were over 65. The GDP of US$ 1300 is very low, and an estimated 45 per cent are living in poverty (defined by the UN as living on under US$1 a day).

Figure 13.55

High population density in Bangladesh

As most of the country is a flat delta, it is prone to frequent and severe flooding. This results from either flooding by the Ganges and Brahmaputra rivers, mainly due to the monsoon rains and to deforestation in the Himalayas, or from tropical cyclones moving up the Bay of Bengal (Places 19, page 148; Places 31, page 238). Most Bangladeshis are farmers (63 per cent) who live in rural communities (76 per cent urban dwellers). There is a shortage of industry, services and raw materials (it has no energy or mineral resources of note) and the transport network is limited. The low level of literacy (54 per cent male, 32 per cent female) has restricted internal innovation and a lack of capital has meant that the country can ill afford to buy overseas technical skills (its trade is valued at US$ 177 per person per year). In 2007 Bangladesh received US$ 9.31 per person in international aid.

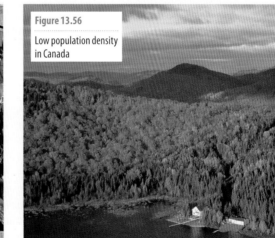

Figure 13.56

Low population density in Canada

Is Canada underpopulated?

Canada, with 33.2 million inhabitants (2008), has one of the world's lowest population densities with just over 3 persons per km² (Figure 13.56). It has a low birth rate (16 per 1000 in 1970, 10 in 2008), a low fertility rate (2.2 per woman in 1970, 1.6 in 2008), a low death rate (7 per 1000 in both 1970 and 2008) and a low infant mortality rate (16 per 1000 in 1970, 5 in 2008) although life expectancy continues to increase (74 years in 1970, 81 years in 2008). Together, these give an extremely low natural increase (1.0 per cent in 1970, 0.8 per cent in 2008). In 2008, only 18 per cent of the population was under 15 years of age but 15 per cent were over 65. The GDP of US$ 38 400 is very high, and less than

10 per cent are said to be living in poverty (that is by Canada's standard, not that of the UN which would be negligible).

Natural disasters, apart from those associated with extreme cold, are rare. Relatively few Canadians are farmers (2 per cent – Places 70, page 486) or live in rural areas: 80 per cent are urban dwellers. Canada has developed industries, services and an efficient transport network, and has utilised its vast energy supplies and mineral resources. The high level of literacy (99 per cent) and national wealth have enabled the country to develop its own technology and to import modern innovations (its trade is valued at US$ 24 954 per person per year). In 2007 Canada gave US$ 93 per person in international aid.

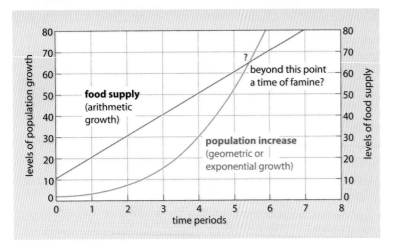

Figure 13.57

Relationships
between population
growth and food
supply (*after* Malthus)

Theories relating to world population and food supply

Malthus

Thomas Malthus was a British demographer who believed that there was a finite optimum population size in relation to food supply and that an increase in population beyond that point would lead to a decline in living standards and to 'war, famine and disease'. He published his views in 1798 and although, fortunately, many of his pessimistic predictions have not come to pass, they form an interesting theory and provide a possible warning for the future. Indeed, his doomsday theory was resurrected in 2007, but due to rising global food prices rather than to food shortages. His theory was based on two principles.

Figure 13.58

Three models
illustrating the
relationships
between an
exponentially
growing population
and an environment
with a limited
carrying capacity

1   Human population, if unchecked, grows at a **geometric** or **exponential rate**, i.e.
    $1 \rightarrow 2 \rightarrow 4 \rightarrow 8 \rightarrow 16 \rightarrow 32$, etc.
2   Food supply, at best, only increases at an **arithmetic rate**, i.e. $1 \rightarrow 2 \rightarrow 3 \rightarrow 4 \rightarrow 5 \rightarrow 6$, etc. Malthus considered that this must be so because yields from a given field could not go on increasing for ever, and the amount of land available is finite.

Malthus demonstrated that any rise in population, however small, would mean that eventually population would exceed increases in food supply. This is shown in Figure 13.57, where the exponential curve intersects the arithmetic curve. Malthus therefore suggested that after five years, the ratio of population to food supply would increase to 16:5, and after six years to 32:6. He suggested that once a ceiling had been reached, further growth in population would be curbed by negative (preventive) or by positive checks.

Preventive (or **negative**) **checks** were methods of limiting population growth and included abstinence from, or a postponement of, marriage which would lower the fertility rate. Malthus noted a correlation between wheat prices and marriage rates (remember that this was the late 18th century): as food became more expensive, fewer people got married.

Positive checks were ways in which the population would be reduced in size by such events as a famine, disease, war and natural disasters, all of which would increase the mortality rate and reduce life expectancy.

The carrying capacity of the environment

The concept of a population ceiling, first suggested by Malthus, is of a saturation level where the population equals the carrying capacity of the local environment. The **carrying capacity** is the largest population of humans/animals/plants that a particular area/environment/ecosystem can carry or support.

Three models portray what might happen as a population, growing exponentially, approaches the carrying capacity of the land (Figure 13.58).

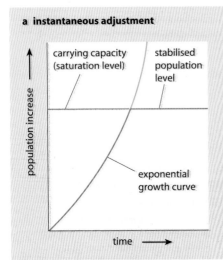

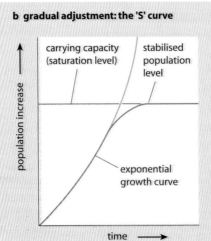

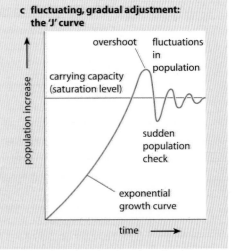

The rate of increase may be unchanged until the ceiling is reached, at which point the increase drops to zero. This highly unlikely situation is unsupported by evidence from either human or animal populations.

More realistically, the population increase begins to taper off as the carrying capacity is approached, and then to level off when the ceiling is reached. It is claimed that populations which are large in size, have long lives and low fertility rates, conform to this 'S' curve pattern.

Here the rapid rise in population overshoots the carrying capacity, resulting in a sudden check – e.g. famine and reduced birth rates. After an initial dramatic fall, the population recovers and fluctuates, then settles down at the carrying capacity. This 'J' curve is more applicable to populations that are small in number, and have short lives and high fertility levels.

Links between population growth, use of resources and economic development

An international team, known collectively as the **Club of Rome**, predicted in 1972, through the use of computers, that if the then rapid trend in population growth and resource utilisation continued, then a sudden decline in economic growth would occur in the next century. Their suggested plans for global equilibrium, few of which have been implemented, included:

- the stabilisation of population growth and the use of resources
- an emphasis on food production and conservation.

At the World Population Conference in Mexico City in 1984, the emphasis was put on taking positive steps to reduce population growth, largely through family planning programmes. The general consensus view articulated the need for population strategies in integration with other development strategies. By 2005, international organisations were suggesting that high population growth rates were a symptom of poverty, not the cause of it. They claimed that all the spending on birth control measures and family planning programmes were having little effect in places where poverty remained the key influence on people's everyday lives.

Ecological footprint

The ecological footprint is a resource management tool that aims to measure the impact of people's lifestyles upon planet Earth. It calculates how much productive land and sea a human population needs to generate the resources it consumes in order to provide all the food, energy, water and raw materials required in people's everyday lives. It also calculates how long it takes to absorb and render harmless the waste that humanity creates or for the ecological balance to renew itself.

Figure 13.59a shows how the ratio between the world's demand and the world's biocapacity has changed over time. Expressed in terms of **'number of planet Earths'**, the biocapacity is always 1 (the horizontal line). The graph shows that whereas in net terms humanity only used about half the planet's biocapacity in 1961, by 2003 this had increased to 1.25 times. The present global ecological deficit of 0.25 represents the world's **ecological overshoot**. This means that as humanity's ecological footprint is 25 per cent more than the planet can regenerate, it now takes one year and three months for the Earth to replace what people use and the waste they create in a single year. By measuring the ecological footprint of a population (a person, a city, a country, and even all humanity) we can assess our overshoot and should, therefore, be able to manage the Earth's ecological resources more carefully.

While the term 'ecological footprint' is now being more widely used and understood, methods of measuring it still vary, although some calculation standards are now emerging. Figure 13.59b lists the countries with the greatest global ecological surplus and the greatest ecological deficit. In 2003, the most recent year for data to be available, the total biocapacity for the world was 2.26 global ha/person. This figure was reached by adding together the global ha/person for each of the following footprints: cropland 0.49, grazing land 0.15, forest 0.23, fishing grounds 0.15, carbon 1.07 (page 638), nuclear 0.09 and built-up land 0.08.

Figure 13.59

The world's ecological footprint
a Human demand and the Earth's biocapacity
b Countries with the greatest global ecological deficit and surplus

a

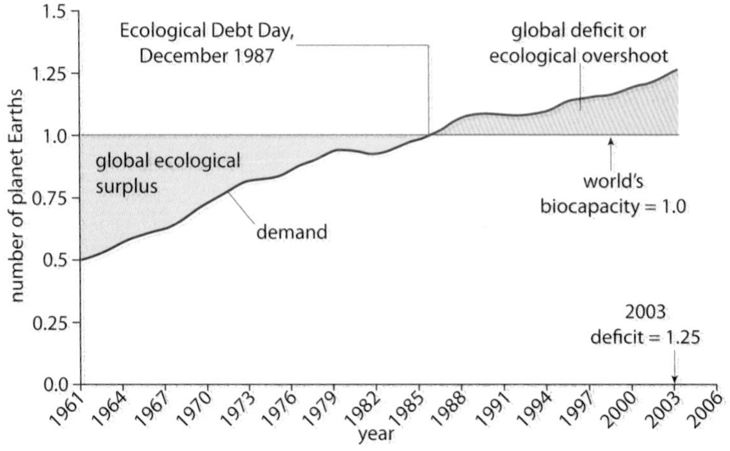

b

Global ecological footprint			
Surplus		Deficit	
1 Gabon	17.8	UAE	−11.0
2 Bolivia	13.7	Kuwait	−7.0
3 New Zealand	9.0	USA	−4.8
4 Mongolia	8.7	Belgium	−4.4
5 Brazil	7.8	Israel	−4.2
6 Congo	7.2	UK	−4.0
7 Canada	6.9	Saudi Arabia	−3.7
8 Australia	5.9	Japan	−3.6

Other selected countries: Germany −2.4, China −0.9, India −0.4, Kenya and Bangladesh −0.2, Ghana +0.3, Malaysia +1.5, Korea, Sweden and Spain each +3.5

China had, in mid-2008, an estimated population of 1.33 billion, which was 20 per cent of the world's total. As in other countries, this population was far from evenly distributed (Figure 13.1), with 95 per cent living on only 40 per cent of the total land area (Figure 13.60). A population density map (Figure 13.61) shows that the highest densities are either in coastal provinces or in the middle and lower Yangtze Basin and the lowest in the mountains and deserts of the north and west. Despite China's large population, the country does not have a particularly high density – only 138 per km^2 (half that of the UK's 273 per km^2).

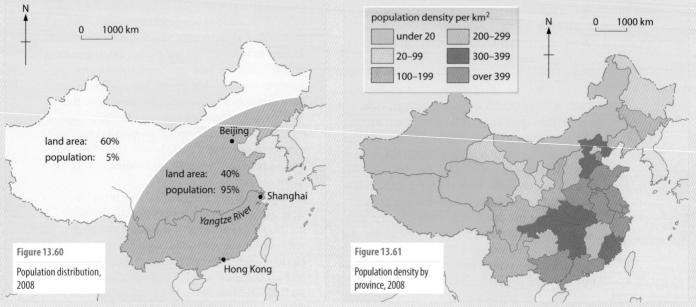

land area: 60%
population: 5%

land area: 40%
population: 95%

Beijing

Shanghai

Yangtze River

Hong Kong

Figure 13.60

Population distribution, 2008

population density per km^2

under 20	200–299
20–99	300–399
100–199	over 399

Figure 13.61

Population density by province, 2008

Figure 13.62 shows the demographic cycle for China since the formation of the People's Republic in 1950; and Figure 13.63 China's age structure based on estimates for mid-2007.

The high birth rate of the 1950s was a response to the state philosophy that 'a large population gives a strong nation', and people were encouraged to have as many children as possible. At the same time, death rates were falling, mainly due to improved food supplies and medical care.

The period between 1959 and 1961 coincided with the 'Great Leap Forward'. It was a time when industrial production had to be increased at all costs, and little attention was paid to farming (Places 63, page 468). The result was a catastrophic famine in which an estimated 20 million people died; infant mortality rates rose and birth rates fell.

During the 1960s, attempts to control population growth were thwarted by the Cultural Revolution. Every three years, China's population increased by 55 million – equal to the UK's total population at that time.

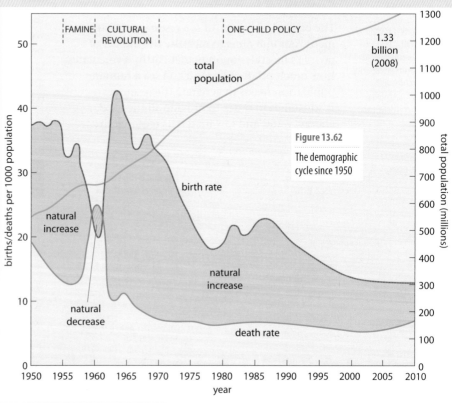

FAMINE CULTURAL REVOLUTION ONE-CHILD POLICY

1.33 billion (2008)

total population

birth rate

natural increase

natural increase

natural decrease

death rate

Figure 13.62

The demographic cycle since 1950

State family planning programmes were introduced in the 1970s and by 1975 the average family size had fallen to three children. The state, considering this to be still too high, began an advertising campaign for *wan-xi-shao* – 'later, longer, fewer' (later marriages, longer gaps and fewer children). Concern grew with the realisation that with millions of couples about to enter the child-bearing age group, the country's population could double within 50 years. A Chinese demographer, on little supportive evidence, calculated that China's optimum population was 700 million (at that time it was already almost 1000 million) and recommended that the state aimed to reduce its population to that figure by 2080. To achieve this, the TFR (page 357) would have to be reduced to a maximum of 1.5.

In 1979 the state decided to 'play safe' and introduced a rigorous carrot-and-stick 'one-child policy' (i.e. to achieve a TFR of 1.0). The 'carrot' for having only one child included free education, priority housing and family benefits, while the 'stick' imposed after the birth of a second child included the loss of these benefits, heavy fines and even forced abortions and threats of sterilisation. The marriageable age was raised and couples had to apply for permission to marry and, later, to have a child. The state did, however, begin to give education on family planning and in a relatively short time over 80 per cent of married women had access to contraception – no mean achievement in view of China's then lack of economic development, its huge size and its mainly rural population. Even so, reports coming out of the country did refer to female infanticide.

The apparently rigid state controlled one-child policy, which has proved successful if the sole aim was to limit population growth, did, however, have many exceptions and loopholes. During 1999, the present author was told, during a month in China researching for the previous edition of this book, that the 'one-child policy was very complicated' and a more recent joint Chinese–American report has claimed that only about 63 per cent of the total population were ever subject to its regulations. The complications, exceptions and loopholes resulted from particular circumstances:

- The Han, who form the ethnic majority (92 per cent of the total population) and were the more likely to live in the expanding urban areas, were the most severely restricted to one child (Figure 13.64) unless their firstborn was mentally or physically handicapped, or died when young.

- Minority groups, of whom there are 56 recognised 'minority nationalities' and whose combined population is now 104 million or 8 per cent of China's total, live mainly in the outlying provinces. They were allowed two children (Figure 13.66) or, if they lived in very remote areas where officials were few in number, possibly up to four children.

- Those Han who lived, often in large numbers, in rural areas where boys were allegedly needed to help work on the farm (the author saw as many girls working in the fields as he did boys) were allowed a second child if the firstborn was a girl (if the second was also a girl, then that was that!).

- The Han who lived in rural areas and who had a second child were often allowed to keep it on payment of a fine. The scale of these fines varied between provinces and often depended on the degree of honesty of local officials.

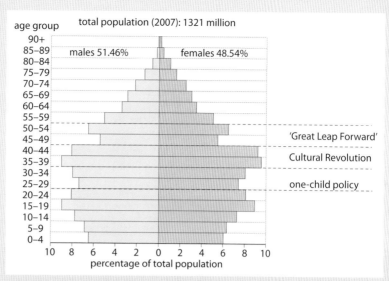

total population (2007): 1321 million

males 51.46%   females 48.54%

'Great Leap Forward'

Cultural Revolution

one-child policy

percentage of total population

Figure 13.63

Age structure of Chinese population, mid-2007

Figure 13.64

One-child family, Kunming, Yunnan

- People working for state firms could be made redundant on the birth of a second child, unlike those who were employed by transnational (overseas) corporations.
- In the event of twins, the state paid the extra costs.

Over a period of time:

- as more married couples received a better education, fewer began to apply even for a single child
- when the first children of one-child families reached marriageable age (women in 2000, men in 2002) then if two 'only' children married they were allowed two children.

In the author's experience, most people, especially those living in urban areas, seem to have accepted the necessity of the policy. Those in their early twenties, who were the first generation to be 'only' children, admitted that while they would have liked to have had a brother or sister, they acknowledged that the policy had helped their family and the local community raise their standard of living. Figure 13.65 gives some of the pros and cons of the one-child policy viewed from a 2008 perspective.

Figure 13.66

Two-boy family, Lijiang, Yunnan

Benefits	Problems
• The birth rate fell from 44 per 1000 in 1950 to 14 in 2008, and the TFR from 6 to 1.5 in the same period (Figure 13.68).	• The birth rate is now lower than the replacement rate (page 358).
• In the first 20 years, only 70 million children, instead of 300 million, were born.	• There is a rapidly ageing population with an increasing number of over 60s (page 383) who are dependent on fewer people in the working/productive age group (page 354).
• It was claimed in 2008 that China's population was 400 million less than it would have been without the policy (the present population of North America is only 338 million).	• China's rapid industrialisation is threatened by a shortage of workers for its factories.
• There is far less pressure on land, water, energy and other resources.	• There is a gender imbalance (page 383) due to, in the early days of the policy, female infanticide; abortion of female foetuses is now illegal. In 2005 the male to female ratio was 118 : 100.
• There has been a greater increase in people's standard of living and, according to the UN, an estimated 120 million people have been lifted out of absolute poverty.	• There is international criticism on grounds of human rights.

Figure 13.65

Consequences by 2008 of the one-child policy

China sticking with 'one-child' policy

Although China's 'one-child' policy and family planning policies have softened over the years, the Minister for the National Population and Family Planning Commission announced that the country's 'one-child-per-couple policy' would not change for at least another decade until the present surge in birth rate subsides. This refuted speculation that officials were contemplating adjustments to compensate for mounting uneven demographic distributions in age and gender. The Minister said that 200 million people would still be entering child-bearing age in the next ten years and that prematurely abandoning the one-child policy could add unwanted volatility to birth rates. With such a large population base [Figure 13.63], this could lead to serious problems and extra pressure on social and economic development.

Figure 13.67

Adapted from *China Daily*, March 2008

As shown in Figure 13.65, China is faced with an imbalance of people in the working age and ageing groups, as well as between the two genders. In the last few years, some provincial authorities, notably Beijing where the replacement rate is not being met, have:

- abolished quotas for child births and replaced them with voluntary family planning programmes that allow a wide choice of contraception types
- relaxed penalties on those having larger families
- allowed more exceptions to the one-child rule, e.g to Sichuan families who lost their children in the earthquake (Places 2, page 11).

Even so, the state announced in 2008 that the one-child policy would be likely to remain in place for at least another decade (Figure 13.67).

Whereas China's major demographic concern in the 1970s and 1980s was population growth, by the beginning of the 21st century it had become that of an ageing population (Figure 13.69) resulting from an increase in life expectancy. Figure 13.70 shows that a person born in 2008 can expect to live 33 years longer than one born in 1950 (men 71.4 years, women 75.2 years). While the problem of ageing is increasingly affecting many developed countries, it is more acute in China than elsewhere. This is partly due to the distortions created by the baby boom encouraged under Chairman Mao in the 1950s and 1960s (notice the 35–54 age groups in China's age structure in Figure 13.63) and partly to the recent improvements in health care. Predictions for the proportion of those aged over 60 are:

1990	4 per cent (of total population)
2008	11 per cent (total of 143 million)
2020	16 per cent
2050	30 per cent (about 430 million).

This means that the ratio of workers to elderly dependants will fall considerably from 10:1 in 2008 to 3:1 by 2030 (Figure 13.68). Also, in time, more single children will have to support up to two parents and four grandparents – the so-called 4-2-1 pattern. This pattern is more common in rural areas, where grandparents still tend to live, as they always have done, with the family – a situation that is less common in large urban areas.

Figure 13.68

Predicted population growth based on different total fertility rates

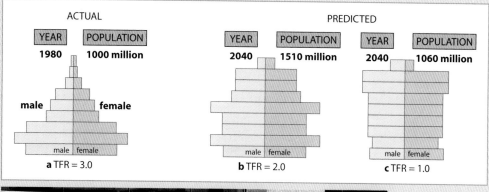

ACTUAL — PREDICTED

YEAR	POPULATION
1980	**1000 million**

male female

| male | female |

a TFR = 3.0

YEAR	POPULATION
2040	**1510 million**

| male | female |

b TFR = 2.0

YEAR	POPULATION
2040	**1060 million**

| male | female |

c TFR = 1.0

Figure 13.69

Ageing Naxi women, Lijiang, Yunnan

Figure 13.70

Increase in life expectancy, 1950–2008

The World Bank claims that China will have an old-age burden of a high-income country such as Japan with only the financial resources of a middle-income economy to shoulder it.

These changes will, according to the *China Business Handbook* (2008), 'have a massive impact on Chinese society and will require urgent reform of the provision of pensions, healthcare and benefits. At present, there is no pension scheme at all for the majority of the population, especially in rural areas, although several pilot schemes have been introduced into cities across the country'. As those same words were expressed in the 1999 edition, it would appear that little progress has been made.

The imbalance between gender, resulting from the traditional preference for boys, is another concern. A recent survey suggests that there are up to 118 newborn Chinese boys to every 100 girls. Statistics, which probably underestimate the ratio, suggest that 99 cities across China have a ratio even in excess of 125:100. These ratios compare with a world average of 105 male births to every 100 female births and a UN recommended ratio of 107:100.

Further reference

Bailey, A. (2005) *Making Population Geography*, Hodder Arnold.

Browning, K. (2002) 'Shanty towns: A South African case study', *Geography Review* Vol 15 No 5 (May).

Clark, J. and Craven, A. (2006) 'Growing grey', *Geography Review* Vol 20 No 2 (November).

Connelly, M. (2008) *Fatal Misconception: The struggle to control world population*, Harvard University Press.

Dahl, T. *et al*. (2006) 'Rural–urban migration in China', *Geography Review* Vol 19 No 5 (May).

Johnston, R. and Poulsen, M. (2007) 'London's changing ethnic Geography', *Geography Review* Vol 21 No 1 (September).

Storey, D. (2006) 'People on the move: Refugees and asylum seekers', *Geography Review* Vol 19 No 5 (May).

2001 Census UK:
www.statistics.gov.uk/Census2001/population_data.asp

Ageing population:
www.un.org/esa/population/publications/WPA2007/wpp2007.htm

China Population Information and Research Center:
www.cpirc.org.cn/en/eindex.htm

CIA World Fact Book:
www.cia.gov/library/publications/the-world-factbook/

Population:
www.popact.org

www.populationconcern.org.uk

Population Reference Bureau:
www.prb.org/Datafinder.aspx

Union of Concerned Scientists (UCS), population/environment:
www.ucsusa.org/

United Nations (UN):
www.un.org/Pubs/CyberSchoolBus/index.html

www.unfpa.org/

UN High Commissioner for Refugees:
www.unhcr.ch/

UN Population Division, population reports:
http://esa.un.org/unpp/index.asp

UN Population Division, searchable database:
www.un.org/esa/population/unpop.htm

US Census Bureau, International Programs Center:
www.census.gov/ipc/www/

World Bank:
www.worldbank.org/data&Statistics

Activities

1 a What do the following terms mean:
   i   total fertility rate   (2 marks)
   ii natural increase of population   (2 marks)
   iii annual growth rate of population?   (2 marks)

   b Study Figure 13.11 on page 351. What statistical change marks the move from:
   i   Stage A (high, fluctuating) to Stage B (early expanding)
   ii Stage B to Stage C (late expanding)
   iii Stage C to Stage D (low, fluctuating)?   (3 marks)

   c Explain how social and/or economic changes could have brought about each of the moves described in **b**. (10 marks)

   d Suggest how the total population of the UK might change over the next 50 years.
      Give reasons for your suggestions.   (6 marks)

2 Study the two population pyramids in Figure 13.71.
   a i   What do you understand by the term 'dependency ratio'?
                                                                 (2 marks)
      ii Suggest, with reasons, what stage of the demographic transition is represented by each of the pyramids.
                                                                 (4 marks)

   b Choose **one** country that has a population structure similar to the one shown in pyramid A.
      i   Suggest two problems that are likely to arise in that country as a consequence of the large proportion of the population in the 0–15 age group.   (4 marks)
      ii How is the country attempting to manage these problems?   (7 marks)

   c Choose **one** country that has a population structure similar to that shown in pyramid B.
      Describe two problems that might arise in future as a consequence of the ageing population structure, and suggest how these problems might be managed.   (8 marks)

3 Study the two tables of data in Figure 13.72.
   a i   Describe how the education of women has affected fertility rates in Morocco.   (3 marks)
      ii Suggest reasons for the changes you have described.
                                                                 (4 marks)

a Average number of children per Moroccan women according to age and educational level, 1994

Figure 13.72

Age group	No education	Primary	Secondary	Higher
20–24	1.6	1.1	0.7	0.5
25–29	2.7	1.7	1.3	0.9
30–34	3.9	2.6	1.9	1.3
35–39	5.1	3.4	2.4	2.2
40–44	6.1	4.2	3.2	2.7
45–49	6.7	4.9	3.4	2.5

b Fertility indices (average number of children born to each woman) of Maghreb women in country of origin and country of residence*

Year	Algerian women in Algeria	Algerian women in France	Moroccan women in Morocco	Moroccan women in France	Tunisian women in Tunisia	Tunisian women in France
1977	7.47	4.73	5.93	5.75	5.84	5.05
1981	6.39	4.35	5.92	5.84	5.19	–
1985	6.24	4.24	–	4.47	4.53	4.67
1987	5.29	3.95	4.46	4.09	4.10	4.49
1989	4.72	3.66	3.95	3.71	3.40	4.30
1991	–	3.35	–	3.25	3.34	3.88
1992	–	3.27	3.28	2.99	3.36	3.56

* 'The Maghreb' is the western part of North Africa. It was colonised by France, gaining independence in the 1950s and 1960s. These countries still have close ties with France, and there has been much migration from Maghreb to France.

   b i   Describe the changes to the fertility indices of women from the Maghreb who have migrated to France.
                                                                 (3 marks)
      ii Suggest reasons for these changes.   (4 marks)
      iii Suggest how the emigration from the Maghreb may have affected the fertility rate of women who have remained in those countries.   (5 marks)

   c What lessons can be learned from these figures by development workers in countries suffering from pressure caused by rapidly increasing population?
                                                                 (6 marks)

4 a What do you understand by:
   i   birth rate   (1 mark)
   ii life expectancy   (1 mark)
   iii overpopulation?   (2 marks)

   b Figure 13.73 illustrates Malthus's view of the relationship between population and food supply in a typical country or region.
   i   Describe what the diagram shows.   (2 marks)
   ii Explain what Malthus thought would be the consequences of the changes shown in the model.
                                                                 (2 marks)

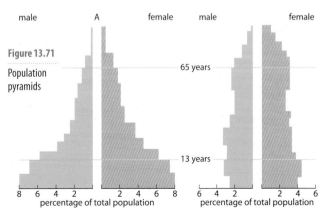

Figure 13.71

Population pyramids

Growth of population	1	2	4	8	16	etc.
Growth of food supply	1	2	3	4	5	etc.
Time periods	→	→	→	→	→	→

Figure 13.73

Malthus's view of population and food supply

iii Suggest why Malthus's predictions did not come true in England following publication of his ideas in the early 18th century. *(4 marks)*

iv Describe the views of Boserup on the balance between population and resources, and explain how these are different from the views of Malthus. *(5 marks)*

Exam practice: basic structured questions

5 a Study the map of Brazil's population distribution (Figure 13.5 on page 347).

i Describe the distribution of areas of dense population. *(2 marks)*

ii The area marked 1B is the tropical rainforest. Suggest why this area, or any other area of tropical rainforest that you have studied, has a very sparse population. *(6 marks)*

b Study the map of population density in London (Figure 13.7 on page 348).

i Describe the distribution of population shown on this map, and explain why this pattern has developed. *(7 marks)*

ii During the 20th century there was a large movement of people out from central London into the suburbs and beyond. Explain why people wished to move, and how changing technology allowed them to make the move. *(10 marks)*

c Choose **one** country that has attempted to manage its population by introducing laws that it hopes will affect birth rate.

i Explain how the population policy was intended to operate.

ii Discuss the consequences of the policy, mentioning both its successes and its failures. *(8 marks)*

6 The period following the Second World War saw some of the biggest international migrations that the world has ever known.

a Name one major international migration that took place during this period. Refer to the source and the destination of the migrants. *(1 mark)*

b Explain the causes of the migration. Refer to pushes from the source and pulls to the destination. *(6 marks)*

c Discuss the consequences of the migration for:

i the source country

ii the host country

iii the migrants themselves. *(12 marks)*

d Suggest why large international migrations have been so common in the period since the Second World War. *(6 marks)*

Exam practice: structured questions

7 Study Figure 13.73 above.

a Outline the theory developed by Malthus to explain the relationship between population increase and the increase in food supply. *(5 marks)*

b Malthus wrote in the early 18th century. He predicted that population growth could soon cause widespread famine and other disasters in England. His predictions have not come true. Explain why. *(10 marks)*

c In recent years views described as neo-Malthusian have become common. Explain why these ideas have developed. Contrast the neo-Malthusian view with the more optimistic view of population growth developed by Boserup. *(10 marks)*

8 a Study Figure 13.22 on page 359.

i Describe the range of life expectancy figures shown by this table and comment on the changes shown over time. *(6 marks)*

ii Choose **one** country in the table with an increasing life expectancy and account for the changes that have been observed and that are predicted. *(7 marks)*

b Name a country that you have studied, where life expectancy has fallen in the last 10–20 years. Explain the causes and the consequences of this fall. *(12 marks)*

9 a At what stage of the demographic transition model is population growth most rapid? Give the reasons for this rapid growth. *(5 marks)*

b Many demographers say that the key to reducing the birth rate in less economically developed countries lies in changing the educational and economic status of women. Discuss this view, with reference to one or more countries that you have studied. *(10 marks)*

c Name **one** country that has adopted policies designed to deal with a rapidly growing population.
Describe the policies and evaluate their success. *(10 marks)*

10 Study the two population pyramids in Figure 13.71 on page 384.

a i Compare and contrast the shapes of the two pyramids. *(3 marks)*

ii Account for the differences between the two pyramids. *(4 marks)*

iii Suggest what population problems are likely to be met in these two countries during the next 20 years or so. *(8 marks)*

b i Name a country that has adopted policies to help it to manage its total population and its rate of population change.

ii Describe its population policies and assess how successful these policies have been. *(10 marks)*

11 **a i** Name a country where immigration has led to the development of a 'multicultural society'. Name the main cultural groups that make up that country's population.
(2 marks)

   ii Explain the causes of the immigration into the country.
(5 marks)

   iii Outline the pattern of distribution of **one** group or people who recently migrated into the country.
(5 marks)

   b Discuss some of the geographical issues caused by the development of a multicultural society in the country named in **a i**.
(13 marks)

Exam practice: essays

12 Many countries in Africa, Asia and Latin America have experienced very rapid population growth since 1950, but now rates of population growth are slowing down in many of these countries.

Discuss the factors affecting the rate of population change in a range of countries in Africa, Asia and Latin America. *(25 marks)*

13 The term 'ecological footprint' is in wide use as a way of assessing the sustainability of the lifestyle of a person, a family, a region or a country.

Explain the meaning of the term, and evaluate the usefulness of the concept of the ecological footprint. *(25 marks)*

Issues Analysis

Population policies – the pros and cons of trying to limit population growth

Across the world and over time countries have adopted a wide range of differing policies to limit their population growth. The characteristics of such policies vary according to the urgency of the situation, the politics of the country and the approach taken. Policies can be compulsory – they tell people what they can and cannot do and use 'sticks' (penalties) to enforce this. Equally, they can lead by example, encouraging people towards certain behaviour with 'carrots' (rewards).

China

With over 1.3 billion people (2005) – one-fifth of global population – China will soon overtake the UK as the world's fourth richest nation. In the early 21st century its economy has grown three times faster than that of the USA (before the 'credit crunch' of 2008, which affects both nations). China has used vast quantities of global raw materials to fuel this unprecedented economic growth and is likely to continue to do so.

People are China's asset and its problem. Without them its labour-intensive industrial growth would not have taken off. Most industry is labour intensive and labour is cheap. However, the government of the 1970s realised that continued rapid growth would create a demographic and economic crisis.

Massive famine was forecast by the end of the end of the 20th century if population growth was not stemmed. This was the reason for the unprecedented 'one-child' policy. Few other countries could have, or ever will have, such a radical policy. Communist governments, such as China's, have great control over people. Democracies vote for their governments, so people have a say in what they do.

Interpretation of the Chinese 'one-child' policy

Read about China's 'one-child' policy above and on pages 381–383 in this chapter.

1 **a** Write down your initial reactions to, and opinions of, this population policy.

   b List the ways in which this policy has been a success.

   c What problems have arisen from the policy, in your opinion?

   d Discuss your ideas with others in your class. Revise any aspects of your answers to (b) and (c), based on these discussions.

2 **a** Summarise the possible future directions of the 'one-child' policy.

   b Which future direction do you consider to be the most practical for China as a country, and why?

   c Which future direction do you consider to be the best for individual Chinese people, and why?

India

'Each year India adds more people to the world's population than any other country.'

Geofile 521, April 2006, by Tim Bayliss and Lawrence Collins, Nelson Thornes

By 2025 India's population (1.1 billion in 2006) will probably overtake that of China. To a limited extent, India has followed similar policies to China to curb its rapid growth, but more free will and persuasion have been employed. Total fertility rate (TFR) has decreased to 3.5 per woman from over 5.0 in the 1970s, suggesting success, but changes are far from even across this large nation. India's growth represents a very different set of challenges from those before the Chinese government.

India's first national family programme began as early as 1952. It has not always been popular; Prime Minister Indira Gandhi's method of encouraging young men to be sterilised in return for something as small as a radio became part of a huge backlash, contributing to the fall of her government. India has great cultural, religious, socio-economic and geographical variations. It might be a case of one policy does not fit all.

Interpretation of India's policies

3 Figure 13.74a shows varying birth rates across India in 2007, while Figure 13.74b shows adult literacy data by region.

   a Describe the patterns shown in the two maps.

   b Is there any correlation between the data in the two maps?

   c Based on the data provided in Figure 13.74a only, summarise the ways Kerala stands out from other regions of India.

4 **a** Research different aspects of India's population control policy. The websites suggested (right) will give you a start.

   b How does Kerala's population policy differ from those followed elsewhere in the country?

Decision-making exercise

5 The population policies of China and India, in particular of Kerala state (Figure 13.75), show very different approaches to the need to limit rapid population growth. Discuss the differing approaches, considering the ways in which people are forced/encouraged to limit family size.

6 Put yourselves in the place of the government of a relatively poor African country with rapid population growth. Assume that this country is a democracy. Devise a population policy suitable for the needs of the country which will not damage its slowly growing economy, yet will also not be too unpopular with the electorate. Consider the best ways to convince your people (who are also your electorate) that this national policy will benefit them as individuals and the country as a whole.

Kerala state in south-west India has one of the lowest birth rates in the country at less than 1 per cent per annum. Women have higher status here than in much of the rest of the country and are encouraged to work outside the home. Education is valued by government and families are encouraged to see that fewer children and better education lead to greater life chances and a better standard of living. The state policy aims to encourage:

- later marriage
- wider spacing of births
- contraception use by married women – the majority do so
- sterilisation once the desired family size is reached
- value placed on education, which can be achieved better in a smaller family.

Spacing of births has not been successful. Parents have tended to have their desired number of children close together and then opted for sterilisation to prevent further births. Abortion is sometimes used, but this is often to achieve the desired gender balance in the family. Between 40 and 75 per cent of abortions are thought to be for this reason. While less than ideal, this does show that people are trying to produce smaller families.

Some 91 per cent of adults are literate – much higher than the national 65 per cent average. This shows that education initiatives are working and perhaps also influences people's reproductive behaviour.

Figure 13.75

Population policy in the state of Kerala

Further reference

http://en/wikipedia.org/wiki/Demographics_of_India

http://news.bbc.co.uk/1/hi/programmes/from_our_own_correspondent/3602862.stm

BNET.com: 'India's Population Reality: Reconciling Change and Tradition'

Geofile numbers 454 (September 2003); 507 (September 2005); 521 (April 2006)

a

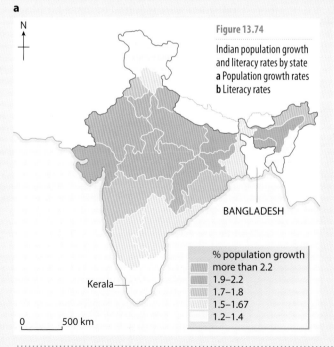

Figure 13.74

Indian population growth and literacy rates by state
a Population growth rates
b Literacy rates

% population growth
- more than 2.2
- 1.9–2.2
- 1.7–1.8
- 1.5–1.67
- 1.2–1.4

BANGLADESH

Kerala

0 500 km

b

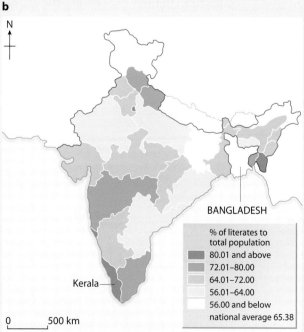

% of literates to total population
- 80.01 and above
- 72.01–80.00
- 64.01–72.00
- 56.01–64.00
- 56.00 and below
national average 65.38

BANGLADESH

Kerala

0 500 km

14

Settlement

• •

'The largest single step in the ascent of man is the change from nomad to village agriculture.'

J. Bronowski, *The Ascent of Man*, 1973

Origins of settlement

About 8000 BC, at the end of the last ice age, the world's population consisted of small bands of hunters and collectors living mainly in sub-tropical lands and at a subsistence level (page 478). These groups of people, who were usually migratory, could only support themselves if the whole community was involved in the search for food. At this time two major technological changes, known as the 'Neolithic revolution', turned the migratory hunter-collector into a sedentary farmer. The first was the domestication of animals (sheep, goats and cattle) and the second the cultivation of cereals (wheat, rice and maize). Slow improvements in early farming gradually led to food surpluses and enabled an increasing proportion of the community to specialise in non-farming tasks.

The evolution in farming appears to have taken place independently, but at about the same time, in three river basins: the Tigris–Euphrates (in Mesopotamia), the Nile, and the Indus (Figure 14.1). These areas had similar natural advantages:

■ hills surrounding the basins provided pasture for domestic animals
■ flat floodplains next to large rivers
■ rich, fertile silt deposited by the rivers during times of flood
■ a relatively dry – but not too dry – climate which maintained soil fertility (i.e. limited leaching) and enabled mud from the rivers to be used to build houses (climatically, these areas were more moist than they are today)
■ a warm subtropical climate, and
■ a permanent water supply from the rivers for domestic use and, as farming developed, for irrigation.

By 1500 BC, larger towns and urban centres had developed with an increasingly wider range of functions. Administrators were needed to organise the collection of crops and the distribution of food supplies; traders exchanged surplus goods with other urban centres; early engineers introduced irrigation systems; and a ruling elite appropriated taxes from the agricultural and trading population to support the military, the priesthood, and 'non-productive' members of society, such as artists, philosophers and astronomers. Craftsmen were required to make farming equipment and household articles – the oldest-known pottery and woven textiles were found at Catal Huyuk in present-day Turkey – and copper

Figure 14.1

Civilisations and cities before 1500 BC

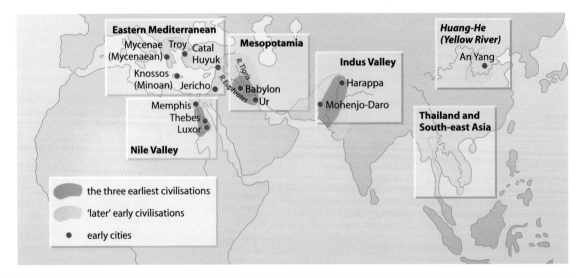

Eastern Mediterranean
Mycenae Troy
(Mycenaean) Catal Huyuk
Knossos
(Minoan) Jericho

Mesopotamia
R.Tigris
R.Euphrates
Babylon
Ur

Huang-He
(Yellow River)
An Yang

Indus Valley
Harappa
Mohenjo-Daro

Thailand and
South-east Asia

Memphis
Thebes
Luxor

Nile Valley

the three earliest civilisations

'later' early civilisations

• early cities

and bronze were being worked by 3000 BC. As towns continued to grow, it became necessary to have a legal system and an army for defence.

Although there is divergence of opinion over the exact dates, Figure 14.2 gives a chronological sequence of early settlements.

Approximate date BC		Near East {Tigris–Euphrates, Nile, Indus}	Rest of world
9000		Hunters and collectors	
8000	8500	First domesticated animals and cereals	Northern Europe recovering from the last ice age
	8300	Jericho: first walled city	
7000			
6000	6250	Catal Huyuk: first pottery and woven textiles; became largest city in world	
5000	5500	Growth of villages in Mesopotamia Growth of many villages in Nile and Indus valleys	
	5000	Early methods of irrigation	Rice cultivation in South-east Asia
4000		Bronze casting	
3000	3500	Invention of the wheel and plough in Mesopotamia, and the sail in Egypt	First Chinese city
	3000	Cities in Mesopotamia	First crops grown in central Africa; bronze worked in Thailand
2000	2600	Pyramids	
	2000	Minoan civilisation in Crete	Metal-working in the Andes
1000	1600	Mycenaean civilisation in Greece	

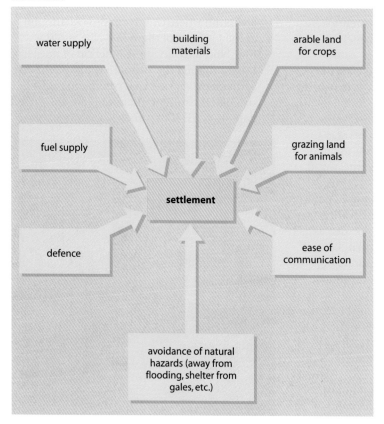

Site and situation of early settlements

Site describes the characteristics of the actual point at which a settlement is located, and was of major importance in the initial establishment and growth of a village or town. **Situation** describes the location of a place relative to its surroundings (neighbouring settlements, rivers and uplands). Situation, along with human and political factors, determined whether or not a particular settlement remained small or grew into a larger town or city (Figure 14.9).

Early settlements developed in a rural economy which aimed at self-sufficiency, largely because transport systems were limited. While the most significant factors in determining the site of a village include those shown in Figure 14.3 and described below, remember that several factors would usually operate together when a choice in the location of a settlement was being made.

Among the most important factors are:

■ **Water supply** A nearby, guaranteed supply was essential as water is needed daily throughout the year and is heavy to carry any distance. In earlier times, rivers were sufficiently clean to give a safe, permanent supply. In lowland Britain, many early villages were located along the spring line at the foot of a

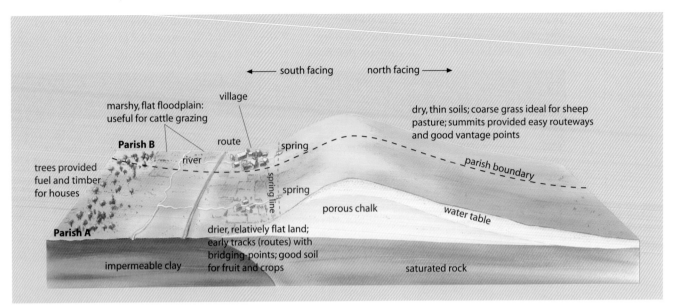

Labels within figure:
south facing ← → north facing

marshy, flat floodplain: useful for cattle grazing

village

dry, thin soils; coarse grass ideal for sheep pasture; summits provided easy routeways and good vantage points

Parish B

route

spring

river

trees provided fuel and timber for houses

spring line

spring

parish boundary

Parish A

porous chalk

water table

drier, relatively flat land; early tracks (routes) with bridging-points; good soil for fruit and crops

impermeable clay

saturated rock

Figure 14.4

A spring-line village at the foot of the North Downs, south-east England

chalk or limestone escarpment (Figures 8.10 and 14.4). In regions where rainfall is limited or unreliable, people settled where the water table was near to the surface (a desert oasis, Figure 14.5) enabling shallow wells to be dug. Such settlement sites are known as **wet-point** or **water-seeking sites**.

■ **Flood avoidance** Elsewhere, the problem may have been too much water. In the English Fenlands, and on coastal marshes, villages were built on mounds which formed natural islands (Ely). Other settlements were built on river terraces (page 82) which were above the flood level and, in some cases, avoided those diseases associated with stagnant water. Such sites are known as **dry-point** or **water-avoiding sites**.

■ **Building materials** Materials were heavy and bulky to move and, as transport was poorly developed, it was important to build settlements close to a supply of stone, wood and/or clay.

■ **Food supply** The ideal location was in an area that was suitable for both the rearing of animals and the growing of crops – such as the scarps and vales of south-east England (page 199). The quality, quantity and range of farm produce often depended upon climate and soil fertility and type.

■ **Relief** Flat, low-lying land such as the North German Plain was easier to build on than steeper, higher ground such as the Alps. However, the need for defence sometimes overruled this consideration.

■ **Defence** Protection against surrounding tribes was often essential. Jericho, built over 10 000 years ago (about 8350 BC), is the oldest city known to have had walls. In Britain, the two best types of defensive site were those surrounded on three sides by water (Durham, Figure 14.6) or built upon high ground with commanding views over the surrounding countryside (Edinburgh). Hilltop sites may, however, have had problems with water supply (Figure 14.7).

■ **Nodal points** Sites where several valleys meet were often occupied by settlements which became **route centres** (Carlisle – Places 49, page 396 – and Paris). **Confluence towns** are found where two rivers join (Khartoum at the junction of the White Nile and the Blue Nile, St Louis at the junction of the Mississippi and the Missouri (Figure 3.59)). Settlements on sites that command routes through the hills or mountains are known as **gap towns** (Dorking and Carcassonne).

Figure 14.5

An oasis: Morocco

Figure 14.6

A settlement within a meander loop: Durham

Figure 14.7

A hilltop defensive site: Andalucia, Spain

■ **Fuel supply** Even tropical areas need fuel for cooking purposes as well as for warmth during colder nights. In most early settlements, firewood was the main source – and still is in many of the least economically developed areas, such as the Sahel.

■ **Bridging-points** Settlements have tended to grow where routes had to cross rivers, initially where the river was shallow enough to be forded (Oxford) and later where the site was suitable for a bridge to be built. Of great significance for trade and transport was the lowest bridging-point before a river entered the sea (Newcastle upon Tyne).

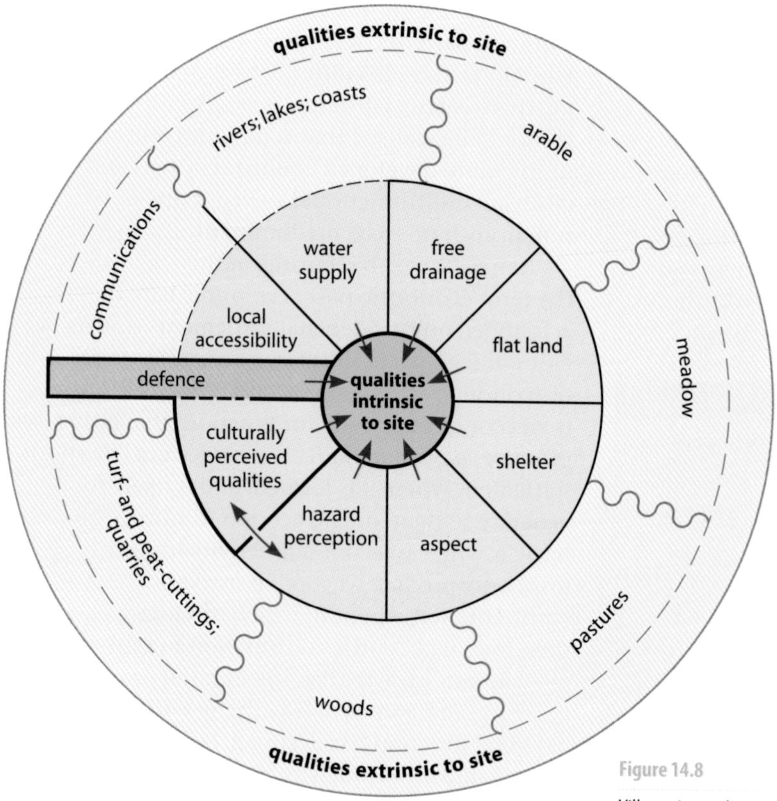

■ **Harbours** Sheltered sea inlets and river estuaries provided suitable sites for the establishment of coastal fishing ports, such as Newquay in Cornwall; later, deep-water harbours were required as ships became larger (Southampton and Singapore, Places 104, page 636). Port sites were also important on many major navigable rivers (Montreal on the St Lawrence) and large lakes (the Great Lakes in North America).

■ **Shelter and aspect** In Britain, south-facing slopes offer favoured settlement sites because they are protected from cold, northerly winds and receive maximum insolation (Torquay).

■ **Resources** Settlements also grew in places with access to specific local resources such as salt (Nantwich, Cheshire), iron ore, coal, etc. Whereas most of the factors listed above were natural, today the choice of a site for a new settlement is more likely to be **political** (Israeli settlements on the West Bank; Brasilia), **social** (some of Britain's new towns) or **economic** (Blaenau Ffestiniog for its slate – Places 78, page 523 – or, in Brazil, Carajas for its iron ore, and Iguaçu for its hydro-electricity).

Roberts has produced a model (Figure 14.8) which draws together not only site and situation factors, but also the perceptions of different settler groups as to the relative importance of the specific factors – e.g. in a desert, water may be perceived to be the most important; in parts of Mediterranean Europe, it may have been defence. The inner circle in Figure 14.8 is concerned with desirable **site** characteristics (intrinsic qualities) and the outer circle with the general **situation** factors (extrinsic qualities). Roberts stresses that each settlement location represents a complex balancing act of all these factors (London, Figure 14.9), with few sites and situations being ideal. You should be aware that 'extrinsic' factors change over time, and that settlements are dynamic in nature.

Figure 14.8

Village site analysis (*after* Roberts)

Figure 14.9

The site and situation of early London

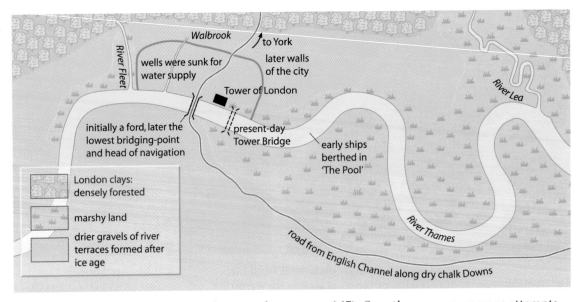

Figure 14.9 shows some of the early site and situation factors that helped determine the original location and early growth of London. As with other settlements, some of these early factors are no longer important, e.g. London now has piped water, has numerous shops to provide food, has bridges and tunnels to cross the river, and no longer needs a castle or city walls for defence.

Functions of settlements

As early settlements grew in size, each one tended to develop a specific function or functions. The **function** of a town relates to its economic and social development and refers to its main activities. There are problems in defining and determining a town's main function and often, due to a lack of data such as employment and/or income figures, subjective decisions have to be made. As settlements are very diverse, it helps to try to group together those with a similar function (Framework 7, page 167). Over the years, numerous attempts have been made to classify settlements based on function, but these tended to refer to places in industrialised countries and are often no longer applicable to post-industrial societies. Further problems arose when the growth of some settlements was based on an activity that no longer exists (the former coalmining villages of north-east England and South Wales), or where the original function has changed over **time** (a Cornish fishing village may now be a holiday resort). As functions change in time, this has a direct bearing on settlement morphology (page 394) and patterns of land use (Chapter 15). Functions may also differ between continents – i.e. there is also a difference over **space**. Finally it should be realised that today, and especially in the more developed countries, towns and cities are multifunctional – even if one or two functions tend to be predominant.

It may be worth referring, at this stage, to the term **economic base.** Economic base theory is founded on the idea that settlements (towns, cities or regions) perform two broad categories of economic activity: basic and non-basic. **Basic** is an economic activity (or function) that either produces a good or markets a service outside the settlement where it is located, and is likely to generate settlement and economic growth. **Non-basic** is when an economic activity (or function) only produces a good or markets a service within the settlement in which it is located and, therefore, makes little contribution to settlement or economic growth. Bearing in mind that the value to geographers of classifying settlements based on function has declined, Figure 14.10 has been included, as much as anything, as a checklist should you wish to conduct personal fieldwork or make an individual study of this topic.

Figure 14.10

Classification of settlement based on function

Rural	Urban	
	Developed countries	**Developing countries**
Market and agricultural	Mining	Administration
Route centre/transport	Manufacturing/industrial	Marketing/agricultural
Small service town	Route centre/transport	Route centre/port
Defensive	Retail/wholesale	Mining
Dormitory/overspill/satellite	Religious/cultural	Commercial
	Trade/commerce/financial	Religious
	Administration	Residential
	Resort/recreation	
	Residential	
	New towns	

Differences between urban and rural settlement

Figure 14.11 shows the commonly accepted types of settlement, but hides the divergence of opinion as to how and where to draw the borders between each type. Several methods have been suggested in trying to define the difference between a village, or rural settlement, and a town, or urban settlement.

■ **Population size** There is a wide discrepancy of views over the minimum size of population required to enable a settlement to be termed a town, e.g. in Denmark it is considered to be 250 people, in Ireland 500, in France 2000, in the USA 2500, in Spain 10 000 and in Japan 30 000. In India, where many villages are larger than British towns, a figure of less than 25 per cent engaged in agriculture is taken to be the dividing point.

■ **Economic** Rural settlements have traditionally been defined as places where most of the workforce are farmers or are engaged in other primary activities (mining and forestry). In contrast, most of the workforce in urban areas are employed in secondary and service industries. However, many rural areas have now become commuter/dormitory settlements for people working in adjacent urban areas or, even more recently, a location for smaller, footloose industries, such as high-tech industries.

■ **Services** The provision of services, such as schools, hospitals, shops, public transport and banks, is usually limited, at times absent, in rural areas (Figure 14.21).

■ **Land use** In rural areas, settlements are widely spaced with open land between adjacent villages. Within each village there may be individual farms as well as residential areas and possibly small-scale industry. In urban areas, settlements are often packed closely together and within towns there is a greater mixture of land use with residential, industrial, services and open-space provision.

■ **Social** Rural settlements, especially those in more remote areas, tend to have more inhabitants in the over-65 age group, whereas the highest proportion in urban areas lies within the economically active age group (page 354) or those under secondary school age.

It has becoming increasingly more difficult to differentiate between villages and towns, especially where urban areas have spread outwards into the **rural fringe**. The term **rural–urban continuum** (page 516) is used to express the fact that in many highly urbanised countries such as Japan and the UK, there is no longer either physically or socially a simple, clear-cut division between town and country. Instead there is a gradation between the two, with no obvious point where it can be said that the urban way of life ends and the rural way of life begins (Figure 17.1). It is, therefore, more realistic to talk about a transition zone from 'strongly rural' to 'strongly urban'. Cloke (1977) devised an **index of rurality** based upon 16 variables taken largely from census data for England and Wales (Figure 17.2). These variables included people aged over 65; proportion employed in primary, secondary and tertiary sectors; population density; population mobility (those moving home in the previous 5 years); proportion commuting; and distance from a large town (Figure 14.20). Cloke then identified four categories (Figure 17.3):

■ extreme rural (parts of south-west England, central Wales, East Anglia and the northern Pennines)
■ intermediate rural
■ intermediate non-rural, and
■ extreme non-rural (mainly suburbanised villages (page 398) around London in Surrey, Cambridgeshire, Hertfordshire and Essex).

Figure 14.11

Type of settlement

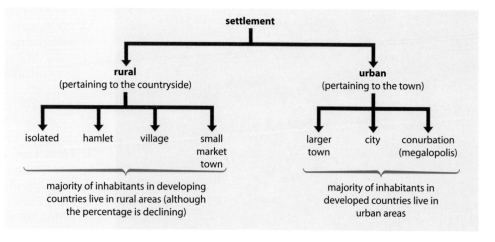

Figure 14.12

Isolated settlement: in the Amazon rainforest

Figure 14.13

Dispersed settlement: in North Yorkshire

Rural settlement

Pattern and morphology

Geographers have become increasingly interested in the **morphology**, i.e. the pattern (numbers 1 and 2 below) and shape (numbers 3–7 below) of settlements. Although village shapes vary spatially in Britain and across the world, it has been – again traditionally – possible to identify seven types (remember that, as in other classifications, some geographers may identify more or fewer categories).

1   **Isolated** This refers to an individual building, usually found in an area of extreme physical difficulty where the natural resources are insufficient to maintain more than a few inhabitants, e.g. the Amazon rainforests where tribes live in a communal home called a *maloca* (Figure 14.12). Isolated houses may also be found in planned pioneer areas such as on the Canadian Prairies where the land was divided into small squares, each with its own farm buildings.

2   **Dispersed** Settlement is described as dispersed when there is a scatter of individual farms and houses across an area; there are either no nucleations present, or they are so small that they consist only of two or three buildings forming a hamlet (Figure 14.13). Each farm or hamlet may be separated from the next by 2 or 3 km of open space or farmland. In the Scottish Highlands and Islands, some communities consist of crofts spaced out alongside a road or raised beach. Hamlets are common in rural areas of northern Britain, on the North German Plain (where their name *urweiler* means 'primeval hamlet') and in sub-Saharan Africa.

3   **Nucleated** Nucleated settlement is common in many rural parts of the world where buildings have been grouped closely together for economic, social or defensive purposes (Figure 14.14). In Britain, where recent evidence suggests that nucleation only took place after the year 1000, villages were surrounded by their farmland, where the inhabitants grew crops and grazed animals in order to be self-sufficient; this led to an unplanned and variable spacing of villages, usually 3–5 km apart. Some villages grew up around crossroads and at T-junctions, as is the case of many villages in India. Many border villages in Britain, hilltop settlements around the Mediterranean Sea, and *kampongs* in Malaysia became nucleated for defensive reasons.

4   **Loose-knit** These are similar to nucleated settlements except that the buildings are more spread out, possibly due to space taken up by individual farms which are still found within the village itself.

Figure 14.14

Nucleated settlement: in Sumatra, Indonesia

5 **Linear**, or **ribbon** Where the buildings are strung out along a main line of communication or along a confined river valley (Figure 14.15), the settlement is described as linear. **Street villages** – planned linear villages – were common in medieval England. Unplanned linear settlements also developed on long, narrow, flood-avoidance sites, e.g. along the raised beaches of western Scotland and on river terraces, as in London. Later, unplanned linear settlement grew up along the floors of the narrow coalmining valleys of South Wales and on main roads leading out of Britain's urban areas following the increase in private car ownership and the development of public transport. In the Netherlands, Malaysia and Thailand, houses have been built along canals and waterways.

6 **Ring and 'green' villages** Ring villages are found in many parts of sub-Saharan Africa and the Amazon rainforest (Figure 14.16). Houses were built around a central area which was left open for tribal meetings and communal life. In Kenya, the Maasai built their houses around an area into which their cattle were driven for protection during most nights. In England, many villages have been built around a central green.

Figure 14.15

Linear (ribbon) settlement: Combe Martin, Devon

Figure 14.16

Ring village: Kraito, in the Amazon rainforest

7 **Planned** Although many early settlements were planned (Pompeii, York), the apparently random shape of many British villages appears to suggest that they were not. More recently, villages surrounding large urban areas in, for example, Britain and the Netherlands, have expanded and become suburbanised, having small and often crescent-shaped estates (Places 49).

If you study maps of village plans, it is very likely that you will find many settlements with a mixture of the above shapes, e.g. a village may have a nucleated centre, a planned estate on its edges and a linear pattern extending along the road leading to the nearest large town (Figure 14.17).

Roberts (1987) suggested a different basis for classification (Figure 14.18). Even so, he concedes there are difficulties in trying to fit a particular village into a specific category, as when determining if a strip of grass is large enough to be called a green, and concludes that many villages are **composite** (or **polyfocal**), incorporating several plans and phases of development.

Figure 14.18

A method of classifying village types in Britain (*after* Roberts)

Basic shape	Plan and morphology		Village green
Linear (in a row)	regular		with
		———	without
	irregular		with
		– – –	without
Agglomerated (more nucleated)	regular grid		with
		⧣	without
	regular radial		with
		✳	without
	irregular grid		with
		-¦- ¦-	without
	irregular agglomerated		with
		●	without

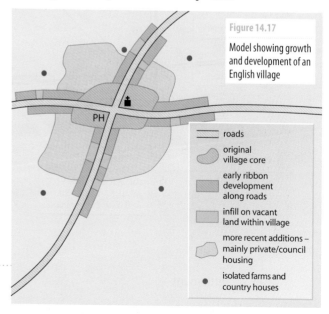

Figure 14.17

Model showing growth and development of an English village

PH

roads

original village core

early ribbon development along roads

infill on vacant land within village

more recent additions – mainly private/council housing

isolated farms and country houses

Figure 14.19

Carlisle: site, morphology and functions, as shown on a map of 1810

Figure 14.19 is a map of Carlisle in 1810. It shows some of the original **site** factors (some of which still applied), the developing **morphology** (pattern) of the city, and some of its initial and subsequent **functions**.

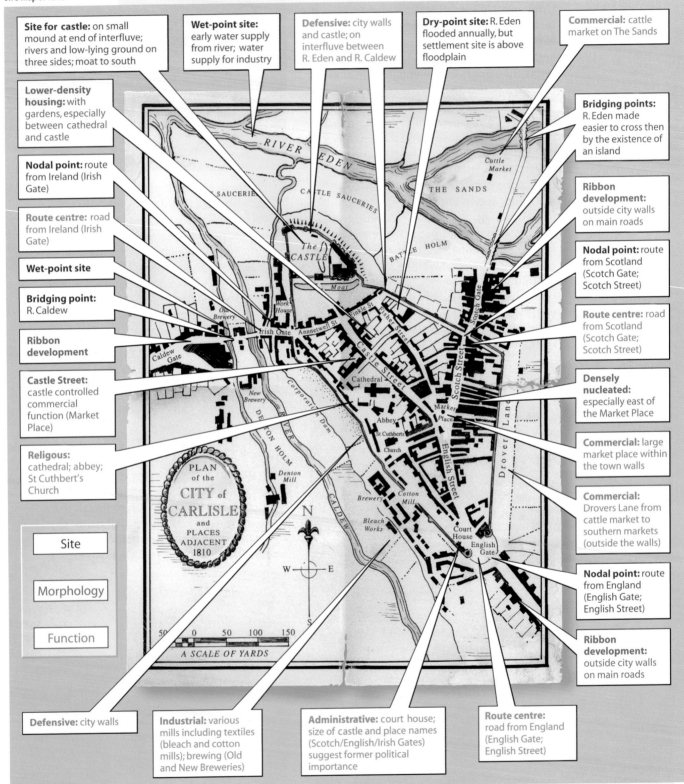

Site for castle: on small mound at end of interfluve; rivers and low-lying ground on three sides; moat to south

Lower-density housing: with gardens, especially between cathedral and castle

Nodal point: route from Ireland (Irish Gate)

Route centre: road from Ireland (Irish Gate)

Wet-point site

Bridging point: R. Caldew

Ribbon development

Castle Street: castle controlled commercial function (Market Place)

Religous: cathedral; abbey; St Cuthbert's Church

Wet-point site: early water supply from river; water supply for industry

Defensive: city walls and castle; on interfluve between R. Eden and R. Caldew

Dry-point site: R. Eden flooded annually, but settlement site is above floodplain

Commercial: cattle market on The Sands

Bridging points: R. Eden made easier to cross then by the existence of an island

Ribbon development: outside city walls on main roads

Nodal point: route from Scotland (Scotch Gate; Scotch Street)

Route centre: road from Scotland (Scotch Gate; Scotch Street)

Densely nucleated: especially east of the Market Place

Commercial: large market place within the town walls

Commercial: Drovers Lane from cattle market to southern markets (outside the walls)

Nodal point: route from England (English Gate; English Street)

Ribbon development: outside city walls on main roads

Site

Morphology

Function

Defensive: city walls

Industrial: various mills including textiles (bleach and cotton mills); brewing (Old and New Breweries)

Administrative: court house; size of castle and place names (Scotch/English/Irish Gates) suggest former political importance

Route centre: road from England (English Gate; English Street)

Dispersed and nucleated rural settlement

Whether settlement is dispersed or nucleated depends upon local physical conditions; economic factors such as the time and distance between places; and social factors which include who owns the land and how the people of the area live and work on it.

Causes of dispersion

The more extreme the physical conditions and possible hardship of an area, the more probable it is that the settlement will be dispersed. Similarly, dispersed settlement develops in areas where natural resources are limited and insufficient to support many people (Figure 13.4). This lack of resources could include a limited water supply (the Carboniferous limestone outcrops of the Pennines); forested areas (the Canadian Shield and the Amazon Basin); and marginal farmland (the Scottish Highlands and the Sahel countries), where pastoral farming is limited by the quality and quantity of available grass. Areas with physical difficulties are also less likely to have good transport networks.

Forms of land tenure can also result in dispersed dwellings, especially in those parts of the world where inheritance laws have meant that the farm is successively divided between several sons. Similar patterns, though with larger farm units, can be found in pioneer areas such as the Canadian Prairies and the Dutch polders.

The 'agrarian revolution' in Britain in the 18th century ended the open-field system, in which strips were owned individually but the crops and animals were controlled by the community. It was replaced by enclosing several fields which were owned by a farmer who became responsible for all the decisions affecting that farm; new farmhouses were sometimes built outside the village.

Two other changes at about the same time increased the incidence of dispersed settlement. The first was the growth of large estates belonging to wealthy landowners. The second was the extension of farming in hilly areas, in the 18th and again in the 19th century, to produce the extra food needed to feed the rapidly growing urban areas. Much moorland in the Pennines was walled; while fenland areas, previously of limited value, were drained and farmed. Areas of downland were also put under the plough. Increased mechanisation reduced labour needs, resulting in overpopulation and, eventually, out-migration.

Finally, settlement was more likely to develop a dispersed pattern where there was less risk of war or civil unrest as there was then less need for people to group together for protection.

Causes of nucleation

The majority of humans have always preferred to live together in groups, as witnessed by the cities of ancient Mesopotamia and Egypt (Figure 14.1), and the present-day conurbations and cities with more than 5 million inhabitants (Figure 15.3). Two major reasons for people to group together have been either a limited or an excess water supply. Settlements have grown up around springs, as at the foot of chalk escarpments in southern England (Figures 14.4 and 8.10), and at waterholes and oases in the desert (Figure 14.5). Settlements have also been built on mounds in marshy fenland regions and on river terraces above the level of flooding (Figures 14.9).

A further cause was the need to group together for defence and protection. Examples of defensive settlements include living in walled cities on relatively flat plains (Jericho and York); behind stockades (African kraals); in hilltop villages in southern Italy and Greece (Figure 14.7); or in meander loops, taking advantage of a natural water barrier (Durham, Figure 14.6).

In Anglo-Saxon England, when many villages had their origin, the feudal open-field system of farming encouraged nucleation: the local lord could better supervise his serfs if they were clustered around him; while the serf, living in the village, was probably equidistant from his fragmented strips of farmland (Places 51, page 400). Today, the more intensive the nature of farming, the more nucleated the settlement tends to be. People like to be as near as possible to services so that the larger and more nucleated the village, the more likely it is to have a wide range of services such as a primary school, shops and a public house (Figure 14.21).

Transport and routeways have always had a major influence on the clustering of dwellings. Buildings tend to be grouped together at crossroads and T-junctions; controlling a gap through hills; at bridging-points (Places 49); and along main roads, waterways and railways. Compact settlement patterns are also found in areas with an important local resource (a Durham coalmining town or a North Wales slate quarry village – Places 78, page 523), or where there was an abundance of building materials. More recently, many governments have encouraged new, nucleated settlements in an attempt to achieve large-scale self-sufficiency. Examples may be found as far afield as the Soviet collective farm, the Chinese commune (Places 63, page 468), the Tanzanian *ujamaa* and the Israeli *kibbutz*.

Changes in rural settlement in Britain

Within the British Isles, there are areas, especially those nearer to urban centres, where the rural population is increasing and others, usually in more remote locations, where the rural population is decreasing (rural depopulation). These population changes affect the size, morphology and functions of villages. Figure 14.20 shows that there is some relationship between the type and rate of change in a rural settlement and its distance from, and accessibility to, a large urban area.

Accessibility to urban centres

As public and private transport improved during the inter-war period (1919–39), British cities expanded into the surrounding countryside at a rapid and uncontrolled rate. In an attempt to prevent this urban sprawl, a **green belt** was created around London following the 1947 Town and Country Planning Act. The concept of a green belt, later applied to most of Britain's conurbations, was to restrict the erection of houses and other buildings and to preserve and conserve areas of countryside for farming and recreational purposes.

Beyond the green belt, **new towns** and **overspill towns** were built, initially to accommodate new arrivals seeking work in the nearby city and, later, those forced to leave it due to various redevelopment schemes. These new settlements, designed to become self-supporting both economically and socially, developed urban characteristics and functions. New towns, overspill and green belts were part of a wider land-use planning process which aimed to manage urban growth (compare Figure 14.22).

Meanwhile, despite the 1968 Town and Country Planning Act, uncontrolled growth also continued in many small villages beyond the green belt. Referred to during the inter-war period as **dormitory** or **commuter villages** (page 375), these settlements have increasingly adopted some of the characteristics of nearby urban areas and have been termed **suburbanised villages**. Figure 14.21 lists some of the changes which occur as a village becomes increasingly suburbanised.

Less accessible settlements

These villages are further in distance from, or have poorer transport links to, the nearest city, i.e. they are beyond commuting range. This makes the journey longer in time, more expensive and less convenient. Though these villages may be relatively stable in size, their social and economic make-up is changing. Many in the younger age groups move out, pushed by a shortage of jobs and social life. They are replaced by retired people seeking quietness and a pleasant environment but who often do not realise that rural areas lack many of the services required by the elderly such as shops, buses, doctors and libraries.

Villages in National Parks and other areas of attractive scenery in upland or coastal areas are being changed by the increased popularity of **second** or **holiday homes** (Figure 14.20 and Places 50). Wealthy urban dwellers, seeking relaxation away from the stress of their own working and living environment, have bought vacant properties at prices that local people cannot afford. While this may improve trade at the village shop and pub during holiday periods, and improve the quality of some buildings, it often means local people cannot afford the inflated house prices, properties standing empty for much of the year, many jobs being seasonal, and an end of public transport.

Figure 14.20

Rural settlements and distance from large urban areas (*after* Cloke, 1977; see page 393)

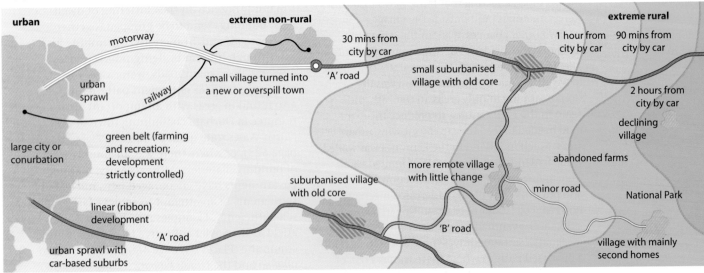

Characteristic	Extreme non-rural (increasingly suburbanised)	Original village	Extreme rural (increasingly depopulated)
Housing	Many new detached houses, semi-detached houses and bungalows; renovated barns and cottages; expensive estates	Detached, stone-built houses/cottages with slate/thatch roofs; some farms, many over 200 years old; barns	Poor housing lacking basic amenities; old stone houses, some derelict, some converted into holiday/second homes
Population structure	Young/middle-aged married couples with children; very few born in village; professional/executive groups; some wealthy retired people	An ageing population; most born in village; labouring/manual groups	Mainly elderly/retired; born and lived all life locally; labouring/manual groups; younger people have moved away
Employment	New light industry (high-tech and food processing); good salaries; many commuters (well-paid); tourist shops	Farming and other primary activities (forestry, mining); low-paid local jobs	Low-paid; unemployment; farming jobs (declining if in marginal areas) and other primary activities; some tourist-related jobs
Transport	Good bus service (unless reduced by private car); most families have one or two cars; improved roads	Bus service (limited); some cars; narrow/winding roads	No public transport; poor roads
Services	More shops; enlarged school; modern public houses/restaurants; garage	Village shop; small junior school; public house; village hall	Shop and school closed; perhaps a public house
Community/social	Local community swamped; division between local people and newcomers; may be deserted during day (commuters absent)	Close-knit community (many are related)	A small community; more isolated
Environment	Increase in noise and pollution, especially from traffic; loss of farmland/open space	Quiet, relatively pollution-free	Quiet; increase in conserved areas (National Parks/forestry)

Remote areas

These areas suffer from a population loss which, by leaving houses empty and villages decreasing in size, adds to the problems of rural deprivation (Figure 14.21 and Places 50). Resultant problems include a lack of job opportunities, fewer services and poor transport facilities. Employment is often limited to the shrinking primary industries which are low-paid and lack future prospects. The cost of providing services to remote areas is high, and there is often insufficient demand to keep the local shop or village school open. With fewer inhabitants to use public transport, bus services may decline or stop altogether, forcing people to move to more accessible areas.

Places 50 Bickington, Devon: a village

Bickington is a village of some 270 residents set on the edge of Dartmoor National Park. Now by-passed by the busy A38 road, it encapsulates most of the problems faced by many small rural settlements. Until recently it was a thriving farming community with its own post office, pub, garage, two churches, a children's nursery and a police house. Today, apart from an ailing village hall and the one remaining church, which has had to advertise for more worshippers, all have gone.

Bickington's location in such an attractive area has meant that property prices have been driven up far beyond the reach of local people and planning restrictions have meant no new affordable housing has been built. Without public transport, inhabitants have become increasingly reliant on the car and, by travelling to supermarkets and other public amenities in nearby Newton Abbot or further afield in Exeter, have caused the closure of the village shop, pub and post office. Meanwhile the nursery group, run in the church hall, was forced to close after government inspectors demanded improvements to the building that the church could not afford. The positive sign in 2008 is that the local community realises the need for radical action and is about to ask for exceptional permission to build affordable homes in the village, covenanted and price-capped so that they can only be sold to local workers, and to group together with five other nearby villages to share facilities.

Bickington's problems are shared by villages across the country. The Commission for Rural Communities claims that in villages each year 800 shops, 400 garages, over 100 churches and 7 primary schools close, while 27 village pubs close each week. Added to this, 95 per cent of village halls are struggling and most of the few remaining village hospitals are under threat. To many villages the death-knell may be the government's decision, in 2008, to close most village post offices, many of which had doubled up as the local shop.

Planned 'eco-towns'

The government has unveiled 15 potential sites for the first 10 of England's 'eco-towns', low-energy, carbon neutral settlements each with between 5000 and 15 000 homes. Of these five will be built by 2016 and the remaining five by 2020. The advantages will be the provision of many new homes to fill the housing shortage with 30 per cent being affordable housing. There will be good transport links with surrounding towns and cities for jobs and services, each settlement will have its own shops, secondary school, business space and leisure facilities and, by being carbon neutral, it will take no more energy from the National Grid than it replaces through renewable power. Opponents point out the likely increase in petrol costs and pressure on existing roads and schools, the location of most being in the south and east and many being on greenbelt sites.

Abridged from The Guardian, *3 April 2008*

Three million new homes by 2028

The government plans to force local councils to allow development on previously protected land in order to achieve their aim of providing 3 million new homes by 2028. Ministers want 33 000 new houses to be built each year, almost 100 every day, in the South East alone for the next 20 years. Of these a high proportion should be either affordable or social homes. Released documents reveal that the green belts around Oxford, Guildford and Woking are to be reviewed and that expansion into London's green belt may also be required. Similar strategies for other areas in the south and east will be released later. Despite growing opposition to their recently announced 10 'eco-towns', the government is set to push ahead with its large house-building programme regardless of the global credit crises and the fact that several of the country's leading house-builders are in difficulty.

Abridged from the Daily Telegraph, *17 July 2008*

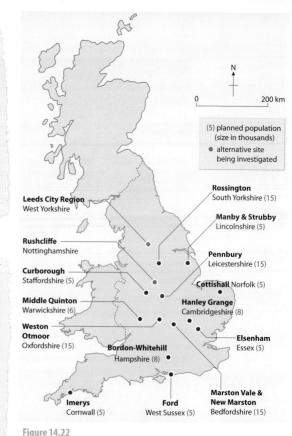

Figure 14.22

Pressure on green belts

Figure 14.23

Villages in the British landscape

Places 51 Britain: evolution of settlement

When Britain's first census was taken in 1801, almost 80 per cent of the population still lived in hamlets and villages. (The corresponding figure in the 1991 census was 7 per cent, rising to an estimated 10 per cent in 1998.) Most people have their own mental image of a 'traditional' hamlet, village, or market town. However, in reality, the development of rural settlement has been so dynamic and complex that, due to differences in site, form (morphology) and function (Places 49), there is no such thing as a 'typical' rural settlement (Figure 14.23) – nor is there a 'typical' urban settlement.

Iron Age settlements

Palaeolithic man left behind flint tools, but few marks on the landscape. The first people to alter their natural surroundings were those of the Neolithic period, the Bronze Age and the Iron Age (Figure 11.18). They began, despite limited technology, to clear woodlands and to leave a legacy of stone circles, tumuli, barrows, hillforts (Figure 14.24) and settlement sites. The hillfort built on the volcanic sill at Drumadoon (Figure 1.37) had a fine panorama of an enemy approaching from the sea, while the steep cliffs prevented any frontal attack. Hillforts may, however, have only been settled during times of attack, as few

There is tendency to think of country life as stable, conservative and unchanging but this is far from the truth. Settlements, like the people who live in them, are mortal. There is, however, no recognisable expected life-span, and a village can survive for twenty or two thousand years depending on its ability to adapt to changing economic and social conditions. In addition to extant village communities there are in Britain thousands of former occupation sites which have been abandoned.

Rural settlement in the past reflected the ever-changing relationship between man and his environment. Human society is never completely static and the settlements which serve it can never remain absolutely still for very long; and before a well-balanced form of settlement becomes generally established, new forces will be at work altering that form. The forces which created our hamlets and villages have involved factors as varied as the pace of technological change, the nature of local authority, inheritance customs, the presence of arable or pasture, and the availability of building materials. Village history tells a story of fluctuating expansion, decline and movement, sometimes reflecting national factors such as pestilence, economic changes and social development, and sometimes purely local events, such as the silting up of a river estuary or the bankruptcy of a local entrepreneur. Such factors have combined to give each village a unique history and plan.

T. Rowley, 1978

had a guaranteed water supply. Not all Iron Age settlements were hillforts; some forts were located in lowland areas, while other settlements may have had a religious or market function as opposed to a military one.

Figure 14.24

Maiden Castle hillfort, Dorset, England

forest, and thatch for the roof from local reeds or straw left over after the harvest. The huts, which were shared with the animals in winter, may have been protected by a stone or wooden wind-break. It was only by late Anglo-Saxon times that larger nucleated villages, with their open fields worked in strips by a heavy plough drawn by oxen, became more commonplace.

Medieval settlements

By medieval times, each village was dominated by a large farm, or manor, house in which the lord of the manor lived. The village would have contained several peasant cottages, built with materials similar to those of Anglo-Saxon homes, a church, a house for the priest, a blacksmith's forge and a mill. Surrounding the village were (usually) three large open fields – open because they had neither hedges nor fences as boundaries. Each field was divided into numerous, long, narrow strips, shared between the peasants. Two of the fields were likely to be growing cereals such as wheat, barley and rye (mainly for bread), while the third was left fallow (allowed to rest). The crops were rotated so that each field was left fallow every third year – the three-field system of crop rotation. When the fields were ploughed, a ridge was formed about 0.3 m above an adjacent furrow. Over many years of ploughing, the ridges built up so that they can still be recognised in our present-day landscape (Figure 4.25).

In the scarp-and-vale areas of south-east England (page 199), the villages were often close together along the spring lines. The parish boundaries were laid out between each village and parallel to each other, so that each individual parish had a long, narrow strip of land extending across the clay vale and over the chalk escarpment (Figure 14.4). This allowed each parish to be self-contained by having a permanent water supply together with land suitable for both rearing animals and growing crops. Although individual parishes no longer need to be self-supporting, the old boundaries still remain.

Romano-British settlements

While the Romans preferred to live in well-planned towns or in large rural villas, it is clear that at the same time many nucleated villages existed in lowland Britain, many of which showed evidence of Roman influence by having well-planned streets. One characteristic feature of Romano-British villages was the presence of small-scale industrial activity – usually pottery production and iron-working.

Anglo-Saxon settlements

Although many English village and town names have Anglo-Saxon origins, it does not prove that they existed during those times. Most Anglo-Saxon settlements were sited in clearings in the natural forest, on 'islands' in marshy areas or near to the coast. Archaeological evidence suggests that most settlements were likely to have consisted of several farms grouped together to form self-contained hamlets. The houses, or rather huts, were rectangular in shape and built from local materials – wood for the frame from the forest, mud and wattle (interlaced twigs and branches) for the walls from the river and

Figure 14.25

Ridge and furrow, south-east Leicestershire

Measuring settlement patterns

Several theories and statistical tests have been put forward to explain and to allow objective comparisons to be made between settlements in different parts of the world, e.g. within a country or between countries.

- **Nearest neighbour analysis** is a statistical test to describe the settlement pattern.
- **The rank–size rule** seeks to find a numerical relationship between the population size of settlements.
- **Central place theory** is concerned with the functional importance of places.
- **Gravity models** seek to determine the interaction (i.e. movement) between places.

Nearest neighbour analysis

Settlements often appear on maps as dots. Dot distributions are commonly used in geography, yet their patterns are often difficult to describe. Sometimes patterns are obvious, such as when settlements are extremely nucleated or dispersed (Figure 14.26). As, in reality, the pattern is likely to lie between these two extremes, then any description will be subjective. One way in which a pattern can be measured objectively is by using nearest neighbour analysis.

This technique was devised by a botanist who wished to describe patterns of plant distributions. It can be used to identify a tendency towards nucleation (clustering) or dispersion for settlements, shops, industry, etc., as well as plants. Nearest neighbour analysis gives a precision that enables one region to be compared with another and allows changes in distribution to be compared over a period of time. It is, however, only a technique and therefore does *not* offer any explanation of patterns.

The formula used in nearest neighbour analysis produces a figure (expressed as *Rn*) which measures the extent to which a particular pattern is clustered (nucleated), random, or regular (uniform) (Figure 14.26).

- **Clustering** occurs when all the dots are very close to the same point. An example of this in Britain is on coalfields where mining villages tended to coalesce. In an extreme case, *Rn* would be 0.
- **Random** distributions occur where there is no pattern at all. *Rn* then equals 1.0. The usual pattern for settlement is one that is predominantly random with a tendency either towards clustering or regularity.
- **Regular** patterns are perfectly uniform. If ever found in reality, they would have an *Rn* value of 2.15 which would mean that each dot (settlement) was equidistant from all its neighbours. The closest example of this in Britain is the distribution of market towns in East Anglia.

Using nearest neighbour analysis

Figure 14.27 shows settlements in part of northeast Warwickshire and south-west Leicestershire, an area of the English Midlands where it might be expected that there would be evidence of regularity in the distribution.

1 The settlements in the study area were located. (The minimum number recommended for a nearest neighbour analysis is 30.) Each settlement was given a number.
2 The nearest neighbour formula was applied. This formula is:

$$Rn = 2\bar{d} \sqrt{\frac{n}{A}}$$

where:

Rn = the description of the distribution
$\bar{d}$ = the mean distance between the nearest neighbours (km)
n = the number of points (settlements) in the study area
A = the area under study (km²).

Figure 14.26

Nearest neighbour values (*Rn*)

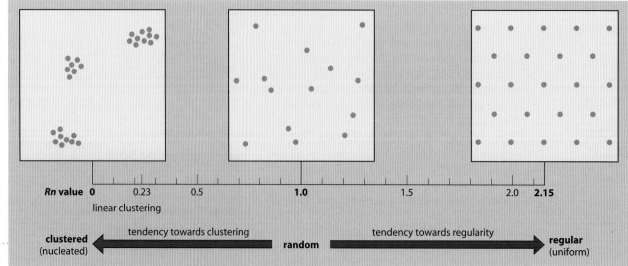

| *Rn* value | 0 | 0.23 | 0.5 | 1.0 | 1.5 | 2.0 | 2.15 |

linear clustering

clustered (nucleated) ← tendency towards clustering — random — tendency towards regularity → regular (uniform)

Figure 14.27

Nearest neighbour analysis: a worked example for part of north-east Warwickshire and south-west Leicestershire

Settle-ment number	Nearest neigh-bour	Distance (km)
1	2	1.0
2	1	1.0
3	4	0.6
4	3	0.6
5	6	1.6
6	5	1.6
7	8	1.8
8	9	1.3
9	8	1.3
10	9	2.1
12	10	2.2
13	13	2.2
14	15	3.3
15	18	1.7
16	17	1.3
17	18	1.0
18	17	1.0
19	16	3.0
20	19	3.2
21	22	1.6
22	21	1.6
23	24	2.1
24	25	1.1
25	24	1.1
26	25	1.5
27	26	1.8
28	27	2.5
29	30	2.2
30	29	2.2
		Σ51.7

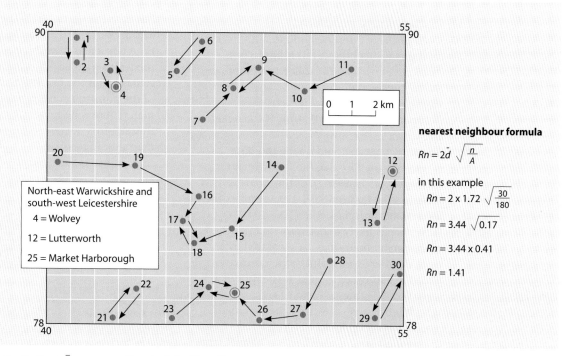

North-east Warwickshire and south-west Leicestershire

4 = Wolvey

12 = Lutterworth

25 = Market Harborough

nearest neighbour formula

$$Rn = 2\bar{d}\sqrt{\frac{n}{A}}$$

in this example

$$Rn = 2 \times 1.72 \sqrt{\frac{30}{180}}$$

$$Rn = 3.44 \sqrt{0.17}$$

$$Rn = 3.44 \times 0.41$$

$$Rn = 1.41$$

3 To find $\bar{d}$, measure the straight-line distance between each settlement and its nearest neighbour, e.g. settlement 1 to 2, settlement 2 to 1, settlement 3 to 4, and so on. One point may have more than one nearest neighbour (settlement 8) and two points may be each other's nearest neighbour (settlements 1 and 2). In this example, the mean distance between all the pairs of nearest neighbours was 1.72 km – i.e. the total distance between each pair (51.7 km) divided by the number of points (30).

4 Find the total area of the map: i.e. 15 km x 12 km = 180 km².

5 Calculate the nearest neighbour statistic, *Rn*, by substituting the formula. This has already been done in Figure 14.27 and gives an *Rn* value of 1.41.

6 Using this *Rn* value, refer back to Figure 14.26 to determine how clustered or regular is the pattern. A value of 1.41 shows that there is a fairly strong tendency towards a regular pattern of settlement.

7 However, there is a possibility that this pattern has occurred by chance. Referring to Figure 14.28, it is apparent that the values of *Rn* must lie outside the shaded area before a distribution of clustering or regularity can be accepted as significant. Values lying in the shaded area at the 95 per cent probability level show a random distribution. (*Note:* with fewer than 30 settlements, it becomes increasingly difficult to say with any confidence that the distribution is clustered or regular.) The graph confirms that our *Rn* value of 1.41 has a significant element of regularity.

How can the nearest neighbour statistic be used to compare two or more distributions? Figure 14.28 shows the *Rn* value for three areas in England, including that for our worked example, the English Midlands. The *Rn* statistic of 1.57 for part of East Anglia shows that the area has a more pronounced pattern of regularity than the Midlands. An *Rn* value of 0.61 for part of the Durham coalfield indicates that it has a significant tendency towards a clustered distribution.

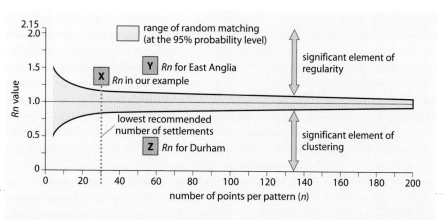

Figure 14.28

Interpretation of *Rn* statistic: significant values

Limitations and problems

As noted earlier, nearest neighbour analysis is a useful statistical technique but it has to be used with care. In particular, the following points should be considered:

1 The size of the area chosen is critical. Comparisons will be valid only if the selected areas are a similar size.

2 The area chosen should not be too large, as this lowers the *Rn* value (i.e. it exaggerates the degree of clustering), or too small, as this increases the *Rn* value (i.e. it exaggerates the level of regularity).

3 Distortion is likely to occur in valleys, where nearest neighbours may be separated by a river, or where spring-line settlements are found in a linear pattern as at the foot of a scarp slope (Figures 8.10 and 14.4).

4 Which settlement sizes are to be included? Are hamlets acceptable, or is the village to be the smallest size? If so, when is a hamlet large enough to be called a village (page 393)?

5 There may be difficulty in determining the centre of a settlement for measurement purposes, especially if it has a linear or a loose-knit morphology.

6 The boundary of an area is significant. If the area is a small island or lies on an outcrop of a particular rock, there is little problem; but if, as in Figure 14.27, the area is part of a larger region, the boundaries must have been chosen arbitrarily (in this instance by predetermined grid lines). In such a case, it is likely that the nearest neighbour of some of the points (e.g. number 20) will be off the map. There is disagreement as to whether those points nearest to the boundary of the map should be included, but perhaps of more importance is the need to be consistent in approach and to be aware of the problems and limitations.

Despite these problems, nearest neighbour analysis forms a useful basis for further investigation into why any clustering or regularity of settlement has taken place.

The rank–size rule

This is an attempt to find a numerical relationship between the population size of settlements within an area such as a country or county. The rule states that **the size of settlements is inversely proportional to their rank**. Settlements are ranked in descending order of population size, with the largest city placed first. The assumption is that the second-ranked city will have a population one-half that of the first-ranked, the third-ranked city a population one-third of the first-ranked, the fourth-ranked one-quarter of the largest city, and so on.

The rank–size rule is expressed by the formula:

$$Pn = Pl \div n \text{ (or } R)$$

where:

Pn = the population of the city

Pl = the population of the largest (primate) city

n (or R) = the rank–size of the city.

For example, if the largest city has a population of 1 000 000, then:

the second-largest city will be 1 000 000 ÷ 2, i.e. 500 000

the third-largest city will be 1 000 000 ÷ 3, i.e. 333 333

the fourth-largest city will be 1 000 000 ÷ 4, i.e. 250 000.

If such a perfect negative relationship actually occurred (Framework 19, page 612), it would produce a steeply downward-sloping, smooth, concave curve on an arithmetic graph (Figure 14.29a). However, it is more usual to plot the rank–size distribution on a logarithmic scale, in which case the perfect negative relationship would appear as a straight line sloping downwards at an angle of 45° (Figure 14.29b). Figure 14.30 shows the rank–size rule applied to Brazil.

Figure 14.29

The rank–size rule

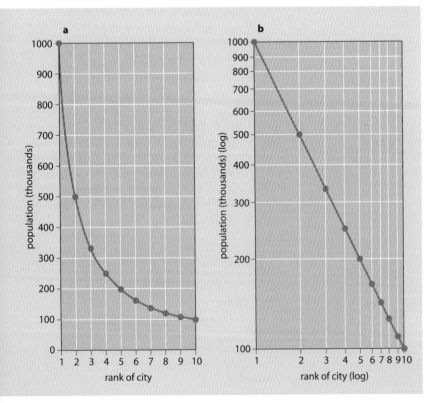

Variations from the rank–size rule

In reality, it is rare to find a close correlation between the city size of a country and the rank–size rule. There are, however, two major variations from the rank–size rule.

1 **Primate distribution** (urban primacy) is found where the largest city, often the capital, completely dominates a country or region (in terms of population size, economic development, wealth, services and cultural activities). In such a case, the primate city will have a population size many times greater than that of the second-largest city (Lima in Figure 14.31). Montevideo in Uruguay is 17 times larger than the second-largest city, and Lima in Peru is 11 times larger than Arequipa.

2 **Binary distribution** occurs where there are two very large cities of almost equal size within the same country: one may be the capital and the other the chief port or major industrial centre. Examples of binary distribution include Madrid and Barcelona in Spain, and Quito and Guayaquil in Ecuador.

It has been suggested (though there are many exceptions) that the rank–size rule is more likely to operate if the country is developed; has been urbanised for a long time; is large in size; and has a complex and stable economic and political organisation. In contrast, primate distribution is more likely to be found (also with exceptions, including France and Austria) in countries which are small in size; less developed; former colonies of European countries; only recently urbanised; and which have experienced recent changes in political organisation and/or boundaries.

Two schools of thought exist concerning the causes of variation in urban primacy. One suggests that as a city begins to dominate a country it attracts people, trade, industry and services at an increasingly rapid rate and at the expense of rival cities (arguably this is more applicable to economically less developed countries). The other claims that as a country becomes more urbanised and industrialised, the growth of several cities tends to be stimulated, thus reducing the importance of the primate city (arguably more applicable to economically more developed countries where some of the largest cities are now experiencing urban depopulation, page 365).

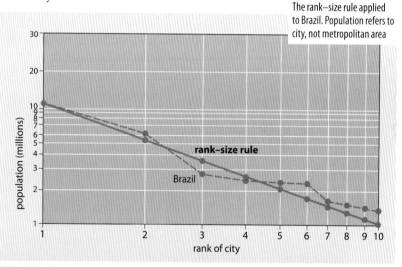

Figure 14.30

The rank–size rule applied to Brazil. Population refers to city, not metropolitan area

Rank 2007	City	Actual population	Estimated population according to rank–size rule (000s)
1	São Paulo	10 239	–
2	Rio de Janeiro	6094	5120
3	Salvador	2891	3413
4	Fortaleza	2431	2560
5	Belo Horizonte	2413	2048
6	Brasilia	2349	1707
7	Curitiba	1797	1463
8	Manaus	1602	1280
9	Recife	1534	1138
10	Belem	1400	1024

Rank	USA 2007	actual population (000s)	Italy 2007	actual population (000s)	Peru 2007	actual population (000s)	Japan 2007	actual population (000s)
1	New York	8275	Roma	2706	Lima	8473	Tokyo	8536
2	Los Angeles	3834	Milano	1303	Arequipa	749	Yokohama	3603
3	Chicago	2837	Napoli	975	Trujillo	683	Osaka	2635
4	Houston	2208	Torino	901	Chiclayo	524	Nagoya	2223
5	Phoenix	1552	Palermo	667	Piura	377	Sapporo	1889
6	Philadelphia	1450	Genova	616	Iquitos	371	Kobe	1529
7	San Antonio	1329	Bologna	373	Cusco	349	Kyoto	1473
8	San Diego	1267	Firenze	366	Chimbote	335	Fukuoka	1414
9	Dallas	1241	Bari	325	Huancayo	323	Kawasaki	1183
10	Detroit	917	Catania	302	Tacna	242	Hiroshima	1158

Figure 14.31

Largest cities in four selected countries.
Population refers to city, not metropolitan area

Figure 14.32

Size, spacing and functions of settlements

Central place	Population	Distance apart (km)	Sphere of influence (km²)	Functions (services)
Hamlet	10–20	2	–	probably none
Village	1 000	7	45	church, post office, shop, junior school
Small town	20 000	21	415	shops, churches, senior school, bank, doctor
Large town	100 000	35	1 200	shopping centre, small hospital, banks, senior schools
City	500 000	100	12 000	shopping complex, cathedral, large hospital, football team, large bus and rail station, cinemas, theatre
Conurbation	1 million	200	35 000	shopping complexes, several CBDs
Capital or primate city	several million	–	whole country	government offices, all other functions

Notes: The distances and service areas have been taken from Christaller's work in southern Germany (1933) with, in some cases, a rounding-off of figures for simplicity. The population figures and functions are more applicable to the UK and the present time. Populations, distances and service areas vary between and within countries and should be taken as comparative and approximate rather than absolute. All places in the hierarchy have all the services of the settlements below them.

Central place theory

A **central place** is a settlement that provides goods and services. It may vary in size from a small village to a conurbation or primate city (Figures 14.32 and 14.33) and forms a link in a hierarchy. The area around each settlement which comes under its economic, social and political influence is referred to as its **sphere of influence**, **urban field** or **hinterland**. The extent of the sphere of influence will depend upon the spacing, size and functions of the surrounding central places.

Functional hierarchies

Four generalisations may be made regarding the spacing, size and functions of settlements:

1 The larger the settlements are in size, the fewer in number they will be, i.e. there are many small villages, but relatively few large cities.

2 The larger the settlements grow in size, the greater the distance between them, i.e. villages are usually found close together, while cities are spaced much further apart.

3 As a settlement increases in size, the range and number of its functions will increase (Figure 14.33).

4 As a settlement increases in size, the number of higher-order services will also increase, i.e. a greater degree of specialisation occurs in the services (Figure 14.32).

Figure 14.33

Settlement hierarchy: the relationship between size and function

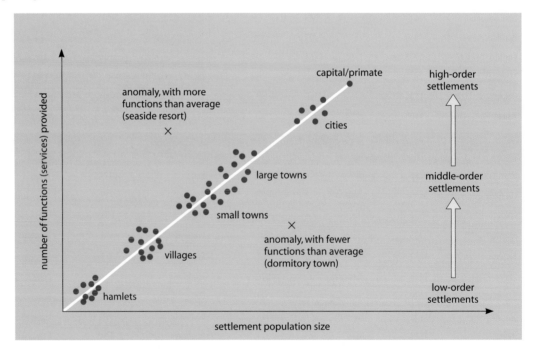

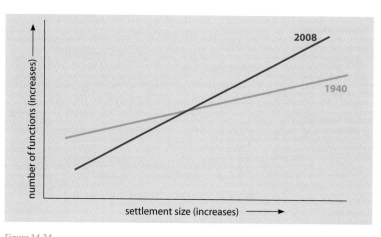

Figure 14.34

Relationship between the number of functions and settlement size in the UK, 1940 and 2008

The range and threshold of central place functions

Central place functions are activities, mainly within the tertiary sector, that market goods and services from central places for the benefit of local customers and clients drawn from a wider hinterland. The **range** of a good or service is the maximum distance that people are prepared to travel to obtain it. It is dependent upon the value of the good, the length of the journey, and the frequency that the service is needed. People are not prepared to travel as far to buy a newspaper (a low-order item), which they need daily, as they are to buy furniture (a high-order item), which they might purchase only once every several years. Low-order functions, such as corner shops and primary schools, need to be spaced closely together as people are less willing and less able to travel far to use them. High-order functions, such as regional shopping centres and hospitals, are likely to be widely spaced as people are more prepared to travel considerable distances to them (page 432).

The **threshold** of a good or service is the minimum number of people required to support it. It is assumed, incorrectly in practice, that people will always use the service located nearest to them (the nearest superstore). As a rule, the more specialised the service, the greater the number of people needed to make it profitable or viable. It has been suggested that, in the UK, about 300 people are necessary for a village shop, 500 for a primary school, 2500 for a doctor, 10 000 for a senior school or a small chemist's shop, 25 000 for a shoe shop, 50 000 for a small department store, 60 000 for a large supermarket, 100 000 for a large department store, and over 1 million for a university. Services locate where they can maximise the number of people in their catchment area and maximise the distance from their nearest rival. Threshold analysis was used by planners of British new towns who equated, for example, 20 000 people with a cinema, 10 000 people with a swimming pool and 100 000 people with a theatre.

Changes in population size and number of functions

Figure 14.34 shows that over the last 50 years in the UK there has been a decrease in the number of services available in small settlements and an increase in the number of functions provided by large settlements. This may be due to many factors, for example:

- Small villages are no longer able to support their former functions (village shop) as the greater wealth and mobility (car ownership) of some rural populations enable them to travel further to larger centres where they can obtain, in a single visit, both high- and low-order goods (Places 50, page 399).
- Domestic changes (deep freezers, convenience foods) mean that rural householders need no longer make use of daily, low-order services previously available in their village.
- As larger settlements attract an increasingly larger threshold population, they can increase the variety and number of functions and, by reducing costs (supermarkets), are likely to attract even more customers.
- In areas experiencing rural depopulation, villages may no longer have a population large enough to maintain existing services.

Christaller's model of central places

Walter Christaller was a German who, in 1933, published a book in which he attempted to demonstrate a sense of order in the spacing and function of settlements. He suggested that there was a pattern in the distribution and location of settlements of different sizes and also in the ways in which they provided services to the inhabitants living within their sphere of influence. Regardless of the level of service provided, he termed each settlement a **central place**. Although Christaller's **central place theory** was based upon investigations in southern Germany, and it was not translated into English until 1966, his work has contributed a great deal to the search for order in the study of settlements.

The two principles underlying Christaller's theory were the **range** and the **threshold** of goods and services. He made a set of assumptions which were similar to those of two earlier German economists, von Thünen (agricultural land use model, page 471) and Weber (industrial location theory, page 557).

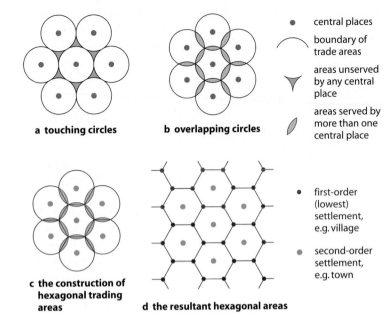

a touching circles	**b** overlapping circles

- central places
- boundary of trade areas
- areas unserved by any central place
- areas served by more than one central place

c the construction of hexagonal trading areas

d the resultant hexagonal areas

- • first-order (lowest) settlement, e.g. village
- • second-order settlement, e.g. town

Figure 14.35

Constructing spheres of influence around settlements (*after* Christaller)

These assumptions were:
- There was unbounded flat land so that transport was equally easy and cheap in all directions. Transport costs were proportional to distance from the central place and there was only one form of transport.
- Population was evenly distributed across the plain.
- Resources were evenly distributed across the plain.
- Goods and services were always obtained from the nearest central place so as to minimise distance travelled, i.e. the assumed rational behaviour that all consumers will minimise their travel in the pursuit of goods and services.
- All customers had the same purchasing power (income) and made similar demands for goods.

Figure 14.36

Christaller's central places and spheres of influence

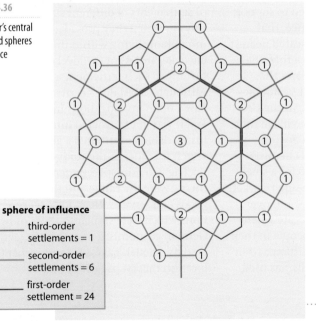

central place	sphere of influence	
③	——	third-order settlements = 1
②	——	second-order settlements = 6
①	——	first-order settlement = 24

- Some central places offered only low-order goods, for which people were not prepared to travel far, and so had a small sphere of influence. Other central places offered higher-order goods, for which people would travel further, and so they had much larger spheres of influence. The higher-order central places provided both higher-order and lower-order goods.
- No excess profit would be made by any one central place, and each would locate as far away as possible from a rival to maximise profits.

The ideal shape for the sphere of influence of a central place is circular, as then the distances from it to all points on the boundary are equal. If the circles touch at their circumferences, they leave gaps which are unserved by any central place (Figure 14.35a); if the circles are drawn so that there are no gaps, they necessarily overlap (Figure 14.35b) – which also violates the basic assumptions of the model. To overcome this problem, the overlapping circles are modified to become touching hexagons (Figure 14.35c). A hexagon is almost as efficient as a circle in terms of accessibility from all points of the plain and is considerably more efficient than a square or triangle (Figure 14.35d). A hexagonal pattern also produces the ideal shape for superimposing the trading areas of central places with different levels of function – the village, town and city of Christaller's hierarchy. Figure 14.36 shows a large trade area for a third-order central place, a smaller trade area for the six second-order central places, and even smaller trade areas for the 24 first-order central places.

By arranging the hexagons in different ways, Christaller was able to produce three different patterns of service or trading areas. He called these $k = 3$, $k = 4$ and $k = 7$, where k is the number of places dependent upon the next-highest-order central place.

The following should be noted at this point.
- Where $k = 3$, the trade area of the third-order (i.e. highest) central place is three times the area of the second-order central place, which in turn is three times larger than the trade area of the first-order (lowest) central place.
- Where $k = 4$, the trade area of the third-order central place is four times the area of the second-order central place, which is four times larger than the trade area of the first-order central place.
- Where $k = 7$, the trade area of each order is seven times greater than the order beneath it.

Figure 14.37

Christaller's $k = 3$

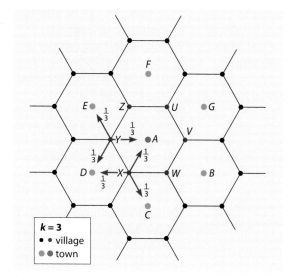

$k = 3$

The arrangement of the hexagons in this case is the same as given in Figure 14.36 and the explanation of how $k = 3$ is reached is shown in Figure 14.37, where:

> A is the central place or third-order settlement
> B, C, D, E, F and G are 6 second-order settlements surrounding A
> U, V, W, X, Y and Z are some of the 24 first-order settlements which lie between A and the second-order settlements.

It is assumed that one-third of the inhabitants of Y will go to A to shop, one-third to D and one-third to E. Similarly, one-third of people living at X will shop at A, one-third at D and one-third at C. This means that A will take one-third of the customers from each of U, V, W, X, Y and Z ($6 \times \frac{1}{3} = 2$) plus all of its own customers (1). In total, A therefore serves the equivalent of three central places (2 + 1).

Christaller based the $k = 3$ pattern on a **marketing principle** which maximises the number of central places and thus brings the supply of higher-order goods and services as close as possible to all the dependent settlements and therefore to the inhabitants of the trade area.

$k = 4$

In this case, the size of the hexagon is slightly larger and it has been re-oriented (Figure 14.38). The first-order settlements, again labelled U, V, W, X, Y and Z, are now located at the mid-points of the sides of the hexagon instead of at the apexes as in $k = 3$. Customers from Y now have a choice of only two markets, A and N, and it is assumed that half of those customers will go to A and half to N. Similarly, half of the customers from X will go to A and the other half to M. A will therefore take half of the customers from each of the six settlements at U, V, W, X, Y and Z ($6 \times \frac{1}{2} = 3$) plus all of its own customers (1) to serve the equivalent of four central places (3 + 1). This pattern is based on a **traffic principle**, whereby travel between two centres is made as easy and as cheap as possible. The central places are located so that the maximum number may lie on routes between the larger settlements.

$k = 7$

Here the pattern shows the same high-order central place, A, but all the lower-order settlements, U, V, W, X, Y and Z, lie within the hexagon or trade area (Figure 14.39). In this case, all of the customers from the six smaller settlements will go to A ($6 \times 1 = 6$), together with all of the inhabitants of A (1). This means that A serves seven central places (6 + 1). As this system makes it efficient to organise or control several places, and as the loyalties of the inhabitants of the lower-order settlements to a higher one are not divided, it is referred to as the **administrative principle**.

Figure 14.38

Christaller's $k = 4$

Figure 14.39

Christaller's $k = 7$

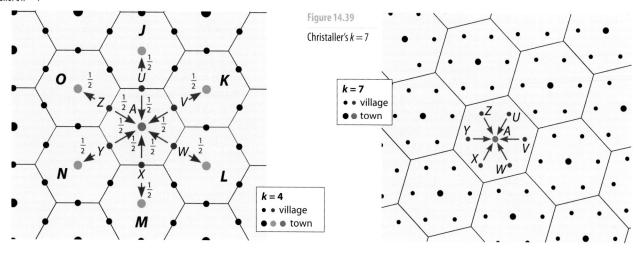

Why, with the possible exception of the reclaimed Dutch polders, can no perfect example of Christaller's model be found in the real world? The answer lies mainly in the basic assumptions of the model.

- Large areas of flat land rarely exist and the presence of relief barriers or routes along valleys means that transport is channelled in certain directions. There is more than one form of transport; costs are not proportional to distance; and both systems and types of transport have changed since Christaller's day.
- People and wealth are not evenly distributed.
- People do not always go to the nearest central place – for example, they may choose to travel much further to a new edge-of-city hypermarket.
- People do not all have the same purchasing power, or needs.
- Governments often have control over the location of industry and of new towns.
- Perfect competition is unreal and some firms make greater profits than others.
- Christaller saw each central place as having a particular function whereas, in reality, places may have several functions which can change over time.
- The model does not seem to fit industrial areas, although there is some correlation with flat farming areas in East Anglia, the Netherlands and the Canadian Prairies.

Christaller has, however, provided us with an objective model with which we can test the real world. His theories have helped geographers and planners to locate new services such as retail outlets and roads.

Interaction or gravity models

These models, derived from Newton's law of gravity, seek to predict the degree of interaction between two places. Newton's law states that:

'Any two bodies attract one another with a force that is proportional to the product of their masses and inversely proportional to the square of the distance between them.'

When used geographically, the words 'bodies' and 'masses' are replaced by 'towns' and 'population' respectively.

The interaction model in geography is therefore based upon the idea that as the size of one or both of the towns increases, there will also be an increase in movement between them. The further apart the two towns are, however, the less will be the movement between them. This phenomenon is known as **distance decay**.

This model can be used to estimate:
1. traffic flows (page 411)
2. migration between two areas
3. the number of people likely to use one central place, e.g. a shopping area, in preference to a rival central place.

It can also be used to determine the sphere of influence of each central place by estimating where the **breaking point** between two settlements will be, i.e. the point at which customers find it preferable, because of distance, time and expense considerations, to travel to one centre rather than the other.

Reilly's law of retail gravitation (1931)

Reilly's interaction breaking-point is a method used to draw boundary lines showing the limits of the trading areas of two adjacent towns or shopping centres. His law states that:

'Two centres attract trade from intermediate places in direct proportion to the size of the centres and in inverse proportion to the square of the distances from the two centres to the intermediate place.'

Unlike Christaller, Reilly suggested that there were no fixed trade areas, that these areas could vary in size and shape, and that they could overlap.

This can be expressed by the formula:

$$Db = \frac{Dab}{1 + \sqrt{\dfrac{Pa}{Pb}}}$$

or similarly

$$djk = \frac{dij}{1 + \sqrt{\dfrac{Pi}{Pj}}}$$

where:

Db (or djk)	= the breaking-point between towns A and B
Dab (or dij)	= the distance (or time) between towns A and B
Pa (or Pi)	= the population of town A (the larger town)
Pb (or Pj)	= the population of town B (the smaller town).

Taking as an example Grimsby–Cleethorpes which has a population of 131 000 and Lincoln, 71 km away, with a population of 75 000, the formula can be written as:

$$Db = \frac{71}{1 + \sqrt{\dfrac{131\,000}{75\,000}}}$$

which means that

Figure 14.40

Reilly's breaking-point
between settlements of
different sizes, applied to
north Lincolnshire

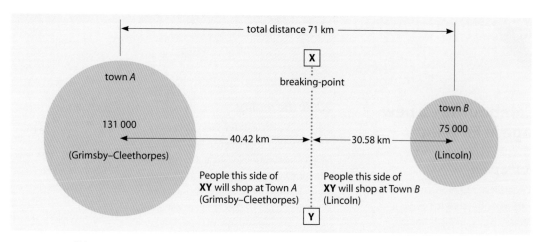

$$Db = \frac{71}{1 + 1.32}$$
$$\therefore Db = 30.58$$

Thus the breaking-point is 30.58 km from Lincoln (town *B*) and 40.42 km from Grimsby–Cleethorpes (town *A*). This is shown in Figure 14.40.

Limitations of Reilly's model

As with other models, Reilly's model is based on assumptions which are not always applicable to the real world. In this case, the assumptions are that:

- the larger the town, the stronger its attraction
- people shop in a logical way, seeking the centre which is nearest to them in terms of time and distance.

These assumptions may not always be true. For example:

- there may be traffic congestion on the way to the larger town and, once there, car parking may be more difficult and expensive
- the smaller town may have fewer but better-quality shops
- the smaller centre may be cleaner, more modern, safer and less congested, and
- the smaller town may advertise its services more effectively.

A variation on Reilly's law of retail gravitation

Like central place theory, Reilly's law seems to fit rural areas better than closely packed, densely populated urban areas. One of several variations on Reilly's law of retail gravitation is based on the drawing power of shopping centres (i.e. the number and type of shops in each) rather than distance between the two towns. (Other variations include retail floorspace and retail sales.)

The version based on the drawing power of shopping centres has the formula:

$$Db = \frac{Dab}{1 + \sqrt{\dfrac{Sa}{Sb}}}$$

where:

Sa = the number of shops in town *A*

Sb = the number of shops in town *B*.

Referring to our original example, suppose Grimsby–Cleethorpes has 800 shops and Lincoln has 300 shops. The formula could then be written:

$$Db = \frac{71}{1 + \sqrt{\dfrac{800}{300}}}$$
$$\therefore Db = 27$$

This means that out of every 71 shoppers, 44 would go to Grimsby–Cleethorpes and 27 to Lincoln.

In reality, the competitive commercial relationships between urban centres can change over a period of time. On Humberside, for example, there have been the effects of the opening of the Humber Bridge on places either side of the estuary, the construction of the M62 and M180 motorways, and the development of new out-of-town shopping centres (pages 433 and 458).

Measuring settlement patterns: conclusion

Nearest neighbour analysis, the rank–size rule, Christaller's central place theory and the interaction models are all difficult to observe in the real world. Their value lies in the fact that they form hypotheses against which reality can be tested – provided you do not seek to *make* reality fit them (Framework 10, page 299)! Also, they offer objective methods of measuring differences between real-world places. When theory and reality diverge, the geographer can search for an explanation for the differences. An important shared characteristic of these approaches is that they aim to find order in spatial distributions.

14 *Case Study*

A Cambourne – a new village in England

1998: the plan

Work began in late 1998 on a new village in South Cambridgeshire to be called Cambourne (Figure 14.41). Eventually 8000 people will live here, in 3300 houses (up to 900 of which will be 'affordable homes'), which are to be built over 12 years. Cambourne, which covers 400 hectares, will be laid out as three distinct villages (Figure 14.43), each with its own central green (Figure 14.42) and separated by two small valleys that will provide open space and leisure amenities. There will also be a church, two primary schools, a library, 18 hectares of playing fields, a multi-purpose sports centre, a health centre, police and fire stations. The developers have agreed to provide funds for a park and ride scheme, cycle tracks and a bus service. The development aims to enhance the environment by including 69 hectares of planted woodland, 56 hectares for a new Country Park, and the construction of a series of lakes. It is hoped that a new 20 hectare business park will eventually create up to 3000 new jobs, many of which, as the village is so close to Cambridge, are likely to be high-tech (Places 86, page 566). In time, the A428 arterial route linking Cambourne to Cambridge will become a dual-carriageway.

2008: the reality

Cambourne has, in some ways, become a unique type of settlement in that the planners have managed to create a village environment with the facilities of a small town. An evaluation by Cambridge Architectural Research Ltd (2007) concluded that the settlement had the advantages of being less congested, polluted and noisy than Cambridge; had cheaper, newer and a wider choice of houses; had easy access to the countryside, a dual-carriageway and mainline stations; and had, despite a bad press, less crime and antisocial behaviour. Residents appreciate the green space and lakes that have been incorporated into the scheme and perceive it to be a safe place to raise a family. In contrast, there is less choice in shops and fewer public transport options than in Cambridge; some residents, especially those without children or a reason to mix, feel isolated; there is less civic pride and an obvious lack of history or a sense of belonging; is not large enough for a secondary school (needs a population of 6000) and – a key issue – there is a lack of local job opportunities.

By early 2008, 2600 houses had been built, of which almost 30 per cent were 'affordable'. By that date, house-building in Great Cambourne should have been completed and the first house in Upper Cambourne should be occupied. Cambourne (Figure 14.43) has primary schools in Lower and Great Cambourne and a day nursery. Morrisons supermarket and several other retail outlets, including a pharmacy,

occupy sites in the High Street (which is in Great Cambourne), along with an estate agent's, a petrol station and the Monkfield Arms pub. The medical practice and public library share Sackville House, and the village also has a dental practice and a new church. The landscaped business park, in the north-west corner near to the interchange with the A428 dual-carriageway, employs over 1000 people and includes the new offices for South Cambridgeshire District Council. Cambourne has a 4-star hotel with 120 bedrooms and a leisure complex, as well as a fully equipped sports centre and community centre, both recently opened. An eco-park has segregated areas for the under 4s and 4–10-year-olds as well as a 'teenage hangout'. There is also a large sports field, skateboard park and golf course. The country park has lakes, which provide opportunities for fishing, a wetland habitat for wildlife, and large wooded areas.

The future

It was planned, before the recent slump in house-building, to complete the last house by 2010. However, an outline planning application for 950 additional homes in Upper Cambourne has been lodged, 40 per cent of which would be affordable. If successful, it would increase the final number of homes in Cambourne to 4250. The application followed a government directive allowing housing densities to increase from 25 to 30 per hectare in an attempt to provide more homes in south-east England (page 400). It is also hoped that a wider range of high street shops will be available by 2010.

Figure 14.41

Lower Cambourne: houses off the village green 2008

Figure 14.42

Upper Cambourne: village green 2008

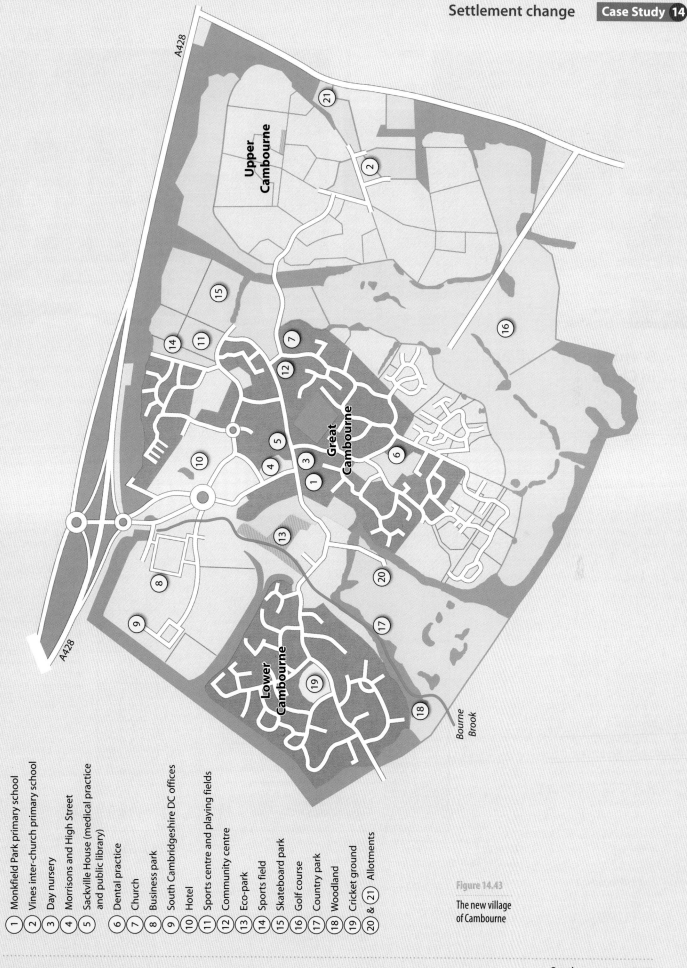

1 Monkfield Park primary school
2 Vines inter-church primary school
3 Day nursery
4 Morrisons and High Street
5 Sackville House (medical practice and public library)
6 Dental practice
7 Church
8 Business park
9 South Cambridgeshire DC offices
10 Hotel
11 Sports centre and playing fields
12 Community centre
13 Eco-park
14 Sports field
15 Skateboard park
16 Golf course
17 Country park
18 Woodland
19 Cricket ground
20 & 21 Allotments

Figure 14.43

The new village of Cambourne

Figure 14.44

Farmhouses around a central courtyard

Figure 14.46

Farmer within the courtyard

B Hua Long – a village in China

Hua Long is situated in the province of Sichuan, 280 km from Chengdu and 180 km from the Yangtze port of Chongqing. Like many other villages in the area it dates from the later Ming period (1550–1644). Between that time and the 1990s, little changed. Today, some 2000 people live in the village, which is fairly small by Chinese standards. Hua Long is linear in shape with most of its buildings strung out along the wide, but poorly maintained, 'main' road which passes through it (Figure 14.47).

Most families in Hua Long are farmers (Sichuan is known as the 'rice bowl of China'), working long hours at little more than a subsistence level (page 477). Many live in farmhouses which are usually grouped together, in typical Chinese fashion, around a central courtyard. Around the courtyard shown in Figure 14.44 are 13 doors, signifying 13 families (Figure 14.45). The address of this group of families – a legacy of the commune days of the 1960 and 1970s (Places 63, page 468), is Group 4 Team 1. Most of the families (Figure 14.46) have lived in these one-roomed houses for several generations. Despite the lack of running water and sewerage, and the presence of several pigs, there is no smell. The wooden or mud-bricked houses have tiled roofs and shutters, or iron bars, across openings that served as windows. There are no chimneys. Central to the courtyard is an area for collecting household waste that can be fed to the pigs. The remainder of the area is used for drying and storing crops, or as a social meeting-place. The houses are soon to be pulled down (the families visited by the author in 1999 wanted copies of the photos shown here as mementoes), and although many of the occupants will be sad to lose their ancestral home, they are looking forward to living in modern, brick-built houses with water and electricity.

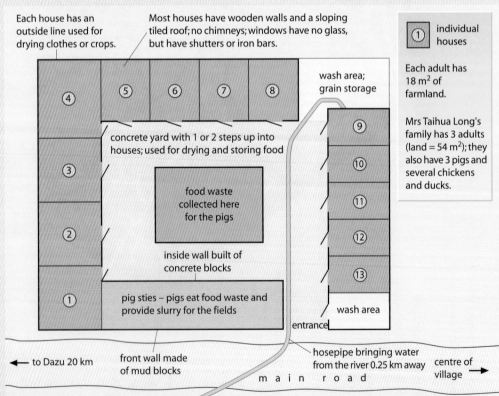

Each house has an outside line used for drying clothes or crops.

Most houses have wooden walls and a sloping tiled roof; no chimneys; windows have no glass, but have shutters or iron bars.

concrete yard with 1 or 2 steps up into houses; used for drying and storing food

food waste collected here for the pigs

inside wall built of concrete blocks

pig sties – pigs eat food waste and provide slurry for the fields

front wall made of mud blocks

← to Dazu 20 km

entrance

wash area; grain storage

wash area

hosepipe bringing water from the river 0.25 km away

m a i n   r o a d

centre of village →

① individual houses

Each adult has 18 m² of farmland.

Mrs Taihua Long's family has 3 adults (land = 54 m²); they also have 3 pigs and several chickens and ducks.

Figure 14.45

Courtyard plan

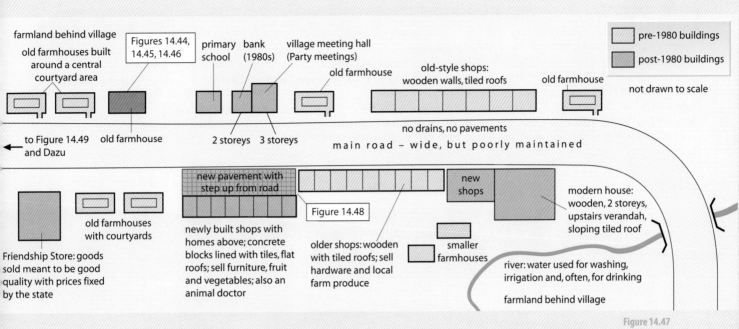

farmland behind village
old farmhouses built around a central courtyard area

Figures 14.44, 14.45, 14.46

primary school

bank (1980s)

village meeting hall (Party meetings)

old farmhouse

old-style shops: wooden walls, tiled roofs

old farmhouse

pre-1980 buildings

post-1980 buildings

not drawn to scale

to Figure 14.49 and Dazu

old farmhouse

2 storeys   3 storeys

no drains, no pavements

main road – wide, but poorly maintained

new pavement with step up from road

Figure 14.48

new shops

modern house: wooden, 2 storeys, upstairs verandah, sloping tiled roof

Friendship Store: goods sold meant to be good quality with prices fixed by the state

old farmhouses with courtyards

newly built shops with homes above; concrete blocks lined with tiles, flat roofs; sell furniture, fruit and vegetables; also an animal doctor

older shops: wooden with tiled roofs; sell hardware and local farm produce

smaller farmhouses

river: water used for washing, irrigation and, often, for drinking

farmland behind village

Several changes have taken place in Hua Long in the last 20 years (Figure 14.47). These include the building of a bank (not needed before 1980 as people were not allowed to earn money), an improved primary school for children aged 6 to 12 (funded by the voluntary Hope Project which aims to improve education in the poorest parts of China) and a Friendship Store. Along the main road are several tile-faced, double-storey buildings (Figure 14.48) where newly rehoused people live above shops and small workshops, and several new detached houses (signs of increasing wealth among a few of the inhabitants).

In contrast, the Yangs live, with their two children, in a large, two-year-old brick farmhouse built on the outskirts of the village (Figure 14.49). Mr Yang is a farmer, but he also operates a trishaw 'taxi' in the nearby town Dazu. The family saved enough money, and borrowed the rest from Mr Yang's cousin, to replace their old wooden farm with a seven-roomed, double-storeyed house (though some rooms are only used for storing crops, and furniture is sparse). The Yangs claim that most people in the village are better off and much happier than they were 20 years ago (Figure 16.8).

Even so, some are likely to have joined China's 150 million migrant workers who have left villages such as Hua Long to seek better-paid jobs in the coastal cities (Case Study 19).

Figure 14.47

Plan of Hua Long

Figure 14.48

New shops and houses

Figure 14.49

A new farmhouse

Further reference

Bradford, M.G. and Kent, W.A. (1977) *Human Geography: Theories and Their Applications*, Oxford University Press.

Roberts, B. (1987) *The Making of the English Village*, Longman.

Wilson, J.G. (1984) *Statistics in Geography for A Level Students*, Schofield and Sims.

 Cambourne:
www.cambourne-uk.com/

Countryside Agency, UK National Parks and regional sites:
www.naturalengland.co.uk

Early civilisation in Crete:
www.dilos.com/region/crete/kn_01.html

Future of rural England:
http://ruralnet.org.uk/

Gretton: a Northamptonshire village:
www.grettonvillage.org.uk/

Milton Keynes:
www.mkweb.co.uk/

Questions & Activities

Activities

1 a What is the meaning of:

   i the 'site' of a settlement *(1 mark)*

   ii the 'situation' of a settlement? *(1 mark)*

  b In the past various factors had to be considered by people seeking a settlement site. Explain what each of the following terms means, and why each type of site was sometimes chosen for settlements:

   i a 'wet point site' *(2 marks)*

   ii a 'dry point site' *(2 marks)*

   iii a 'nodal point'. *(2 marks)*

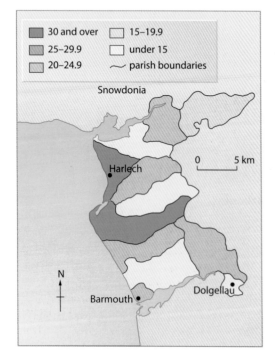

Figure 14.50

Second homes as a percentage of all houses in part of North Wales

  c Many towns and cities in the UK have changed their functions many times since they were first built. This has often caused serious planning problems because the original sites are not suitable for the modern functions of the settlement. Name one town or city in the UK that has problems caused by its original site.

   i Describe the site, and explain why it was originally chosen. *(4 marks)*

   ii Explain why that site causes problems now. *(4 marks)*

   iii Describe how the planners are attempting to tackle the problems caused by the site. *(6 marks)*

   iv To what extent have the planners been successful in tackling the problems? *(3 marks)*

2 a What is meant by:

   i the morphology of a settlement *(1 mark)*

   ii a nucleated settlement *(1 mark)*

   iii dispersed settlement? *(1 mark)*

  b Name an example of each of the settlement types listed below. Describe the main features of each of the settlements that you name. Explain why each of the named settlements developed at that location.

   i linear settlement *(4 marks)*

   ii ring or green village *(4 marks)*

   iii commuter village *(4 marks)*

  c Study Figure 14.50. It shows the development of second homes in a remote area of rural North Wales.

   i Suggest why such a high proportion of houses have become second homes for people who have their main homes elsewhere. *(5 marks)*

   ii Explain why the growth of second home ownership can create problems in areas such as that shown on the map. *(5 marks)*

Exam practice: basic structured questions

Figure 14.51

MetroCentre, Gateshead

3   **a**   Study Figure 14.51 showing the MetroCentre on Tyneside.

    i   What evidence supports the view that this site was chosen because it was:

      (i) accessible to a large number of people   *(5 marks)*

      (ii) built on comparatively cheap land?   *(5 marks)*

    ii   What evidence shows that the MetroCentre has been carefully designed to allow customers to have the easiest possible access to all parts of the complex?   *(5 marks)*

  b   Some modern offices are built as close as possible to city centres, whilst others are located on the rural–urban fringe. Compare the advantages of these two types of location for offices. Refer to specific examples.   *(10 marks)*

Exam practice: structured questions

4   **a**   What do the following phrases mean?

    i   the 'range' of a good or service   *(1 mark)*

    ii   the 'threshold population' for a good or service   *(1 mark)*

  b   When geographers develop models they always make a set of assumptions before they start to describe the model. Explain three of the assumptions that Christaller made before he developed his central place model.   *(6 marks)*

  c   Explain how first, second and third order settlements are distributed in the $k = 3$ version of the Christaller model.   *(7 marks)*

  d   How useful is Christaller's central place model for modern geographers?   *(10 marks)*

5   Study Figure 14.20 on page 398.

  a   Six settlements are shown outside the main conurbation.

    Explain why these settlements have developed in different ways.   *(12 marks)*

  b   Choose a region in which rural settlement changes in nature with distance away from a large urban area. Discuss the extent to which this model helps explain variations in the form of settlements in your chosen region.   *(13 marks)*

6   Choose a town in a more developed country that shows evidence of its evolution through different periods of history.

  a   Describe how the present settlement shows evidence of the form of the settlement in previous periods.   *(12 marks)*

  b   Discuss the problems and benefits which the historical development presents for today's inhabitants of the town.   *(13 marks)*

7   **a**   Study the map of Cambourne on page 413. Referring to map evidence:

    i   Describe how the planners of Cambourne have tried to make Cambourne an ideal place for people of all ages to live.   *(7 marks)*

    ii   Discuss whether they have been successful.   *(8 marks)*

  b   With reference to examples of settlements in less developed countries, discuss why settlement structures have to be adapted as the functions of the settlements change.   *(10 marks)*

8   **a**   **i**   Outline how you would carry out a nearest neighbour analysis for an area of 1000 km². You have been provided with an Ordnance Survey map (or the local equivalent) at a scale of 1:50 000 (or 2 cm = 1 km).   *(7 marks)*

    ii   If, having completed the nearest neighbour calculation, you obtained a figure of $Rn = 1$, what conclusions could you draw?   *(3 marks)*

  b   **i**   What is the 'rank–size rule' of settlements in a country?   *(5 marks)*

    ii   How can the rank–size rule be helpful to geographers who are studying urban patterns in different countries?   *(10 marks)*

Exam practice: essays

9   Study the description of Bickington, Devon on page 399.

To what extent does Bickington illustrate issues that affect all rural villages in the UK at the present time?   *(25 marks)*

10   Name **one** town or city that you have studied. Explain how the growth and development of the town have been influenced by the physical geography of its site.   *(25 marks)*

11   'Settlement morphology is usually a result of an interaction between physical geography, economic geography and cultural development.'

Discuss this statement with reference to a range of settlements that you have studied.   *(25 marks)*

15

Urbanisation

• •

'The invasion from the countryside … is overwhelming the ability of city planners and governments to provide affordable land, water, sanitation, transport, building materials and food for the urban poor. Cities such as Bangkok, Bogota, Bombay, Cairo, Delhi, Lagos and Manila each have over one million people living in illegally developed squatter settlements or shanty towns.'

L. Timberlake, *Only One Earth*, 1987

Urban growth – trends and distribution

Urbanisation is defined as the process by which an increasing proportion of the total population, usually that of a country, lives in towns and cities. Although the process began at least as far back as the fourth millennium BC (Figure 14.2), the number of people living in urban areas formed, until fairly recently, only a small proportion of a country's population. One estimate suggests that in 1800 only 3 per cent of the world's population were urban dwellers, a figure that has risen, according to latest UN estimates, to 50 per cent (2008) and which is predicted to rise to 60 per cent (Figure 15.2) before 2025.

Rapid urbanisation has occurred twice in time and space.

1   During the 19th century, in what are now referred to as the economically more developed countries, industrialisation led to a huge demand for labour in mining and manufacturing centres. Urbanisation was, in these parts of the world, a consequence of economic development.

2   Since the 1950s, in the economically less developed countries, the twin processes of migration from rural areas (page 366) and the high rate of natural increase in population (resulting from high birth rates and falling death rates, Figure 13.10) have resulted in the uncontrolled growth of many cities. Urbanisation is, in the developing countries, a consequence of population movement and growth and is not, as was previously believed, an integral part of development.

In 2008, the UN claimed that 74 per cent of the total population lived in urban areas of the developed countries, and 45 per cent in developing countries (the prediction for 2050 is 86 and 67 per cent respectively) (Figure 15.1).

Simultaneous with urbanisation has been the growth of very large cities. Whereas the only cities in the world with a population exceeding 1 million in 1900 were London and Paris, there were, again according to the UN, 70 in 1950 and 410 in 2005. Of these cities, most of which are in developing countries and including China, 48 had a population of over 5 million with 18 – the so-called **megacities** – exceeding 10 million. Although the largest cities are named and listed in rank-order of size in Figure 15.3, their population is not given due to problems in collecting accurate data, although figures are available from the UN's World Urbanisation Prospects. These problems include:

- the use of different criteria by countries to define the size of an urban area, e.g São Paulo city is quoted as 10.239 million, its urban agglomeration as 18.333 million (2007), while other countries give data for conurbations, e.g. Osaka-Kobe
- problems in collecting accurate census data (e.g. within shanty towns) or accurately estimating natural changes made annually between each 10-year census
- difficulties in obtaining accurate migration figures, especially where refugees and illegal immigrants are involved (page 367).

Figure 15.1

Urban population growth (UN)

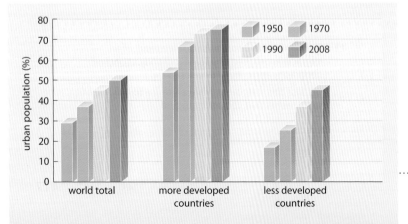

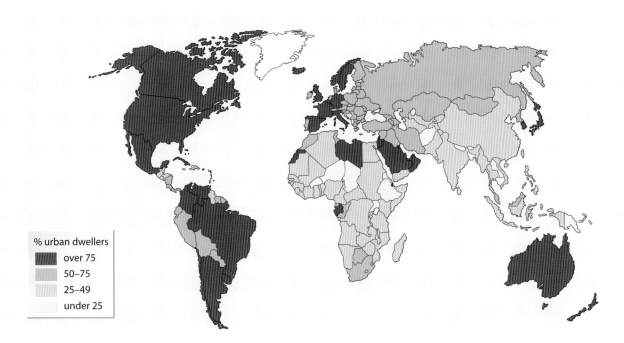

There have been several noticeable trends in the growth of the so-called 'million cities' since the mid-1980s.

- Most of the largest cities are in South-east Asia and Latin America. Of 410 global cities with a population in excess of 1 million in 2005, 117 were in China, 40 in India and 17 in Brazil (the USA had 39).

- Most of the fastest-growing cities are in South-east Asia although in-migration is usually more significant than natural increase.

- Although the rate of growth slowed in many developed countries in the second half of the

20th century, it has increased again, mainly due to immigration, in this century.

- In China, with the most large cities, it is those nearest the coast that have grown most rapidly due to rural–urban migration (Places 41, page 363 and Places 98, page 618).

What affects most people who live in large urban areas is not the actual population size of the city but rather its density. Of the world's 100 largest cities, the 10 with the lowest population density are in developed countries (mainly North America) and the 15 with the highest density are in developing countries.

■ Over 15 million	▲ 10–15 million
1 Tokyo	8 Kolkata
2 Mexico City	9 Dhaka
3 New York-Newark	10 Buenos Aires
4 São Paulo	11 Los Angeles
5 Mumbai	12 Karachi
6 Shanghai	13 Cairo
7 Delhi	14 Rio de Janeiro
	15 Osaka-Kobe
	16 Manila
	17 Beijing
	18 Moscow

● 5–10 million	
19 Paris	35 Bangkok
20 Seoul	36 Bangalore
21 Jakarta	37 Chongqing
22 Chicago	38 Lahore
23 London	39 Hyderabad
24 Lagos	40 Santiago
25 Guangzhou	41 Miami
26 Shenzhen	42 Madrid
27 Lima	43 Philadelphia
28 Tehran	44 Baghdad
29 Tianjin	45 Toronto
30 Bogotá	46 Belo Horizonte
31 Kinshasa	47 Ahmadabad
32 Wuhan	48 Ho Chi Minh City
33 Hong Kong	
34 Chennai	

Models of urban structure

20th century

As cities grew in area and population in the 20th century, geographers and sociologists tried to identify and to explain variations in spatial patterns. These patterns, which may show differences and similarities in land use and/or social groupings within a city, reflect how various urban areas evolved economically and socially (culturally) in response to changing conditions over a period of time. While each city had its own distinctive pattern, or patterns, studies of other urban areas showed that they too often exhibited similar patterns. As a result several models which tried to describe and explain the then urban structure were put forward.

21st century

Before looking at the basic assumptions of four such models, together with the theory behind them, their value at the time and their limitations, it should be pointed out that, to many present-day geographers, urban models belong to the realm of 'historic geography'. Urban models, like all models (Framework 12, page 352), have limitations and have always been open to criticism. It is, therefore, understandable why models put forward at a particular time (early to mid-20th century and before the advent of the post-industrial city), for a particular place (Western Europe and North America), and using criteria and referring to processes that may have changed (increased mobility and migration) should be, to some, ready for the 'recycling bin'. Yet perhaps it is only by understanding the early structure, both physical and social, of an urban area that we can appreciate the changing processes that are shaping our cities of today.

1 Burgess, 1924

Burgess attempted to identify areas within Chicago based on the outward expansion of the city and the socio-economic groupings of its inhabitants (Places 52).

Basic assumptions

Although his main aim was to describe residential structures and to show processes at work in a city, geographers subsequently made further assumptions:

- The city was built on flat land which therefore gave equal advantages in all directions, i.e. morphological features such as river valleys were removed.
- Transport systems were of limited significance being equally easy, rapid and cheap in every direction.
- Land values were highest in the centre of the city and declined rapidly outwards to give a zoning of urban functions and land use.
- The oldest buildings were in, or close to, the city centre. Buildings became progressively newer towards the city boundary.
- Cities contained a variety of well-defined socio-economic and ethnic areas.
- The poorer classes had to live near to the city centre and places of work as they could not afford transport or expensive housing.
- There were no concentrations of heavy industry.

Burgess's concentric zones

The resultant model (Figure 15.4) shows five concentric zones:

1 The **central business district** (CBD) contains the major shops and offices; it is the centre for commerce and entertainment, and the focus for transport routes.
2 The **transition** or **twilight zone** is where the oldest housing is either deteriorating into slum property or being 'invaded' by light industry. The inhabitants tend to be of poorer social groups and first-generation immigrants.
3 Areas of **low-class housing** are occupied by those who have 'escaped' from zone 2, or by second-generation immigrants who work in nearby factories. They are compelled to live near to their place of work to reduce travelling costs and rent. In modern Britain, these zones are equated with the inner cities.
4 **Medium-class housing** of higher quality which, in present-day Britain, would include inter-war private semi-detached houses and council estates.
5 **High-class housing** occupied by people who can afford the expensive properties and the high cost of commuting.

The model's limitations are listed in Figure 15.15.

Figure 15.4

The Burgess concentric model

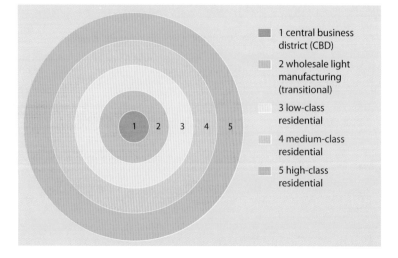

1 central business district (CBD)

2 wholesale light manufacturing (transitional)

3 low-class residential

4 medium-class residential

5 high-class residential

Burgess, in producing his model, was influenced by the emerging science of plant ecology at the University of Chicago. He made analogies with such ecological processes as the **invasion** of an area by competing groups, **competition** between the invaders and the natural groups, and the eventual **dominance** of the area by the invaders which allowed them to **succeed** the natural groups.

Relating this to urban geography, Burgess suggested that people living in the inner zone were **invaded** by newcomers and, in face of this **competition** by immigrants who became **dominant** there, **succeeded** to the next outer zone – a process also referred to as **centrifugal movement**. The energy to maintain this dynamic system came from a continual supply of immigrants to the centre, and existing groups being forced (or choosing) to move towards the periphery.

Chicago lies on the shores of Lake Michigan, with its CBD, known as the 'Loop', facing the lake. Surrounding the CBD, the city's housing developed a distinctive pattern (Figure 15.5). The initial migrants, from north-western Europe, settled around the CBD. In time, they were replaced by newer immigrants from southern Europe (especially Italy) and by Jews who were, in turn, replaced by blacks from the American south (Figure 15.6). This led to the creation of a series of income, social and ethnic zones radiating outwards from the centre. These zones showed:

1 That wealth, as seen by the quality of housing, increased towards the outskirts of the city. People with the highest incomes lived in the newest property (on the north-west fringe) while those with the lowest incomes occupied the poorest housing next to the CBD.

2 That people in their early twenties or over 60 tended to live close to the CBD, while middle-aged people and families with young children tended to live nearer to the city boundary.

3 That areas of ethnic segregation existed, with the early white immigrants – whose wealth had tended to increase in relation to the length of time they had lived in the city – living towards the outskirts, and non-white groups living nearer to the city centre, e.g. in China Town and the black belt.

Figure 15.5

Urban areas of Chicago (*after* Burgess)

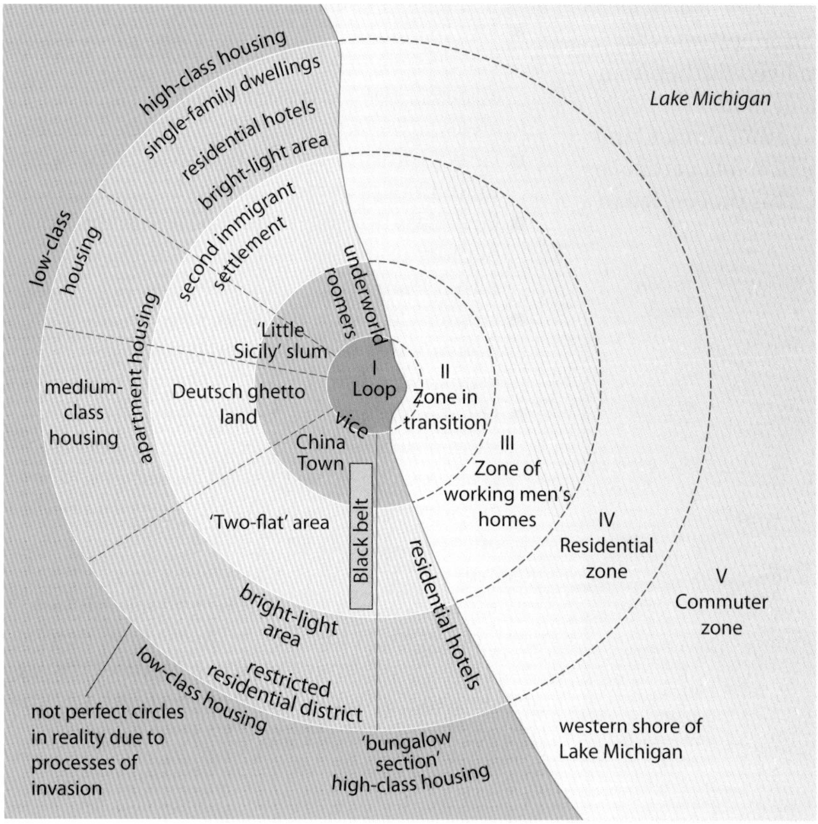

high-class housing
single-family dwellings
residential hotels
bright-light area
second immigrant settlement
low-class housing
underworld
roomers'
'Little Sicily' slum
Deutsch ghetto land
China Town
vice
medium-class housing
apartment housing
'Two-flat' area
Black belt
bright-light area
restricted residential district
low-class residential housing
not perfect circles in reality due to processes of invasion
'bungalow section' high-class housing

I Loop
II Zone in transition
III Zone of working men's homes
IV Residential zone
V Commuter zone
residential hotels
western shore of Lake Michigan
Lake Michigan

Figure 15.6

Centrifugal movement in Chicago

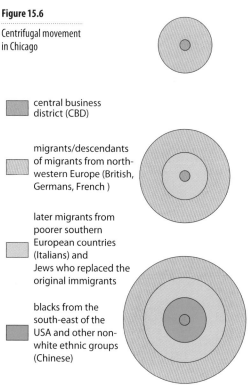

central business district (CBD)

migrants/descendants of migrants from north-western Europe (British, Germans, French)

later migrants from poorer southern European countries (Italians) and Jews who replaced the original immigrants

blacks from the south-east of the USA and other non-white ethnic groups (Chinese)

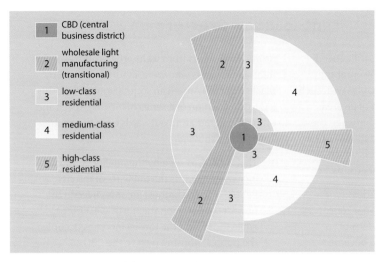

Figure 15.7

The Hoyt sector model

2 Hoyt, 1939

Hoyt's model was based on the mapping of eight housing variables for 142 cities in the USA. He tried to account for changes in, and the distribution of, residential patterns.

Basic assumptions

Hoyt made the same implicit assumptions as had Burgess, with the addition of three new factors:

- Wealthy people, who could afford the highest rates, chose the best sites, i.e. competition based on 'ability to pay' resolved land use conflicts.
- Wealthy residents could afford private cars or public transport and so lived further from industry and nearer to main roads.
- Similar land uses attracted other similar land uses, concentrating a function in a particular area and repelling others. This process led to a 'sector' development.

Figure 15.8

Growth of Sheffield

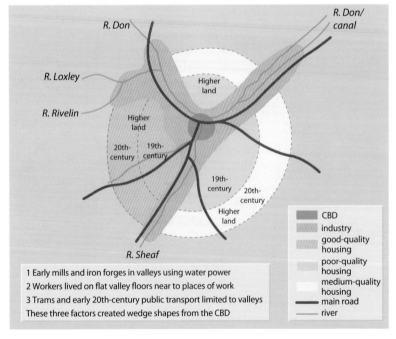

1 Early mills and iron forges in valleys using water power
2 Workers lived on flat valley floors near to places of work
3 Trams and early 20th-century public transport limited to valleys
These three factors created wedge shapes from the CBD

Hoyt's sector model

Hoyt suggested that areas of highest rent tended to be alongside main lines of communication and that the city grew in a series of wedges (Figure 15.7). He also claimed that once an area had developed a distinctive land use, or function, it tended to retain that land use as the city extended outwards, e.g. if an area north of the CBD was one of low-class housing in the 19th century, then the northern suburbs of the late 20th century would also be likely to consist of low-class estates. Calgary, in Canada, is the standard example of this model. The model's limitations are listed in Figure 15.15.

3 Mann, 1965

Mann tried to apply the Burgess and Hoyt models to three industrial towns in England: Huddersfield, Nottingham and Sheffield (Figure 15.8). His compromise model (Figure 15.8) combined the ideas of Burgess's concentric zones and Hoyt's sectors. Mann assumed that because the prevailing winds blow from the south-west, the high-class housing would be in the south-western part of the city and industry, with its smoke (this was before Clean Air Acts), would be located to the north-east of the CBD. His conclusions can be summarised as follows.

- The twilight zone was not concentric to the CBD but lay to one side of the city which allowed, elsewhere, more wealthy residential areas.
- Heavy industry was found in sectors along main lines of communication.
- Low-class housing should be called the 'zone of older housing' (age-based classification, rather than social).
- Higher-class or, in Hoyt's terms, 'modern' housing was usually found away from industry and smoke.
- Local government (politics) played a role in slum clearance and gentrification. This led to large council estates which took the working class/low incomes to the city edge (opposite of the Burgess model).

Robson (1975) applied Mann's model to a north-eastern industrial town, Sunderland (Figure 15.10), and to Belfast. Mann's model does show, despite its small sample, that a variety of approaches are possible to the study of urban structures. Its limitations are listed in Figure 15.15.

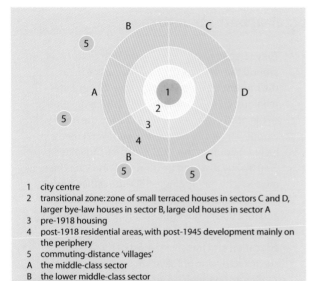

1 city centre
2 transitional zone: zone of small terraced houses in sectors C and D, larger bye-law houses in sector B, large old houses in sector A
3 pre-1918 housing
4 post-1918 residential areas, with post-1945 development mainly on the periphery
5 commuting-distance 'villages'
A the middle-class sector
B the lower middle-class sector
C the working-class sector and main municipal housing areas
D industry and lowest working-class sector

Figure 15.9

Mann's model of urban structure

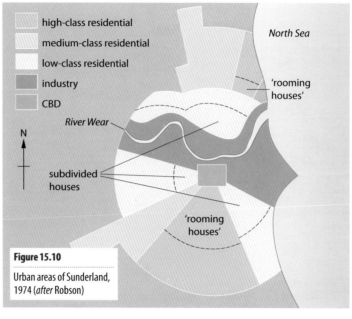

high-class residential
medium-class residential
low-class residential
industry
CBD

North Sea

'rooming houses'

River Wear

subdivided houses

'rooming houses'

Figure 15.10

Urban areas of Sunderland, 1974 (*after* Robson)

4 Ullman and Harris, 1945

Ullman and Harris set out to produce a more realistic model than those of Burgess and Hoyt but consequently ended with one that was more complex (Figure 15.11) – and more complex models may become descriptive rather than predictive if they match reality too closely in a specific example (Framework 12, page 352).

Basic assumptions

- Modern cities have a more complex structure than that suggested by Burgess and Hoyt.
- Cities do not grow from one CBD, but from several independent nuclei.
- Each nucleus acts as a growth point, and probably has a function different from other nuclei within that city. (In London, the City is financial; Westminster is government and administration; the West End is retailing and entertainment; and Dockland was industrial.)
- In time, there will be an outward growth from each nucleus until they merge as one large urban centre (Barnet and Croydon now form part of Greater London; Figure 13.7).
- If the city becomes too large and congested, some functions may be dispersed to new nuclei. (In Greater London, edge-of-city retailing takes place at Brent Cross and new industry has developed close to Heathrow Airport/M25/M4.)

Multiple nuclei developed as a response to the need for maximum accessibility to a centre, to keep certain types of land use apart, for differences in land values and, more recently, to decentralise (Places 53). The model's limitations are listed in Figure 15.15.

Urban structure models: conclusions

The four models described were put forward to try to explain differences in structure within cities in the developed world. It must be remembered that:

- each model will have its limitations (Figure 15.15)
- if you make a study of your local town or city, you must avoid the temptation of saying that it *fits* one of the models – at best it will show characteristics of one or possibly two; each city is unique and will have its own structure – a pattern not necessarily derived according to any existing model (Framework 12, page 352).

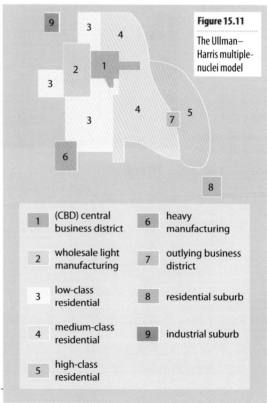

Figure 15.11

The Ullman–Harris multiple-nuclei model

1	(CBD) central business district	6	heavy manufacturing
2	wholesale light manufacturing	7	outlying business district
3	low-class residential	8	residential suburb
4	medium-class residential	9	industrial suburb
5	high-class residential		

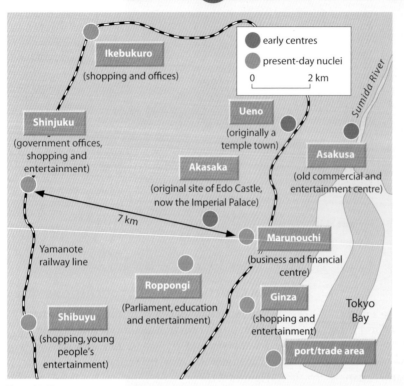

early centres

present-day nuclei

0   2 km

Ikebukuro
(shopping and offices)

Shinjuku
(government offices,
shopping and
entertainment)

Ueno
(originally a
temple town)

Asakusa
(old commercial and
entertainment centre)

Akasaka
(original site of Edo Castle,
now the Imperial Palace)

7 km

Yamanote
railway line

Marunouchi
(business and financial
centre)

Roppongi
(Parliament, education
and entertainment)

Ginza
(shopping and
entertainment)

Tokyo
Bay

Shibuyu
(shopping, young
people's
entertainment)

Sumida River

port/trade area

Figure 15.12

Multiple nuclei in
Tokyo, 1994

Tokyo began to grow in the late 16th century
around the castle of the Edo Shogunate (near
the present Imperial Palace, Figure 15.12). Later
religious, cultural and financial districts developed
to the north-east. Over the centuries, the mainly
wooden-built city was destroyed several times,
including during the 1923 Kanto earthquake
(140 000 deaths) and by US aircraft in 1945. The
modern city has no single CBD but, rather, has
several nuclei each with its own specialist land use
and functions – government offices (Figure 15.13),
shopping (Figure 15.14), finance, entertainment,
education and transport. Most of these nuclei
are linked by one of Tokyo's many railways, the
Yamanote line, which forms a circle with a diameter
of 7 km.

Figure 15.14

The Ginza shopping district

Figure 15.13

The Shinjuko
business district

Figure 15.15

Limitations/criticisms of
the four urban models

	Burgess	Hoyt	Mann	Ullman–Harris
1	zones, in reality, are never as clear-cut as shown on each model			
2	each zone usually contains more than one type of land use/housing			
3	no consideration of characteristics of cities outside USA and north-west Europe			
	based on 1 USA city	based on 142 USA cities	based on 3 English cities (in north and Midlands)	based on cities in economically more developed world
4	redevelopment schemes and modern edge-of-city developments are not included (most of the models pre-date these developments)			
5	based mainly on housing: other types of land use neglected		industry not always to north-east of British cities	
6	cities not always built upon flat plains			
7	tended to ignore transport			

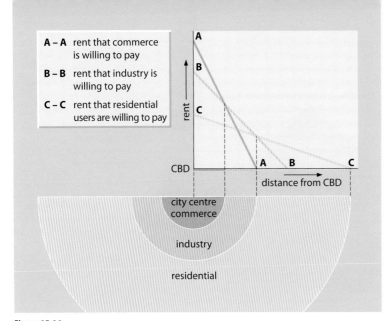

A – A rent that commerce is willing to pay

B – B rent that industry is willing to pay

C – C rent that residential users are willing to pay

Figure 15.16

Bid–rent curves

The land value model or bid–rent theory

This model is the urban equivalent of von Thünen's rural land use model (page 471) in that both are based upon locational rent. The main assumption is that in a free market the highest bidder will obtain the use of the land. The highest bidder is likely to be the one who can obtain the maximum profit from that site and so can pay the highest rent. Competition for land is keenest in the city centre. Figure 15.16 shows the locational rent that three different land users are prepared to pay for land at various distances from the city centre.

The most expensive or 'prime' sites in most cities are in the CBD, mainly because of its accessibility and the shortage of space there. Shops, especially department stores, conduct their business using a relatively small amount of ground-space, and due to their high rate of sales and turnover they can bid a high price for the land (for which they try to compensate by building

upwards and by using the land intensively). The most valuable site within the CBD is called the **peak land value intersection** or PLVI – a site often occupied by a Marks and Spencer store! Competing with retailers are offices which also rely upon good transport systems and, traditionally, proximity to other commercial buildings (this concept does not have the same relevance in centrally planned economies).

Away from the CBD, land rapidly becomes less attractive for commercial activities – as indicated by the steep angle of the bid–rent curve (A–A) in Figure 15.16. Industry, partly because it takes up more space and uses it less intensively, bids for land that is less valuable than that prized by shops and offices. Residential land, which has the flattest of the three bid–rent curves (C–C), is found further out from the city centre where the land values have decreased due to less competition. Individual householders cannot afford to pay the same rents as shopkeepers and industrialists.

The model helps to explain housing (and population) density. People who cannot afford to commute have to live near to the CBD where, due to higher land values, they can only obtain small plots which results in high housing densities. People who can afford to commute are able to live nearer the city boundary where, due to lower land values, they can buy much larger plots of land, which creates areas of low housing density. Figure 15.17 shows the predicted land use pattern when land values decrease rapidly and at a constant rate from the city centre. The resultant pattern is similar to that suggested by Burgess (Figure 15.4).

One basis of this model is 'the more accessible the site, the higher its land value'. Rents will therefore be greater along main routes leading out of the city and along outer ring roads. Where two of these routes cross, there may be a secondary or subsidiary land value peak (Figure 15.18). Here the land use is likely to be a small suburban shopping parade or a small industrial estate. The 'retail revolution' of the 1980s (page 432), which led to the development of large edge-of-city shopping complexes (MetroCentre in Gateshead, Places 55, page 433, Bluewater in Kent and Brent Cross in north London), has altered this pattern. Similarly, large industrial estates and science parks (Places 86, page 566) have been located near to motorway interchanges.

Figure 15.17

Urban land use patterns based on land values

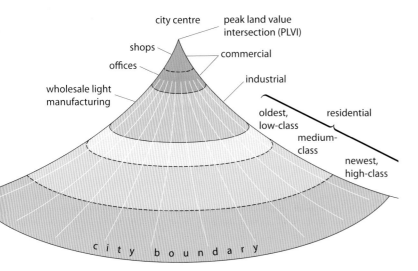

Figure 15.18

Secondary land
value peaks

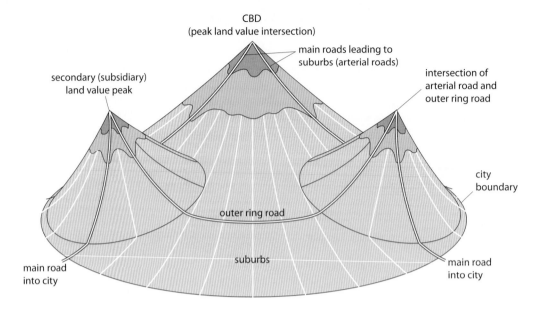

Functional zones within a city

Different parts of a city usually have their own specific functions (Figure 15.12). These functions may depend upon:

■ the age of the area: buildings usually get older towards the city centre except that most CBDs and many old inner-city areas have been redeveloped and modernised
■ land values: these increase rapidly from the city boundary in towards the CBD (Figure 15.16)
■ accessibility: some functions are more dependent on transport than others.

While each urban area will have its own unique pattern of functional zones and land use, most British cities exhibit similar characteristics. These characteristics have been summarised and simplified in Figure 15.19 where:

Zone A = the CBD (shops and offices)
Zone B = old inner city (including, before redevelopment, 19th-century/ low-cost/low-class housing, industry and warehousing and, after redevelopment/regeneration, modern low-cost housing and small industrial units
Zone C = inter-war (medium-class housing)
Zone D = suburbs (modern/high-cost/high-class housing, open space, new industrial estates/science and business parks, shopping complexes and office blocks).

The central business district (CBD)

The CBD is regarded as the centre for retailing, office location and service activities (banking and finance). It contains the principal commercial streets and main public buildings and forms the **core** of a city's business and commercial activities. Some large cities, such as London and Tokyo (Figure 15.12) may have more than one CBD. Other types of city-centre land use, such as government and public buildings, churches and educational establishments, are classed as non-CBD functional elements.

The delimitation of the CBD

Most of you are likely to have relatively easy access to a town or city centre. If so, your geography group may be able to make one or more visits to that CBD with the aim of trying to delimit its extent. Bearing in mind possible dangers, such as from moving traffic, your group could attempt one or more of several methods, based on the pioneer work of Murphy and Vance in North America, and described in Places 54, page 430. Ideally you should:

1 formulate one (or more) hypothesis before you begin your fieldwork (Framework 10, page 299)
2 collect, as a group, the relevant data
3 determine how you will record that data (i.e. using which geographical techniques)
4 discuss – again as a group – your findings.

One of several dangers that may result from putting forward geographical models and from making generalisations is that of creating stereotypes. For example:

a Urban models have the tendency to suggest that some areas are 'better' than others, e.g. that all housing in inner city areas is low-class/low-income and that only the elderly and single-parent families live here in a zone lacking open space, whereas wealthy families only reside in the 'tree-lined' suburbs.

b Different groups of people tend to develop their own customs and ways of life. By putting such characteristics together, we make mental pictures and develop preconceptions of different groups of people, i.e. we create stereotypes.

The following unsupported, emotive statements may not only be grossly inaccurate, they may also be considered, by many, to be offensive.

- The Germans, on holiday, are always first to the swimming pool and dining room.
- All Italians drive cars dangerously.
- All Chinese and Japanese are small.
- *Favelas* are shanty settlements whose residents have no chance of improving their living conditions and who can only survive by a life of crime (see below).
- The Amazon Amerindian way of life remains undeveloped as the people are lazy and unintelligent (see below).

The following accounts are based on the author's experiences in Brazil.

Example One

'According to books which I had read in Britain and advice given to me by guides in São Paulo, *favelas* were to be avoided at all costs (Places 57, page 443). Any stranger entering one was sure to lose his watch, jewellery and money and was likely to be a victim of physical violence.

With this in mind, I set off in a taxi to take photographs of several *favelas*. On reaching the first *favela*, to my horror the driver turned into the settlement and we bumped along an unmade track. He kept stopping and indicating that I should take photographs. Expecting at each stop that

the car would be attacked and my camera stolen, I hastily took pictures – which turned out to be over-exposed because, not daring to open windows, I took them through the windscreen and looking into the sun!

Suddenly the taxi spluttered and stopped. In one movement, I had hidden my camera and was outside trying to push the car. I raised my eyes to find three well-built males helping me to push the car. Which one would hit me first? I smiled and they smiled. I pointed to each one in turn and called him after one of Brazil's football players and then referred to myself as Lineker. Huge smiles, big pats on the back and comments like *Ingleesh amigo* were only halted by the car re-starting. As we drove away, I began to question my original stereotyped view of a *favela* inhabitant.'

Example Two

'I was surprised to find, on landing at Manaus airport in the middle of the Amazon rainforest, that our courier was an Amerindian. He dashed around quickly getting our party organised and our luggage collected. (He certainly did not seem to be slow or lazy.) He later admitted, and proved, that he could speak in seven languages (hardly the sign of someone unintelligent – how many can *you* speak?). I asked him why so few Amerindians appeared to have good jobs and why he kept talking about returning to the jungle. His reply was simple: "to avoid hassle". He considered that the Indian lifestyle was preferable to the Western one with its quest for material possessions. Had he returned to the jungle, he would have rejoined his family and become a shifting cultivator living in harmony with the environment (Places 66, page 480). Is that traditional way of life really less demanding of intelligence than that imposed by invading timber and beefburger transnationals engaged in the destruction of large tracts of rainforest?'

From these examples, we can see how easy it is to accept stereotypes without realising we are doing so, and also how seeing a situation for ourselves may lead us to question our original picture. Should geographers take a role in overcoming the problems of stereotyped images (on the basis of which planning decisions, for example, may be made) by helping to provide relatively unbiased information to improve knowledge and understanding?

A CBD

A1

Figure 15.19

Functional zones
in a British city

A1 Indoor shopping mall (St Enoch's Centre, Glasgow)
A2 High-rise office development (the City of London)

B1 An inner-city corner shop (Leeds)
B2 19th-century terraced housing (Lancashire)
B3 Inner-city redevelopment (London)
B4 19th-century industry and transport (Manchester)

B Inner city

B1

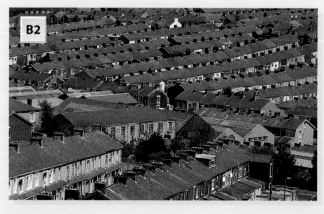

B2

C Inter-war areas

C1

C2

D Edge of city

D1

D2

1   Shopping types

2   Residential styles

C1 A suburban shopping parade
C2 Inter-war semi-detached private housing (Enfield)
C3 Inter-war council housing estate (Carlisle)
C4 Public open space (Brockwell Park, London)

D1 Edge-of-city shopping complex (Lakeside shopping centre, Dartford)
D2 A modern private housing estate (Wirral)
D3 Post-war edge-of-city council housing estate (Kenton, Newcastle upon Tyne)
D4 Business/science park (Guildford)

2   Residential styles

3   Other land uses

The main characteristics of the CBD

1 The CBD contains the major retailing outlets. The principal department stores and specialist shops with the highest turnover and requiring largest threshold populations compete for the prime sites (Figure 15.19 A1).

2 It contains a high proportion of the city's main offices (Figure 15.19 A2).

3 It contains the tallest buildings in the city (more typical in North America), mainly due to the high rents which result from the competition for land (Figure 15.16).

4 It has the greatest number and concentration of pedestrians.

5 It has the greatest volume and concentration of traffic. The city centre grew at the meeting point of the major lines of communication into the city and therefore had the greatest accessibility.

6 It has the highest land values in the city (Figure 15.17).

7 It is constantly undergoing change, with new shopping centres, taller office blocks and traffic schemes. Some of the grandiose schemes of the early 1960s are now viewed as out of date and unattractive (Birmingham's Bull Ring; London's Paternoster Square at St Paul's). Many have since been demolished and rebuilt.

Recent studies have shown that the CBD of many cities is advancing in some directions (**zone of assimilation**) and retreating in others (**zone of discard**). The zone of assimilation is usually towards the higher-status residential districts whereas the zone of discard tends to be nearer the industrial and poorer-quality residential areas (Figure 15.20). There has also been a trend in many CBDs for retailing to be static, or even declining (due to competition from out-of-town developments), while offices, banks and insurance companies are increasing in terms of space taken and income generated.

Mapping the characteristics of the CBD

The following fieldwork methods may be used to evaluate the seven characteristics described above.

1 **Land use mapping of shops**

  a Plot the location of all the shops. Where the ratio of shops to other properties is more than 1:3, count that area as being within the CBD (based on evidence that over 33 per cent of buildings in the CBD are connected with retailing).

  b An alternative method is to include within the CBD all shops that are within 100 m (or any agreed distance) of adjacent shops. This may produce a central 'core' and several smaller groupings.

  c A third possibility is to take the mean frontage (in metres) of, for example, the middle five buildings or shop units in a block. Shop frontages are likely to be greatest near to the PLVI where most department stores are located.

2 **Land use mapping of offices** Method **1a** above could be repeated using offices instead of shops, and a ratio of 1:10. This recognises that, at ground-floor level, offices are less numerous than shops. Include banks and building societies in your count.

3 **Height of buildings** Plot the height (i.e. the number of storeys) of individual buildings, or the mean of a group of buildings in the centre of a block. Most cities tend to have a sharp decline in building height at the edge of the CBD.

4 **Number of pedestrians** This is a group activity – the more groups the better! Each group counts the number of pedestrians passing a given point at a given time (e.g. 1100–1115 hours). The greater the number of sites (ideally chosen by using random numbers, Framework 6, page 159), the greater the accuracy of the survey. Define a pedestrian as someone of school age and over, walking into, out of or past a shop on your side of the street. These criteria may be altered as long as they are applied by all the groups.

5 **Accessibility to traffic** This is similar to the previous survey except that here vehicles are counted. Make sure all groups have the same definition of a vehicle, e.g. do you include a bicycle and/or pram?

Figure 15.20

The core and frame concept for the CBD

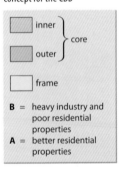

inner ⎫
         ⎬ core
outer ⎭

frame

B = heavy industry and poor residential properties
A = better residential properties

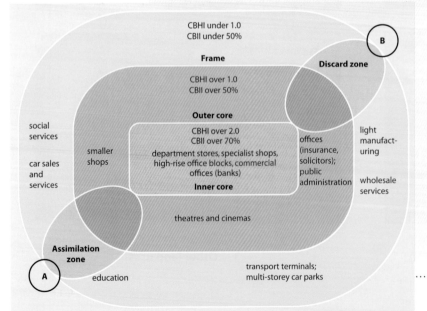

CBHI under 1.0
CBII under 50%

B

Frame

Discard zone

CBHI over 1.0
CBII over 50%

Outer core

CBHI over 2.0
CBII over 70%
department stores, specialist shops, high-rise office blocks, commercial offices (banks)
Inner core

offices (insurance, solicitors); public administration

light manufacturing

wholesale services

social services

car sales and services

smaller shops

theatres and cinemas

Assimilation zone

A

education

transport terminals; multi-storey car parks

6 **Land values** These might be expected to decline outwards at a fairly uniform rate. Providing rateable values can be obtained (try the rates office) and there is the time to process them (or a sample of the total), this is often a good indicator of the CBD. It may be useful to take the PLVI point and call this 100 per cent, and then convert the rateable index for all other properties as a percentage of the PLVI. It has been suggested that a figure of 20 per cent delimits the CBD for a British city.

7 **Changing land use and functions** This is a mapwork exercise using old maps of the central area (shopping maps are produced by GOAD plans) and superimposing onto them present-day land uses. Look for evidence of zones of assimilation and discard (Figure 15.20).

• **Central business index** This is probably the best method as it involves a combination of land use characteristics, building height and land values. The problem is in obtaining the necessary data, i.e.

  a the total floor area of all central or CBD functions

  b the total ground floor area (central and non-central functions)

  c the total floor area (upstairs floor area as well as the ground floor).

You may laboriously work this out from a large-scale plan, or choose to compromise by taking the mean of a sample of buildings in each block. From these data, two indices can be derived:

• The **central business height index**, or **CBHI**, which is expressed as:

$$\text{CBHI} = \frac{\text{total floor area of all CBD functions}}{\text{total ground floor area}}$$

• The **central business intensity index**, or **CBII**, which is expressed as:

$$\text{CBHI} = \frac{\text{total floor area of all CBD functions}}{\text{total ground floor area}} \times \frac{100}{1}$$

To be considered part of the CBD, the CBHI of a plot should be over 1.0 and the CBII over 50 per cent (Figure 15.21).

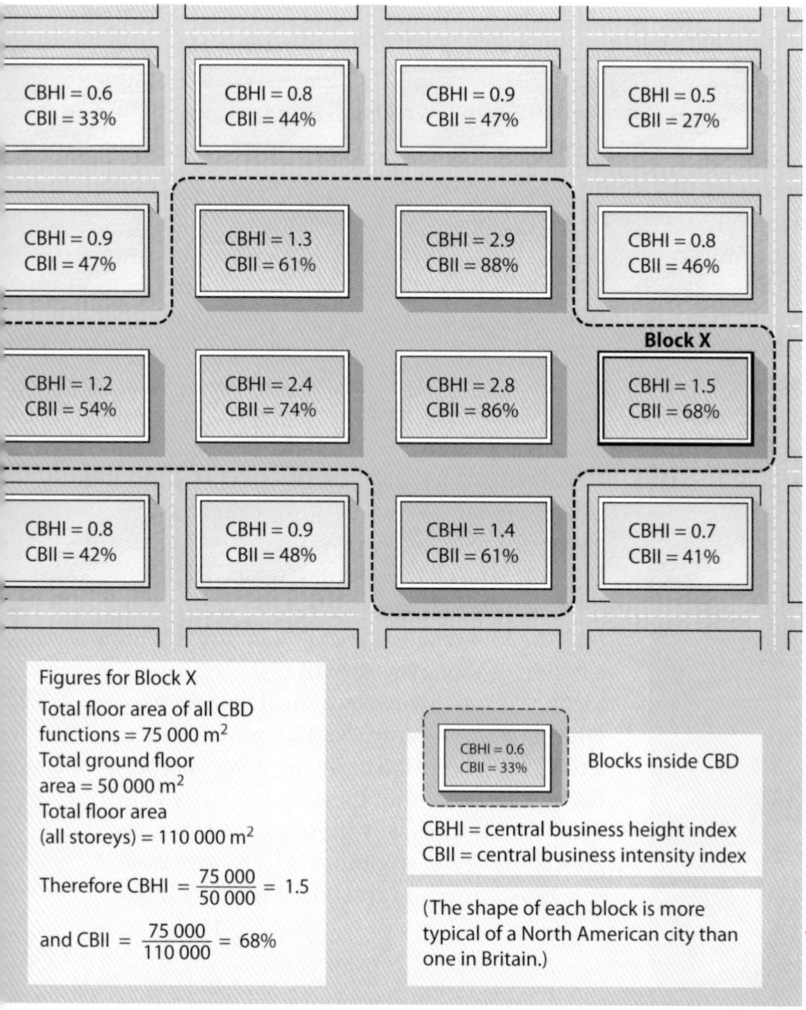

Plotting the data

Careful consideration should be given as to which cartographic technique is best applied to each set of collected data. You may wish to use one or several of the following: land use maps, isolines, choropleths, flow graphs, histograms, bar graphs, scattergraphs and transects. Alternatively, you may be able to devise a technique of your own. You may save time and produce results that are easier to compare by using tracing overlays and/or a computer.

Delimiting the CBD: conclusions

If you have carried out your own survey, your report might include comments on the following questions:

1 What problems did you encounter in collecting and refining the data?

2 Which of the methods used in collecting the data appeared to give the most, and the least, accurate delimitation of the CBD?

3 In your town, was there an obvious CBD; did you find an inner and an outer core (Figure 15.20)? Was there evidence of zones of assimilation and discard? Were there any specific functional zones other than shops and offices? Was the area of the CBD similar to your mental map (your preconceived picture) of its limits?

4 What refinements would you make to the techniques used if you had to repeat this task in a different urban area?

Figure 15.21

The central business index (CBI)

CBHI = 0.6 CBII = 33%	CBHI = 0.8 CBII = 44%	CBHI = 0.9 CBII = 47%	CBHI = 0.5 CBII = 27%
CBHI = 0.9 CBII = 47%	CBHI = 1.3 CBII = 61%	CBHI = 2.9 CBII = 88%	CBHI = 0.8 CBII = 46%
CBHI = 1.2 CBII = 54%	CBHI = 2.4 CBII = 74%	CBHI = 2.8 CBII = 86%	**Block X** CBHI = 1.5 CBII = 68%
CBHI = 0.8 CBII = 42%	CBHI = 0.9 CBII = 48%	CBHI = 1.4 CBII = 61%	CBHI = 0.7 CBII = 41%

Figures for Block X

Total floor area of all CBD functions = 75 000 m²

Total ground floor area = 50 000 m²

Total floor area (all storeys) = 110 000 m²

Therefore CBHI $= \frac{75\ 000}{50\ 000} = 1.5$

and CBII $= \frac{75\ 000}{110\ 000} = 68\%$

CBHI = 0.6 CBII = 33% → Blocks inside CBD

CBHI = central business height index
CBII = central business intensity index

(The shape of each block is more typical of a North American city than one in Britain.)

Retailing

Traditional shopping patterns

Traditionally, as neatly summarised by Prosser, 'Retailing in British cities has been based upon a well-established hierarchy, from the CBD or "High Street" at the top, through major district centres, local suburban centres, to neighbourhood parades and the local corner shop. Using numbers of outlets, floor space, type and range of goods, for example, as measures of size or "mass", Christaller's central place (page 407) and gravity models (page 410) have been applied to the hierarchical structure, relating mass to spatial distribution of shopping centres and their spheres of influence.'

Within this hierarchy were two main types of shop:

1   Those selling **convenience** or **low-order goods** which are bought frequently, usually daily, and are not sufficiently high in value to attract customers from further than the immediate catchment area, e.g. newsagents and small chain stores.
2   Those selling **comparison** or **high-order goods** which are purchased less frequently but which need a much higher threshold population, e.g. goods found in department stores and specialist shops.

The preferred location of these two types of shop was usually determined by the frequency of visit, their accessibility and the cost of land and, therefore, rent (page 425).

Convenience shops are commonly located in housing estates, both in the inner city and the suburbs, and in neighbourhood units so as to be within easy reach of their customers – often within walking distance. With a lower turnover of goods than retail units in the CBD, they may have to charge higher prices but their rent and rates are lower. Ideally, they are located along suburban arterial roads or at a crossroads for easier access and, possibly, to encourage impulse buying by motorists driving into the CBD (Figure 15.18).

Convenience shops are also located in inner cities where the corner shop (Figure 15.19 B1) caters for a population that cannot afford high transport costs; in suburban shopping parades (Figure 15.19 C1) where the inhabitants live a long way from the central shopping area; and along side-streets in the CBD where they take advantage of lower rents to provide daily essentials for those who work in the city centre.

Comparison shops need a large threshold population (page 407) and therefore have to attract people from the whole urban area and beyond. As they bid for a central location, they must have a high turnover in order to pay the high rents. This central area has traditionally afforded the greatest accessibility for shoppers, with public transport competing with the private motorist. Large department stores and specialist shops usually locate within the CBD (Figure 15.19 A1), although comparison shops may also locate in the more affluent suburbs.

The retailing revolution

Since the 1970s there has been such a revolution in retailing that, by 2007, it provided 8 per cent of the UK's GDP and employed 11 per cent of its total workforce. It began with the growth of superstores, often in then traditional city centre shopping areas, and hypermarkets, locating on new edge-of-city sites (Figure 15.19 D1). The 1980s saw a growth in both non-food retail parks and, at MetroCentre (Places 55), the first of the now dominant out-of-town regional shopping centres.

Town centres

Many city centres have undergone constant change either to try to attract new customers or, as is more usual, to restrict losses of existing shoppers to the regional shopping centres or to internet shopping. Most city centres contain covered malls, where shoppers can compare styles and prices while staying warm and dry, and which are either traffic free or have access limited to delivery vehicles and public transport. Many local councils have allowed an extension of land use to include places for eating, drinking and entertainment and have improved the quality of the shopping environment (Figure 15.22).

A report by the New Economics Foundation (August 2004) claimed that Britain was becoming a nation of 'clone towns with high streets having identical shops owned by a small number of powerful chains'. The only variation was how smart a town is perceived to be by the stores' market researchers – that this, there is a hierarchy in the quality of shop: usually the larger the town, the greater the degree of cloning. The report claimed that local businesses are suffocated by identikit

Figure 15.22

An environmentally improved city centre shopping area in Sheffield

chain stores that have marketing budgets, political contacts and resources that give them an unfair economic advantage. The only real recent gainers have been coffee shops, pub chains, mobile phones and charity shops.

Despite attempts, both locally and nationally, to try to restrict further shopping developments on edge-of-city sites, an increasing number of the smaller city-centre retailers are still being forced to close. Initially these were mainly food, clothing and other specialist shops but, as 2008 has shown with the previously unforeseen closure of banks, nothing in the CBD is immune to an economic downturn. These recent events may well buck the trend by which city centres have responded, often successfully, to the challenges of the out-of town centres through considerable re-branding and updating.

Out-of-town shopping centres

An increasing number of shopping outlets began locating on the edge of towns and cities to take advantage of economies of scale, lower rents, and a more pleasant and planned environment. Superstores in particular were built on cheaper land at, or beyond, city margins (Figure 15.17), which allowed them space for immediate use, future expansion and essential large car parking areas. The ideal location is also near to a motorway interchange facilitating access for both customers and delivery drivers.

Developments have included the following:

■ The present 'big four' supermarkets of Tesco, Sainsbury's, ASDA and Morrisons – there is much concern about these as they continue, between them, increasingly to dominate Britain's retailing industry:

2007	Stores in existence	Applications	
		New	Extensions
Tesco	1819	37	26
Sainsbury's	751	6	9
ASDA	302	21	6
Morrisons	370	10	2
Total	3242	74	43

■ Retail parks, which have also been attracted to inner city brownfield sites, tend to concentrate on the sale of non-food items (e.g. B&Q, Comet and Homebase).

■ Regional shopping centres not only sell 'everything' under one roof, but often include restaurants, children's play areas and cinemas. The earliest such centres, each covering over 100 000 m² , were Gateshead MetroCentre (Places 55), Sheffield Meadowhall, Dudley Merry Hill, Lakeside and Dartford Bluewater. They were controversial in that they not only took a large amount of business from local city-centres, they also attracted literally coachloads of shoppers from places up to 150 km away. While the volume of trade in city centres has been increasing by less than 1 per cent annually in recent times, that of the regional shopping centres has seen a growth of over 20 per cent a year. New ones are still being developed as at Liverpool One (2007) and Cabot Circus in Bristol (2008). However, Westfield in west London (2008 – page 458) and Stratford City in east London (planned) are attempts to keep retail spending within the capital (at the expense of Bluewater). However, predictions are that, following any economic downturn, it will be the regional centres, not the city centres, that will be the first to recover.

Figure 15.23

Aerial view of the MetroCentre site, Gateshead: to the left is the A1 (Western by-pass) and to the right the Newcastle–Carlisle railway, with station, and the River Tyne

Places 55 Gateshead: the MetroCentre

Location

The MetroCentre, on the edge of Gateshead, opened in 1986, and was the prototype for a new concept in retailing in Britain: out-of-town shopping. After several upgrades and extensions, it still remains Europe's largest single shopping centre. Before development the site was marshland, which meant that a large amount of land was available and relatively cheap to buy.

Access

The site is adjacent to the western by-pass which now forms part of the main north–south trunk road, the A1, which avoids central Newcastle and Gateshead. It has 10 000 free car parking spaces

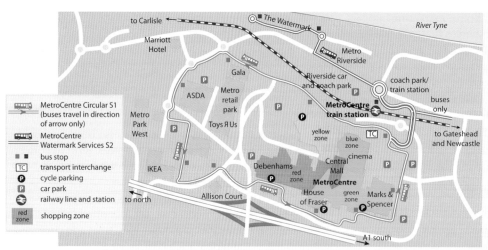

Figure 15.24

Layout of the MetroCentre

with special facilities for the disabled motorist, 100 buses per hour, and 69 trains daily. The centre, which has its own bus and railway stations, is less than 70 minutes' drive away for 2.6 million people (Figures 15.23 and 15.24).

Shopping environment

In 2008 there were 342 shops with Marks & Spencer, House of Fraser, Debenhams and BHS forming the anchor stores. The shops, grouped into four colour-coded zones (Figure 15.24), are set on two levels in a pleasant environment which includes tree-lined malls, air conditioning, 1 km² of glazed roof to let in natural light (supplemented by 'old world' lamps), numerous seats for relaxing, escalators and, for the disabled, lifts (Figure 15.25). A street atmosphere is created by traders selling from stalls and there are over 50 places for eating and drinking – many staying open long after the shops close.

Additional amenities

Leisure has always been a vital part of the scheme. Sir John Hall, whose idea the MetroCentre was, believed that shopping should be an enjoyable occasion for the whole family (Figure 15.24). There is a 'children's village', a crèche, a 10-screen cinema, a 'space city' for computer enthusiasts, and various theme areas such as The Forum, The Village and Garden Court. Opening in the Yellow Mall in 2009 will be bowling, pool, soft play and electronic games in a family entertainment centre, together with a new cinema and more restaurants. This could increase the present working population of 7000 by another 1000. The MetroCentre complex also includes a 150-bedroom hotel, an office block and a petrol station, while the adjacent Metro Retail Park contains IKEA, DFS, ASDA, Toys Я Us and Harry Ramsden's.

Visitors

MetroCentre's sphere of influence extends as far as York, Cumbria and Scotland. Over 24 million people, 70 per cent of whom are female, visit annually with 84 per cent arriving by car and 16 per cent by public transport (Figure 15.26).

Figure 15.25

Inside the MetroCentre

Significant changes have taken place in the top five ranking of the UK's leading shopping centres, according to the 2006 shopping centre analysis carried out by Trevor Wood Associates. The most notable is that Gateshead's MetroCentre has regained top spot from Bluewater by which it had been replaced six years earlier. MetroCentre gained maximum scores in many categories in the report in which every shopping centre was ranked by overall attractiveness to shoppers, retailers and investors. Categories included quality and mix of tenants, gross lettable retail area, whether it is open or closed to the elements and whether it has a food court and a crèche. The ranking was achieved by checking details, including tenants, for over 850 shopping centres, parks and factory outlets for schemes covering over 4600 m². However, it was the opening of the Red Mall in 2006, with Debenhams and 23 other retail units, which raised MetroCentre's floor area to 220 000 m², that made it, once again, Europe's largest shopping centre.

Figure 15.26

Adapted from the *Newcastle Journal*, November 2007

Financial institutions and offices

Financial institutions employ large numbers of people, especially in world centres such as New York, Tokyo, Hong Kong and London. These institutions, which include banking, insurance and accountancy, operate within offices. Traditionally offices have vied with shops for city centre locations regardless of the country's level of development (compare Tokyo, Figure 15.13; London, Figure 15.19 A2; Hong Kong, Figure 15.27; and Nairobi, Figure 15.36). However, whereas shops offer assistance to local individuals, offices form part of an agglomeration of businesses usually served by, and in close association with, a myriad consultants, media, hospitality and recreational establishments.

Company head offices and major institutions such as the stock exchange locate in the capital city. As offices use land intensively, they compete with shops for prime sites within city centres (Figure 15.17). Increasingly, due to high land values, they have had to locate in ever-taller office blocks. Elsewhere in city centres, offices may locate above shops in the main street or on ground-floor sites in side-streets running off the main shopping thoroughfares. Banks can afford prime corner sites, while building societies and estate agents vie for high-visibility locations. A city centre office location may have been desirable for prestige reasons, for ease of access for clients and staff, for proximity to other **functional links** (banks, insurance and entertainment) and sources of data and information, and for face-to-face contact.

Taking London as an example, it can be seen that demands for office space and location change over time. In the late 1940s, some firms re-located to the then New Towns. Later the decentralisation of government offices saw a movement often to areas where there was high unemployment (DHS to Newcastle, DVLA to Swansea and Giro to Bootle) and, until the early 1990s, to smaller towns where rents were lower, more space was available and the quality of life perceived to be higher. Since then there has been a remarkable reversal, with a huge demand for space within the capital itself resulting from London's increasing status as a global city. The Docklands can now be considered to be a CBD in it own right (Ullman and Harris, page 423), while media and advertising companies have put real pressure on commercial space in Soho and central London. It will be interesting to see how the global financial crisis of 2008 affects future growth and location.

New technology has allowed the easier transfer of data and has reduced the need for face-to-face contact, while increased computerisation has often led to a reduced workforce (banking) but one that is more highly skilled. Many new office locations are on purpose-built business, office or science parks (Figure 15.19 D4).

Figure 15.27

Office development on Hong Kong Island

Industrial zones

Industry within urban areas has changed its location over time. In the early 19th century, it was usually sited within city centres, e.g. textile firms, slaughter houses and food processing. However, as the Industrial Revolution saw the growth in size and number of factories, and later when shops began to compete for space in the city centre, industry moved centrifugally outwards into what today is the inner city (Places 52, page 421). Inner-city areas could provide the large quantity of unskilled labour needed for textile mills, steelworks and heavy engineering. The land was cheaper and had not yet been built upon. Factories were also located next to main lines of communication: originally, rivers and canals, then railways and finally roads (Figure 15.19 B4). Firms including bakeries, dairies, printing (newspapers) and furniture, which have strong links with the city centre, are still found here.

Between the 1950s and the 1980s this zone increasingly suffered from industrial decline as older, traditional industries closed down and others moved to edge-of-city sites. In Britain, recent changes in government policy have led to attempts to regenerate industry in these areas through initiatives such as Enterprise Zones, derelict land grants and Urban Development Corporations (page 439). Even so, the replacement industries are often on a small scale and compete for space with warehouses and DIY shops.

Most modern industry is 'light' and clean in comparison to that of the last century and has moved to greenfield sites near to the city boundary (Figure 15.19 D4). Industrial estates and modern business and science parks are located on large areas of relatively cheap land where firms have built new premises, use modern technology and, by being near to local housing estates, can satisfy the need for a wider range of skills and the increased demand for female labour (Places 86, page 566). Most industries are 'footloose' and include high-tech, electronics, IT software houses, media/news companies, food processing and distribution firms and those providing services such as waste recycling.

Residential zones

The Industrial Revolution also led to the rapid growth in urban population and the outward expansion of towns. Long, straight rows of terraced houses (Figure 15.19 B2) were constructed as close as possible to the nearby factories where most of the occupants worked. The closeness was essential as neither private nor public transport was yet available. Houses and factories competed for space. As a result, houses were small, sometimes with only one room upstairs and one downstairs or they were built 'back to back'. The absence of gardens and public open space added to the high housing density.

By the 1950s, many of these inner-city areas, the low-class/low-income houses of the urban models, had become slums. Wholesale clearances saw large areas flattened by bulldozers and redeveloped with high-rise blocks of flats (Figure 15.19 B3). Within 20 years, the previously unforeseen social problems of these flats led to a change in policy where, under urban renewal, older housing was improved, rather than replaced, by adding bathrooms, kitchens, hot water and indoor toilets. The tower blocks and estates, mainly due to the action of housing associations, are themselves being replaced on an ambitious scale.

Some inner-city areas have undergone a process known as **gentrification**. This is where old, substandard housing is bought, modernised and occupied by more wealthy families. In some Inner London districts, like Chelsea, Fulham and Islington, such properties are much sought-after and have become very expensive. The process is partly triggered by the proximity of employment and services in the city centre and partly through the availability of improvement grants. Once begun, it is often maintained by the perception of social prestige derived from living in such areas. More recently, inner-city areas with a waterfront location, as in London, Bristol, Manchester, Liverpool and Newcastle, have undergone a renaissance which has also seen them becoming fashionable and expensive (Figure 15.30).

The outward growth of the city continued both during the inter-war period when, aided by the development of private and public transport, large estates of semi-detached houses were built (the medium-class houses of the urban models, Figure 15.19 C2 and C3), and after the 1950s. Many of the present edge-of-city estates consist of low-density private housing. Due to low land values (Figure 15.17), the houses are large, and have gardens and access to open space (Figure 15.19 D2). Other estates were created by local councils in an attempt to rehouse those people forced to move during the inner-city clearances. These estates, a mixture of high-rise and low-rise buildings (Figure 15.19 D3), have a high density and, like some older inner-city areas, are now experiencing extreme social and economic problems (page 441).

Existing A2/AS syllabuses state as one of their aims: 'improve as critical and reflective learners, aware of the importance of attitudes and values, including their own'. This is not a new aim: since the early 1970s geography teachers have been trying to encourage their students to develop and clarify their own values and attitudes, a process by which geographers do not simply measure and quantify but confront some of the questions and concepts that arise from those measurements, e.g. inequalities and deprivation (page 438). It is not a case of teachers 'passing on' their own values but getting their students to enquire, for example, why there are inequalities and how they have developed.

The present author has tried, rightly or wrongly, to maintain a 'neutral' stance. Some would claim that what has been included in this book has been influenced by the author's own values and attitudes, e.g. a belief in the fundamental role of physical geography in an understanding of environmental problems; a preference for living in a semi-rural area rather than an inner city. Criticism could also be levelled for using personal experiences as exemplars in some **Places** and **Case Study** sections. What the author has tried to do is to present readers with information in the hope that they may become more aware of their own values in relation to the behaviour of others, and to enable them to discuss, with fewer prejudices and preconceptions, the foundations of their own values.

This may be illustrated with reference to the following section on inner cities which is structured as follows:

1   **The problem of inner cities** Will these issues be seen differently by the inhabitant of an inner-city area and a person living in a rural environment?

2   **The image of an inner-city area** Will a description of inner-city problems give a negative picture of the quality of life in those environments and in doing so perpetuate the problems, or could it help in the understanding and tackling of them?

3   **Possible solutions to the inner-city problem** Would solutions proposed by inner-city residents be similar to those suggested and implemented by the government or the local authority?

4   **What successes have government schemes had?** Your answer to this may depend upon your own political views. Before the 1997 general election, Conservatives pointed out the many achievements of the period 1980–97; Labour, the Liberal Democrats and other opposition parties claimed that little had been done. Who, if either, was correct? Presumably since that election, which led to a reversal of roles, the two main parties will be changing their attitudes!

Issues in Britain's inner cities

Tremendous changes have taken place in inner-city areas since the last edition of this book was published a decade ago. In many areas these changes, which include land use and social composition, have replaced the largely negative picture that was described in the late 1990s (the dangers of stereotyping, Framework 13, page 427) and, with an increasing mix in the types of building and in population structure (Hampstead and Brixton are both inner-city locations), it has become impossible to make broad, accurate generalisations such as those suggested by 20th-century urban models and textbooks.

1991

The widest definition of an inner city at that time was 'an area found in older cities, surrounding the CBD, where the prevailing economic, social and environmental conditions pose severe problems'. This definition was, intentionally or otherwise, reinforced by the 'Small-area Census' of that year which listed the characteristics of inner-city London, and of other inner-city areas, as:

- a lack of basic household amenities (1 million without a bathroom, WC or hot water)
- high densities in high-rise flats, overcrowding in houses
- lower life expectancies and a greater incidence of illness
- a predominance of lower-income, semi-skilled manual workers
- a higher incidence of single-parent families and the elderly
- a concentration of ethnic minorities.

Even if it was true at the time, these indicators only tended to reinforce the concept of the inner cities being areas of poverty and deprivation (Figures 15.28 and 15.29).

2008

Certain London boroughs have seen considerable regeneration and most areas have seen improvements, to a greater or lesser degree, in housing, transport, employment and the provision of amenities. The biggest transformation has occurred in the former Docklands where – as in similar locations in places like Liverpool, Bristol and Newcastle – derelict land

and unused buildings have been cleared, poor-quality housing has been upgraded and former warehouses converted into expensive accommodation. Transport links have been improved by the construction of a new light railway and the extension of the Jubilee line. Numerous new jobs, often office-based, have been created, as at Canary Wharf, together with improvements in leisure amenities, shopping and the environment (Figure 15.30).

Elsewhere in London, Brixton Market and the Notting Hill Carnival are examples of events where local people and visitors from a wider area come together. In 1996, Tower Hamlets (Figure 13.7) recorded only 11 per cent of its students obtaining 5 GCSEs, but ten years later that figure was 44 per cent. New shopping centres are appearing and more are planned (Stratford City, page 433). The biggest change of all is beginning in east London with the regeneration in preparation for the 2012 Olympics (Places 56). Perhaps the best indicator of all of London's successful transformation is the large number of people, mainly in the 20s and 30s age groups and from overseas, moving here to be part of a multi-ethnic, global city.

Yet to many people living in parts of inner London, this now positive view of the city is either unrecognisable or remains beyond their reach. There are still too many pockets of poverty, especially in some of the boroughs towards east London. However, whereas 30 years ago it was the result of industrial decline, especially in the former docklands, now it often results from poor housing and social conditions. Canning Town, quoted as the poorest ward in the poorest borough (Newham), owes much of its poverty to (a) a housing policy that led to the selling of large tracts of public (council) housing and which resulted in an increasing accumulation of deprived families and individuals and (b) the low level of educational attainment compared with other London boroughs.

Indicators of welfare and deprivation

The Department of the Environment describes deprivation as: 'when an individual's well-being falls below a level generally regarded as a reasonable minimum for Britain today' and it is measured by several economic, social, housing and environmental indicators (Figure 15.28). In 2007, despite a determined government effort, over 7 million people were living in households which received less than the national annual income, and up to one-quarter of children born each year are born into poverty (Figure 15.29).

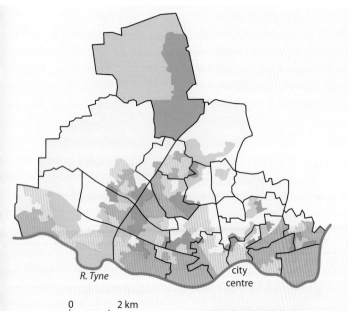

Figure 15.28

Deprivation index by wards, Newcastle upon Tyne, 2007

R. Tyne

city centre

0   2 km

IMD 2007
National Deprivation

▨ in 10% most deprived
10–20%
20–30%
30–50%
50–100% least deprived
— ward boundary
IMD = indicators of multiple deprivation

Deprivation indicators
Economic stress
– unemployment
– low-income families
Social stress
– all dependants in house: no family earner
– lone-parent families
– families without a car
– crime/overt delinquency
– racial tension
Housing stress
– lacking one or more basic amenities (WC, running water, bathroom)
– overcrowding (more than 1 person per room)
– no central heating
Environmental stress
– noise pollution
– derelict land

Figure 15.29

Cycle of poverty, or deprivation

Cycle of poverty, or deprivation

This is a concept that is largely, though not exclusively, linked to inner-city problems. It offers some explanation of how the problems have arisen. The cycle of poverty involves a continuous process which transmits relative poverty from one generation to another and which makes escape from deprivation very difficult. Certain occupational groups earn very low incomes, which makes for a low standard of living, including poor housing (since they cannot afford any better). The poor environment may produce stresses and strains in the household, and poor health amongst household members. In turn this affects the educational and other prospects of younger members in the family. The school and neighbourhood may lack the resources and skilled people needed to improve conditions for the young who are caught in this cycle of poverty. They tend to leave school early with insecure job prospects.

Poor conditions and poor prospects encourage criminal activity and lack of interest in the neighbourhood environment, discouraging outside investment and incentives to improve it. On the contrary, the neighbourhood becomes even more run-down and an adverse image of it is created, discouraging inward movement of all but the desperate households who have nowhere else to go.

The cycle of poverty is thus characteristic of the underclasses and is also increasingly concentrated in particular areas of the city and in certain housing estates on the edge of some cities.

Source: Material adapted from Department of the Environment

Government policies for the inner cities

Innumerable inner city initiatives have been introduced by various governments since 1945. These have sought to try to achieve one or more of the following:

- enhance job prospects and re-train local people to compete for them
- bring derelict land and buildings back into use
- improve housing conditions and local services
- encourage private sector investment
- encourage community co-operation and involvement to improve the social fabric
- improve the quality of the environment.

Since the 1980s many schemes have proved to be short-lived and to have had only limited effect, e.g. Urban Development Grants, Derelict Land Grants, Inner City Task Force, City Challenge, Urban Task Force and Neighbourhood Renewal Units. The two most successful and longest lasting initiatives operated throughout most of the 1980s and 1990s.

1 **Enterprise Zones (EZs)** tried to stimulate economic activity in areas of high unemployment by lifting certain tax burdens, e.g. exemption from paying rates for the first ten years; 100 per cent grants for machinery and new buildings; and the relaxing or speeding up of planning applications. Included in the 26 EZs that affected inner cities were Gateshead's MetroCentre (Places 55), the cleaning up of the Lower Swansea Valley, and the opening of the independent television studios at Limehouse in London's Isle of Dogs.

2 **Urban Development Corporations (UDCs)** were introduced to spearhead the then government's attempts to regenerate areas that contained large amounts of derelict, unused land or buildings. UDCs were given the power to acquire, reclaim and service land; to restore buildings to effective use; to promote new industrial activity and housing developments; and to support local community facilities. Financed by private-sector investment, the first two, the London Dockland Development Corporation (LDDC, Fig 15.30) and the Merseyside Development Corporation (MDC), were set up in 1981. By 1993 there were 13 – 12 in England and 1 in Wales. Most of these schemes changed the face of the areas in which they operated, for example the LDDC (which transformed London's former docklands and included the pulling down of the Limehouse television studios (see above) and replacing them with Canary Wharf); the MDC (which revitalised Liverpool's Albert Dock); Trafford Park DC in Manchester; Cardiff Bay DC; and Sheffield DC (which regenerated the Lower Don Valley). The UDCs in England were all wound up by 1998, and Cardiff Bay DC in 2000.

Present schemes

- **Urban Regeneration Companies (URCs)** are local partnerships with the task of achieving radical physical, economic and social transformation of towns and cities in declining urban areas. Launched in 1999, with three pilot companies in Liverpool, east Manchester and Sheffield, they now operate in 22 areas, including one in each of Wales and Northern Ireland.
- **New Deal for Communities (NDC)** operates in 39 of England's most deprived areas including in Lambeth and Hackney in London, as well as in Bradford, Manchester, Leicester, Oldham, Hull and Middlesbrough. Its aim was to deliver real improvements to people's lives and to narrow the gap between the most deprived areas and the rest of the country by, among other factors, reducing crime and improving education, health and a community spirit.

It is difficult to generalise on the overall success of so many wide-ranging schemes introduced over such a long period. There have been many positive improvements, especially to the environment, but social and economic problems still remain, with some former inner city areas experiencing above the national average in terms of unemployment, amounts of poor-quality housing and levels of crime, while standards in education and health care are often below it.

Figure 15.30

Canary Wharf and London Docklands, 2008

The East End of London would appear to have had its fair share of government inner city initiatives, with parts having been in the Isle of Dogs EZ and under the London Docklands DC (page 439). These schemes resulted in many improvements, especially in housing, job opportunities, transport links and the environment. Yet, as was noted on page 438, Canning Town, in the borough of Newham and just a short distance from the prestigious Canary Wharf development (Figure 15.30) was, according to statistics in the UK 2001 census, the poorest and most deprived area in the country not only for standards of housing but also for people employed, having a limiting illness or disability, and lacking educational qualifications or job skills. One meaningful comment comes from Bob Digby who wrote: 'a tube journey along the Jubilee line between Westminster and the Stratford terminus in east London links two areas with nine years' difference in life expectancy – one year for every station'.

A major reason for London being granted the 2012 Olympics Games was its plan to use the event as a way of regenerating deprived areas such as Canning Town and Stratford (Figure 15.31). London's bid was made on certain basic principles: that the long-term benefits of the Olympics would outweigh the total costs; that London is a global city with one of the world's most culturally and ethnically mixed populations; and that by portraying children as the ones who would be likely to benefit the most, this could help link, through sport, the nations of the world. The specific site, alongside the River Lea, is at present a mixture of industrial estates with many firms in the service sector, university halls of residence, low-cost housing and large tracts of waste land that create an eyesore. However, this area has the advantage of being near to Stratford which is a major transport 'hub' (page 637) with nine surface and underground rail links and, opening in 2009, an international station on the high-speed Channel Tunnel rail link, bringing the site within two hours of Paris and Brussels. The construction of

the Olympic site, with its village and stadiums, will mean relocating existing factories, students and permanent residents, and cleaning up the environment. After the Olympics, the plan is to re-model the village, where 17 000 athletes and officials will have stayed, into 3500 mainly affordable homes; to construct up to a further 9000 new houses of which 50 per cent will be affordable; and to be left with an improved transport system, a new primary healthcare centre and an academy school. Also, once some of the sporting facilities have either been dismantled or re-located, such as the multi-sports arena at Hackney Wick, the area will have a large urban park extending alongside the River Lea, with protected wildlife and cleared river and canal channels. The reality, though, could be that the region, like Barcelona, Atlanta, Sydney and Athens after previous Olympics, will struggle to create permanent jobs, have sporting amenities unused and many of the houses (especially in 2008's financial climate), remaining unsold.

Figure 15.31

The Lea Valley area before redevelopment

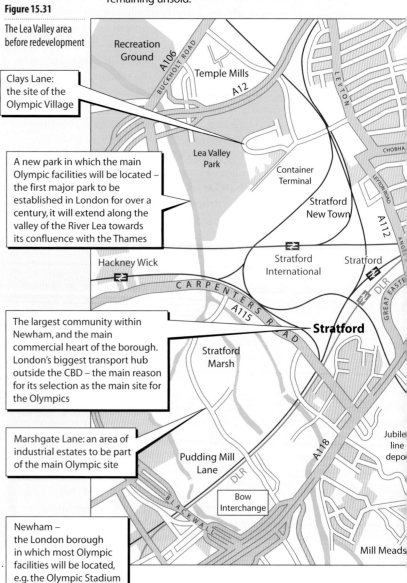

Clays Lane: the site of the Olympic Village

A new park in which the main Olympic facilities will be located – the first major park to be established in London for over a century, it will extend along the valley of the River Lea towards its confluence with the Thames

The largest community within Newham, and the main commercial heart of the borough. London's biggest transport hub outside the CBD – the main reason for its selection as the main site for the Olympics

Marshgate Lane: an area of industrial estates to be part of the main Olympic site

Newham – the London borough in which most Olympic facilities will be located, e.g. the Olympic Stadium

Issues in Britain's council estates

Whereas most government policies and funding have been focused on the inner cities, there is increasing evidence that poverty, unemployment, crime and social stress may be even higher on council estates.

1   **Inter-war estates** have the most acute problems. One such estate, in Newcastle upon Tyne's West End, which was passed by the city council to private builders in the mid-1980s, was revitalised with government Urban Development Grants. The flats, over 170 in total, were modernised for aspiring home-owners, and the local parade of shops was regenerated. Despite this, many flats have proved difficult to sell, even at a very affordable price. For the people still living there, crime is a constant threat, and finding work is difficult. It has been suggested that the scheme failed because it was an 'oasis' and that, in future, the redevelopment of brownfield sites must neither be as isolated nor as small.

2   **Edge-of-city estates** (Figure 15.19 D3), built on greenfield sites during the 1950s and 1960s, were created to house people forced to move by inner-city redevelopment schemes. Today they exhibit several common features:
   - The physical fabric of the buildings, many originally built using cheap materials and methods, is deteriorating rapidly. Local councils, without the financial help given to the inner cities, are trying to upgrade selected estates as and when they can.
   - Many estates include high-rise buildings which, as in the inner cities, have created feelings of isolation and stress-related illnesses. Flats and maisonettes have not proved popular under the 'right-to-buy' schemes and have been too expensive for most of the occupants to consider buying.
   - The low level of car ownership and high bus fares have increased the feeling of isolation from jobs, shops and entertainment.
   - Levels of unemployment often exceed 30 per cent. There are also many low-income families; many elderly living on small pensions; and up to two-thirds of households may be receiving housing benefit.
   - The environmental quality of the estates is poor, often with a lack of open space.
   - The estates tend to have high levels of problems, drug-taking, petty crime and vandalism, and low levels of academic attainment and aspiration.

Brownfield and greenfield sites

In 2008, the government announced that 3 million new homes would have to be built by 2020 (Figure 14.22) to accommodate the predicted rise in households by that date (mainly due to an increase in both single person households, from 5.8 million to 8.7 million, and in immigrants). The intention is that 60 per cent of the new houses will be built on **brownfield** sites, i.e. on land within urban areas, and 40 per cent on **greenfield** sites, i.e. in the countryside.

- Why greenfield sites? Developers claim that most British people want their own home, complete with garden, set in a rural, or semi-rural, location. As evidence they quote that, at present, for every three people moving into cities, five move out. Greenfield sites are cheaper to build on than brownfield sites as they are likely to have lower land values and are less likely to be in need of clearing-up operations than former industrial locations.
- Why brownfield sites? Groups such as the Council for the Protection of Rural England and Friends of the Earth argue that there are already three-quarters of a million unoccupied houses in cities which could be upgraded, while a further one and a quarter million could be created by either subdividing large houses or using empty space above offices and shops. They quote the database which showed that one-third of a million homes could be built on vacant and derelict land and another one-third of a million by re-using old industrial and commercial buildings. They also argue that urban living reduces the use of the car and maintains services, especially retailing, within city centres. Arguably, of course, many of those people wanting to protect the green belt and build in cities are probably already living in rural areas themselves.

However, the National Data Base shows a mismatch between:
- the South East of England where brownfield sites are limited but where most homes are needed (1996–2021 has a projected increase in households in the South of 24.2 per cent)
- the Midlands and the North where more brownfield sites are available but where demand for new properties is likely to be less (Midlands a 16.1 per cent increase and the North an 11.4 per cent increase in households).

There are two considerations as to how sustainable urban development can take place in the South East:
1   Settlements should become self-contained for work, living and leisure (see Case Study 14A, page 412).
2   Public transport needs to be improved for the resulting longer journeys to work.

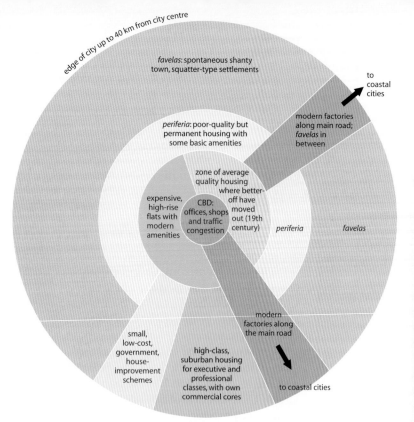

edge of city up to 40 km from city centre

favelas: spontaneous shanty town, squatter-type settlements

periferia: poor-quality but permanent housing with some basic amenities

to coastal cities

modern factories along main road; *favelas* in between

zone of average quality housing where better-off have moved out (19th century)

CBD: offices, shops and traffic congestion

expensive, high-rise flats with modern amenities

periferia

favelas

small, low-cost, government, house-improvement schemes

high-class, suburban housing for executive and professional classes, with own commercial cores

modern factories along the main road

to coastal cities

Figure 15.32

Model showing land use and residential areas in Brazilian cities (excluding Brasilia): the zoning of housing, with the more affluent living near to the CBD and the poorest further from the centre, is typical of cities in developing countries (*after* Waugh, 1983)

Figure 15.33

Living conditions in Howrah, Kolkata

Cities in developing countries

Cities in economically less developed countries, which have grown rapidly in the last few decades (page 419), have developed different structures from those of older settlements in developed countries. Despite some observed similarities between most developing cities, few attempts have been made to produce models to explain them. Clarke has proposed a model for West African cities, McGhee for South-east Asia, and the present author (based on two television programmes on São Paulo and Belo Horizonte together with some limited fieldwork) for Brazil (Figure 15.32).

Functional zones in developing cities

The CBD is similar to those of 'Western' cities except that congestion and competition for space are even greater (São Paulo, Cairo, and Nairobi, Places 58).

Inner zone In pre-industrial and/or colonial times, the wealthy landowners, merchants and administrators built large and luxurious homes around the CBD. While the condition of some of these houses may have deteriorated with time, the well-off have continued to live in this inner zone – often in high-security, modern,

high-rise apartments, sometimes in well-guarded, detached houses.

Middle zone This is similar to that in a developed city in that it provides the 'in between' housing, except that here it is of much poorer quality. In many cases, it consists of self-constructed homes to which the authorities *may* have added some of the basic infrastructure amenities such as running water, sewerage and electricity (the *periferia* in Figure 15.32 and the 'site and service' schemes on page 449 and in Figure 15.41).

Outer zone Unlike that in the developed city, the location of the 'lower-class zone' is reversed as the quality of housing decreases rapidly with distance from the city centre. This is where migrants from the rural areas live, usually in shanty towns (the *favelas* of Brazil and *bustees* in Kolkata, Places 57 and Figure 15.33) which lack basic amenities. Where groups of better-off inhabitants have moved to the suburbs, possibly to avoid the congestion and pollution of central areas, they live together in well-guarded communities with their own commercial cores.

Industry This has either been planned within the inner zone or has grown spontaneously along main lines of communication leading out of the city.

Kolkata's *bustees*

Although over 100 000 people live and sleep on Kolkata's streets, one in three inhabitants of the city lives in a *bustee* (Figure 15.33). These dwellings are built from wattle, with tiled roofs and mud floors – materials that are not particularly effective in combating the heavy monsoon rains. The houses, packed closely together, are separated by narrow alleys. Inside, there is often only one room, no bigger than an average British bathroom. In this room the family, often up to eight in number, live, eat and sleep. Yet, despite this overcrowding, the interiors of the dwellings are clean and tidy. The houses are owned by landlords who readily evict those *bustee* families who cannot pay the rent.

Rio de Janeiro's *favelas*

A *favela* is a wildflower that grew on the steep *morros*, or hillsides, which surround and are found within Rio de Janeiro. Today, these same *morros* are covered in *favelas* or shanty settlements (Figure 15.34). A *favela* is officially defined as a residential area where 60 or more families live in accommodation that lacks basic amenities. The *favelados*, the inhabitants, are squatters who have no legal right to the land they live on. They live in houses constructed from any materials available – wood, corrugated iron, and even cardboard. Some houses may have two rooms, one for living in and the other for sleeping. There is no running water, sewerage or electricity, and very few local jobs, schools, health facilities or forms of public transport. The land upon which the *favelas* are built is too steep for normal houses. The most favoured sites are at the foot of the hills near to the main roads and water supply, although these may receive sewage running in open drains downhill from more recently built homes above them. Often there is only one water pump for hundreds of people and those living at the top of the hill (with fine views over the tourist beaches of Copacabana and Ipanema!) need to carry water in cans several times a day. When it rains, mudslides and flash floods occur on the unstable slopes (Places 8, page 49; page 55). These can carry away the flimsy houses (over 200 people were killed in this way in February 1988).

Almost 1.1 million people – nearly one-fifth of the total population – live in Rio's estimated 750 *favelas*. The two largest, Roçinha and Morro de Alemao, each have a population in excess of 100 000. Living conditions are improving and UN figures say that 95 per cent of *favela* residents now have access to clean water and 76 per cent to improved sanitation. The Brazilian government has pledged $1.7 billion on further improvements including dealing with the major problem which, in over half the *favelas*, is the influence of powerful drug gangs.

Figure 15.34

A *favela* in Rio de Janeiro

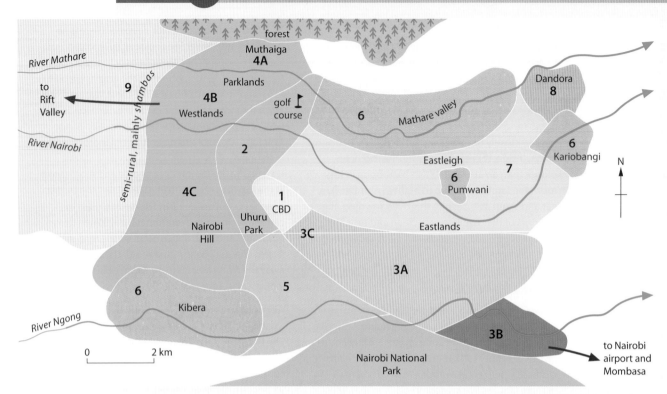

River Mathare

to Rift Valley

River Nairobi

forest

Muthaiga **4A**

Parklands

4B

Westlands

golf course

semi-rural, mainly shambas

9

Mathare valley

6

Dandora **8**

6 Kariobangi

N

Eastleigh

6 Pumwani

7

2

4C

Nairobi Hill

Uhuru Park

1 CBD

3C

Eastlands

3A

6

Kibera

5

3B

to Nairobi airport and Mombasa

River Ngong

0   2 km

Nairobi National Park

Figure 15.35

Functional zones and residential areas in Nairobi

In 1899 a railway, being built between Mombasa, on the coast, and Lake Victoria, reached a small river which the Maasai called *enairobi* (meaning 'cool'). The land that surrounded the river was swampy, malarial and uninhabited. Despite these seemingly unfavourable conditions, a railway station was built and, less than a century later, the settlement at Nairobi had grown to over 1.5 million people. The present-day functional zones (Figure 15.35) show the early legacy of Nairobi as a colonial settlement and the more recent characteristics associated with a rapidly growing city in an economically developing country.

1 **CBD** This is the centre for administration; it includes the Parliament Buildings, the prestigious Kenyatta International Conference Centre, commerce and shopping (Figure 15.36). Also located here are large hotels and, in the north, the University and the National Theatre.

2 **Open space** Immediately to the west and north of Nairobi's CBD (unlike in developed cities), are several large areas of open space. These include Uhuru (Freedom) Park and several other parks, sports grounds and a golf course. Other areas of open space, notably the Nairobi National Park to the south and the Karura Forest to the north, lie outside the city boundary.

Figure 15.36

The CBD (zone 1)

Figure 15.37

Higher-income housing (zone 5)

Figure 15.38

Shanty settlement, Mathare Valley (zone 6)

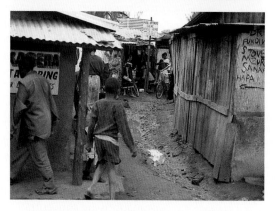

Figure 15.39

Inside a shanty settlement, Kibera (zone 6)

3 Industrial zone Early industry, much of which is formal, grew up in a sector that borders the railway linking Nairobi with the port of Mombasa (**3A** in Figure 15.35). The main industries, most of which are formal (Figure 19.34), include engineering, chemicals, clothing and food processing. A modern industrial area (**3B**) extends alongside the airport road and contains many well-known transnational firms. This zone includes (**3C**) the *Jua kali* workshops (Places 89, page 575).

4 High-income residential Wealthy European colonists and, later, immigrant Asians lived on ridges of highland to the north and west of the CBD where they built large houses above the malarial swamps (Figure 15.37). Today, Europeans

Figure 15.40

Low-income, council-built housing (zone 7)

Figure 15.41

Dandora 'site and services scheme' (zone 8)

tend to concentrate in Muthaiga (**4A**) and the Asians and more wealthy Africans in Parklands and Westlands (**4B**). Westlands, with its shops and restaurants, forms a small secondary core while several large hotels are located on Nairobi Hill (**4C**). Many of the largest private properties have their own security guards.

5 Middle-income residential The southern sector was originally built for Asians who worked in the adjacent industrial zone. The estates, which were planned, are now mainly occupied by those Africans who have found full-time employment.

6 Shanty settlements As in other developing cities, shanty settlements have grown up away from the CBD on land that had previously been considered unusable – in Nairobi, this was on the narrow, swampy floodplains of the Rivers Mathare and Ngong. The two largest settlements are those that extend for several kilometres along the Mathare valley (Figure 15.38) and in Kibera (Figure 15.39). Estimates suggest that over 100 000 people, almost exclusively African, live in each area. They find work in informal industries (page 574).

7 Low-income residential These areas include flats, 3–5 storeys in height and council-built (Figure 15.40), and former shanty settlements to which the council has added a water supply, sewerage and electricity.

8 Self-help housing Under this scheme (page 449), the council provided basic amenities and, at a cheap price, building materials. In Dandora (Figure 15.41), which has over 120 000 residents, relatively wealthy people bought plots of land and built up to six houses around a central courtyard. The council then installed a tap and a toilet in each courtyard and added electricity and roads to the estate. The 'owner' is able to sell or rent the houses that are not needed by his/her own family.

In 1993, an article in Nairobi's daily newspaper *The Nation* stated that 'Kenya has been hailed as Africa's leading example of multi-racial harmony, yet one has only to tour its residential districts to see a form of "apartheid". Despite a façade of racial harmony, people live according to colour and status and, unlike in the UK or USA, do not feel they have to mix with each other.' This example of global harmony was unexpectedly shattered in December 2007 by post-election violence, mainly between two powerful ethnic groups, the Kalenjin and the Kikuyu, which led to over 1000 deaths and the displacement of over 600 000 people.

Problems resulting from rapid growth

The 'pull' and rapid growth of cities in the developing world has led to serious problems in providing housing, basic services and jobs – problems accentuated by a much wider gulf between the minority rich and the majority poor than exists in the developed world. (Remember that developing cities do have positive as well as negative features.)

Housing

Despite some promising initiatives, most authorities have been unable to provide adequate shelter and services for the rapidly growing urban population and so the majority of the poor have to fend for themselves and to survive by their own efforts. Estimates suggest that one-third of the urban dwellers in developing countries either cannot afford or cannot find accommodation that meets basic health and safety standards. Consequently, they are faced with three alternatives: to sleep on pavements or in public places; to rent a single room if they have some resources; or to build themselves a shelter, possibly with the help of a local craftsman, on land which they do not own and on which they have no permission to build (Figure 15.38 and Places 57 and 58).

In time, some squatter settlements may develop into residential areas of 'adequate' standards (the *periferia* in Figure 15.32 and Dandora in Places 58). Rather than trying to build new housing, city councils find it cheaper and easier to add water supplies, sewerage systems, electricity and public services (refuse disposal, street lighting) to existing shanties, and to allow occupants to obtain legal tenure of the land (pages 448–49).

Services

Only small areas within many developing cities have running water and mains sewerage. Rubbish, dumped in the streets, is rarely collected. When heavy rains fall, especially in the monsoon countries, the drains are inadequate to carry the surplus water away. The lack of electricity hinders industrial growth and affects the material standard of living in homes. There is a shortage of schools and teachers, and of hospitals, doctors and nurses. Police, fire and ambulance services are unreliable. Shops may sell only essentials, and food may be exposed to heat and infection-carrying flies.

Pollution and health

Drinking water is often contaminated with sewage which may give rise to outbreaks of cholera, typhoid and dysentery. The uncollected rubbish is an ideal breeding-ground for disease. Many children have worms and suffer from malnutrition as their diet lacks fresh vegetables, protein, calories and vitamins. Local industry is rarely subjected to pollution controls and so discharges waste products into the air which may cause respiratory diseases, and/or into water supplies. The constant struggle for survival often causes stress-related illnesses. It is not surprising that in these rapidly growing urban areas infant mortality is high and life expectancy is low.

Unemployment and underemployment

New arrivals to a city far outnumber the jobs available and so high unemployment rates result. As manufacturing industry is limited, full-time occupations are concentrated in service industries such as the police, the army, cleaning, security guards and the civil service. The majority of people who do work are in the informal sector, i.e. they have to find their own form of employment (page 574). Informal jobs may include street trading (selling food or drinks), food processing, services (shoe-cleaning) and local crafts (making furniture and clothes, often out of waste products). Most of these people are underemployed and live at a subsistence level.

Transport

Relatively few developing cities can afford an elaborate public transport system. This means that the road network is likely to be unable to deal with the large volume of traffic. This traffic will, at the best, consist mainly of old cars, vans, trucks, overcrowded minibuses and buses and, at the worst and depending upon the individual city, an added complication of rickshaws, bullock carts, donkeys, matatus, tuc-tucs and bicycles (Figure 15.42). Apart from congestion, there is likely to be severe air pollution and a high accident rate. As countries develop, the main city may consider building a subway system, or metro, as a means of relieving pressure on the roads, e.g. Hong Kong (Places 106, page 640), São Paulo, Singapore and Seoul in the NICs (page 578) and, more recently in an emerging country, Shanghai (Case Study 15B).

Figure 15.42

Overcrowding in Chennai

Figure 15.43

Apartments in the medieval centre of Cairo

Figure 15.44

The 'City of the Dead'

In 1996, when the author was taken on an eye-opening journey through the back alleys and markets of old Cairo, the population of the city was given, according to the census of that year, as 6.801 million (see Framework 15). At the time, compared with other developing cities, Cairo had relatively few squatter settlements. Most newcomers to the city disappeared into the medieval centre of the old town to live either in:

- overcrowded two-roomed apartments within tall blocks of flats (Figure 15.43)

- roof-top slums (the flat roofs are suitable for the desert climate and allow the later addition, often illegally, of an extra storey

- the 'City of the Dead', a huge Muslim cemetery where, according to one estimate, up to 3 million people actually live in the tombs because they are cleaner and give more shelter than the city apartments, even though they are a kilometre from water (Figure 15.44).

Cairo's narrow streets were not built for the volume of its present noisy and air-polluting traffic. Pollution also comes from a dilapidated early 20th-century sewerage system and numerous small factories located in backyards, within houses and on rooftops, that emit their waste both into the air and onto the streets. Donkey carts take rubbish

to waste dumps on the edge of the city where it is sorted by people looking for bottles, plastic and paper that can be recycled in local factories.

A return visit in 2009 showed how the city authorities have tried to overcome these problems by extending and improving the sewerage system in what became one of the world's largest public-health engineering schemes; widening roads and building a 10-lane ring road; opening an efficient underground 'metro' system with two lines operating and a third planned; organising refuse collection and converting one of the largest tips into a large urban park that overlooks the city; erecting numerous high-rise apartment buildings; and creating low-cost housing in several 'new towns' that have sprung up in the desert that surrounds the city (one of which, the Sixth of October, already has a population of 2.6 million). Even so, the Cairo authorities are struggling to keep pace with population growth, which results from a combination of high fertility rates (3.1 per family) and rural–urban migration, and which has all but doubled in the three decades since 1975. One consequence has been the rapid growth of informal settlements that now encircle the city, including that of Ezbet El Haggana, a shanty in the north-east of Cairo with over 1 million inhabitants.

2001 UK census

Accurate and reliable statistics are often difficult to obtain, even for developed countries. Some of the least reliable figures are for population, and those presented in this book, e.g. fertility rates and urban populations, should be used with some caution, even though the most reliable resources were used to acquire them. It is suggested that Britain's 2001 census could have had a margin of error of 1 per cent either way, even with the use of the latest available technology and with supposedly high levels of refinement. This was partly because many people failed to complete and return the relevant forms (over 20 per cent in 10 inner London boroughs), and partly due to a rise in illegal immigration which, by its very nature (page 367), means that people arriving in a country do not want to be recorded and so do not appear in official figures. Later, the (House of) Commons Public Accounts Committee questioned the accuracy of the census, after claims were made that the population of England and Wales was 900 000 lower than previously predicted (it was

1.92 million lower at the 1991 census). This may have been due to a failure to record people who leave the country permanently or who were away from their homes (e.g. on holiday) on census night.

What is Cairo's population?

For 2005–06 the population of Cairo has variously been given as:

- Egypt State Information Service – Cairo: 6.8 million

- UN population division – Cairo governorate: 7.786 million

- Collin's Atlas: 9.462 million

- UN World Urbanisation – Cairo agglomeration (world's 13th largest): 11.487 million

- World Bank – Cairo region and its new towns: 15.2 million

- Rough Guide – Cairo region: about 18 million

These discrepancies may arise from organisations using different criteria, notably:

- the land area covered can vary, from the city itself to urban developments such as El Giza that have sprawled beyond the city's 'official' limits, or the Cairo region including the new town settlements

- except for the actual year of the census, the population has to be estimated from birth and death rates (assuming that these are always recorded) and migration – but there is no effective means of counting in-coming migrants from rural areas or from overseas countries.

In a city like Cairo, with its growing shanties and the 'City of the Dead', it is highly unlikely that every resident was consulted during the census, and even for those who were it is unlikely all were able to read, and then to complete, the census forms.

2001 India census

India, the second most populous nation in the world, has begun its mammoth task of conducting its first census for a decade, a year after the population officially exceeded 1 billion. Several states, including Jammu and Kashmir, have already been surveyed, while recording in Gujarat, where the authorities are struggling to deal with the aftermath of the earthquake [Places 5, page 20], has been delayed. Elsewhere, the exercise will take to the end of the month and will involve around 2 million census workers who will visit 5000 towns and cities and more than 600 000 villages. One of India's most publicised revolutions, the greater use of computers, has enabled the authorities to promise 98 per cent accuracy.

Source: Adapted from BBC News Online, 9 February 2001

Government housing

Upgrading and self-help schemes

A policy of wholesale demolition of squatter settlements, as was attempted in Rio de Janeiro (Places 57) and South Africa (Places 45, page 372), is often a mistaken one. Squatters have shown that they are capable of constructing cheap accommodation for themselves, but that they cannot provide the essential basic services. In Latin America, and less successfully in Africa and South-east Asia, governments have, albeit reluctantly, at times accepted that shanties are

permanent and that it is cheaper and easier to improve them by adding basic amenities than it is to build new houses.

The concept of '**site and services**', funded by the World Bank and several voluntary organisations, encourages local people to become involved in self-help projects. This approach seems to be most appropriate in the poorer countries whose governments cannot afford large rehousing schemes. One such scheme, in Dandora in Nairobi (Figure 15.41), was briefly described in Places 58.

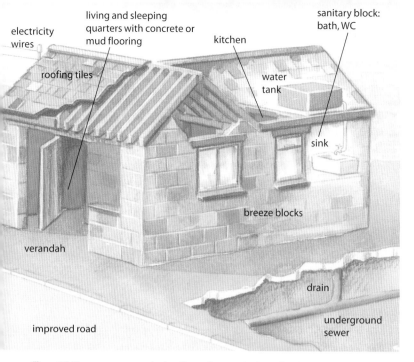

living and sleeping quarters with concrete or mud flooring

electricity wires

roofing tiles

kitchen

sanitary block: bath, WC

water tank

sink

breeze blocks

verandah

drain

improved road

underground sewer

Figure 15.45

A 'site and services' scheme, São Paulo, Brazil

Figure 15.46

A self-help scheme in Kenya linked to Practical Action

A similar scheme in Lusaka (Zambia) encourages about 25 individuals to group together. They are given a standpipe and 8 hectares of land. If the group digs ditches and foundations then, with the money saved, the authorities will lay water and drainage pipes and construct the houses. Moreover, if local craftsmen are prepared to build the shells of the houses, the group will be supplied with low-priced building materials and the extra money saved by the authorities may be used to add electricity and to tarmac the roads. In some cases, a small clinic and school may be added.

Several schemes in São Paulo's *periferia* (Figure 15.45) have enabled running water, main drains and electricity to be added to houses, with street lighting and improved roads if there was any surplus money. The result over a lengthy period of time has been an upgrading of living conditions, and the introduction of some shops and small-scale industry, although the people are still poor.

Elsewhere in Brazil, an estimated 62 per cent of Recife's population (Figure 13.5) live in *favelas*. Here, following over a decade of popular organisation and collective negotiation, the city's Plan for the Regularisation and Urbanisation of Special Zones of Social Interest (PREZEIS) became law. It meant that urban services such as sewers and paved streets would be forthcoming and that *favela* residents would be protected from eviction (or from being ignored as if they did not exist). Each *favela* elected two representatives who met weekly with officials to develop and carry out urbanisation schemes. By 2008, living conditions in many *favelas* had improved dramatically (Places 57), mainly due to the enthusiasm of local people, whereas in others, where less interest has been shown, limited progress had been made.

Self-help schemes can create a community spirit, can improve the skills of local people and can result in cheap-to-erect accommodation. Yet their success often depends upon the motivation and skills of the local people and the use of appropriate and cheap building materials under expert guidance.

Practical Action and 'materials for shelter'

Practical Action (Places 90, page 577) helps people in Africa, Asia and South America to develop and use technologies and methods that give them more control over their lives and which contribute to the long-term development of their communities. Several of Practical Action's projects involve investigating, developing and promoting a range of building materials suitable and affordable for self-help schemes (Figures 15.46 and 15.47). A Practical Action-sponsored scheme in India prolongs the lives of thatch roofs by coating them with a waterproof compound of copper sulphate and cashew nut resin. In Kenya, the Maasai are under increasing internal pressure to give up their semi-nomadic way of life and settle in permanent houses. Practical Action has responded to this situation by working closely with the Maasai in helping to modify their traditional houses by adding a concrete mix to the cow-dung roof (which always seemed to leak), inserting a small chimney to remove smoke (all cooking is done inside the house), improving lighting (previously each house had only one minute opening as a 'window'), and using chicken wire as a framework for the walls. It also provides, in several parts of the world, technical assistance in the mining, quarrying and processing of local raw materials which can be used for building.

Figure 15.47

Production of low-cost roofing tiles in Kenya

Relocation housing and new towns

Some of the more wealthy developing countries, such as Venezuela with its oil revenue and the NICs of South Korea, Hong Kong and Singapore with their income from trade and finance, have made considerable efforts to provide new homes to replace squatter settlements. In most cases, high-rise blocks of flats have been built on sites as close as possible to the CBD or in new towns beyond the city boundary (Places 60).

Places 60 Singapore: a housing success story

Figure 15.48

Early high-rise flats on the edge of China Town, Singapore

Faced with a large and rapidly increasing number of slum dwellers, and an overcrowded, unplanned, central area, the Singapore government set up, in 1960, the Housing and Development Board (HDB). The HDB cleared old property near to the CBD, especially in the Chinese, Arab and Indian ethnic areas (Figure 15.48), and created purpose-built estates (with 10 000–30 000 people) within a series of 23 new towns, each with up to 250 000 people and all within 25 km of the CBD.

In both cases, the HDB constructed housing units of 1–3 rooms in closely packed high-rise flats (Figure 15.49). The flats were initially for low-income families and rents were kept to a minimum. However, one-quarter of every wage-earner's salary is automatically deducted and individually credited by the government into a central pension fund (CPF). Western-style welfare benefits are regarded as an anti-work ethic, but Singaporeans can use their CPF capital to buy their own apartment or flat. Since 1974 the HDB have built many 4-room and 5-room units for the average and higher-income groups who have then been expected to buy their own property. In 2008, 81 per cent of Singaporeans lived in government-built housing, with 79 per cent of them having managed to buy their own home.

Figure 15.49

Blocks of 1960s high-rise flats

The large estates are functional in design and were developed on the neighbourhood concept of British new towns. Each estate contains much greenery and is well provided with amenities such as shops, schools, banks, medical and community centres. Where several estates are in close proximity, better services are provided such as department stores and entertainment facilities. All the new towns have been linked to, and are within half an hour of, the city centre by the MRT (mass rapid transport railway). Each estate has its own light industries producing, usually, clothing, food products and high-tech goods. As everywhere else in Singapore, the estates are models of cleanliness with the buildings constantly being painted, grass areas cut and where there is an absence of litter and graffiti (the state has always imposed heavy fines for litter). By 1999, when over 825 000 flats had been built, the HDB had set out to provide every householder with a minimum of three rooms. This was achieved by pulling down and replacing some of the earliest apartment blocks, merging adjacent flats to make them larger, and building more architect-designed estates in specially designated 'new towns' (Figures 15.50 and 15.51). To ensure that all Singaporeans had a home, the HDB bought three-bedroomed flats on the open market and then sold them at a discount price to low-income families as well as introducing their 'Rent and Purchase Scheme'. This scheme allowed families who had a minimum of four members and who had previously only been eligible for a one-room or two-room flat, initially to rent a three-roomed flat from the HDB and then, subsequently, to buy it.

The government also continued its Selective Enbloc Redevelopment Scheme under which all estates were extensively modernised once they were 17 years old (providing that 75 per cent of the occupants agreed). This included allowing owners to apply for a larger flat and/or to relocate to a newer estate as well as refurbishing the interior and decorating the exterior of existing flats. A corresponding improvement in public utilities and services included the addition, or upgrading, of communal facilities, improving roads and planting more trees and shrubs. This has meant that these estates, unlike those elsewhere in the world, show little or no sign of decay.

By the end of 2007, the HDB had built 99 320 flats in which 81 per cent of Singaporeans lived (3 million out of Singapore's total population of 3.6 million). Under the Home Ownership for the People Scheme, whereby most residents had bought their own home, special help had always been being given to assist low-income families. The HDB, under their Build-to-Order system, now offer new two-room and three-room flats to families, initially after 2004 with a monthly household income of under $3000 and, since 2006, to those with an income of under $2000. Also by the end of 2006, various renewal schemes had seen the continual improvement and upgrading of all estates, especially the earlier ones.

Visitors from the West unjustly and incorrectly compare living in one of these 'boxes in the sky' with the often poor-quality high-rise projects found in places like the UK and the USA. However, set in a self-sufficient 'new town' with its own commercial, shopping and leisure facilities and in a clean and increasingly green environment, HDB flats have become very much part of the Singapore way of life and the country's estates are studied by planners from around the world, who consider them a model of success. In 2008 the HDB won a UN public service award for its home ownership programme.

Figure 15.50

An early 1990s estate in Bishan

Figure 15.51

A late 1990s estate

A Los Angeles

Physical hazards

For several generations, southern California was seen as America's promised land. Now it seems that this part of the 'sunshine state' is cursed by natural disasters such as earthquake, fire, fog, drought and flood – disasters which, in part, are created or exacerbated by the lifestyle and economic activities of its inhabitants. The Los Angeles agglomeration, with a population in excess of 12 million people, has become known as 'hazard city'.

Earthquakes

Not only does the San Andreas Fault, marking the conservative boundary between the Pacific and North American Plates, cross southern California (Places 6, page 21), but Los Angeles itself has been built over a myriad transform faults (Figure 15.52). Although the most violent earthquakes are predicted to occur at any point along the San Andreas Fault between Los Angeles and San Francisco, earth movements frequently occur along most of the lesser-known faults. The most recent of 11 earthquakes to affect Los

Angeles since 1970 occurred in January 1994. It registered 6.7 on the Richter scale, lasted for 30 seconds, and was followed by aftershocks lasting several days. The quake killed 60 people, injured several thousand, caused buildings and sections of freeways to collapse, ignited fires following a gas explosion, and left 500 000 homes without power and 200 000 without water.

Tsunamis

Tsunamis are large tidal waves triggered by submarine earthquakes which can travel across oceans at great speed. The 1964 Alaskan earthquake caused considerable damage in several Californian coastal regions. Although Los Angeles has escaped so far, it is considered to be a tsunami hazard-prone area.

Sinking coastline

The threat of coastal flooding has increased due to crustal subsidence. Although this may, in part, be due to tectonic processes, the main cause has been the extraction of oil and, to a much lesser extent, subterranean water. Parts of Long Beach have sunk by up to 10 m since 1926. Although this sinking has now been checked, parts of

the harbour area lie below sea-level and are protected from flooding by a large sea wall.

Landslides and mudflows

Landslides and mudflows occur almost annually during the winter rainfall season within the city boundary of Los Angeles. They have increased in number and frequency due to effects of urbanisation such as the removal of vegetation from, and the cutting of roads through, steep hillsides and by channelling rivers (Figure 3.8). In 1994, winter storms buried parts of the Pacific Coast Highway to a depth of over a metre in mud, trapped hundreds of people in their cars and houses, and threatened the Malibu homes of film and TV stars. Landslides are frequent along coastal cliffs, and the 1994 earthquake caused several thousand of them in the hills surrounding the city.

Heavy rain

Winter storms bring rain and strong winds. These are especially severe during an El Niño event (Figure 15.53 and Case Study 9A). Although most rivers in the Los Angeles basin are short in length and seasonal, they can transport large volumes of water during times of flood. Deforestation

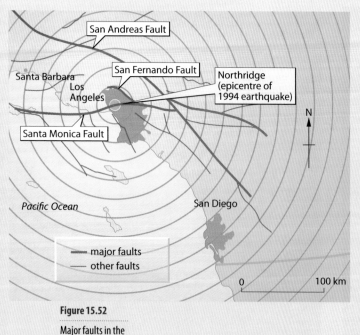

Figure 15.52

Major faults in the Los Angeles basin

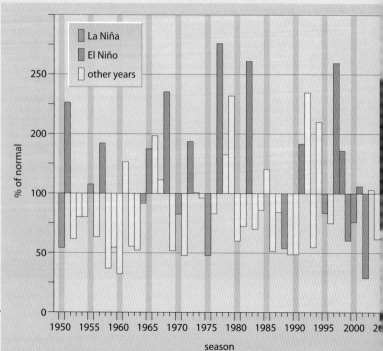

Figure 15.53

Los Angeles rainfall, 1950–2006

and brush fires on the steep surrounding hillsides, and rapid urbanisation (page 63), have increased surface runoff. Large dams have been built to try to hold back flood-water but even so the flood risk remains. In February 1992 (during an El Niño event) eight people died and dozens of cars and caravans were swept out to sea when, fol-lowing two days of torrential rain, floodwa-ters poured through a caravan park to the south of Malibu. Heavy rain also triggers landslides and mudflows.

El Niño and La Niña events

El Niño events seem to coincide with years of above-average rainfall, and La Niña events with periods of drought, though to a lesser extent (Figure 15.53). In February 1998, parts of southern California were declared a disaster area. El Niño was blamed for the serious floods, mudflows, landslides, storms and, in the mountains, heavy snowfalls.

Drought

The long, dry summers associated with the Mediterranean climate may be ideal for tourists but, as the population of Los Angeles continues to grow, they put tre-mendous pressure on the limited water resources. Much of the city's water comes, via the Colorado aqueduct, from the River Colorado 400 km to the east. So much water is now extracted from the river that,

in very dry years, it almost dries up before reaching the sea. Droughts are expected to increase with global warming.

Brush fires

Much of the Los Angeles basin is covered in drought-resistant (xerophytic) chap-paral, or brush vegetation (page 324). By the autumn, after six months without rain, this vegetation becomes tinder-dry. The Santa Ana is a hot, dry wind that owes its high temperature to adiabatic heating as it descends from the mountains. The heat and extreme low humidity of Santa Ana winds cause discomfort to humans and increase the dryness in vegetation. A care-less spark or an electrical storm can prove sufficient to set off serious fires.

In October 2007, over 500 000 Californians were forced to flee from brush fires that extended from north of Los Angeles down to the Mexican border. Worst-hit was San Diego county, south of Los Angeles, and the 'celebrity' enclave of Malibu, to the north of the city (Figure 15.54). The fires caused the deaths of seven people, destroyed over 2000 homes and thousands of hectares of vegetation. Fire-fighters, fifty of whom were injured while on duty, worked around the clock for several days trying to control flames that were fuelled by a Santa Ana wind gusting up to 160 km/hr over parched vegetation in temperatures of 38°C.

Fog and smog

Advection fog (page 222) occurs when cool air from the cold offshore Californian current drifts inland where it meets warm air. Fog can form most afternoons between May and October as the strength of the sea-breeze increases (page 240). This event can cause a temperature inversion (page 217), where warm air becomes trapped under cold air. When many pollutants from Los Angeles' traffic, power stations and industry are released into the air, the result is smog (Figure 9.25) and, when they return to Earth, acid rain. Smog in Los Angeles can be a major health problem (Figure 15.55). It has been confirmed that there is a correlation between fog and hospital admissions. For each 10 microgram increase in airborne particulate concentrations, admissions jumped 7 per cent for chronic respiratory patients and 3.5 per cent for cardiovascular disease patients. According to another recent study reported in the *Los Angeles Times*, local residents show lung damage that might be expected of someone who smoked half a pack of cigarettes every day.

Latest figures suggest that 9000 Californians die annually from diseases caused or aggravated by air pollution, more than half from the south of the state, and that one in every 15 000 are at risk of con-tracting cancer from breathing chemicals in the air.

Figure 15.54

Brush fire near Malibu

Figure 15.55

Smog over Los Angeles

Social contrasts

Living in Los Angeles presents great contrasts in lifestyle and opportunity. The census data for Compton (Figures 15.56 and 15.58), an area between downtown Los Angeles and the docks, contrasts with the idealised picture given by films and TV of expensive Beverly Hills located to the north-west and life in the more distant southern district of Mission Viejo in Orange County (Figure 15.57).

Population

In 2006, the population of Los Angeles city was given as 3.834 million; Los Angeles county as 9.948 million; and the Los Angeles-Long Beach-Santa Ana urban agglomeration as 12.307 million (the 11th largest in the world – Figure 15.3).

Immigration

Over 36 per cent of people living in Los Angeles county were born outside the USA, and 58 per cent do not speak English at home. Classified by race (the term used by the US Census Bureau), 45 per cent are Hispanics (mainly from Mexico and Latin America), 31 per cent are white, 9 per cent are black and most of the remaining 15 per cent are Asian (mainly from China, Japan and Korea).

Most of the Hispanics are men of working age, which includes a high proportion who entered the country as illegal immigrants. Most are young, have little money and limited qualifications or skills. They are attracted to California's wealthy image (Stereotypes – Framework 13), but the reality they face on arrival is often very different. Until they can obtain a Green Card from the Department of Immigration, they may not work legally nor can they receive welfare. They are therefore forced to take very low-paid jobs, often in the informal (page 574) or hidden sectors (page 367). Low educational standards, a lack of qualifications and poor health and housing characterise some of the black American-African communities such as Compton, although this relatively small ethnic group is likely to have migrated here from elsewhere in the USA rather than from overseas. The city authorities, as well as the state and federal government, are making attempts to improve housing, health and education for both the Hispanic and black communities.

	Beverly Hills	Mission Viejo	Compton
Average household income	$70 945	$78 248	$31 819
Households below poverty level	7.9%	2.3%	25.5%
Households earning over $150 000 a year	25.2%	3.8%	1.9%
Unemployment rate	3.2%	2.4%	7.1%

Figure 15.56

Contrasts in Los Angeles
Source: US 2000 Census

Of the growing number of Asian immigrants who have settled in Los Angeles, those from Japan and Korea include highly educated professional and business people who are improving low- and medium-cost housing, creating many new jobs and helping to provide services for their own community and for the city. In districts where they have settled, such as Norwalk, neighbourhood schools have improved, house prices have risen and violence has decreased.

However, there does appear to be an increasing re-location within Los Angeles of earlier immigrants, on the basis of social class, as the more successful, especially those from Asia and South America, move to more affluent areas. In contrast many Mexicans, who still are often forced to take the poorer jobs, remain in the least desirable districts.

Figure 15.57

Los Angeles

Greater Los Angeles sprawl 115 km east to west and c an area of 1166 km²

- Los Angeles urban area
- major route
- upland area (over 500 m)
- highest-income areas
- poorest districts

0   20 km

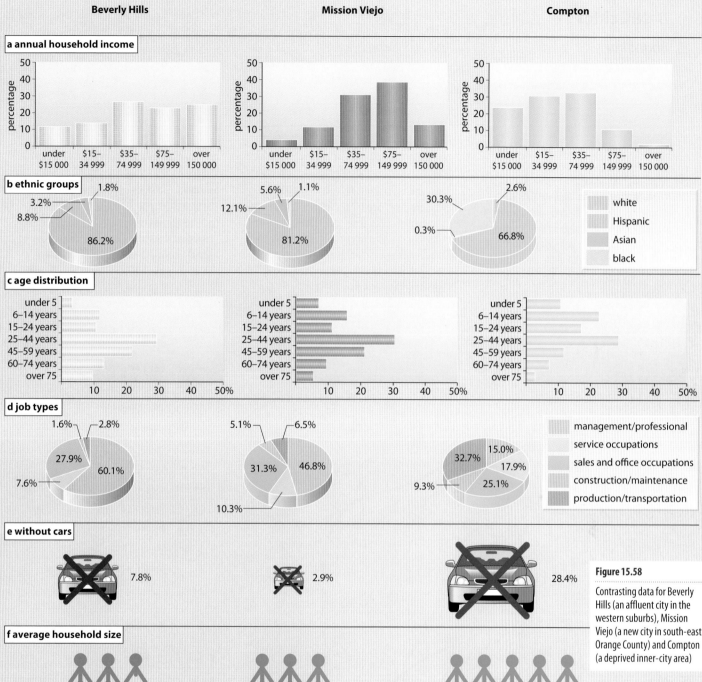

Beverly Hills | Mission Viejo | Compton

a annual household income

b ethnic groups

white
Hispanic
Asian
black

c age distribution

d job types

management/professional
service occupations
sales and office occupations
construction/maintenance
production/transportation

e without cars

Figure 15.58

Contrasting data for Beverly Hills (an affluent city in the western suburbs), Mission Viejo (a new city in south-east Orange County) and Compton (a deprived inner-city area)

f average household size

Housing

Increased migration has led to a lack of affordable housing in districts near to downtown Los Angeles. Many immigrants earn less than $5 an hour and so can neither afford to buy their own home nor pay the high rents. Twenty-five per cent of Asians and over 60 per cent of Hispanics live in overcrowded conditions, yet still pay over one-third of their income on housing in areas that are disadvantaged socially, economically and environmentally. Should

the newcomers establish themselves with a reliable job, then movement away from the poor housing conditions becomes a possibility, continuing the processes of centrifugal movement (Places 52, page 421) and urban sprawl. Migrants from the same country tend to group together, often maintaining the cultures of their place of origin as in Chinatown and Koreatown.

The Los Angeles agglomeration now extends eastwards for over 115 km inland from the Pacific coast and, as it continued

to grow, a number of **edge-cities**, such as Mission Viejo in Orange County, have sprung up. Edge-cities have large, modern houses, new schools and hospitals, and large shopping centres all set in a pleasant environment. However, they often lack both the types of work that the wealthy inhabitants seek and an adequate public transport system. This has resulted in over 80 per cent of Mission Viejo's working population becoming long-distance commuters.

B NIC cities

Shanghai

Shanghai is the industrial, commercial, financial and fashion centre of China and, with a population of 15.789 million, is the world's seventh largest urban agglomeration. Industrially, it has more than 400 000 firms in the private sector, and over 31 000 foreign-invested companies and is the regional headquarters for 130 transnational corporations. Despite China's huge domestic market, its rapid emergence as a major world economic power – perhaps *the* major world power – depends on its ability to trade and to increase its overseas links. One of the first major developments took place at Pudong directly across the River Huangpu which, in 1989, was an area of farmland reached only by ferry. Within ten years Pudong (Figure 19.43) was a city in its own right, with Shanghai's stock exchange, numerous industrial areas and a large resident population.

Shanghai's development during the 1980s was handicapped by congestion on its roads and a lack of port facilities. Drastic measures were taken. By 1999, the city had three bridges and three tunnels (two metro and one road) linking it with Pudong; a 47 km long outer ring road which only took three years to build; an east–west and a north–south three-lane elevated freeway, each of which cut straight through the existing city regardless of what lay in its path (Figure 15.60); two underground (metro) rail lines; and the beginnings of a new international airport (Figure 15.59).

By 2008, the international airport had two terminals and three runways in use with both a maglev railway and an eight-lane expressway linking it with Shanghai. The city also had eight metro lines and the first stage of a 1318 km rail track to Beijing with trains that will run at speeds of up to 350 km/hr. Deep-harbour facilities at the mouth of the Yangtze River make Shanghai the largest port in the world (according to 2008 figures based on the weight of goods) and the third largest for containers (50 new container berths are at present under construction).

Figure 15.59

Shanghai's transport development

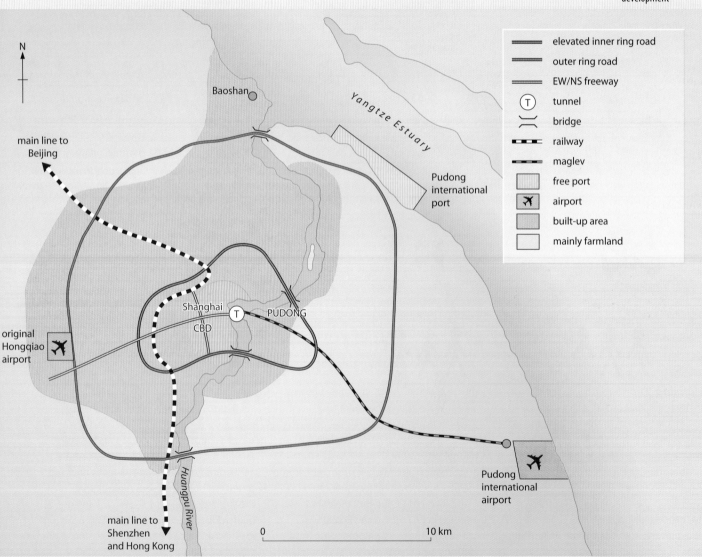

Figure 15.60

Elevated freeways, Shanghai

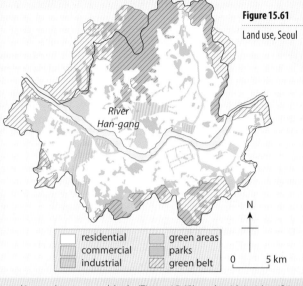

Figure 15.61

Land use, Seoul

River Han-gang

N

residential   green areas
commercial   parks
industrial   green belt

0   5 km

Seoul

In the early 1950s Seoul, like most of the country and its economy, lay in ruins after the Korean War. It had a population of under 2 million and no industries capable of competing in the global market. Today the city has a population of over 10 million and its buildings, office blocks and transport system are as modern as those anywhere in the world.

Figure 15.61 shows the present land use in this very modern city. The commercial centre (Figure 15.62) lies east–west and in part has been re-created by opening up a previously concreted-over river. Alongside this 'buried' river used to be countless small family-run businesses, many of which have been re-located in a major redevelopment scheme. Along with numerous larger, newer industries, they are now grouped together along part of the south bank of the River Han-gang. Much of the city is covered in high-rise residential blocks (Figure 15.63), each with its identifying number visible from some distance. The seemingly endless blocks of flats mean that low-quality housing has all but been replaced. Just south of the CBD, and creating a large area of open space, is Namsan Park in which the Seoul Tower caps a hill 262 m high, while surrounding the city itself are vast areas of parkland and woods that form part of an extensive green belt.

Figure 15.62

Seoul CBD from Namsan Park

Figure 15.63

High-rise flats in Seoul

Further reference

Agnew, J. (2004) 'International migration: the view from Los Angeles', *Geography Review* Vol 17 No 4 (March).

Ashby, M. (2008) 'Megacities, migration and Manila', *Geography Review* Vol 21 No 4 (April).

Barke, M. and O'Hare, G. (1991) *The Third World*, Oliver & Boyd.

Bradford, M.G. and Kent, W.A. (1982) *Human Geography: Theory and Applications*, Hodder & Stoughton.

Digby, B. (2007) 'Progress on the London Olympics', *Geography Review* Vol 21 No 1 September.

Digby, B. (2007) 'Regeneration in east London', *Geography Review* Vol 21 No 2 November.

Smith, D. (2005) '"Clone town" surveys', *Geography Review* Vol 19 No 1 September.

Southern California Earthquake Center, Los Angeles, Northridge earthquake: www.data.scec.org/chrono_index/ northreq.html

UN Population Division World Urbanisation Prospects: 2007 Revision Population Database: http://esa.un.org/unup/

UN urbanisation: www.unep.org/geo2000

Urban regeneration in the UK: www.urcs-online.co.uk

US Census Bureau: www.census.gov/hhes/www/poverty. html

USGS Response to an Urban Earthquake: http://pubs.usgs.gov/of/1996/ ofr-96-0263/

World urbanisation: http://cities.canberra.edu.au/ publications/OECDpaper/World_ urbanisation.htm

The Westfield Centre, Shepherd's Bush

In October 2008, the Westfield shopping centre opened in Shepherd's Bush in west London. From the time that it was first planned it has been controversial. Why should this be? Retail centres and shopping malls have opened throughout the UK; what's different about this one?

What is the Centre like?

The Westfield Centre (Figure 15.64) is the UK's third largest shopping centre after Gateshead's MetroCentre and Bluewater in Kent. It covers a shopping area of 149 000 m^2 – the same as 30 football pitches. The Centre is owned by the Westfield Group, a multinational Australian company which owns shopping centres in Australia, the USA, New Zealand and the UK. It resembles American-style malls more than British high streets.

The core of the Centre is the shopping complex, but one with a difference. Eighty per cent of the stores are high-value, upmarket fashion outlets; of its 265 shops, the Centre has 40 luxury brands including Louis Vuitton, Mulberry and Prada. Mainstream chain stores include Marks & Spencer, Debenhams, Next, and one supermarket, Waitrose.

What makes the Centre different is that it brings luxury, high street and supermarket functions together on one site. That is unusual for such centres, which normally try to attract mass shopping.

It is also unusual for London, with its upmarket stores in Chelsea and Knightsbridge, and mass-market stores in Oxford Street.

However, the Centre is not simply about shopping. It aims to attract customers to stay longer and spend more. Its facilities include 50 restaurants and a 14-screen multiplex cinema. In its bid to be upmarket, it has barred KFC, McDonald's and plastic cutlery, offering instead upmarket choices such as the Square Pie Company. With this range of businesses on site, Westfield estimates that 21 million people will visit annually. What Westfield really wants is high spending per customer.

Where is it located?

The Centre is located 4 km west of London's main shopping areas in Oxford Street, Knightsbridge and Chelsea (Figure 15.65). Access is good: close by is the Westway, the branch of the main A40 heading west to Oxford, and it is a short distance from the start of the M4 to Heathrow and the west.

The Centre is a regeneration project designed to 're-brand' the area, just as the

Figure 15.64

The Westfield Centre from the air, looking north

Olympics are expected to re-brand east London. On Westfield's doorstep is the White City estate, one of London's most deprived areas. The Centre is actually built on land formerly owned by London Underground, before which it was the site of the 1908 Franco-British Exhibition. Close by is the BBC's Television Centre, itself a regeneration project from the mid-1980s, on the site of the former White City Stadium, where London's 1908 Olympics were held (Figure 15.66).

The demand for retail space in London

Until the 2008 credit crunch, demand for retail space in London was considerable. Incomes in London are higher than anywhere in the UK; the average weekly income per worker was £619 in 2008, 40 per cent above the UK average. In addition, some of London's wealthiest suburbs are on the Centre's doorstep, such as Holland Park and Notting Hill. In the London Borough of Hammersmith and Fulham, in which the Centre is located, 33 per cent of the population work in managerial or professional jobs, compared with 26 per cent for London as a whole. Not far away are the riverside suburbs of Chiswick (52 per cent of the population in such occupations), Kew and Richmond (56 per cent).

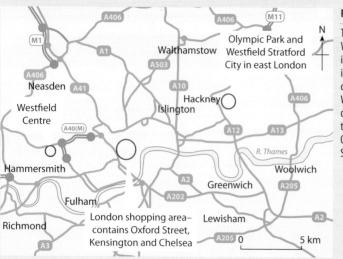

Figure 15.65

The location of the Westfield Centre in west London, in relation to central London and Westfield's next development on the edge of the new Olympic Park at Stratford City

Figure 15.66

The layout and transport links close to the Westfield Centre

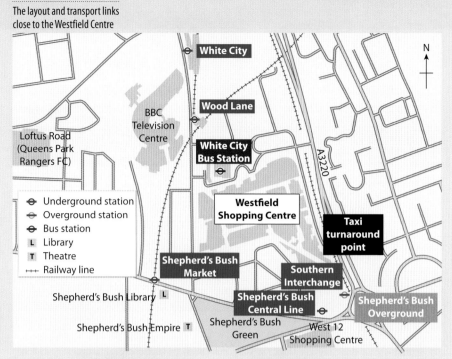

However, London's problem is that so much money earned within the city is spent outside it. Commuters are as likely to do their spending in Essex (at Lakeside) or Kent (at Bluewater). Westfield Centre is an attempt to get Londoners to spend more of their cash in London.

What are the issues?

1 Diverting trade from local shops

One of the problems with any new shopping centres is to what extent it diverts trade from other shops. Close to Westfield is Shepherd's Bush Green, where there is a small shopping centre and supermarket, cinema and gym, and several small, mostly independent shops, many of which cater for local ethnic minority communities. Most shop owners believe that the Westfield Centre will bring increased trade for them. However, others in the area are less certain, especially in Oxford Street, a few kilometres away. Some feel that the two areas are not competing, and that the new Centre will actually bring new money into London. But in November 2008, the number of customers in Oxford Street and Regent Street fell by 25 per cent compared with figures for a year earlier. However, it is difficult to know whether this was due to the Centre, or a result of reduced consumer spending during the credit crunch.

2 Accessibility

Compared with shops in central London, Westfield is less accessible, and is badly affected by traffic congestion. Westfield has invested £170 million into local transport improvements (Figure 15.66). These include
- a new Underground station, Wood Lane, on the Hammersmith and City Line, linking to central and east London
- a new Shepherd's Bush overground station on the line between East Croydon and Milton Keynes, giving the Westfield Centre a potential sphere of influence up to 80 km north of London.

But road traffic is a concern; there are only 4500 parking spaces in the Centre. Local residents and businesses claim that traffic jams and parking shortages on local streets have become worse. Westfield estimate that 60 000 visitors per day will visit the Centre. They claim that public transport will bring 60 per cent of its visitors, i.e. 36 000. The shortage of parking can only worsen if the remaining 24 000 visitors are fighting for 4500 spaces.

3 Impact on the local area

Westfield estimates that the Centre will create 7000 new jobs, and claims that 1000 of these have gone to local residents. In the local area, Shepherd's

Bush Green has had a £3 million revamp, with another £4 million spent on 24-hour policing, a new library and 78 affordable homes.

What local people have to say

'I think parking is an issue. If there isn't enough in the Westfield complex something has to be done so that those cars don't come to the local area. It's not about Westfield; it's about residents.'

Jamie Bishop, 35, shopowner

'Very few people are against the regeneration. The site hasn't been properly used since the 1908 Olympics. But the council has done nothing to look at parking or congestion.'

Andrew Slaughter, local Labour MP

'We might have to introduce residents-only parking with some kind of visitors' scheme. There's no doubt that something of this magnitude isn't going to come without pain but there is huge economic benefit and social regeneration for local people.'

Stephen Greenhalgh, Leader of Conservative-run Hammersmith and Fulham Council

'TfL's own research proves Hammersmith and Fulham already has the most clogged-up streets in London with 7.6 million hours lost in traffic every year.'

Spokesperson for Hammersmith and Fulham Council

Activities

1 Summarise the need for economic and social regeneration in this area.
2 Analyse the benefits and problems of locating a centre of this size in this part of west London.
3 Draw a table to show the economic, social and environmental benefits and problems brought by the Westfield Centre.
4 Justify whether you think that the Centre a represents or b does not represent a good example of sustainable development.

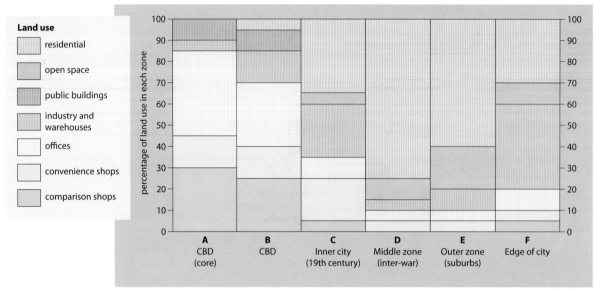

Figure 15.67

Land use in a British city: an idealised transect from the CBD to the city boundary

Land use
- residential
- open space
- public buildings
- industry and warehouses
- offices
- convenience shops
- comparison shops

percentage of land use in each zone

A CBD (core)
B CBD
C Inner city (19th century)
D Middle zone (inter-war)
E Outer zone (suburbs)
F Edge of city

Activities

1 Study Figure 15.67.

a i Describe and account for the differences between Zone A and Zone B, which are both described as part of the CBD. *(5 marks)*

ii Which two types of land use occupy the most area in Zone C? Explain why this is so. *(4 marks)*

iii How would you expect the appearance of the housing areas in Zone D to be different from those in Zone E? *(4 marks)*

iv Explain why Zone F has more industry and warehouses, offices and comparison shops than Zones D and E. *(5 marks)*

b Name a city in the UK that you have studied. Assess how closely it matches the idealised city shown in the diagram. Make specific reference to named areas within your chosen city. *(7 marks)*

2 There are many factors that 'push' people away from rural areas in less economically developed countries and make them migrate to cities. These include poverty, shortage of land, famine and natural disasters, and the lack of opportunity.

a Explain what the main 'pull' factors are that attract people to move to the cities. *(4 marks)*

b Many of the newcomers in the cities find themselves living in 'squatter settlements' on the outskirts of the city.

i Why do many newcomers end up living in such settlements? *(2 marks)*

ii Why are such settlements often found on the edges of cities? *(2 marks)*

iii Describe the main features of a squatter settlement in a named city that you have studied. *(5 marks)*

c With reference to a named example, explain why traffic congestion can be a problem in cities in less economically developed countries. *(4 marks)*

d Name a city in a less economically developed country. Explain how that city is tackling the problem of housing its growing population, and show how successful it has been. *(8 marks)*

Exam practice: basic structured questions

3 **a** What is the meaning of:

i urbanisation *(2 marks)*

ii gentrification *(2 marks)*

iii brownfield development? *(2 marks)*

b With reference to one or more inner city areas in the UK, explain what is meant by the 'cycle of deprivation'. *(5 marks)*

c Choose **one** of the following policies for inner city redevelopment that have been tried in the UK:

- Urban Development Corporations (UDCs)
- Enterprise Zones (EZs)
- Urban Regeneration Companies (URCs)
- New Deal for Communities (NDCs)
- English Partnership (EP) agreements.

Describe how your chosen scheme has affected one area in which it has been tried, and assess its success. *(14 marks)*

4 **a** Describe the main features of the Burgess model of urban structure, and explain why the model is useful to geographers. *(5 marks)*

   b Select **one** of the following models of urban development:
   - the Hoyt model
   - the Mann model
   - the Ullman and Harris model.

   i Describe your chosen model, and explain how it is different from the Burgess model of urban development. *(5 marks)*

   ii Discuss the limitations of the model. *(5 marks)*

   c With reference to a named city, describe the structure of the city and discuss the extent to which any of the models of urban structure fit that city. *(10 marks)*

Exam practice: structured questions

5 Study Figure 15.67.

   a Describe and explain the changes in land use along the transect. *(15 marks)*

   b Draw an idealised transect from the CBD to the city boundary for a typical city in a less economically developed country. Add notes below your transect to explain some of the key features of your diagram. *(10 marks)*

6 Study Figure 15.68.

   a Describe and compare the rates of urbanisation shown in the table. *(7 marks)*

   b Choose one of the less economically developed regions shown in the table. Explain why that region is experiencing rapid urbanisation. *(9 marks)*

   c Choose one of the more economically developed regions shown in the table. Explain why the rate of urbanisation was comparatively slow in the last 30 years of the 20th century. *(9 marks)*

Area	Urban population (percentage)				
	1950	1970	1990	2000	2030 (estimate)
World	29.2	37.1	45.2	48.2	61.9
Europe and Russia	56.3	66.7	73.4	73.5	80.6
North America	63.9	73.8	74.3	77.4	84.6
Oceania	61.3	70.8	71.3	74.2	72.2
Latin America	41.0	57.4	75.1	75.3	84.1
Asia (excl. Russia)	16.4	24.1	28.2	37.5	54.1
Africa	15.7	22.5	33.9	48.2	52.9

Figure 15.68

The proportion of world population living in urban areas

Exam practice: essays

7 Study Figure 15.68 above. Compare and contrast the rates of urbanisation in a range of regions at different stages of economic development. Suggest reasons for the differences that you have observed. *(25 marks)*

8 Several different schemes have been developed by UK governments since 1979 to improve conditions in declining inner city areas.

Choose any **two** of these schemes. Describe the aims and methods of each of the schemes. Assess the successes and failures of each scheme, with reference to one or more cities where the schemes were put into practice. *(25 marks)*

16

Farming and food supply

● ●

'But of all the occupations by which gain is secured, none is better than agriculture, none more profitable, none more delightful, none more becoming to a free man.'
Cicero, *De Officiis*, 1.51

'Behold, there shall come seven years of great plenty throughout all the land of Egypt: and there shall arise after them seven years of famine; and all the plenty shall be forgotten in the land of Egypt; and the famine shall consume the land …'
The Bible, Genesis 41: 29, 30

'He who slaughters his cows today shall thirst for milk tomorrow.'

Muslim proverb

The location of different types of agriculture at all scales depends on the interaction of physical, cultural and economic factors (Figure 16.25). Where individual farmers in a market economy (capitalist system) or the state in a centrally planned economy have a knowledge, or understanding, of these three influences, then decisions may be made. How these decisions are reached involves a fourth factor: the behavioural element.

Environmental factors affecting farming

Although there has been a movement away from the view that agriculture is controlled solely by physical conditions, it must be accepted that environmental factors do exert a major influence in determining the type of farming practised in any particular area. Increasingly, the environment is seen to be an input converted into monetary terms, e.g. yields and slopes.

In 1966 McCarty and Lindberg produced their optima and limits model, an adaptation of which appears in Figure 16.1. They suggested that there was an optimum or ideal location for each specific type of farming based on climate, soils, slopes and altitude. The optimum is defined as where the total cost of production per unit output (TCP) is minimised for that crop or live-stock. As distance increases from this optimum, conditions become less than ideal, i.e. too wet or dry; too steep or high; too hot or cold; or a less suitable soil. Consequently, the profitability of producing the crop or rearing animals is reduced, and the **law of diminishing returns** operates when either the output decreases or the cost of maintaining high yields becomes prohibitive. Eventually a point is reached where physical conditions are too extreme to permit production on an economically viable scale, and later at even a subsistence level (page 477). McCarty and Lindberg applied their model to the cotton belt of the USA (Figure 16.2), but it can equally be adapted to account for the growth of spring wheat on the Canadian Prairies (Figure 16.3).

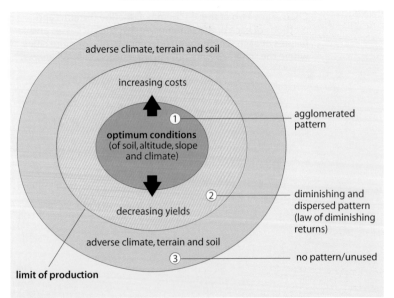

Figure 16.1

The optima and limits model (*after* McCarty and Lindberg)

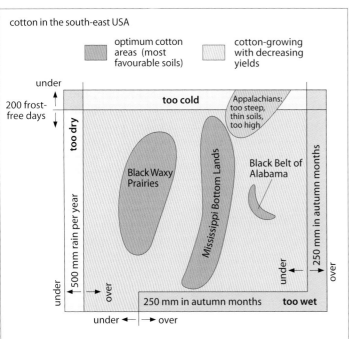

cotton in the south-east USA

optimum cotton areas (most favourable soils)

cotton-growing with decreasing yields

under 200 frost-free days | too cold

Appalachians: too steep, thin soils, too high

too dry

under 500 mm rain per year

Black Waxy Prairies

Mississippi Bottom Lands

Black Belt of Alabama

250 mm in autumn months

under ← → over

250 mm in autumn months   too wet

under ← → over

Figure 16.2

Optima and limits model applied to the former cotton belt in south-eastern USA

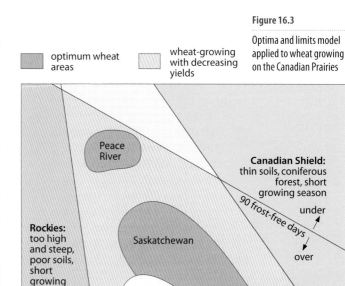

Figure 16.3

Optima and limits model applied to wheat growing on the Canadian Prairies

optimum wheat areas

wheat-growing with decreasing yields

Peace River

Canadian Shield: thin soils, coniferous forest, short growing season

90 frost-free days

under

over

Rockies: too high and steep, poor soils, short growing season

Saskatchewan

too dry

300 mm

over ← → under

400 mm

under ← → over

United States of America

Temperature

This is critical for plant growth because each plant or crop type requires a minimum growing temperature and a minimum growing season. In temperate latitudes, the critical temperature is 6°C. Below this figure, members of the grass family, which include most cereals, cannot grow – an exception is rye, a hardy cereal, which may be grown in more northerly latitudes.

In Britain, wheat, barley and grass begin to grow only when the average temperature rises above 6°C, which coincides with the beginning of the growing season. The growing season is defined as the number of days between the last severe frost of spring and the first of autumn. It is therefore synonymous with the number of frost-free days that are required for plant growth. Figures 16.2 and 16.3 show that cotton

needs a minimum of 200, and spring wheat 90. Barley can be grown further north in Britain than wheat, and oats further north than barley because wheat requires the longest growing season of the three and oats the shortest. Frost is more likely to occur in hollows and valleys. It has beneficial effects as it breaks up the soil and kills pests in winter, but it may also damage plants and destroy fruit blossom in spring.

Within the tropics there is a continuous growing season, provided moisture is available. As well as decreasing with distance from the Equator, both temperatures and the length of the growing season decrease with height above sea-level. This produces a succession of natural vegetation types according to altitude, although many have been modified for farming purposes (Figure 16.4).

Figure 16.4

The effect of altitude on farming and vegetation

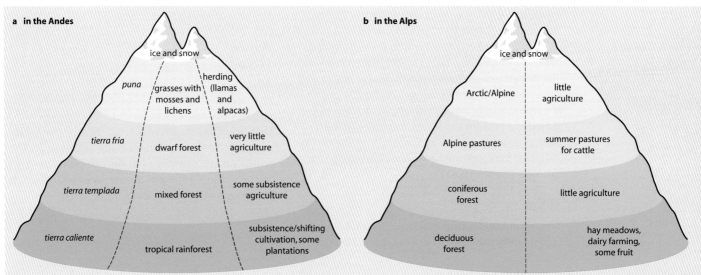

a in the Andes

ice and snow

puna / grasses with mosses and lichens

herding (llamas and alpacas)

tierra fria — dwarf forest — very little agriculture

tierra templada — mixed forest — some subsistence agriculture

tierra caliente — tropical rainforest — subsistence/shifting cultivation, some plantations

b in the Alps

ice and snow

Arctic/Alpine — little agriculture

Alpine pastures — summer pastures for cattle

coniferous forest — little agriculture

deciduous forest — hay meadows, dairy farming, some fruit

Precipitation and water supply

The mean annual rainfall for an area determines whether its farming is likely to be based upon tree crops, grass or cereals, or irrigation. The relevance and effectiveness of this annual total depends on temperatures and the rate of evapotranspiration. Few crops can grow in temperate latitudes where there is less than 250 mm a year or in the tropics where the equivalent figure is 500 mm. However, the seasonal distribution of rainfall is usually more significant for agriculture than is the annual total. Wheat is able to grow on the Canadian Prairies (Places 70, page 486) because the summer rainfall maximum means that water is available during the growing season. The Mediterranean lands of southern Europe have relatively high annual totals, yet the growth of grasses is restricted by the summer drought. Some crops require high rainfall totals during their ripening period (maize in the American corn belt), whereas for others a dry period before and during harvesting is vital (coffee).

The type of precipitation is also important (page 62). Long, steady periods of rain allow the water to infiltrate into the soil, making moisture available for plant use. Short, heavy downpours can lead to surface runoff and soil erosion and so are less effective for plants. Hail, falling during heavy convectional storms in summer in places such as the Canadian Prairies, can destroy crops. Snow, in comparison, can be beneficial as it insulates the ground from extreme cold in winter and provides moisture on melting in spring. In Britain we tend to take rain for granted, forgetting that in many parts of the world amounts and occurrence are very unreliable (Figure 9.28). India depends upon the monsoon; if this fails, there is drought and a risk of famine (page 502). Even in the best of years, the Sahel countries receive a barely adequate amount of moisture. The ecosystem is so fragile that should rainfall decrease even by a small amount (and in several years recently no rain has fallen at all), then crops fail disastrously – an event which appears to be occurring with greater frequency. In Britain, we would barely notice a shortfall of a few millimetres a year: in the Sahel and sub-Saharan Africa, an equivalent fluctuation from the mean can ruin harvests and cause the deaths of many animals (Figure 16.61).

Wind

Strong winds increase evapotranspiration rates which allows the soil to dry out and to become vulnerable to erosion. Several localised winds have harmful effects on farming: the *mistral* brings cold air to the south of France (Figure 12.22); the *khamsin* is a dry, dust-laden wind found in Egypt; Santa Ana winds can cause brush fires in California (Case Study 15A); and hurricanes, typhoons and tornadoes can all destroy crops by their sheer strength. Other winds are beneficial to agriculture: the *föhn* and *chinook* (page 241) melt snows in the Alps and on the Prairies respectively, so increasing the length of the growing season.

Altitude

The growth of various crops is controlled by the decrease in temperature with height. In Britain few grasses, including those grown for hay, can give commercial yields at heights exceeding 300 m, whereas in the Himalayas, in a lower (warmer) latitude, wheat can ripen at 3000 m. As height increases, so too does exposure to wind and the amounts of cloud, snow and rain, while the length of the growing season decreases. Soils take longer to develop as there are fewer mixing agents; humus takes longer to break down and leaching is more likely to occur. Those high-altitude areas where soils have developed are prone to erosion (Case Study 10).

Angle of slope (gradient)

Slope (see catena, page 276) affects the depth of soil, its moisture content and its pH (acidity, page 269), and therefore the type of crop that can be grown on it. It influences erosion and is a limitation on the use of machinery. Until recently, a 5° slope was the maximum for mechanised ploughing but technological improvements have increased this to 11°. Many steep slopes in Southeast Asia have been terraced to overcome some of the problems of a steep gradient and to increase the area of cultivation (Figure 16.29).

Aspect

Aspect is an important part of the microclimate. **Adret** slopes are those in the northern hemisphere that face south (Places 28, page 213). They have appreciably higher temperatures and drier soils than the **ubac** slopes which face north. The adret receives the maximum incoming radiation and sunshine, whereas the ubac may be permanently in the shade. Crops and trees both grow to higher altitudes on the adret slopes.

Soils (edaphic factors)

Farming depends upon the depth, stoniness, water-retention capacity, aeration, texture, structure, pH, leaching and mineral content of the soil (Chapter 10). Three examples help to show the extent of the soil's influence on farming:

1 Clay soils tend to be heavy, acidic, poorly drained, cold, and give higher economic returns under permanent grass.
2 Sandy soils tend to be lighter, less acidic, perhaps too well-drained, warmer and more suited to vegetables and fruit.
3 Lime soils (chalk) are light in texture, alkaline, dry, and give high cereal yields.

Although soils can be improved, e.g. by adding lime to clay and clay to sands, and by applying fertiliser, there is a limit to the increase in their productivity – i.e. the law of diminishing returns operates.

Global warming

Despite uncertainty as to the exact effects of global warming, scientists agree that the greenhouse effect will not only lead to an increase in temperature but also to changes in rainfall patterns. The global increase in temperature will allow many parts of the world to grow crops which at present are too cold for them: wheat will grow in more northerly latitudes in Canada and Russia, while maize, vines, oranges and peaches may flourish in southern England (Case Study 9B). Of greater significance will be the changes in precipitation, with some places becoming wetter and more stormy (Australia and South-east Asia) while others are likely to become drier (the wheat-growing areas of the American Prairies and the Russian Steppes).

Places 61 Northern Kenya: precipitation and water supply

The Rendille tribe live on a flat, rocky plain in northern Kenya where the only obvious vegetation is a few small trees and thorn bushes. Their traditional way of life has been to herd sheep, goats and camels, moving about constantly in search of water. (See Places 65, page 479 and Figure 16.5.)

'On the government map of Kenya, the realities of the Rendille's land are summarised in a few words: "Koroli Desert", it says, and just above this is the warning "Liable to Flood". There are two rainy seasons here: the long rains in April and May and the short rains in November. But the word "season" suggests that the rains are much more predictable and steady than they are in reality. Add together rainfall from the long and the short rains and you arrive at only 150 mm on the Rendille's central plains in an average year. But the word average means nothing here, because "normal variation" from that average can bring only 35 mm of rain one year and 450 mm the next. Variation from place to place is even more erratic than variation from year to year. Rains can be heavy when they do come, and water often rushes off the baked ground in flash floods; thus the apparent contradiction of a flood-prone desert.

It may suddenly rain in a valley for the first time in ten years; and it may not rain there for another decade. Therefore, the Rendille do not so much follow the rains as chase them, rushing to get their animals on to new grasses, which are more easily digested and converted into milk than are the drier, older shoots.'

L. Timberlake, *Only One Earth*, p. 92

Figure 16.5

Rendille herders at a shallow hand-dug well

Although the former Soviet Union is the largest country in the world, physical controls of climate, relief and soils have restricted farming to relatively small parts of the country. Of the land area of 22.27 million km², only 27 per cent was farmed in 1989 (10 per cent arable and 17 per cent pastoral), mainly in the deciduous forest belt, where the land had been cleared, and on the Steppes. The remaining 73 per cent (non-farmed) consisted of forest (42 per cent), tundra, desert and semi-desert (Figure 16.6).

After the Second World War, farmers were offered incentives to exceed their production targets. This task was most difficult for those farmers who were 'encouraged' by state directives rather than by financial incentives to develop the 'virgin lands' (Figure 16.6), in such states as Kazakhstan, by ploughing up the natural grassland in order to grow wheat and other cereals. Unfortunately, the unreliable rainfall, with totals often less than 500 mm a year, did not guarantee reliable crop yields. Later, to help cereal production, irrigation schemes were begun. These have since been extended into semi-desert areas where cotton is now grown. This necessitated the Soviets constructing large-scale transfer schemes by which water from rivers in the wetter parts of the country was diverted to areas suffering a deficiency.

Future water-transfer schemes are even more ambitious and may never reach fruition, as they involve diverting water from the northward-flowing Pechora, Ob and Yenisei rivers towards the south. Apart from the cost, environmentalists fear that this could result in the saline Arctic Ocean receiving less cold river water and then being warmed up sufficiently to cause the pack ice to melt and sea-levels to rise.

Figure 16.6

Physical controls on farming in the former Soviet Union

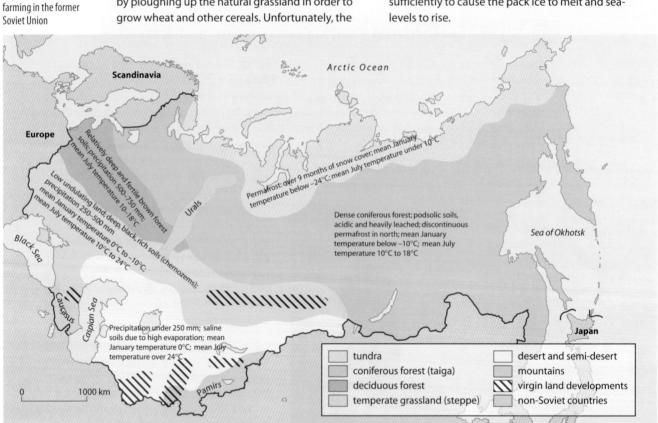

Cultural (human) factors affecting farming

Land tenure

Farmers may be owner-occupiers, tenants, landless labourers or state employees on the land which they farm. The **latifundia** system is still common to most Latin American countries.

The land here is organised into large, centrally managed estates worked by peasants who are semi-serfs. Even in the mid-1980s it was estimated that in Brazil 70 per cent of the land belonged to 3 per cent of the landowners. Land is worked by the landless labourers among the peasantry who sell their labour, when conditions permit, for substandard wages on the large estates or commercial plantations.

Other peasant farmers in Latin America have some land of their own held under insecure tenure arrangements. This land may be owned by the farmer, but it is more likely to have been rented from a local landowner or pawned to a moneylender. This latter type of tenancy takes two forms: cash-tenancy and share-cropping. **Cash-tenancy** is when farmers have to give as much as 80 per cent of their income or a fixed pre-arranged rent to the landowner. If the farmer has a short-term lease, he tends to overcrop the area and cannot afford to use fertiliser or to maintain farm buildings. If the lease is long-term, the farmer may try to invest but this often leads to serious debt. **Share-cropping** is when farmers have to pay, as a form of 'rent-in-kind' for occupying the land, part of their crop or animal produce to the landowner. As this fraction is usually a large one, the farmer works hard with little incentive and remains poor. This system operated in the cotton belt of the USA following the abolition of slavery and still persists in places. Both forms of tenancy, together with that of lati-fundia, resemble feudal systems found in earlier times in western Europe. The **plantation** is a variant form of the large estate system in that it is usually operated commercially, producing crops for the world market rather than for local use as in latifundia. On some plantations (oil palm in Malaysia – Places 68, page 483), the labourers are landless but are given a fixed wage; on others (sugar in Fiji), they are smallholders as well as receiving a payment.

In economically more developed, capitalist countries, many farmers are owner-occupiers, i.e. they own, or have a mortgage on, the farm where they live and work. Such a system should, in theory, provide maximum incentives for the farmer to become more efficient and to improve his land and buildings. Tenant farmers have been and still are, albeit in reduced numbers, an important part of land tenure in developed countries as well as in developing countries.

In sharp contrast to the neo-feudalist (lati-fundia, cash-tenancy and share-cropping) and capitalist systems of land tenure is the socialist system. In the former USSR, the individual farmer and the company-run estates were replaced by the *kolkhoz* (collective farm) and *sovkhoz* (state farm) system of organisation and management. Other forms of socialist tenure include the **commune** system which operated during the early years of communism in China (Places 63) and the *kibbutz*, which is a form of communal farming in Israel.

Inheritance laws and the fragmentation of holdings

In several countries, inheritance laws have meant that on the death of a farmer the land is divided equally between all his sons (rarely between daughters). Also, dowry customs may include the giving of land with a daughter on marriage. Such traditions have led to the sub-division of farms into numerous scattered and small fields. In Britain, fragmentation of land parcels may also result from the legacy of the open-field system (Places 51, page 400) or, more recently, from farmers buying up individual fields as they come onto the market. Fragmentation results in much time being wasted in moving from one distant field to another, and may cause problems of access. It may, however, be of benefit as it can enable a wider range of crops to be grown on land of different qualities.

Farm size

Inheritance laws, as described above, tend to reduce the size of individual farms so that, often, they can operate only at subsistence level or below. In most of the EU and North America, the trend is for farm sizes to increase as competitive market capitalism leads to the demise of small farms, and their land being purchased for enlargement by larger and more efficient, economically successful farms (page 493). Capital-intensive farms use much machinery, fertiliser, etc. and have a wide choice in types of production.

In South-east Asia and parts of Latin America and Africa, the rapid expansion of population is having the reverse effect. Farms, already inadequate in size, are being further divided and fragmented, making them too small for mechanisation (even if the farmers could afford machines). They are increasingly limited in the types of production possible, and output in certain areas, such as sub-Saharan Africa, is falling. Although farms of only 1 ha can support families in parts of South-east Asia where intensive rice production occurs and several croppings a year are possible, the average plot size in many parts of Taiwan, Nepal and South Korea has fallen to under 0.5 ha (about the size of a football pitch). In comparison, farms of several hundred hectares are needed to support a single family in those parts of the world where farming is marginal (upland sheep farming in Britain, cattle ranching in northern Australia).

Pre-1949

Before the establishment of the People's Republic in 1949, farming in China was typical of South-east Asia, i.e. it was mostly intensive subsistence (page 481). Farms were extremely small and fragmented, with the many tenants having to pay up to half of their limited produce to rich, often absentee landlords. Cultivation was manual or using oxen. Despite long hours of intensive work, the output per worker was very low. The need for food meant that most farmland was arable, with livestock restricted to those kept for working purposes or which could live on farm waste (chickens and pigs).

People's communes, 1958

After taking power in 1949, the communists confiscated land from the large landowners and divided it amongst the peasants. However, most plots proved too small to support individual farmers. After several interim experiments, the government created the 'people's communes'. The communes, which were meant to become self-sufficient units, were organised into a three-tier hierarchy with communist officials directing all aspects of life and work (Figure 16.7). Members of the commune elected a people's council, who elected a subcommittee to ensure that production targets, set by the Central Planning Committee (the government) in a series of Five Year Plans, were met. The committee was also responsible for providing an adequate food supply to make the unit self-supporting (crops, livestock, fruit and fish), for providing small-scale industry (mainly food processing and making farm implements), organising housing and services (hospital, schools) and for flood control and irrigation systems. Most communes had a research centre which trained workers to use new forms of machinery, fertiliser and strains of seed correctly (Green Revolution, page 504). By pooling their resources, farmers were able to increase yields per hectare.

Responsibility system, 1979

The introduction in 1979 of this more flexible approach, which encouraged farming families to become more 'responsible', preceded the abolition of the commune system in 1982. Under it, individual farmers were given rent-free land in their own village or district. They then had to take out contracts with the government, initially for 3 years but now extended to 30, to deliver a fixed amount of produce. To help meet their quota, individual farmers were given tools and seed. Once farmers had fulfilled their quotas, they could sell the remainder of their produce on the open market for their own profit. The immediate effect, due to farmers working much harder, was an increase in yields by an average of 6 per cent per year throughout the 1980s. Rural markets thrived and some farmers have become quite wealthy. Profits were used to buy better seed and machinery and to create village industry. Although most farmers have improved their standard of living, admittedly from an extremely low base, those living near to large cities (large nearby market) and in the south of the country (climatic advantages) have benefited the most.

1999

Hua Long (Case Study 14B) was one of several villages where the residents claimed that both their standard of living and quality of life had improved considerably over the last 20 years (Figure 16.8). Even the more rural villages were showing signs of an improvement in services and amenities (Figures 14.47 and 14.48), while the more efficient and prosperous farmers were able to save money and to invest it in new homes (Figure 14.49) and machinery. Farmers were now able to sublet land, hire labour, own machinery and make agricultural decisions.

50 families	= 1 production team	(300 people, 20 ha)	Responsible for own finances and payment of taxes for welfare services
10 production teams	= 1 brigade	(3000 people, 200 ha)	Responsible for overall planning, although they left the details to the production team
5 brigades	= 1 commune	(15 000 people, 1000 ha)	Responsible for ensuring that production targets set by the state were met

Figure 16.7

The structure of a former Chinese commune

Figure 16.8

Group 4 Team 1 in Hua Long village

Large farms are often	extensive on more marginal land	commercial in the EU and North America	animal grazing (sheep, cattle ranching); plantations; and temperate cereals (wheat)	further from large cities	areas of low population density and/or under-populated	increasing in size and efficiency due to amalgamation and mechanisation
Small farms are often	intensive on flat, fertile land	subsistence in Asia, Latin America and Africa	tropical crops (rice); and market gardening	nearer large cities	areas of high population density and/or overpopulated	decreasing in size and efficiency due to fragmentation and hand labour

Figure 16.9

Reasons for spatial variations in farm size

Bearing in mind the dangers of making generalisations (Framework 11, page 347), Figure 16.9 gives some of the spatial variations, and reasons for these variations, between large and small farms. Differences in farm size also affect other types of land use and the landscape.

Economic factors affecting farming

However favourable the physical environment may be, it is of limited value until human resources are added to it. Economic man – a term used by von Thünen (page 471) – applies resources to maximise profits. Yet these resources are often available only in developed countries or where farming is carried out on a commercial scale.

Transport

This includes the types of transport available, the time taken and the cost of moving raw materials to the farm and produce to the market. For perishable commodities, like milk and fresh fruit, the need for speedy transport to the market demands an efficient transport network, while for bulky goods, like potatoes, transport costs must be lower for output to be profitable. In both cases, the items should ideally be grown as near to their market as possible.

Markets

The role of markets is closely linked with transport (perishable and bulky goods). Market demand depends upon the size and affluence of the market population, its religious and cultural beliefs (fish consumed in Catholic countries, abstinence from pork by Jews), its preferred diet, changes in taste and fashion over time (vegetarianism) and health scares (BSE and GM foods).

Capital

Most economically developed countries, with their supporting banking systems, private investment and government subsidies, have large reserves of readily available finance, which over time have been used to build up **capital-intensive** types of farming (Figure 16.24) such as dairying, market gardening and mechanised cereal growing. Capital is often obtained at relatively low interest rates but remains subject to the law of diminishing returns. In other words, the increase in input ceases to give a corresponding increase in output, whether that output is measured in fertiliser, capital investment in machinery, or hours of work expended.

Farmers in developing countries, often lacking support from financial institutions and having limited capital resources of their own, have to resort to **labour-intensive** methods of farming (Figure 16.24). A farmer wishing to borrow money may have to pay exorbitant interest rates and may easily become caught up in a spiral of debt. The purchase of a tractor or harvester can prove a liability rather than a safe investment in areas of uncertain environmental, economic and political conditions.

Technology

Technological developments such as new strains of seed, cross-breeding of animals, improved machinery and irrigation may extend the area of optimal conditions and the limits of production (Green Revolution, page 504). Lacking in capital and expertise, developing countries are rarely able to take advantage of these advances and so the gap between them and the economically developed world continues to increase.

The state

We have already seen that in centrally planned economies it is the state, not the individual, that makes the major farming decisions (Places 62 and 63). In the UK, farmers have been helped by government subsidies. Initially, organisations such as the Milk and Egg Marketing Boards ensured that British farmers got a guaranteed price for their products. Today, most decisions affecting British farmers are made by the EU. Sometimes EU policy benefits British farmers (support grants to hill farmers) and sometimes it reduces their income (reduction in milk quotas). Certainly countries in the EU have improved yields, evident by their food surpluses (pages 487 and 493), and have adapted farming types to suit demand. Increasingly farmers in the UK are being pressurised by the demands of supermarkets and in developing countries by those of transnational companies.

Figure 16.10

Farming in China: the relationship between precipitation and farming type

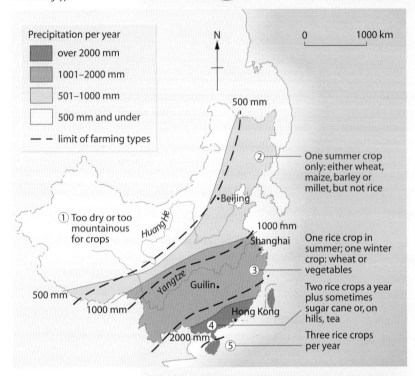

Precipitation per year

- over 2000 mm
- 1001–2000 mm
- 501–1000 mm
- 500 mm and under
- – – limit of farming types

① Too dry or too mountainous for crops

② One summer crop only: either wheat, maize, barley or millet, but not rice

③ One rice crop in summer; one winter crop: wheat or vegetables

④ Two rice crops a year plus sometimes sugar cane or, on hills, tea

⑤ Three rice crops per year

It is very difficult to generalise (Framework 11, page 347) about farming in a country that is the world's third largest in terms of area (40 times that of the UK) and largest in population. An atlas will show more accurately that, in general, the height of the land decreases, while temperatures and rainfall together with the length of the growing and rainy season (the monsoon, page 239), increase from the deserts and mountains of north and west China to the subtropics of the south-east. The type of farming – i.e. the type of crop grown and the number of croppings per year – shows a close correlation with such physical factors as the length of the growing season and the amount and distribution of annual rainfall (Figure 16.10).

Although there has been a population movement towards the towns, increasingly since 1979, especially to those near to the coast, and an increase in employment in the manufacturing and service sectors, 56 per cent of Chinese still live in rural areas and 44 per cent are farmers. Despite many improvements both in farming and in rural settlements (Case Study 14B), most farmers still have a very hard life and live at, or only a little above, subsistence level (page 477). Many work in their fields from daylight to dusk and have to rely upon hand labour (Figure 16.11). Although machinery is increasingly being used on the larger, flatter fields and the bigger farms of north-east China (Figure 16.12), animals such as the water buffalo are better suited to the smaller fields and farms found towards the south of the country where every conceivable piece of land is intensively used (Figure 16.13). Pastoral farming is practised in the higher, drier lands to the north and west (Figure 16.14).

Most farmers are still short of capital, although since the introduction of the responsibility system (Places 63) they now have the freedom to grow those crops or rear the animals they choose, together with the incentive to produce more and to diversify, as they can now sell any surplus. (The creation of wealth was not allowed during the first 30 years of the People's Republic, which coincided with a time when food shortages caused the deaths of millions of people.) As a result farmers across the country now claim that their standard of living, their quality of life and the country's food supply are better than they have been in living memory (Case Study 14B).

Figure 16.11

Intensive farming: planting rice near Kunming

Figure 16.12

Extensive farming: wheat and oilseed rape near Xi'an

Figure 16.14

Pastoral farming:
northern China

Figure 16.13

Use of animals: water
buffalo near Dazu

Von Thünen's model of rural land use

Heinrich von Thünen, who lived during the early 19th century, owned a large estate near to the town of Rostock (on the Baltic coast of present-day Germany). He became interested in how and why agricultural land use varied with distance from a market, and published his ideas in a book entitled *The Isolated State* (1826). To simplify his ideas, he produced a model in which he recognised that the patterns of land use around a market resulted from competition with other land uses. Like other models, von Thünen's makes several simplifying assumptions. These include:

- The existence of an isolated state, cut off from the rest of the world (transport was poorly developed in the early 19th century).
- In this state, one large urban market (or central place) was dominant. All farmers received the same price for a particular product at any one time.
- The state occupied a broad, flat, featureless plain which was uniform in soil fertility and climate and over which transport was equally easy in all directions.
- There was only one form of transport available. (In 1826 this was the horse and cart.)
- The cost of transport was directly proportional to distance.
- The farmers acted as 'economic men' wishing to maximise their profits and all having equal knowledge of the needs of the market.

In his model, von Thünen tried to show that with increasing distance from the market:

a the intensity of production decreased, and
b the type of land use varied.

Both concepts were based upon **locational rent** (*LR*) which von Thünen referred to as **economic rent**. Locational rent is the difference between the revenue received by a farmer for a crop grown on a particular piece of land and the total cost of producing and transporting that crop. Locational rent is therefore the profit from a unit of land, and should not be confused with **actual rent**, which is that paid by a tenant to a landlord.

Since von Thünen assumed that all farmers got the same price (revenue) for their crops and that costs of production were equal for all farmers, the only variable was the cost of transport, which increased proportionately with distance from the market. Locational rent can be expressed by the formula:

$$LR = Y(m - c - td)$$

where:

LR = locational rent
Y = yield per unit of land (hectares)
m = market price per unit of commodity
c = production cost per unit of land (ha)
t = transport cost per unit of commodity
d = distance from the market.

Since Y, m, c and t are constants, it is possible to work out by how much the *LR* for a commodity decreases as the distance from the market increases. Figure 16.15 shows that *LR* (profit) will be at its maximum at **M** (the market), where there are no transport costs. *LR* decreases from **M** to **X** with diminishing returns, until at **X** (the **margin of cultivation**) the farmer ceases production because revenue and costs are the same – i.e. there is no profit.

Farming and food supply   **471**

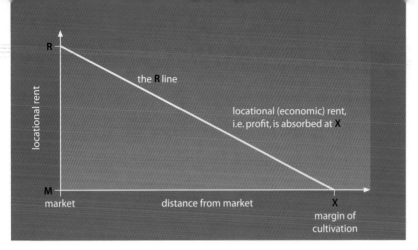

Figure 16.15

The relationship
between locational (eco-
nomic) rent and distance
from the market

Details of von Thünen's theory

Von Thünen tried to account for the location
of several crops in relation to the market. He
suggested that:

a   bulky crops, such as potatoes, should be
    grown close to the market as their extra
    weight would increase transport costs

b   perishable goods, such as vegetables and
    dairy produce, should also be produced as
    near as possible to the market (he wrote
    before refrigeration had been introduced)

c   intensive crops should be grown nearer to the
    market than extensive crops (Figure 16.16).

Consequently, bulky, perishable and intensive
crops (or commodities) will have steep *R* lines
(Figures 16.15, 16.16 and 16.17).

Figure 16.16 shows the result of two crops,
potatoes and wheat, grown in competition. The
two *R* lines, showing the locational rent or profit
for each crop, intersect at **Y**. If a perpendicular
is drawn from **Y** to **Z**, locational rent can be

translated into land use. Potatoes, an intensive, bulky crop, are grown near to the market
(between **M** and **Z**) as their transport costs are
high. Wheat, a more extensively farmed and less
bulky crop, is grown further away (between **Z**
and **X**) because it incurs lower transport costs.

What happens if three crops are grown in
competition? This is the combination of von
Thünen's two concepts: variation of intensity
and type of land use, with distance from market.
Let us suppose that wool is produced in addition
to potatoes and wheat (Figure 16.17).

Potatoes give the greatest profit if grown
at the market, and wool the least. However, as
potatoes cost £10 to transport every kilometre,
after 7 km their profit will have been absorbed in
these costs (£70 profit – £70 transport = £0). This
has been plotted in Figure 16.18a which is a **net
profit graph**. Wheat costs £3/km to transport
and so can be moved 15 km before it becomes
unprofitable (£45 profit – £45 transport = £0).
Wool, costing only £1/km to transport, can be
taken 30 km before it, too, becomes unprofitable.
Figure 16.18 also shows that although potatoes
can be grown profitably for up to 7 km from the
market, at point **A**, only 3.5 km from the market,
wheat farming becomes equally profitable and
that, beyond that point, wheat farming is more
lucrative. Similarly, wheat can be grown up to
15 km from the market, but beyond 7.5 km it is
less profitable than, and is therefore replaced by,
wool. The point at which one type of land use
is replaced by another is called the **margin of
transference**.

The types of land use can now be plotted
spatially. Figure 16.18b shows three concentric
circles, with the market as the common central
point. As on the graph, potatoes will be grown
within 3.5 km of the market. This is because
competition for land, and consequently land
values, are greatest here so only the most intensive farming is likely to make a profit. The plan
also shows that wheat is grown between 3.5 and
7.5 km from the market, while between 7.5 and
30 km, where the land is cheaper, farming is
extensive and wool becomes the main product.
Von Thünen's land use model is therefore based
on a series of concentric circles around a central
market.

The formula for locational rent (page 471)
assumed that market prices (*m*), production costs
(*c*) and transport costs (*t*) were all constant. What
would happen to a crop's area of production if
each of these in turn were to alter?

If the market price falls or the cost of production increases, there is a decrease in both the

Figure 16.16

Locational rents for
two crops grown in
competition

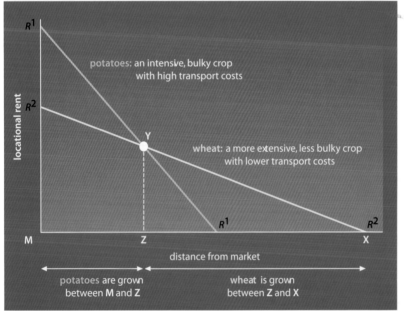

Farm product	Market price per unit of commodity	Production costs per unit of land (ha)	Transport costs per unit of commodity	Profit if grown at market
Potatoes	100	30	10	70
Wheat	65	20	3	45
Wool	45	15	1	30

Figure 16.17

Locational rents for three
commodities in competition

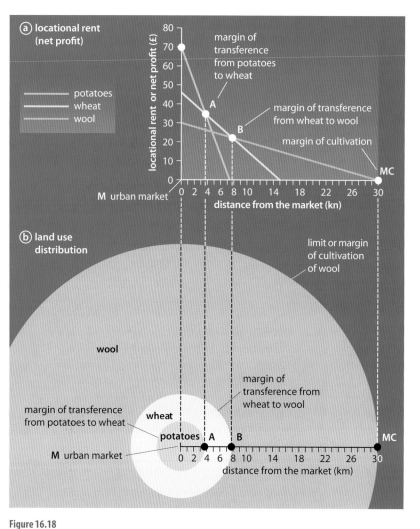

Figure 16.18

Locational rent (net profit) and land use of three commodities produced in competition

profits would rise, leading to an extension in the margin of cultivation. Changes in transport costs will not affect any farm at the market (Figure 16.19c) but an increase in transport costs reduces profits for distant farms, causing a decrease in the margin of cultivation. Conversely, a fall in transport costs makes those distant farms more profitable and enables them to extend their margin of cultivation.

Von Thünen's land use model

Von Thünen combined his conclusions on how the intensity of production decreased and the type of land use varied with distance from the market, to create his model (Figure 16.20a). He suggested six types of land use which were located by concentric circles.

1 Market gardening (horticulture) and dairying were practised nearest to the city, due to the perishability of the produce. Cattle were kept indoors for most of the year and provided manure for the fields.
2 Wood was a bulky product much in demand as a source of fuel and as a building material within the town (there was no electricity when von Thünen was writing). It was also expensive to transport.
3 An area with a 6-year crop rotation was based on the intensive cultivation of crops (rye, potatoes, clover, rye, barley and vetch) with no fallow period.
4 Cereal farming was less intensive as the 7-year rotation system relied increasingly on animal grazing (pasture, rye, pasture, barley, pasture, oats and fallow).

profit and the margin of cultivation of that crop (Figure 16.19a and b). Conversely, if the market price rises or the costs of production decrease,

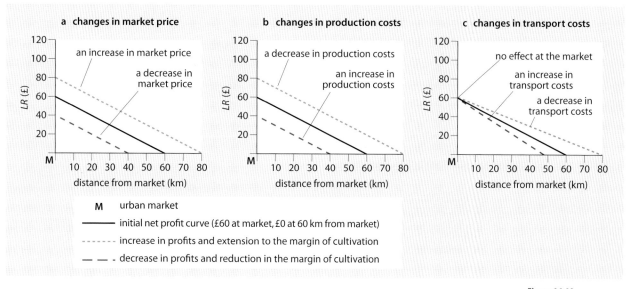

M   urban market

—— initial net profit curve (£60 at market, £0 at 60 km from market)

········ increase in profits and extension to the margin of cultivation

— — · decrease in profits and reduction in the margin of cultivation

Figure 16.19

Some causes of variation in locational rent

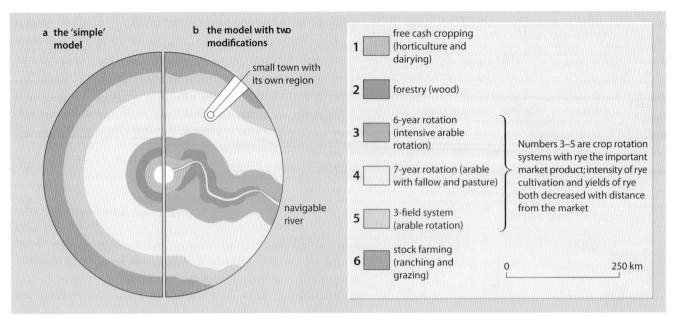

a the 'simple' model

b the model with two modifications

small town with its own region

navigable river

1 free cash cropping (horticulture and dairying)

2 forestry (wood)

3 6-year rotation (intensive arable rotation)

4 7-year rotation (arable with fallow and pasture)

5 3-field system (arable rotation)

6 stock farming (ranching and grazing)

Numbers 3–5 are crop rotation systems with rye the important market product; intensity of rye cultivation and yields of rye both decreased with distance from the market

0   250 km

Figure 16.20

The von Thünen land use model

5 Extensive farming based on a 3-field crop rotation (rye, pasture and fallow). Products were less bulky and perishable to transport and could bear the high transport costs.

6 Ranching with some rye for on-farm consumption. This zone extended to the margins of cultivation, beyond which was wasteland.

Modifications to the model

Later, von Thünen added two modifications in an attempt to make the model more realistic (Figure 16.20b). This immediately distorted the land use pattern and made it more complex. The inclusion of a navigable river allowed an alternative, cheaper and faster form of transport than his original horse and cart. The result was a linear, rather than a circular, pattern and an extension of the margin of cultivation. The addition of a secondary urban market involved the creation of a small trading area which would compete, in a minor way, with the main city.

Later still, von Thünen relaxed other assumptions. He accepted that climate and soils affected production costs and yields (though he never moved from his concept of the featureless plain) and that, as farmers do not always make rational decisions, it was necessary to introduce individual behavioural elements.

Why is it difficult to apply von Thünen's ideas to the modern world?

Models, in order to represent the totality of reality, rely upon the simplifying of assumptions (Framework 12, page 352). These simplifications can, in turn, be subject to criticisms which in the case of von Thünen's model can be grouped under four headings:

a Oversimplification There are very few places with flat, featureless plains, and where such landscapes do occur they are likely to contain several markets rather than one. As large areas with homogeneous climate and soils rarely exist, certain locations will be more favourable than others. Similarly, the 'isolated state' is rarely found in the modern world – Albania may be nearest to this situation – and there is much competition for markets both within and between countries. Von Thünen accepted that while his model simplified real-world situations, the addition of two variables immediately made it more complex (Figure 16.20b).

b Outdatedness As the model was produced 170 years ago, critics claim it is out-dated and of limited value in modern farming economics. Certainly since 1826 there have been significant advances in technology, changing uses of resources, pressures created by population growth, and the emergence of different economic policies. The invention of motorised vehicles, trains and aeroplanes has revolutionised transport, often increasing accessibility in one particular direction and making the movement of goods quicker and relatively cheaper. Milk tankers and refrigerated lorries allow perishable goods to be produced further from the market (London uses fresh milk from Devon) and stored for longer (the EU's food mountains). The use of wood as a fuel in developed countries has been replaced by gas and electricity and so trees need not be grown so near to the market, while supplies of timber in developing countries are being rapidly consumed and not always replaced. Improved farming techniques using fertilisers and irrigation have improved yields and extended the margins of cultivation.

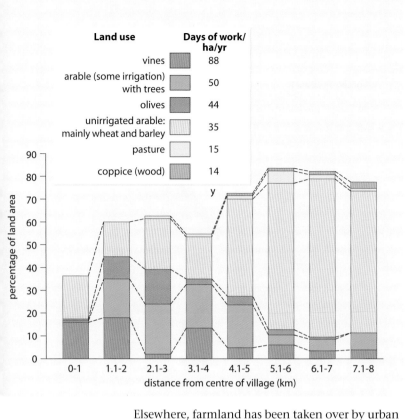

d Failure to include behavioural factors

Von Thünen has been criticised for assuming that farmers are 'rational economic men'. Farmers do not possess full knowledge, may not always make rational or consistent decisions, may prefer to enjoy increased leisure time rather than seeking to maximise profits and may be reluctant to adopt new methods. Farmers, as human beings, may have different levels of ability, ambition, capital and experience and none can predict changes in the weather, government policies or demand for their product.

How relevant is von Thünen's theory to the modern world?

It is pointed out on pages 411 and 557 that although theories are difficult to observe in the real world, they *are* useful because reality can be measured and compared against them. In the case of von Thünen's model:

a Figure 16.21 takes, at a **local** level, a relatively remote, present-day hill village in the Mediterranean lands of Europe. Many villages in southern Italy, Spain and Greece have hilltop sites (in contrast to von Thünen's featureless plain) where, usually, transport links are poor, affluence is limited and the village provides the main – perhaps the only – market (Figure 14.7). As the distance from the village increases, the amount of farmland used, and the yields from it, decrease. Two critical local factors are the distance which farmers are prepared to travel to their fields and the amount of time, or intensity of attention, needed to cultivate each crop.

b Figure 16.22 shows, at the **national** level, the spatial pattern of land use in Uruguay. The capital city, Montevideo, is located on the coast, and Fray Bentos is on the navigable Rio Uruguay: a situation similar in some respects to von Thünen's modified model (Figures 16.20b and 16.35).

Elsewhere, farmland has been taken over by urban growth or used by competitors who obtain higher economic rents.

c Failure to recognise the role of government

Governments can alter land use by granting/ reducing subsidies and imposing/removing quotas. The EU (page 493) has recently reduced milk quotas and paid farmers to take land out of production (set-aside). Centrally planned economies, as in the former USSR and in the early years of the People's Republic of China (Places 63, page 468), directly control the types and amount of production rather than manipulating market mechanisms.

Conclusions

Von Thünen's land use model still has some modern relevance, particularly at the local level, provided its limitations are understood and accepted. His concept of locational rent, which is useful in studying urban as well as rural land use (page 425), is still applicable today, as conceptually the land use providing the greatest locational rent will be the one farmed. However, cheap and efficient transport systems, powerful retailers, variable regulatory and planning frameworks, and uneven patterns of wealth now severely limit the model's application in the modern world.

Figure 16.22

Land use patterns in Uruguay

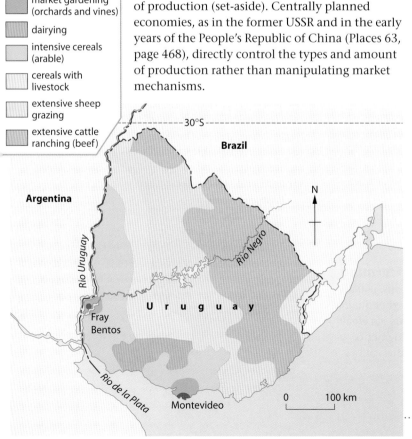

Farming and food supply **475**

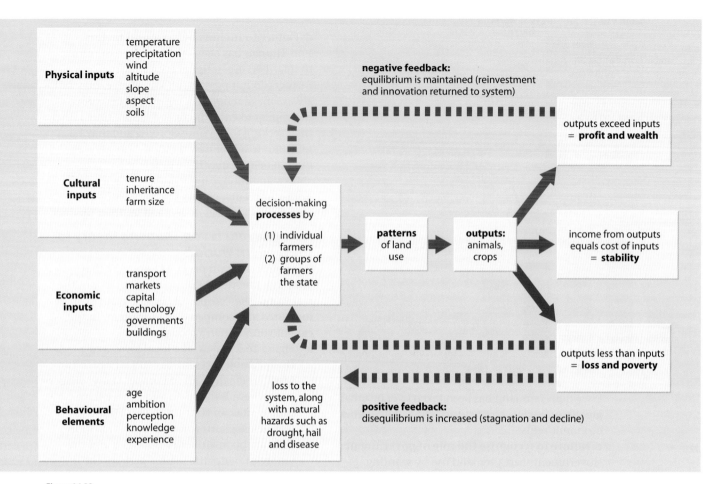

The farming system

Farming is another example of a system, and one which you may have studied already (Framework 3, page 45). The system diagram (Figure 16.23) shows how physical, cultural, economic and behavioural factors form the inputs. In areas where farming is less developed, physical factors are usually more important but as human inputs increase, these physical controls become less significant. This system model can be applied to all types of farming, regardless of scale or location. It is the variations in inputs that are responsible for the different types and patterns of farming.

Types of agricultural economy

The simplest classifications show the contrasts between different types of farming.

1 Arable, pastoral and mixed farming

Arable farming is the growing of crops, usually on flatter land where soils are of a higher quality. It was the development of new strains of cereals which led to the first permanent settlements in the Tigris–Euphrates, Nile and Indus valleys (Figure 14.1). Much later, in the mid-19th century, the building of the railways across the Prairies, Pampas and parts of Australia led to a rapid increase in the global area 'under the plough' (page 485). Today, there are few areas left with a potential for arable farming. This fact, coupled with the rapid increase in global population, has led to continued concern over the world's ability to feed its present and future inhabitants, a fear first voiced by Malthus (page 378). Already, there has been a decrease in the amount of arable land in some parts of the world, especially those parts of Africa affected by drought and soil erosion (Places 75, page 503).

Pastoral farming is the raising of animals, usually on land which is less favourable to arable farming (i.e. colder, wetter, steeper and higher land). However, if the grazed area has too many animals on it, its carrying capacity is exceeded or the quality of the soil and grass is not maintained, and then erosion and desertification may result (Case Study 7).

Mixed farming is the growing of crops and the rearing of animals together. It is practised on a commercial scale in developed countries, where it reduces the financial risks of relying upon a single crop or animal (monoculture), and at a subsistence level in developing countries, where it reduces the risks of food shortage.

2 Subsistence and commercial farming

Subsistence farming is the provision of food by farmers only for their own family or the local community – there is no surplus (Places 67, page 481). The main priority of subsistence farmers is self-survival which they try to achieve, whenever possible, by growing/rearing a wide range of crops/animals. The fact that subsistence farmers are rarely able to improve their output is due to a lack of capital, land and technology, and not to a lack of effort or ability. They are the most vulnerable to food shortages.

Commercial farming takes place on a large, profit-making scale. Commercial farmers, or the companies for whom they work, seek to maximise yields per hectare. This is often achieved – especially within the tropics – by growing a single crop or rearing one type of animal (Places 68, page 483). Cash-cropping operates successfully where transport is well developed, domestic markets are large and expanding, and there are opportunities for international trade (Places 69 page 484, and 70 page 486).

3 Shifting and sedentary farming

Many of the earliest farmers moved to new land every few years, due to a reduction in yields and also reduced success in hunting and gathering supplementary foods. **Shifting cultivation** is now limited to a few places where there are low population densities and a limited demand for food; where soils are poor and become exhausted after three or four years of cultivation (Places 66, page 480); or where there is a seasonal movement of animals in search of pasture (Places 65, page 479). However, farming over most of the world is now **sedentary**, i.e. farmers remain in one place to look after their crops or to rear their animals.

4 Extensive and intensive cultivation

These terms have already been used in describing von Thünen's model (Figure 16.16). **Extensive farming** is carried out on a large scale, whereas **intensive farming** is usually relatively small-scale. Farming is extensive or intensive depending on the relationship between three factors of production: labour, capital and land (Figure 16.24). Extensive farming occurs when:

- Amounts of labour and capital are small in relation to the area being farmed. In the Amazon Basin (Places 66, page 480), for example, the yields per hectare and the output per farmer are both low (Figure 16.24a).
- The amount of labour is still limited but the input of capital may be high. In the Canadian Prairies (Places 70, page 486), for example, the yields per hectare are often low but the output per farmer is high (Figure 16.24b).

Intensive farming occurs when:

- The amount of labour is high, even if the input of capital is low in relation to the area farmed. In the Ganges valley (Places 67, page 481), for example, the yields per hectare may be high although the output per farmer is often low (Figure 16.24c).
- The amount of capital is high, but the input of labour is low. In the Netherlands (Places 71, page 487), for example, both the yields per hectare and the output per farmer are high (Figure 16.24d).

Figure 16.24

Extensive and intensive farming (*after* Briggs)

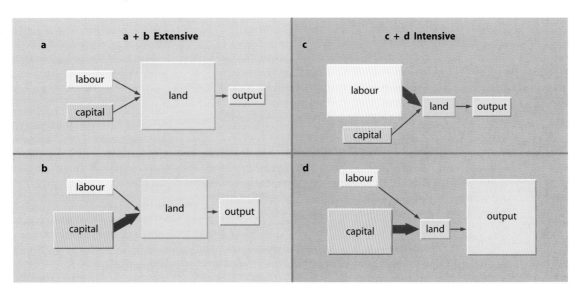

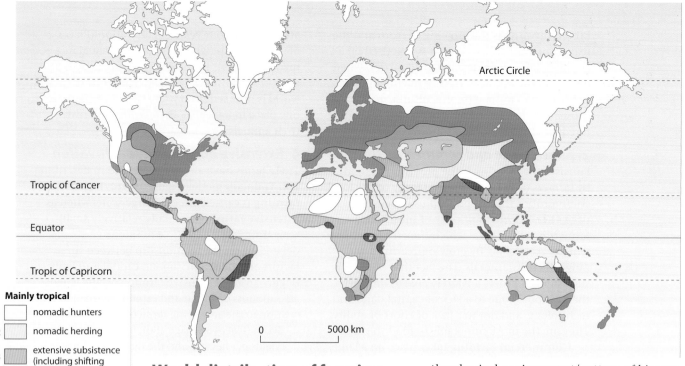

Mainly tropical

1   nomadic hunters

2   nomadic herding

3   extensive subsistence (including shifting cultivation)

4   intensive subsistence agriculture

5   commercial plantation agriculture

Mainly temperate

6   livestock ranching (commercial pastoral)

7   cereal cultivation (commercial grain)

8   intensive commercial (mixed)

9   Mediterranean agriculture

10   irrigation

11   unsuitable for agriculture

Figure 16.25

Location of the world's major farming types

World distribution of farming types

There is no widely accepted consensus as to how the major types of world farming should be classified or recognised (Framework 7, page 167). There is disagreement over the basis used in attempting a classification (intensity, land use, tropical or temperate, level of human input, the degree of commercialisation); the actual number and nomenclature of farming types; and the exact distribution and location of the major types.

You should be aware that:

1 Boundaries between farming types, as drawn on a map, are usually very arbitrary.

2 One type of farming merges gradually with a neighbouring type; there are few rigid boundaries.

3 Several types of farming may occur within each broad area, e.g. in West Africa, sedentary cultivators live alongside nomadic herdsmen.

4 A specialised crop may be grown locally, e.g. a plantation crop in an area otherwise used by subsistence farmers.

5 Types of farming alter over time with changes in economies, rainfall, soil characteristics, behavioural patterns and politics.

Figure 16.25 suggests one classification and shows the generalised location and distribution of farming types based upon the four variables described in the previous section. On a continental scale, this map demonstrates a close relationship between farming types and the physical environment/pattern of biomes (page 306). It disguises, however, the important human–economic factors that operate at a more local level.

The following section describes the main characteristics of each of these categories of farming together with the conditions favouring their development. A specific example is used in each case (which should be supplemented by wider reading) together with an account of recent changes or problems within that agricultural economy.

1 Hunters and gatherers

Some classifications ignore this group on the grounds that it is considered to be a relict way of life, with the original lifestyle now largely, or totally, destroyed by contact with the outside world. Others feel that even if it did exist then it does not constitute a 'true' farming type, as no crops or domesticated animals are involved. It is included here as, before the advent of sedentary farming, all early societies had to rely upon hunting birds and animals, catching fish, and collecting berries, nuts and fruit in order to survive … which is surely why we rely upon farming today. There are now very few hunter-gatherer societies remaining – the Bushman of the Kalahari, the Pygmies of central Africa, several Amerindian tribes in the Brazilian rainforest, and the Australian Aborigines. All have a varied diet resulting from their intimate knowledge of the environment, but each group need an extensive area from which to obtain their basic needs.

2 Nomadic herding

In areas where the climate is too extreme to support permanent settled agriculture, farmers become **nomadic pastoralists**. They live in inhospitable environments where vegetation is sparse and the climate is arid or cold. The movement of most present-day nomads is determined by the seasonal nature of rainfall and the need to find new sources of grass for their animals, e.g. the Bedouin and Tuareg in the Sahara and the Rendille and Maasai in Kenya (Places 65 and Case Study 12A). The indigenous Sami of northern Scandinavia have to move when their pastures become snow-covered in winter, while the Fulani in West Africa may migrate to avoid the tsetse fly.

There are two forms of nomadism. **Total nomadism** is where the nomad has no permanent home, while **semi-nomads** may live seasonally in a village. There is no ownership of land and the nomads may travel extensive distances, even across national frontiers, in search of fresh pasture. There may be no clear migratory pattern, but migration routes increase in size under adverse conditions, e.g. during droughts in the Sahel. The animals are the source of life. Depending upon the area, they may provide milk, meat and blood as food for the tribe; wool and skins for family shelter and clothing; dung for fuel; mounts and pack animals for transport; and products for barter. Just as sedentary farmers will not sell their land unless they are in dire economic difficulty, similarly pastoralists will not part with their animals, retaining them to regenerate the herd when conditions improve.

Places 65 Northern Kenya: nomadic herders

Rainfall is too low and unreliable in northern Kenya to support settled agriculture (Places 61, page 465). Over the years, the Rendille have learned how to survive in an extreme environment (Figure 16.26). All they need are their animals (camels, goats and a few cattle): all their animals need is water and grass. The tribe are constantly on the lookout for rain, which usually comes in the form of heavy, localised downpours. Once the rain has been observed or reported, the tribe pack their limited possessions onto camels (a job organised by the women) and head off, perhaps on a journey of several days, to an area of new grass growth. In the past, this movement prevented overgrazing, as grazed areas were given time to recover. Camels, and to a much lesser extent goats, can survive long periods without water by storing it within their bodies or by absorbing it from edible plants – food supply is as important as water. Humans, who can go longer than animals without food but much less long without water, rely upon the camels for milk and blood, and the goats for milk and occasional meat. Indeed, the main diet of blood and milk avoids the necessity of cooking and the need to find firewood.

But the Rendille way of life is changing. Land is becoming overpopulated and resources overstretched as the numbers of people and animals increase and as water supplies and vegetation become scarcer. Consequently, as the droughts of recent years continue, pastoralists are forced to move to small towns, such as Korr. Here there is a school, health centre, better housing, jobs, a food supply and a permanent supply of water from a deep well (Figure 16.26). The deep well waters hundreds of animals, many of which are brought considerable distances each day. However, the increase in animal numbers has resulted in overgrazing, and the increase in townspeople has led to the clearance of all nearby trees for firewood. This has resulted in an increase in soil erosion, creating a desert area extending 150 km around the town (desertification, Case Study 7). Although attempts are being made to dig more wells to disperse the population, travelling shops now take provisions to the pastoralists, and the tribespeople have been shown how to sell their animals at fairer prices, many Rendille are still moving to Korr to live. There the children, having been educated, remain, looking for jobs, with the result that there are fewer pastoralists left to herd the animals.

Figure 16.26

Rendille camels and goats at a waterhole

3 Shifting cultivation (extensive subsistence agriculture)

Subsistence farming was the traditional type of agriculture in most tropical countries before the arrival of Europeans, and remains so in many of the less economically developed countries and in more isolated regions. The inputs to this system are extremely limited. Relatively few labourers are needed (although they may have to work intensively), technology is limited (possibly to axes), and capital is not involved. Over a period of years, extensive areas of land may be used as the tribes have to move on to new sites. Outputs are also very low with, often, only sufficient being grown for the immediate needs of the family, tribe or local community.

The most extensive form of subsistence farming is shifting cultivation which is still practised in the tropical rainforests (the *milpa* of Latin America and *ladang* of South-east Asia) and, occasionally, in the wooded savannas (the *chitimene* of central Africa). The areas covered are becoming smaller, due to forest clearances, and are mainly limited to less accessible places within the Amazon Basin (Places 66), Central America, Congo and parts of Indonesia. Shifting cultivation, where it still exists, is the most energy-efficient of all farming systems as well as operating in close harmony with its environment.

Places 66 Amazon Basin: shifting cultivation

With the help of stone axes and machetes, the Amerindians clear a small area of about 1 ha in the forest (Figures 16.27 and 16.28). Sometimes the largest trees are left standing to protect young crops from the sun's heat and the heavy rain; so also are those which provide food, such as the banana and kola nut. After being allowed to dry, the felled trees and undergrowth are burnt – hence the alternative name of 'slash and burn' cultivation. While burning has the advantage of removing weeds and providing ash for use as a fertiliser, it has the disadvantage of destroying useful organic material and bacteria. The main crop, manioc, is planted along with yams (which need a richer soil), pumpkins, beans, tobacco and coca. The Amerindian diet is supplemented by hunting, mainly for tapirs and monkeys, fishing and collecting fruit.

The productivity of the rainforest depends on the rapid and unbroken recycling of nutrients (Figure 12.7). Once the forest has been cleared, this cycle is broken (Figure 12.8). The heavy, afternoon, convectional rainstorms hit the unprotected earth causing erosion and leaching. With the source of humus removed, the loss of nutrients within the harvested crop, and in the absence of fertiliser and animal manure, the soil rapidly loses its fertility. Within four or five years, the decline in crop yields and the re-infestation of the area by weeds force the tribe to shift to another part of the forest. Although shifting cultivation appears to be a wasteful use of land, it has no long-term adverse effect upon the environment as, in most places, nutrients and organic matter can build up sufficiently to allow the land to be re-used, often within 25 years.

The traditional Amerindian way of life is being threatened by the destruction of the rainforest. As land is being cleared for highways, cattle ranches, commercial timber, hydro-electric schemes, reservoirs and mineral exploitation, the Amerindians are pushed further into the forest or forced to live on reservations. Recent government policy of encouraging the in-migration of landless farmers from other parts of the country, together with the development of extensive commercial cattle ranching, has meant that sedentary farming is rapidly replacing shifting cultivation. After just a few years, as should have been foreseen, large tracts of some cattle ranches and many individual farms have already been abandoned as their soils have become infertile and eroded.

Figure 16.27

'Slash and burn': a shifting cultivator clearing the rainforest

Figure 16.28

Crops gown in *chagras* (fields) around the *maloca* (communal house)

4 Intensive subsistence farming

This involves the maximum use of the land with neither fallow nor any wasted space. Yields, especially in South-east Asia, are high enough to support a high population density – up to 2000 per km² in parts of Java and Bangladesh. The highest-yielding crop is rice which is grown chiefly on river floodplains (the Ganges and Figure 16.30) and in river deltas (the Mekong and Irrawaddy). In both cases, the peak river flow, which follows the monsoon rains, is trapped behind bunds, or walls (Places 67). Where flat land is limited, rice is grown on terraces cut into steep hillsides, especially those where soils have formed from weathered volcanic rock as in Indonesia and the Philippines (Figure 16.29). Upland rice, or dry padi, is easier to grow but, as it gives lower yields, it can support fewer people. Rice requires a growing season of only 100 days, which means that the constant high temperatures of South-east Asia enable two, and sometimes even three, croppings a year (Figure 16.10).

The high population density, rapid population growth and large family size in many South-east Asian countries mean that, despite the high yields, there is little surplus rice for sale. The farms, due to population pressure and inheritance laws (page 467), are often as small as 1 hectare. Many farmers are tenants and have to pay a proportion of their crops to a landlord. Labour is intensive and it has been estimated that it takes 2000 hours per year to farm each 1 hectare plot. Most tasks, due to a lack of capital, have to be done by hand or with the help of water buffalo. The buffalo are often overworked and their manure is frequently used as a fuel rather than being returned to the land as fertiliser. Poor transport systems hinder the marketing of any surplus crops after a good harvest and can delay food relief during the times of food shortage which may result from the extremes of the monsoon climate: drought and flood.

Figure 16.29

Rice cultivation on terraced hillsides, Bali

Places 67 The Ganges valley: intensive subsistence agriculture

Rice, with its high nutritional value, can form up to 90 per cent of the total diet in some parts of the flat Ganges valley in northern India and western Bangladesh. Padi, or wet rice, needs a rich soil and is grown in silt which is deposited annually by the river during the time of the monsoon floods. The monsoon climate (page 239) has an all-year growing season but, although 'winters' are warm enough for an extra crop of rice to be grown, water supply is often a problem. During the rainy season from July to October, the *kharif* crops of rice, millet and maize are grown. Rice is planted as soon as the monsoon rains have flooded the padi fields and is harvested in October when the rains have stopped and the land has dried out. During the dry season from November to April, the *rabi* crops of wheat, barley and peas are grown and harvested. Where water is available for longer periods, a second rice crop may be grown.

Figure 16.30

Rice harvesting on the flood-plain of the River Ganges

Rice growing is labour intensive with much manual effort needed to construct the bunds (embankments); to build irrigation channels; to prepare the fields; and to plant, weed and harvest the crop (Figure 16.30). The bunds between the fields are stabilised by tree crops. The tall coconut palm is not only a source of food, drink and sugar, but also acts as a cover crop protecting the smaller banana and other trees which have been planted on the bunds. The flooded padi fields may be stocked with fish which add protein to the human diet and fertiliser to the soil.

5 Tropical commercial (plantation) agriculture

Plantations were developed in tropical areas, usually where rainfall was sufficient for trees to be the natural vegetation, by European and North American merchants in the 18th and 19th centuries. Large areas of forest were cleared and a single bush or tree crop was planted in rows

In 1964, many Indian farmers and their families were short of food, lacked a balanced diet and had an extremely low standard of living. The government, with limited resources, made a conscious decision to try to improve farm technology and crop yields by implementing Western-type farming techniques and introducing new hybrid varieties of rice and wheat – the so-called Green Revolution (page 504). Although yields have increased and food shortages have been lessened, the 'Green Revolution' is not considered to be, in this part of the world, a social, environmental or political success (Figure 16.63).

(Figure 16.31 and 16.32) – hence the term mono-culture (page 280). This so-called **cash crop** was grown for export and was not used or consumed locally (Places 68).

Plantations needed a high capital input to clear, drain and irrigate the land; to build estate roads, schools, hospitals and houses; and to bridge the several years before the crop could be harvested. Although plantations were often located in areas of low population density, they needed much manual labour. The owners and managers were invariably white. Black and Asian workers, obtained locally or brought in as slaves or indentured labour from other countries, were engaged as they were prepared, or forced, to work for minimum wages. They were also capable of working in the hot, humid climate. Today, many plantations, producing most of the world's rubber, coffee, tea, cocoa, palm oil, bananas, sugar cane and tobacco, are owned and operated by large transnational companies (Figure 16.32).

Plantations, large estates and even small farmers are being increasingly drawn into making commercial contracts to supply fruit and vegetables to consumers in the developed world. Although such contracts may help some developing countries to provide jobs and to pay off their international debts, it also means they have to import greater volumes of staple foods to make up for the land switched from staples to export crops (page 501).

Figure 16.31

A rubber plantation in Malaysia

Figure 16.32

The advantages and disadvantages of plantation agriculture

Advantages	Disadvantages
Higher standards of living for the local workforce	Exploitation of local workforce, minimal wages
Capital for machines, fertiliser and transport provided initially by colonial power, now the transnational corporations	Cash crops grown instead of food crops: local population have to import foodstuffs
Use of fertilisers and pesticides improves output	Most produce is sent overseas to the parent country
Increases local employment	Most profit returns to Europe and North America
Housing, schools, health service and transport provided, also often electricity and a water supply	Dangers of relying on monoculture: fluctuations in world prices and demand
	Overuse of land has led, in places, to soil exhaustion and erosion

A plantation is defined in Malaysia as an estate exceeding 40 ha in size. Many extend over several thousand hectares. The first plantations were of coffee, but these were replaced at the end of the 19th century by rubber. Rubber is indigenous to the Amazon Basin, but some seeds were smuggled out of Brazil in 1877, brought to Kew Gardens in London to germinate and then sent out to what is now Malaysia. The trees thrive in a hot, wet climate, growing best on the gentle lower slopes of the mountains forming the spine of the Malay peninsula. Rubber tends not to be grown on the coasts where the land is swampy, but near to the relatively few railway lines and the main ports. The 'cheap' labour needed to clear the forest, work in the nurseries, plant new trees and tap the mature trees was provided by the poorer Malays and immigrants from India (Figure 16.31).

The Malaysian government has now taken over all the large estates, formerly run by such trans-nationals as Dunlop and Guthries, having seen them as a relict of colonialism. In the early 1970s, the Federal Land Development Authority (FELDA) was set up. Initially its job was to clear areas of forest, divide the land into 5 ha plots and to plant young rubber trees. After four years, smallholders were put in charge of the trees but FELDA still provided fertiliser and pesticide and, later, bought and marketed the crop.

The world demand for rubber steadily declined after the 1950s, mainly due to competition from synthetic rubber – apart from the years immediately after the first AIDS scare (page 622) which saw an increased demand for contraceptives. By 2000, the income of one-quarter of smallholders was said to be below the poverty line. Official figures suggested that half the country's smallholders, each with an average of four dependants, were totally reliant on rubber which, by 2001, hit a low price of RM1 per kg (RM = Malaysian ringgit, the local currency). Since then it has risen sharply, reaching over RM7 in

Figure 16.33
Oil palm

2006, which was high enough even for trees felled by storms to be tapped. However, the high price is predicted to be a short-term trend.

The Malaysian plantation industry is now heavily dependent on just one crop, oil palm (Figure 16.33). Oil palm, which covers over 80 per cent of the country's plantations, has many advantages over rubber including higher yields, higher prices, lower production costs and a less intensive use of labour (Figure 16.34). It is also more versatile because, apart from providing an edible oil and being used in a wide range of foodstuffs, it is also used in the oleochemical industry in the manufacture of soap, cosmetics and paint. Since 2000, further large areas have been converted into oil palm plantations where the crop is grown as a source of biofuel, mainly in the EU (page 543). Palm oil currently accounts for 6 per cent of Malaysia's GDP.

Although oil palm fruits have still to be harvested manually (the fronds get in the way of machines) and the fruits have to be harvested within a short period of time (otherwise the oil is lost), the spraying of herbicides, the application of fertiliser and transportation have all been mechanised.

Figure 16.34

The changing importance of rubber and oil palm

Production		Rubber	Oil palm
	1950 (thousand tonnes)	722	49
	1995 (thousand tonnes)	1089	7810
	2007 (thousand tonnes)	1200	15 823
	Tonnes per ha	2	24–26
	Years for trees to mature	6–7	4–5
	Labour intensive	Higher	Lower
	Price	1990s very low; since 2003 rising	Higher, rising

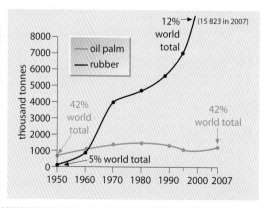

6 Extensive commercial pastoralism (livestock ranching)

Livestock ranching returns the lowest net profit per hectare of any commercial type of farming. It is practised in more remote areas where other forms of land use are limited and where there are extensive areas of cheaper land with sufficient grass to support large numbers of animals. It is found mainly in areas with a low population density and aims to give the maximum output from minimum inputs – i.e. there is a relatively small capital investment in comparison with the size of the farm or ranch, but output per farmworker is high. This type of farming includes commercial sheep farming (in central Australia, Canterbury Plains in New Zealand, Patagonia, upland Britain) and commercial cattle ranching (Places 69), mainly for beef (in the Pampas, American Midwest, northern Australia and, more recently, Amazonia and Central America). It corresponds, therefore, to the outer land use zone of von Thünen's model (Figure 16.20) and does not include commercial dairying which, being more intensive, is found nearer to the urban market (Places 71, page 487).

The raising of beef cattle is causing considerable environmental concern. It is a cause of deforestation (uses 40 per cent of the cleared forest in Amazonia), desertification and soil erosion (overgrazing) and global warming (release of methane). It also takes more water and feed to produce 1 kg of beef than the equivalent amount of any other food or animal product.

Places 69 The Pampas, South America: extensive commercial pastoralism

The Pampas covers Uruguay and northern Argentina. The area receives 500–1200 mm of rainfall a year – enough to support a temperate grassland vegetation. During the warmer summer months the water supply has to be supplemented from underground sources, while in the cooler, drier winter much of the grass dies down. Temperatures are never too high to dry up the grass in summer, nor low enough to prevent its growth in winter. The relief is flat and soils are often of deep, rich alluvium, deposited by rivers such as the Parana which cross the plain (Figure 16.35). The grasses help to maintain fertility by providing humus when they die back (Figure 11.29b).

Many ranches, or estancias, exceed 100 km² and keep over 20 000 head of cattle. Most are owned by businessmen or large companies based in the larger cities, and are run by a manager with the help of cowboys or gauchos. Several economic improvements have been added to the natural physical advantages. Alfalfa, a leguminous, moisture-retaining crop, is grown to feed the

Figure 16.35

Land use on the South American Pampas, an area with a zonation similar to that suggested by von Thünen

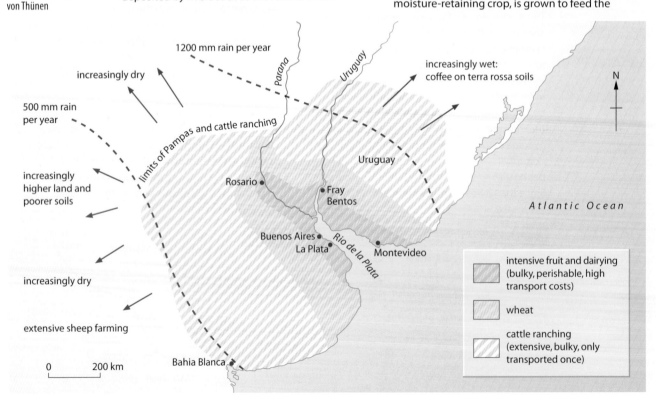

cattle when the natural grasses die down in winter. Barbed wire, for field boundaries, was essential where rainfall was insufficient for the growth of hedges. Pedigree bulls were brought from Europe to improve the local breeds and later British Hereford cattle were crossed with Asian Brahmin bulls to give a beef cow capable of living in warm and drier conditions. Initially, due to distances from world markets, cattle were reared for their hides. It was only after the construction of a railway network, linking places on the Pampas to the stockyards (*frigorificos*) at the chief ports of Rosario, Buenos Aires, La Plata and Montevideo on the Rio de la Plata (Figure 16.35), that canned products such as corned beef became important. Later still, the introduction of refrigerated wagons and ships meant that frozen beef could be exported to the more industrialised countries.

7 Extensive commercial grain farming

As shown on the map of the Pampas (Figure 16.35) and in the von Thünen model (Figure 16.20), cereals utilise the land use zone closer to the urban market than commercial ranching. Grain is grown commercially on the American Prairies (Places 70), the Russian Steppes (Figure 16.6) and parts of Australia, Argentina and north-west Europe (Figure 16.25). In most of these areas, productivity per hectare is low but per farmworker it is high.

It was the introduction and cultivation of new strains of cereals that led to the first permanent settlements (Figure 14.1) and, later, it was a reliance upon these cereals to provide a staple diet which allowed steady population growth in Europe, Russia and South-east Asia (Figure 16.36). A demand for increased cereal production came, in the mid-19th century, from those countries experiencing rapid industrialisation and urban growth. This demand was met following the building of railways in Argentina, Australia and across North America (Figure 16.36). More recent demands have, so far, been met by the Green Revolution in South-east Asia (page 504) and increases in irrigation and mechanisation.

Figure 16.36

Changes in the world's arable areas, 1870–2005

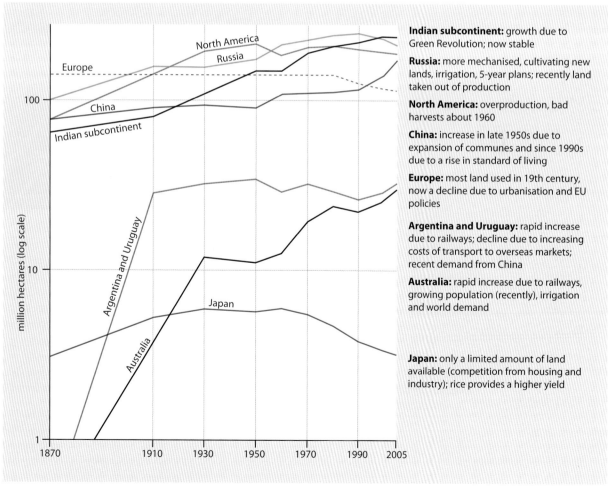

Indian subcontinent: growth due to Green Revolution; now stable

Russia: more mechanised, cultivating new lands, irrigation, 5-year plans; recently land taken out of production

North America: overproduction, bad harvests about 1960

China: increase in late 1950s due to expansion of communes and since 1990s due to a rise in standard of living

Europe: most land used in 19th century, now a decline due to urbanisation and EU policies

Argentina and Uruguay: rapid increase due to railways; decline due to increasing costs of transport to overseas markets; recent demand from China

Australia: rapid increase due to railways, growing population (recently), irrigation and world demand

Japan: only a limited amount of land available (competition from housing and industry); rice provides a higher yield

The Prairies have already been referred to in the optima and limits model (Figure 16.3). Although this area has many favourable physical characteristics, it also has disadvantages (Case Study 12B). Wheat, the major crop, ripens well during the long, sunny, summer days, while the winter frosts help to break up the soil. However, the growing season is short and in the north falls below the minimum requirement of 90 days. Precipitation is low, about 500 mm, but though most of this falls during the growing season there is a danger of hail ruining the crop, and droughts occur periodically. The winter snows may come as blizzards but they do insulate the ground from severe cold and provide moisture on melting in spring. The chinook wind (page 241) melts the snow in spring and helps to extend the growing season, but tornadoes in summer can damage the crop. The relief is gently undulating, which aids machinery and transport. The grassland vegetation has decayed over the centuries to give a black (chernozem – page 327) or very dark brown (prairie) soil (page 328). However, if the natural vegetation is totally removed, the soil becomes vulnerable to erosion by wind and convectional rainstorms.

Figure 16.37

Extensive commercial cereal farming on the Canadian Prairies

When European settlers first arrived, they drove out the local Indians, who had survived by hunting bison, and introduced cattle. The world price for cereals increased in the 1860s and demand from the industrialised countries in Western Europe rose. The trans-American railways were built in response to the increased demand (and profits to be made) and vast areas of land were ploughed up and given over to wheat. The flat terrain enabled straight, fast lines of communication to be built (essential as most of the crop had to be exported) and the land was divided into sections measuring 1 square mile (1.6 km^2). In the wetter east, each farm was allocated a quarter or a half section; while in the drier west, farmers received at least one full section.

The input of capital has always been high in the Prairies as farming is highly mechanised (Figure 16.37). Mechanisation has reduced the need for labour although a migrant force, with combine harvesters, now travels northwards in late summer as the cereals ripen. Seed varieties have been improved, and have been made disease-resistant, drought-resistant and faster-growing. Fertilisers and pesticides are used to increase yields and the harvested wheat is stored in huge elevators while awaiting transport via the adjacent railway.

In the last three decades, spring wheat has become less of a dominant crop and the area on which it is grown has decreased considerably, with many farms diversifying into canola (second biggest crop), barley, sugar beet, dairy produce and beef (Case Study 12B).

8 Intensive commercial (mixed) agriculture

This corresponds with von Thünen's inner zone where dairying, market gardening (horticulture) and fruit all compete for land closest to the market. All three have high transport costs, are perishable, bulky, and are in daily demand by the urban population. Similarly, all three require frequent attention, particularly dairy cows which need milking twice daily, and market gardening. Although this type of land use is most common in the eastern USA and north-west Europe (Places 71), it can also be found around every large city in the world. Intensive commercial farming needs considerable amounts of capital to invest in high technology, and numerous workers: it is labour intensive. The average farm size used to be under 10 ha but recently this has been found to be uneconomic and amalgamations have been encouraged by the American government and the EU in order to maximise profits. This type of farming gives the highest output per hectare and the highest productivity per farmworker.

Food surpluses

As farming in the more developed countries of North America and the EU continued to become more efficient, output increased. Farmers were paid subsidies, or a guaranteed minimum price, for their produce. The result was the overproduction of certain commodities for the American and European markets. Although food surpluses are needed for trade to take place, the issue is the unit cost at which the surpuses are produced. The problem arises when surpluses are produced at costs and subsidised prices that are above world market prices. This means that the commodities can only be sold on at a further subsidised price which then either distorts world markets or makes them inaccessible to developing countries. During the 1980s and 1990s, the EU introduced a variety of measures to try to limit the production of surplus products. These included fixed quotas on milk production, with penalties for overshoots; limits on the area of crops or number of animals for which the farmer could claim subsidies; and – voluntary at first but later compulsory – set-aside, which obliged farmers to leave a proportion of their land uncultivated (page 493). Since 2000, the EU has also encouraged farmers to restructure their farms, to diversify and to improve their product marketing. Farmers are, therefore, no longer paid just to produce food. Today's CAP is demand driven and takes into account consumers' and taxpayers' concerns while still giving farmers the freedom to produce what the market needs but not to receive subsidies on products that are overproduced. In addition, farmers have to respect environmental (page 496), food safety and animal welfare standards. These initiatives have meant that over time the EU has managed to reduce its use of export subsidies while managing to maintain, and in some cases even increase, its agricultural exports. Even so the EU remains a net importer of farm products.

Places 71 The western Netherlands: intensive commercial farming

Most of the western Netherlands, stretching from Rotterdam to beyond Amsterdam, lies 2–6 m below sea-level. Reclaimed several centuries ago from the sea, peat lakes or areas regularly flooded by rivers, this land is referred to as the **old polders**. Today, they form a flat area drained by canals which run above the general level of the land. Excess water from the fields is pumped (originally by windmills) by diesel and electric pumps into the canals. With 469 persons per km^2 in 1998 (compared with only 360 in 1975), the Netherlands has the highest population density in Europe. Consequently, with farmland at a premium, the cost of reclamation so high and the proximity of a large domestic urban market, intensive demands are made on the use of the land (Figure 16.38).

There are three major types of farming on the old polders.

- Dairying is most intensive to the north of Amsterdam, in the 'Green Heart' and in the south-west of Friesland. It is favoured by mild winters, which allow grass to grow for most of the year; the evenly distributed rainfall, which provides lush grass; the flat land; and the proximity of the Randstad conurbation. Most of the cattle are Friesians. Some of the milk is used fresh but most is turned into cheeses (the well-known Gouda and Edam) and butter. Most farms have installed computer systems to control animal feeding.

- The land between The Hague and Rotterdam (Figure 16.38) is a mass of glasshouses where **horticulture** is practised on individual holdings averaging only 1 ha. The cost of production is exceptionally high. Oil and natural gas-fired central heating maintain high temperatures and sprinklers provide water. Heating, moisture and ventilation are all controlled by computerised systems. Machinery is used for weeding and removing dead flowers, and the soils are heavily fertilised and manured. Sometimes plants are grown through a black plastic mulch (heat-absorbing) which has the effect of advancing their growth and thus extending the cropping season to meet market demand for fresh produce. Several crops a year can be grown in the glasshouses, i.e. cut flowers in spring, tomatoes and cucumbers in summer, and lettuce in autumn and winter.

- The sandier soils between Leiden and Haarlem are used to grow **bulbs**. Tulips, hyacinths and daffodils, protected from the prevailing winds by the coastal sand dunes, are grown on farms averaging 8 ha (Figure 16.39). The flowers form a tourist attraction, especially in spring, and bulbs are exported all over Europe from nearby Schipol Airport.

Figure 16.38

Agricultural land use in the western Netherlands

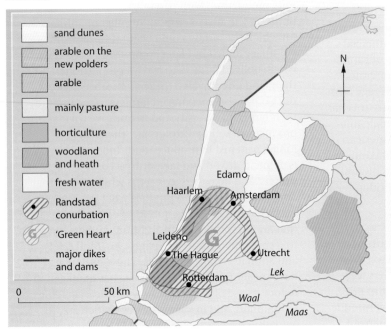

Figure 16.39

Intensive farming on reclaimed polder land in the western Netherlands

9 Mediterranean agriculture

A distinctive type of farming has developed in areas surrounding the Mediterranean Sea. Winters are mild and wet, allowing the growth of cereals and the production of early spring vegetables or *primeurs*. Summers are hot, enabling fruit to ripen, but tend to be too dry for the growth of cereals and grass. As rainfall amounts decrease and the length of the dry season increases from west to east and from north to south, irrigation becomes more important. River valleys and their deltas (the Po, Rhône and Guadalquivir) provide rich alluvium, but many parts of the Mediterranean are mountainous with steep slopes and thin rendzina soils (page 274). Due to earlier deforestation, many of these slopes have suffered from soil erosion. Frosts are rare at lower levels, though the cold mistral and bora winds may damage crops (Figure 12.22).

Farming tends to be labour intensive but with limited capital. There are still many absentee landlords (latifundia, page 466) and outputs per hectare and per farmworker are usually low. Most farms tend to be small in size.

Land use (Figure 16.21) shows the importance of tree crops such as olives, citrus and nuts, while land use frequently illustrates that crops which need most attention are grown nearest to the farmhouse or village, and that land use is more closely linked to the physical environment than controlled by human inputs (Places 72). Many village gardens and surrounding fields are devoted to citrus fruits, such as oranges, lemons and grapefruit, as these have thick waxy skins to protect the seeds and to reduce moisture loss. These fruits are also grown commercially where water supply is more reliable, e.g. oranges in Spain around Seville and on *huertas* (irrigated farms) near Valencia, lemons in Sicily and grapefruit in Israel. Recently there has been a rapid increase in the use of polythene, especially in south-east Spain, where the area around Almeria has become known as the 'Costa del Polythene', and in Israel. The polythene, which is stretched across 3 m high poles, creates a hothouse environment suitable for the growth of tomatoes and other crops such as melons, green beans, peppers and courgettes. The crops are harvested twice yearly, usually when they are out of season in more northerly parts of Europe.

Vines, another labour-intensive crop, and olives, the 'yardstick' of the Mediterranean climate, are both adapted to the physical conditions. They tolerate thin, poor, dry soils and hot, dry summers by having long roots and protective bark. Wheat may be grown in the wetter winter period in fields further from the village as it needs less attention, while sheep and goats are reared on the scrub and poorer-quality grass of the steeper hillsides. Grass becomes too dry in summer to support cattle and so milk and beef are scarce in the local diet.

Apart from central Chile, other areas experiencing a Mediterranean climate have developed a more commercialised type of farming based upon irrigation and mechanisation. Central California supports agribusiness based on a large, affluent, domestic market which is, in terms of scale, organisation and productivity, the ultimate in the capitalist system. Southern Australia produces dried fruit to overcome the problem of distance from world markets. All Mediterranean areas have now become important wine producers.

Places 72 The Peloponnese, Greece: Mediterranean farming

Figure 16.40

Land use and farming types in the Peloponnese (not to scale)

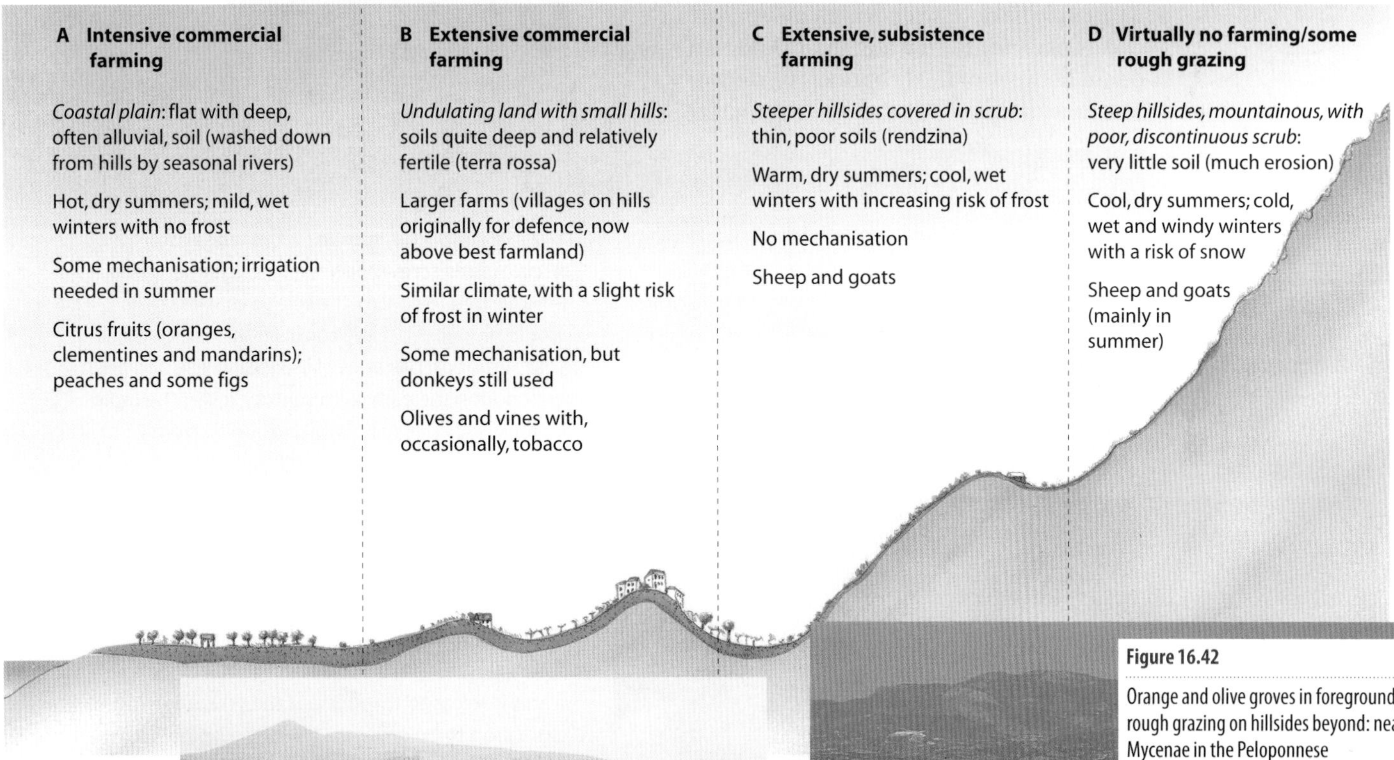

Figure 16.40 is a transect, typical of the Peloponnese and many other Mediterranean areas, showing how relief, soils and climate affect land use and farming types. The area next to the coast, unless taken over by tourism, is farmed intensively and commercially (Figure 16.41). As distance from the coast increases, farming becomes more extensive and eventually, before the limit of cultivation, at a subsistence level (Figure 16.42).

A Intensive commercial farming	B Extensive commercial farming	C Extensive, subsistence farming	D Virtually no farming/some rough grazing
Coastal plain: flat with deep, often alluvial, soil (washed down from hills by seasonal rivers)	*Undulating land with small hills*: soils quite deep and relatively fertile (terra rossa)	*Steeper hillsides covered in scrub*: thin, poor soils (rendzina)	*Steep hillsides, mountainous, with poor, discontinuous scrub*: very little soil (much erosion)
Hot, dry summers; mild, wet winters with no frost	Larger farms (villages on hills originally for defence, now above best farmland)	Warm, dry summers; cool, wet winters with increasing risk of frost	Cool, dry summers; cold, wet and windy winters with a risk of snow
Some mechanisation; irrigation needed in summer	Similar climate, with a slight risk of frost in winter	No mechanisation	Sheep and goats (mainly in summer)
Citrus fruits (oranges, clementines and mandarins); peaches and some figs	Some mechanisation, but donkeys still used	Sheep and goats	
	Olives and vines with, occasionally, tobacco		

Figure 16.42

Orange and olive groves in foreground, rough grazing on hillsides beyond: near Mycenae in the Peloponnese

Figure 16.41

Intensive fruit, mainly oranges, next to the coast, settlement on higher, less fertile land in mid-distance, and deforested hills in the background: near Navplion in the Peloponnese

10 Irrigation

Irrigation is the provision of a supply of water from a river, lake or underground source to enable an area of land to be cultivated (Figure 16.43). It may be needed where:

1 rainfall is limited and where evapotranspiration exceeds precipitation, i.e. in semi-arid and arid lands such as the Atacama Desert in Peru (Places 24, page 180) and the Nile valley (Places 73)
2 there is a seasonal water shortage due to drought, as in southern California with its Mediterranean climate (Case Study 15A)
3 amounts of rainfall are unreliable, as in the Sahel countries (Figure 9.28)
4 farming is intensive, either subsistence or commercial, despite high annual rainfall totals, e.g. the rice-growing areas of Southeast Asia.

In economically more developed countries, large dams may be built from which pipelines and canals may transport water many kilometres to a dense network of field channels (Case Study 17). The flow of water is likely to be computer-controlled. Unfortunately, it is the economically less developed countries, lacking in capital and technology, that suffer most severely from water deficiencies. Unless they can obtain funds from overseas, most of their schemes are extremely labour intensive as they have to be constructed and operated by hand.

Figure 16.43

Boom irrigation in Saudi Arabia

Places 73 The Nile valley: irrigation

Figure 16.44

Landsat photo of the Nile delta: the River Nile, Suez Canal and Mediterranean Sea are shown in black, the irrigated lands in magenta, and Cairo and other settlements in pale blue

From the time of the Pharaohs until very recently, water for irrigation was obtained from the River Nile by two methods. First, each autumn the annual floodwater was allowed to cover the land, where it remained trapped behind small bunds until it had deposited its silt. Second, during the rest of the year when river levels were low, water could be lifted

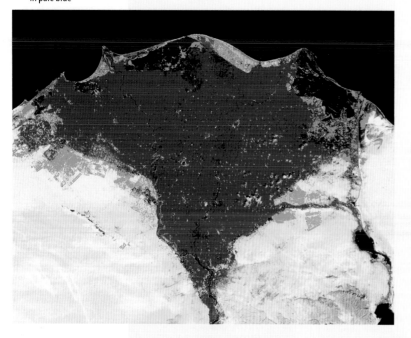

1–2 m by a *shaduf*, *saquia* (*sakia*) **wheel** or **Archimedes screw**. However, the Egyptians had long wished to control the Nile so that its level would remain relatively constant throughout the year. Although barrages of increasing size had been built during the early 20th century, it was the rapid increase in population (which doubled from 25 to 50 million between 1960 and 1987) and the accompanying demand for food that led to the building of the Aswan High Dam (opened 1971) and several new schemes to irrigate the desert near Cairo (late 1980s).

The main purpose of the High Dam was to hold back the annual floodwaters generated by the summer rains in the Ethiopian Highlands. Some water is released throughout the year, allowing an extra crop to be grown, while any surplus is saved as an insurance against a failure of the rains. The river regime below Aswan is now more constant, allowing trade and cruise ships to travel on it at all times. Two and sometimes three crops can now be grown annually in the lower Nile valley (Figure 16.44). Yields have increased and extra income is gained from cash crops of cotton, maize, sugar cane, potatoes and citrus fruits. The dam incorporates a hydro-electric power station which provides Egypt with almost a third of its energy needs for domestic and industrial purposes. Lake Nasser is important for fishing and tourism.

Following the construction of the Aswan High Dam (Figure 16.45), Egypt has modernised its methods of irrigation. Electricity is now used to power pumps which, by raising water to higher levels, allow a strip of land up to 12 km wide on both sides of the Nile to be irrigated. **Drip irrigation** utilises plastic pipes in which small holes have been made; these are laid over the ground and water drips onto the plants in a much less wasteful manner, as less evaporates or drains away. Between the Nile and the Suez Canal, **boom irrigation** has been introduced (Figure 16.43 shows this method in use in Saudi Arabia), creating fields several hectares in diameter. However, the Dam has created several problems.

Environmental

The cessation of the summer Nile flood has also meant the ending of the annual deposition of fertile silt on the fields, which in turn means that fertiliser now has to be added. Without its supply of sediment, which included silt and sand, the delta has begun to retreat, causing a loss of tourist beaches, the threat of saltwater contaminating existing irrigation schemes

and – not envisaged when the Dam was originally built – less protection against the rising sea-level of the Mediterranean which is resulting from global warming. The number of bilharzia snails has also risen due to the greater number of irrigation channels, while moisture in the air, caused by evaporation over Lake Nasser, is affecting ancient buildings.

Economic and social

Farmers have been encouraged to grow cash crops for export instead of providing a better diet for themselves, and their costs have increased due to the need to buy fertiliser. The area of land under irrigation has actually decreased since the Dam was built due to the increased effects of salinisation (page 273 and Figure 16.53) and clay is less easy to find for making sun-baked bricks in the traditional manner. More recently, great concern has been raised over rising damp and the deposition of salt (salt crystallisation, page 40), resulting from the constantly higher river level, in the foundations of temples and other buildings dating from the time of Ancient Egypt.

Figure 16.45

The Nile: sources and uses of water

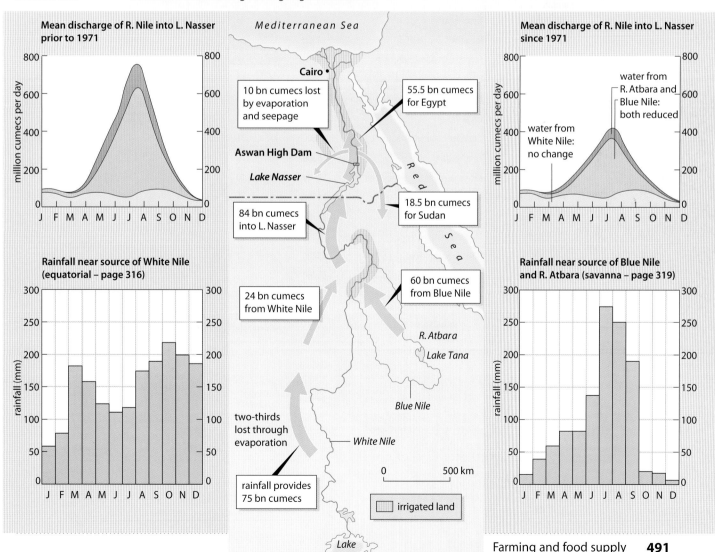

Farming types and economic development

Throughout the section on types of farming, several fundamental assumptions have been made. These include the generalisations that: 'the poorest countries are those which, because they have the lowest inputs of capital and technology, have the lowest outputs'; and 'the wealthier countries are those which can afford the highest inputs, giving them the maximum yield, or profit, per person'. Is it really possible to make a simple correlation between wealth (the standard of living) and the type of agriculture?

Figure 16.46 shows 15 countries selected (not chosen randomly) as representing the main types of farming and used as examples in the previous section. Using the five variables A–E, it is possible to postulate four hypotheses:

1 The less developed a country (i.e. the lower its GDP/GNP per capita – see pages 604 and 606), the greater the percentage of its population involved in agriculture.

2 The less developed a country, the greater the percentage of its GDP/GNP is made up from agriculture.

3 The less developed a country, the less fertiliser it will use.

4 The less developed a country, the less mechanised will be its farming (fewer tractors per head of population, for example).

Figure 16.47 shows schematically possible links between farming and economic development. As with all data, there are considerations which you should remember when drawing conclusions from Figure 16.46.

- The countries were selected with some bias in order to cover all the main types of farming.
- GDP/GNP is not the only indicator of wealth or development (pages 607–608).
- GDP/GNP figures are not necessarily accurate and may be derived from different criteria (page 606).
- There may be several different types of farming in each country.

Figure 16.46

Types of farming, GDP and agricultural data for selected countries

	Country	Major farming type	A	B	C	D	E
1	Ethiopia	Nomadic herding	800	81	46	12	0.3
2	Bangladesh	Intensive subsistence	1300	52	21	88	0.7
3	Kenya	Nomadic herding/subsistence	1700	74	27	31	2.8
4	India	Intensive subsistence	2700	58	20	101	15.7
5	China	Intensive subsistence/centrally planned	5300	54	16	278	7.0
6	Egypt	Irrigation	5500	31	15	434	30.7
7	Uruguay	Extensive commercial ranching	1600	12	11	94	24.2
8	Malaysia	Commercial plantation	13 300	16	9	683	24.1
9	Argentina	Extensive commercial grain/ranching	13 500	9	10	27	10.7
10	Russia	Centrally planned	14 700	9	5	12	4.8
11	Greece	Mediterranean	29 700	15	7	149	94.5
12	Spain	Mediterranean	30 100	6	4	157	68.7
13	UK	Intensive commercial	35 100	2	1	311	88.3
14	Canada	Extensive commercial/grain	38 400	2	2	57	16.0
15	Netherlands	Intensive commercial/mixed	38 500	3	2	478	163.9

A Gross domestic product (GDP) per capita in US$
B Percentage of population engaged in agriculture
C Percentage of GDP derived from agriculture
D Kg of fertiliser used per hectare of agricultural land
E Number of tractors per 1000 ha of land

Figure 16.47

An alternative method of showing links between types of farming and levels of economic development (as devised by a group of A-level Geography students)

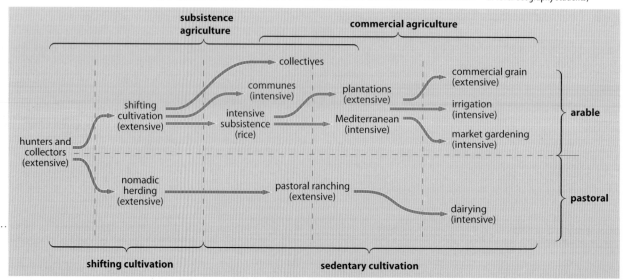

The EU – an example of a supra-national agricultural policy

Member countries of the EU are meant to implement the Common Agricultural Policy (CAP) by which most major decisions affecting farming are made in Brussels and not by individual countries or by individual farmers. The five basic aims of the CAP were to:

1 increase agricultural productivity and to improve self-sufficiency
2 maintain jobs on the land, preferably on family farms
3 improve the standard of living (income) of farmers and farmworkers
4 stabilise markets
5 keep consumer food prices stable and reasonable.

Although many of these aims had been fulfilled by the early 1990s, there was increasing concern over both the running and the effects of the CAP.

■ 70 per cent of the EC's (as it was then) budget was spent supporting farming when agriculture only provided 5 per cent of the EC's total income.
■ As farmers were encouraged, and were helped by improved technology, to produce as much as possible, large surpluses were created (page 487).
■ Imports were subject to duties to make them less competitive with EC prices. This handicapped the economically less developed countries.
■ EC farmers were granted generous subsidies to maintain prices. This helped restrict imports from non-EC economically developed countries.
■ As EC farms became larger and more efficient, it was the more prosperous farmers who benefited, often at the expense of those farming in upland areas and on the periphery, especially in southern Europe.
■ There was insufficient regard for the environment.

Since 1992 the CAP has undergone a series of reforms in order to solve some of these problems and has introduced policies aimed at encouraging the de-intensification of farming and the protection of the environment.

■ **Subsidies** guaranteed farmers a minimum price and an assured market for their produce. Farmers tended, therefore, to overproduce (hence the EU surpluses), and the payment of subsidies became a drain on EU finances. Since the 1990s, steps have been taken to limit the production of surplus products either by reducing subsidies for them or, in some cases, imposing penalties (page 487). In the early 1990s, the EU began a programme of progressive reductions of subsidies in cereal, beef and other commodities which has led, over time, to the elimination of the so-called 'mountains and lakes' surpluses of agricultural products. Even so, in 2006 the CAP still accounted for 45 per cent of the EU's total budget.
■ **Quotas** were introduced in 1984 to reduce milk output. These, like subsidies, have been gradually phased out and, as announced in 2006, will end by 2015. To try to reduce the impact on dairy farmers (Figure 16.48), the EU has proposed five annual quota increases between 2008 and 2013.
■ **Set-aside** was initially introduced on a voluntary basis, but later enforced, to try to reduce overproduction of arable crops. Farmers who took 20 per cent of their cultivated land out of production (pasture and fallow were not included) were given £20 a hectare, provided that the land was either left fallow, turned into woodland (under the Farm Woodland Management Scheme) or diversified into other non-agricultural land uses such as golf courses, nature trails, wildlife habitats and caravan parks. By the early 2000s, there was little surplus production and so when 2007–08 saw a rapid global increase in food prices, the EU fixed the set-aside rate at zero. This meant that British farmers could bring up to 5 million hectares back into production.
■ **Environmentally friendly farming** is a new EU approach by which, instead of paying farmers to produce more food, they are given payments if they meet environmental and animal standards and keep their land in good condition – the so-called '**health check**'. This health check is an attempt to streamline and modernise the CAP and to encourage farmers to be 'guardians of the countryside'.
■ The **World Trade Organisation (WTO)** has been trying, with minimum success (pages 627–629), to encourage the EU, and other well-off trade blocs, to reduce tariffs, quotas and subsidies so as to help the developing countries.

The CAP reforms of the early 2000s did not anticipate the increased global needs to fight climate change (page 256), to improve water management and supply (page 610), to satisfy the growing demand for biocrops as a source for renewable energy (page 543), or the rise in food prices. In 2008 it was claimed that even within the EU itself, 43 million people were at risk of food poverty – that is, they had less than one meal in two days that included meat, chicken or fish.

Figure 16.48

Problems for dairy
farmers, 2008

UK dairy farmers on brink of collapse

UK dairy farmers lose an average of 4.7p on every pint of milk they produce, giving the average dairy farm an annual loss of £37,600, new figures show.

The figures from First Milk, a farmer-owned dairy business that supplies more than 1.8 bn litres of milk a year, lay bare the desperate plight of the UK dairy industry.

According to a report out today, the average price paid to a farmer for a litre of milk over the year to March 31 2007 was 17.5p. However, the cost of producing this milk was 22p. This 4.7p loss multiplied by the 800,000 litres that the average farm produces each year equates to £37,600.

The UK dairy industry has been shrinking rapidly since 2000. Around 11pc of the national herd has disappeared in the past five years, while farmer numbers are reducing at the rate of 6.5pc a year.

First Milk is calling for the introduction of a new formula to calculate a 'consistently fair price' for milk. The formula should take into account rising production costs, labour costs on the farm and should include a profit margin so that farmers can reinvest in their businesses.

It says that farmers should be paid 29.6p a litre for their milk this year.

The report – called *The Real Price of Milk* – details the rising costs that have been absorbed by farmers in recent years, including animal feed, fertiliser and fuel. These rising costs were equivalent to an extra £36,000 a year since 2006.

Daily Telegraph, 3 March 2008

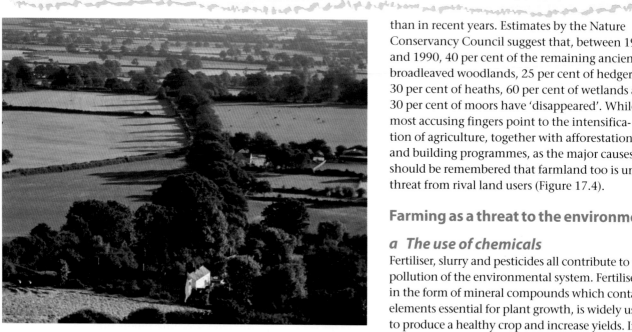

Figure 16.49

A rural landscape with trees and hedges, Dorset

Figure 16.50

How eutrophication can upset the ecosystem

Farming and the environment

Numerous pressure groups are claiming that the traditional British countryside is being spoilt, yet the countryside of today is not 'traditional' – it has always been changing. The primeval forests, regarded as Britain's climatic climax vegetation (page 286), were largely cleared, initially for sheep farming and later for the cultivation of cereals. Although there is evidence that hedges were used as field boundaries by the Anglo-Saxons, it was much later that land was 'enclosed' by planting hedges and building dry-stone walls (page 397). It is this 18th- and 19th-century landscape which has become looked upon, incorrectly, as the traditional or natural environment (Figure 16.49). However, the rate of change has never been faster than in recent years. Estimates by the Nature Conservancy Council suggest that, between 1949 and 1990, 40 per cent of the remaining ancient broadleaved woodlands, 25 per cent of hedgerows, 30 per cent of heaths, 60 per cent of wetlands and 30 per cent of moors have 'disappeared'. While most accusing fingers point to the intensification of agriculture, together with afforestation and building programmes, as the major causes, it should be remembered that farmland too is under threat from rival land users (Figure 17.4).

Farming as a threat to the environment

a The use of chemicals

Fertiliser, slurry and pesticides all contribute to the pollution of the environmental system. Fertiliser, in the form of mineral compounds which contain elements essential for plant growth, is widely used to produce a healthy crop and increase yields. If too much nitrogenous fertiliser or animal waste (manure) is added to the soil, some remains unabsorbed by the plants and may be leached to contaminate underground water supplies and rivers. Where chemical fertiliser accumulates in lakes and rivers, the water becomes enriched with nutrients (eutrophication) and the ecosystem is upset (Figure 16.50). In parts of north-west Europe, levels of nitrates in groundwater are above EU safety limits and over 80 per cent of lowland areas in the UK are said to be affected.

In Britain, the Water Authorities claim that slurry (farmyard effluent) is now the major pollutant of, and killer of life in, rivers. After several decades in which the quality of river water had improved, the last few years have seen levels of pollution again increasing, especially in farming areas.

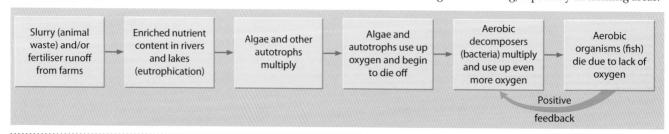

Figure 16.51

The case for and against hedgerows and ponds in a rural area

For	Against
Hedgerows	
Form part of the attractive, traditional British landscape	Are not traditional and were initially planted by farmers
Form a habitat for wildlife: birds, insects and plants (Large Blue butterfly is extinct, 10 other species are endangered)	Harbour pests and weeds
Act as windbreaks (and snowbreaks)	Costly and time-consuming to maintain
Roots bind soil together, reducing erosion by water and wind	Take up space which could be used for crops
	Limit size of field machinery (combine harvesters need an 8 m turning circle)
Ponds	
Form a habitat for wildlife: birds, fish and plants	Take up land that could be used more profitably
Add to the attractiveness of the natural environment	Stagnant water may harbour disease
i.e. Concern is environmental	**i.e. Concern is economic**

Pesticides and herbicides are applied to crops to control pests, diseases and weeds. Estimates suggest that, without pesticides, cereal yields would be reduced by 25 per cent after one year and 45 per cent after three. The Friends of the Earth claim that pesticides are injurious to health and, although there have been no human fatalities reported in Britain in the last 15 years, there are many incidents in developing countries resulting from a lack of instruction, fewer safety regulations and faulty equipment. A UN report claims that 25 million agricultural workers in developing countries (3 per cent of the total workforce) experience pesticide poisoning each year. Pesticides are blamed for the rapid decrease in Britain's bee and butterfly populations, and an up to 80 per cent reduction in 800 species of fauna in the Paris basin. Pesticides can dissipate in the air as vapour, in water as runoff, or in soil by leaching to the groundwater.

b The loss of natural habitats

The most emotive outcries against farmers have been at their clearances of hedges, ponds and wet-lands. These clearances mean a loss of habitat for wildlife and a destruction of ecosystems, some of which may have taken centuries to develop and, being fragile, may never recover or be replaced. As stated earlier, over 25 per cent of British hedge-rows were removed between 1949 and 1990 – in Norfolk, the figure was over 40 per cent. Figure 16.51 lists some of the arguments for and against the removal of hedgerows and the drainage of ponds/wetlands. Figures 16.49 and 16.52 show the contrast between a landscape with trees and hedges, and one where they have been removed.

Farming can increase soil erosion. The rate of erosion is determined by climate, topography, soil type and vegetation cover (Case Study 10), but it is accelerated by poor farming practices (overcropping and overgrazing) and deforestation. In Britain, wind erosion (Figures 7.8 and 10.34) tends to be restricted to parts of East Anglia and the Fens where the natural vegetation cover, including hedges, has been removed and where soils are light or peaty. Water erosion (page 62) is most likely to occur after periods of prolonged and heavy rainfall, on soils with less than 35 per cent clay content, in large and steeply sloping fields and where deep ploughing has exposed the soil.

Arable farming, especially when ploughing is done in the autumn, removes the protective vegetation cover, increasing surface runoff. The intensification of farming, and overcropping, in areas of highly erodible soils in the USA have led to a decrease in yields and an estimated loss of one-third of the country's topsoil – much of it from the Dust Bowl during the 1930s. Deforestation in tropical rainforests, mountainous and semi-arid areas – Brazil, Nepal and the Sahel, respectively – also accelerates soil erosion.

Figure 16.52

An agricultural land-scape without trees or hedges, Cambridgeshire

Figure 16.53

Salinisation in California

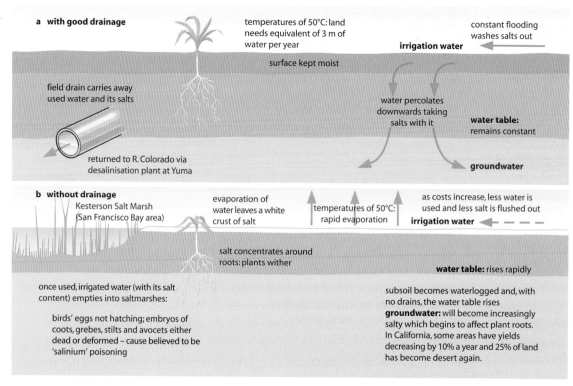

a with good drainage

temperatures of 50°C: land needs equivalent of 3 m of water per year

constant flooding washes salts out

irrigation water

surface kept moist

field drain carries away used water and its salts

water percolates downwards taking salts with it

water table: remains constant

returned to R. Colorado via desalinisation plant at Yuma

groundwater

b without drainage
Kesterson Salt Marsh (San Francisco Bay area)

evaporation of water leaves a white crust of salt

temperatures of 50°C: rapid evaporation

as costs increase, less water is used and less salt is flushed out

irrigation water

salt concentrates around roots: plants wither

once used, irrigated water (with its salt content) empties into saltmarshes:

birds' eggs not hatching; embryos of coots, grebes, stilts and avocets either dead or deformed – cause believed to be 'salinium' poisoning

water table: rises rapidly

subsoil becomes waterlogged and, with no drains, the water table rises
groundwater: will become increasingly salty which begins to affect plant roots. In California, some areas have yields decreasing by 10% a year and 25% of land has become desert again.

Irrigation (Places 73, page 490) also needs the surplus water to be drained away. Without this careful, and often expensive, management, the soil can become increasingly saline and waterlogged (Figure 16.53). As the water table rises it brings, through capillary action (page 261), dissolved salts into the topsoil. These affect the roots of crops, which are intolerant of salt, so that over a period of time they die. Where water is brought to the surface and then evaporates, a crust of salt is left on the surface and the area may revert to desert. To date, only rough estimates have been made of the amount of irrigated land now affected by salinisation, but figures suggest that it may be as high as 40 per cent in Pakistan and Egypt, and 30 per cent in California.

Attempts by farming to improve the environment

a Environmental improvement schemes
The EU and the British government introduced several schemes in which financial incentives were offered to farmers who tried to improve their environment, e.g. set-aside, woodland management and Environmentally Sensitive Area (ESA) schemes (page 493).

Many parts of Britain benefited from set-aside because, when this was in operation, soils that were left under either permanent or rotational fallow with its protective vegetation cover were given the time to improve their humus content, while other areas saw the restoration of ponds, wetlands and other wildlife habitats. The woodland management scheme increased the number of trees and small woods, while the Countryside Commission and the Nature Conservancy Council looked at areas where it was considered that farming landscapes were under threat from changing farming practices. These two parties originally looked at 46 'search' areas which targeted chalk and limestone grasslands, lowland heath, river valleys, coasts, uplands and historic landscapes. From these, 22 were eventually to be designated, at four different stages, as Environmentally Sensitive Areas (ESAs) because of 'their high landscape, wildlife or historic value' (Figure 16.54). Farmers living in ESAs were then invited to join the scheme at one of two levels: a lower level paid on condition that they maintained the existing landscape; and a higher level if they made environmental improvements such as replanting hedges or restoring ponds and traditional farm buildings. This, the Countryside Stewardship Scheme (CSS), was superseded in 2005 by the Environmental Stewardship Scheme (ESS).

Environmental Stewardship is a joint farming–environmental initiative that builds on the success of the former ESAs and Countryside Stewardship schemes. Its primary objectives are to:
- conserve wildlife (biodiversity)
- maintain and enhance landscape quality and character
- protect the historic environment and natural resources

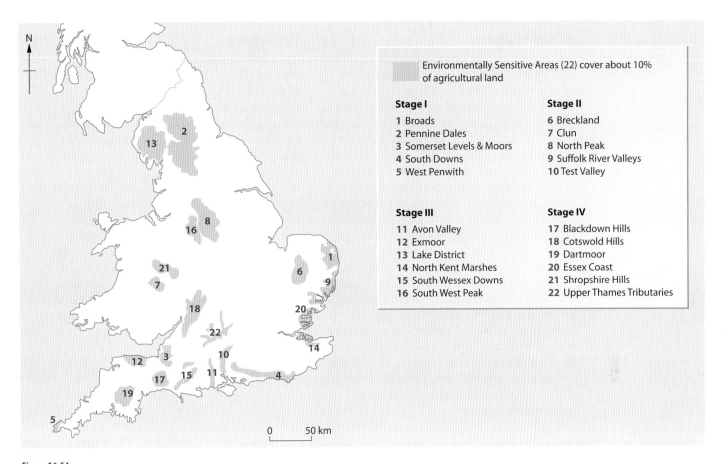

Figure 16.54

Environmentally Sensitive Areas (ESAs) in England and Wales

- promote public access and understanding of the countryside.

Within these primary objectives were the secondary aims of:

- genetic conservation and
- flood management.

There are three levels of stewardship:

- Entry Level Stewardship (ELS) is open to all farmers and landowners and provides a straightforward approach to land management for which payment is £30 per hectare.
- Organic Entry Level Stewardship (OELS) is similar but is geared to organic farming with payments of £60 per hectare (Case Study 16B).
- Higher Level Stewardship (HLS) is designed to build on the first two in that it aims to achieve a wide range of environmental benefits across the farm. As it also concentrates on more complex types of management where landowners need advice and support and where agreements need to be tailored to suit local circumstances, then payments are less rigid.

b Organic farming

Since the mid-1980s there has been a small but increasing number of farmers in Britain and elsewhere who have turned to organic farming (Case Study 16B). Organic farming aims to produce food of high nutrient quality by using management practices that avoid the use of agro-chemical inputs and which minimise damage to the environment and to wildlife. As such, it is both self-supporting and an example of sustainable development (Framework 16, page 499). For any food to qualify for the organic label it must adhere to a strict set of rules enforced by a regulatory body, such as the Soil Association.

Figure 16.55 describes both the advantages of organic farming and some of its problems. In the last few years, more British shoppers have been prepared to pay the higher prices asked for organic produce, believing it to be healthier than conventionally produced food. This, together with a greater range of organic brands, has persuaded the giant supermarkets that it is worth their while to stock organic products. However, it will be interesting to see if these shoppers continue to buy 'organic' at a time of rising global food prices and during the 'credit crunch'.

Figure 16.55

Advantages and problems of organic farming

Advantages

Compared with conventional farming, organic farming is self-sustaining in that it produces more energy than it consumes and it does not destroy itself by misusing soil and water resources (Framework 16, page 499). It rules out the use of artificial (chemical) fertiliser, herbicides and pesticides, favouring instead only animal and green manures (compost) and mineral fertilisers (rock salt, fish and bonemeal). These natural fertilisers put organic matter back into the soil, enabling it to retain more moisture during dry periods and allowing better drainage and aeration during wetter spells. Organic farming involves the intensive use of both land and labour. It is a mixed farming system which involves crop rotations and the use of fallow land. It is less likely to cause soil erosion or exhaustion as the soils contain more organic material (humus), earthworms and bacteria than soil in non-organic farms. It is also less likely to harm the environment as there will be no nitrate runoff (no eutrophication in rivers) and less loss of wildlife (no pesticides to kill butterflies and bees).

Problems

If organic farming replaces a conventional farming system, yields can drop considerably in the first two years, when artificial fertiliser is no longer used, although they soon rise again as the quality of the soil improves. Also, during the conversion period, farmers cannot market any goods as 'organic': they must wait until they meet the regulatory body's standards before receiving its label guaranteeing the authenticity of their produce. Weeds can increase without herbicides, and may have to be controlled by hand labour or by being covered with either mulch or polythene. This means that, although organic farming is helpful to the environment and, arguably, less harmful to human health, its produce is more expensive to buy. Producers, processors and importers must all be registered and are subject to regular inspections.

GM crops

The growing of genetically modified (GM) crops is an issue of global concern that has led to the extreme polarisation of opinions held by those in favour and those against. GM crops are a result of a deliberate attempt, using biotechnology, to alter the genetic make-up of a plant with the intention of increasing yields by making it resistant to either disease, pests or a climatic extreme such as drought. At present, nearly all the world's GM crops being grown are in the USA, Argentina, Canada and China (where the world's first GM crop was planted in 1992). In the USA, around 70 per cent of all packaged foods already contain GM material. Of about 40 million hectares of GM crops at present being grown worldwide, most are soya and maize (corn):

	1998	2007 (million ha)
Soya	14	20
Maize	8	12
Oilseed rape	3	4
Sugar beet	1	2
Potatoes	>1	1

The production of GM crops is dominated by several large transnational corporations. They claim that GM crops are essential in order to feed the world's growing population and to combat the rise in global food prices which, with the effects of climate change, they believe is the main cause of the increasing food shortages, especially in sub-Saharan Africa (Places 74 and 75). The TNCs claim that after 30 years of growing GM soya and maize in the USA, there appear to be no ill-effects either to people's health or to the environment, although recent reports suggest that, instead of improving yields of those crops, output has actually fallen by up to 10 per cent. The TNCs also suggest that, apart from reducing hunger, GM crops will reduce the use of weedkillers and insecticides and will provide both cheaper and higher-nutrient food.

But the production of GM crops is opposed by virtually all the main environmental groups, which claim that the crops remain untested and that such crops are not a solution to food shortages as, so far, being grown intensively in developed countries, they seem inappropriate to the needs and demands of up to 400 million subsistence smallholders in many of the world's poorest countries. The environmental groups claim that governments, including that of Britain, are being misled if they believe GM crops will end food shortages, as they neither increase yields nor tackle the fundamental problem of poverty. They also fear that pollen from GM crops is adversely affecting insect wildlife, especially bees and butterflies. This debate is far from over, with DEFRA claiming (2008) that 'while tests in Britain are continuing, no GM crop will be released if there is any doubt about its impact'.

The concept of sustainable development dominated the environmental agenda during the 1990s and, following the 1992 Earth Summit at Rio de Janeiro, has been embraced by governments at all levels of development. The term is not, however, easy to explain; Dobson, in 1996, claimed that there were over 300 different definitions and interpretations. Of these, the most widely used is that taken from the Brundtland Report (The World Commission on Environment and Development, 1987) which claims that sustainable development 'meets the needs of the present without compromising the ability of future generations to meet their own needs'. This definition, according to Munton and Collins (*Geography*, 1998), 'highlights the socio-economic rather than the environmental basis of sustainable development and, unlike earlier understandings of the term "environmental sustainability", it gives absolute primacy to improving human conditions and not to environmental limits'.

Put more simply, sustainable development should lead to an improvement in people's:

- quality of life, allowing them to become more content with their way of life and the environment in which they live
- standard of living, enabling them, and future generations, to become better off economically.

This may be achieved in a variety of ways:

- by encouraging economic development at a pace that a country can both afford and manage so as to avoid that country falling into debt
- by developing technology that is appropriate to the skills, wealth and needs of local people irrespective of the country's level of development, and developing local skills so that they may be handed down to future generations
- by using natural resources without spoiling the environment, developing materials that will use fewer resources, and using materials that will last for longer – ideally, once a resource is used, it should either be renewed, recycled or replaced.

Sustainable development needs careful planning and, increasingly as it involves a commitment to conservation, the co-operation of groups of countries and, under extreme conditions, global agreement.

Sustainable development is a theme that keeps re-appearing throughout this book. It is a concept that, from a geographer's point of view, can be studied:

- through a selection of physical and human environments
- at a variety of levels of development

in the context of people and food supply, resources, and natural and human created/adapted environments. Examples referred to in this book, with chapter numbers in brackets, include the following:

- **People and environments**
  - world biomes and fragile environments such as the tropical rainforest (11 and 12) and the tundra (5)
  - smaller-scale ecosystems including wetlands (16) and sand dunes and saltmarshes (6 and 11)
  - effects of economic development on scenic areas and the wildlife of coastal and mountainous areas (6, 17 and 20)
- **People and resources**
  - finite resources of fossil fuels (18) and minerals (17)
  - renewable resources, providing that they are carefully managed, including soils (10); fresh and reliable water supply (3 and 21); forests (11 and 17); crops and food supply (16); energy (18); recycled materials (19); and the atmosphere (9)
  - ecological footprint (13) and carbon credits (21)
- **Socio-economic**
  - population growth and family planning (13)
  - urban growth/loss of countryside (15)
  - housing materials (15 and 19)
  - development of skills and levels of education (21).

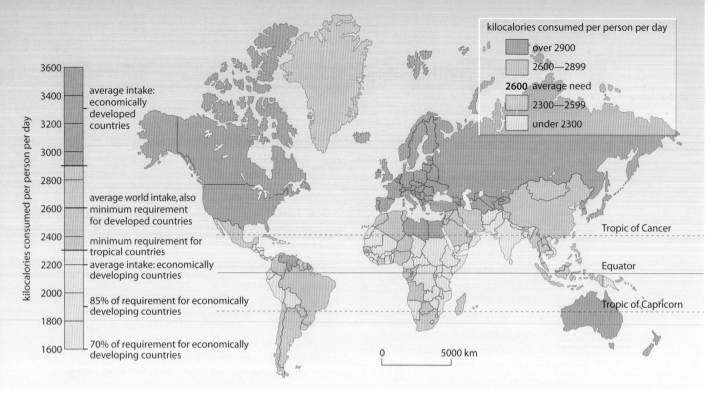

Food supplies

Diet and health

It is over 200 years since Malthus expressed his fears that world population would outstrip food supply (page 378). Today, despite assurances from various international bodies such as the Food and Agriculture Organisation (FAO) that there is still sufficient food for everyone, it is estimated that three-quarters of the world's population is inadequately fed, and that the majority of these live in less economically developed countries. The problem is, therefore, the unevenness in the distribution of food supplies: surpluses still exist in North America and the EU; and there are shortages in many developing countries.

This uneven distribution is reflected in Figure 16.56 which shows variations in kilocalorie intake throughout the world. Dieticians calculate that the average adult in temperate latitudes requires 2600 kilocalories a day, compared with 2300 kilocalories for someone living within the tropics. The FAO reports that the actual average intake for the economically more developed world is 3300 kilocalories, but only 2200 kilocalories in less developed countries. However, the quantity of food consumed is not always as important as the quality and balance of the diet. A good diet should contain different types of food to build and maintain the body, and to provide energy to allow the body to work. A balanced diet should contain:

- **proteins**, such as meat, eggs and milk, to build and renew body tissues
- **carbohydrates**, which include cereals, sugar, fats, meat and potatoes, to provide energy, and
- **vitamins** and **minerals**, as found in dairy produce, fruit, fish and vegetables, which prevent many diseases.

Malnutrition and undernutrition, often caused by poverty, affect many people including even a surprisingly high number in developed countries. Malnutrition may not be a primary cause of death, but by reducing the ability of the body to function properly, it reduces the capacity to work and means that people, and especially children, become less resistant to disease and more likely to fall ill. Nutritional diseases, which include rickets (vitamin D deficiency), beri-beri (vitamin B1 deficiency and common in rice-dependent China), kwashiorkor (protein deficiency) and marasmus (shortage of protein and calories), can reduce resistance to intestinal parasitic diseases, malaria and typhoid. In contrast, people in developed countries are at risk from over-eating and from an unbalanced diet which often contains too many animal fats which can cause heart disease. Malnutrition, a Millennium Development Goal (page 609), is believed to be the underlying cause for almost half of all child deaths worldwide. Figure 16.57 shows the proportion of children aged under 5 who are underweight. More than one-quarter of all under 5s living in the developing countries are underweight, about 143 million in total, with the highest levels in South Asia and sub-Saharan Africa. In these countries children living in rural areas are twice as likely to be underweight as those living in urban areas. There is no gender difference.

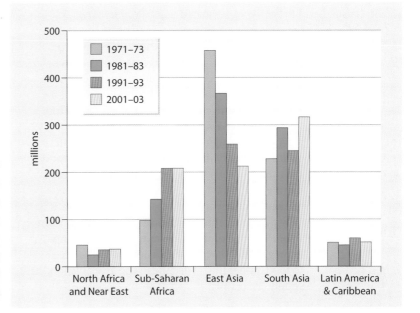

Figure 16.57

Percentage of children under 5 who are underweight

Trends in food supply

Since the early 1950s world food output has usually increased more rapidly than has world population. This increase has been more rapid in the developing countries, albeit from a very low base, than in developed countries and, during the 1960s and 1970s, was attributed mainly to the 'Green Revolution' (page 504). There were, however, exceptions to still this trend. The main exception was sub-Saharan Africa where, in several countries, food output per person actually fell and has continued to fall. A second exception was short-term periodic declines in either global or, more often, regional rainfall. During the late 1980s, for example, many places received below average totals which resulted in an estimated fall in the world's food reserves from 101 days to 54 days, and when up to 35 per cent of the world's total population was left living at or below starvation level. At that time there was much pessimism about future food prospects, and Malthus's gloomy predictions were, for a time, revived (page 378).

However, food production did once again begin to exceed the rate of population growth, this time in the 1990s, although there were disparities on a continental scale. While there was 5 kg of food per person for North America, 3.5 kg for Oceania and 2 kg for Western Europe, there was only 1 kg for Latin America and South-east Asia and less than 0.5 kg for Africa. Even so, at the beginning of the 21st century the WHO was able to report that 'new farming techniques are improving output, nutrition seems to be improving, life expectancy is increasing due to a better diet and global food supplies are in a relatively good shape with surpluses in certain areas'.

However, the WHO report did highlight several areas of considerable concern.

- There was a continued decline of food production in Africa (Places 75) and an inability of several of the countries located there to afford to buy sufficient to satisfy their shortages and, therefore, they were forced to rely increasingly on food aid (page 632).

- Although production was increasing in most regions, there had been a global decline in the yields of the three staple food crops of wheat, rice and maize.

- Throughout history, whenever extra food was needed, people simply cleared more land for crops. Today, most high-quality land is already in use, or, increasingly, has been built upon. Much of the remaining areas have soils that are less productive and more fragile, i.e. less sustainable (Framework 16, page 499).

- There is a lack of **food security** – a term used by the WHO which means a lack of nutritious food needed to keep people alive and healthy. Although numbers have dropped, some 20 per cent of the world's population are still thought to experience chronic undernutrition (Figure 16.58).

- There is increasing globalisation of food production, with transnational corporations and large supermarkets in developed countries sourcing more of their food from developing countries. Smallholders in less well-off countries are being drawn into contracts to supply fruit and vegetables to markets in the developed world which is resulting in a decline in the growth of staple foods for their own domestic consumption.

Famine

Famines were once considered to be an inevitable occurrence but increasingly they can be seen to result from human mismanagement of the environment, or localised wars leading to the displacement of people. The notion that famine means a total food shortage (as implied in the introductory quote to this chapter) has been challenged, as recent studies suggest that it only affects certain groups in society (the poorest, the least skilled and the rural dwellers). Even during the worst times of famine, some food still appears in local markets – but at a price beyond the reach of most people. It is now widely accepted that most famines result from a combination of natural events and human mismanagement together with a decline in the access to food, rather than a decline in the available food supply (Places 74).

Rising food prices

The year 2008 saw an unprecedented rise in global food prices which resulted from a combination of factors (e.g. wheat £70 per tonne in 2006, £180 per tonne in 2008). These factors included:

- a change in diet, especially in India and China where greater affluence has led to a rising demand for meat products which in turn means more grain is needed to feed the extra number of reared animals
- more land and more cereal crops being used to produce biofuels (e.g. one-third of the USA's maize crop) in an attempt to provide more renewable energy
- climate change causing more erratic rainfall patterns, e.g. drought in the cereal growing areas of Australia and northern India, floods in the American mid-west
- growth in the world's population
- the reduction of subsidies to American and EU farmers, meaning that less food is held in storage
- encouragement of developing countries to grow cash crops rather than cereal crops.

Places 74 Niger: famine

In 2005, Niger was on the brink of a famine with over 3 million of its inhabitants (one in every three) suffering from severe hunger. To many people living in remote rural areas – which is the majority of the population – the only food available until that year's crop was ready was a watery-looking porridge look-alike. The often quoted causes of famine and food shortages are poverty and overpopulation, but this is too simplistic an answer. While it was true that many people were, at that time, unable to afford what food was available from within the country, and Niger itself was too poor to buy much from other countries, especially at a time when global food prices were beginning to rise, the real cause of the threatened famine was a combination of environmental, economic, social, cultural and political factors.

Niger had experienced two natural disasters in 2004: drought (it is a Sahel country – Case Study 7 and page 280), and a locust infestation. These were exacerbated by social causes that included a growing population that needed to be fed and, within that population, a considerable unevenness in the distribution of wealth. The country's limited development is shown by a lack of technology in farming, in which most of the population is engaged. Also, there is a limited amount of land suitable for agriculture, and what there is lies on the fringes of the Sahara Desert and so is 'marginal' (zone 3 on Figure 16.1, and Figure 16.59), with nutrient-deficient soils and a lack of water (rain or irrigation).

The lack of development (notice its position at the foot of the HDI table on page 607) is partly due to a lack of resources (other than uranium), which means that with little to export, Niger has a balance of trade deficit (page 624) which places it on the wrong side of the development gap (page 605). With limited money for investment and a legacy of colonialism, the country lacks a basic infrastructure. This includes a poorly developed transport network, which makes it difficult to distribute food internally at times of shortage and limits links with the outside world (the only long-haul flight is to Paris); it also has a poorly developed banking system.

Figure 16.59

Niger

Severe drought, civil strife and economic security have displaced large numbers of people and disrupted food production (Figure 16.60). Food shortages at present affect 26 countries in sub-Saharan Africa (Figure 16.61).

The population of this region is growing faster than anywhere else in the world. With over 70 per cent of its labour force in agriculture and 66 per cent living in rural areas, the income, nutrition and health of most Africans is closely tied to farming. In an area where, due to limited capital and technology, the use of new seeds, fertiliser, pesticides, machinery and irrigation is the lowest in the world, agriculture is almost totally reliant upon an environment that is not naturally favourable. The soils often have fertility constraints, a low water-holding capacity and limited nutrients, making them vulnerable to erosion. High evapotranspiration rates harm crops while the unreliable rains which may cause flooding one year may then fail for several that follow. Periods of drought are getting longer and more frequent with experts arguing as to whether this is part of a natural climatic cycle, less moisture in the air due to deforestation, or the effects of global warming.

With increases in population, fallow periods have been reduced and the land has been overgrazed or overcropped which, together with destruction of trees for fuelwood, has accelerated soil erosion and desertification (Case Studies 7 and 10). The region has limited money for investment in agriculture and when overseas aid has been given it has often been channelled into unsuitable projects such as promoting monoculture, growing crops for export instead of domestic consumption, increasing the size of animal herds on marginal land and ploughing fragile soils that would have been better left under a protective vegetation cover.

Financial aid from overseas can also increase the debt of the recipient country (page 632). People's diet often lacks sufficient calories or protein and, with many living in extreme poverty (page 609), they cannot afford the inflated food prices at times of shortage. Animals may be attacked by tsetse fly, crops in the field by locusts, and crops in storage by rats and fungi. To add to these difficulties, several countries are, or have been recently, torn by civil strife resulting in the problem of internally displaced persons and refugees. This, together with administrative corruption, interrupts farming and the distribution of relief supplies. Last, but by no means least, is the effect of the HIV/AIDS pandemic which, even when not fatal to individual farmers, considerably reduces their ability to work (Places 100, page 623).

Figure 16.60

Children awaiting food aid: Somalia

Figure 16.61

Countries with exceptional shortfalls in food production supplies

Country	Reasons
Burundi–Rwanda–Uganda	Civil strife, IDPs
Central African and Congo Reps	IDPs
Chad	Civil strife, refugees
Congo Democratic Republic	Civil strife, refugees, IDPs
Eritrea–Somalia	Civil strife, drought, IDPs
Ethiopia	Drought, IDPs
Ghana	Flood and drought
Guinea	Civil strife, refugees
Kenya	Civil strife, drought, pests
Lesotho–Swaziland	Drought, HIV/AIDs
Liberia–Sierra Leone	Civil strife, refugees, IDPs
Malawi–Zambia	Drought, HIV/AIDs
Mali–Niger–Burkina Faso	Drought, locusts
Mauritania	Drought, locusts
Mozambique	Floods, drought, HIV/AIDs
Sudan	Civil strife, IDPs, drought
Zimbabwe	Economic crisis, HIV/AIDs

IDPs = internally displaced persons

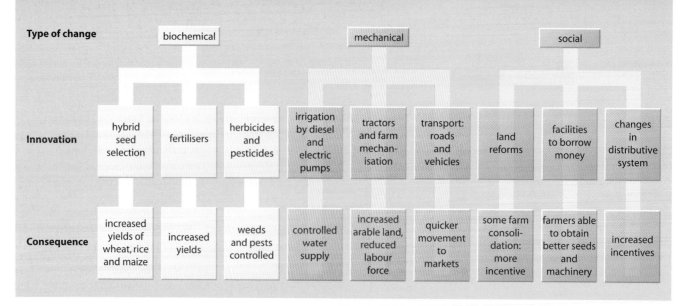

Type of change	biochemical			mechanical			social		
Innovation	hybrid seed selection	fertilisers	herbicides and pesticides	irrigation by diesel and electric pumps	tractors and farm mechan- isation	transport: roads and vehicles	land reforms	facilities to borrow money	changes in distributive system
Consequence	increased yields of wheat, rice and maize	increased yields	weeds and pests controlled	controlled water supply	increased arable land, reduced labour force	quicker movement to markets	some farm consoli- dation: more incentive	farmers able to obtain better seeds and machinery	increased incentives

Figure 16.62

The Green Revolution

What might be done to improve food supplies in developing countries?

As most areas with an average or high agricultural potential have already been used, future extension of cropland can only take place on marginal land where the threats of soil erosion and desertification are greatest. The solution is not, therefore, to extend the cultivated area but to make better use of those areas already farmed.

Land reform can help to overcome some inefficiencies in the use of land and labour. The redistribution of land has been tackled by such methods as the expropriation of large estates and plantations and distributing the land to individual farmers, landless labourers or communal groups; the consolidation of small, fragmented farms; increasing security of tenure; attempting new land colonisation projects; and state ownership. The success of these schemes has been mixed. Not all have increased food production, although many farms in China have seen an increase in yields since the transference of farming decisions to individual farmers under the responsibility system (Places 63, page 468; Places 64, page 470).

The **Green Revolution** refers to the application of modern, Western-type farming techniques to developing countries. Its beginnings were in Mexico when, in the two decades after the Second World War, new varieties or hybrids of wheat and maize were developed in an attempt to solve the country's domestic food problem. The new strains of wheat produced dwarf plants capable of withstanding strong winds, heavy rain and diseases (especially the 'rusts' which had attacked large areas). Yields of wheat and maize tripled and doubled respectively, and the new seeds were taken to the Indian subcontinent. Later, new varieties of improved rice were developed in the Philippines.

The most famous, the IR-8 variety, increased yields sixfold at its first harvest. Another 'super rice' increased yields by a further 25 per cent (1994). Further improvements have shortened the growing season required, allowing an extra rice crop to be grown, and new strains have been developed that are tolerant of a less than optimum climate.

In 1964 many farmers in India were short of food, lacked a balanced diet and had an extremely low standard of living. The government, with limited resources, was faced with the choice (Figure 16.62) of attempting a land reform programme (redistributing land to landless farmers) or trying to improve farm technology. It opted for the latter. Some 18 000 tonnes of Mexican HYV (high-yielding varieties) wheat seeds and large amounts of fertiliser were imported. Tractors were introduced in the hope that they would replace water buffalo; communications were improved; and there was some land consolidation. The successes and failures of the Green Revolution in India are summarised in Figure 16.63. In general, it has improved food supplies in many parts of the country, but it has also created adverse social, environmental and political conditions. The question now being asked in India is: 'How green was the Green Revolution?' For the first time in four decades, population growth is outstripping food production. This is due to high birth rates, longer life expectancy, more land being devoted to commercial crops and a mass rural–urban migration caused by India's rapidly emerging economy. At the same time there are growing health concerns with fertiliser and pesticides, leached into water supplies, blamed for a rapid increase in cancers, birth defects and other illnesses. A small but growing number of farmers are turning away from a reliance on chemicals to a more organic-type of farming.

Successes	Failures
Wheat and rice yields have doubled	HYV seeds need heavy application of fertiliser and pesticides, which has increased costs, encouraged weed growth and polluted water supplies
Often an extra crop per year	Extra irrigation is not always possible; it can cause salinisation and a falling water table
Rice, wheat and maize have varied the diet	HYVs not suited to waterlogged soils
Dwarf plants can withstand heavy rain and wind and photosynthesise more easily	Farmers unable to afford tractors, seed and fertiliser have become relatively poorer
Farmers able to afford tractors, seed and fertiliser now have a higher standard of living	Farmers with less than 1 ha of land have usually become poorer
Farmers with more than 1 ha of land have usually become more wealthy	Farmers who have to borrow are likely to get into debt
The need for fertiliser has created new industries and local jobs	Still only a few tractors, partly due to cost and shortage of fuel
Some road improvements	Mechanisation has increased rural unemployment
Area under irrigation has increased	Some HYV crops are less palatable to eat
Some land consolidation	Fertiliser and pesticides have contaminated water supplies causing health problems
Conclusions	
A production and economic success which has lessened but not eliminated the threat of food shortages	Social, environmental and political failure: bigger gap between rich and poor

Figure 16.63

An appraisal of the Green Revolution in the Indian sub-continent

Appropriate technology (Case Study 18) is needed to replace the many, often well-intentioned schemes that involved importing capital and technology from the more developed countries. Appropriate technology, often funded by non-governmental organisations such as the British-based Practical Action (Places 90, page 577), seeks to develop small-scale, sustainable projects which are appropriate to the local climate and environment, and the wealth, skills and needs of local people. This means:

- *Not* large dams and irrigation schemes, but more wells so that people do not migrate to the few existing ones, drip irrigation as this wastes less water, stone lines (Figures 10.40 and 16.64) and check-dams (Figure 10.43). For stone lines, stones are laid down, following the contours, even on gentle slopes in Burkina Faso, while small dams built of loess are constructed across gulleys in northern China. In both cases, surface

runoff is trapped giving water time to infiltrate into the soil and allowing silt to be deposited behind the barriers. These simple methods, taking up only 5 per cent of farmland, have increased crop yields by over 50 per cent.
- *Not* chemical fertiliser, but cheaper organic fertiliser from local animals (which can also provide meat and milk in the diet). Unfortunately, in many parts of Africa dung is needed as fuel instead of being returned to the fields.
- *Not* tractors, but simple, reliable, agricultural tools made, and maintained, locally.
- *Not* cash crops (often monoculture) on large estates, but smallholdings where both cash crops (income) and subsistence crops (food supply) can be grown. Mixed farming and crop rotation are less likely to cause soil erosion and exhaustion. Intercropping can protect crops and increase yields (smaller plants protected by tree crops).

Figure 16.64

Stone lines in Burkina Faso

Farming

A Farming in western Normandy

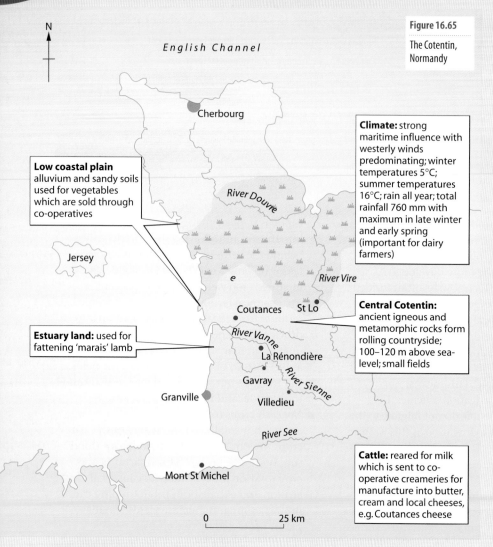

Figure 16.65

The Cotentin, Normandy

English Channel

Cherbourg

Low coastal plain alluvium and sandy soils used for vegetables which are sold through co-operatives

River Douvre

Jersey

e

River Vire

Coutances   St Lo

Climate: strong maritime influence with westerly winds predominating; winter temperatures 5°C; summer temperatures 16°C; rain all year; total rainfall 760 mm with maximum in late winter and early spring (important for dairy farmers)

Estuary land: used for fattening 'marais' lamb

River Vanne

La Rénondière

Central Cotentin: ancient igneous and metamorphic rocks form rolling countryside; 100–120 m above sea-level; small fields

Gavray

River Sienne

Granville

Villedieu

River See

Mont St Michel

Cattle: reared for milk which is sent to co-operative creameries for manufacture into butter, cream and local cheeses, e.g. Coutances cheese

0   25 km

The Cotentin lies between the Vire estuary and Mont St Michel Bay (Figure16.65). It is mainly an agricultural region, although tourism is also important. The maritime climate, with rain (760 mm per year) occurring at all seasons and reaching a maximum in the late winter and spring months, is important for the farming. The maximum occurs just as temperatures are rising and the grass is starting to grow. This has been the basis of the successful dairy farming industry. Cattle are reared for their milk from which Normandy butter is made in addition to many local cheeses and cream. Most farms also produce fodder for their cattle, either in the form of silage in the late spring or as crops of corn in the late summer.

La Rénondière is a typical Cotentin dairy farm (Figure 16.66). It lies at 71 m above sea-level in a small valley whose stream flows into the River Vanne 0.75 km to the north. The land slopes very gently; fields are small and bounded by dense hedges; and most of the farm can be ploughed except for a small area in the valley bottom which becomes very wet. The Normandy-style farmhouse of grey stone covered in creeper, with white shutters, faces south. It is sheltered from the westerly winds, as are most of the buildings grouped around it.

The farm is 44 ha in area. This is large for Normandy, where the average size is between 15 and 24 ha. Cattle are kept on 4 ha close to the farm; the rest of the

land is used for producing fodder for the animals. The present herd consists of 52 cattle – mainly Friesian, with some traditional Normandy cows. The black-and-white Friesians have high milk yields, but the Normandy cows have better-quality milk with a high cream content. They are kept outdoors all year round, with some protection in the winter. The cattle in milk are brought to the dairy twice a day and they produce on average 116 litres per cow per day (Figure 16.67). The small milking parlour is similar to many in the region. It holds eight cows at a time, and is simpler than large dairies in the English Midlands or on dairy farms close to Paris. The milk is kept under refrigeration on the farm until it is collected by the creamery lorry – each day in summer, but every two days at other times of the year (Figure 16.68).

The cows are artificially inseminated and produce one calf a year. Bull calves are sold in Gavray market for veal, and female calves are sold or used to replenish the herd. They are carefully checked for yield and as this drops off they are replaced. They are kept as long as possible, as the return from cull cows is not high.

The present farmer has been on the farm for over 20 years, but it was farmed earlier by his parents and grandparents. All the work is done by the farmer, his wife (she is in charge of the dairy) and his father. Neighbours help during silage making. There is a strong tradition of dairy farming in the region.

On the western side of the Cotentin, there is a low-lying plain approximately 15–60 m above sea-level. It contains areas of sandy soils which are important for producing vegetables, including carrots, leeks, sweet corn, lettuce and tomatoes. These vegetables are marketed through co-operatives in the larger towns of the region, as well as in Paris and the UK.

The lowlands along the estuary of the Sienne and the Vanne are used as grazing land for the 'marais lamb'; large flocks of sheep are fattened on the marshes, providing yet another income for the farmers of the region.

As income from farming declines, farmers across the EU are having to diversify. In

Figure 16.66

A typical Normandy farmhouse

Figure 16.67

A small milking parlour

Figure 16.68

A co-operative creamery in Normandy

addition to their regular enterprises, many Normandy farmers breed and train trotting ponies – making regular visits to the long open sandy beaches to train them at low tide. As in Britain, bed and breakfast accommodation during the short tourist season from June to the end of August provides an additional source of income.

A major issue facing farmers in this part of France is the steady loss of people from the land. Many small farmers are going out of business, leaving houses empty. As in other peripheral regions of Europe, young people are moving to the cities. There is evidence that one or two wealthier large farmers are buying up vacant land. Some of the villages contain summer homes, owned by Parisians, with a number of British residents both in holiday and permanent homes. Prices for some houses without land have been low, encouraging overseas buyers. Villages still contain their bakery and shop, often with a butcher, but children are being forced to travel increasing distances to school. These features of rural life are common to many remoter areas within the EU.

The impact of EU regulations can be seen. Milk quotas in line with EU rulings have been set by the government (page 493). They are generally higher than in the UK, perhaps due to the political strength of the farmers, and are an established part of the farm economy. However, they are generally unpopular with local farmers. Perhaps they will not be too disappointed when milk quotas are phased out by 2015 (page 493).

Subsidies for lamb encourage the producer to maintain flocks. Demand for lamb is high, as is shown by the high prices in the supermarkets.

From 1988, EU farmers were paid subsidies if they left parts of their land uncropped. Payments for this set-aside land ended in 2008 when the rise in global food prices forced the EU to encourage farmers to bring back into production former crop-growing areas and to introduce new policies by which farmers will only get subsidies if they keep their land in good condition – the so-called 'health check' (page 493).

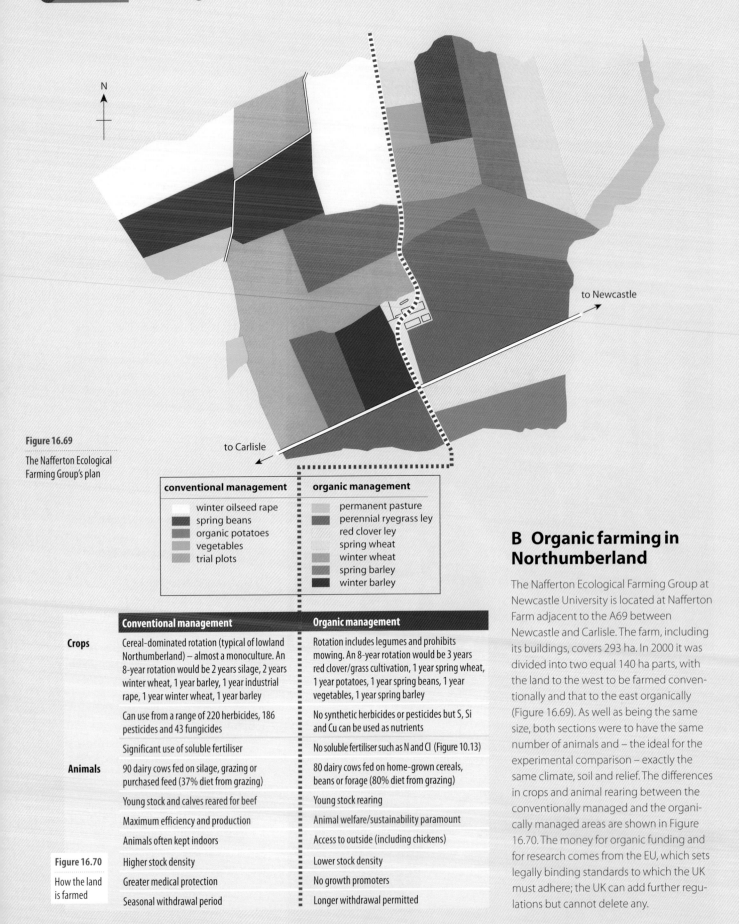

Figure 16.69

The Nafferton Ecological Farming Group's plan

conventional management	organic management
winter oilseed rape	permanent pasture
spring beans	perennial ryegrass ley
organic potatoes	red clover ley
vegetables	spring wheat
trial plots	winter wheat
	spring barley
	winter barley

N

to Newcastle

to Carlisle

Figure 16.70

How the land is farmed

	Conventional management	Organic management
Crops	Cereal-dominated rotation (typical of lowland Northumberland) – almost a monoculture. An 8-year rotation would be 2 years silage, 2 years winter wheat, 1 year barley, 1 year industrial rape, 1 year winter wheat, 1 year barley	Rotation includes legumes and prohibits mowing. An 8-year rotation would be 3 years red clover/grass cultivation, 1 year spring wheat, 1 year potatoes, 1 year spring beans, 1 year vegetables, 1 year spring barley
	Can use from a range of 220 herbicides, 186 pesticides and 43 fungicides	No synthetic herbicides or pesticides but S, Si and Cu can be used as nutrients
	Significant use of soluble fertiliser	No soluble fertiliser such as N and Cl (Figure 10.13)
Animals	90 dairy cows fed on silage, grazing or purchased feed (37% diet from grazing)	80 dairy cows fed on home-grown cereals, beans or forage (80% diet from grazing)
	Young stock and calves reared for beef	Young stock rearing
	Maximum efficiency and production	Animal welfare/sustainability paramount
	Animals often kept indoors	Access to outside (including chickens)
	Higher stock density	Lower stock density
	Greater medical protection	No growth promoters
	Seasonal withdrawal period	Longer withdrawal permitted

B Organic farming in Northumberland

The Nafferton Ecological Farming Group at Newcastle University is located at Nafferton Farm adjacent to the A69 between Newcastle and Carlisle. The farm, including its buildings, covers 293 ha. In 2000 it was divided into two equal 140 ha parts, with the land to the west to be farmed conventionally and that to the east organically (Figure 16.69). As well as being the same size, both sections were to have the same number of animals and – the ideal for the experimental comparison – exactly the same climate, soil and relief. The differences in crops and animal rearing between the conventionally managed and the organically managed areas are shown in Figure 16.70. The money for organic funding and for research comes from the EU, which sets legally binding standards to which the UK must adhere; the UK can add further regulations but cannot delete any.

The Ecological Farming Group researches the effects of soil, crop and livestock management on food quality and safety, environmental impact, soil health and biological activity, biodiversity, and the economic viability of the two types of farming system. It has confirmed that the organic management area:

- by using less fertiliser, produces less CO_2 and has a smaller ecological footprint (page 379)
- by using less nitrogen, reduces eutrophication (page 494)
- by using compost to bind the soil together, reduces soil erosion (page 495)
- has a greater biological activity (e.g. earthworms)
- despite not adding fertiliser, which increases crop yields, has outputs similar to those of conventional methods of the 1980s
- produces milk that is both better in quality and healthier than that produced conventionally (Figure 16.71).

Being a commercial venture, what the research centre actually grows can be influenced by market demand – so long as this demand fits into the rotation system. For example, if the market price for wheat increases, then more wheat might be planted that year. The centre does sell some of its own produce but cereals are sent to a grain merchant for processing before being sent to shops and supermarkets.

Figure 16.72

Organically grown cabbages are smaller but less affected by disease than those grown conventionally

Figure 16.73

Trial beds showing crops with fertiliser added (darker green) and vegetables being grown with and without protective netting

Organic milk has more healthy benefits

A study of organic milk, conducted by Professor Carlo Leifert of Newcastle University, has shown that drinking organic milk has greater health benefits than drinking normal milk. The study showed that organic milk contained 67 per cent more antioxidants and vitamins than ordinary milk and 60 per cent more of a healthy fatty acid called conjugated linoleic acid (CLA9) which tests have shown can shrink tumours. Similar levels of vaccenic acid, which has been shown to cut the risk of heart disease, diabetes and obesity, were also found as was an extra 39 per cent of the fatty acid Omega-3 which has also been shown to cut the risk of heart disease.

Gillian Butler, the livestock project manager, pointed out the health benefits even if consumers did not switch completely to organic milk. She pointed out that organic milk is more expensive to produce, as you get less milk per unit of land, and to buy, but because it is higher in all these beneficial compounds you do not need to buy as much to get health benefits.

Adapted from *Daily Telegraph*, 28 May 2008

Figure 16.71

Findings on organic milk

C Banana cultivation in South and Central America

Bananas are the main fruit in international trade and the most edible in the world. In terms of volume they are the first export fruit while in value they rank second after citrus fruits. The banana industry is a very important

Countries	% total
a World producers	
India	23
Brazil	9
Ecuador	9
China	8
Philippines	8
Rest of world	43
Countries	**% total**
b World exporters	
Ecuador	29
Costa Rica	14
Philippines	12
Colombia	10
Guatemala	6
Rest of world	29

Figure 16.74

World producers and exporters of bananas

source of income, employment and export earnings for several major exporting countries, mainly in Latin America and the Caribbean as well as in Asia and Africa.

Over half the world's bananas are grown in just five countries (Figure 16.74a) and 98 per cent in developing countries. Despite this, only one in five bananas enters the export market and of these 70 per cent come from five countries (Figure 16.74b). Although they are the major export of Ecuador and Costa Rica, the highest levels of dependence can be found in the Windward Islands of St Lucia (50 per cent of its exports), St Vincent and the Grenadines, Dominica and Grenada.

World trade in bananas is dominated by two groups of producers, the ACP (Africa, Caribbean and Pacific) producers and the 'dollar producers' of the Central American republics Colombia and Ecuador (controlled by large American transnationals). Over 80 per cent of bananas entering the EU come from the Caribbean where they are grown on small family-owned farms by people who are almost totally reliant on this single crop as a source of income (Figure 16.75). Bananas are grown on plantations in the Ivory Coast and Cameroon which are also members of ACP. Each country is given a quota based on

the amount it exports. In 1998, bananas were at the centre of a major trade dispute between the EU and the USA.

Bananas are cultivated under tropical conditions where the temperatures are high and rainfall exceeds 120 mm per month. In some tropical plantation conditions where evapotranspiration is high, irrigation may be used. Drip irrigation is more effective and produces a better bunch weight of bananas than basin irrigation. In order to meet the demands of the marketing companies, the bunches (or hands) of bananas must be over 270 g in weight. Bananas grown for local consumption are mainly cultivated on small landholdings, whilst those produced for export are grown on large plantations (Figure 16.74a). In most Caribbean countries, bananas are grown on small family-run plots. The crop requires a high labour input, which in the Caribbean islands is mainly provided by the smallholder's family. Suckers taken from a mother plant are rooted and grow well in the deep volcanic soils. Weeds growing between the plants need to be kept down until the plant is tall enough to outgrow them. It is common to see plants being supported by props so that the weight of the bunch does not pull the plant over. Fruit has to be protected from bruising and scarring. Each bunch may be covered by a large plastic bag until it is ready for harvest. This takes place about 10 months after the plant is established. The fruit is cut when it is still green and hard, and then it is taken to the processing plant. Here it is packed and refrigerated before being sold or shipped overseas (Figure 16.74b).

On the Caribbean islands marketing is done through transnationals such as Fyffes. The small farmers rely on the banana industry to provide their basic needs of food, shelter and education.

These small-scale farmers are also the ones who suffer most from hurricane damage as in 1998 when Hurricane Mitch destroyed much of the plantation area of Nicaragua and Honduras, and in 2005 when Grenada's crop was devastated (Figure 16.75).

Figure 16.75

Banana production in the Caribbean and Central America, 2005

Country	Production (tonnes)	Export (tonnes)	% total exported	Export (value £'000s)
Belize	76 000	64 891	85	21 353
Colombia	1 764 501	1 621 746	92	464 959
Costa Rica	1 875 000	1 775 519	95	483 492
Dominica	16 000	12 732	80	6 800
Dominican Republic	547 433	163 510	29	44 640
Ecuador	6 118 425	4 764 193	78	1 068 659
Grenada*	0	0	0	0
Guatemala	1 150 200	1 129 477	98	238 100
Honduras	887 072	545 527	61	134 698
Jamaica	125 000	11 713	93	4 693
Mexico	2 250 041	70 166	31	25 342
Nicaragua	49 915	45 532	91	11 579
Panama	439 228	352 480	80	96 517
St Lucia	45 000	30 630	68	15 542
St Vincent & Grenadines	50 000	24 470	55	12 815
Trinidad & Tobago	7000	39	6	23
World total	69 644 923	15 946 146	23	5 651 321

*Grenada lost all its crop in 2005 through hurricane damage

The influence of the large transnational companies is strong in the Central American countries where the bananas are grown on the rich alluvial soils found on the coastal lowlands, providing high yields per hectare for large plantations owned by transnationals. Labour is hired and often low-paid. Land is carefully cultivated and more mechanisation is used than on smaller farms. There is intensive use of fertiliser and pesticides which is having cumulative environmental effects. One of the most serious of these is the damage to the coral reefs off the Costa Rican coast, where 90 per cent are now dead as a result of pesticide runoff from banana plantations.

Bananas were to become one of the first products to be traded internationally under the Fairtrade label (Figure 21.44) and also, in places, to be grown organically. Under Fairtrade, farmers in South and Central America are getting a fairer price for their produce, enabling them to improve their standard of living (Figures 16.77b and 21.45).

Following years of expansion because of increased demand for the fruit, there is now a problem of oversupply. Economies such as those of St Vincent and St Lucia depend on the crop for survival. There is a need to diversify into food crops and other cash crops to reduce the dependency on one major export.

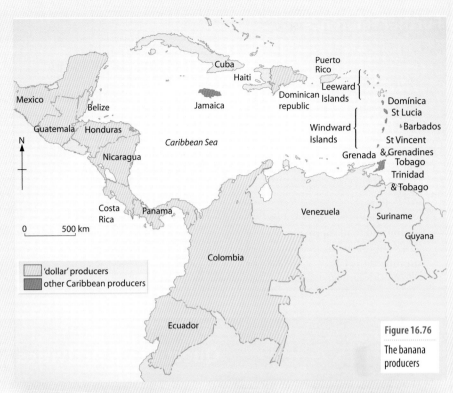

'dollar' producers
other Caribbean producers

Figure 16.76

The banana producers

Figure 16.77

The banana industry:
a Bananas on the tree
b Preparing Fairtrade bananas for export

Further reference

Barke, M. and O'Hare, G. (1991) *The Third World*, Oliver & Boyd.

Gee, N. (2005) 'Farm diversification', *Geography Review* Vol 19 No.2 (November).

O'Riordan, T. (2007) 'Agriculture and the environment', *Geography Review* Vol 21 No 11 (September).

Timberlake, L. (1987) *Only One Earth*, Earthscan/BBC Books.

CAP Policy:
www.sustainweb.org/news.php?id=93

Famine and food supply:
www.ifpri.cgiar.org

Farming in the UK:
www.defra.gov.uk/environment/statistics

Sustainable development:
www.defra.gov.uk/sustainable/government/

UK Department for Environment, Food and Rural Affairs (DEFRA):
www.defra.gov.uk/

Union of Concerned Scientists (UCS):
www.ucsusa.org/global_warming/

UN Food and Agriculture Organisation (FAO):
www.fao.org/

UN FAO Compendium of Food and Agriculture Indicators: (searchable by country)
www.fao.org/ES/ess/compendium_2006/list.asp

UN FAO Statistics, land/agriculture: (searchable by country or region)
http://faostat.fao.org/site/377/default.aspx#ancor

UN World Food Programme (WFP):
www.wfp.org/english/

US Department of Agriculture (USDA):
www.usda.gov/

World Resources Institute: Feeding the World:
www.igc.org/wri/wri/wri/wr-98-99/feeding.htm

Questions & Activities

Activities

1 Study the map in Figure 16.78. It shows the general pattern of intensity of farming in Europe.

   a i Describe the location of the areas where average intensity of farming is 75 per cent of the average, or lower. *(2 marks)*

   ii Choose a named location within the area described in **i** and explain why physical geography makes farming difficult in that area. *(3 marks)*

   b i Describe the location of the area with average intensity 50 per cent or more above average. *(2 marks)*

   ii Explain how market forces have affected the development of the area of intensive farming you have described in **i**. *(4 marks)*

   c Name **one** area of intensive farming that is found within the peripheral area of Europe.

   i Describe the type of farming.

   ii Explain why this area of intensive farming has developed there. *(7 marks)*

   d Name **one** area of low-intensity farming found within the farming core.

   i Describe the type of farming.

   ii Explain why this area of low-intensity farming has developed, despite the favourable market conditions. *(7 marks)*

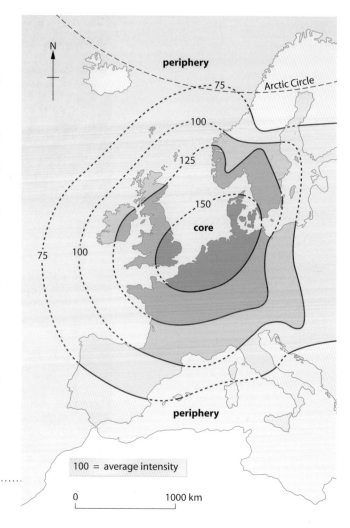

Figure 16.78

Intensity of agriculture in Europe (*after* van Valkenburg and Held, 1952)

2 Study Figure 16.79.

a Complete a copy of the table below. *(4 marks)*

b Moorland and woodland both produce low returns for farmers.

    i Using information from your table, suggest what is the main physical type of land in this sample that is left as:
- moorland
- woodland. *(2 marks)*

   ii Suggest why each of these types of land is not used for a type of farming that produces better returns. *(6 marks)*

c i What do the following terms mean:
- extensive farming
- capital-intensive farming
- labour-intensive farming? *(3 marks)*

   ii Name **one** area where capital-intensive farming has developed. Explain how market conditions in that area have encouraged the development of this type of farming. *(5 marks)*

  iii Name one area where labour-intensive farming with low capital inputs has developed. Explain how physical and social conditions have encouraged the development of this type of farming. *(5 marks)*

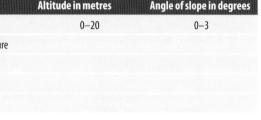

	Altitude in metres	Angle of slope in degrees
Arable	0–20	0–3
Improved pasture		
Rough pasture		
Woodland		
Moorland		

Figure 16.79

Relationships between land use, altitude and slope in south-east Arran

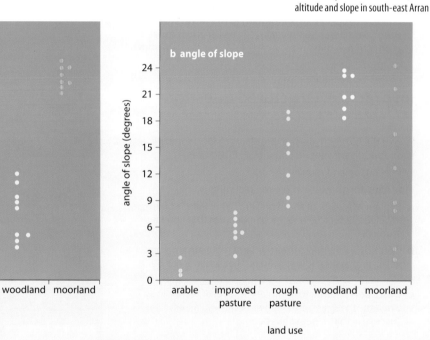

3 **a** Modern farming practices threaten the environment in many ways. Describe one problem that can result from each of the following practices:

    i increasing use of chemicals on the land *(4 marks)*

   ii increasing the size of fields *(4 marks)*

  iii draining wetlands. *(4 marks)*

b Choose **one** of the problems that you described in **a**. Explain how changes in the management of the land can reduce this problem. *(6 marks)*

c 'I would like to manage my farm in a more eco-friendly way, but I feel that I must farm as intensively as modern scientific techniques will allow. Farmers like me must produce maximum possible yields in order to feed the starving millions in poor countries throughout the world.'

Imagine that a farmer who ran a very intensive farm in East Anglia made the statement above. How might you reply if you wanted to convince him that he ought to consider a less intensive form of farming? *(7 marks)*

Exam practice: basic structured questions

4 a Physical controls have an important effect on the type of farming in most agricultural areas. Choose **two** of the following physical factors. For each of your chosen factors, explain how it influences farming. Illustrate each part of your answer with reference to a named area.

       i temperature
       ii precipitation
       iii soil. *(8 marks)*

   b The use of technology can reduce the farmer's dependence on physical factors. Explain how this has happened in:

       i a named farming region in a more economically developed country *(5 marks)*
       ii a less economically developed country where intermediate technology has been used. *(5 marks)*

   c Explain what is meant by 'organic farming' and explain why it has grown in importance in recent years. Illustrate your answer by reference to one or more case studies. *(7 marks)*

5 a Name **one** region where commercial grain production makes an important contribution to the world's food supply and describe the main features of agricultural production in that area. *(7 marks)*

   b Name **one** region where farming mainly for subsistence is still important. Outline the main features of the farming system and explain why subsistence farming is still important there. *(8 marks)*

   c With reference to **one or more** crops, discuss the strengths and weaknesses of the plantation system of agriculture. *(10 marks)*

6 a Name a less economically developed country (LEDC) that has suffered / is suffering from famine. Explain the causes of the famine. You should refer to both natural and human causes. *(10 marks)*

   b 'Famine and food shortage are likely to increase in future.' Give two reasons why this is likely. *(5 marks)*

   c i With reference to one or more named case studies, explain how land reform can improve total food production in LEDCs. *(5 marks)*

       ii With reference to one or more case studies, explain how appropriate technology (intermediate technology) can help increase agricultural yields in LEDCs. *(5 marks)*

Exam practice: structured questions

7 Two of the biggest causes of problems of food supply in less economically developed countries (LEDCs) are:

   • the need for land reform

   • the need for access to improved technology.

   a Explain why each of these presents problems for farmers in LEDCs. Refer to one or more examples that you have studied. *(9 marks)*

   b Describe a scheme to improve land tenure in a named LEDC, and assess how successful that scheme has been. *(8 marks)*

   c Describe a scheme to improve the level of technology available to farmers in a named LEDC, and assess how successful that scheme has been. *(8 marks)*

8 a i Outline three of the basic aims of the European Union's (EU) Common Agricultural Policy (CAP). *(3 marks)*

       ii Why did the CAP lead to overproduction and surpluses in the 1980 and 1990s? *(5 marks)*

   b Recent reforms of the CAP have led to the introduction of a number of schemes that are designed to improve the rural environment.

       i Describe the policy of 'set-aside' and explain its role in improving the environment. *(5 marks)*

       ii Explain how improvements in the rural environment in the UK can be brought about by **either** the introduction of Environmentally Sensitive Areas (ESAs) **or** Stewardship schemes. *(12 marks)*

9 Study the photographs in Figure 16.80. They were both taken near Guilin in China.

Figure 16.80

Agriculture in the Li Valley, near Guilin

a i Describe evidence in photograph B which shows that farming is intensive in this area. *(4 marks)*

ii Two crops per year can be taken from farmland in photograph A. Suggest how the land is kept fertile, even though the people cannot afford inputs of artificial fertiliser. *(4 marks)*

iii Land in the background of photograph A is not farmed. Suggest why not. *(4 marks)*

b • Before the revolution in 1949, farming in this part of China was mostly subsistence farming. Farms were small and fragmented and tenants had to give up to half their produce to absentee landlords.

• After the revolution, land was divided amongst the peasants, but most plots were too small to support the families who worked them.

• After several experiments the government created 'people's communes' in which around 15 000 people pooled their land and labour to run the farm.

• Since 1979 individual farmers have been given more responsibility, and now they are allowed to sell surplus crops at local markets, and to keep the profits.

Suggest why yields are higher under the present system than they have been under any of the previous systems. *(13 marks)*

10 a On a copy of Figure 16.81 add:

i net profit curves for dairying and wheat when locational rent for:
• dairying is £120 at the market and £0 at 60 km
• wheat is £80 at the market and £0 at 80 km. *(6 marks)*

ii labels to show:
• the margin of transference from market gardening to dairying
• the margin of transference from dairying to wheat
• the margin of cultivation for wheat. *(3 marks)*

b Explain why land use changes at the margins of transference. *(4 marks)*

c i Explain why von Thünen's model is difficult to apply to agricultural patterns in the modern world.

ii In what ways is von Thünen's model still useful to an understanding of modern agricultural geography? *(12 marks)*

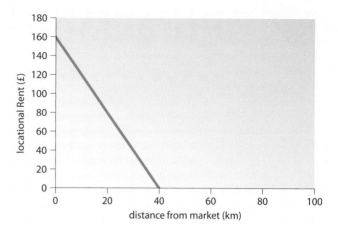

Figure 16.81

Net profit curve for market gardening around a town on a uniform plain

11 Study Figure 16.82.

a Explain how the Common Agricultural Policy of the European Community (now the European Union) led to the development of surpluses like those shown in the table. *(6 marks)*

b Explain how these surpluses were reduced during the period from 1986. *(7 marks)*

c Increasing intensification of farming in the UK and other parts of the European Union has damaged the environment in several ways.

Evaluate methods that have been introduced by the EU and the UK government to encourage the sustainable development of farming. *(12 marks)*

Commodity	January 1986	January 1992
	(figures in thousand tonnes unless otherwise stated)	
Butter	1400	300
Skimmed mild powder	800	0
Beef	500	800
Cereals	15 000	7 000
Wine/alcohol	4 000 (hectolitres)	2 500 (hectolitres)

Figure 16.82

EU food surpluses

Exam practice: essays

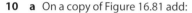

12 'Modern agri-business is not a sustainable form of farming.'
Discuss this statement using the following headings:
• What is the nature of modern agri-business?
• Is modern agri-business sustainable?
• Can agri-business be made less damaging to the environment? *(25 marks)*

13 'As farming becomes more modernised the influence of economic factors increases while the influence of physical factors decreases.'
Discuss this statement with reference to farming in regions at varying levels of development. *(25 marks)*

14 'Since the 1950s increased food production has meant increased food security for most of the world's people, but there are exceptions to this pattern. Moreover, food production cannot go on increasing for ever.'
Discuss this statement, with reference to countries at different stages of development. *(25 marks)*

15 Evaluate the outcomes of the Green Revolution and consider how the lessons from this should influence the introduction of modern developments such as GM crops. *(25 marks)*

17 Rural land use

'*Nor rural sights alone, but rural sounds, exhilarate the spirit.*'

William Cowper

'*I see the rural virtues leave the land.*'

Oliver Goldsmith

The term **rural** refers to those less densely populated parts of a country which are recognised by their visual 'countryside' components. Areas defined by this perception will depend upon whether attention is directed to economic criteria (a high dependence upon agriculture for income), social and demographic factors (the 'rural way of life' and low population density) or spatial criteria (remoteness from urban centres). Usually it is impossible to give a single, clear definition of rural areas as, in reality, they often merge into urban centres (the rural–urban fringe) and differ between countries. Although generalisations may lead to over-simplifications (Framework 11, page 347), it is useful to identify three main types of rural area.

1 Where there is relatively little demand for land, certain rural activities can be carried out on an extensive scale, e.g. arable farming in the Canadian Prairies and forestry on the Canadian Shield.

2 In many areas, especially in economically developing countries, there is considerable pressure upon the land which results in its intensive use. Where human competition for land use becomes too great to sustain everyone, the area is said to be overpopulated (page 376). This often leads to rural depopulation, e.g. the movement to urban centres in Latin American countries (page 366).

3 In many economically developed countries, competition for land is greater in urban than in rural areas. The resultant high land values and declining quality of life are leading to a repopulation of the countryside (urban depopulation), e.g. migration out of New York and London (page 365).

The urban–rural continuum

It is now unusual to find a clear distinction between where urban settlements and land use end and rural settlements and land use begin. Instead, there is usually a gradual gradation showing a decrease in urban characteristics with increasing distance from the city centre Figure 17.1). This is known as the **urban–rural continuum** (page 393).

Figure 17.1

The urban–rural continuum

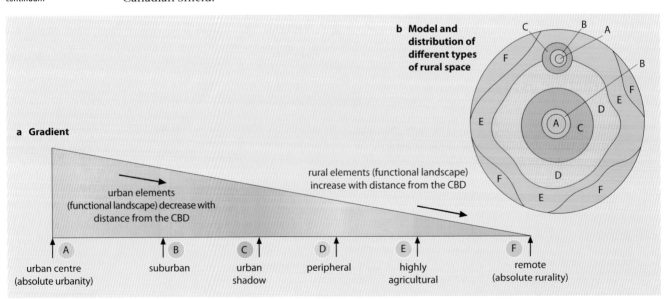

a Gradient

urban elements (functional landscape) decrease with distance from the CBD

rural elements (functional landscape) increase with distance from the CBD

A urban centre (absolute urbanity)
B suburban
C urban shadow
D peripheral
E highly agricultural
F remote (absolute rurality)

b Model and distribution of different types of rural space

The urban–rural continuum includes the rate at which rural settlements expand or decrease as people move out of or into nearby cities; changes in the socio-economic base as services and other functions are transferred to the countryside; and changes in land use resulting from increased pressure exerted on rural areas by nearby urban areas.

Figure 17.3

Rurality in England and Wales

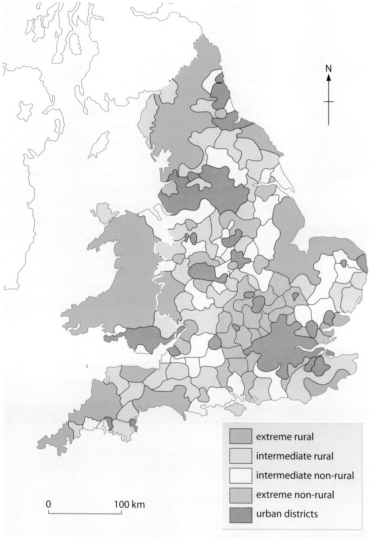

extreme rural

intermediate rural

intermediate non-rural

extreme non-rural

urban districts

0   100 km

Figure 17.2

An index of rurality for England and Wales (*after* Cloke, 1977)

Indices	Characteristics in rural areas
Population per ha	Low
% change in population	Decrease
% total population: over 65 years	High
% total population: male 15–45 years	Low
% total population: female 15–45 years	Low
Occupancy rate: % population at 1.5 per room	Low
Households per dwelling	Low
% households with exclusive use of (a) hot water (b) fixed bath (c) inside WC	High
% in socio-economic groups: 13/14 farmers	High
% in socio-economic group: 15 farmworkers	High
% residents in employment working outside the rural district	Low
% population resident < 5 years	Low
% population moved out in last year	Low
% in-/out-migrants	Low
Distance from nearest urban centre of 50 000	High
Distance from nearest urban centre of 100 000	High
Distance from nearest urban centre of 200 000	High

There are a number of measures of the intensity of change over distance, of which the best known is Cloke's **index of rurality** (Figure 17.2). The index is obtained by combining a range of socio-economic measures or variables, with absolute urbanity at one extreme and absolute rurality at the other. Using his index of rurality, Cloke then produced a map with a five-fold classification to show rurality in England and Wales (Figure 17.3).

Figure 17.4 shows some of the major competitors for land in a rural area. In many parts of the world, farming takes up the majority of the land and, especially in developing countries, employs most of the population.

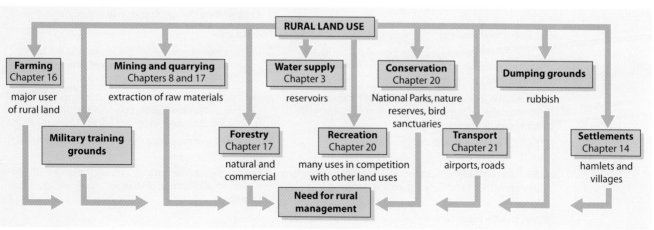

Forestry

In Britain

Neolithic farmers began the clearance of Britain's primeval forests about 3000 years ago. Aided by the development of axes, some clearances may have been on a scale not dissimilar to that in parts of the tropical rainforests of today. In 1919, with less than 4 per cent of the UK covered in trees, the Forestry Commission was set up to begin a controlled replanting scheme. Since then the policy has been to look towards an economic profit over the long term and to try to protect the environment. By 2008, 11 per cent of the UK was classified as woodland, but this still remained one of the lowest proportions in the EU.

Deciduous trees are more suited to England where the relief is lower and the location more southerly, whereas conifers are better adapted to Scotland with its higher relief and more northerly latitude.

	Coniferous	Deciduous
England	32%	68%
Scotland	78%	22%
UK	58%	42%

Much of Britain's surviving, established woodland is deciduous while most of the 20th-century forest planted by the Forestry Commission was coniferous. This is mainly because conifers, being softwoods, have a much greater commercial value than deciduous trees. This is partly due to their greater range of use, ranging from paper to furniture, and partly because, regardless of whether they are grown naturally or have been replanted, as there are fewer species within a given area than in a deciduous woodland, selection and felling of trees is made easier.

Softwoods growing in the poor soils and harsh climate of northern Britain take between 40 and 60 years to mature and so afforestation was always looked upon as an investment for future generations. Most of the pre-1980 plantations were neither attractive for human recreation nor as a habitat for wildlife (Figure 17.5). Since then, a strong conservation lobby has ensured that modern plantations are carefully landscaped while a more sustainable forest management aims to provide social and environmental advantages, to maintain an economically viable forestry sector and to protect woodlands for future generations. In the last two or three decades, the previously all-important economic factor has given way to a broader range of objectives that include amenity landscaping, wildlife management and recreation, while forest operations have moved towards smaller-scale practices that are environmentally and aesthetically more sensitive (Figure 17.6). Such has been the movement away from what had been virtually a monoculture, with perhaps only one, or two at the most, species of conifer being planted over a large area, that between 2004 and 2008, 84 per cent of newly planted trees within the UK were deciduous.

Figure 17.5

The case for and against forestry in Britain (*after* Warren, *Geography Review*, March 1998)

Advantages	Disadvantages
Socio-economic	**Landscape**
■ National timber needs – the UK supplies only 13% of its own timber and has a large annual import bill for wood products.	■ Early plantations were visually intrusive with their rigid geometric patterns, and with no regard for natural features.
■ Provides employment, especially as located in those rural areas where jobs are in short supply.	■ Often a 'blanket afforestation', using just one species of tree, created a monoculture with a uniformity of height and colour.
■ A positive method of using set-aside land.	■ They transformed the landscape and obliterated views. Concern over the speed and scale of replanting.
Non-market/environmental	**Environmental**
■ Trees are a renewable resource if carefully managed and, by planting in the UK, reduces pressures on tropical forests (sustainable development).	■ Introduction of non-native species, such as the North American Sitka spruce and lodgepole pine, as they were faster-growing than indigenous species.
■ Trees replace oxygen in the atmosphere and so help counterbalance the increase of carbon dioxide and its effects on global warming.	■ Destruction of valued environments such as the Flow Country wetlands of Caithness and Sutherland, and moorlands elsewhere in upland Britain.
■ Forests reduce water runoff (page 63).	■ Adverse impacts on flora and fauna, e.g. moorland birds and plants.
■ Forests contribute to biodiversity, providing habitats for a range of fauna and flora, e.g. red deer and red squirrels.	■ Concerns over water quality as afforestation led to increased acidification of lakes and rivers, and disrupted runoff.
■ Forests offer opportunities for recreation, and trees make an aesthetic contribution to the countryside.	
■ Some people argue that forests are part of Britain's traditional landscape.	

summits left clear for heather moorlands which provide a habitat for grouse and golden eagles

mature woodland forms a habitat for tawny owls and provides food for short-eared owls

trees planted at different times as differences in height are scenically more attractive

a variety of species and a lower density of trees replanted: helps to encourage more bird life which feeds on insects and so reduces the need for pesticide spraying

only small areas cleared at one time to reduce 'scars'

grassland provides a habitat for short-eared owls and food for tawny owls

winding forest road

land beside roads/tracks cleared to a width of 100 m and left as grass or planted with attractive deciduous trees

ponds created

cleared forest: branches left to rot; it takes 10 years for the nutrients to be returned to the soil

land next to river left clear for migrating animals such as deer

Figure 17.6

Managing an upland British forest (Kielder)

In developing countries

Commercial forestry is a relatively new venture in the tropics. It is usually controlled by trans-nationals based overseas which look for an imme-diate economic profit and have little thought for the long-term future or the environment. The UN suggests that over half of the world's forests were cleared during the last millennium and that the present rate of clearance is 102 000 km² annually. Of this, 94 000 km² is in developing countries located in the tropical areas of Africa, Latin America and South-east Asia where rates of replanting are often minimal.

The underlying causes of deforestation in devel-oping countries are varied. Key issues, according to the World Wide Fund for Nature, include unsustainable levels of consumption; the effects of national debt; pressure for increased trade and development; poverty; patterns of land ownership; and growing populations and social relationships. It is also usual to blame forest destruction on the poor farmers of these countries rather than on the resource-consuming developed countries.

During commercial operations the forest is totally cleared by chainsaw, bulldozer and fire: there is no selection of trees to be felled. The sec-ondary succession (page 318) is of poorer-quality trees, as little restocking is undertaken. Where afforestation of hardwoods does take place, there is often insufficient money for fertiliser and pes-ticide. The hope for the future may lie in **agro-forestry**, where trees and food crops are grown alongside each other. Forest soils, normally rated unsuitable for crops, can be improved by growing leguminous tree species. Commercial forestry is more difficult to operate in developing coun-tries as they are distant from world markets, the demand for hardwood is less than for softwood, and although there are several hundred species in a small area only a few are of economic value.

The threat of the destruction of the rainfor-ests has become a major global concern. Some of the consequences of deforestation are described in Places 76 and Case Study 11.

Ethiopia

Early in the 20th century, 40 per cent of Ethiopia was forested. Today the figure is 11 per cent. In 1901, a traveller described part of Ethiopia as being 'most fertile and in the heights of commercial prosperity with the whole of the valleys and lower slopes of the mountains one vast grain field. The neighbouring mountains are still well-wooded. The numerous springs and small rivers give ample water for domestic and irrigation purposes, and the water meadows produce an inexhaustible supply of good grass the whole year.' A century later, the same area was described as 'a vast barren plain with eddies of spiralling dust that was once topsoil. The mountains were bare of vegetation and the river courses dry.' As the trees and bushes were cleared, less rainfall was intercepted and surface runoff increased, resulting in less water for the soil, animals and plants. There has been little attempt to treat the forest as a *sustainable* resource.

Amazonia

The clearance of the rainforests means a loss of habitat to many Indian tribes, birds, insects, reptiles and animals. Over half of our drugs, including one from a species of periwinkle which is used to treat leukaemia in children, come from this region. It is possible that we are clearing away a possible cure for AIDS and other as yet incurable diseases. (Despite the rainforests being the world's richest repository of medical plants, only 2 per cent have so far been studied for potential health properties.)

Without tree cover, the fragile soils are rapidly leached of their minerals, making them useless for crops and vulnerable to erosion (Figure 12.8). The Amazon forest supplies one-third of the world's oxygen and stores one-quarter of the world's fresh water – both would be lost if the region was totally deforested. The burning of the forest not only reduces the amount of oxygen given off, but increases the release of carbon dioxide (a contributory cause of global warming). It has also been suggested that the decrease in evapotranspiration, and subsequently rainfall, caused by deforestation could also have serious climatic repercussions – could the Amazon Basin become another Ethiopia? There is a much greater need for *sustainable* logging.

Malaysia: a model for the future?

Malaysia has several thousand species of tree, mainly hardwoods, with timber and logs being the country's third-largest export. However, the government has imposed strict controls, and the Forestry Department 'manages the nation's forest areas to ensure sufficient supply of wood and other forest produce and manages and implements forest activities that would help to sustain and increase the productivity of the forest' (*Malaysia Official Year Book*, 2007). The Department insists that trees reach a specific height, age and girth before they can be felled (Figures 17.7 and 17.8). Logging companies are given contracts only on agreement that they will replant the same number of trees as they remove. Many newly planted hardwoods are ready for harvesting within 20–25 years due to the favourable local growing conditions. Further experiments are being made with acacias and rattan, both of which grow even faster. Consequently, half of Malaysia is still forested and as most of the remaining third is under tree crops such as rubber, oil palm and coca (Places 68, page 483) stocks are being successfully maintained. Even so, Malaysia's rapid industrialisation (Places 91, page 578) is causing increased deforestation, especially around the capital of Kuala Lumpur. Attempts have been made to make logging *sustainable*.

Figure 17.7

Logging operations in Malaysia

Figure 17.8

Timber lorry in the Malaysian rainforest

From September 1997 to June 1998, much of South-east Asia was blanketed by a thick smoke haze, in reality smog, caused by thousands of uncontrolled forest fires, mainly in Sumatra and Borneo (Figure 17.9). At its peak the smoke haze covered an area the size of western Europe and caused visibility to be reduced to 50 m. Its effects were various:

- **Human** The Air Pollution Index on Sarawak reached 851 (300 is considered 'hazardous' for human life), children and high-risk groups already suffering from respiratory or cardio-vascular diseases (Places 99, page 621) were prone to major health problems, and schools on Sumatra were closed.

- **Economic** Airports throughout the region were closed (an airline crash in Sumatra and a ship collision in the Strait of Malacca were both attributed to the haze), logging operations were suspended and farm crops destroyed.

- **Environmental** An estimated 90 per cent of canopy trees were lost in Sumatra and Borneo, and the rate of secondary succession would be slow; soils were seriously degraded; and wildlife habitats were lost (including those for such endangered species as the orang-utan, Sumatran rhinoceros and Sumatran tiger, and an irreparable loss in biodiversity).

Many Indonesians, accustomed to the humid climate and with little experience of dry weather, still adhere to fire-using traditions. Fire has long been used as a quick and cheap method of land clearance by farmers, and by plantation and forestry-concession owners. In 1997 the monsoon rains failed and the resultant prolonged drought, believed to have been triggered by the El Niño event (Case Study 9A), together with the prevailing land use and land management conditions, proved ideal conditions for the spread of forest fires on an unprecedented scale. The remoteness of the fires and the lack of resources, organisation and expertise combined to make fire-control impossible. Satellite imagery suggested that, although the blame for most of the fires was apportioned to the many small farmers, 80 per cent of the fires were due to large companies. By the time the rains did come, in May 1998, 10 million ha of forest had been burnt. Lessons were not learned, however, and fires and the resultant smoke haze kept returning each year until, in 2006, the consequences were almost as bad as in 1997–98. As in 1997, the fires followed a summer drought associated with an El Niño event (Case Study 9A). Most of the out-of-control fires were, as in previous years, on the Indonesian islands of Sumatra and Kalimantan (Indonesian Borneo). Government officials accused the many small farmers who clear their land annually by fire, whereas environmentalists claimed 80 per cent of the fires were begun by large companies clearing land on big plantations, timber estates and protected areas. By July over 100 fires were spotted by satellite, by which time many people were already experiencing breathing difficulties. During the first week of October, visibility in Pontianak (Kalimantan) was reduced to less than 50 m for several days, and many flights from the town's airport were either delayed or cancelled. Air pollution was said to be at a 'dangerous' level and people were advised to wear protective face masks if they went out of doors. Schools remained closed. A thick haze, blown by a strong wind from Sumatra, prompted Singapore to warn people against vigorous outside activities, while in adjacent Malaysia, Kuala Lumpur recorded 'unhealthy air quality'. The event lasted several months.

In 2007, the Indonesian government pledged to reduce forest fires while admitting to its neighbours that it might be incapable of totally eradicating them. With Malaysian co-operation, personnel were being trained in fire prevention, fire control and public education.

Figure 17.9

Maximum extent of smoke haze in 1997 and 2006

Mining and quarrying

Even since the Neolithic (when flint was excavated from chalk pits), Bronze and Iron Ages, quarrying and mining have been an integral part of civilisation. It was through the extraction and processing of minerals that many of today's 'developed' countries first became industrialised, while to some 'developing' countries the export of their mineral wealth provides the only hope of raising their standard of living. The modern world depends upon 80 major minerals, of which 18 are in relatively short supply, including lead, sulphur, tin, tungsten and zinc.

Minerals are a finite, non-renewable resource which means that, although no essential mineral is expected to run out in the immediate future, their reserves are continually in decline. **Resources** are the total amount of a mineral in the Earth's crust. The quantity and quality are determined by geology. **Reserves** are the amount of a mineral that can be economically recovered.

Although many items in our daily lives originated as minerals extracted from the ground, no mineral can be quarried or mined without some cost to local communities and the environment. Extractive industries provide local jobs and create national wealth, but they also cause inconvenience, landscape scars, waste tips, loss of natural habitats, and various forms and levels of pollution.

The most convenient methods of mining are **open-cast** and **quarrying**. In open-cast mining, all the vegetation and topsoil are removed, thus destroying wildlife habitats and preventing other types of economic activity such as farming (Places 79). Sand and gravel are extracted from depressions which, although shallow, often reach down to the water table, as in the Lea valley in north-east London. Coal and iron ore are often obtained from deeper depressions using drag-line excavators which are capable of removing 1500 tonnes per hour (Figure 17.10). Often, the worst scars (eyesores) result from quarrying into hillsides to extract 'hard rocks' such as limestone and slate (Figure 17.11 and Places 78). There is usually greater economic and political pressure for open-cast coalmining than to quarry any other resource: it is the cheapest method of obtaining a strategic energy resource, but none generates greater social and environmental opposition. The increased demand for aggregates for road building and cement manufacture has led to the go-ahead being given for superquarries to be opened up in many different parts of the world, including that at Dehra Dun in northern India (Case Study 8).

Mining involves the construction of either horizontal **adit mines**, where the mineral is exposed on valley sides, or vertical **shaft mines**, where seams or veins are deeper.

Deep mining still affects local communities and the environment either by the piling up of rock waste to form tips – of coal in South Wales valleys (Aberfan, Case Study 2B) and china clay in Cornwall, for example – or by causing surface subsidence – as in some Cheshire saltworkings. Waste can also be carried into rivers where it can cause flooding by blocking channels and, when it contains poisonous substances, can kill fish and plants and contaminatse drinking water supplies. This was highlighted in early 1992 when floodwaters from Cornwall's last working tin mine, Wheal Jane, flowed into rivers and to the coast, carrying with them arsenic and cadmium.

Figure 17.10

Opencast mining for coal, West Virginia, USA

Figure 17.11

The Dinorwic slate quarries, Gwynedd, North Wales

The Oakley slate quarries were first worked in 1818. By the 1840s, the most easily obtained slate had been won and mining began. The introduction of steam power and the building of the Ffestiniog railway led to the export of 52 million slates from Porthmadog in 1873. At the quarry's peak productivity, 2000 men and boys were employed on seven different levels. Each level was steeply inclined into the hillsides and was worked to a depth of 500 m. Apart from farming, the slate mines were the sole providers of employment,

and Blaenau Ffestiniog's population peaked at 12 000. Working in candlelight in damp and dusty conditions for up to 12 hours a day, and with rock falls common (pressure release, page 41) the life expectancy of miners was short. By the turn of the century, the manufacture of clay roof tiles heralded the beginning of the industry's decline and in 1971 the mine at Blaenau closed. A decade later, renamed Gloddfa Ganol, the underground galleries were re-opened to tourists, some of whom arrive via the narrow-gauge Ffestiniog railway.

As the mines closed, people became either unemployed or were forced to move to seek work – the present population of Blaenau is under 500. Today the slate mines are a tourist attraction and have again become the town's largest employer. Above the rows of the former miners' cottages tower the large and unsightly spoil heaps (Figure 17.12) as for every tonne of usable slate, ten tonnes of waste was created – though these spoil heaps seem more stable than the coal tips which affected Aberfan (Case Study 2B). Some of the old buildings have been restored as tourist attractions and there is little evidence of subsidence as in other mining areas.

Figure 17.12

Spoil heaps above Blaenau Ffestiniog, Gwynedd

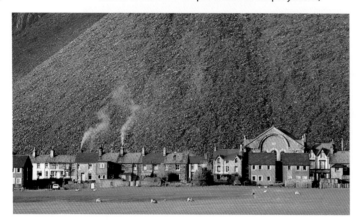

Malaysia (2008) is the world's seventh major producer of tin ore but whereas it was the world leader until 1993 and in 1970 was producing 40 per cent of the world's output, now its contribution is only 1 per cent. Early tin mining was typical of the colonial trade period (page 624). British settlers brought in the capital, machinery and technology; supervised the mining; and organised the export of tin for refining. Malaya, as it was then known, received few advantages. Most tin was obtained by opencast methods and the use of hydraulic jets.

After independence, when the mines were nationalised and operated under the Malaysian government, tin played a major role in the country's economic development and its emergence as one of the 'Asian tigers' (page 578). However, since then the

total output and the number of workers have fallen rapidly due to the depletion of reserves (especially those of the highest-quality), the low market prices and the rising costs of extraction. Many of the former mines have been left as land either covered in mining spoil (Figure 17.13) and polluted lakes or with abandoned overhead 'railways', machinery and buildings. There is talk of re-opening some of the mines in Perak in the north-west of the country due to a resurgence in world prices.

One of the largest abandoned mines lies 15 km south of Kuala Lumpur in an area of rapidly growing housing and high-tech industry. It has been converted into a theme park with the world's longest aerial ropeway, together with water slides and various water sports (Figure 17.14) .

Figure 17.13

Disused tin mine, Malaysia

Figure 17.14

Sunway Lagoon Theme Park, near Kuala Lumpur, Malaysia

Having completed any sampling exercise (Framework 6, page 159), it is important to remember that patterns exhibited may not necessarily reflect the parent population. In other words, the results may have been obtained purely by chance. Having determined the mean of the sample size, it is possible to calculate the difference between it and the mean of the parent population by assuming that the parent population will conform to the normal distribution curve (Figure 6.37). However, while the sample mean must be liable to some error as it was based on a sample, it is possible to estimate this error by using a formula which calculates the **standard error of the mean** (SE).

$$SE\,\bar{x} \;=\; \frac{\sigma}{\sqrt{n}}$$

where: $\bar{x}$ = mean of the parent population

σ = standard deviation of parent population

$\sqrt{n}$ = square root of number of samples

We can then state the reliability of the relationship between the sample mean and the parent mean within the three confidence levels of 68, 95 and 99 per cent (Framework 6). Unfortunately, when sampling, the standard deviation of the parent population is not available and so to get the standard error we have to use the standard deviation of the sample, i.e. using s rather than σ. Although this introduces a margin of error, it will be small if n is large (n should be at least 30).

For example: a sample of 50 pebbles was taken from a spit off the coast of eastern England. The mean pebble diameter was found to be 2.7 cm and the standard deviation 0.4 cm. What would be the mean diameter of the total population (all the pebbles) at that point on the spit?

$$SE \;=\; \frac{0.4}{\sqrt{50}} \;=\; \frac{0.4}{7.07} \;=\; 0.06 \text{ (to two decimal places)}$$

This means we can say:

1 with 68 per cent confidence, that the mean diameter will lie between 2.7 cm ± 0.06 cm, i.e. 2.64 to 2.76 cm

2 with 95 per cent confidence, that the mean diameter will lie between 2.7 cm ± 2 x SE (2 x 0.06 = 0.12 cm), i.e. 2.58 to 2.82 cm

3 with 99 per cent confidence, that the mean diameter will lie between 2.7 cm ± 3 x SE (3 x 0.06 = 0.18 cm), i.e. 2.52 to 2.88 cm.

If we wanted to be more accurate, or to reduce the range of error, then we would need to take a larger sample. Had we taken 100 values in the above example, we would have had:

$$SE \;=\; \frac{0.4}{\sqrt{100}} \;=\; \frac{0.4}{10.00} \;=\; 0.04 \;\;\text{(to two decimal places)}$$

which means we can now say with 68 per cent confidence that the mean diameter size will lie between 2.7 cm ± 0.04 cm (i.e. 2.66 to 2.74 cm). Of course, this also means there is a 32 per cent chance that the mean of the parent population is *not* within these values. This is why most statistical techniques in geography require answers at the 95 per cent confidence level.

This standard error formula is applicable only when sampling actual values (**interval** or **measured data**). If we wish to make a count to discover the frequency of occurrence where the data are **binomial** (i.e. they could be placed into one of two categories), we have to use the **binomial standard error**. For example, we may wish to determine how much of an area of sand dune is covered in vegetation and how much is *not* covered in vegetation. When using binary data, the sample population estimates are given as percentages, not actual quantities – i.e. x per cent of points on the sand dune *were* covered by vegetation; x per cent of points on the sand dune *were not* covered by vegetation.

The formula for calculating standard error using binomial data is:

$$SE \;=\; \sqrt{\frac{p \times q}{n}}$$

where: p = the percentage of occurrence of points in one category

q = the percentage of points not in that category

n = the number of points in the sample.

A random sample of 50 points was taken over an area of sand dunes similar to those found at Morfa Harlech (Figure 6.33). Of the 50 points, 32 lay on vegetation and 18 on non-vegetation (sand) which, expressed as a percentage, was 64 per cent and 36 per cent respectively. How confident can we be about the accuracy of the sample?

$$SE \;=\; \sqrt{\frac{64 \times 36}{50}} \;=\; \sqrt{46.08} \;=\; 6.79$$

As the sample found 64 per cent of the sand dunes to be covered in vegetation and knowing the standard error to be ± 6.79, we can say:

1   with 68 per cent confidence, that the vegetated area will lie between 64 per cent ± 6.79, i.e. between 57.21 and 70.79 per cent

2   with 95 per cent confidence, that it will lie between 64 per cent ± 2 x *SE* (2 x 6.79 = 13.58), i.e. between 50.42 and 77.58 per cent will be vegetated

3   with 99 per cent confidence, that it will lie between 64 per cent ± 3 x *SE* (3 x 6.79 = 20.37), i.e. between 43.63 and 84.37 per cent.

Minimum sample size

It seems obvious that the larger the size of the sample, the greater is the probability that it accurately reflects the distribution of the parent population. It is equally obvious that the larger the sample, the more costly and time-consuming it is likely to be to obtain. There is, however, a method to determine the minimum sample size needed to get a satisfactory degree of accuracy for a specific task, e.g. to find the mean diameter of pebbles on a spit, or the amount of vegetation cover on sand dunes. This is achieved by reversing the two standard error calculations.

For measured data: Imagine you wish to know the mean diameter of pebbles at a given point on a spit to within ± 0.1 cm at the 99 per cent confidence level.

The 99 per cent confidence level is 3 x *SE*.

$$3\ SE\ =\ \frac{3s}{\sqrt{n}}$$

i.e. $3s\ =\ 0.1\ \sqrt{n}$

$$3s$$

i.e. $\dfrac{}{0.1}\ =\ \sqrt{n}$

We determined earlier that s (standard deviation of the sample) for the pebble size was 0.4, and so by substitution we get:

$$\frac{1.2}{0.1}\ =\ \sqrt{n}$$

i.e. $12\ =\ \sqrt{n}$

$$n\ =\ 12^2$$

$$n\ =\ 144$$

We would need, therefore, to measure the diameter of 144 pebbles to get an estimate of the parent population at the 99 per cent confidence level.

For binomial data: How many sample values are needed to estimate the area of sand dunes which is vegetated, with an accuracy which would be within 5 per cent of the actual area (i.e. at the 95 per cent confidence level)?

$$n\ =\ \frac{p \times q}{(SE)^2}$$

Again by substitution we get:

$$n\ =\ \frac{64 \times 36}{(5)^2}$$

$$n\ =\ 92.16$$

We would therefore have to take a sample of 93 values to achieve results within 5 per cent of the parent population.

The need for rural management

As was shown on Figure 17.4, there is often considerable competition for land in most rural areas and, therefore, there is a need, in most people's opinion, for careful management. In Britain, this management may be the task of national, local or voluntary organisations such as the Department of the Environment, the various National Parks Planning Boards (Places 92, page 592) and the Council for the Protection of Rural England (CPRE). Pressures on rural areas increase towards large urban areas where there is a greater demand for housing, shopping, business parks and recreational facilities (Figure 14.20 and pages 433 and 567). Pressure on

the land may be even greater in economically less developed countries where the need to improve people's basic standard of living is likely to take preference over management schemes.

One attempted management scheme in a developing country is described in Places 80. It draws together several topics discussed in this book, i.e. an island (Chapter 6) with interrelated ecosystems (Chapter 11) offering alternative, rural land use possibilities (Chapter 17), where the population is increasing (Chapter 13) and wishing to improve its standard of living (Chapter 21), thus putting pressure on natural resources (Chapter 17).

Over two-thirds of Tanzania's 900 km long coastline consists of three fragile ecosystems – a fringing coral reef, separated from mangrove swamps on the mainland by a lagoon. Mafia Island, where the coral reaches above sea-level, is a national marine park.

An island management plan was put forward in the 1990s to try to maintain economic development, to conserve resources for future generations and to avoid conflict between different land uses and users. Some of the various economic activities threatening the fragile island ecosystems include the following:

- **Coral mining** The removal of live coral for the tourist and curio trade, and of fossilised coral rock for building purposes (Places 37, page 302). For lime, the rock is burnt over fires made from locally collected wood.

- **Fisheries** At all scales from subsistence to commercial, taking fin-fish, octopus, crayfish and edible shellfish.

- **Dynamite fishing** The illegal use of dynamite to stun and kill fish. This destroys the physical structure of the reef and kills virtually every organism within 15 m of the blast.

- **Seaweed farming** Important as a means of diversifying income but suffers from the problems associated with cash crops and could lead to biodiversity loss through the creation of monoculture (page 501).

- **Salt production** By evaporation: hyper-saline seawater is boiled using local mangrove wood for fuel, a crude process that can cause the denudation of large areas of mature trees.

- **Tourism** A rapidly growing industry and one that the government is keen to promote. Coastal tourism includes game-fishing, 'sea-safaris', diving, snorkelling and beach activities. Tourists, per capita, are major consumers of resources (drinking water, fuel and foods), can damage the natural environment (new hotels, destruction of the reef) and can cause cultural conflicts (dress code in a Muslim country).

- **Off-shore gas extraction** From the small Songo-Songo gasfield.

- **Farming** Pesticides entering the lagoon behind the roof are killing coral.

A new management approach

This aims to satisfy economic, social and environmental objectives in order to ensure:

- the maximum sustainable economic benefit from the long-term use of natural resources

- the maintenance of the conditions and productivity of the natural environment

- the allocation of resources between competing uses and users.

These aims are often seen as contradictory, and the main problem is how to cope with the diverse requirements of the different user-groups, especially those who utilise finite resources.

Developing a management model

To achieve an understanding of the nature and conditions of the resources in a proposed management area, the following considerations should be explored:

- **Political factors** What is the scale and structure of the area? Is it stable? Who will pay? Can it provide finance or secure funding? Who will advise? Are there powerful interest groups either for or against?

- **Physical factors** What are the main physical features? Are these stable? Are there any natural hazards?

- **Biological factors** What biological communities exist? In what condition are they? Are there records of change or overuse over a period of time? Are there species of endangered, cultural or commercial importance?

- **Socio-economic** What are the current uses of the area? Who uses it? Are they traditional or local uses? Have they a commercial interest? How are the resources exploited? Are these practices sustainable or destructive?

Once an area has been chosen, four stages can lead to a practical plan for its creation as a multi-user management scheme:

1 The definition of management goals – normally including conservation, sustainable resource use and economic development (Framework 16, page 499).

2 The establishment of an administrative authority – the process of human representation.

3 The formulation of a management strategy and objectives – an assessment of the physical and human characteristics of the whole area and, within it, sub-zones.

4 The development of legislation – to achieve the objectives.

But remember – no plan should be considered as final – it is simply an improvement on what was done before.

Figure 17.15

National Parks and Recreation
Areas in south-western USA

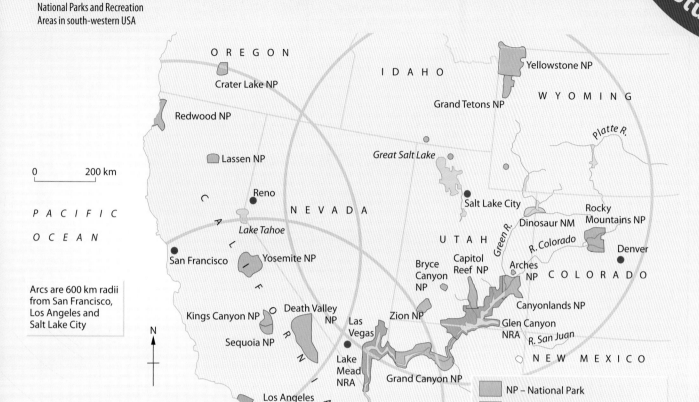

Figure 17.16

Old Faithful,
Yellowstone
National Park

The scenery of the mountain states of western USA is spectacular and varied. It includes some of the country's highest peaks, as well as extensive desert and wild river scenery. Yellowstone was set up in 1872, and is arguably the world's best known National Park with its variety of mountain and volcanic scenery, geysers, hot springs, canyons, lava flows and wildlife including bear, elk and buffalo (bison). Tourists from all over the world now flock to the region to visit the large number of designated National Parks, Monuments and Recreation Areas. Figure 17.15 indicates the accessibility of the most popular attractions for visitors from major cities and international airports. Over 30 million people live within 500 km of the major National Parks and Recreation Areas.

The National Parks were set up to preserve and protect the environment for future generations. Visitors are encouraged to stay, but apart from workers associated with the National Park there are no permanent residents within the parks. (This is a major difference from National Parks in the UK where farmers and other residents live on the land throughout the year.) Lodges, hotels, tourist villages and regulated camping grounds are provided, together with well-made roads and tracks to the different scenic attractions. This contributes to visitor pressure on 'honeypot' sites such as Old Faithful in Yellowstone where people wait for the geyser to blow once every 85 minutes (Figure 17.16). There are traffic jams as cars stop to watch animals such as bison grazing or herds of elk close to the road (Figure 17.17). Roads are closed by rangers if pressure is considered too great.

Visitor numbers to the Parks and Recreation Areas in the mountain states have continued to increase. Many come in private cars, camper vans and buses. Park authorities are working to provide better traffic management, which includes vehicle restriction. The most popular 'round tour' of parks includes those located in Utah and neighbouring Arizona. Of these, by far the most popular are the Grand Canyon National Park with 4.4 million visitors in 2007 (Figure 7.19) and the Glen Canyon National Recreation Area with 1.9 million visitors (Figures 17.22 and 17.23). Both of these areas stretch along the sides of the Colorado River but their access and uses are very different, as most visitors to the Grand Canyon travel to the North or South Rim to look down at the river flowing 1.6 km below them, whereas at Glen Canyon people have access to Lake Powell which was created by damming the river.

At present most visitors to the Grand Canyon go to the South Rim, mainly because it has easier access, more facilities and better panoramic vistas although the North Rim, which is closed by snow in winter, is becoming increasingly popular. The Canyon itself continues to attract rafting and canoeing enthusiasts but their number is strictly limited to protect the natural habitat along the river banks. At the western (downriver) end of the National Park is the Lake Mead National Recreation Area which has taken advantage of the lake created by the construction of the Hoover Dam (Figure 17.18). This dam, known as the Boulder Dam when it was built in the 1930s, has created a lake which has a shoreline of over 1100 km and whose water is used for irrigation, to provide hydro-electricity for the local area and recreation opportunities for lake cruising, boating and swimming. The dam is only half an hour away from the 'bright-lights' of Las Vegas.

The numbers visiting Utah's five National Parks of Arches (Figure 17.19), Bryce Canyon, Canyonlands, Capitol Reef and Zion doubled between 1982 and 2007. These parks, together with smaller protected areas such as Goosenecks State Park (Figure 17.20) and, straddling the border with Arizona, Monument Valley Navajo Tribal Park (Figure 7.25), offer some of the world's most spectacular desert and river-eroded (canyon) scenery. Visitors using motor caravans, 4x4s or tourist buses can manage to visit all of these attractions in a week, but for those who are more energetic, twice that time is preferable. At places like St George, near to Zion National Park, and Moab, close to Arches National Park, many holiday

Figure 17.17

Yellowstone National Park: bison grazing on the verge; the vegetation is recovering from a brush fire

Figure 17.18

The Hoover Dam and Lake Mead

lodges and small hotels have been built, together with housing for 'retirees' wishing to move away from large urban areas.

Rainfall is low in the Basin and Range province lying between the Sierra Nevada, which forms the border with California to the west, and the Rocky Mountains in the east. Although the states in this region now rely upon tourism as their major source of employment, the rural economy also depends on ranching, irrigation and mining.

In an area that is naturally short of water, and where the problem has been accentuated by recent droughts, it is not surprising that there should be conflicts over its use. In the last three or four decades there has

been rapid in-migration to the region, particularly to the larger urban areas of Salt Lake City, Phoenix and Las Vegas. These, and other smaller towns, are growing as more people decide to move here partly for the climate and partly as they choose to work from home using computers and other electronic equipment.

Increasing amounts of expensive water are taken by canal and pipeline to fill the swimming pools of new houses and to work the fountains of Las Vegas. This extra water means there is less for agriculture, yet farming itself needs more as the extra fruit and vegetables demanded by both new residents and tourists can only be grown

Figure 17.19

Arches National Park

Figure 17.20

Gooseneck State Park, Utah

under irrigation. Irrigation is also necessary close to cities where good-quality pasture land is needed if dairy cows are to be reared for their milk, butter and cheese, as well as in more remote areas where it helps produce the silage for beef cattle (Figure 17.21). There are also water disputes between individual states.

Traditionally, several of the south-western states were important mineral producers, especially of copper, silver and gold. Much of the easily obtained and higher-quality ores have already been used, while falling world prices and rising extraction costs have forced the closure of all but the most profitable of mines, leaving scars on the landscape.

Figure 17.21

Irrigation on rangelands near Salt Lake City

Glen Canyon National Recreation Area

This Recreation Area is based on Lake Powell (named after Major Powell who led the first expedition down 1600 km of rapids in the Colorado River), which is the country's second largest artificial reservoir. Despite a hard fight by conservationists, the Glen Canyon Dam was begun in 1956, completed in 1963 and the reservoir had filled by 1972. When full the lake – which accounts for only 13 per cent of the total Recreation Area – is 300 km long and has

a shoreline of 3000 km due to its zigzag-ging through 96 major canyons. High water, reached in the mid-1980s when the lake was over 125 m deep, is marked by a white ring etched into the red canyon walls. Since then the onset of numerous drought years has resulted in a drop in the lake level, by 2005 of over 30 m, and its volume has decreased by one-third. The fall in level revealed petroglyphs (indian carvings, compare Figure 7.7) and enabled visitors to walk (rather than visiting by boat)

into some of the tributary canyons. It also meant the exposure of huge areas of mud at the head of the drawn-down reservoir and the closure of marinas, as at Hite. Since that time the wet winter of 2005 and record snowfalls in 2008 have seen the levels of Lake Powell rise by 15 m – equal to half its previous fall.

The lake is ideal for water sports such as canoeing, water boarding, water-skiing, wind-surfing, scuba diving and fishing. Most tourists hire houseboats whose

Rural land use **529**

lengths range from 15 to 25 m and which can cost up to $14 000 a week to hire. The latest houseboats, which can sleep 8 to 12, come with a hot tub, a wet bar and a 120cm flat-screen TV. Around the shores of Lake Powell are six marinas (Figures 17.22 and 17.23) with the names of Bullfrog, Hite, Wahweap, Hall's Crossing, Dangling Rope and (the latest) Antelope Point. On land there are wilderness trails and back-country roads which can only be used by four-wheel-drive vehicles, but which give access to isolated geological, historical and archaeological sites, such as the Rainbow Arch Monument Park.

Environmental damage is evident along the busiest stretches of shoreline. 'Adopt-a-Canyon' has become a slogan, encouraging visitors to take out everything they take in. Water quality is constantly tested and water-skiiers are designated to specific areas. Summers can be extremely hot – up to 43°C – while winters, when fishing is almost the sole recreational activity, are very cold.

A survey of visitors in 2007 showed that 48 per cent were aged in the 41–65 age group; 78 per cent had visited before; people came from 48 states and 21 countries; and most came for either the scenery or for motorised boating. Over 85 per cent found the quality of services, facilities and recreational opportunities as 'very good' or 'good'.

Figure 17.22

Glen Canyon National Recreation Area: Bullfrog Ferry, Lake Powell

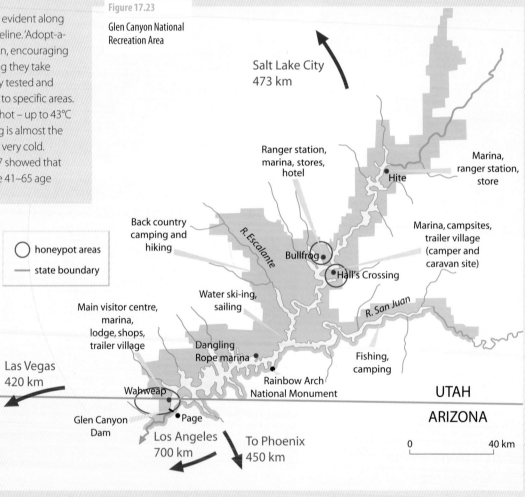

Figure 17.23

Glen Canyon National Recreation Area

Further reference

Pickering, K.T. and Owen, L.A. (1997) *An Introduction to Global Environmental Issues*, Routledge.

Wilson, J. (1984) *Statistics in Geography for A Level Students*, Schofield & Sims.

Council for the Protection of Rural England:
www.cpre.org.uk

Finnish Forest Association:
www.metla.fi/forestfin/intro/eng.index.htm

Forestry Commission of Great Britain:
www.forestry.gov.uk/

ForestWorld:
http://forestworld.com/

Natural England:
www.naturalengland.org.uk

UN Food and Agriculture Organisation Forestry:
www.fao.org/forestry/home/en/

Activity

1 **a** 'Forestry is not usually economically viable in developed countries unless supported by the state with subsidies.'

Explain the advantages of forestry in rural areas of the United Kingdom, giving:

    i two socio-economic advantages   *(4 marks)*

    ii two environmental advantages.   *(4 marks)*

  b **i** Explain why some people think that commercial forestry plantations have caused environmental damage in some parts of the United Kingdom. *(4 marks)*

    ii 'In the last decade ... forest operations have been transformed, with a shift towards smaller-scale practices which are more environmentally and aesthetically sensitive.' Explain how the changes in forest management referred to above have improved the rural environment in parts of the United Kingdom. *(5 marks)*

  c With reference to a named tropical country, explain how commercial forestry in the rainforest can be a form of sustainable development. *(8 marks)*

Exam practice: basic structured question

Figure 17.24

2 Study Figure 17.24. It shows how conflicts may arise in a rural area where recreation and tourism are important.

  a Name a rural area in a more economically developed country where recreation and tourism are important, and where their development has caused conflicts with local people and conservationists. Describe conflicts in that area between:

    i tourists and the local community

    ii tourists and conservation

    iii the local community and conservation. *(8 marks)*

  b Explain how management of the area is attempting to reduce the conflicts described in **a** above. *(7 marks)*

  c Can tourism ever lead to sustainable development in rural areas in less economically developed countries? Illustrate your answer with reference to one or more case studies. *(10 marks)*

Exam practice: structured question

3 **a** 'In the last decade ... forest operations have been transformed, with a shift towards smaller-scale practices which are more environmentally and aesthetically sensitive.'

With reference to a named area of forestry in the United Kingdom, explain how the changes referred to above have altered forest management practices. Explain how this has benefited the environment. *(10 marks)*

  b **i** How has commercial forestry caused environmental damage in tropical regions? *(7 marks)*

    ii Suggest how commercial forestry in tropical regions can be managed so that it is a sustainable form of development. *(8 marks)*

Exam practice: essay

4 Study Figure 17.24.

'The development of the tourist industry can bring both benefits and problems for communities and the environment of rural areas.'

Discuss this statement with reference to examples from more economically developed and less economically developed countries. *(25 marks)*

Energy resources

· ·

'If each Indian were to start consuming the amount of commer-cial energy a Briton does, that would mean the world finding the equivalent of an extra 3190 million tonnes of oil each year. Imagine what consuming that would do to the greenhouse effect, not to mention its effect on oil and other reserves.'

Mark Tully, *No Full Stops in India*, 1991

What are resources?

Resources have been defined as commodities that are useful to people although the value and impor-tance of individual resources may differ between cultures. Although the term is often taken to be synonymous with **natural resources**, geographers and others often broaden this definition to include **human resources** (Figure 18.1). Natural resources can include raw materials, climate and soils. Human resources may be subdivided into people and capital. A further distinction can be made between **non-renewable resources**, which are finite as their exploitation can lead to the exhaus-tion of supplies (oil), and **renewable resources**, which, being a 'flow' of nature, can be used over and over again (solar energy). As in any classifica-tion, there are 'grey' areas. For example, forests and soils are, if left to nature, renewable; but, if used carelessly by people, they can be destroyed (defor-estation, soil erosion).

Reserves are known resources which are considered exploitable under current economic and technological conditions. For example, North Sea oil and gas needed a new technology and high global prices before they could be brought ashore; in contrast, tidal power still lacks the technology, and often the accessibility to markets, that are needed to allow it to be devel-oped on a widespread, commercial scale.

Energy resources

The sun is the primary source of the Earth's energy. Without energy, nothing can live and no work can be done. Coal, oil and natural gas, which account for an estimated 88 per cent of the global energy consumed in 2007 (Figure 18.2) compared with 85.5 per cent in 1996, are forms of stored solar energy produced, over thousands of years, by photosynthesis in green plants. As these three types of energy, which are referred to as **fossil fuels**, take long periods of time to form and to be replenished, they are classified as non-renewable. As will be seen later, these fuels have been relatively easy to develop and cheap to use, but they have become major polluters of the environment. Nuclear energy is a fourth non-renewable source but, as it uses uranium, it is not a fossil fuel.

Figure 18.1

A classification of resources

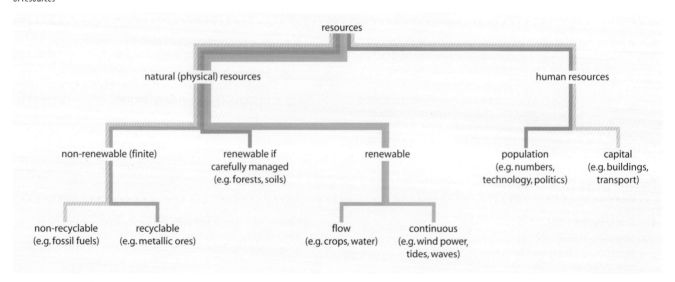

resources

natural (physical) resources

human resources

non-renewable (finite)

renewable if carefully managed (e.g. forests, soils)

renewable

population (e.g. numbers, technology, politics)

capital (e.g. buildings, transport)

non-recyclable (e.g. fossil fuels)

recyclable (e.g. metallic ores)

flow (e.g. crops, water)

continuous (e.g. wind power, tides, waves)

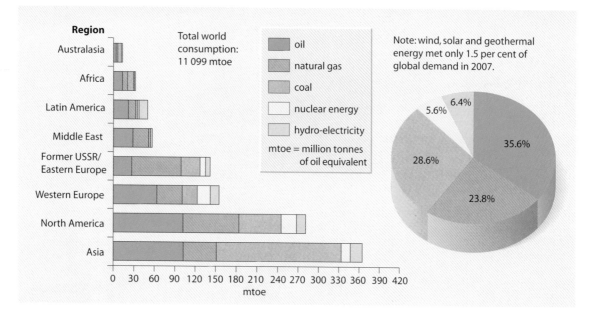

Figure 18.2

World energy consumption: by region, 2007

Note: wind, solar and geothermal energy met only 1.5 per cent of global demand in 2007.

Renewable sources of energy are mainly forces of nature which can be used continually, are sustainable and cause minimal environmental pollution. They include running water, waves, tides, wind, the sun, geothermal, biogas and biofuels. At present, with the exception of running water (hydro-electricity), the wind and biomass, there are economic and technical problems in converting their potential into forms which can be used.

World energy producers and consumers

It has been estimated that, annually, the world consumes an amount of fossil fuel that took nature about 1 million years to produce, and that the rate of consumption is constantly increasing. This consumption of energy is not evenly distributed over the globe (Figure 18.2). At present, the 83 per cent of people living in the 'developing' countries consume only 47 per cent of the total energy supply.

Although recently the consumption of energy in 'developed' countries has begun to slow down, due partly to industrial decline and environmental concerns, it has been increasing more rapidly in 'developing' countries with their rapid population growth and aspirations to raise their standard of living (China's energy consumption doubled between 1997 and 2007). This led to a conflict of interest between groups of countries at the 1992 Rio Earth Summit conference. The 'industrialised' countries, with only 17 per cent of the world's population yet consuming 53 per cent of the total energy, wished to see resources conserved and, belatedly, the environment protected. The 'developing' countries, which blame the industrialised countries for most of the world's pollution and depletion of resources, considered that it was now their turn to use energy resources, often regardless of the environment, in order to develop economically and to improve their way of life.

The world's reliance upon fossil fuels (Figures 18.2 and 18.3) is likely to continue well into this century. However, while the economically recoverable reserves of coal remain high (Figure 18.4), the similar life expectancies of oil and natural gas are much shorter (coal: about 200–400 years; oil: about 50 years; natural gas: about 120 years). The distribution of recoverable fossil fuels is spread very unevenly across the globe, with the former USSR being well endowed with coal and natural gas; North America and parts of Asia with coal; and the Middle East with oil and natural gas (Figure 18.5). As these producers are not always major consumers, there is a considerable world movement of, and trade in, fossil fuels (Figure 18.6).

Figure 18.3

World energy consumption: by type, 1982–2007

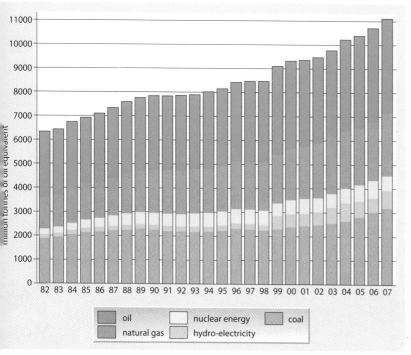

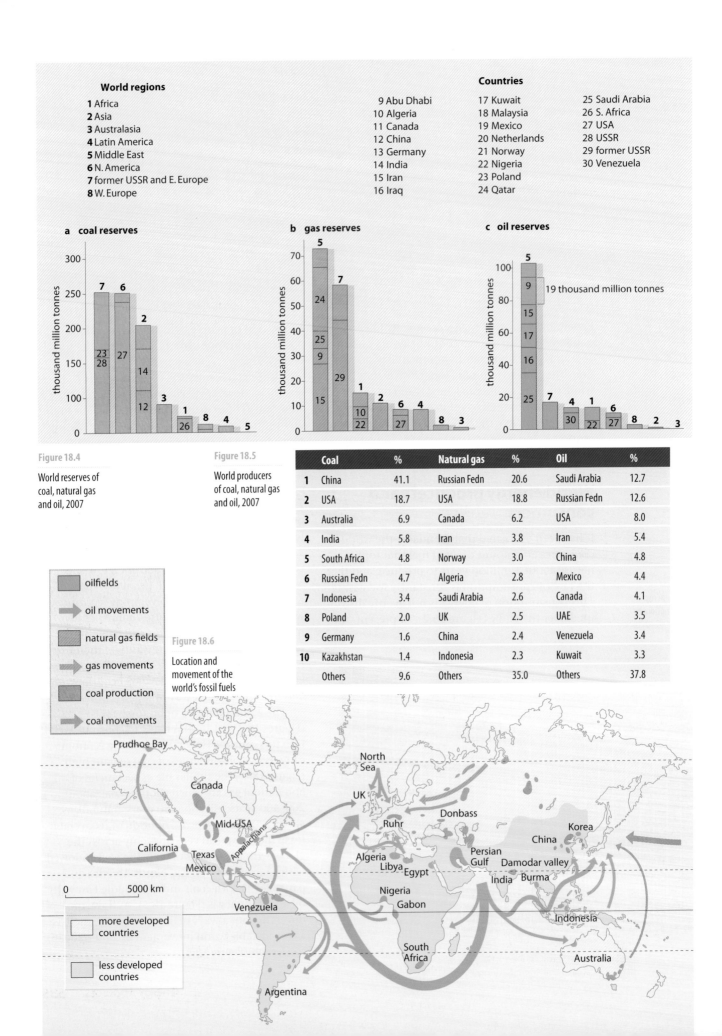

World regions

1 Africa
2 Asia
3 Australasia
4 Latin America
5 Middle East
6 N. America
7 former USSR and E. Europe
8 W. Europe

Countries

9 Abu Dhabi
10 Algeria
11 Canada
12 China
13 Germany
14 India
15 Iran
16 Iraq
17 Kuwait
18 Malaysia
19 Mexico
20 Netherlands
21 Norway
22 Nigeria
23 Poland
24 Qatar
25 Saudi Arabia
26 S. Africa
27 USA
28 USSR
29 former USSR
30 Venezuela

a coal reserves

b gas reserves

c oil reserves

19 thousand million tonnes

Figure 18.4

World reserves of coal, natural gas and oil, 2007

Figure 18.5

World producers of coal, natural gas and oil, 2007

	Coal	%	Natural gas	%	Oil	%
1	China	41.1	Russian Fedn	20.6	Saudi Arabia	12.7
2	USA	18.7	USA	18.8	Russian Fedn	12.6
3	Australia	6.9	Canada	6.2	USA	8.0
4	India	5.8	Iran	3.8	Iran	5.4
5	South Africa	4.8	Norway	3.0	China	4.8
6	Russian Fedn	4.7	Algeria	2.8	Mexico	4.4
7	Indonesia	3.4	Saudi Arabia	2.6	Canada	4.1
8	Poland	2.0	UK	2.5	UAE	3.5
9	Germany	1.6	China	2.4	Venezuela	3.4
10	Kazakhstan	1.4	Indonesia	2.3	Kuwait	3.3
	Others	9.6	Others	35.0	Others	37.8

oilfields

oil movements

natural gas fields

gas movements

coal production

coal movements

Figure 18.6

Location and movement of the world's fossil fuels

more developed countries

less developed countries

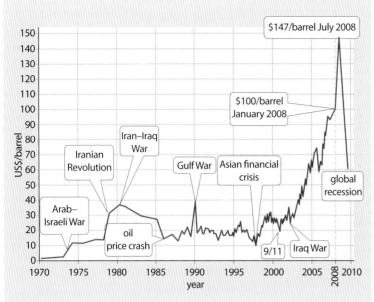

Figure 18.7

Crude oil prices, 1970–2008

UK energy consumption

The UK has always been fortunate in having abundant energy sources. In the Middle Ages, fast-flowing rivers were used to turn water-wheels while, in the early 19th century, the use of steam, from coal, enabled Britain to become the world's first industrialised country. Just when the accessible and cheapest supplies of coal began to run short, natural gas (1965) and oil (1970) were discovered in the North Sea, and improvements in technology enabled the controversial production of nuclear power. Looking ahead to a time when the UK's reserves of fossil fuels become less available and their use environmentally unacceptable, Britain's seas and weather have the potential to provide renewable sources of energy using the wind, waves and tides. Even so Britain is, for the first time, having to rely on energy imports.

The total energy consumption in the UK rose from 152.3 mtoe (million tonnes of oil equivalent, a standard measure for comparing energy consumption) in 1960 to 233.5 mtoe in 2004, since when it has fallen back a little, to 226 mtoe in 2007. Of that, 97.5 per cent still came from fossil fuels and nuclear energy and only 2.5 per cent from renewables, including hydro-electricity and waste, despite pledges to increase renewables to 20 per cent by 2020 (Figure 18.8).

Energy consumption by final user continues to see a decline by industry (34 per cent in 1980 to 21 per cent in 2007), with domestic (28 per cent) and services (12 per cent) remaining fairly steady, and a rise by transport (25 per cent in 1980 to 39 per cent in 2007).

Recent global trends

Energy consumption rose by an average of nearly 3 per cent per annum for the decade up to 2008 (Figure 18.3). The Asia-Pacific region accounted for two-thirds of this total growth, with China averaging over 8 per cent (Places 82, page 544) and India recently exceeding 6 per cent. In comparison, North America had only a slight rise, while Japan and the EU saw a decrease. Of the five main sources of primary energy, coal again, despite its contribution to climate change, saw the biggest growth. The year 2008 may be remembered as the year when the price of oil doubled that of its previous peak (Figure 18.7) before falling almost as rapidly with the onset of a global recession.

Figure 18.8

The UK's changing sources of energy, 1950–2007

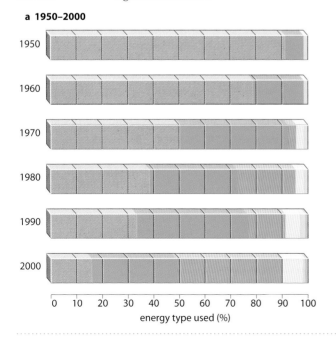

a 1950–2000

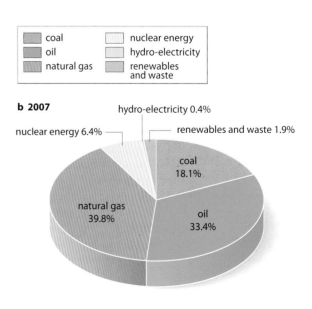

b 2007

Sources of energy

Decisions by countries as to which source, or sources, of energy to use may depend upon several factors. These include:

- Availability, quality, lifetime and sustainability of the resource.
- Cost of harnessing, as well as transporting (importing or within the country), the source of energy: some types of energy, such as oil, may be too expensive for less wealthy countries; while others, such as tides, may as yet be uneconomical to use.
- Technology needed to harness a source of energy: like costs, this may be beyond some of the less developed countries (nuclear energy), or may yet have to be developed (wave energy).
- Demands of the final user: in less developed countries, energy may be needed mainly for domestic purposes; in more developed countries, it is needed for transport, agriculture and industry.
- Size, as well as the affluence, of the local market.
- Accessibility of the local market to the source.
- Political decisions: for example, which type of energy to utilise or to develop (nuclear), or whether to deny its sale to rival countries (embargoes).
- Competition from other forms of available energy.
- Environment: this may be adversely affected by the use of specific types of energy, such as coal and nuclear; it may only be protected if there are strongly organised local or international conservation pressure groups such as Friends of the Earth and Greenpeace. More recently there has been growth in carbon trading (page 639) and the concept of our ecological footprint (page 379).

These factors will be considered in the next section, which discusses the relative advantages and disadvantages of each available or potential source of energy.

Non-renewable energy

Coal

Coal provided the basis for the Industrial Revolution in Britain, Western Europe and the USA. Despite its exploitation for almost two centuries, it still has far more economically recoverable reserves than any of the other fossil fuels (Figure 18.4). Improved technology has increased the output per worker (Figure 17.10), has allowed deeper mining with fewer workers, and has made conversion for use as electricity more efficient.

In Britain, both production and employment reached a peak in the early 1950s. Between then and 2007, the number of deep mines decreased from 901 to 6 (plus 25 opencast), the number employed from 691 000 to 6000 (4000 in deep mines) and production from 206 million tonnes to 20 million tonnes (9.6 million from deep mines). The social and economic consequences, especially in former single-industry coal-mining villages, were devastating, although people in these areas a generation or two later seem to have little desire to return to those earlier times (Figure 18.9a). Similar problems were created in other old mining communities such as in Belgium, the Ruhr (Germany) and the Appalachians (USA).

There are many reasons for this decline. The most easily accessible deposits have been used up, and many of the remaining seams are dangerous, due to faulting, and uneconomic to work. Costs have risen due to expensive machinery and increased wages. The demand for coal has fallen for industrial use (the decline of such heavy industries as steel), domestic use (oil- and gas-fired central heating) and power stations (now preferring gas). British coal has had to face increased competition from cheaper imports (USA and Australia), alternative methods of generating electricity (gas-fired) and cleaner forms of energy. Political decisions have seen subsidies paid to the nuclear power industry and a greater investment in gas rather than coal-fired power stations (the 'dash for gas' policy – page 538). Green pressures have also led to a decline in coal mining, which creates dust and leaves spoil tips; and in the use of coal to produce electricity, as this releases sulphur dioxide and carbon dioxide which are blamed for acid rain and for global warming (Case Study 9B). However, coal may still have an important future (Figure 18.9b) as alternative sources of energy run out – globally there are an estimated 155 years of high-quality coal left whereas oil and natural gas only have 45 and 65 years respectively. In the short term, coal is seen by emerging countries, such as China and India with their large reserves (Figure 18.5), as the main source for their increased energy consumption (Places 82, page 544); in the long term some countries will be dependent on the development of 'clean coal technologies' (as in Germany) and coal will be in competition with renewables.

New Opencast Mine Bid

Anger erupted after planners backed a new opencast pit in Northumberland – 12 months after saying it should be rejected. Hundreds of people have opposed UK Coal's bid to dig up 250 hectares of countryside on the edge of Ashington, in a six-year operation, only a year after the county council recommended the scheme be thrown out 'as opencasting would harm vital regeneration efforts'. Ashington, once one of the largest colliery towns in Britain, lies in an official 'constraint area' where county council policy says 'there is a strong presumption against opencast mining close to towns'. The planners appear to have changed their minds after the government gave permission for an opencast mine to be developed a few kilometres away at Cramlington, which is also in a 'constraint area', and since fears were raised that Britain may not, in the near future, have sufficient energy to 'keep its lights on'.

Abridged from the Ashington Journal, October 2008

Kingsnorth Coal-fired Power Station

The Cabinet is split over whether or not to approve a controversial plan for a £1 bn coal-fired power station at Kingsnorth on the Thames estuary in Kent. If the scheme went ahead it would be the first coal-fired station to be built in Britain for many years. The issue is, on one hand, the need to safeguard Britain's electricity supplies in the near future and, on the other, a test of the government's green credentials and assurances to reduce emissions.

E.ON UK, which has made the application, is hoping that the EU will choose this power station as part of their carbon capture experiment under which carbon emissions would be 'captured' and stored under the sea. A decision on the successful applicants is not expected for nine months. Environmental protest groups are not convinced with the assurance that, even if not selected, emissions from the power station would be lower than existing coal-fired stations.

Abridged from the The Independent, September 2008

Figure 18.9

What is the future for coal?

Oil

Oil is the world's largest business, with commercial and political influence transcending national boundaries. Indeed, several of the largest transnational enterprises are oil companies. Oil, like other fossil fuels, is not even in its distribution (Figure 18.6), and is often found in areas that are either distant from world markets or have a hostile environment, e.g. the Arctic (Alaska), tropical rainforests (Nigeria and Indonesia), deserts (Algeria and the Middle East) or under stormy seas (North Sea). This means that oil exploration and exploitation is expensive, as is the cost of its transport by pipeline or tanker to world markets. Oil, with its fluctuating prices, has been a major drain on the financial reserves of many developed countries and has been beyond the reach of most developing countries. Countries where oil is at present exploited can only expect a short 'economic boom' as, apart from several states in the Middle East where production may continue for a little longer, most world reserves are predicted to become exhausted within 45 years.

Figure 18.10

Milford Haven oil refinery

New technology has had to be developed to tap less accessible reserves. Before oil could be recovered from under the North Sea, large concrete platforms, capable of withstanding severe winter storms, had to be designed and constructed. Each platform, supported by four towers, had to be large enough to accommodate a drilling rig, process plant, power plant, helicopter landing-pad and living and sleeping quarters for its crew. The towers may either be used to store oil or may be filled with ballast to provide extra anchorage and stability. Two 90 cm trunk pipelines were laid, by a specially designed pipe-laying barge, over an uneven sea-bed to Sullom Voe on Shetland. Since then, production has spread northwards to even deeper and stormier waters west of Shetland. In 2007 the UK had 211 offshore oilfields although production from these had decreased by over 40 per cent since 1997.

On a global scale, oil production and distribution are affected by political and military decisions. OPEC (Figure 21.34) is a major influence in fixing oil prices and determining production – although even it is helpless in the face of international conflicts such as Suez (1956), the Iran–Iraq War (early 1980s) and the Gulf War (1991). Closer to home, recent British and EU fuel policies have favoured the gas and nuclear industries at the expense of oil. Oil is used in power stations, by industry, for central heating and by transport. Although it is considered less harmful to the environment than coal, it still poses many threats. Oil tankers can run aground during bad weather (*Braer*, 1993) releasing their contents which pollute beaches and kill wildlife (*Exxon Valdez*, 1989) or be hijacked by pirates (Somalia, 2008), while explosions can cause the loss of human life (Alpha Piper rig, 1988). To try to reduce the dangers of possible spillages and explosions, oil refineries have often been built on low-value land adjacent to deep, sheltered tidal estuaries, well away from large centres of population (Figures 18.10 and 18.12). Oil production is becoming increasingly concentrated in a few countries, all of whose current levels of production have been affected by geopolitics, both internal and external (Indonesia, Iran, Iraq, Kuwait, Mexico, Nigeria, Russia and Venezuela).

Natural gas

Natural gas has become the fastest-growing energy resource. It provides an alternative to coal and oil and, in 2007, it comprised almost a quarter of the world's primary energy consumption. Latest estimates suggest that global reserves will last another 65 years. Gas is often found in close proximity to oilfields (Figure 18.6) and therefore experiences similar problems in terms of production and transport costs and requirements for new technology. In 2008, Russia, Iran and Qatar announced an OPEC-type cartel that will control 60 per cent of the world's gas, a decision not welcomed by the EU which feared it could lead to a price rise and a means of achieving political goals. The UK had, in the latter part of the last century, a surplus of gas from its North Sea fields which resulted in the so-called 'dash for gas' by the electricity companies. By 2007, North Sea gas production had halved in ten years and the UK is now a net importer, some via a pipeline from Norway (2006) and an increasing amount from Russia. At present natural gas is considered to be the cheapest and cleanest of the fossil fuels.

Nuclear energy

During the 1950s, nuclear energy, with its slogan 'atoms for peace', was seen by many to be a sustainable, inexpensive and clean energy resource and by others as a potentially dangerous military weapon and a threat to the environment.

Nuclear energy uses uranium as its raw material. Compared with fossil fuels, uranium is only needed in relatively small amounts (50 tonnes of uranium a year, compared with 500 tonnes of coal per hour for coal-fired power stations). Uranium has a much longer lifetime than coal, oil or gas and can be moved more easily and cheaply. However, the development of nuclear energy, with its new technology, specially designed power stations and essential safety measures, has been very expensive. As a result, it has generally been adopted by the more wealthy countries, and even then only by those (Figure 18.11) lacking fossil fuels (Japan) or with significant energy deficits (UK, USA and France). As a source of electricity, it is fed into the National Grid; but, as a source of energy, it cannot be used by transport or for heating. The decision to develop nuclear power has been, universally, a political decision.

At its peak in the early 1990s, nuclear energy provided the UK with 30 per cent of its energy needs from 22 power stations. Since then plans for new stations have been dropped, and the plan for a fast-breeder reactor has been abandoned. By 2008, only ten stations remained open (Figure 18.12) and, by 2015, these will be reduced to four. While their closure will please the anti-nuclear lobby, it does not explain from where Britain is to get the replacement energy. This question has led to the government having to review its nuclear policy, especially when, with global warming and climate change so high on the environmental agenda, the nuclear industry can claim that the energy it produces is 'clean' and that improvements in technology have made it 'safer and more affordable' than in the past. In 2008, the government, partly also in an attempt to reduce Britain's increasing reliance on imported energy, opened the way for up to ten new stations to be built by 2020, the first by 2017.

Although nuclear power stations produce fewer greenhouse gases than thermal (coal-, oil- and gas-fired) power stations, they do present potential risks in three main areas: routine emissions of radioactivity, waste disposal, and radioactive contamination accidents. Routine emissions have been linked – without proven evidence – with clusters of increased leukaemia, especially in children, around several power stations (notably Sellafield and Dounreay). Radioactive waste has to be stored safely, either deep underground or at Sellafield. Every radioactive substance has a 'half-life', i.e. the time it takes for half of its radioactivity to die away.

Figure 18.11

Major users of nuclear energy, 2007

Amount used (mtoe)		Proportion of total energy use (%)	
USA	192	France	76.8
France	100	Lithuania	64.4
Japan	63	Slovakia	54.3
Russia	36	Belgium	54.0
Germany	32	Ukraine	48.1
South Korea	32	Sweden	46.1
Canada	22	Armenia	43.5
Ukraine	21	Slovenia	41.6
Sweden	15	Switzerland	40.0
China	14	Hungary	36.8
UK	14	South Korea	35.3
Spain	13	Bulgaria	32.1

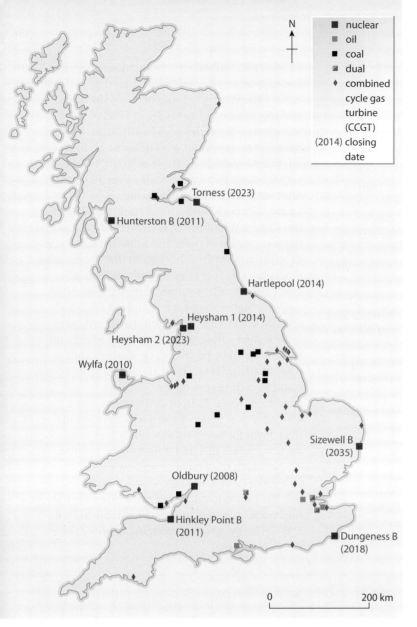

Figure 18.12

Power stations in the UK with 50 MW or more capacity, 2008

Legend (map):
- ■ nuclear
- ■ oil
- ■ coal
- ■ dual
- ♦ combined cycle gas turbine (CCGT)
- (2014) closing date

Map labels:
Torness (2023)
Hunterston B (2011)
Hartlepool (2014)
Heysham 1 (2014)
Heysham 2 (2023)
Wylfa (2010)
Sizewell B (2035)
Oldbury (2008)
Hinkley Point B (2011)
Dungeness B (2018)

0   200 km

Iodine, with a half-life of 8 days, becomes 'safe' relatively quickly. In contrast, plutonium 239, produced by nuclear reactors, has a half-life of 250 000 years and may still be dangerous after 500 000 years. The two worst radioactive accidents resulted from the melt-down of reactor cores at Three Mile Island in the USA (1979) and at Chernobyl in Ukraine (1986). Fortunately there was no such leak when the world's largest nuclear power plant was forced to close following an earthquake in Japan in 2007. It was mainly for economic and safety reasons that British nuclear power stations (Figure 18.12) were built on coasts and estuaries where there is water for cooling and cheap, easily reclaimable land well away from major centres of population. However, the British government had to agree in 1998, following renewed calls from several EU countries, to make a large reduction in discharges into the Irish Sea from Sellafield.

Renewable energy

With the depletion of oil and gas reserves during the early years of the 21st century and the unfavourable publicity given to all types of fossil fuels, especially regarding their contribution towards global warming, renewable energy resources are likely to become increasingly more attractive. They are likely to become more cost-competitive, offer greater energy diversity, and allow for a cleaner environment. As shown in Figure 18.1, there are two types of renewable energy:

- ■ **Continuous sources** are recurrent and will never run out. They include running water (for hydro-electricity), wind, the sun (solar), tides, waves and geothermal.
- ■ **Flow sources** are sustainable providing that they are carefully managed and maintained (Framework 16, page 499). Biomass, including the use of fuelwood, is sustainable in that it has a maximum yield beyond which it will begin to become depleted.

Hydro-electricity

Hydro-electricity is the most widely used commercially produced renewable source of energy (fuelwood is used by more people and in more countries). Its availability depends on an assured supply of fast-flowing water which may be obtained from rainfall spread evenly throughout the year, or by building dams and storing water in large reservoirs. The initial investment costs and levels of technology needed to build new dams and power stations, to install turbines and to erect pylons and cables for the transport of the electricity to often-distant markets, are high. However, once a scheme is operative, the 'natural, continual, renewable' flow of water makes its electricity cheaper than that produced by fossil fuels.

Although the production of hydro-electricity is perceived as 'clean', it can still have very damaging effects upon the environment. The creation of reservoirs can mean large areas of vegetation being cleared (Tucurui in Amazonia), wildlife habitats (Kariba in Zimbabwe) and agricultural land (Volta in Ghana) being lost, and people being forced to move home (Aswan in Egypt and the Three Gorges Dam in China – Places 82, page 544). Where new reservoirs drown vegetation, the resultant lake is likely to become acidic and anaerobic. Dams can be a flood risk if they collapse or overflow (Case Study 2B), have been linked to increasing the risk of earthquake activity (Nurek Dam in Tajikistan) and can trap silt previously spread over farmland (Nile valley, Places 73, page 490). Despite these negative aspects, many countries rely on large, sometimes prestigious, schemes or, increasingly in less developed countries, on smaller projects using more appropriate levels of technology (Case Study 18).

Wind

Wind is the most successful of the new renewable technologies. Wind farms are best suited to places where winds are strong, steady and reliable and where the landscape is either high or, as on coasts, exposed. Although expensive to build – wind farms cost more than gas or coal-fired power stations – they are cheap and safe to operate. Most of Britain's new wind farms are to be located offshore where, although more costly to construct, winds are more reliable than on land. As wind farms are mainly pollution free, they do not contribute to global warming or acid rain and they should significantly contribute to world commitments to reduce carbon dioxide emissions by 60 per cent by 2050. Winds, especially in Western Europe and California, are strongest in winter when demand for electricity is highest. Wind farms can provide extra income for farmers who could earn more from them than they could from growing a crop on the same-sized plot. Wind farms also create extra jobs for people living in rural areas and in the electricity generation supply chain. As fossil fuels become less available, countries will have to become increasingly dependent on renewables such as wind.

However, British environmentalists are now less supportive of wind power than they originally were. This is partly because many of the actual and proposed wind farms are in areas of scenic attraction, where they are visually intrusive, or too close to important wildlife habitats. In an attempt to make them more efficient, turbines are becoming increasingly tall – over 50 m on onshore wind farms and even higher on those located offshore, where some could be taller than the Canary Wharf tower. Elsewhere, local residents complain of noise and impaired radio and TV reception, while others claim that the rotating blades are a danger to birds, the turbines can affect airport radar systems, and that electricity costs are higher than for power from fossil fuels and nuclear energy. As yet, electricity companies cannot store surplus power for times when wind power cannot be produced, i.e. during calms or when the wind is less than about 15 km/hr which could be during very cold winter anticyclonic conditions (page 234); or during gales when winds are over 55 km/hr and wind farms must shut down for safety reasons. Both eventualities are times when demand is likely to be greatest.

Although the first large-scale wind farms were located in California (Figure 18.13) and the USA still has over one-quarter of the world's capacity, the fastest growth is in the EU, notably in Spain and Germany, and the emerging countries of China and India.

Places 81 California and the UK: wind farms

California

Most wind farms in the USA have been developed by private companies. The developers, who use either their own or leased land, sell electricity to electric utilities. At present, 90 per cent of the USA's capacity comes from California. California's wind farms are in an ideal location mainly because peak winds occur about the same time of year as does peak demand for electricity in the large cities nearby. Approximately 16 000 turbines within the state produce enough electricity to supply a city the size of San Francisco. The three largest wind farms are at Altamont Pass (east of San Francisco), Tehachapi (between the San Joaquin Valley and the Mojave Desert) and San Gorgonio (north of Palm Springs). The Altamont Pass, with 7000 turbines, is one of the largest wind farms in the world (Figure 18.13). The average wind speed averages between 20 and 37 km/hr. The land is still used for cattle grazing as there is only one turbine for every 1.5–2 ha.

Figure 18.13

Wind farm at Altamont Pass, California

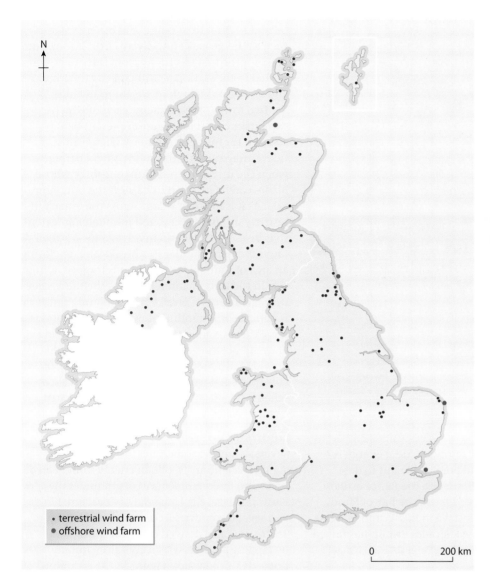

0   200 km

The UK

Britain's first wind farm was opened in 1991 near Camelford in Cornwall. The farm, on moorland 250 m above sea-level and where average wind speeds are 27 km/hr, generates enough electricity for 3000 homes. In 2008, Britain had 188 operational wind farms, 7 of which were offshore, but these in total provided less than 1 per cent of the country's energy needs (Figure 18.14). With another 43 under construction (8 offshore) and 130 projected (8 offshore), the government hopes that, by 2010, 10 per cent of Britain's energy will come from renewables (60 per cent from the wind); and, by 2020, 20 per cent. To achieve this, another 4000 onshore turbines and 3000 offshore wind farms (with 11 000 turbines) will be required.

Figure 18.14

Wind farms in the UK, 2007

Solar energy

The sun, as stated earlier, is the primary source of the Earth's energy. Estimates suggest that the annual energy received from the sun (insolation) is 15 000 times greater than the current global energy supply. Solar energy is safe, pollution-free, efficient and of limitless supply. Unfortunately, it is expensive to construct solar 'stations', although many individual homes have had solar panels added, especially in climates that are warmer and sunnier than in Britain. It is hoped, globally, that future improvements in technology will result in reduced production costs. This would enable many developing countries, especially those lying within the tropics, to rely increasingly on this type of energy. In Britain, the solar energy option is less favourable partly due to the greater amount of cloud cover and partly to the long hours of darkness in winter when demand for energy is at its highest.

In 2008, South Korea opened the world's largest solar power plant. It covers the equivalent of 93 football stadiums and provides electricity for 100 000 homes.

Wave power

Waves are created by the transfer of energy from winds which blow over them (page 140). In western Europe, winter storm waves from the Atlantic Ocean transfer large amounts of energy towards the coast where it has the potential to generate the same amount of energy for the UK as wind does now. At present there are two experimental schemes off the Scottish coast, making it ten years behind wind power. The first, LIMPET, is a 500 kW shoreline oscillating water column in Islay; the second is the 750 kW Pelamis sea snake – a hinged contour device – in Orkney (the Portuguese are now using Pelamis commercially).

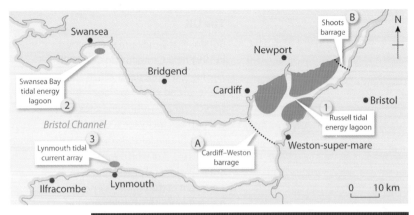

Figure 18.15

A Severn barrage or tidal lagoons?

	Barrages		Lagoons
	Cardiff–Weston	**Shoots**	**Russell**
Cost	£15 bn	£1.5 bn	est. £10 bn
UK's energy supply	5%	0.75%	7%
Generation cost	5p kW/hr	?	2–2.5p kW/hr
Impounded areas	480 km^2	very little	300 km^2
Length of barrage/walls	16 km	4 km	100 km
Environment	low carbon; loss of feeding grounds for up to 50 000 birds	low carbon; less loss of feeding grounds	low carbon; little loss of feeding grounds

Tidal energy

Of all the renewable resources, tidal energy is the most reliable and predictable but to date major schemes are limited to the Rance estuary in north-west France (1960s), the Bay of Fundy in eastern Canada, Kislaya in Russia and Jiangxia in China. For over two decades Britain has talked about (and is still debating) erecting a barrage across the Severn estuary, which has the world's second highest tidal range, and other estuaries such as the Mersey and Solway Firth. It took until 2008 for the first electricity from tides to be connected to the National Grid (at Eday in Orkney and Strangford Lough in Northern Ireland).

Two forms of technology, each of which exploits the tidal range, are at present being assessed for the Severn estuary (Figure 18.15): tidal barrage and tidal lagoon. A tidal barrage is when the incoming tide turns a turbine whose blades can be reversed to harness the outgoing tide. As it is in effect a dam across the estuary, it restricts shipping access and inundates an extensive area. A tidal lagoon involves a rock-walled impoundment, similar to a breakwater, enclosing an area of shallow water. Water is trapped at high tide in the lagoon and released as the tide recedes through a bank of electricity-generating turbines within the impoundment walls. This method is less extensive, less environmentally damaging, does not obstruct shipping access and would provide both more and cheaper electricity. If constructed, it would be the world's first such scheme (it is favoured by Friends of the Earth Cymru (Wales)).

Geothermal energy

Several countries, especially those located in active volcanic areas, obtain energy from heated rocks and molten magma at depth under the Earth's surface, e.g. Iceland, New Zealand, Kenya, and several countries in central America (Figure 18.16). It is also derived from hot springs and by tapping aquifers which contain naturally hot water. Cold water (Figure 18.17) is pumped downwards, is heated naturally and is then returned to the surface as steam which can generate electricity. Geothermal energy does pose

Figure 18.16

Wairakei geothermal power station, New Zealand

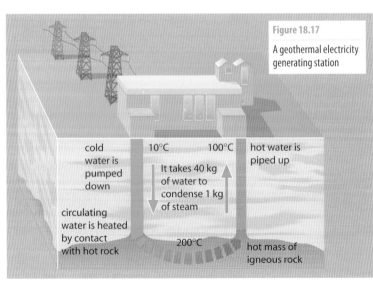

Figure 18.17

A geothermal electricity generating station

cold water is pumped down

circulating water is heated by contact with hot rock

10°C   100°C

It takes 40 kg of water to condense 1 kg of steam

200°C

hot water is piped up

hot mass of igneous rock

environmental problems as carbon dioxide and hydrogen sulphide emissions may be high, the water supply can become saline, and earth movements can damage power stations.

Biomass

Biomass, also known as biofuels and bioenergy, is the dominant form of energy for most of the world's population who are living in extreme poverty (page 609 and Figure 18.19) and who use it for cooking and, when necessary and if sufficient is available, for heating. It is obtained from organic matter, i.e. crops, plants and animal waste, of which the most important to those living in the least developed countries, especially in Africa, is **fuelwood**. Trees are a sustainable resource, providing that those cut down are replaced, which costs money, or allowed to regenerate, which takes time – but money and time are what these people do not have. As nearby supplies are used up, collecting fuelwood becomes an increasingly time-consuming task; in extreme cases, it may take all day (Figure 18.18 and Case Study 18). Many of these countries have a rapid population growth, which adds greater pressure to their often meagre resources, and lack the capital and technology to develop or buy alternative resources. In places

where the demand for fuelwood outstrips the supply, and where there is neither the money to replant nor the time for regeneration, the risk of desertification and irreversible damage to the environment increases – i.e. the cycle of environmental deprivation (Figure 18.18).

The use of biomass is generally considered to be a 'carbon neutral' process as the carbon dioxide released in the generation of energy balances that absorbed by plants during their growth. This is not, however, applicable to those parts of Africa where animal dung is allowed to ferment to produce methane gas. While the methane, a greenhouse gas, provides a vital domestic fuel, it means that the dung cannot be spread as a fertiliser on the fields.

Biomass can also be used to produce **biofuels** (**bioenergy**), the first being used in Brazil where sugar cane was allowed to ferment to produce bio-ethanol which was then used as a vehicle fuel that was cheaper than petrol. More recently, and more widely used, biodiesel comes from oil palm, a use that has led to increased forest clearances for that crop in Malaysia; in the EU oilseed rape, and in the USA maize, are being grown for the same purpose. Governments are viewing the use of this renewable resource as a way of reducing their carbon emissions without foreseeing that their increased growth, at the expense of food crops, will lead to food shortages and rising global food prices.

It has become apparent that the sustainable use of bioenergy requires a balancing of several factors, including the competition between food and energy security, the effects on rural development and on agricultural markets and food prices, as well as the effects on the environment. In 2007, biomass accounted for 82 per cent of the UK's renewable energy sources (wind 9 per cent, and hydro-electricity 8 per cent), the majority of which was derived from landfill gas and waste combustion. It is also the fastest-growing renewable in the EU while its use in the USA is said to be equal to the output of ten nuclear power stations.

Hydrogen

Hopes are high for the development of a fuel cell in which a chemical reaction takes place that generates electricity from hydrogen. The reaction produces clean, efficient energy in a process that releases nothing more damaging to the environment than water vapour. Although developed countries see the petrol-free hydrogen car as a major breakthrough in transport, fuel cells hold potential for developing countries too as they are equally economic on a small scale and require little maintenance.

Figure 18.18

Collecting fuelwood, sub-Saharan Africa

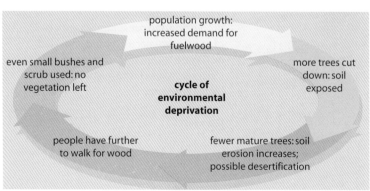

population growth: increased demand for fuelwood

even small bushes and scrub used: no vegetation left

more trees cut down: soil exposed

cycle of environmental deprivation

people have further to walk for wood

fewer mature trees: soil erosion increases; possible desertification

Figure 18.19

The cycle of environmental deprivation

Energy conservation through greater efficiency

One of the UN's main objectives since the first 'Earth Summit' on the environment (Rio de Janeiro 1992) has been to try to get countries to agree to set global limits and timescales in reducing harmful emissions from vehicles, factories and power stations and to seek methods of greater energy efficiency. Since then, progress has been limited partly due to opposition from vested interests and, until more recently, a lack of political will. However, there have been some successes. Several industries, including the steel industry, have improved their techniques, reducing the amount of energy needed; factories have made savings by reducing to a single source of electricity needed for heating, lighting and operating machinery; and the number of coal- and oil-fired power stations has been reduced in favour of gas (a cleaner greenhouse gas). At home, heat loss has been reduced through roof and wall insulation and double glazing, while for lighting the EU is trying to enforce the use of energy-saving bulbs.

Places 82 China: changes in energy production and consumption

Figure 18.20

Energy consumption in China, 1992–2007

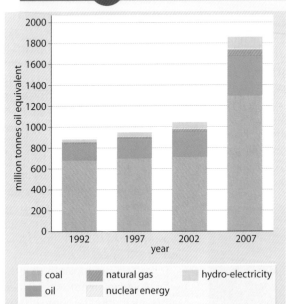

legend:
- coal
- oil
- natural gas
- nuclear energy
- hydro-electricity

In the early 1990s, China's energy industry was dominated by coal (Figure 18.20), which was not surprising since the country was producing nearly two-fifths of the world's total (Figure 18.5). Coal is mined in most parts of the country, although production is lower to the south of the Yangtze River and least in the mountains to the west (Figure 18.21). With industry, transport and homes all so reliant upon the burning of coal in one form or another, many Chinese cities experienced severe atmospheric pollution (Figure 18.22), and China was blamed for releasing annually 10 per cent of the world's greenhouse gases. The remainder of China's energy budget was made up from oil, in which it was self-sufficient, and hydro-electricity. The fact that the country could provide sufficient energy for its needs was because, despite having such a large population, the country's standard of living and consumption of energy was low and it had yet to embark on the rapid economic development seen in the late 1990s and early 21st century. Even so, in 1995 China was ranked second in the world for generated energy, mainly from its thermal and hydro-electric power stations, and generating capacity.

Since the turn of this century and as China has become, economically, the world's most rapidly emerging country, there has been a huge increase in both its energy production and consumption, especially of coal, hydro-electricity and imported oil (Figure 18.20). This emergence (Chapter 21) has had a major effect on the global economy.

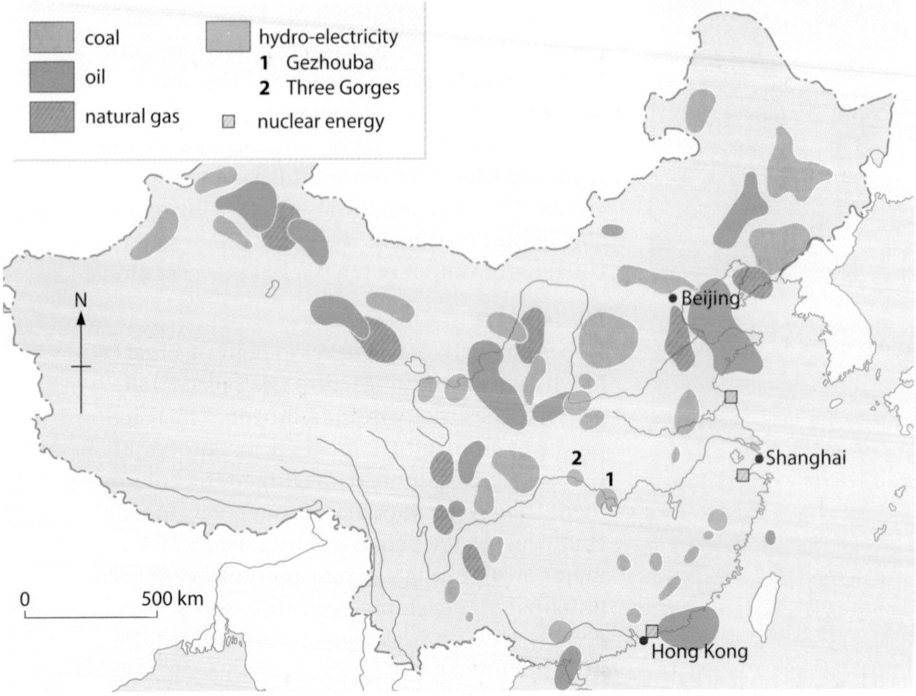

Figure 18.21

Energy resources in China

Map legend:
- coal
- oil
- natural gas
- hydro-electricity
  1 Gezhouba
  2 Three Gorges
- nuclear energy

N

0   500 km

Beijing
Shanghai
Hong Kong

Figure 18.22

Atmospheric pollution over Chengdu

Coal still accounts for 70 per cent of China's energy consumption (Figure 18.20), despite its known effect on global warming and the pollution it causes in Chinese cities. Production almost doubled between 2002 and 2007 to meet the growing demand, despite a continually high rate of mining accidents and resultant deaths (Figure 18.23a). As the country's road and air transport systems develop, an increasing amount of oil has to be imported, mainly from the Middle East. This increase was partly to blame for a world shortage of oil in 2008 that led to the record high global price per barrel (Figure 18.7).

a

China's dependence on coal continues amid the incidence of accidents – 16 miners were killed and 46 injured in a mine blast last week – and claims of inefficient mining methods and high levels of pollution. In the last few years more than 18 000 small mines have either been closed, or merged with larger ones, but 14 000 are still operating. This number will be reduced further to 10 000 by 2010. Two effects of the merging of small mines have been a doubling in coal production and a drop in fatalities. In 2007, when 2900 died – 450 fewer than in the previous year – for every million tonnes of coal produced, the death toll at small mines was eight times that of the larger state-owned ones as the latter are believed to pay more attention to safety.

October 2008

b

China, now the world's second biggest gas consumer, plans to boost its own production by 50 per cent by 2010 by which time gas will have increased its share of the nation's total energy consumption from 3.5 per cent to 5.3 per cent. However, China will still have to import a significant amount which it will do through a second west–east pipeline at present being built to connect the Central Asian countries, notably Turkmenistan, with the energy-thirsty eastern and southern regions that include Shanghai and Guangdong.

November 2007

c

The first nuclear power reactor to be built in China (early 1990s) was at Qinshan in Zhejiang Province to the south of Shanghai. Since then six more reactors have been added to the complex. A second site is at Daya Bay (2002) in Guangdong Province where two reactors now provide energy for Shenzhen and Guangzhou, while a third at Lianyungang (2007) in Jiangsu Province, equidistant between Shanghai and Beijing, also has two reactors. All these reactors are second-generation, but work has just begun near Qinshan on a new third-generation type reactor.

June 2008

d

The last generator of China's Three Gorges Dam went online yesterday, meaning that the world's largest hydropower plant has become fully operational – five years after the first of the 26 turbines in the project's original plan began producing energy. The Three Gorges is now, in 2008, producing 58 per cent of the country's total hydro-electricity. The original plan has since been expanded to include six more generators which will be completed by 2012.

October 2008

Figure 18.23

Changes in coal, natural gas, nuclear power and hydro-electric power, adapted from *China Daily*

In comparison, consumption of natural gas and nuclear power is small but both show an increase (Figure 18.23b and c). Of the renewable sources of energy, hydro-electricity is by far the most important and is expected to become even more so as fossil fuels, as they run out, cannot go on satisfying China's rising needs for energy and as the country looks for cleaner options. Schemes such as the Three Gorges (Figures 18.23d and 18.24) are predicted to account for 28 per cent of China's total power generation by 2015.

Figure 18.24

The Three Gorges Dam on the Yangtze River

Development and energy consumption

To many people, especially in developed countries, economic development is linked to the wealth of a country, with wealth being measured by GDP per capita (page 606). Of several other variables that can be used to measure development, one is energy consumption per capita – i.e. how much energy, often given in tonnes of coal or oil equivalent, that each person in a country uses per year. Consequently a correlation between the wealth of a country and energy consumption might be expected (Framework 19, page 612).

The log-log graph in Figure 18.25 seems to show that there is a good, positive correlation between the two variables, i.e. as the wealth of a country increases, so too does its energy consumption. The huge gap in energy consumption between the developed and the developing world is shown in Figure 18.26. Note also that those countries above the line in Figure 18.25 tend to have more natural energy reserves (Russia – gas, Saudi Arabia – oil; Zambia – hydro-electricity) than those below the line (Italy, Peru).

Energy is the driving force behind most human activities, so it is fundamental to development. Energy allows people to make greater use of the resources that they have. According to Practical Action (Case Study 18 and Places 90, page 577):
'reliable, accessible and affordable energy supplies can play an important role in improving living conditions in the developing world. They provide light and heat for homes, and power workshops that create jobs and generate wealth. Poor people in developing countries face particular problems in securing energy for their daily needs. This is a pressing issue in rural areas where most people live. More than half the world's population relies upon biomass fuel (usually wood but also charcoal, crop residues or animal dung). Poor people cannot afford alternative fuels such as gas or kerosene. National grids mainly serve urban areas, or large industrial operations; it is prohibitively expensive to extend them far into the countryside. In places where there are renewable energy resources such as the sun, wind and water, communities often lack the knowledge, expertise and capital needed to install the most appropriate system.'

Figure 18.25

Correlation between GDP (US$) and energy consumption, 2008

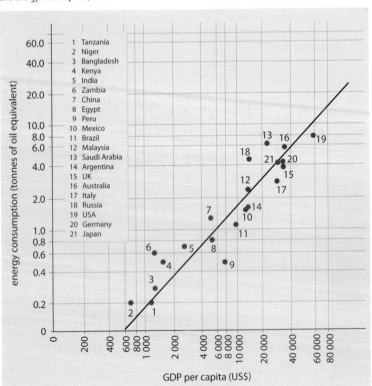

Figure 18.26

Energy consumption per capita in the developed and developing worlds

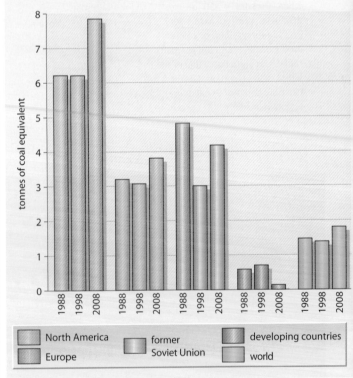

'An Appropriate Technology is exactly what it says – a technology appropriate or suitable to the situation in which it is used [page 576]. If that situation is a highly industrialised urban centre the appropriate technology may well be "high tech". If, however, the situation is a remote Nepalese village "appropriateness" will be measured in the following terms:
• Is it culturally acceptable?
• Is it what people really want?
• Is it affordable?
• Is it cheaper or better than alternatives?
• Can it be made and repaired with local material, by local people?
• Does it create new jobs or protect existing ones?
• Is it environmentally sound?

For many decades "Aid" meant sending out the same large-scale, expensive, labour-saving technologies that we use: huge hydro-electric schemes, coal-fired power stations, diesel-powered generators. In some cases, for example towns and industrial areas, these have been appropriate. But such schemes do not reach the poorer communities in the rural areas. What was needed was some way of using local resources appropriately, and best of all some way of using renewable resources to decrease the need for reliance on outside help. Wind, solar and biogas energy are possibilities, but another resource widely available and already in use for thousands of years is water. Water is attracting much attention in the search for renewable sources of

energy. However, despite continuing public outrage at the devastating impact of large hydro-electric schemes on people's livelihoods and the environment [page 539 and Places 82], vast sums of money continue to be pumped into big dams and other inappropriate power generation plans. On the other hand, the intermediate approach, through small-scale hydro, has no negative impact on the environment, offers positive benefits to the local community, and uses local resources and skills.'

Practical Action

Practical Action and micro-hydro in Nepal
'The small Himalayan kingdom of Nepal ranks as one of the ten poorest countries in the world. Around 90 per cent of its 19 million people earn their living from farming, often at a subsistence level. The Himalaya mountains offer Nepal one vast resource – the thousands of streams which pour down from the mountains all year round. Nepali people have harnessed the power in these rivers for centuries, albeit on a small scale [Figure 18.27].

About 20 years ago, two local engineering workshops began to build small, steel, hydro-power schemes for remote villages. These turbines have the advantage of producing more power than the traditional mills, as well as being able to run a range of agricultural processing machines [Figure 18.28]. Practical Action first became involved in Nepal's micro-hydro sector in the late 1970s when the local manufacturers asked for help in using their micro-hydro schemes to generate electricity.

In the mid-1980s, Practical Action ran two training courses on micro-hydro power aimed at improving the technical ability of the nine new water turbine manufacturers that had been established in Nepal. These courses were very successful and prompted an agreement between Practical Action and the Agricultural Development Bank (the agency which funds micro-hydro power in Nepal) to collaborate on the development of small water turbines for rural areas. This work not only improved and extended the range and number of micro-hydro schemes in Nepal, but also established Practical Action as a leader in the field. In 1990 Practical Action was included in a government task force investigating the whole area of rural electrification; and in 1992 Practical Action was asked by the government to help establish an independent agency to promote all types of appropriate energy in rural areas of the country.'

Practical Action

Figure 18.27

Cross-section of a traditional Nepali water mill

1 chute delivering the water to the paddles of the wheel
2 grain hopper (basket)
3 device to keep the grain moving
4 metal piece to lock top of shaft in upper millstone
5 grinding stones
6 metal shaft
7 thick wooden hub
8 wooden horizontal wheel, with obliquely set paddles attached to hub
9 metal pin and bottom piece
10 lifting device to adjust gap between millstones

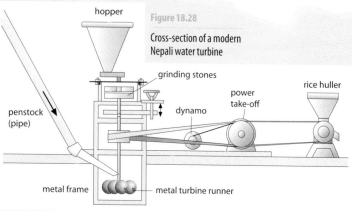

Figure 18.28

Cross-section of a modern Nepali water turbine

Life before power

'Water power, harnessed using water wheels or *ghattas*, has been used for centuries for grinding corn. The micro-hydro system in the village now has improved the efficiency of milling, so that what used to take a woman four hours to grind by *ghatta* can be done in fifteen minutes. The power can also be used for dehusking rice and extracting oil from sunflower seeds [Figure 18.29].

The mechanical power produced by Ghandruk's micro-hydro system is also converted to electric power, which is distributed to every house in the village. Apart from the obvious benefit of lighting, many households are starting to use electric cookers or *bijuli dekchis*, which work like slow cookers [Figure 18.29].

Women are turning to *bijuli dekchis* because they reduce smoke levels in the kitchen, they save time by reducing the amount of firewood the family needs to collect, and they are more convenient and cook faster than traditional stoves. In a country ravaged by deforestation – villagers spend up to 12 hours on a round trip to collect wood – fuel saving is becoming more and more important.

Micro-hydro schemes like the one in Ghandruk work because the community has "ownership" of the scheme by participating in its planning, installation and management; because the machinery needed can be made and maintained by local manufacturers using local materials available in the country; and because production and consumption are linked within a community.

The lives of villagers all over Nepal are literally being lit up by micro-hydro schemes, and the country could serve as a model for decentralised, sustainable energy production. Already, 700 mechanical and 100 electrical schemes have been installed.

Much of the impetus for the development of hydro in Nepal initially stemmed from the absence of fossil fuel reserves to exploit. However, if the Government can resist the temptations of big dam schemes and the dollars being thrown at them by the big, international donor agencies, it could have the last laugh watching the rest of the world scrabble for the last of fossil fuel reserves.'   Practical Action

Crop processing

A woman grinds corn for the family meal.

An elderly Nepalese woman and young girl hull rice with a traditional footpowered *dhiki*.

Cooking

Cooking on an open fire.

Living

Figure 18.29

Practical Action's work in Nepal

Light comes from a single kerosene lamp.

Power for life

The grinder in the mill-house.

Grinding enough corn to feed a family for just 3 days takes 15 hours when it is done by hand.

By taking corn to the grinder in the mill-house – usually a popular meeting place for villagers – 3 days' worth of corn can be ground in just 15 minutes.

Rice being hulled mechanically in a 3 kW mill.

For thousands of women, the supply of power releases them from the many labour-intensive and time-consuming tasks they previously had to carry out by hand.

Villagers can now hull their rice mechanically with this 3 kW mill, driven by a micro-hydro turbine. Time is saved and quality and productivity increased.

Low-wattage electric cookers.

Cooking on an open fire burns up a great deal of wood (which is becoming increasingly scarce) and gives off a lot of thick smoke. As a result, the villagers not only have to walk long distances to collect their fuel, but many women and children suffer from serious lung disorders.

Practical Action is helping to develop two low-wattage electric cookers which have been specifically designed to make use of 'off-peak' electricity. The *bijuli dekchi* heats water during off-peak times for use in cooking later on, while the heat storage cooker stores the energy available during off-peak periods and releases it at mealtimes for cooking. Both save fuelwood and help to reduce deforestation.

A Nepalese family enjoying the benefits of electric light.

Kerosene lamps are costly to run, and those who can afford them have to collect fuel in cans from towns which are usually several days' walk away.

With electric light, children and adults can improve their education by learning to read and write in the evenings. Electric light is also cheaper, cleaner and brighter than kerosene.

Further reference

Blades, H. (2007) 'Canada's black gold', *Geography Review* Vol 20 No 3 (January).

Clark, N.A. (2006) *Tidal Barrages and Birds*, Ibis.

Middleton, N.J. (2008) *The Global Casino: An Introduction to Environmental Issues*, Hodder Arnold.

Patterson, W. (2007) *Keeping the Lights On: towards sustainable electricity*, Earthscan.

Pickering, K.T. and Owen, L.A. (1997) *An Introduction to Global Environmental Issues*, Routledge.

Sinden, G. (2007) 'Wind power', *Geography Review* Vol 21 No 1 (September).

BP Amoco Statistical Review of World Energy:
www.bp.com/productlanding.do?
categoryId=6929&contentId=7044622

British Wind Energy Association:
www.bwea.com

Energy consumption:
www.un.org/Pubs/CyberSchoolBus/
index.shtml

www.worldenergy.org

International Atomic Energy Agency (IAEA):
www.iaea.org/

International Energy Agency (IEA), World Energy Outlook:
www.worldenergyoutlook.org/

Renewable energy:
www.berr/gov.uk/energy/index.html

www.bbc.co.uk/climate/adaptation/
renewable_energy.shtml

UK Energy policies and statistics:
www.berr.gov.uk/whatwedo/energy/
index.html

US Department of Energy's Efficiency and Renewable Energy Network:
www.eere.energy.gov/

US Energy Information Administration:
www.eia.doe.gov/

World Commission on Dams:
www.dams.org/

World Energy Council:
www.worldenergy.org/

Questions & Activities

Activities

1 **a** **i** What are 'natural resources'? *(1 mark)*

ii What is the difference between renewable and non-renewable resources? *(2 marks)*

iii Name a renewable source of energy that is used commercially. State where it is produced and explain why conditions in that area are suitable. *(3 marks)*

iv Explain what will happen to the amount of reserves of a fuel such as natural gas if:
- the market price of gas goes up
- new technology is developed, allowing deeper wells to be drilled. *(4 marks)*

Figure 18.30

Rate of world energy usage in terawatts (TW)

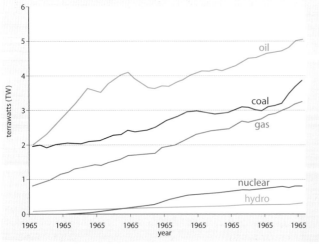

b Study Figure 18.30.

i Describe the main trends shown by the graph. *(4 marks)*

ii During the 1990s the use of energy resources by the more economically developed countries did not increase, and may even have fallen. At the same time the amount used by less economically developed countries increased.

Explain this situation. *(4 marks)*

c Describe the main features of the world trade in any one fuel. *(7 marks)*

2 **a** Fuelwood is an important source of power in many remote regions in less economically developed countries (LEDCs). Name an example of a region where fuelwood is widely used and:

i explain why people in that region rely on fuelwood. *(3 marks)*

ii describe some of the problems caused for the economy and the environment by the reliance on fuelwood. *(5 marks)*

b Large hydro-electric power schemes are seen as the solution to the energy shortages of many LEDCs.

i Suggest why some people see such schemes as a welcome development for that country. *(5 marks)*

ii Suggest why other people see such schemes as being unwelcome. *(5 marks)*

c Recent conferences on global warming have concluded that more economically developed countries should share their technological knowledge with the LEDCs. How might such sharing help to reduce global warming in future? *(7 marks)*

Exam practice: basic structured questions

3   **a**   In many less economically developed countries fuelwood is the main source of energy for heating and cooking. Explain how this can cause:

       i   damage to the environment

       ii   social problems.   *(10 marks)*

   b   **i**   What does 'appropriate technology' mean?   *(2 marks)*

       ii   Appropriate technology can be used by poor people in remote areas to harness energy supplies. Describe one such scheme in a named region of the world.   *(5 marks)*

       iii   Explain how the scheme described in **b ii** brings social and economic benefits to the people who use it.   *(8 marks)*

4   **a**   What is meant by the term 'fossil fuel'?   *(2 marks)*

   b   Choose **one** country that has important reserves of coal.

       i   Describe the distribution of coal reserves in that country.   *(4 marks)*

       ii   Explain the economic factors that are influencing decisions about whether those reserves should be exploited at the present time.   *(9 marks)*

       iii   Name **one** environmental problem caused by the use of coal as a fuel. Describe the problem. Explain how good management can reduce the problem.   *(10 marks)*

Exam practice: structured questions

5   Lack of a suitable power supply is holding back development in many remote areas of the world. For a named area:

   a   explain how shortage of power has caused economic and social problems.   *(12 marks)*

   b   explain how the problems are being reduced by provision of an appropriate power supply.   *(13 marks)*

6   Study Figure 18.31a.

   a   **i**   Describe the major changes in the UK's energy mix between 1971 and 2005.   *(5 marks)*

       ii   Account for the decline in the use of coal and the increase in the use of natural gas over this period.   *(10 marks)*

   b   Should the UK increase its use of nuclear energy over the next 10 years?

       Justify your answer.   *(10 marks)*

7   Study the two graphs in Figure 18.31.

   a   Describe the major changes in France's energy supply between 1971 and 2005.   *(6 marks)*

   b   Compare France's energy mix in 2005 with the energy mix of the UK.   *(4 marks)*

   c   Which of the two countries has the better mix in terms of:

       • energy security

       • minimising environmental damage?

       Justify your answer.   *(15 marks)*

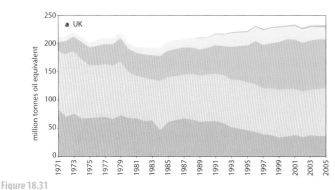
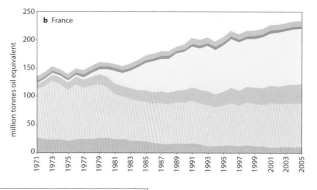

Figure 18.31

Total energy supply in the UK and France, 1971–2005

coal   oil   gas   nuclear   hydro   combined renewables & waste

Exam practice: essays

8   Evaluate the arguments for and against the development of new coal-fired power stations, such as the one that is proposed at Kingsnorth.   *(25 marks)*

9   Discuss the benefits and problems that would be involved in an increased reliance on biomass as a major source of energy supply.   *(25 marks)*

10   Choose any **two** of the following sources of renewable energy.

   • wind   • solar

   • waves   • tidal power

   • geothermal

   Discuss the economic, environmental and technological issues that are involved in the development of each of your chosen sources.   *(25 marks)*

19

Manufacturing industries

• •

'Science finds, industry applies, man confirms.'

Anon, Chicago World Fair 1933

'We need methods and equipment which are cheap enough so that they are accessible to virtually everyone; suitable for small-scale production; and compatible with man's need for creativity. Out of these three characteristics is born non-violence and a relationship of man to nature which guarantees permanence. If one of these three is neglected, things are bound to go wrong.'

E. F. Schumacher, *Small is Beautiful*, 1974

What is meant by industry? In its widest and more traditional sense, the word industry is used to cover all forms of economic activity: **primary** (farming, fishing, mining and forestry); **secondary** (manufacturing and construction); **tertiary** (back-up services such as administration, retailing and transport); and **quaternary** (high-technology and information services/knowledge economy). In this chapter, the use of the term 'industry' has been confined to its narrowest

definition, i.e. manufacturing. Manufacturing industry includes the processing of raw materials (iron ore, timber) and of semi-processed materials (steel, pulp), together with – where necessary – the assembling of these products (cars, computers).

It needs to be pointed out, however, that while this definition may be convenient, it does create several major problems. Not the least of these problems has been the unprecedented transformation of the global economy in the last 20 or so years. This change has included rapid deindustrialisation and a growth of the service sector which has caused some advanced economies to view 'manufacturing' as almost peripheral compared with their increasing reliance, until the shockwaves of 2008, on banking and finance. At present, only some 27 per cent of the UK's working population are employed in manufacturing, a trend that is repeated across most of the developed market economies. This shift from an industrial to a post-industrial society is shown in Figure 19.1. In reality, it is also unrealistic to draw boundaries between 'manufacturing' and 'services'. Not only are the two integrated in reality through linkages (page 568 and Figure 19.2), buyer–supplier relations, etc., but many people who are officially classified as working in the manufacturing sector also have occupations that are service based (salespeople, administrators, accountants and financial advisers as well as those in research and development) within 'manufacturing' sector firms. It can be argued, with much justification, that it is conceptually (and empirically) unrealistic to sever manufacturing from services. This distinction becomes particularly problematic when discussing, for example, high-tech developments along the M4 (Places 86, page 566) as, by their nature, many firms are 'information-intensive' and knowledge based rather than production or materials based; or when describing the differences between the 'formal' and 'informal' sectors in less economically developed, less industrialised countries (page 574). Finally, the world financial events of 2008 showed countries, regardless of their level of economic development, just how interdependent the process of globalisation has made them (page 605).

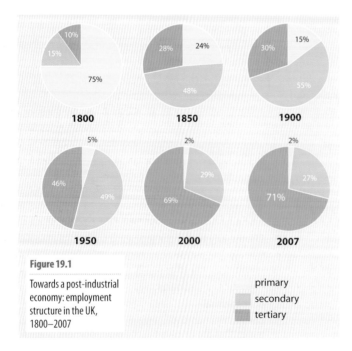

Figure 19.1

Towards a post-industrial economy: employment structure in the UK, 1800–2007

■ primary
■ secondary
■ tertiary

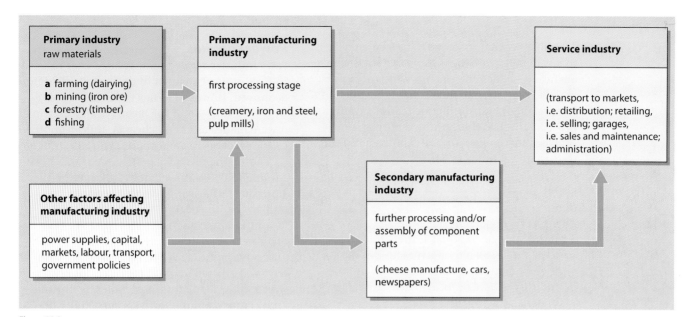

Traditional locations of industry

The processes which contribute to determine the location and distribution of industry are more complex and dynamic than those affecting agriculture. This means that the making of generalisations becomes less easy and the dangers of stereotyping increase. Reasons for this complexity include:

- Some locations were chosen before the Industrial Revolution and many more during it. Initial factors favouring a location may no longer apply today. For example, the original raw materials may now be exhausted (iron ore and coal in South Wales – Places 87, page 570) or replaced by new innovations (cotton by synthetic fibres) and sources of energy (water power by electricity).
- New locational factors which were not applicable last century include cheaper and more efficient transport systems, the movement of energy in the form of electricity, automation and new technologies such as email, the Internet and mobile phones.
- Some industries have developed from older industries and are linked to these former patterns of production even when the modern product is different (in Japan the Mazda Car Corporation began as a cork-making and then a machine-tools firm).
- Before the 20th century, industry was usually financed and organised by individual **entrepreneurs** who initiated and organised, usually for a profit, an enterprise or business; this included risk-taking, deciding what goods would be produced or services provided, the scale of production, and marketing. Nowadays

these decisions are often taken far away from the site of a factory, originally by the state, now usually by **transnationals** (**multinationals**, page 573).
- Many factories now produce a single component and therefore are a part of a much larger organisation which they supply.
- The sites of some early factories were chosen by individual preference or by chance, i.e. the founder of a firm just happened to live at, or to like, a particular location (Unilever at Port Sunlight and Rowntree at York).

Factors affecting the location of manufacturing industry

Raw materials

Industry in 19th-century Britain was often located close to raw materials (ironworks near iron ore), sources of power (coalfields) or ports (to process imports), mainly due to the immobility of the raw materials which were heavy and costly to move when transport was then expensive and inefficient. In contrast, today's industries are rarely tied to the location of raw materials and so are described as **footloose** (see post-Fordism, page 561). There is now a greater efficiency in the use of raw materials; power is more mobile; transport of raw materials, finished products and the workforce is more efficient and relatively cheaper; components for many modern, and especially high-tech, industries are relatively small in size and light in weight; and some firms may simply rely on assembling component parts made elsewhere. A location close to markets, labour supply or other linked firms has become increasingly important.

Industries that still need to be located near to raw materials are those using materials which are heavy, bulky or perishable; which are low in value in relation to their weight; or which lose weight or bulk during the manufacturing process. **Alfred Weber**, whose theory of industrial location is referred to later, introduced the term **material index** or **MI**.

$$MI = \frac{\text{total weight of raw materials}}{\text{total weight of finished product}}$$

There are three possible outcomes.

1 If the MI is greater than 1, there must be a weight loss in manufacture. In this case, the raw material is said to be **gross** and the industry should be located near to that raw material, e.g. iron and steel:

$$MI = \frac{\text{6 tonnes raw material}}{\text{1 tonne finished steel}} = 6.0$$

2 If the MI is less than 1, there must be a gain in weight during manufacture. This time the industry should be located near to the market, e.g. brewing:

$$M1 = \frac{\text{1 tonne raw material}}{\text{5 tonnes beer}} = 0.2$$

3 Where the MI is exactly 1, the raw material must be **pure** as it does not lose or gain weight during manufacture. This type of industry could therefore be located at the raw material, the market or any intermediate point.

Industries that lose weight during manufacture include food processing (butter has only one-fifth the weight of milk, refined sugar is only one-eighth the weight of the cane), smelting of ores (copper ore is less than 1 per cent pure copper, iron ore has a 30–60 per cent iron content; Places 84, page 563) and forestry (paper has much less mass than trees; Places 83, page 562). Industries that gain weight in manufacture include those adding water (brewing and cement), and those assembling component parts (cars, Places 85, page 565; and electrical goods, Places 86, page 566). In these cases, the end product is more bulky and expensive to move than its many smaller constituent parts.

Power supplies

Early industry tended to be located near to sources of power, which in those days could not be moved. However, as newer forms of power were introduced and the means of transporting it were made easier and cheaper, this locational factor became less important (Figure 19.3).

During the medieval period, when water was a prime source of power, mills had to be built alongside fast-flowing rivers. When steam power took over at the beginning of the Industrial Revolution in Britain, factories had to be built on or near to coalfields, as coal was bulky and expensive to move. When canals and railways were constructed to move coal, new industries were located along transport routes. By the mid-20th century, oil (relatively cheap before the 1973 Middle East War) was being increasingly used as it could be transported easily by tanker or pipeline. This began to free industry from the coalfields and to offer it a wider choice of location (except for such oil-based industries as petrochemicals). Today, oil, coal, natural gas, nuclear and hydro-electric power can all be used to produce electricity to feed the National Grid. Electricity, in addition to its cleanliness and flexibility, has the advantage that it can be transferred economically over considerable distances either to the long-established industrial areas, where activity is maintained by **geographical inertia**, or to new areas of growth.

Transport

Transport costs were once a major consideration when locating an industry. Weber based his industrial location theory on the premise that transport costs were directly related to distance (compare von Thünen's assumptions, page 471). Since then, new forms of transport have been introduced, including lorries (for door-to-door delivery), railways (preferable for bulky goods) and air (where speed is essential). Meanwhile, transport networks have improved, with the building of motorways, and methods of handling goods have become more efficient through containerisation. For the average British firm, transport costs are now only 2–3 per cent of their total expenditure. Consequently, raw materials can be transported further and finished goods sold in more distant markets without any considerable increase in costs. The increasing reliance, since the late 1990s, on emails, the Internet and mobile phones (page 642) has speeded up the transfer of data, including orders and payments, both within and between firms – a major factor in the process of globalisation (page 605).

Period	Source of power	Examples of location
early iron industry	charcoal	wooded areas (the Weald, the Forest of Dean)
later iron industry	waterwheels	fast-flowing rivers (River Don, Sheffield)
early steel industry	coal	coalfields (South Wales, north-east England)
present-day steel industry	electricity	coastal (Port Talbot)

Figure 19.3

Power supply and the location of iron and steelworks

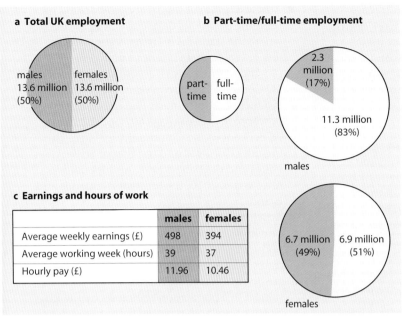

a Total UK employment

males 13.6 million (50%) / females 13.6 million (50%)

b Part-time/full-time employment

part-time / full-time

males: 2.3 million (17%) / 11.3 million (83%)

females: 6.7 million (49%) / 6.9 million (51%)

c Earnings and hours of work

	males	females
Average weekly earnings (£)	498	394
Average working week (hours)	39	37
Hourly pay (£)	11.96	10.46

Figure 19.4

UK employment data, 2007

The functioning of the present world depends upon a range of **space-shrinking technologies** that connect firms, workers, governments and consumers (i.e. commodity chains, page 643).

Two important types of space-shrinking technologies are **transport systems**, examples of which are commercial jet aircraft and containerisation (page 636), and **communication systems**, which include satellite and optic-fibre technology, the Internet, mobile phones and the electronic mass media (radio and TV).

Today, the pull of a large market is more important than the location of raw materials and power supplies; indeed, it has been suggested that flexibility and rapid response to changing market signals are perhaps the most important determinants of location.

Industries will locate near to markets if:
- the product becomes more bulky with manufacture or there are many linkage industries involved (the assembling of motor vehicles)
- the product becomes more perishable after processing (bread is more perishable than flour); it is sensitive to changing fashion (clothes); or it has a short life-span (daily newspapers)
- the market is very large (north-eastern states of the USA, south-east England, or global)
- the market is wealthy
- prestige is important (publishing).

Labour supply

Labour varies spatially in its cost, availability and quality.

In the 19th century, a huge force of semi-skilled, mainly male, workers operated in large-scale 'heavy' industries doing manual jobs in steelworks, shipyards and textile mills. Today, there are fewer semi-skilled and more highly skilled workers operating in small-scale 'light' industries which increasingly rely upon machines, computers and robots. The cost of labour, especially in EU countries, can be high, accounting for 10–40 per cent of total production costs. Three consequences of this have been the introduction of mechanisation to reduce human inputs, the exploitation of female labour, and the use of 'cheap' labour in developing countries.

Traditionally, labour has been relatively immobile. Although there was a drift to the towns during the Industrial Revolution, since the First World War British people have expected respective governments to bring jobs to them rather than they themselves having to move for jobs. Certain industries often located in a specific area or place to take advantage of local skills (cutlery at Sheffield, electronics around Cambridge) but as transport improvements allowed greater mobility, firms were able to locate, and their workforce to travel, more freely. In emerging economies industries often locate in large coastal cities where they can attract large numbers of unskilled workers from surrounding rural areas (in China, Places 41, page 363), while in more developed countries space-shrinking technologies have allowed a greater number of people, both employees and self-employed, to work from home.

Similarly, the roles of women and trade unions have both changed. At the turn of the 21st century, half of Britain's workforce were women (Figure 19.4a), with an increasing number either seeking career jobs or prepared to work part-time (Figure 19.4b), even flexi-time, although many still have to accept a lower salary than males (Figure 19.4c). The role of trade unions has declined significantly as their membership numbers have fallen with the decline of the large 'heavy' industries.

Capital

Capital may be in three forms.
1 **Working capital** (money) which is acquired from a firm's profits, shareholders or financial institutions such as banks. Money is mobile and can be used within and exchanged between countries. Location is rarely constrained by working capital unless money is to be borrowed from the government which might direct industry to certain areas (see below). In Britain, capital is more readily available in the City of London, where most of the financial institutions are based.
2 **Physical** or **fixed capital** refers to buildings and equipment. This form of capital is not mobile, i.e. it was invested for a specific use.

3 **Social capital** and cultural amenities are linked to the workforce's out-of-work needs rather than to the factory or office itself. Houses, hospitals, schools, shops and recreational amenities are social capital which may attract a firm, particularly its management, to an area.

Government policies

Government policies attempt to even out differences in employment, income levels and investment. In Britain, this was initially by the British government; now it is through the EU. At present, areas can only receive financial aid if they conform to EU guidelines. Under the latest guidelines, which came into effect in 2007, the proportion of the UK population covered by the **Assisted Areas** will be 23.9 per cent (Figure 19.5) compared with the previous 30.9 per cent. This reduction in coverage reflects partly the EU's objective to reduce areas of state aid amongst longer-standing members so as to help new, poorer member states, and partly due to the UK's own sustained (until 2008) economic success. As Figure 19.5 shows, the Assisted Areas in the British Isles can be divided into two groups.

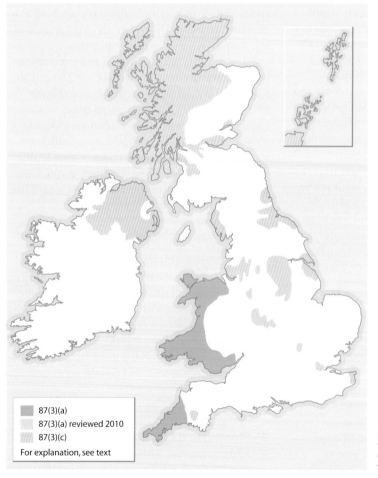

- 87(3)(a)
- 87(3)(a) reviewed 2010
- 87(3)(c)

For explanation, see text

■ Article 87(3)(a) of the Treaty provides aid to promote economic development in areas where the standard of living is abnormally low or where there is serious underemployment. These areas, according to the EU Commission, have a Gross Domestic Product per capita, measured in purchasing power parity terms (page 607), of less than 75 per cent of the Community average. This includes Cornwall and the Isles of Scilly, West Wales and the Valleys and, until its status is reviewed in 2010, the Scottish Highlands and Islands.

■ Article 87(3)(c) permits aid to certain regions providing that it does not affect the working of the single market. In this instance, member states can designate regions that they feel are disadvantaged within their own country, subject to a population ceiling set by the EU. Northern Ireland, together with smaller locations in England, Scotland and Wales, fit this category.

Land

In the 19th century, extensive areas of flat land were needed for the large factory units. Today, although modern industry is usually smaller in terms of land area occupied, it prefers cheaper land, less congested and cramped sites and improved accessibility, as are to be found on greenfield sites on the edges of cities and in smaller towns. Now, partly due to pressure from environmental and influential local groups, attempts are being made to attract new industry, including service industries, to derelict and under-used brownfield sites (page 441), or to former industrial premises (page 439), where existing infrastructures still exist.

Environment

The latter part of the 20th century saw an increasing demand by both managers and employees to live and work in an attractive environment. This led to firms moving away from large urban areas and relocating either in smaller towns that have easy access to the countryside, or on new science and business parks with landscaped green areas and ornamental lakes (Places 86, page 566).

Figure 19.5

Assisted Areas in the UK, 2007–13

Changing approaches to industrial geography

The term 'manufacturing industries' does not refer to a discrete, bounded, measurable entity, but is dynamic with profound cultural, social and political dimensions. This can be illustrated through the following brief evolution of industrial geography.

1 Location theory and the neoclassical approach

The 1960s was the era of the 'quantitative revolution' when established deductive theories about the world (mainly German) were tested to see if they could be used to accurately predict spatial patterns, e.g von Thünen's rural land use (put forward in 1826 – page 471), Weber's industrial location (in 1909 – below) and Christaller's central place (in 1933 – page 407). This approach has been classically illustrated by the study of Henry Ford's car industry (hence the term 'Fordism', Figure 19.13) with its assembly-line organisation. It has regained some credence since the late 1990s ('McDonaldisation').

2 Behaviouralism (late 1960s to early 1980s)

This examined the role of cognitive information and personal choice in determining decision-making and locational outcomes. While still focusing on locational issues and spatial behaviour, it concentrated more on detailed surveys and avoided the mathematical modelling that dominated the neoclassical approach.

3 Political economy (mid-1970s)

Writers such as David Harvey focused less on the idealised assumptions of rational economic man and perfect knowledge and more on how global political and economic forces (capitalism) shaped the space economy. This saw a shift in the main focus away from spatial patterns of industrial location towards structures of social relations. By 1990, this approach manifested itself in the **post-Fordism** debate (Figure 19.13) at a time when industry was showing greater flexibility both in production techniques and between institutions and industrial districts.

4 Cultural economic interpretations (post mid-1990s)

An even more recent 'cultural turn' has focused on hitherto neglected dimensions, such as gender (Places 96, page 608), which has placed a greater emphasis on the meanings of terms such as production, industry and labour.

Space here only permits a study of one industrial location model – Weber's – and a comparison between Fordism and post-Fordism.

Theories of industrial location

It has already been seen that 'models form an integral and accepted part of present-day geographical thinking' (Framework 12, page 352). Models, as in many branches of geography, have been formulated in an attempt to try to explain, in a generalised, simplified way, some of the complexities affecting industrial location. The most commonly quoted is that by Weber (1909) who based his model on the industrialist who seeks the lowest-cost location (LCL). Before looking more closely at this model, it must be remembered that it is an abstract framework which may be difficult to observe in the real world, but against which reality can be tested. It should also be pointed out that this model has a 'traditional' approach to industrial location applicable to a particular moment in time (history) and that there have, on other occasions, been alternative theories.

Weber's model of industrial location

Alfred Weber was a German 'spatial economist' who, in 1909, devised a model to try to explain and predict the location of industry. Like von Thünen before him and Christaller later, Weber tried to find a sense of order in apparent chaos, and made assumptions to simplify the real world in order to produce his model. These assumptions were as follows:

- There was an isolated state with flat relief, a uniform transport system in all directions, a uniform climate, and a uniform cultural, political and economic system.
- Most of the raw materials were not evenly distributed across the plain (this differs from von Thünen). Those that were evenly distributed (water, clay) he called **ubiquitous materials**. As these did not have to be transported, firms using them could locate as near to the market as was possible. Those raw materials that were not evenly distributed he called **localised materials**. He divided these into two types: gross and pure (page 554).
- The size and location of markets were fixed.
- Transport costs were a function of the mass (weight) of the raw material and the distance it had to be moved. This was expressed in tonnes per kilometre (t/km).
- Labour was found in several fixed locations on the plain. At each point it was paid the same rates, had equivalent skills, was immobile and in large supply. Similarly, entrepreneurs had equal knowledge, related to their industry, and motivation.
- Perfect competition existed over the plain (i.e. markets and raw materials were unlimited)

which meant that no single manufacturer could influence prices (i.e. there was no monopoly). As revenue would therefore be similar across the plain, the best site would be the one with the minimal production costs (i.e. the least-cost location or LCL).

Possible least-cost locations

Weber produced two types of locational diagram. A straight line was sufficient to show examples where only one of the raw materials was localised (it could be pure or gross). However, when two localised raw materials were involved, he introduced the idea of the **locational triangle**. Figure 19.6 summarises the nine possible variations based on the type of raw material involved.

1 One gross localised raw material. As there is weight loss during manufacture (the material index for a gross raw material is more than 1) then it is cheaper to locate the factory at the source of the raw material – there is no point in paying transport costs if some of the material will be left as waste after production (Figure 19.7a).

2 **a** One ubiquitous raw material or **b** one pure localised raw material gaining weight on manufacture (MI less than 1). If the raw material is found all over the plain (ubiquitous) then transport is unnecessary as it is already found at the market. If a pure material gains mass on manufacture then it is cheaper to move it rather than the finished product and so again the LCL will be at the market (Figure 19.7b).

3 One pure localised raw material. If this neither gains nor loses weight during manufacture (MI = 1), the LCL can be either at the market, at the location of the raw material, or at any intermediate point (Figure 19.7c).

4 Two ubiquitous (gross or pure) raw materials. As these are found everywhere, they do not have to be transported and so the LCL is at the market.

5 Two raw materials: one ubiquitous and one pure and localised. The LCL is at the market because the ubiquitous material is already there and so only the pure localised material has to be transported (Figure 19.8a). It will be cheaper to move one raw material than the more cumbersome final product.

6 Two raw materials: one ubiquitous and the other gross and localised. The ubiquitous material is available at every location. As the gross material loses weight, the LCL could theoretically be at any intermediate point between its source and the market. However, if the mass of the product is greater than that

Figure 19.6

Least-cost locations dependent on types of raw material

Type(s) of raw material (RM) MI = material index	LCL at raw material		LCL at any intermediate point		LCL at market
1 one gross localised RM MI > 1	■				
2 one RM gaining weight or one ubiquitous RM MI < 1					■
3 one pure localised RM MI = 1			■		
4 two ubiquitous RMs (pure or gross)					■
5 two RMs (one ubiquitous, one pure)					■
6 two RMs (one ubiquitous, one gross)			(could be any site, according to amount of weight loss)		
7 two RMs (both pure)					■
8 two RMs (one pure, one gross)	■	(if big weight loss)	■	(if a small weight loss)	
9 two RMs (both gross)	■	(at RM with greatest weight loss)	■	(equal weight loss)	

Figure 19.7

Least-cost locations with one raw material

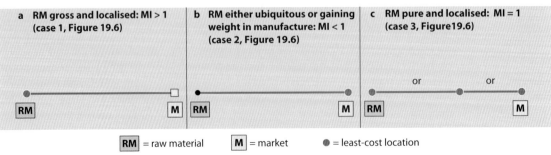

a RM gross and localised: MI > 1 (case 1, Figure 19.6)	**b** RM either ubiquitous or gaining weight in manufacture: MI < 1 (case 2, Figure 19.6)	**c** RM pure and localised: MI = 1 (case 3, Figure19.6)

RM = raw material   M = market   ● = least-cost location

of the localised raw material, the LCL is at the market; if it is less, the LCL is at the location of the raw material; and if it is the same, the LCL is at the mid-point (Figure 19.8b).

7 Two raw materials: both localised and pure. In the unlikely event of the two raw materials lying to the same side of and in line with the market, the LCL will be at the market. If the materials do not conform with this arrangement but form a triangle with the market (Figure 19.9), the LCL is at an intermediate point near to the market. This is because the weight and therefore the transport costs of the raw material are the same as, or less than, those of the product.

8 Two localised raw materials: one pure and one gross. In this case, the industry will locate at an intermediate point (Figure 19.10a). The greater the loss of weight during production, the nearer the LCL will be to the source of the gross material.

9 Two raw materials: both localised and gross. If both raw materials have an equal loss of weight, the LCL will be equidistant between these two sources but closer to them than to the market (Figure 19.10b1). However, if one raw material loses more mass than the other, the industry is more likely to be located closer to it (Figure 19.10b2).

Weber claimed that four factors affected production costs: the cost of raw materials and the cost of transporting them and the finished product, together with labour costs and agglomeration/deglomeration economies (page 560).

Spatial distribution of transport costs

As transport costs lay at the heart of his model, Weber had to devise a technique that could both measure and map the spatial differences in these costs in order to find the LCL. His solution was to produce a map with two types of contour-type lines which he called isotims and isodapanes. An **isotim** is a line joining all places with equal transport costs for moving either the raw material (Figure 9.11a) or the product (Figure 9.11b). An **isodapane** is a line joining all places with equal total transport costs, i.e. the sum of the costs of transporting the raw material and the product (Figure 19.11c).

Figure 19.11a shows the costs of transporting 1 tonne of a raw material (R) as concentric circles. In this example, it will cost 5 t/km (tonne/kilometres) to transport the material to the market. Figure 19.11b shows, also by concentric circles, the cost of transporting 1 tonne of the finished product (P). The total cost of moving the product from the market to the source of the raw material is again 5 t/km. By superimposing these two maps it is possible to show the total transport costs (Figure 19.11c).

If a factory were to be built at **X** (Figure 19.11c), its transport costs would be 7 t/km (i.e. 2 t/km for moving the raw material plus 5 t/km for the product). A factory built at **Y** would have lower transport costs of 6 t/km (4 t/km for the raw material plus 2 t/km for the product). However, the LCL in this case may be at the source of the raw material, the market or any intermediate point in a straight line between the two because all these points lie on the 5 t/km isodapane.

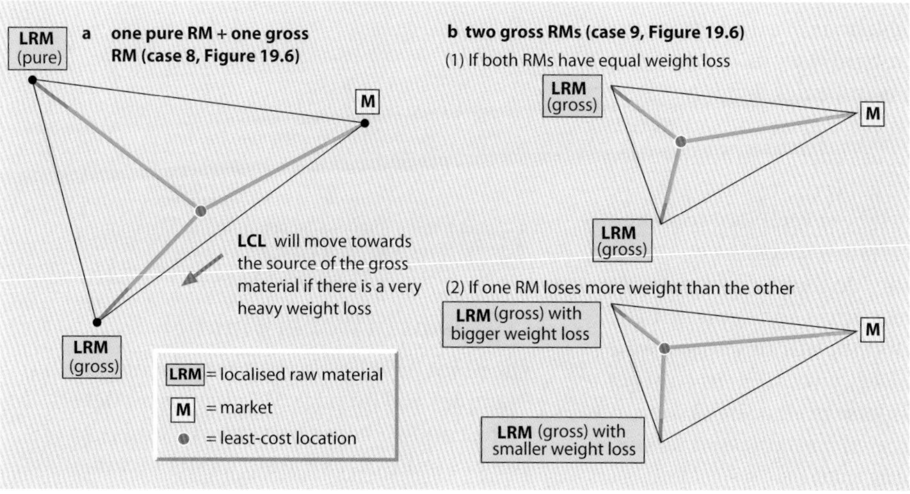

Figure 19.8

Least-cost locations with two raw materials, one of which is ubiquitous

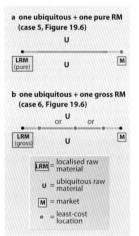

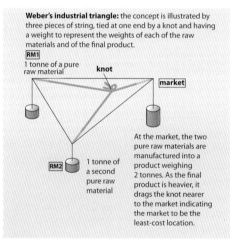

Figure 19.9

Least-cost locations with two localised pure raw materials, illustrating Weber's industrial triangle (case 7, Figure 19.6)

Figure 19.10

Least-cost locations with two localised raw materials, illustrating Weber's industrial triangle

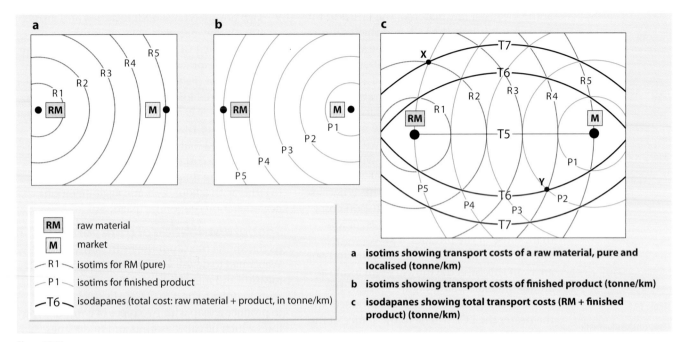

a   isotims showing transport costs of a raw material, pure and localised (tonne/km)

b   isotims showing transport costs of finished product (tonne/km)

c   isodapanes showing total transport costs (RM + finished product) (tonne/km)

Legend:

RM	raw material
M	market
R1	isotims for RM (pure)
P1	isotims for finished product
T6	isodapanes (total cost: raw material + product, in tonne/km)

Figure 19.11

Isotims and isodapanes

The effects of labour costs and agglomeration economies

It has been stated that Weber considered that four factors affected production costs: we have seen the effects of the costs of raw materials and transport – let us now look at labour costs and agglomeration economies.

- **Labour costs** Weber considered the question of whether any savings made by moving to an area of cheaper or more efficient labour would offset the increase in transport costs incurred by moving away from the LCL. He plotted isodapanes showing the increase in transport costs resulting from such a move. He then introduced the idea of the **critical isodapane** as being the point at which savings made by reduced labour costs equalled the losses brought about by extra transport costs. If the cheaper labour lay within the area of the critical isodapane, it would be profitable to move away from the LCL in order to use this labour.

- **Agglomeration economies** Agglomeration is when several firms choose the same area for their location in order to minimise their costs. This can be achieved by linkages between firms (where several join together to buy in bulk or to train a specialist workforce), within firms (individual car component units) and between firms and supporting services (banks and the utilities of gas, water and electricity). Deglomeration, in contrast, is when firms disperse from a site or area, possibly due to increased land prices or labour costs or a declining market.

Figure 19.12 shows the **critical isodapane** for three firms. It would become profitable for all the firms to locate within the central area formed by the overlapping of all three critical isodapanes. It may be slightly more profitable for firms **A** and **B**, but less profitable for firm **C**, to locate within the purple area. However, it would not be additionally profitable for any firm to move if none of the isodapanes overlapped. Agglomeration is now considered by many to be probably the most important single factor in the location of a firm or industry.

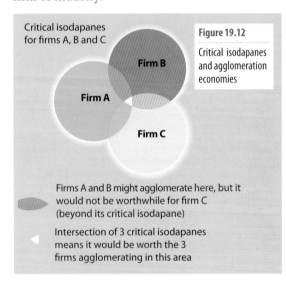

Critical isodapanes for firms A, B and C

Figure 19.12

Critical isodapanes and agglomeration economies

Firm B

Firm A

Firm C

Firms A and B might agglomerate here, but it would not be worthwhile for firm C (beyond its critical isodapane)

Intersection of 3 critical isodapanes means it would be worth the 3 firms agglomerating in this area

Criticisms of Weber's model

The point has already been made with previous examples and on page 557 that no model is perfect and all have their critics. Criticisms of Weber's industrial location model include:

- It no longer relates to modern conditions – such as the present extent of government intervention (grants, aid to Enterprise Zones), improvements in and reduced costs of transport, technological advances in processing raw materials, the development of new types of industry other than those directly involved in the processing of raw materials, the increased mobility of labour and the increased complexity of industrial organisation (transnationals instead of single-product firms).
- Each country evolves its own industrial patterns and may be in different stages of economic development (pages 604–608).
- There are basic misconceptions in his original assumptions. For example, there are changes over time and space in demand and price; there are variations in transport systems; perfect competition is unreal as markets vary in size and change over a period of time; and decisions made by industrialists (who do not all have the same knowledge) may not always be rational (von Thünen's 'economic man', page 471).
- Weber's material index was a crude measure and applicable only to primary processing or to industries with a very high or very low index.
- Dr L. Crewe (2008) claimed that 'traditional locational theories, such as that of Weber, are becoming increasingly less significant. Although labour costs and agglomeration factors *are* important in determining the location of an economic activity, they are handled far too simplistically in our ever increasingly complex world. Traditional models often cannot cope with the volatile global system in which we now live, and in which technological change is both rapid and endemic. A whole range of organisational and institutional forces shape economic change within a global economy. The real problem is the interconnectedness and complexity of the various processes at work.'

Production process technologies

According to Coe, Kelly and Yeung (2007), there are three different kinds of industrial system co-existing in the present global economy (Figure 19.13). These are:

1 **Fordism** – where scale economies remain crucial, e.g. food processing and electronic components. Traditionally this was associated with mass production (Henry Ford's 1910s car factory in Detroit, USA) and today more likely with sweatshops (Case Study 21).

2 **Post-Fordism** – where the chief characteristic is flexibility which, by allowing the use of information technology and computerisation in machines and their operation, gives more control over the production process. It can be sub-divided into:

   a flexible specialisation, when skilled workers use flexible machinery to provide a wider range of product to suit the smaller volumes of high-value or specialist goods that they produce, e.g. shoes and jewellery

   b flexible production, originating in Japan, which combines information technologies with the flexible organisation of either workers or commodity chains (page 643).

Figure 19.13

Present-day industrial systems

Characteristic	Fordism	Flexible	
	Mass production	Specialisation	Production
Labour force	Division of labour: a few skilled organisers/managers. Large number of semi-/unskilled workers doing repetitive jobs.	Highly skilled.	Multi-skilled, flexible workers all with some responsibility. Work in teams.
Technology/machines	Complex but single-purpose. Hard to change product. Machines in a sequence linked by conveyor belt. Standardised products.	Simple, flexible machines. Non-standardised products.	Highly flexible methods. Relatively easy to change products.
Supplier relationship	Arm's length. Stocks held in factory to ensure supply, i.e. 'just-in-case'.	Close contact with customers and suppliers.	Very close links with suppliers. No stored stock, i.e. 'just-in-time' delivery.
Product (volume, value and variety)	Very high volume. Small range/single product. Low value.	Low volume. Wide variety. High value.	Very high volume. Wide range. High value.

Industrial location: changing patterns

Four different types of industry have been selected as exemplars to try to demonstrate how the importance of different factors affecting the location of industry have changed through time. Their choice may reinforce the generalisation, by no means true in every case, that the more important locational factors in the 19th century were physical, while in modern industry they tend to be human and economic. They also show that while Weber's theory may have had some relevance in accounting for the location of older industries (remembering that it was put forward in 1909), it has less when explaining the location of contemporary industry.

The four industries are:

1 A primary manufacturing industry where, due to weight loss, the presence of raw materials and sources of energy is more important than the market and other economic factors (Places 83).
2 A secondary manufacturing industry initially tied to raw materials and sources of energy but in which economic and political factors have become increasingly more important (Places 84). This is an example of Fordism with its conveyor belt/assembly line production.
3 A secondary manufacturing industry where the nearness of a market and labour supply is more important than the presence of raw materials and sources of energy (Places 85). This illustrates flexible production (just-in-time).
4 Modern secondary (quaternary) manufacturing industries where human and economic factors are the most important (Places 86). This is an example of flexible specialisation (a footloose industry).

Places 83 Sweden: wood pulp and paper

There are three stages in this industry: the felling of trees, the processing of wood pulp (primary processing), and the manufacture of paper (secondary processing). In Sweden, most pulp and paper mills (Figure 19.14) are located at river mouths on the Gulf of Bothnia (Figure 19.15). Timber is a gross raw material which loses much of its weight during processing; it is bulky to transport; and it requires much water to turn it into pulp. Towns such as Sundsvall and Kramfors are ideally situated (Figure 19.15): the natural coniferous forests provide the timber; the fast-flowing Rivers Ljungan, Indals and Angerman which initially provided cheap water transport for the logs are a source of the necessary and cheap hydro-electricity; and the Gulf of Bothnia provides an easy export route. Paper has a higher value than pulp and it is convenient and cheaper to have integrated mills.

Weber's agglomeration economies, together with Fordism's mass production techiniques, seem to operate with the clustering of so many mills.

Figure 19.14

Pulp mill on the Gulf of Bothnia

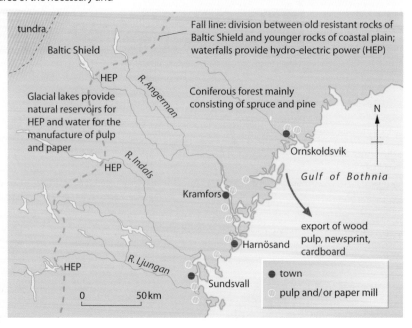

Figure 19.15

Location of wood pulp and paper factories in central Sweden

tundra

Baltic Shield

HEP

Glacial lakes provide natural reservoirs for HEP and water for the manufacture of pulp and paper

HEP

R. Angerman

Fall line: division between old resistant rocks of Baltic Shield and younger rocks of coastal plain; waterfalls provide hydro-electric power (HEP)

Coniferous forest mainly consisting of spruce and pine

N

Ornskoldsvik

Gulf of Bothnia

R. Indals

Kramfors

export of wood pulp, newsprint, cardboard

Harnösand

HEP

R. Ljungan

Sundsvall

0   50km

● town

◎ pulp and/or paper mill

Although the early iron and later steel industries were tied to raw materials, modern integrated iron and steelworks have adopted new locations as the sources of both ore and energy have changed.

- **Before AD 1600** Iron-making was originally sited where there were surface outcrops of iron ore and abundant wood for use as charcoal (the Weald, the Forest of Dean, Figure 19.16a). Locations were at the source of these two raw materials as they had a high material index, were bulky and expensive to transport, had a limited market and could not be moved far owing to the poor transport system.

- **Before AD 1700** Local ores in the Sheffield area were turned into iron by using fast-flowing rivers to turn waterwheels as water provided a cheaper source of energy.

- **After AD 1700** In 1709, Abraham Derby discovered that coke could be used to smelt iron ore efficiently. At this time, it took 8 tonnes of coal and 4 tonnes of ore to produce 1 tonne of iron, and so new furnaces were located on coalfields. One of the first areas to develop was South Wales where bands of iron ore (blackband ores) were found between seams of coal. The advantages possessed by South Wales at that time are shown in Figure 19.17a. Later, the industry extended into other British coalfields. When local ores became exhausted, the industry continued in the same locations because of geographical inertia, a pool of local skilled labour, a local market using iron as a raw material, improved techniques reducing the amount of coal needed (2 tonnes per 1 tonne of final product), improved and cheaper transport systems (rail and canal) which brought distant mined iron ore, and the beginnings of agglomeration economies.

- **After 1850** Until the 1880s, the low ore and high phosphorus content of deposits found in the Jurassic limestone, extending from the Cleveland Hills to Oxfordshire, had not been touched. After 1879, the **Gilchrist–Thomas process** allowed this ore to be smelted economically. As iron ore now had a higher material index than coal it was more expensive to move. As a result, new steelworks were opened on Teesside, near to the Cleveland Hills deposits, and at Scunthorpe and Corby, on the ore fields. However, the major markets remained on the coalfields.

Figure 19.16

Location of iron and steelworks in England, Scotland and Wales

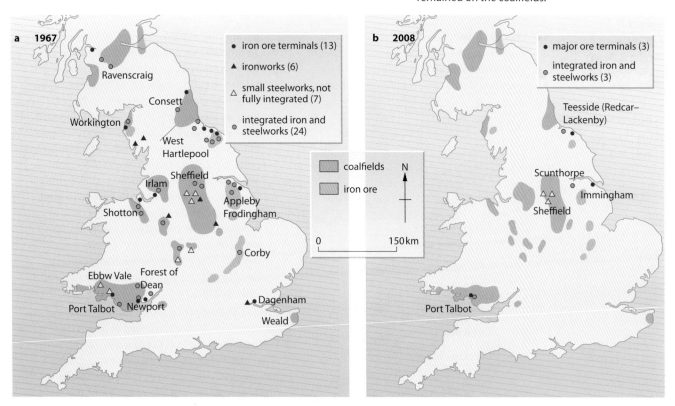

Period of time			a Location of early 19th-century iron foundries in South Wales (e.g. Ebbw Vale)	b Disadvantages of these early locations by 1960 (e.g. Ebbw Vale)	c Location of only remaining integrated steelworks in 2008 (Port Talbot)
Physical	Raw materials	Coal	mined locally in valleys	older mines closing	little now needed; imported
		Iron ore	found within the Coal Measures	had to be imported: long way from coast	imported from N Africa and N America
		Limestone	found locally	found locally	found locally
		Water	for power and effluent: local rivers	insufficient for cooling	for cooling: coastal sites
	Energy/fuel		charcoal for early smelting, later rivers to drive machinery; then coal	electricity from National Grid	electricity from National Grid using coal, oil, natural gas, nuclear power
	Natural routes		materials local; export routes via the valleys	poor; restricted by narrow valleys	coastal sites
	Site and land		narrow valley floor locations	cramped sites; little flat land	large areas of former sand dunes
Human and economic	Labour		large quantities of semi-skilled labour	still large numbers of semi-skilled workers	still relatively large numbers but with a higher level of skill; fewer due to high-tech/mechanisation
	Capital		local entrepreneurs	no investment	government and EU incentives
	Markets		local	difficult to reach Midlands and ports	tin plate industry (Llanelli) and the Midland car industry
	Transport		little needed; some canals; low costs	poor; old-fashioned; isolated	M4; purpose-built port
	Geographical inertia		not applicable	not strong enough	tradition of high-quality goods
	Economies of scale		not applicable	worked against the inland sites	one large steelworks more economical than numerous small iron foundries
	Government policy		not applicable	Ebbw Vale kept open by government help	having the capital, governments can determine locations and closures and provide heavy investment
	Technology		small scale: mainly manual	out of date	high-technology: computers, lasers, etc.

Figure 19.17

Growth, decline and changing location of iron and steelworks in South Wales

- **After 1950** With iron ore still the major raw material (less than 1 tonne of coal was now needed to produce 1 tonne of steel), but with deposits in the UK largely exhausted, Britain became increasingly reliant on imported ores. This meant that new **integrated steelworks** were located on coastal sites while those inland tended to close (Figure 19.16). Since the 1950s three new elements, unforeseen by Weber, became increasingly important in the location of new steelworks: government intervention, improved technology and reduced transport costs. It is a now a government/EU decision as to where any new steelworks (unlikely in the present economic climate) will be located, and which existing works will either close or remain open; improved technology has seen a reduction in raw materials consumed and workers needed; while lower transport costs have aided both imports of raw materials and exports of finished goods.

Even so, the industry still uses complex machines set out in a sequence and linked by a conveyer belt system. At the Port Talbot works, raw materials enter one end of the factory, pass through several processing stages, all highly computerised, to finally emerge, several kilometres away, as a standardised end product. The steelworks is also part of a value added chain in a global industry.

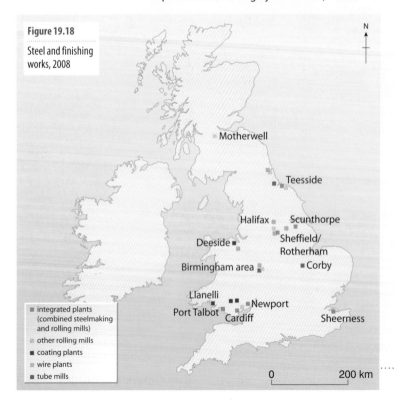

Figure 19.18

Steel and finishing works, 2008

Motherwell

Teesside

Halifax Scunthorpe

Deeside Sheffield/ Rotherham

Birmingham area Corby

Llanelli Newport
Port Talbot Cardiff Sheerness

N

- integrated plants (combined steelmaking and rolling mills)
- other rolling mills
- coating plants
- wire plants
- tube mills

0   200 km

Japan's production of 8.6 million cars in 2006, which was 20.8 per cent of the world's total, kept it as the world leader ahead of Germany (5.1 m) and the USA (5.0 m). This has been achieved despite a lack of basic raw materials.

Japan has very limited energy resources for, although it produces hydro-electricity and nuclear energy, it has to import virtually all its coal, oil and natural gas requirements. Similarly, most of the iron ore and coking coal needed to manufacture steel also has to be imported. The result has been the location of the major steelworks on tidal sites found around the country's many deep and sheltered natural harbours. As only 17 per cent of the country is flat enough for economic development (for homes, industry and agriculture), most of the population also has to live in coastal areas and around the harbours. The five major conurbations, linked by modern communications, provide both the workforce and the large, affluent, local markets needed for such steel-based products as cars (Figure 19.19). Within these conurbations are numerous firms engaged in making car component parts. This agglomeration of firms limits transport costs and conforms with Weber's concept that industries gaining weight through processing (car assembly) are best located at the market. As many of the smaller, older and original firms have amalgamated into large-scale companies, the extra space required for their factories has

had to come from land reclaimed from the sea (Figure 19.20). These new locations, despite the high costs of reclamation, make excellent sites from which to export finished cars to all parts of the world. The large local labour force contains both skilled and semi-skilled workers who, as well as being educated and industrious, are very loyal to their firm. The car industry, which has received considerable government financial assistance, has an organisation which centres around teamworking, worker involvement, total-quality management, and 'just-in-time' production (this is when various component parts arrive just as they are needed on the assembly line, thus avoiding the need to store or to overproduce). The Japanese car industry has a high level of automation and uses the most modern technology: it produces three times the number of cars per worker as does western Europe. The assembled cars are reliable and universally acceptable in design which means, together with the shift from mass production to lean, or flexible, production, that the Japanese have gained strong footholds in world markets. To expand further into these markets, the Japanese have either built overseas assembly plants or have amalgamated with local companies so that more cars can be produced close to the large urban markets within western Europe and the USA, e.g. Honda at Swindon, Nissan at Sunderland, and Toyota at Burnaston and Deeside in the UK.

Figure 19.19

Major industrial areas in Japan

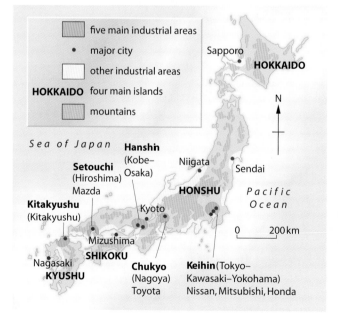

Figure 19.20

Mazda's Hofu car plant, built on land reclaimed from the sea

The term **high-technology** refers to industries that have developed within the last 35 years and whose processing techniques often involve micro-electronics, but may include medical instruments, biotechnology and pharmaceuticals. These industries, which collectively fit into the **quaternary** sector (page 552), usually demand high inputs of information, expertise and research and development (R&D). They are also said to be **footloose** (the modern term is **flexible specialisation**) in that, not being tied to raw materials, they have a free choice of location (Figure 19.13). However, they do tend to occur in clusters in particular areas, forming what Weber would have called 'agglomerated economies', such as along the M4 and M11 corridors in England (also Silicon Glen in Scotland, Silicon Valley in California and Grenoble and the Côte d'Azur in France). By locating close together, high-tech firms can exchange ideas and information and share basic amenities such as connecting motorways.

Two of the major concentrations of high-tech industries in Britain are along the M4 westwards (Sunrise Valley) from London to Reading, Newbury ('Video Valley'), Bristol (Aztec West) and into South Wales; and the M11 northwards to Cambridge (Figure 19.21). Transport is convenient due to the proximity of several motorways and mainline railways, together with the four main London airports. Transport costs are, in any case, relatively insignificant as the raw materials (silicon chips) are lightweight and the final products (computers) are high in value and small in bulk. Even so, it has been argued that two of the main reasons for high-tech development in this part of Britain were:

1 the presence of government-sponsored research establishments at Harwell and Aldermaston and of government aerospace contractors in the Bristol area

2 its attractive environment, e.g. the valley of the Thames and the nearby upland areas of the Cotswolds, Chilterns and Marlborough Downs (Figure 19.21), and its proximity to cultural centres, e.g. London, Oxford and Cardiff.

Figure 19.21

The M4 and M11 Corridors

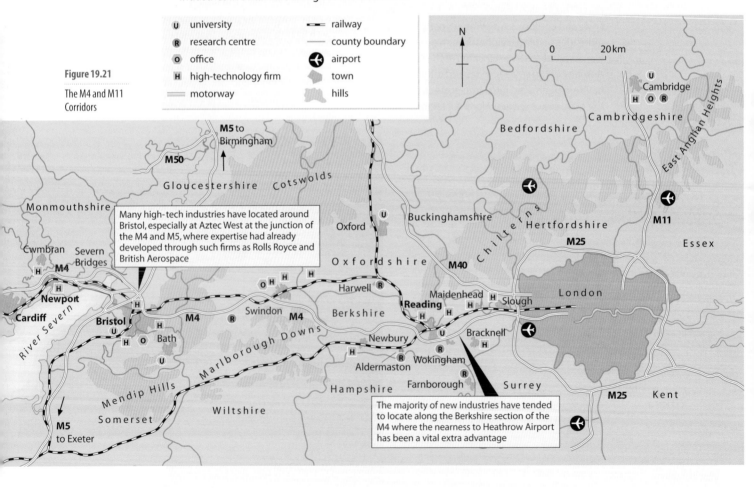

Most firms that have located here claim that the major factor affecting their decision was the availability of two types of labour:

- Highly skilled and inventive research scientists and engineers, the majority of whom were university graduates or qualified technicians. These specialists, whose abilities were in short supply, could often dictate areas where they wanted to live and work, i.e. areas of high environmental, social and cultural quality. The proximity of several universities (Figure 19.21) provided a pool of skilled labour and facilities for R&D.

- Female workers who either tended to be plentiful as an increasing number of career-minded women were among those who had recently moved out of London and into new towns and suburbanised villages (page 398), or were prepared to accept part-time/flexi-time jobs (Figure 19.4).

Science parks are often joint ventures between universities and local authorities. They are usually located adjacent to universities on edge-of-town greenfield sites where, because the land is of lower value, there is plenty of space for car parking, landscaping (ornamental gardens and lakes) and possible future expansion.

The Cambridge Science Park (Figure 19.22) has been developed in conjunction with Trinity College, Cambridge. Opened in 1972, the success of early firms soon attracted more (agglomeration economies), so that by 2008 there were 109 companies employing about 5000 people. Existing companies can be divided into those making electronics, scientific instruments, drugs and pharmaceuticals (biomedicinal), with a strong emphasis on scientific R&D. Only selected firms, using the high-quality, flexible buildings for specific purposes, are allowed to locate in the business park. Almost one-quarter of these firms are medium-sized, each employing between 20 and 49 workers. Some 70 per cent of the park, which covers 62 ha, is left as open space with trees, grass and ornamental gardens with lakes (Figure 19.23). As this, and other business and science parks in the Cambridge area, continue to develop, new housing has to be provided, e.g. at Cambourne (Case Study 14A), and building pressure increases on the surrounding transport system and countryside (Figure 14.22).

Figure 19.22

The Cambridge Science Park

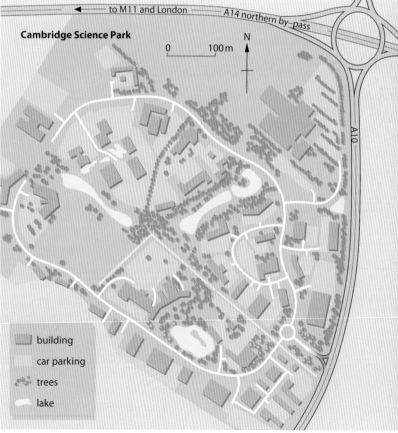

Cambridge Science Park

to M11 and London · A14 northern by-pass

A10

0   100 m   N

- building
- car parking
- trees
- lake

Figure 19.23

Layout of the Cambridge Science Park

Industrial linkages and the multiplier

When Weber introduced the term 'agglomeration economies', he acknowledged that many firms made financial savings by locating close to, and linking with, other industries. The success of one firm may attract a range of associated or similar-type industries (cutlery in Sheffield), or several small firms may combine to produce component parts for a larger product (car manufacture in Coventry). **Industrial linkages** may be divided into **backward linkages** and **forward linkages**:

backward linkages		**forward linkages**
to firms providing raw materials or component parts	← FACTORY →	to firms further processing the product or using it as a component part

A more detailed classification of industrial linkages is given in Figure 19.24. The more industrially advanced a region or country, the greater is the number of its linkages. Developing countries have few linkages, partly because of their limited number of industries and partly because few industries go beyond the first stage in processing – the simple chain in Figure 19.24a. Industrial linkages may result in:

- energy savings
- reduced transport costs
- waste products from one industry forming a raw material for another
- energy given off by one process being used elsewhere
- economies of scale where several firms buy in bulk or share distribution costs
- improved communications, services and financial investment
- higher levels of skill and further research
- a stronger political bargaining position for government aid (the securing of EU funding now depends upon having a network of linked organisations).

Louise Crewe has stressed the 'increasingly critical importance of local linkages in ensuring competitive success, and the need to emphasise how agglomeration is becoming an increasingly important factor in explaining industrial location'. In the fashion quarter of Nottingham's Lace Market, for example, 85 per cent of all firms are linked to others, e.g. supplier links, manufacturers, retailers, local intelligence, and so on. Other examples of linkages and industrial location include the Motor Sport valley in Oxfordshire and car assembly in the West Midlands, together with both the fashion and jewellery agglomerations and the semiconductor clusters in California and the UK (Places 86).

Figure 19.24

Types of industrial linkage

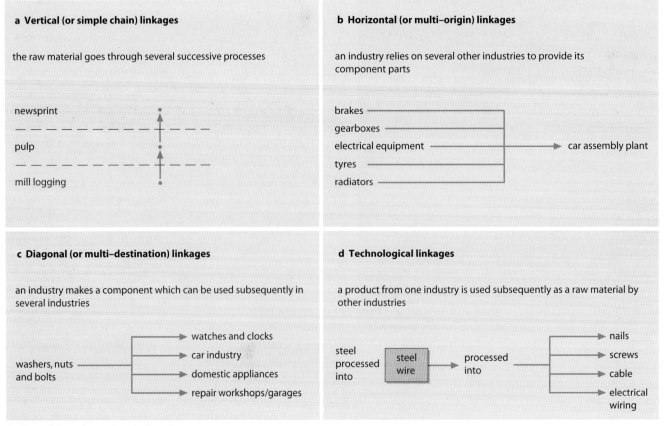

a Vertical (or simple chain) linkages

the raw material goes through several successive processes

newsprint

pulp

mill logging

b Horizontal (or multi–origin) linkages

an industry relies on several other industries to provide its component parts

brakes
gearboxes
electrical equipment → car assembly plant
tyres
radiators

c Diagonal (or multi–destination) linkages

an industry makes a component which can be used subsequently in several industries

washers, nuts and bolts →
watches and clocks
car industry
domestic appliances
repair workshops/garages

d Technological linkages

a product from one industry is used subsequently as a raw material by other industries

steel processed into → steel wire → processed into →
nails
screws
cable
electrical wiring

The multiplier effect and Myrdal's model of cumulative causation

If a large firm, or a specialised type of industry, is successful in an area, it may generate a **multiplier effect**. Its success will attract other forms of economic development creating jobs, services and wealth – a case of 'success breeds success'. This circular and cumulative process was used by **Gunnar Myrdal**, a Swedish economist writing in the mid-1950s, to explain why inequalities were likely to develop between regions and countries. Figure 19.25 is a simplified version of his model.

Myrdal suggested that a new or expanding industry in an area would create more jobs and so increase the spending power of the local population. If, for example, a firm employed a further 200 workers and each worker came from a family of four, there would be 800 people demanding housing, schools, shops and hospitals. This would create more jobs in the service and construction industries as well as attracting more firms linked to the original industry. As **growth poles**, or points, develop there will be an influx of migrants, entrepreneurs and

Figure 19.25

A simplified version of Myrdal's model to show development of an industrial region

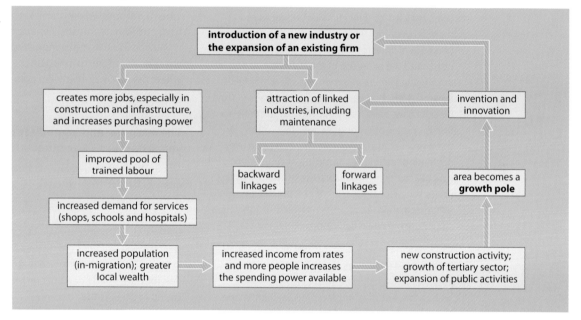

capital, together with new ideas and technology. Myrdal's multiplier model may be used to explain a number of patterns.

1 The growth of 19th-century industrial regions (South Wales and the Ruhr) and districts (cutlery in Sheffield, guns and jewellery in Birmingham and clothing in Nottingham).
2 The development of growth poles (page 617) in developing countries (São Paulo in Brazil and the Damodar Valley in India), where increased economic activity led, in turn, to multiplier effects, agglomeration economies and an upward spiral resulting in core regions (Places 87 and Places 98, page 618). At the same time, cumulative causation worked against regions near the **periphery** where Myrdal's **backwash effects** included a lack of investment and job opportunities.
3 The creation of modern government regional policies which encourage the siting of new, large, key industries in either peripheral, less developed (Trombetas and Carajas in Amazonia) or high unemployment (Nissan and Toyota in England) areas in the hope of

stimulating economic growth. This policy is more likely to succeed if the industries are labour intensive.

Industrial regions

Much of Britain's early industrial success stemmed from the presence of basic raw materials and sources of energy for the early iron, and the later iron and steel, industries; the mass production of materials using the processed iron and steel; and the development of overseas markets. During the 19th century it was the coalfields, especially those in South Wales, northern England and central Scotland, which became the core industrial regions. However, as the initial advantages of raw materials (which became exhausted), specialised skills and technology (no longer needed as the traditional heavy industries declined) and the ability to export manufactured goods (in the face of growing overseas competition) were lost, these early industrial regions have become more peripheral. Recent attempts to revive their economic fortunes have met with varying success (Places 87).

Pre-1920: industrial growth creating a core region

The growth of industry in South Wales was based on readily obtainable supplies of raw materials (Figure 19.17a). Coking coal and blackband iron ore were frequently found together, exposed as horizontal seams outcropping on steep valley sides. Their proximity to each other meant that the area around Merthyr Tydfil and Ebbw Vale (Figure 19.26) was ideally suited for industrial development (Weber's least-cost location for two gross raw materials, Figure 19.10b). Added to this was the presence of limestone only a few kilometres to the north, and the expertise of the local population in iron-making where waterwheels, driven by fast-flowing rivers, had earlier been used to power the blast furnace bellows. By the 1860s there were 35 iron foundries operating in the Welsh valleys. By the time the more accessible coal had been used up, mining techniques had improved sufficiently to allow shafts to be sunk vertically into the valley floors. When local supplies of iron ore became exhausted, there were ports nearby through which substitute ore could be imported.

'Thus began the spread of the well-known industrial landscape of the Valleys. Pits crammed themselves into the narrow valley bottoms, vying for space with canals, housing and, later, railways and roads. Housing began to trail up the valley sides, line upon line of terraces pressed against the steep slopes [Figure 19.27]. The opening-up of the underground coal seams resulted in massive immigration, much of it from rural areas. Working conditions, living conditions and wages were deplorable while health and safety standards underground were poor. Housing was overcrowded as the provision of homes, financed by the local entrepreneur ironmasters, lagged far behind the supply of jobs.'

The rapid increase in coalmining and iron-working partly resulted from the growth of large overseas markets as both products were mainly exported. Transport to the Welsh ports first involved simply allowing trucks to run downhill under gravity. Later, canals and then railways were used to move the bulky materials. While Barry, Cardiff and Newport developed as exporting ports, Swansea and Neath grew as 'break of bulk' ports smelting the imported ores of copper, nickel and zinc. **Break of bulk** is when a transported product has to be transferred from one form of transport to another – a process that involves time and money. It was easier and cheaper, therefore, to have had the smelting works where the raw materials were unloaded, rather than transporting them inland.

The inter-war and immediate post-war years: depression and industrial decline

Just as the existence of raw materials and overseas markets had led to the growth of local industry, so did their loss hasten its decline. Iron ore had long since been exhausted and it increasingly became the turn of coal, even though there were still over 500 collieries employing 260 000 miners in 1925. The steelworks which had replaced the iron foundries had been built on the same inland, cramped sites; as they became less competitive mainly due to rising transport costs, so they became increasingly dependent on government support (Figure 19.17b). Overseas markets were lost as rival industrial regions with lower costs and more up-to-date technology were developed overseas. The difficulties of an economy reliant on a narrow industrial base, dependent on an increasingly out-of-date infrastructure, and unable to compete with overseas competition, led to major economic, social and environmental problems.

Figure 19.26

Early industrial development in South Wales

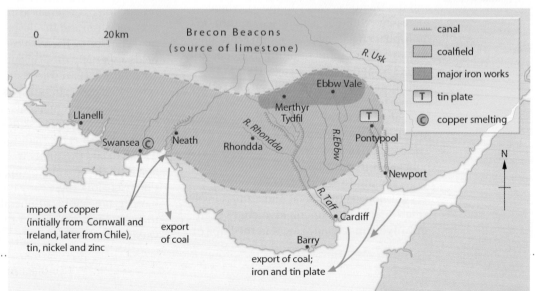

Figure 19.27

Industry, communications and terraced housing strung along the valley floor and lower valley sides: Rhondda Fawr, looking towards Treorchy, mid-Glamorgan

Towards the present: industrial diversification in a peripheral area

Steel-making and non-ferrous metal smelting have been maintained, partly due to geographical inertia, despite a significant fall in output and workers. As the centre of gravity for steel-making moved to coastal sites, so too did the location of the two South Wales integrated works, to Llanwern (closed 2001) and Port Talbot (Figures 19.17c and 19.28). Tin plate, using local steel, is produced at Trostre near Llanelli (the Felindre works near Swansea closed in 1989), while the Mond nickel works near Swansea is the world's largest (Figure 19.28).

The major factor to have affected industry in the region in the last 50 years has been government intervention (or lack of it, depending on your

Figure 19.28

Recent industrial development in South Wales

political views). The Special Areas Act of 1934 saw the first government assistance which set up industrial estates at Treforest, Merthyr Tydfil and Rhondda (Figure 19.28), while Cwmbran became one of Britain's first new towns (1949). Much of the former coalfield remains an Assisted Area (Figure 19.5). The last NCB colliery closed in 1994, although the Tower Colliery, near Merthyr Tydfil, reopened privately between 1995 and 2008. At present coal comes from seven opencast mines, and a current planning application, if successful, would make one of those – Ffos-y-Fran, also near Merthyr Tydfil – the largest in Europe (there is strong local opposition to the scheme).

Two local areas of exceptionally high unemployment, Swansea and Milford Haven Waterway, were designated two of Britain's 27 Enterprise Zones (page 439). The Swansea EZ included five parks – the Enterprise (commerce and light industry), Leisure (recreation facilities), Riverside (heritage and environmental schemes), City (retailing) and Maritime (housing and cultural) Parks. The Ford Motor Company took advantage of government incentives to build two plants in the region, one of which, at Bridgend, has been expanded. It was government policy that built an integrated steelworks at Ebbw Vale in 1938, and which closed it in 1979. The future of Port Talbot is also in government hands. A policy to decentralise some government departments has seen vehicle licensing moved to Swansea and the Royal Mint to Llantrisant (Figure 19.28). Improvements in communications have included the M4, the Heads of Valleys Road, the InterCity rail link, and Cardiff international airport – some of which were financed by EU funds.

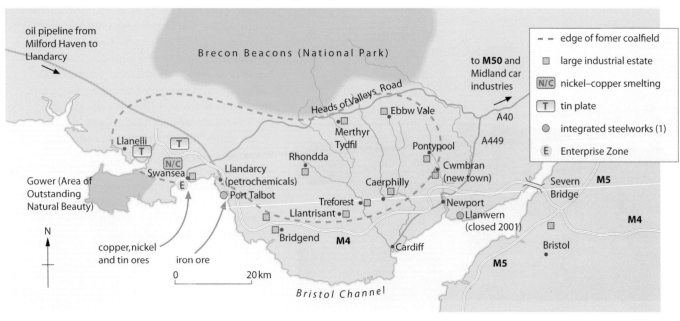

Figure 19.29

Swansea Enterprise Park, west Glamorgan

Money has also been spent on landscaping old industrial areas which had been scarred either by metal-smelting industries (lower Swansea Valley) or by slag (Ebbw Vale) and colliery waste tips (Aberfan – Case Study 2B). The Ebbw Vale Garden Festival (1992), sited on part of the former steelworks, was part of a larger scheme aimed at creating new jobs, improving housing, renovating old properties and improving the local environment (page 439). Other schemes, some funded by the WDA (see below), include tourist and cultural facilities such as the Welsh Industrial and Maritime Museum in Cardiff's newly created Marina area and the international sports village in Cardiff Bay.

The Cardiff Bay project, environmentally controversial, was aimed at improving transport and housing as well as providing jobs and retailing and leisure opportunities.

Figure 19.30

Sony's CTV European headquarters at Pencoed, Bridgend, occupies a 25 ha site

The Welsh Development Agency (WDA) was set up in 1976 'to attract high-quality investment, to help the growth of Welsh businesses and to improve the environment' (Figure 19.29). It saw as its main advertising points:

- a workforce that was skilled (although it needed retraining for the new-style high-tech industries)
- low labour costs, high productivity and good labour relations
- a well-developed transport infrastructure with modern road, rail and air links
- the availability of advanced factory sites with quality buildings at competitive rates
- a local market, and access to a national and the international market
- low rates and rents for firms wishing to locate in either the Development or Intermediate Areas (Figure 19.5)
- lower house prices and cost of living than south-east England
- the University of Wales with its five separate colleges
- the Welsh countryside, including the Pembrokeshire Coast and Brecon Beacons National Parks and 500 km of Heritage Coastline (including the Gower Peninsula), and the Pembrokeshire Coast footpath.
- the Welsh culture, including music, the performing arts and sport.

At the beginning of the 21st century, South Wales had a more varied and broad economic base than it had ever had before, with both manufacturing and inward investment growing at a faster rate than anywhere else in the UK. Of nearly 500 international companies that had located here, 150 were from North America (Ford and General Electric), 60 were German (Bosch) and 50 were Japanese (Sony, Figure 19.30; Aiwa, Matsushita and Hitachi). Other companies have come from France, Italy, Singapore, South Korea and Taiwan. The major types of new industry include aerospace and defence (six of the world's top ten companies including Airbus and BAE systems), car assembly (Bridgend), chemicals, electronics, medical devices, optical equipment, pharmaceuticals and telecommunications. A recent addition has been the Amazon (books) distribution centre at Swansea, which is expected to employ 1200 full-time and 1500 seasonal staff.

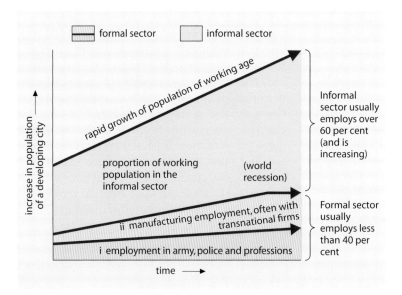

Figure 19.31

Growth in the informal sector

Industry in economically less developed countries

In cities in economically less developed countries, the number of people seeking work far outweighs the number of jobs available. As these cities continue to grow, either through natural increase or in-migration, the job situation gets continually worse. The UN estimates that in developing countries, on average, only about 40 per cent of those people with jobs work in the **formal sector** (Figure 19.31). These jobs, which are permanent and relatively well paid, include those offered by the state (police, army and civil service) or by overseas-run **transnational (multinational) corporations**, which are a major feature of globalisation (Chapter 21). The remaining 60 per cent – a figure which the UN claims is rising – have to seek work in the **informal sector**. The main differences between the formal and informal sectors are listed in Figure 19.34.

Transnational (multinational) corporations

A transnational, or multinational, corporation is one that operates in many different countries regardless of national boundaries. The headquarters and main factory are usually located in an economically more developed country. Although, at first, many branch factories were in economically less developed countries, increasingly there has been a global shift to the more affluent markets of Europe, North America, Japan and South Korea. Transnationals (TNCs) are believed to directly employ nearly 50 million people worldwide and to indirectly influence an even greater number. It is estimated that the largest 300 TNCs control over 70 per cent of

Advantages to the country	Disadvantages to the country
Brings work to the country and uses local labour	Numbers employed small in comparison with amount of investment
Local workforce receives a guaranteed income	Local labour force usually poorly paid and have to work long hours
Improves the levels of education and technical skills of local people	Very few local skilled workers employed
Brings inward investment and foreign currency to the country	Most of the profits go overseas (outflow of wealth)
Companies provide expensive machinery and introduce modern technology	Mechanisation reduces the size of the labour force
Increased gross national product/personal income can lead to an increased demand for consumer goods and the growth of new industries and services	GNP grows less quickly than that of the parent company's headquarters, widening the gap between developed and developing countries
Leads to the development of mineral wealth and new energy resources	Raw materials are usually exported rather than manufactured locally, and energy costs may lead to a national debt
Improvements in roads, airports and services	Money possibly better spent on improving housing, diet and sanitation
Prestige value (e.g. Volta Project)	Big schemes can increase national debt (e.g. Brazil)
Widens economic base of country	Decisions are made outside the country, and the firm could pull out at any time
Some improvement in standards of production, health control, and recently in environmental control	Insufficient attention to safety and health factors and the protection of the environment

Figure 19.32

Advantages and disadvantages of transnational (multinational) corporations

world trade (compared with only 20 per cent in 1960) and produce over half of its manufactured goods. The largest TNCs have long been car manufacturers and oil corporations but these have, more recently, been joined by electronic and high-tech firms. Several of the largest TNCs have a higher turnover than all of Africa's GNP in total.

Transnationals, with their capital and technology, have the 'power' to choose what they consider to be the ideal locations for their factories. This choice will be made at two levels: the most suitable country, and the most suitable place within that country. The choice of a country usually depends on political factors. Most governments, regardless of the level of economic development within their country, are prepared to offer financial inducements to attract transnationals which they see as providers of jobs and a means of increasing exports. (Sony, Figure 19.30, was reputed to have been offered better inducements to locate at Bridgend rather than in Barcelona.) Many governments of economically less developed countries, due to a greater economic need, are prepared to impose fewer restrictions on transnationals because they often have to rely on them to develop natural resources, to provide capital and technology (machinery, skills, transport), to create jobs and to gain access to world markets (Places 88). Despite political independence, many poorer countries remain economically dependent (neo-colonialism) on the large transnationals (together with international banks and foreign aid). Some of the advantages and disadvantages of transnational corporations to developing countries are listed in Figure 19.32.

Transnationals, having selected a country, then have to decide where to locate within that country. If the country is economically developed, the location is likely to be where financial inducements are greatest, land values are low, transport is well developed, and levels of skill and unemployment are high (Japanese companies in South Wales, page 572). If the country is economically less developed, the location is more likely to be in the primate city (page 405), especially if that city is also the capital or the chief port. A capital city location, with an international airport, allows quick access to the companies' overseas headquarters; and a port location enables easier export of manufactured goods. Should several transnational companies locate in the same area, the multiplier effect (page 569) is likely to result in the development of a core region (Places 98, page 618).

Places 88 Pune, India: a hub for transnationals

Figure 19.33

Location of Pune in India

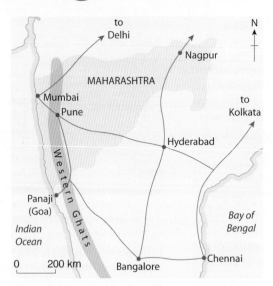

Pune, a city of 5 million inhabitants, lies 150 km south-east of Mumbai in the western coastal state of Maharashtra. It is known as the 'Oxford of the East', as it has nine universities, and 'The Detroit of India' due to the presence of numerous global car TNCs. Its rapid industrial growth has partly been due to congestion, pollution, lack of space and exceptionally high property prices in nearby Mumbai (population 18.2 million), as well as to its own advantages.

Pune has good transport links, especially with the port and financial centre of Mumbai. It is also on the 'Golden Quadrilateral', a four-lane expressway that links Bangalore, Chennai, Kolkata, Delhi and Mumbai, as well as being on a main rail line and having its own airport. Its climate is healthier than

that of Mumbai, being 650 m above sea-level, which makes it less humid, and, lying in the rain shadow to the east of the Western Ghats, it receives only 650 mm of rain a year compared with Mumbai's 2200 mm (Figure 9.57).

Pune's universities produce large numbers of skilled graduates and Mercedes-Benz founded an international school for professional people from overseas. The state of Maharashtra is viewed positively as a manufacturing and commercial centre as it is less prone to industrial strikes and corruption which affect other parts of India. Other favourable factors that are important when trying to attract TNCs include its good health care service and a reliable supply of water and electricity.

Pune has also benefited from the setting up, in 1960, of the Maharashtra Industrial Development Corporation (MIDC) which offers business incentives that include exemptions from electricity duty and stamp duty, refund on Octroi (a tax applied to goods entering and leaving an area) and special financial help, together with interest rate subsidies for the textile industry – incentives that are not available in Mumbai. Among the TNCs that have located in and around Pune are automotive corporations (Daimler-Chrysler, Fiat, General Motors, Mercedes-Benz, Skoda, Tata Motors and Volkswagen), electrical companies (Panasonic, Philips, Siemens and Whirlpool), technology centres (Barclays, HSBC and John Deere) and outsourcing call centres (Next and British Gas).

The informal sector

A large and growing number of people with work in developing countries have found or created their own jobs in the informal sector (Figure 19.34).

'This sector covers a wide variety of activities meeting local demands for a wide range of goods and services. It contains sole proprietors, cottage industries, self-employed artisans and even moonlighters. They are manufacturers, traders,

	Formal	Informal
Description	Employee of a large firm	Self-employed
	Often a transnational	Small-scale/family enterprise
	Much capital involved	Little capital involved
	Capital-intensive with relatively few workers; mechanised	Labour-intensive with the use of very few tools
	Expensive raw materials	Using cheap or recycled waste materials
	A guaranteed standard in the final product	Often a low standard in quality of goods
	Regular hours (often long) and wages (often low)	Irregular hours and uncertain wages
	Fixed prices	Prices rarely fixed and so negotiable (bartering)
	Jobs done in factories	Jobs often done in the home (cottage industry) or on the streets
	Government and transnational help	No government assistance
	Legal	Often outside the law (illegal)
	Usually males	Often children and females
Type of job	Manufacturing: both local and transnational companies	Distributive (street peddlers and small stalls)
	Government-created jobs such as the police, army and civil service	Services (shoe cleaners, selling clothes and fruit)
		Small-scale industry (food processing, dress-making and furniture repair)
Advantages	Uses some skilled and many low-skilled workers	Employs many thousands of low-skilled workers
	Provides permanent jobs and regular wages	Jobs may provide some training and skills which might lead to better jobs in the future
	Produces goods for the more wealthy (food, cars) within their own country so that profits may remain within the country	Any profit will be used within the city: the products will be for local use by the lower-paid people
	Waste materials provide raw materials for the informal sector	Uses local and waste materials

Figure 19.34

Differences between 'formal' and 'informal' sectors

transporters, builders, tailors, shoemakers, mechanics, electricians, plumbers, flower-sellers and many other activities. The [Kenyan] government have recognised the importance of these small-scale *jua kali* enterprises [Places 89] and a few commercial banks are beginning to extend loans to these new entrepreneurs who are themselves forming co-operatives. There are many advantages in developing these concerns. They use less capital per worker than larger firms; they tend to use and recycle materials that would otherwise be waste; they provide low-cost, practical on-the-job training which can be of great value later in more formal employment; and, as they are flexible, they can react quickly to market changes. Their enterprising spirit is a very important national human resource.'

Central Bank of Kenya

The governments of several developing countries now recognise the importance of such local ventures as Kenya's *jua kali* which, apart from creating employment, provide goods at affordable prices. India, for example, encourages the growth of co-operatives to help family concerns, under the 'Small Industries Development Organisation', by setting up district offices that offer technical and financial advice. Under its Development Plans, the manufacture of 600 products will be exclusively reserved for small firms and family enterprises.

Children, many of whom may be under the age of 10, form a significant proportion of the informal-sector workers. Very few of them have schools to go to and, from an early age, they go onto the streets to try to supplement the often meagre family income. They may try to earn money by shining shoes or selling items such as sweets, flowers, fruit and vegetables.

Places 89 Nairobi, Kenya: *jua kali* workshops

Jua kali means 'under the hot sun'. Although there are many smaller *jua kali* in Nairobi, the largest is near to the bus station where, it is estimated, over 1000 workers create jobs for themselves (Figure 19.34). The plot of land on which the metal workshops have been built measures about 300 m by 100 m. The first workshops were spontaneous and built illegally as their owners did not seek permission to use the land, which did not belong to them. As more workshops were set up and the site developed, the government was faced with the option of either bulldozing the temporary buildings, as governments had done to shanty settlements in other developing countries, or encouraging and supporting local initiative.

Realising that the informal workshops created jobs in a city where work was hard to find, the government opted to help. The Prime Minister himself became personally involved by organising the erection of huge metal sheds which protected the workers from the hot sun and occasional heavy rain.

Groups of people are employed touring the city collecting scrap. The scrap is melted down, in charcoal stoves, and then hammered into various shapes including metal boxes and drums, stoves and other cooking utensils, locks and water barrels, lamps and poultry water troughs (Figure 19.35). Most of the workers are under 25 and have had at least some primary education. The technology

Figure 19.35

Jua kali workshops

they use is appropriate and sustainable, suited to their skills and the availability of raw materials and capital. Most of the products are sold locally and at affordable prices.

It is estimated that there are approximately 600 000 people engaged in 350 000 small-scale *jua kali* enterprise units in Kenya. This figure needs to be compared with the 180 000 recorded as employed in large-scale manufacturing and the 2.2 million total in all areas of the non-agricultural economy. *Jua kali* form, therefore, a most significant part of the total employment picture.

Intermediate (appropriate) technology

Dr E.F. Schumacher developed the concept of **intermediate technology** as an alternative course for development for poor people in the 1960s. He founded the Intermediate Technology Development Group (ITDG) in 1966, now renamed Practical Action, and published his ideas in a book, *Small is Beautiful* (1973). Schumacher himself wrote:

'If you want to go places, start from where you are.

If you are poor, start with something cheap.

If you are uneducated, start with something relatively simple.

If you live in a poor environment, and poverty makes markets small, start with something small.

If you are unemployed, start using labour power, because any productive use of it is better than letting it lie idle.

In other words, we must learn to recognise boundaries of poverty.

A project that does not fit, educationally and organisationally, into the environment, will be an economic failure and a cause for disruption.'

In 1988 the ITDG stated that:

'Essentially, this alternative course for development is based on a local, small-scale rather than the national, large-scale approach. It is based on millions of low-cost workplaces where people live – in the rural areas – using technologies that can be made and controlled by the people who use them and which enable those people to be more productive and earn money.'

These ideas challenged the conventional views of the time on aid. Schumacher said:

'The best aid to give is intellectual aid, a gift of useful knowledge … The gift of material goods makes people dependent, but the gift of knowledge makes them free – provided it is the right kind of knowledge, of course.'

To illustrate this he quoted an old proverb:

'Give a man a fish and you feed him for a day; teach him how to fish and he can feed himself for life.'

The first part of this might be seen as the traditional view of aid where 'giving' leads to dependency. The second part, 'teaching', is a move in the direction of self-sufficiency and self-respect. Schumacher added a further dimension to the proverb by saying: 'teach him to make his own fishing tackle and you have helped him to become not only self-supporting but also self-reliant and independent'.

In most developing countries, not only are high-tech industries too expensive to develop, they are also usually inappropriate to the needs of local people and the environment in which they live. Examples of intermediate, or **appropriate technology** as it is now known (Places 90), include:

- labour-intensive projects; since, with so many people already being either unemployed or underemployed, it is of little value to replace workers by machines
- projects encouraging technology that is sustainable and the use of tools and techniques designed to take advantage of local resources of knowledge and skills
- the development of local, low-cost schemes using technologies which local people can afford, manage and control rather than expensive, imported techniques
- developing projects that are in harmony with the environment.

Practical Action (formerly known as ITDG – see page 576) is a British charitable organisation that works with people in developing countries, especially those living in rural areas, by helping them to acquire the tools and skills needed if they are to raise themselves out of poverty and meet the UN's Millennium Development Goals (page 609). Practical Action helps people to meet their basic needs of food, clothing, housing, energy and jobs. It also uses, and adds to, local knowledge by providing technical advice, training, equipment and financial support so that people can become, in Schumacher's words, 'more self-sufficient and independent' (page 576). Although Practical Action operates globally, the following examples are taken from Kenya. They are all:

- suitable for the local environment (local raw materials and climate)
- appropriate to the wealth, skills and needs of the local people.

1 Improved building materials include roofing tiles that are made from a mix of cement, sand and water (and sometimes a pigment if a different colour is required). They are left in their moulds for a day to cure (but not to dry), placed in a reservoir of water for a week and finally covered with plastic, as a protection against the hot sun, and allowed to dry slowly for three weeks. They are cheaper than commercially produced tiles, as they do not need firing, and lighter (Figure 19.36).

2 In another scheme, lime and natural fibres are added to soil to produce 'soil blocks'. Soil is important because it can be obtained locally, can easily be compressed and, once heated, retains its heat. Soil blocks are replacing the more expensive concrete blocks and industrially produced bricks.

3 Other projects have helped to improve ventilation and lighting in existing houses. Traditionally, most Kenyan women cooked on wooden stoves in houses that had no chimneys and few windows. The result was a smoky and unhealthy atmosphere. To reduce reliance on wood and charcoal, which may be difficult and/or expensive to obtain, and to improve living conditions, Practical Action has helped to train potters to produce two types of improved cooking stoves (Figure 19.37): the *mandaleo* for wood-burning stoves in rural areas, which are made from ceramic; and the *jiko* for charcoal-burning stoves in urban areas, which are made from recycled scrap metal, often in *jua kali* workshops (Places 89), to which potters add a ceramic lining. The new stoves, based on traditional designs, reduce smoke, improve women's health and pay for themselves within a month. They also reduce the amount of time rural families have to spend collecting firewood (page 543) and the cost that urban families have to pay for charcoal, and help to conserve a rapidly declining natural resource.

4 Practical Action has also helped the Maasai improve their houses. This has been done by adding a thin layer of concrete reinforced with chicken wire over the old mud roof; adding a gutter and downpipe which leads to a water barrel (saving a likely long trek to the nearest river, Figure 21.11); and adding a small window and chimney cowl to make the inside of the house lighter and less smoky, which improves health.

Figure 19.36

Roofing tiles

Figure 19.37

New cooking stoves

Newly industrialised countries (NICs)

Newly industrialised countries (NICs) is a term applied to a select group of developing countries that, over the last three or four decades, have sustained a high rate of economic growth (Figure 19.38). They have out-performed all the more developed countries, mainly due to their competitive edge in manufacturing. Although Brazil and Mexico were among the first NICs, most are located in eastern Asia. Encouraged by Japan's success, governments in other countries in Asia's Pacific Rim set out to improve their standard of living by:

- encouraging the processing of primary products, as this added value to their exports
- investing in manufacturing industry, initially by developing heavy industries such as steel and shipbuilding, and later by concentrating on high-tech products
- encouraging transnational firms to locate within their boundaries (many countries now have their own TNCs)
- grouping together to form ASEAN (Figure 21.34) to promote, among other aims, economic growth
- having a dedicated workforce that was reliable and, initially, prepared to work long hours for relatively little pay
- long-term industrial planning.

The term 'tiger economies' was first given to Hong Kong, Singapore, South Korea and Taiwan because of their ferocious growth after 1970. This growth continued during the 1980s at a time when economic growth in the developed world was slowing down. Since then, Malaysia (the most successful, Places 91), Thailand and, to a lesser extent, Indonesia and the Philippines, have also emerged as NICs. The latest, and likely to be the largest if it maintains its present unprecedented rate of growth, is China (Case Study 19). China and India together have become known as the **emerging countries**.

Figure 19.38

Annual economic growth rate (%) – NICs in eastern Asia and the emerging countries

		1981–90	1991–2000	2001–05	2007
NICs	Hong Kong	7.1	4.7	4.3	6.8
	Indonesia	6.3	4.3	4.7	5.6
	Japan	4.1	1.3	1.4	2.7
	Malaysia	5.1	7.2	4.5	5.9
	Philippines	1.1	3.0	4.7	5.4
	Singapore	7.0	7.7	4.0	7.9
	South Korea	10.1	6.1	4.5	5.0
	Taiwan	5.2	6.5	3.2	4.7
	Thailand	7.6	4.5	5.1	5.0
Emerging countries	China	9.5	10.5	9.5	10.7
	India	3.1	4.5	5.1	5.0

Places 91   Malaysia: a newly industrialised country

Until the 1980s, Malaysia's economy was based on primary products such as rubber and palm oil (Places 68, page 483), timber (Places 76, page 520), tin (Places 79, page 523) and oil (Figure 19.39a). The government at that time proclaimed its vision of Malaysia becoming a fully developed and industrialised nation by the year 2020. Since then the country has emerged as the leader of the second wave of Asian 'tiger economies', averaging – between 1990 and mid-1997 – an annual growth rate of 8 per cent. This allowed the World Bank to classify Malaysia as an 'upper middle income country', no longer a developing country. This was achieved without high inflation or unmanageable foreign aid.

Malaysia's economic development was based on its pivotal position as a gateway to ASEAN (Figure 21.34), it being a springboard to eastern Asia, its affordable land and liberal investment rules,

Figure 19.39

Malaysia's changing exports, 1970 and 2008

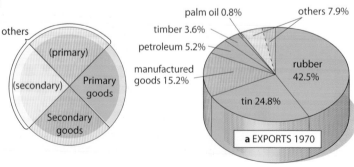

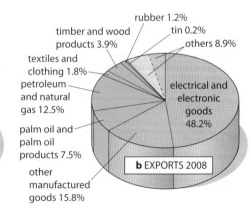

Figure 19.40

Car assembly (Proton)

and its encouragement, through tax incentives, for transnationals to locate there. The country's industrial strategy emphasised the development of high-value goods for the domestic market and export (e.g. cars) and the encouragement of high-tech industries (e.g. electronics – Figure 19.39b). In 1985, the government founded the Proton car company, initially in conjunction with Mitsubishi (Figure 19.40) and in 1995, the Perodua company, in partnership with Daihatsu.

During the 1980s and 1990s, industry was confined to specifically designed areas such as the new town of Shah Alam. This policy was good environmentally as only certain tracts of primary forest or farmland were used, but had the social disadvantage of concentrating jobs and development within a few core places (page 617 and Places 98). As firms newly locating in Shah Alam need not pay taxes for 10 years, then many of the world's better-known transnationals, together with Proton, located there.

The government had also, during the early 1990s, invested less money in industries that required large workforces and more in those where the emphasis was on technology. Its Technology Action Plan covered micro-electronics, biotechnology and information technology (Figure 19.41).

The Second Industrial Plan, which operated between 1996 and 2005, focused on the manufacturing sector and R&D (research and development), together with the integration of support industries. The plan concentrated on the production of electrical and electronic goods (including IT and multimedia), oleochemicals (from palm oil, timber and rubber), chemicals (petrochemicals and pharmaceuticals), transport equipment, machinery, and high-value textiles and clothing.

Figure 19.44

New high-tech industry, Penang

By the mid-2000s, an extensive road system linked Kuala Lumpur (the financial and commercial centre), Putrajaya (the new seat of government), Shah Alam (the industrial new town), Port Klang (the chief port) and Westport (with its new deepwater ocean terminal), Sepang (the new international airport) and Subang (the old airport now mainly used for domestic flights). The country had also completed a series of expensive 'prestige' projects including, in Kuala Lumpur, the twin Petronas Towers (1998), a three-line light rail transit system (LRT) and a 'linear city' (a 2 km long, 10 storey high structure comprising shopping malls, hotels, restaurants, apartments and offices).

The Third Industrial Master Plan (IMP3) is to operate between 2006 and 2020. Its main objective is 'to achieve long-term global competitiveness through transformation and innovation of the manufacturing and service sectors'. The government has targeted:

- six non-resource based manufacturing industries – electrical and electronics, medical devices, textiles and apparel, machinery and equipment, metals and transport equipment

- six resource-based manufacturing industries – petrochemicals, pharmaceuticals, wood-based, rubber-based, oil palm based and food processing

- eight services sub-sectors – logistics, business and professional, ICT, distributive trade, construction, education and training, healthcare and tourism.

The plan was introduced with a predicted average economic growth of 6.3 per cent per annum (compare Figure 19.38) and, during that period, a threefold increase in trade – but that was before the global recession of 2008.

Opening up to the outside world

In 1979 the Chinese government made several monumental decisions including replacing the commune system with the responsibility system, initially in farming (Places 63, page 468) and then in industry, together with the implementation of both the one-child policy (Case Study 13) and the 'open-door' policy which allowed trade with the outside world. The following year, China established five **Special Economic Zones (SEZs)** in Shenzhen, Zhuhai and Shantou in Guangdong Province, Xiamen in Fujian Province and the whole of Hainan island (Figure 19.42). According to the *China Business Handbook*, the SEZs: 'integrate science and industry with trade, and benefit from preferential policies and special economic managerial systems intended to facilitate exports. The SEZs also offer preferential conditions to foreign investors by granting them more favourable rates than in inland areas, and relaxing entry and exit procedures for business people. SEZs aim to attract foreign investment, to import advanced techniques,

keep up to date with trends and activity in international markets, to expand export trade, to stimulate foreign exchange earnings, to facilitate participation in international economic and technological co-operation, and to provide a training ground for scientific and technological personnel specialising in international economics and trade'.

In 1984, China opened 14 coastal cities to overseas investment. These **open cities** (Figure 19.42), as they are known, were given the dual role of being 'windows' opening to the outside world and 'radiators' spreading economic development inland in an export-oriented economy. The economic and technological development zones that were set up within these open cities became such hotspots for overseas investment that in 1985 the state decided to expand the SEZs and open cities to form one continuous coastal belt. Five years later, several additional open cities were created along the Yangtze River, as far as Chongqing (Figure 21.24). When the Pudong New Zone was established in 1990 it meant that, with Pudong acting as the 'dragon's head' (reflecting the shape of the river), a chain of open cities extended up the Yangtze Valley.

Pudong

The development of Pudong, along with Shenzhen, must rank as the world's fastest-growing area, with huge commercial, industrial and residential zones together with a modern transport system (Case Study 15B and Figure 19.44). Since Pudong's development was first announced in 1990, when it was little more than an area of padi fields, it has been a **New Open Economic Development Zone**. It has emerged, in less than two decades, as China's financial and commercial hub, being home to the Shanghai World's Financial Centre and the Shanghai Stock Exchange as well as the Lujiazui Trade and Finance Zone, the Waigaoqiao Free Trade Zone, the Jinqiao Export Processing Zone and the Zhangjiang Hi-tech Park. Perhaps the most spectacular feature of Pudong's growth, at least to the visitor, be it for business or pleasure, is the skyline viewed across the river from Shanghai itself (Figure 19.43). In 1990 the Chinese saw Pudong as the engine pulling Shanghai into position as a major international economic, financial and trade centre – a vision that seems to have been fulfilled.

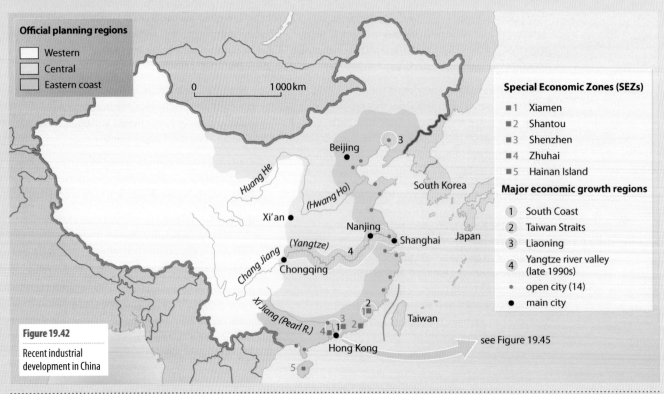

Figure 19.42 Recent industrial development in China

Figure 19.43

Pudong's new skyline

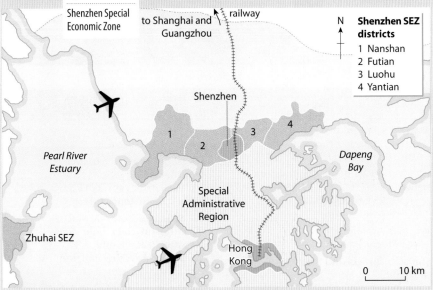

Figure 19.44

Development in Pudong

Figure 19.45

Shenzhen Special Economic Zone

to Shanghai and Guangzhou   railway

N   **Shenzhen SEZ districts**
1 Nanshan
2 Futian
3 Luohu
4 Yantian

Shenzhen

Pearl River Estuary

Dapeng Bay

Special Administrative Region

Zhuhai SEZ

Hong Kong

0   10 km

Shenzhen

Before 1989, when it became a Special Economic Zone, Shenzhen was a group of small fishing villages surrounded by padi fields (Figure 19.47). It had a population of 20 000. By 1997, this population had risen to 3.8 million and by 2006 it was given as 8.46 million, making it the world's 27th largest city (Figure 19.48). The average age is under 30. The workforce can be divided into two polarised groups: those who have had a high level of education (20 per cent of China's PhDs are said to work here), and the majority, many of whom are migrants from surrounding rural provinces, with little education. Added to this are over 7000 daily commuters from Hong Kong.

Shenzhen was chosen as an SEZ due to its coastal location for trade, its deep natural harbours, its proximity to the financial and commercial centre of Hong Kong (then still a British colony and with which it has a similar culture), its plentiful supply of labour (which is adaptable but cheaper than in other Asian NICs) and its low land values (rents are half those in Hong Kong). It has benefited from financial incentives offered by the Chinese government and from over US$30 billion invested by overseas TNCs for the building of factories and in forming joint ventures. The SEZ comprises four of the seven districts (397 km^2) that make up Shenzhen (Figure 19.45). Nanshan (164 km^2), with its Science and Technology Park, is the focus for high-tech industries and foreign companies. Futian (78 km^2) is the trading centre and includes the Stock Exchange and the municipal government building. Luohu (79 km^2) is the financial and commercial centre with the new People's Bank of China. Yantian (76 km^2) is the centre for logistics as well as being China's second biggest and the world's fourth largest deepwater container terminal.

Since its inception, Shenzhen has focused on selective industries which include computer software, IT, microelectronics and components, video and audio products and electro-mechanical integration. More recently, new industries, such as pharmaceuticals, medical equipment and biotechnology, have grown rapidly. At present, electronics and telecoms equipment is the largest industry with, for example, over 100 million handsets for mobile phones being manufactured in 2007. There are over 200 R&D

organisations within the SEZ, many having strong links with inland universities. TNCs located here include Sanyo, Hitachi, Matsushita (all Japanese), IBM (American), Siemens (German) and Great Wall (China) together with, from the retail sector, over 5000 companies producing goods for Wal-Mart (Figure 19.46). Shenzhen has the largest manufacturing base in the world as well as being a powerhouse in the economy of China – and all in less than 20 years! But success rarely comes without its problems and Shenzhen has these in the form of an unreliable electricity supply, insufficient clean water, difficult disposal of waste and uneasy labour relations.

Figure 19.47

Shenzhen (1999) – this was all farmland in 1980

Figure 19.46

Wal-Mart's Shenzhen base

Wal-Mart is the world's largest retailer by far. In 2004, the company had 4900 stores worldwide and its 1.6 million sales assistants sold goods to some 138 million customers. But where do the products it sells come from? For many of the non-perishable consumer goods on the store shelves, such as toys, clothes and electronics, the answer is increasingly likely to be China. In 2004, Wal-Mart sourced US$18 billion worth of goods from China, representing 3 per cent of that country's exports. The huge sourcing operation is run from Wal-Mart's overseas procurement office located in Shenzhen in the southern Guangdong province, from which the retailer has established ongoing supply relations with over 5000 local companies. Individual companies can do huge amounts of business with Wal-Mart. Guangdong's Yili Electronics Group, for example, started supplying hi-fi systems in 1995, and now supplies Wal-Mart with over US$200 million worth of goods each year, accounting for half its sales.

Wal-Mart sources its goods from China because labour costs there are just 4 per cent of those in the USA. This means that a product can be manufactured in China, packaged, shipped around the world, sold to American or European consumers and still return a decent profit for both manufacturer and then retailer [page 643].

Adapted from *Economic Geography* (Blackwell, 2007)

Figure 19.48

Today, Shenzhen is a city of tower blocks

Further reference

Barke, M. and O'Hare, G. (1991) *The Third World*, Oliver & Boyd.

Coe, N.M., Kelly P.F. and Yeung, H.W.C. (2007) *Economic Geography*, Blackwell.

Malaysia Official Yearbook 2008.

Schumacher, E.F. (1993) *Small is Beautiful*, Vintage.

Assisted Areas in the UK:
www.berr.gov.uk/whatwedo/regional/assisted-areas/index.html

CIA World Fact Book, employment structures:
www.odci.gov/cia/publications/factbook/index.html

Practical Action:
www.itdg.org

Statistics Bureau and Statistics Centre of Japan:
www.stat.go.jp/english/index.htm

Statistics Singapore:
www.singstat.gov.sg/

UK labour market statistics, manufacturing:
www.statistics.gov.uk/CCI/SearchRes.asp?term=manufacturing

UK Office for National Statistics (NOMIS), official labour market statistics: (searchable)
www.nomisweb.co.uk/Default.asp

UK steel statistics:
www.eef.org.uk/uksteel/publications/steel/data/public/UK_Steel_Key_Statistics_2008.htm

Questions & Activities

Activities

1 a i What is 'manufacturing industry'? *(1 mark)*

    ii 'With the shift from an industrial to a post-industrial society it is sometimes unrealistic to try to draw clear boundaries between "manufacturing" and "services"'.

    Explain the problems that led to this statement. *(2 marks)*

    iii Explain why the proportion of the UK's population in secondary employment has fallen so sharply in recent years. *(6 marks)*

  b The number of people employed in manufacturing has not fallen evenly across the country.

    i Name an area where a loss of manufacturing jobs has caused a serious local unemployment problem. *(1 mark)*

    ii Explain what caused the loss of manufacturing jobs in that area. *(3 marks)*

    iii Describe a strategy that has been used to create new employment opportunities in that area. Assess the success of the strategy. *(6 marks)*

  c The gender structure of the workforce in the UK has changed rapidly since 1960. Describe and account for the changes. *(6 marks)*

2 a i What are 'high-tech industries'? *(2 marks)*

    ii It has been noted that firms involved in high-tech industries often have two quite distinct parts to their operations. These are:

    • research and development

    • mass manufacturing.

    Suggest why these two separate parts of the industry often locate in different places. *(3 marks)*

  b i Name **one** area where a concentration of research and development centres for high-tech industry has developed. *(1 mark)*

    ii Explain why the area you named in **i** is attractive to this industry. *(9 marks)*

  c i Name **one** area where mass production for the high-tech industry has developed quite separately from research and development. *(1 mark)*

    ii Explain why the area you named in **c i** is attractive to this industry. *(9 marks)*

3 a Study the diagram below. It shows some of the factors that influence the location of manufacturing industry.

Give **one** example of an industry where the most important factor influencing its location is:

  i raw materials

  ii power supply

  iii labour supply

  iv access to market.

For each example you have given, explain why that factor is so important. *(12 marks)*

  b i What is meant by:

    • footloose industry?

    • greenfield site?

    ii Name an example of a footloose manufacturing industry that has located on a greenfield site. Suggest why that site was a suitable location for that industry. *(5 marks)*

  c i What is 'inward investment'? *(1 mark)*

    ii Choose an example of a factory that has been built by a foreign-based company investing in the UK. Suggest why that company chose to invest in the UK, and why that particular location was suitable for its investment. *(7 marks)*

Exam practice: basic structured questions

4 a i Name a region in the UK that has suffered unemployment as a result of the decline of its traditional manufacturing industry. *(1 mark)*

    ii Explain why the traditional industry developed in that area and then declined. *(8 marks)*

    iii Describe the other social and economic problems that are found in that area as a result of the unemployment that followed the decline of the traditional industry. *(6 marks)*

  b The government has developed several policies to try to attract new industry into regions that have suffered the loss of their traditional industry.

  Choose any such government initiative, and describe how it has affected any one area. Assess how successful the initiative has been in attracting new industry. *(10 marks)*

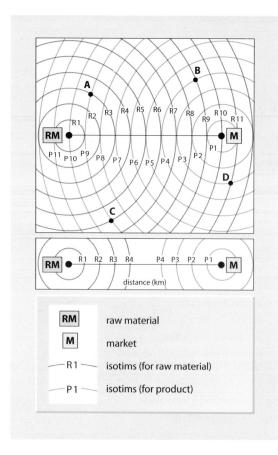

Site	Total transport costs if the raw material is pure	Total transport costs if the raw material is gross and loses 50% of its weight during manufacture
A		
B		
C		
D		

Figure 19.49

RM raw material

M market

R 1 isotims (for raw material)

P 1 isotims (for product)

5
a In the Weber model of industrial location, what is the 'least cost location'? Explain why it is important. *(3 marks)*
b Show, with the aid of diagrams, how the least cost location in Weber's locational triangle may be:
i near to the raw material
ii near to the market
iii midway between the raw material and the market. *(6 marks)*
c Study Figure 19.49, which shows an area with one raw material and one market. Assume two different situations:
X – the raw material is pure
Y – the raw material loses 50 per cent of its weight during manufacture.
i Complete a copy of the table above to show the total transport costs (in tonne km) for an industry located at each of the sites A – D. *(4 marks)*
ii For each situation, X and Y, describe the least cost location.
Give reasons for your answer. *(6 marks)*

d How useful is Weber's model to an understanding of modern industrial location? Justify your answer. *(6 marks)*

6
a i Describe **three** differences between the formal and the informal economic sectors in economically less developed countries. *(6 marks)*
ii Explain why *jua kali* workshops are very important in Kenya's economy. *(6 marks)*
iii 'If governments wish to encourage development that will benefit the poorer sections of the population, one of the most important actions they can take is to reduce the rules and regulations which hinder the development of the informal economy.'
Suggest why this is seen to be important. *(4 marks)*
b Discuss the advantages and disadvantages for less economically developed countries of investment by transnational companies. You should make specific references to one or more countries that you have studied. *(9 marks)*

7
a In Myrdal's model of industrial location he referred to 'cumulative causation' which is also sometimes called 'the multiplier effect'. What does this term mean? *(4 marks)*
b With reference to the industrialisation of **either** South Wales in the late 19th and early 20th centuries **or** any other region with which you are familiar, explain how cumulative causation helped to cause the development of industry. *(10 marks)*
c Myrdal realised that as some areas industrialise this may cause a 'backwash effect' on other regions in the country. He named the areas that were affected 'the periphery'. Explain:
i the 'backwash effect' *(3 marks)*
ii the 'periphery' of a country or region. *(3 marks)*
d Name an area which could be regarded as part of the economic periphery in a country or region that you have studied.
Describe the features that make this area part of the economic periphery. *(5 marks)*

Exam practice: structured questions

8
a Discuss the problems that have been caused by a high concentration of employment in a small number of industries in the UK. *(10 marks)*
b Explain how one or more government initiatives have been used to try to broaden the base of employment. *(15 marks)*

9 Study Figure 19.50.
a Referring to Figure 19.50 and your own knowledge, explain why India has become a major centre of automobile manufacture. *(10 marks)*
b Discuss the extent to which the recent growth of India's manufacturing and service economy have depended on investment by transnational corporations (TNCs). *(15 marks)*

The market for cars in India is growing so quickly that it seems likely that it will overtake China's sales totals soon. Sales of passenger cars increased by 12.17% to 1.5 million in the year to March 2008.

India's car industry is concentrated in the region around Pune, Maharashtra. Plans are that the Pune region will employ 25,000 people in car making in two years.

Volkswagen, General Motors, Tata Motors, Mercedes-Benz, Fiat and Peugeot already have plants there and the local Development Corporation is in discussions with four or five other major international companies seeking land for new factories. The cost of building a factory here is cheaper than almost anywhere else in the world.

But huge savings are made on manpower – with manual workers in India paid about £1.30 a day. As a result, major car makers are considering using their India plants for export, both for finished cars and components. GM has said it wants to make India an export hub for small and mid-sized cars destined to be sold in other emerging markets and Hyundai plans to make India the sole production centre for its new I20 model, even though it will not be sold domestically.

Figure 19.50

10 Study Figure 19.51. It shows details of solar cookers, an example of appropriate technology.

   a Describe the solar cookers shown here and explain why they are good examples of appropriate technology for use in developing countries of Africa and Asia.
   (10 marks)

   b With reference to one or more examples, explain how appropriate technology can be used to improve the quality of housing in developing countries. *(15 marks)*

11 a Describe the main features of Myrdal's theory of cumulative causation. *(10 marks)*

   b With reference to a named peripheral region in a country outside the UK:

      i explain the problems that have been caused by its peripheral position *(5 marks)*

      ii describe one scheme that has been tried in an attempt to overcome these problems, and evaluate its success. *(10 marks)*

Figure 19.51
Solar cookers

Box cookers
Box cookers cook at moderate to high temperatures and often accommodate multiple pots. Worldwide, they are the most widespread. There are several hundred thousand in India alone.

Curved concentrator cookers
Curved concentrator cookers cook fast at high temperatures, but require frequent adjustment and supervision for safe operation. Several hundred thousand exist, mainly in China.

Panel cookers
Panel cookers are simple and relatively inexpensive to buy or produce. Solar Cookers International's 'CooKit' is the most widely used combination cooker.

Exam practice: essays

12 Changes in technology during the past 30 years have had a major effect on industrial location throughout the world.

Describe the major changes. Explain why they have taken place and how they have affected the location of industry. *(25 marks)*

13 Assess the importance of transnational corporations in the development of the global pattern of industrialisation in the late 20th and early 21st centuries. You should refer to their effect in both more and less economically developed countries. *(25 marks)*

14 Account for the development of the 'tiger economies' of South-east Asia and discuss the extent to which they can be seen as models for the development of the economies of other developing countries. *(25 marks)*

20

Tourism

• •

'In the Middle Ages people were tourists because of their religion whereas now they are tourists because tourism is their religion.'

Robert Runcie, former Archbishop of Canterbury

'Travel broadens the mind.'

Proverbs

Tourism is an example of a service industry and as such falls into the tertiary sector, one of the four major sectors into which economies of all countries may, for convenience, be divided (page 552). Individual services may be grouped as follows:

1 **Public services**, e.g. electricity and water companies.
2 **Producer services** help businesses carry out their activities, e.g. banking, law and transport.
3 **Consumer services** are those that have direct contact with the consumer, e.g. retailing (Chapter 15) and leisure, recreation and **tourism**.

Leisure, recreation and tourism

Leisure is a broad term associated with 'time, free from employment, at one's own disposal'.

In developed countries, with shorter working weeks and earlier retirement, many people have an increasing amount of 'free time' which allows them to participate in recreational activities.

Recreation refers to activities, events and pursuits that are undertaken though choice, e.g. sport, gardening, fireworks displays, bird watching, video games. An increase in leisure time generates the demand for additional recreational amenities such as golf courses, country parks, swimming pools and night clubs. **Tourism** involves travel away from home to visit friends and relations or different places. The official UK definition is 'a stay away from one's normal place of residence which includes at least one night but is less than a year'. The World Tourism Organization (UNWTO), however, does not stipulate the 'one night away' so its definition includes day visitors as tourists, as well as 'business tourism'.

The UK travel and tourist industry consists of a wide variety of commercial and non-commercial organisations that interact to supply products and services to tourists. This often makes it difficult to differentiate leisure and tourism from other forms of employment, e.g. a fish and

Figure 20.1

Types and location of various leisure and tourist facilities

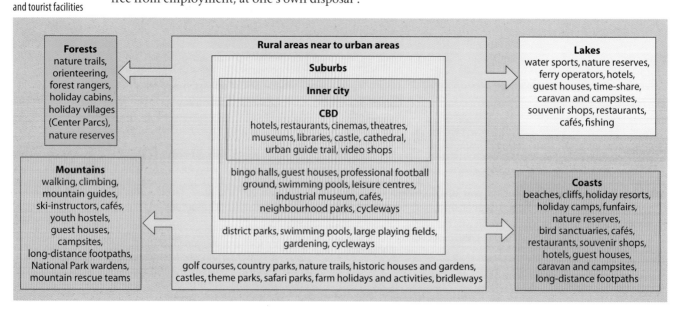

Forests	Rural areas near to urban areas	Lakes
nature trails, orienteering, forest rangers, holiday cabins, holiday villages (Center Parcs), nature reserves	**Suburbs** — **Inner city** — **CBD**: hotels, restaurants, cinemas, theatres, museums, libraries, castle, cathedral, urban guide trail, video shops; bingo halls, guest houses, professional football ground, swimming pools, leisure centres, industrial museum, cafés, neighbourhood parks, cycleways; district parks, swimming pools, large playing fields, gardening, cycleways; golf courses, country parks, nature trails, historic houses and gardens, castles, theme parks, safari parks, farm holidays and activities, bridleways	water sports, nature reserves, ferry operators, hotels, guest houses, time-share, caravan and campsites, souvenir shops, restaurants, cafés, fishing
Mountains walking, climbing, mountain guides, ski-instructors, cafés, youth hostels, guest houses, campsites, long-distance footpaths, National Park wardens, mountain rescue teams		**Coasts** beaches, cliffs, holiday resorts, holiday camps, funfairs, nature reserves, bird sanctuaries, cafés, restaurants, souvenir shops, hotels, guest houses, caravan and campsites, long-distance footpaths

chip shop proprietor in Blackpool sells to both tourists and residents, while farmers on a Greek or West Indian island sell their produce to both local people and hotels.

People with limited income, access to transport or leisure time tend to seek recreational amenities and activities that are near their homes. As the majority of British people live in towns and cities, then most amenities are located within or near to urban areas (Figures 20.1 and 20.2). People with more leisure time tend to travel further afield to scenic rural areas, especially those with added amenities (coasts, mountains and National Parks), to large urban areas (historical towns and cultural centres), and to places outside the UK.

As in other areas of their subject, geographers have tried to classify aspects of tourism (Framework 7, page 167). One suggested classification is:

- by nature of attraction, e.g. coastal, mountains, rivers and lakes, climate, woodland, flora and fauna, historic heritage buildings and sites, cruises, retailing, activity centres, urban and rural
- by length of stay, e.g. weekend break, annual two-week holiday
- by travel within or beyond national, borders, e.g. domestic and international
- by type of transport, e.g. caravan, bicycle, canal boat, cruise liner
- by type of accommodation, e.g. camping, safari lodge, beach village.

The growth in tourism

The Romans must rank amongst the earliest tourists, as many of their most wealthy families used to move to their country villas during the hot, dry summers. By the 18th and 19th centuries, affluent British people were either visiting spa towns within England or making the 'Grand Tour' of Classical Europe, while the less well-off were beginning to popularise local seaside resorts. Today tourism has become part of everyday life and a major source of employment in many developed countries. Here, the rapid growth of the tourist industry in the last half-century can be linked to numerous factors such as greater affluence (wealth), increased mobility, improvements in accessibility and transport, more leisure time, paid holidays, product development and innovations, improvements in technology, changes in lifestyles and fashion, an increased awareness of other places and, more recently, the need for 'green' (sustainable) tourism (page 597). These factors are summarised in Figure 20.3.

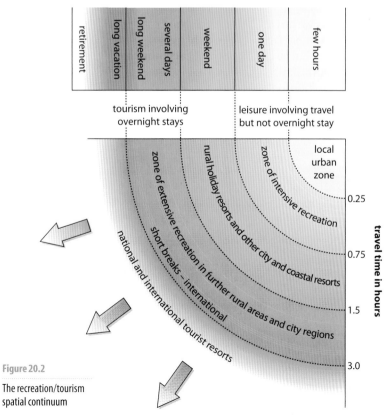

consumer's available leisure time

retirement | long vacation | long weekend | several days | weekend | one day | few hours

tourism involving overnight stays

leisure involving travel but not overnight stay

local urban zone

zone of intensive recreation

rural holiday resorts and other city and coastal resorts

zone of extensive recreation in further rural areas and city regions

short breaks – international

national and international tourist resorts

travel time in hours

0.25
0.75
1.5
3.0

Figure 20.2

The recreation/tourism spatial continuum

Greater affluence	– People in employment earn high salaries and their disposable income is much greater than it was several decades ago. – People in full-time employment also receive holiday with pay, allowing them to take more than one holiday a year and to travel further.	
Greater mobility	– The increase in car ownership has given people greater freedom to choose where and when they go for the day, or for a longer period. In 1951, only 1 UK family in 20 had a car. By 2008, 75 per cent had at least one car. – Chartered aircraft have reduced the costs of overseas travel; wide-bodied jets can carry more people and can travel further, bringing economies of scale.	
Improved accessibility and transport facilities	– Improvements in roads, especially motorways and urban by-passes, have reduced driving times between places and encourage people to travel more frequently and greater distances. – Improved and enlarged international airports (although many are still congested at peak periods). Faster trains, e.g. Eurostar. Reduced air fares. Package holidays.	
More leisure time	– Shorter working week (although the UK's is still the longest in the EU) and longer paid holidays (on average 3 weeks a year, compared with 1 week in the USA). – Flexi-time, more people working from home, and more firms (especially retailing) employing part-time workers. – An ageing population, many of whom are still active.	
Technological developments	– Jet aircraft, computerised reservation systems, use of the Internet.	
Product development and innovation	– Holiday and beach villas, long-haul destinations, package tours.	
Changing lifestyles	– People are retiring early and are able to take advantage of their greater fitness. – People at work need longer/more frequent rest periods as pressure of work seems to increase. – Changing fashions, e.g. health resorts, fitness holidays, winter sun.	
Changing recreational activities	– Slight decline in the 'beach holiday' – partly due to the threat of skin cancer. – Increase in active holidays (skiing, water sports) and in self-catering. – Most rapid growth since mid-1990s has been in cruise holidays. – Importance of theme parks, e.g. Alton Towers, Thorp Park, Center Parcs. – Large number of city breaks.	
Advertising and TV programmes	– Holiday programmes, film and TV sets, magazines and brochures promote new and different places and activities.	
'Green' or sustainable tourism	– Need to benefit local economy, environment and people without spoiling the attractiveness and amenities of the places visited (ecotourism).	

Figure 20.3

Factors causing growth in tourism

Global tourism

In 2008, the travel and tourism industry accounted for 8.4 per cent (238 million) of the world's total employment and contributed 9.9 per cent of its GDP. Of total tourist receipts, 71 per cent was earned by countries in North America and Europe (Figure 20.4), although this only gave them a very small **travel account surplus**. In contrast, the travel account balance for developing countries has shown a persistently high, and widening, surplus (unlike their trade balance, page 624), mainly because they are visited by wealthy tourists from developed countries whereas few of their residents can afford holidays in developed countries (Figure 20.5).

Figure 20.4

Growth in global tourism, 1960–2020

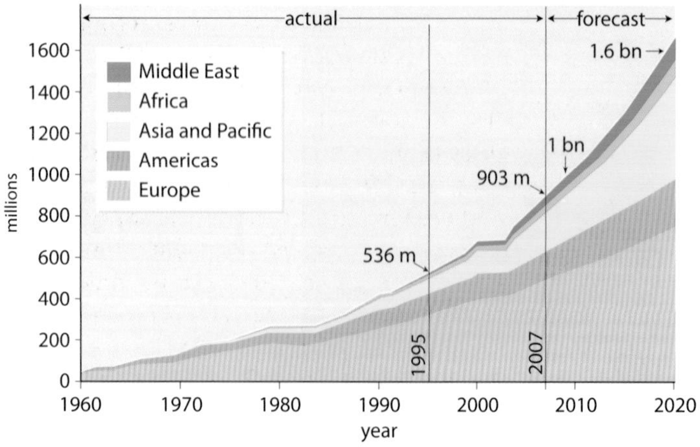

		Arrivals (millions)	% world total
1	France	81.9	9.1
2	Spain	59.2	6.6
3	USA	56.0	6.2
4	China	54.7	6.1
5	Italy	43.7	4.8
6	UK	30.7	3.4
	World	903.0	

		Earners (US$ million)	% world total
1	USA	96.7	11.3
2	Spain	57.8	6.8
3	France	54.2	6.3
4	Italy	42.7	5.0
5	China	41.9	4.9
6	UK	37.6	4.4
	World	856.0	

		Spenders (US$ million)	% world total
1	Germany	82.9	9.7
2	USA	76.2	8.9
3	UK	72.3	8.4
4	France	36.7	4.3
5	China	29.8	3.4
6	Italy	27.3	3.2
	World	856.0	

Figure 20.5

Leading tourist countries, 2007

The travel and tourism industry is dynamic, having to change continually to meet consumer demands and perceptions. Its key features at present include the following:

■ It has a complex structure consisting of a wide variety of interrelated commercial and non-commercial organisations (Figure 20.6).
■ It is predominantly private-sector led.
■ It is dominated by relatively few large, often transnational, firms, e.g. tour operators (Kuoni, Going Places Leisure Travel, Thomas Cook, Thomson), hotel chains (Marriot, Sheridan, Holiday Inn), theme parks (Disney) and air operators (BA, American Airlines). Despite this, the majority of enterprises are small and medium-sized, often catering for the local market.
■ There is an extensive use of new technologies including data handling, advertising, advance bookings and the Internet.
■ There was an increase in the number and range of destinations between 1950, when the top 15 attracted 98 per cent of international arrivals and were mainly based on 'sun, sand and sea', and 2007, when the top 15 destinations only received 57 per cent of arrivals. This reflects the emergence of new locations, especially in developing countries, and a demand for a greater range of activities and experiences. At present, the fastest emerging tourist areas are China and the Middle East.
■ It is vulnerable to external pressures such as currency fluctuations, fuel charges, government legislation and international terrorism.

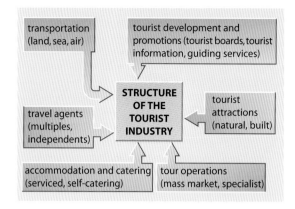

Figure 20.6

Structure of the tourist industry

■ It has both a positive and a negative effect on host communities (economic, social and cultural) and local environments (Figure 20.8).

UK tourism

■ **Number of tourists**. In 2007, Britain received 32.4 million visitors from overseas while at the same time 66.4 million UK residents took their holidays, or a break, outside the country. Over one-third of British tourists still went to Spain (14.4 million) and France (10.9 million) but this proportion continues to decline as people look for different places to visit and activities to do. The same year saw Britain receiving it highest ever number of tourists and business visits. Nearly 80 per cent of these came from the EU, although the USA remained the largest single country of origin of visitors. A record number also came from Poland – presumably friends and relatives of migrant workers (Places 44, page 369).
■ **Consumer spending**. In 2007, UK residents spent £72.3 million overseas (£7.3 million in 1987) compared with overseas residents who spent £16.0 million in the UK (£6.3 million in 1987) – a deficit on the travel account of £56.3 million (£1.0 million in 1987).
■ **Number employed and type of job**. Official figures show that 1.45 million people were directly employed in the tourist industry in Britain in 2007, with an estimated further half million engaged indirectly. The wide range of jobs included hotel and catering, travel agents, coach operators, in entertainment and as tour guides. Approximately 132 000 of these jobs were classified as self-employed.

Figure 20.7

Factors affecting the growth of the holiday industry in the UK

Factor	Specific examples	Example of area or resort
1 Transport and accessibility	• Early resorts (stage-coach), spa towns • Water transport (18th century) • Railways • Car and coach • Plane	Bath Margate Blackpool, Brighton Cornwall, Scottish Highlands Channel Islands
2 Scenery	• Sandy coasts • Coasts of outstanding beauty • Mountains, lakes and rivers	Margate, Blackpool Pembroke, Antrim Lake District, Snowdonia
3 Weather	• Hot, dry, sunny summers • Snow	Margate Aviemore
4 Accommodation	• Hotels and boarding house resorts • Holiday camps • Caravan parks and campsites	Margate, Blackpool Minehead, Pwllheli National and forest parks
5 Amenities	• Culture and historic (castles, cathedrals, birthplaces) • Active amenities (sailing, golf, water-skiing) • Passive amenities (shopping, cinemas) • Theme parks	York, Edinburgh Kielder, St Andrews Most resorts Alton Towers, Chessington
6 Ecotourism and sustainability	• Wildlife conservation areas • Heritage sites	SSSIs, nature reserves York

Positive effects/benefits	Negative effects/problems
Economic	
Increases gross domestic product directly and indirectly via the multiplier effect (see Myrdal, page 569).	May divert government expenditure from other needy areas of the economy.
Taxes on tourism increase government revenue.	Requires government expenditure on tourism.
Increased foreign exchange earnings.	Over-dependence on outside agencies and some external control on the economy.
Foreign investment.	Income reduced by external leakages or outflows, e.g. imported food for tourists.
Creates employment, including in unskilled occupations; labour-intensive.	Profits may go overseas.
Helps fund new infrastructure, i.e. roads, airports and facilities which local people can also use.	Overstretches infrastructure.
Stimulates and diversifies economic activity in other sectors – local craft revival, manufacturers, services and agriculture (the multiplier effect).	Spread effects limited and may therefore increase regional inequalities between tourist growth areas and less developed periphery (page 617).
May act as a seedbed for entrepreneurship, with spin-offs into other sectors.	Diverts labour and resources away from non-tourist regions and may (particularly) affect peripheral areas, leading to out-migration to tourist resort opportunities (Places 42, page 366).
Improves balance of payments through increased trade.	Labour unskilled and seasonal.
	Foreign personnel and firms dominate managerial and higher-paid posts, reducing opportunities for local people.
	Inflated prices for land, housing, food and clothes.
Social	
Cultural exchange stimulated with broadening of horizons and reduction of prejudices amongst tourist visitors and host population.	May cause polarisation between population in advancing tourist regions and less developed areas, creating a 'dual society'.
May enhance role and status of women in society, as opportunity for goals in tourism is created and outlook widened.	Increases rift between 'rich' and 'poor'. Breakdown of traditional family values creates material aspirations.
Encourages education.	Breakdown of families due to stress between younger generation, who are affected by imported culture, and older members of household – called the negative demonstration effect.
Encourages travel, mobility and social integration.	
Improves services (electricity and health), transport (new roads, airports) and widens range of shops and leisure amenities.	Social pathology, including an increase in prostitution, drugs and petty crime.
	Increases health risk, e.g. AIDS.
Cultural	
May save aspects of indigenous culture due to tourist interest in them.	Impact of commercialisation may lead to pseudo-cultural activities to entertain tourists and, at extreme, may cause disappearance or dilution of indigenous culture – known as 'commodification'.
Contact with other cultures may enrich domestic culture through new ideas and customs being introduced.	Mass tourism may create antagonism from host population who are concerned for traditional values, e.g. dress, religion.
Encourages contact and harmonious relations between people of different cultures.	Westernisation of culture, food (McDonalds) and drink (Coca-Cola).
Increases international understanding.	
Environment	
Improved landscaping and architectural standards in resort areas, including increased local funding for improvement of local housing, etc.	Destruction of natural environment and wildlife habitat – marine, coastal and inland.
Promotes interest in monuments and historic buildings, and encourages funding to conserve and maintain them.	Excessive pressure leads to air, land, noise, visual and water pollution, and breakdown in water supplies, etc.
May induce tighter environmental legislation to protect environment, i.e. landscape, heritage sites, wildlife.	Traffic congestion and pollution.
Establishment of nature reserves and National Parks; growing tourist interest and awareness protects areas from economic and building encroachment.	Clearance of natural vegetation, loss of ecosystems.
Poor building and infrastructure development – tourist complexes do not integrate with local architecture.	

Figure 20.8

Adapted from a World Tourism Organization classification

Positive and negative effects of mass global tourism

Tourism and the environment

As the demands for recreation and tourism increase, so too will their impact on other socio-economic structures in society, scenic areas and wildlife habitats. Tourists will compete for space and resources with:

- local people living and working in the area, e.g. farmers, quarry workers, foresters, water and river authority employees (Figure 17.4)

- other visitors wishing to pursue different recreational activities, e.g. water skiers, wind-surfers, anglers and bird watchers all visiting the same lake.

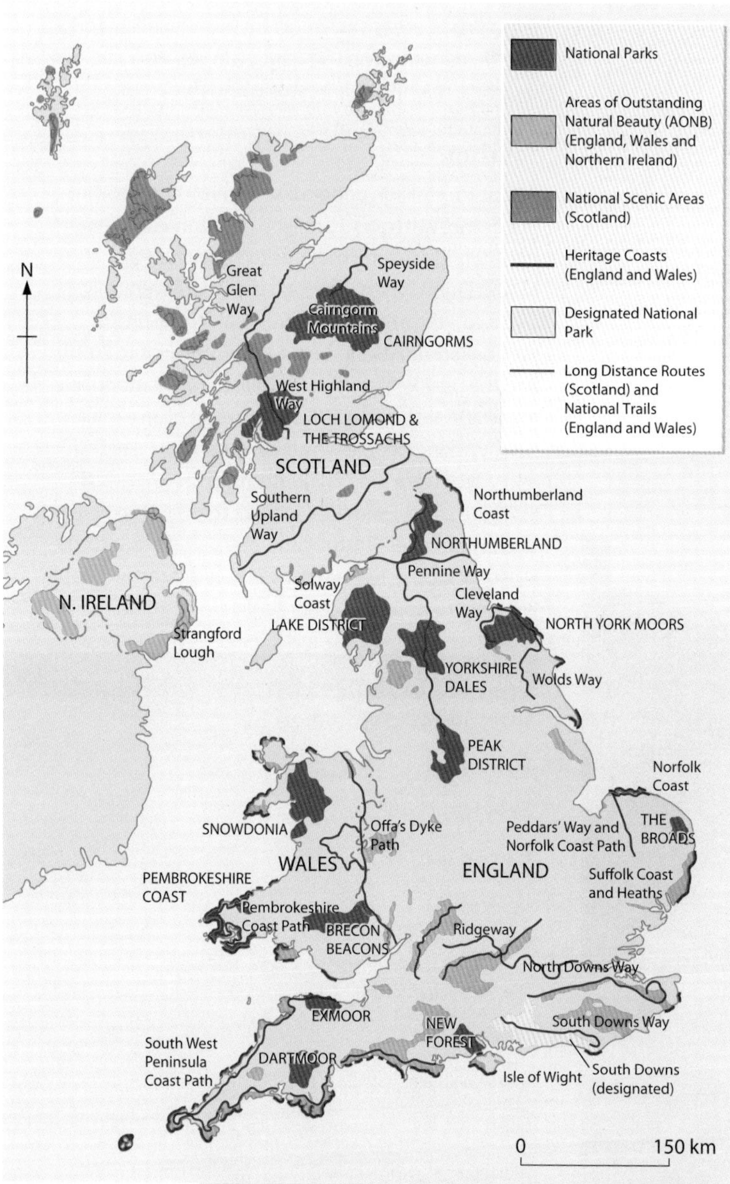

The development of recreation and tourist facilities creates pressure on specific places and environments in both urban and rural areas. Places with special interest or appeal that are very popular with visitors and which tend to become over-crowded at peak times are known as **honeypots**. Honeypots may include, in urban areas, concert halls (Albert Hall), museums (Madame Tussaud's), and historic buildings (Tower of London); and, in rural areas, places of attractive scenery (Lake District), theme parks (Alton Towers), and places of historic interest (Stonehenge). The problem of overcrowding within certain American National Parks (Yellowstone), together with congestion on access roads, has become so acute that permits are needed for entry and quotas are imposed on areas that are ecologically vulnerable (Case Study 17).

Sometimes planners encourage the development of honeypots, especially in British National Parks and African safari parks, to ensure that such sites have adequate visitor amenities (car parks, picnic areas, toilets, accommodation). It is now widely accepted that leisure amenities and tourist areas need to be carefully managed if the maximum number of people are to obtain the maximum amount of enjoyment and satisfaction (Figure 20.9).

It is possible to identify three levels of recreation and tourism in rural areas.

1. High-intensity areas where recreation is the major concern (theme parks such as Alton Towers, honeypots such as at Bowness on Windermere, and resorts such as Aviemore).

2. Average-intensity areas where there needs to be a balance between tourism and other land users, and between recreation and conservation (Peak District National Park, Places 92).

3. Low-intensity areas, usually of high scenic value, where conservation of the landscape and wildlife is given top priority (upland parts of Snowdonia and the Cairngorms – Places 94, page 595).

Recently there has been a growth in **ecotourism** (Places 95, page 598), which aims at safeguarding both natural and built environments, being sustainable (Framework 16, page 499), and enabling local people to share in the economic and social benefits.

Figure 20.9

Protected areas in the UK

12	National Parks + 1 designated in England and Wales, 2 in Scotland
36	AONBs in England, 9 in Northern Ireland, 4 in Wales, 1 in England and Wales
45	National Scenic Areas in Scotland
15	National Trails in England and Wales + 4 Long Distance Routes in Scotland
32	Heritage Coasts in England and Wales

Following the passing of The National Parks and Access to the Countryside Act in 1949, the Peak District became the first National Park in 1951 (Figure 20.9). The Environment Act of 1995, which set up a National Park Authority to administer the affairs of each of the National Parks, defined the purposes of National Parks as:

- conserving and enhancing the natural beauty, wildlife and cultural heritage, and
- promoting opportunities for the understanding and enjoyment of their special qualities.

National Parks must also foster the economic and social well-being of local communities. They are also required to pursue a policy of sustainable development by which they must aim to improve the quality of people's lives without destroying the environment (Framework 16, page 499). Despite their often spectacular scenery, National Parks are not owned by the nation nor managed purely for their landscapes and wildlife. They are, rather, mainly farmed areas where many people live (38 000 within the Peak District) and work.

Land use

There are some 800 farms in the Peak District National Park (PDNP), most of them under 40 ha. Some are owned by the National Trust, water companies and large landowners, with 70 per cent run by farmers who need income from a second job. The PDNP manages 4580 ha of woodland. There are 55 reservoirs, which supply water to large urban areas such as Manchester, Leeds and Sheffield located on the Park's fringes, and 10 quarries, mainly for limestone and fluorspar.

Tourism

About 15.7 million people live within 100 km of the PDNP and, with over 30 million day visits each year, it is the world's second most visited National Park (after Mount Fuji in Japan). The Park is divided, scenically, into two – the attractive 'White Peak' consisting of Carboniferous limestone (page 196 and Figure 20.10) and the Dark Peak, which is a Millstone Grit moorland (page 201 and Figure 20.11). The result of an earlier survey asking people why they visited the Peak District and which were their favourite attractions, is given in Figure 20.12. Estimates suggest that tourism here directly provides 500 full-time, 350 part-time and 100 seasonal jobs, as well as many others indirectly (people working in shops and other service industries).

Reason	(%)
Scenery/landscape/sightseeing	61
Outdoor activities/walking	56
Enjoyed previous visit	39
Peace and quiet	31
Easy to get to	26
New place to visit	17
Specific event/attraction	16
Come every year	9
Own second home/caravan in area	6
Others	14

Most popular areas of the Peak District National Park are:
- **Bakewell**, with interesting buildings and a busy market.
- **Chatsworth**, home of the Duke of Devonshire.
- **Dovedale**, a spectacular limestone dale.
- **Hartington village** and **Eyam**, the plague village.
- **Goyt Valley** and its reservoirs.
- **Hope Valley** and the village of Castleton.
- **Upper Derwent** and the Ladybower and Derwent Reservoirs.

Figure 20.12

Why people visit the Peak District National Park

Figure 20.10

The White Peak: Lathkill Dale Nature Reserve

Figure 20.11

The Dark Peak: Kinder Scout

Conservation

National Parks were set up with the specific purpose of protecting areas of natural beauty in the countryside. Today, although facilities for suitable types of recreation (walking, climbing and fishing) are an important part of the National Parks, the aims of conservation have to take priority. By conservation, the National Parks mean 'keeping and protecting a living and changing environment', which, in the case of the Peak District, is:

- *The Nearly Natural Landscapes* which include the gritstone moorland of the Dark Peak and the limestone heaths and dales of the White Peak. These areas include Sites of Special Scientific Interest (SSSIs), which cover 35 per cent of the National Park, and National Nature Reserves (NNRs), both managed by English Heritage, as well as Environmentally Sensitive Areas (ESAs) which are supervised by DEFRA (Figure 16.54), and farms engaged in the Environmental Stewardship Scheme (ESS) (page 496).

- *The Not So Natural Landscapes* which have resulted from farming and mineral extraction.

- *The Built Landscape* which includes villages, hamlets, listed buildings and archaeological sites. The PDNP Authority has control over the erection of new properties, the range of building materials and the ability to create Conservation Areas in villages that include places of historic or architectural interest (Figure 20.13).

Figure 20.13

Enhancement Project in Eyam Square

The PDNP has identified four main land use conflicts to which it has suggested ways forward:

- conservation and farming – farmers to manage land in traditional ways and be given grants for conservation work

- conservation, water supply and recreation – limit fishing, sailing and other activities to specific reservoirs

- conservation and tourism – more robust footpaths and use of former railway tracks; new footpaths, cycle tracks and bridleways; siting of car parks to spread visitors over a wider area

- conservation and mineral extraction – screening and restoration to be part of the mining process.

The latest PDNP Management Plan is for 2006–11. Its vision is underpinned by two main principles:

- partnership working
- sustainable development.

The headings and sub-headings for this plan, which can be seen in full on the PDNP website, are listed in Figure 20.14.

A	**Social drivers**
	a The need to build cohesive communities
	b Listening to, involving and engaging communities
	c The need for people to have decent and affordable homes
	d Being proactive in providing opportunities for recreation
	e The need for people to adopt healthier lifestyles
	f The need for sustainable improvements to travel
B	**Technological drivers**
	a Mobile phone operators and consumer demand
C	**Environmental drivers**
	a UK and local Biodiversity Plans
	b Climate change
	c The changing historical nature of air pollution
	d Mineral extraction
	e The changing patterns of land use and ownership
D	**Economic drivers**
	a Changes to farm payments system
	b Delivering the outcomes of the Peak District Rural Action Zone
	c Changing nature of sources of funding
	d Developing a sustainable tourism economy
E	**Political drivers**
	Local government and legislative changes

Figure 20.14

The PDNP Management Plan, 2006–11

The tourist resort/area life-cycle model

Despite some of the obvious disadvantages of tourism, the nightmare scenario for any tourist-dependent country, region or resort, is that people will find somewhere else to visit and to spend their money. New resorts develop; old resorts may become run-down; fashions change; places may receive a bad press; economic recessions occur; currency rates alter and new activities are designed. To survive, tourist places have to keep re-inventing themselves by, for example, including new attractions or changing their orientation to a wider or new client group. Places that fail, such as some older British seaside resorts and spa towns, begin to wither away. Places that manage to adapt, such as Blackpool, continue to be successful. On this basis, Butler produced a useful life-cycle model (Framework 12, page 352) for tourist resorts (Figure 20.15); this may also be applied more widely to tourist regions (Places 93).

Figure 20.15

Tourist area/resort life-cycle model (*after* Butler)

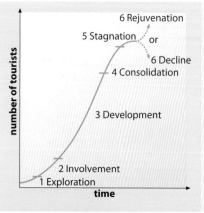

1 **Exploration**: small number of visitors attracted by natural beauty or cultural characteristics — numbers are limited and few tourist facilities exist, e.g. Chile.

2 **Involvement**: limited involvement by local residents to provide some facilities for tourists — recognisable tourist season and market areas begin to emerge, e.g. Guatemala.

3 **Development**: large numbers of tourists arrive, control passes to external organisations, and there is increased tension between local people and tourists, e.g. Florida.

4 **Consolidation**: tourism has become a major part of the local economy, although rates of visitor growth have started to level off and some older facilities are seen as second-rate, e.g. earlier Mediterranean coastal resorts.

5 **Stagnation**: peak numbers of tourists have been reached. The resort is no longer considered fashionable and turnover of business properties tends to be high, e.g. Costa del Sol (Places 93).

6 **Decline or rejuvenation**: attractiveness continues to decline, visitors are lost to other resorts, and the resort becomes more dependent on day visitors and weekend recreationalists from a limited geographical area — long-term decline will continue unless action is taken to rejuvenate the area and modernise as a tourist destination, e.g. Blackpool, British spa towns and older coastal resorts.

Places 93 The Spanish 'costas': the life-cycle of a tourist area

In the 1950s, Benidorm on the Costa Blanca was still a small fishing village (compare the Costa del Sol in Figure 20.16). During the 1960s, the introduction of cheap air travel began to attract visitors from northern Europe and enabled resorts to develop, with their sandy beaches, warm seas and hot, dry, sunny summers. By the 1970s it had turned into a sprawling modern resort with high-rise hotels and all the amenities expected by mass tourism. By the 1980s it had reached Butler's stage of consolidation, when the carrying capacity was reached. By the early 1990s it had begun to stagnate and to decline. Since then the Spanish government has tried to rejuvenate the area by encouraging the refurbishment of hotels, reducing VAT in luxury hotels and ensuring that both beaches and the sea have become cleaner (Spain has the most 'Blue Flag' beaches in the EU).

Figure 20.16

Life-cycle of a holiday area: tourists from the UK to the Costa del Sol, 1960s–2000s

	1960s	1970s	1980s	1990s and 2000s
Tourists from UK to Spain	1960 = 0.4 million	1971 = 3.0 m	1984 =6.2 m 1988 = 7.5 m	1990 = 7.0 m 2000 = 7.3 m
State of, and changes in, tourism	very few tourists	rapid increase in tourism; government encouragement	carrying capacity reached; tourists outstrip resources, e.g.water supply and sewerage	decline (world recession); prices too high; cheaper upper-market hotels elsewhere; government intervention to rejuvenate tourism
Local employment	mainly in farming and fishing	construction workers; jobs in hotels, cafes, shops; decline in farming and fishing	mainly tourism: up to 70 per cent in some places	unemployment increases as tourism declines (30 per cent); farmers use water for irrigation
Holiday accommodation	limited accommodation; very few hotels and apartments; some holiday cottages and campsites	large hotels built (using breeze blocks and concrete); more apartment blocks and villas	more large hotels built, also apartments, time-share and luxury villas	older hotels looking dirty and run down; fall in house prices; only high-class hotels allowed to be built; government oversees the refurbishment of hotels
Infrastructure (amenities and activities)	limited access and few amenities; poor roads; limited streetlighting and electricity	some road improvements but congestion in towns; bars, discos, restaurants and shops added	E340 opened: 'the Highway of Death'; more congestion in towns; marinas and golf courses built	bars/cafés closing; Malaga by-pass and new air terminal opened; re-introduction of local foods and customs
Landscape and environment	clean, unspoilt beaches; warm sea with relatively little pollution; pleasant villages; quiet with little visual pollution	farmland built upon; wildlife frightened away; beaches and sea less clean	mountains hidden behind hotels; litter on beaches; polluted seas (sewage); crime (drugs,vandalism, muggings); noise from traffic and tourists	attempts to clean up beaches and sea (EU Blue Flag beaches); new public parks and gardens opened; nature reserves

At a conference on 'sustainable mountain development', one speaker claimed: 'Mountains are suffering an unprecedented environmental crisis. Wherever you go in the world, you can find the same symptoms – landscapes wrecked by roads; forests cleared for, and slopes shredded by, skiing; vegetation worn away by walkers; and litter left by tourists.

Places 94 The Cairngorms: a mountainous area under threat

The Cairngorm range, which includes four of Britain's five highest mountains, became part of the Cairngorms National Park in 2003. The arctic-alpine plateau is a fragile ecosystem which includes mosses, lichen and dwarf shrubs (page 333) and which provides an irreplaceable habitat for rare birds such as the golden eagle, ptarmigan, snow bunting and dotterel (Figure 20.17). It includes three SSSIs (page 593) and a National Nature Reserve. It also receives the heaviest, and longest lying, snowfall in Britain, making it ideal for downhill and cross-country skiing as well as other winter sports. These advantages have led to conflict between developers and environmentalists.

In the 1990s, the Cairngorm Chairlift Company, now Cairngorm Mountain Ltd, having twice failed to get planning permission to extend its skiing facilities into nearby Lurcher's Gully, put forward a plan which included a 2 km funicular railway that would go to within 150 m of the summit. The plan also included a new chairlift, three new ski tows and four additional ski runs. At the top, the underground terminus to the railway would give access to a 250-seater restaurant, an interpretative exhibition and a retail outlet. The railway would get visitors to the summit in three minutes, would be able to operate, unlike the old chairlift, in high winds and could increase the number of summer visitors from 60 000 to 225 000. It was this increase in the prospect of the extra number of feet trampling the fragile summit plateau during the short growing and nesting season that caused most alarm to conservationists. They feared plants would be crushed, birds disturbed and the landscape eroded.

What swung the decision the developers' way was their proposal to operate a 'closed system' which would confine everyone to the visitor centre with its indoor viewing area. This meant an end to the 50 000 visitors who, until then, could trample without restriction over the summit area. The funicular railway began running at Christmas 2001 and the visitor centre was formally opened the following May. At that opening, it was said that the funicular project demonstrated how it is possible to balance environmental concerns with projected economic benefits.

Meanwhile skiers in the Cairngorms face a greater threat – global warming is reducing both the amount of snowfall and the period of snow cover.

Figure 20.17

The Cairngorm arctic/alpine environment

Other types of tourism

Heritage

According to the World Heritage Convention (WHC), created by UNESCO, 'Heritage is our legacy from the past, what we live with today, and what we pass on to future generations. Our cultural and natural heritage are both irreplaceable sources of life and inspiration.' **Cultural heritage** includes monuments, groups of buildings and sites such as the Pyramids, the Acropolis, the Taj Mahal (Figure 20.18a), Machu Picchu, Chichen Itza (Figure 20.23) and The Great Wall of China. **Natural heritage** includes landscape and wildlife sites such as the Barrier Reef and Tanzania's Serengeti National Park (page 311). There are, at present, over 800 World Heritage Sites.

Theme parks and purpose-built resorts

Theme parks and purpose-built resorts have become centres of mass tourism in the last two or three decades. They include Disney World (Florida – Figure 20.18b), Disneyland (Paris), Legoland (Denmark), Seaworld (Queensland) and Alton Towers (England).

Wildlife

There has been a steady increase in the number of people wishing to see wildlife in its natural environment. The most popular is the African 'safari' in which tourists are driven around, usually in small minibuses with adjustable roofs to allow for easier viewing (Figure 20.18c). Kenya, South Africa, Tanzania and Zimbabwe are all able to capitalise on their abundance of wildlife. Other tourists may go whale-watching (New Zealand), visit marine reserves (Places 80, page 526), view threatened wildlife such as the giant panda and the mountain gorilla, or go to places with a unique ecosystem (Madagascar and the Galapagos Islands).

Wilderness holidays

These are popular in America: one or two people set off into largely uninhabited areas such as Alaska to 'live and compete with nature' (Figure 20.18d).

City breaks

Globally more people take city breaks – often lasting just a few days – than any other type of holiday. In Britain in 2007, 87 per cent of adults visited a city for at least one day, the vast majority – over 11 million – travelling to London to take advantage of its cultural amenities (the National Gallery), theatres (Drury Lane), historic buildings (St Paul's Cathedral), sporting venues (Wembley Stadium), shops (Oxford Street) and businesses (Canary Wharf). Eight of the top ten most visited destinations in Britain are cities (including over 2 million visits a year to Manchester and Birmingham) while many other tourists take city breaks in Europe and beyond.

Religious centres

Religious centres to which people make a pilgrimage include Mecca, The Vatican, Jerusalem, Salt Lake City and Varanasi (Benares).

Figure 20.18

Types of tourism
a Heritage:
    Taj Mahal
b Theme parks:
    Disney World
c Wildlife parks:
    Botswana
d Wilderness:
    Mt McKinley
    in Dynali
    National Park

Cruises

Cruising has been the fastest-growing section of the world's tourist industry for two decades. More, and larger, liners are being built each year (Figure 20.19) while the number of passengers has increased from under 4 million in 1990 to almost 13 million in 2008. Cruise holidays are often an excuse for people to relax and enjoy the sun and the life aboard ship, as seen by over one-third of all passengers opting for the Caribbean (Figure 20.20). Other tourists may take a cruise that follows a theme, such as visiting historical/archaeological sites (Mediterranean), capital cities (Baltic), scenic coasts (Norway, Figure 20.19) or whale-watching (Alaska). While the scores of passengers create jobs for tour guides and shop assistants and generate income for bus companies, taxi drivers, and local craft industries, they rarely spend large amounts of money while on land as they eat and sleep on board ship. Also, their large numbers – up to 3600 on the latest super cruise liners – may swamp local communities and disrupt their way of life.

Certain rivers are also popular for cruising – with the added bonus of calm water! People sail along the Nile (to see ancient temples), the Mississippi (on paddle boats), the Yangtze (Three Gorges), the Amazon, Rhine and Danube. Canal holidays are a self-catering form of cruising.

Figure 20.19

Cruise liners in the Geiranger Fiord, Norway

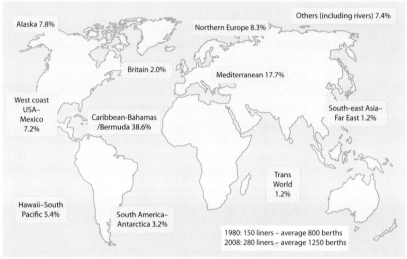

Figure 20.20

Cruise destinations, 2007

Ecotourism

Ecotourism, sometimes known as 'green tourism', is a sustainable form of tourism (Framework 16, page 499) that is more appropriate to developing countries than the mass tourism associated with Florida and certain Mediterranean areas. Ecotourism includes:

- visiting places in order to appreciate the natural environment, ecosystems (page 295), scenery and wildlife, and to understand their culture
- creating economic opportunities (jobs) in an area while at the same time protecting natural resources (scenery and wildlife) and the local way of life.

Compared with mass tourists, ecotourists usually travel in small groups (low-impact/low-density tourism), share in specialist interests (bird-watching, photography), are more likely to behave responsibly and to merge and live with local communities, and to appreciate local cultures (rather than to stop, take a photo, buy a souvenir and then move on). They are likely to visit National Parks and game reserves where the landscape and wildlife which attracted them there in the first place is protected and managed. Places visited include Brazil (rainforests), the east coast of Belize and Mexico (coral reefs – Places 95), Nepal (mountains), Burundi (mountain gorillas) and the Arctic (polar bears).

Even so, ecotourists usually pay for most of their holiday in advance (spending little in the visited country), are not all environmentally educated or concerned, can cause local prices to rise, congregate at prime sites (honeypots), and may still cause conflict with local people. There is a real danger that tour operators, by adding 'eco' as a prefix, give certain holidays unwarranted respectability.

The Xcaret Eco-archaeological Park (Figure 20.21) in Mexico's Yucatan Peninsula won, back in 1999, the Sunday Times Readers' Award for what they considered to be the most successful project in protecting, or improving, the quality of a local environment. Xcaret is located (Figure 20.22) 70 km south of the mass tourist resort of Cancun (Mexico's answer to Miami) and 270 km east of the former Mayan settlement of Chichen Itza, now a World Heritage Site (Figure 20.23).

In 1990, five families set up the Xcaret venture with two aims: to support Mexican research programmes into biodiversity and to encourage ecotourism, the latter by allowing visitors to relax in beautiful surroundings and to learn, almost by accident, the value of the ecosystem on show together with the scientific work being carried out. Visitors are encouraged to travel by bus or taxi, not car, and, on arrival, are asked to hand in any suntan lotion (which pollutes seawater) and, in return, are given a bottle of eco-friendly lotion (though less effective as a sunblock). The inlet, with its warm, crystal-clear water that is home to thousands of multicoloured fish and contains a sea-turtle reserve, is ideal for swimming and snorkelling. Two underground rivers, lit by sunlight streaming through openings in rock holes, allow tourist to explore underground channels. First being warned that touching coral can kill it (Places 80, page 526), people are taken to offshore reefs where they can swim with bottlenose dolphins. At night, a show in the open-air theatre ends with a performance of a famed folkloric ballet.

Figure 20.21

Xcaret

Xcaret also has a wild-bird breeding centre that caters for endangered species, a butterfly pavilion, a botanical garden and a coral reef aquarium. However, the venture is not without its critics, some of whom cite the fact that the underground rivers were blasted and remodelled while others point to the threat that snorkelling poses to the reef and the presence of a mock Mayan village. Yet generally, and by banning high-rise beach complexes, Xcaret has shown that it is possible for people to enjoy themselves without harming the environment.

Jane Dove, in *Geography Review*, describes two more sustainable examples of ecotourism nearby. At the Sian Ka'an Biosphere, visitors are taken on walking tours to see lagoons, mangroves and tropical rainforest. They sleep in tents, use composting toilets and obtain water that is heated by solar and wind power. The Mayan village of Pac Chen limits access to 80 tourists a day. Here they are served local food, are shown a swallow hole in the limestone (page 196) and visit a Mayan ruin (the Mayan civilisation was between AD 990 and 1200). The income generated has helped to build a local school and a clinic.

Figure 20.22

Mexico's tourist sites

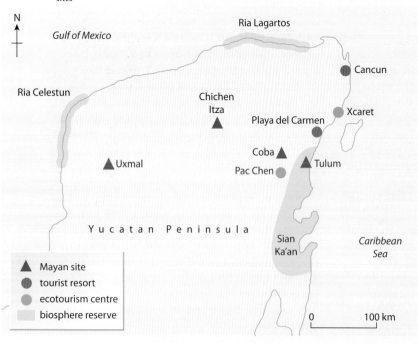

N

Gulf of Mexico

Ria Lagartos

Ria Celestun

Cancun

Chichen Itza

Playa del Carmen

Xcaret

Coba

Uxmal

Pac Chen

Tulum

Yucatan Peninsula

Sian Ka'an

Caribbean Sea

▲ Mayan site
◑ tourist resort
● ecotourism centre
▨ biosphere reserve

0   100 km

Figure 20.23

Chichen Itza

The personal investigative study, or enquiry, is an important part of the examination assessment for AS and A2 Geography. It provides an opportunity for you to develop your individual interests in a particular part of the specification, to make use of fieldwork and to become an 'expert' on a small investigation.

Choosing your study

- Choose a topic in which you have personal interest. This will make it easier to study.
- Check in local papers for current issues which could prove a useful topic to investigate. Collect as much background information as you can before setting up your topic.
- Studies can involve combined fieldwork undertaken at field centres. However, your conclusions must be individual, even though the data collection may have been done as a group.
- Choosing a topic covering a human/environmental theme may allow work in different sections or modules of the course to be linked to an investigation.
- Avoid a topic that will mean travelling long distances to collect data and to do fieldwork. This can be expensive in terms of both time and money, and it will not be easy to make return visits.
- Careful planning is essential – particularly the Schedule for your enquiry.

Collecting your data

- It is important to begin preliminary collection of ideas and materials as early as possible.
- Primary data is the basis of a study, and collecting the data has to be carefully planned, involving surveys, questionnaires, interviews, use of annotated photographs and map construction. Make sure that you choose appropriate dates and times for your fieldwork.
- Questionnaires need to be succinct and to the point – you need to know the types of answers that you require. You should make sure that you have a large sample in order to have well-founded results.
- Take as many photographs as possible of the study area. Carefully annotate and label them, and make sure they are relevant to your enquiry. Bear in mind that you will need to select only the most relevant photographs in your final report.
- If you are visiting an organisation or requesting information it is always useful to write a polite letter beforehand, outlining what you wish to find out and giving time for an answer. People are always busy, so be prepared to wait a few days before telephoning to make an appointment.

- Secondary data collection will mean visits to local libraries, researching newspapers for background, and using the Internet (see Framework 1, page 22). Old maps will show conditions at previous times (page 396). Keep a detailed record of all your sources.

Writing your report

- Plan the structure of your report before you start writing. You may find the following outline useful: Introduction – Aims – Data Collection – Data Analysis – Evaluation – Conclusion.
- Data that you collect will have to be analysed and displayed in maps and statistical form. Although bar charts and pie charts are clear and easy to display, try to use a variety of forms of presentation. Most exam boards require that you are able to use statistical methods effectively in your studies. This helps you to evaluate and then to explain the results of your investigation. Some of the most useful methods are Spearman's rank correlation and chi-squared (see Framework 19, page 612).
- Careful detailed analysis of your statistics and diagrams is vital – do not assume that the examiner or moderator will automatically understand what is set out.
- Thoughtful and detailed evaluation is a very important part of your study. You may have collected the opinions of a number of different groups in your investigations and you must set these out clearly and balance up the different values which may be apparent. Do not forget to include your own ideas.
- An extended conclusion will complete the study, drawing together the different opinions and values, weighing up the options and probably putting forward any alternative proposal you may consider to have value.

Remember...

- Presentation is important. Make your report look good – use ICT where possible.
- Diagrams may be computer-generated but maps should be hand-drawn and not photocopied.
- Check that all maps, diagrams and photographs are labelled and annotated.
- Acknowledge any quotations and draw up a clear bibliography of your references, including any material sourced from the Internet.
- Number all the pages and where necessary cross-reference diagrams and text.

'Half way down India's west coast is the tiny state of Goa. A unique blend of Indian and Portuguese cultures with miles of long, sandy beaches, emerald-green paddy fields and gleaming, white-washed Portuguese-style churches peering out over extensive palm groves.' This is how the former tiny Portuguese enclave of Goa, which became part of India in 1962 and an independent state in 1987, is described in a Kuoni travel brochure.

Goa has become a major tourist centre for both domestic and international visitors.

As Figure 20.24 shows, domestic arrivals increased by 35 per cent between 1986 and 2006 and international arrivals by 26 per cent during that same period, with the number of domestic visitors doubling since 2000 and international visitors since 1998. Especially since the increase in internal low-cost airlines, Goa has been popular with Indian tourists from the large cities of Mumbai and Delhi, and, more recently, Bangalore. Whereas Goa is 12 hours by road or rail from Mumbai with

its population of 16 million, it is only a one-hour flight away (400 km). Most international arrivals arrive by air on charter flights, which have increased from 25 in 1986 to 720 in 2006. Of these recent arrivals, 42 per cent came from the UK, followed by 8.5 per cent from Russia and 6.2 per cent from Germany. However, in the last decade and with the increasing popularity of cruising (page 597), more visitors have been arriving by sea – 18 cruise ships in 1996 and 72 in 2006.

Figure 20.24

Goa's domestic and international arrivals, 1986–2006

	Domestic arrivals	International arrivals	Total
1986	736 548	97 533	834 081
1996	888 914	237 216	1 126 130
2006	2 098 654	380 414	2 479 068

Goa's beaches

Goa's beach resorts can roughly be divided into four types from north to south (Figure 20.25).

The extreme north

The most northerly beaches at Keri and Arambol are, by and large, undeveloped, and tend to attract day visitors and those wishing to find cheap accommodation, food and drink. They are only reached along narrow winding roads by infrequent local buses and, until a year or so ago, Keri had no accommodation at all and only a few beach shacks that sold simple refreshments (Figure 20.26a). Arambol has become more accessible since the opening of a road bridge over the river estuary to the south but has insufficient accommodation to cater for those wishing to stay for longer than a day (Figure 20.26b).

The northern beaches

Vagator and Anjuna, being nearer the state capital of Panaji, are more popular. They have small hotels as well as bars and restaurants, many of which are still family owned. Most of the shops fall into the informal sector (page 574), some only open seasonally.

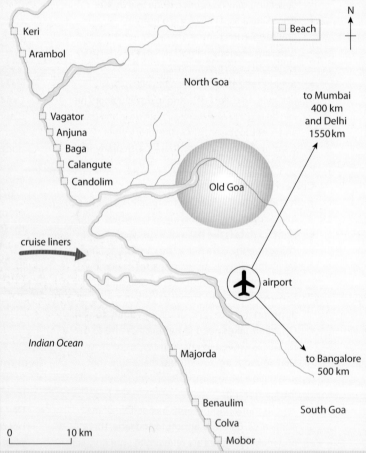

Figure 20.25

Goa's beaches

Figure 20.26

Arambol, one of the less-developed resorts on Goa's northern beaches

The main northern beaches

The long stretches of sand continue south-wards to the beaches of Baga, Calangute and Candolim. Being nearer both the airport and the mainline railway station, these are the places for those arriving from Mumbai and Delhi or by charter from Europe. Even so, many of the older hotels, bars, restaurants and shops are relatively small and family owned (Figure 20.26c). Back from the beach are the larger hotels, with more being built. This rapid development has already caused considerable damage to the sand dune eco-system that runs behind the beaches.

The southern beaches

This is the area for the large five-star beach resort complexes which have opened up at Benaulim, Colva and Mobor (Figure 20.27). These are more likely to attract an older group of overseas and package holiday-maker and the better-off, professional Indian worker. The beach resorts are set in large grounds full of coconut palms, tropical plants and shrubs, each with their own gardens, swimming pools, bars and restaurants, sporting amenities and stretch of beach.

Benefits and problems

Tourism is concentrated mainly along a narrow coastal zone where it has had a number of positive benefits including higher incomes, increased employment, improved local transport and greater foreign exchange earnings. However, tourism has also created socio-economic and environmental problems due to a largely uncontrolled, unplanned development, much employment being seasonal, drug dealing, the concentration and subsequent congestion of people and attractions along a narrow strip, and the destruction of local ecosystems.

Figure 20.27

From a Kuoni travel brochure

GOA RENAISSANCE

Location: Set in 23 acres of gardens on the southern coast is the Goa Renaissance. The hotel is 75 minutes' drive from the airport with gardens leading down over the sand dunes to the wide expanse of Colva beach.

Facilities: A spacious, open-plan lobby with attractive lobby bar overlooks the gardens; there is also a main restaurant, an informal coffee shop and an outdoor barbecue terrace with regular live entertainment. The hotel has a large freeform swimming pool with swim-up bar and a fitness centre with a sauna, jacuzzi, massage and beauty parlour. Windsurfing from the beach, and there is table tennis, chess, floodlit tennis courts as well as a 9-hole pitch and putt course.

Accommodation: 202 rooms

LEELA PALACE

Location: In south Goa some 90 minutes' drive from the airport, this superb hotel is set in 75 acres of coconut groves in grounds full of tropical plants and shrubs, lagoons and waterways leading down to the soft sands of beautiful Mobor beach.

Facilities: A large swimming pool with poolside bar, children's pool, tennis courts, 9-hole golf course, gym, shops and health spa. A uniquely decorated open lounge bar is popular for pre- or post-dinner drinks and dining is a gastronomic delight with a choice of traditional Indian, international and Italian cuisines. From its own watersports centre on the beach is sailing, parasailing and water skiing.

Accommodation: 137 rooms

Further reference

Chapman, R. (2007) 'Sustainable tourism', *Geography Review* Vol 20 No 3 (January).

Chapman, R. (2007) 'Ecotourism', *Geography Review* Vol 20 No 4 (March).

Dove, J. (2004) *Tourism and Recreation*, Hodder & Stoughton.

Dove, J. (2007) 'Tourism: impact on the Yucatan Peninsula', *Geography Review* Vol 20 No 5 (May).

Wyne, M. (2007) 'Modelling tourism', *Geography Review* Vol 20 No 3 (January).

Kenya's Tourism and National Parks: www.tourism.go.ke/wildlife_ministry. nsf/ministryparks

Peak District National Park: www.peakdistrict.org

Sustainable tourism: www.peopleandplanet.net/doc/ php?id=1110

www.ecotourism.org
www.geog.nau.edu/tg/

UK National Parks: www.nationalparks.gov.uk

UN World Tourism Organization (WTO), global tourism facts: www.unwto.org/facts/eng/highlights.htm

Activities

1 **a** What is tourism? *(1 mark)*

   b i Give four factors that have helped cause the growth of world tourism since 1960. *(4 marks)*

   ii For each of your answers in **i**, explain why this factor led to a growth of tourism. *(4 marks)*

   c With reference to a named resort or tourist area that you have studied, explain how the growth of tourism has brought both:

   i benefits and

   ii problems to the people who live in the area. *(10 marks)*

   d If tourism starts to decline in an area it can cause serious economic problems. Name a tourist area where the industry has started to decline. Describe how the area has adapted to try to stop the decline. *(6 marks)*

2 Study Figure 20.28.

   a For each photograph:

   i Describe the attractions of the area that make it a suitable tourist destination. *(6 marks)*

   ii Suggest which sector of the holidays market this area will particularly appeal to. *(3 marks)*

   iii Suggest how tourism has brought advantages and disadvantages to the people of the area. *(6 marks)*

   b In many tourist areas the natural environment is a major attraction for tourists. Unfortunately the pressure of tourism threatens to destroy the natural environment.

   For a named tourist area, explain how management strategies have been, are being, or could be developed to allow tourism to continue without destroying the environment. *(10 marks)*

3 **a i** What is 'ecotourism'? *(1 mark)*

   ii Name an example of a place in a less economically developed country where ecotourism has been developed. *(1 mark)*

   iii Describe the attractions for ecotourists of the area that you named in **ii**. *(4 marks)*

   iv Explain how ecotourism has brought specific benefits to the people and the environment in the area. *(6 marks)*

   b With reference to the Cairngorms or another mountainous area in the UK that is being damaged by increased tourist pressure:

   i explain why the number of tourists has increased in recent years *(4 marks)*

   ii explain how the tourist pressure is damaging the environment *(4 marks)*

   iii describe one management strategy that aims to reduce the damage being done, and explain how the strategy is intended to work. *(5 marks)*

Figure 20.28

a Spain, **b** Nepal, **c** Greece

Exam practice: basic structured questions

4 Study Figure 20.28.

  a Describe the tourist attractions of each of the areas shown in the photographs. *(6 marks)*

  b Butler's model of the life cycle of a tourist resort shows the following stages:
  - exploration
  - involvement
  - development
  - consolidation
  - stagnation
  - rejuvenation or decline.

  Suggest, with reasons, which stage has been reached by each of the tourist areas shown in the photographs. *(12 marks)*

  c Name a tourist resort that has reached the later stages of the model, and explain what is being done to rejuvenate the tourist industry there. *(7 marks)*

5 **a** Refer to Figure 20.29. Name the most popular destinations for tourists from the UK in:

  i Europe

  ii regions outside Europe. *(2 marks)*

  b **i** With reference only to holidays taken in Europe by residents of the UK, describe and account for the distribution of the main holiday destinations. *(5 marks)*

  ii The number of UK residents taking holidays in Europe in February is fairly small. Suggest, with reasons, how the distribution of holiday destinations is likely to be different from that shown on Figure 20.29. *(6 marks)*

  c Study Figure 20.30. Describe and explain the patterns shown by the data of tourism from the different world regions. *(12 marks)*

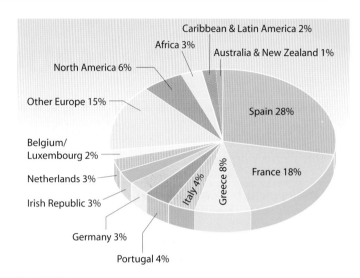

Figure 20.29

Foreign holidays taken by UK residents

Region of origin	Millions	% change 2005/06
Africa	24.5	12.1
Americas	142.2	3.7
Asia & Pacific	166.5	7.7
Europe	473.7	4.7
Middle East	24.8	8.9
World	846.0	5.4

Figure 20.30

International arrivals to UK, 2005/06

Exam practice: structured questions

6 The Peak Park Authority issued a revised Management Plan for 2006–11.

The two main principles underpinning that plan are:
- partnership working
- sustainable development.

Referring to the Peak District National Park or to any other tourist area that you have studied:

  a Describe how conflicts can arise between different groups and individuals who use the land in the Park. *(7 marks)*

  b Discuss how the aims and principles of the Management Plan for the Peak Park could help to manage and reduce land use conflict.

You must make reference to specific conflicts in named places. *(18 marks)*

7 Study the table below Figure 20.9 on page 591.

  a There has recently been an increase in the number of protected areas in the UK. Explain why. *(10 marks)*

  b With reference to one or more such areas, explain how the development of protected areas is affecting tourism. *(15 marks)*

Exam practice: essays

8 Explain how tourism can bring both advantages and disadvantages to the people and environment in areas where it develops. Make reference to countries at different stages of development. *(25 marks)*

9 Can the development of tourism lead to sustainable development in poor, remote areas of the world?

Discuss this with reference to examples that you have studied. *(25 marks)*

10 Account for the recent rapid growth of tourism in Goa (or in any other tourist resort in a less developed country that you have studied). *(25 marks)*

21

Development and globalisation

● ●

'One world, one dream.'
Beijing Olympics, 2008

'Development is more than mere economics.'
Mark Tully, *No Full Stops in India*, 1991

The concept of economic development

Frequent references have been made in earlier chapters to the inequalities in world development and prosperity. Gilbert, in his book *An Unequal World*, began by stating that:

'Few can deny that the world's wealth is highly concentrated. The populations of North America and Western Europe eat well, consume most of the world's fuel, drive most of the cars, live in generally well serviced homes and usually survive their full three score years and ten. By contrast, many people in Africa, Asia and Latin America are less fortunate. In most parts of these continents a majority of the population lack balanced diets, reliable drinking water, decent services and adequate incomes. Many cannot read or write, many are sick and malnourished, and too many children die before the age of five.'

Figure 21.1

Terms used in relation to world development

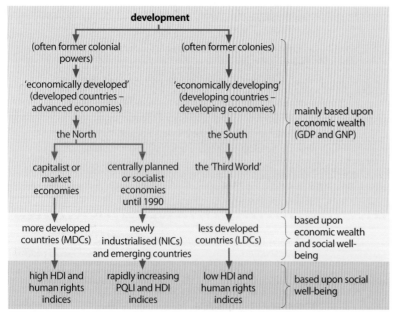

Note: For consistency, the terms 'economically more developed' or 'developed', and 'economically less developed' or 'developing' are mainly used in this book.

Definition of terms

Terms such as 'developed' and 'developing' have been used for several decades to indicate the economic conditions of a group of people or a country. By the 1980s, the term 'developing' had come to be regarded as a stigma and was replaced by the concept of the 'South' (**Brandt Report**, 1980) and, with increasing popularity, the 'Third World' (Figure 21.1). By the 1990s, with the growing realisation and appreciation that poverty is relative, not absolute, the terms **more economically developed countries (MEDCs)** or 'advanced economies', and **less economically developed countries (LEDCs)** or 'developing economies' became increasingly acceptable. Even more recently the nations that had, a decade or two earlier, been grouped together as belonging to the 'developing economies' had now shown among themselves a widening spread of wealth and living standards, for example the growing gap between the NICs (newly industrialised counties, page 578) and, today, the emerging countries (BRIC – Brazil, Russia, India and China) with those of sub-Saharan Africa.

All these definitions (summarised in Figure 21.1) were based on, and overemphasised, economic growth. To those living in a Western, industrialised society, economic development tends to be synonymous with wealth, i.e. a country's material standard of living. This is measured as the **gross domestic product (GDP)** per capita and is obtained by dividing the monetary value of all the goods and services produced in a country by its total population. When trade figures for 'invisibles' (mostly financial services and deals) are included, the term **gross national product (GNP)** is used. It is possible to use either term – GDP is preferred by the EU, and GNP by the UN (our usual source of data) and the USA – as both aim to measure the wealth of a country and to show the differences in wealth between countries. GDP and GNP figures need to be treated cautiously due to problems with exchange rates, differences between countries in their methods of calculation, and difficulties in evaluating services.

Recently, an increasing number of definitions, often involving cultural development, social well-being and political rights, have been suggested as alternatives to those previously based solely upon economic criteria – i.e. they emphasise 'quality of life' in contrast to 'standard of living'. In the early 1990s, the UN introduced the term **Human Development Index (HDI)** – see page 606.

Development is not just the difference between the developed, rich and powerful countries and those that are less developed, poor and subordinate. Each country has areas of prosperity and poverty; contains people with different standards of living based on variations in job opportunities (Shanghai and Sichuan in China), race or tribe (Hutu and Tutsi in Rwanda), religion (Sunnis and Shi'ites in Iraq), language (Dutch-speaking Flemings and French-speaking Walloons in Belgium), or social class (caste in India). Taken a step further, it is also possible to identify differences in development within cities (Places 52 and 58) and inequality between genders.

The difference in wealth and standard of living between the world's richest and poorest countries is referred to as the **development gap**. Despite some attempts to the contrary (aid – page 632), this gap continues to widen (debt – page 608), particularly as **globalisation** puts increasing power into the hands of the most wealthy countries and organisations.

What is meant by globalisation?

Globalisation is a relatively new term. It has a wide range of meanings but generally refers to processes that extend globally to affect or integrate people across the entire world. From a geographer's point of view, it includes any process of change that occurs at a world scale and which has world-wide effects. These processes may be considered to be physical (e.g. rising sea-level), human and economic (e.g. trade) or a combination of both (e.g. global warming) but they are considered to be essentially geographical in that they affect the Earth's environment and its people.

The links created by globalisation are increasing both in range and scale, and are developing at an ever increasing pace. These links, which may be considered to be environmental, economic, technological, cultural, sociological or political, can have – often depending on your own viewpoint – either beneficial or detrimental effects. Some would argue that globalisation spreads wealth, knowledge and personal contacts across the world; others that it is creating an unfair world in which rich countries and large organisations exploit the world's poorest peoples which increases, rather than reduces, the 'development gap'. Figure 21.2 is one of several possible schematic diagrams showings topics related to globalisation that appear within this text, e.g. trade, transport, tourism, migration, aid, health, finance and technology.

Development v. globalisation

It is not straightforward to try to link development and globalisation. Development has conventionally been understood as something that happens, or fails to happen, to countries. Globalisation is increasingly being regarded as a process that disintegrates national economies and constitutes new spatial patterns, e.g. trading blocs, innovative regions, international banking.

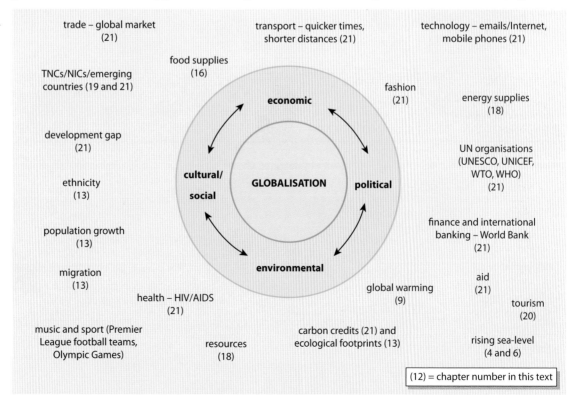

Figure 21.2

References to globalisation

Criteria for measuring the 'development gap'

1 Economic wealth

To many people living in developed countries, economic development has been associated with a growth in wealth based on GDP (or GNP). This implies that the GDP (or GNP) of a country has to increase if its standard of living and quality of life are to improve. An economic growth rate of 8 to 10 per cent, which is the highest, has been achieved in China and Ireland in recent years, and by several South-east Asian countries over the past decade or two (Figure 19.38). A rate of 1 per cent is considered disappointing.

Although GDP/GNP figures are easier to measure and to obtain than other development indicators such as social well-being, there are limitations to their use and validity. They are more accurate in countries that have many economic transactions and where goods, services and labour can be measured as they pass through a market place – hence the term 'market economies'. Where markets are less well developed, and trading is done informally or through bartering, and where much production takes place in the home for personal subsistence, GDP figures are less reliable. In the former centrally planned, socialist economies, with their relatively small role in international trade and with few services, GDP figures were difficult to calculate and interpret.

Comparison of GDP requires the use of a single currency, generally US dollars, but currency exchange rates fluctuate. The size and growth of GDP may prove to be poor long-term economic indicators and fail to take into consideration human and natural resources. GDP per capita is a crude average and hides extremes and uneven distribution of income between regions and across socio-economic groups, especially in less developed countries where there may be very few extremely wealthy people and a large majority living at subsistence level. Despite these limitations, GDP and GNP are still regarded as relatively good indicators of development and good measures for comparing differences between countries (Figure 21.3). Notice that it is the advanced economies and several of the oil-producing states that have the highest GDP per capita and the developing economies that have the lowest, although the fastest-growing are China and several others in South-east Asia. The World Bank now produces figures for income inequality within some countries, e.g. Brazil.

2 Social, cultural and welfare criteria

Human development has changed the purpose of development to that of meeting human needs, and away from the old style of economic development based on changes in a country's economy and wealth. The UN Development Programme's Human Development Index (HDI) gives every

Figure 21.3

World GDP
Source: The UC Atlas of Global Inequality

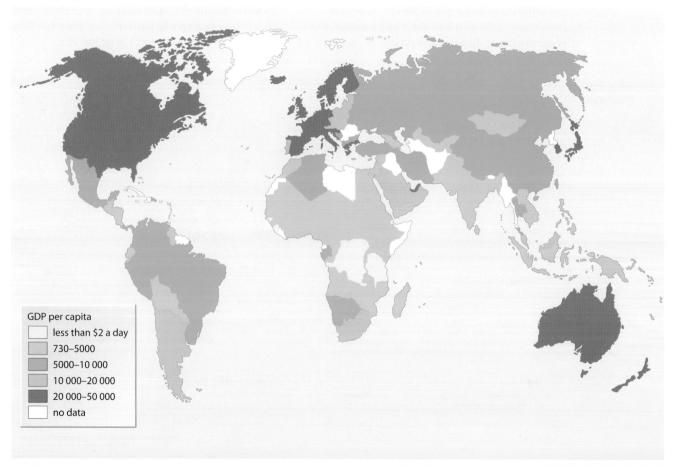

GDP per capita

- less than $2 a day
- 730–5000
- 5000–10 000
- 10 000–20 000
- 20 000–50 000
- no data

country a score between 0 and 1, based on its citizens' longevity, education and income. The three factors are given equal weight. Longevity is measured by average life expectancy at birth – the most straightforward measure of health and safety. Education is derived from the adult literacy rate and the average number of years of schooling. Income is based on GDP per capita converted to 'purchasing power parity dollars' (PPP) and is adjusted according to the law of diminishing returns, i.e. what an actual income will buy in a country. The HDI value for a country shows the distance that it has already travelled towards the maximum possible value of 1, and also allows comparisons with other countries (Figure 21.4). The difference between the value achieved by a country and the maximum possible value shows the country's shortfall, i.e. how far the country has to go. Finding ways of reducing this shortfall is a major challenge for each country.

As the table on the right shows, it is countries in Scandinavia that now top the HDI list and countries in the Sahel of sub-Saharan Africa that tend to be at its foot – an interesting latitude effect.

Countries with a score of over 0.9 correspond closely with the economically more developed countries while those with less than 0.5 equate closely with the least economically developed countries (compare Figures 21.3 and 21.4).

Yet should the similarities between GDP and HDI really be that surprising? Longevity, a good education and a high purchasing power all depend fairly directly on a country's wealth.

A major criticism of the HDI is that it contains no measure of human rights or freedom. Although the UNDP did produce a separate **Human Freedom Index (HFI)** in 1991, it has not done so since, arguing that 'freedom is difficult to measure and is too volatile, given military coups and the whims of dictators'. The issue of personal and political rights has become increasingly important since then.

Perhaps the main point about HDI is that it enables you to spot anomalies, e.g. countries that have a better (Canada, Sri Lanka and Tanzania) or worse (Saudi Arabia and other oil-producing countries) level of well-being than might be expected from their GNP. HDI can serve a purpose if it identifies where poverty is greatest (between countries, within a country or between groups of people in a country) or if it stimulates debate and action as to where aid, trade and debt alleviation needs to be focused.

Year	Top two		Bottom two	
1990	Canada	0.93	Niger	0.28
	Japan	0.92	Mali	0.30
1995	Norway	0.94	Niger	0.30
	Canada	0.94	Mali	0.32
2000	Norway	0.96	Sierra Leone	0.31
	Sweden	0.95	Niger	0.32
2005	Iceland	0.97	Sierra Leone	0.34
	Norway	0.97	Burkina Faso	0.37

Figure 21.4

The UN Human Development Index (2005) World development

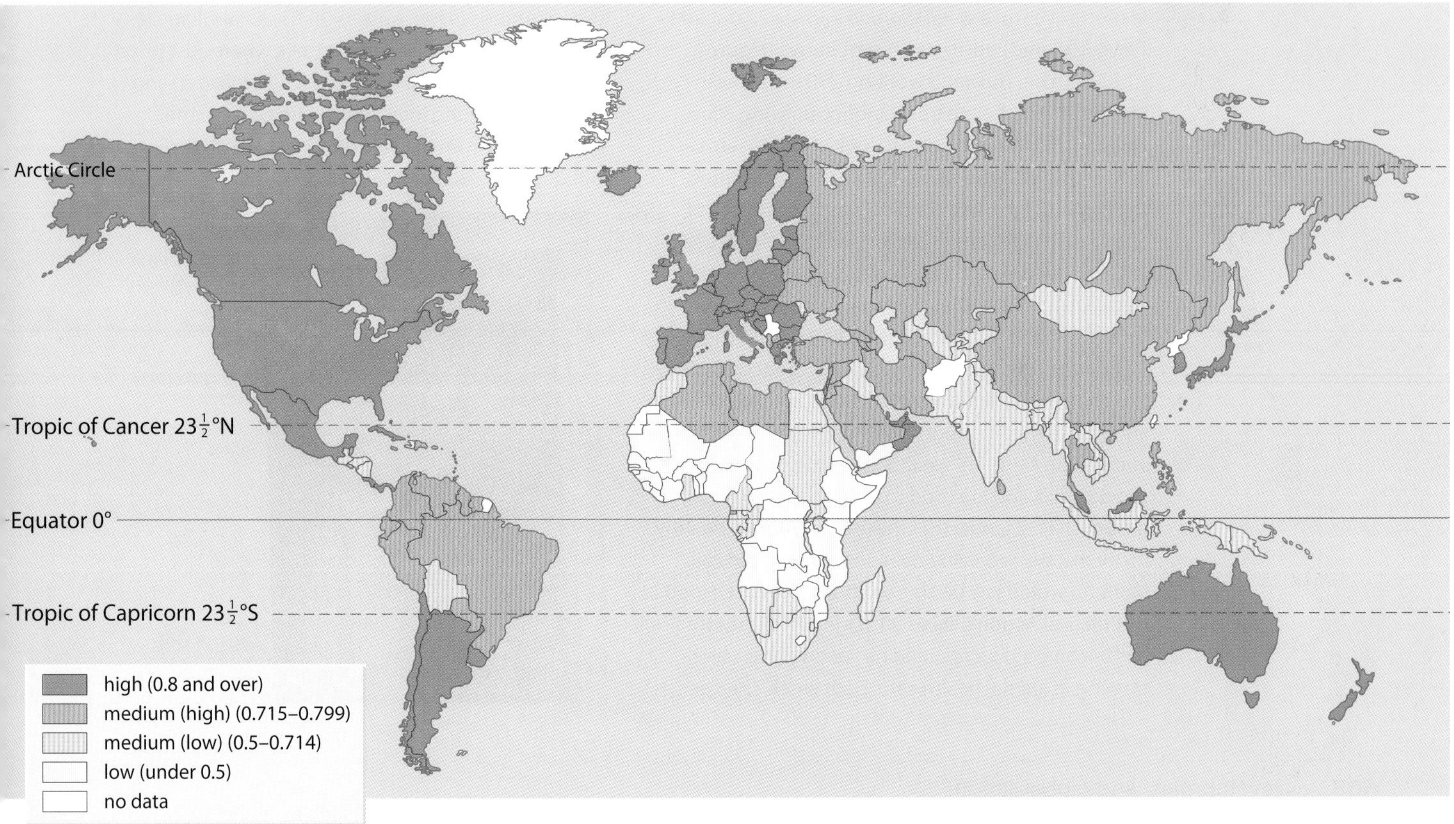

- high (0.8 and over)
- medium (high) (0.715–0.799)
- medium (low) (0.5–0.714)
- low (under 0.5)
- no data

Arctic Circle

Tropic of Cancer 23½°N

Equator 0°

Tropic of Capricorn 23½°S

3 Other criteria for measuring the 'development gap'

Further criteria have also been used to measure the quality of life as an indicator of levels of, or stages in, development. Several are linked to population as, in developing countries, birth rates are generally high, the natural increase is rapid, life expectancy is shorter and a high percentage of the population is aged under 15 (Figures 13.15 and 13.21). Higher death and infant mortality rates reflect the inadequacy of nutrition, health and medical care. In many developing countries, the prevalence of disease may result from an unbalanced diet, a lack of clean water and poor sanitation – a situation often aggravated by the limited numbers of doctors and hospital beds per person. The major-ity of people live in rural areas and are dependent upon farming, while in the country as a whole only a small percentage of the population is likely to find employment in manufacturing or service industries. Many jobs are at a subsistence level, in the informal sector (page 574) and the amount of energy consumed within the country is low (Figure 18.25). Economically less developed countries often import manufactured goods, energy supplies and sometimes even foodstuffs, especially grain. In return, they may export raw materials for processing in the developed world (Figure 21.36), accumulate a trade deficit and get increasingly into debt (page 624). High rates of illiteracy reflect a shortage of schools and trained teachers. The density of communication networks, circulation of newspapers and numbers of cars, telephones and television sets per household or per capita have also been used as indicators of development.

Social and economic development

An often neglected factor in social and economic development is gender, and in particular the role of women. Places 96 describes the lifestyle of a Kenyan woman who, like many other women across the world, is the principal support of her family and local community. It is women like these who form the mainstay of the family, of women's groups, the community and, indeed, of a nation's development. Yet their role as providers and generators of wealth is not matched in most societies by their status or influence. Women (and not just in developing countries) are often:

- denied ownership of property (including land), access to wealth, education and family planning (page 357) and equality in justice and employment
- kept subordinate by being granted lowly positions or given menial tasks which are often poorly paid or even unpaid (farming) or are heavy, tedious and time-consuming (collecting firewood and water)
- subject to violence, both physical and mental
- denied political influence.

Places 96 Kenya: women and development

Marietta lives on a small *shamba* (farm) just outside Tsavo National Park in south-east Kenya (Figure 21.5). With her husband working 250 km away in Mombasa, and her nearest neighbour living 3 km away, Marietta is left alone to look after the farm and her seven children. Her day begins by sharpening the machete needed to collect the daily supply of dead wood (living trees are left for animal grazing), as this is her only source of energy (page 543), and by preparing a meal for the family. The eldest girls, before walking to school, collect water from the river 1 km away. Much of Marietta's day is spent collecting firewood and looking after her crops (maize, beans and sorghum). Although owning a few chickens and goats, Marietta's 'wealth' is her two cows which provide milk and are used to plough the hard ground. It is essential that these cows remain healthy for even if the vet, living over 50 km away, did call, Marietta would not be able to afford the bill. Helped by Practical Action (Places 90, page 577), Marietta has become a *wasaidizi* and has been given basic training in animal health care. Each week, she spends two mornings in her 'surgery' in the local village (a four-hour round walk along a track where, just prior to the author's visit, a lion had killed a villager) and other days visiting local farms. She earns a small commission from the sale of vaccines and medicine, but does not receive a salary.

Figure 21.5

Marietta at her 'surgery'

Living in extreme poverty

At the beginning of the 21st century, the UN claimed that nearly 1 billion people lived in **extreme (or absolute) poverty**, which meant that 1 person in every 6 of the world's population was struggling for survival. Poor countries were finding themselves falling further and further behind the richer countries and the 'development gap' was continuing to grow. As this gap widened, people in the poorest countries became caught up in the so-called **'cycle of poverty'** (Figure 21.6), which leaves successive generations in a 'poverty trap' from which there appears little hope of escape.

At the Millennium Summit of 2000, world leaders committed their nations to a new global partnership aimed at reducing extreme poverty. They set out a series of targets which have become known as the **Millennium Development Goals** (MDGs) and which they hoped would be achieved by 2015 (Figure 21.7). Within five years significant progress had been made in many parts of the world. The number in extreme poverty had declined by an estimated 130 million, average overall incomes had increased by 21 per cent, infant mortality had fallen from 103 per 1000 live births to 88, life expectancy had risen from 63 years to 65, and an extra 8 per cent of the developing world's people had access to clean water and 15 per cent to improved sanitation.

A report by the UN Millennium Project secretariat team in 2006 concluded, however, that this progress had not been uniform and that there were still huge disparities not only between countries but especially between rural areas, where extreme poverty is often still increasing, and urban areas. The team said that sub-Saharan Africa was at the

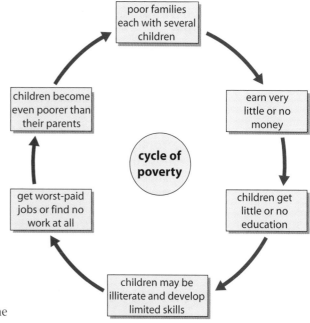

Figure 21.6

The cycle of poverty

centre of the crisis, with continuing food insecurity (page 503), extremely high child and maternal mortality, large numbers living in sub-standard accommodation and a widespread shortfall for most of the MDGs. According to the Human Poverty Index (HPI), the world's six poorest countries were, in descending order, Sierra Leone, Niger, Ethiopia, Burkina Faso, Mali and, at the foot, Chad. Asia was the region with the fastest progress, but even there thousands of people remained in extreme poverty and even the fastest improving countries still failed to meet non-income goals.

Figure 21.7

Millennium Development Goals (MDGs) and basic human rights

MDGs need to address:		**MDGs should between 1990 and 2015:**
• income poverty	• hunger	• reduce by two-thirds the under-5s mortality rate
• lack of adequate shelter	• disease	• have halted and begun to reverse the spread of HIV and AIDS
• lack of clean water	• exclusion	
MDGs need to promote:		• reduce by three-quarters the maternal mortality rate
• gender equality	• education	• aim to halve the number of people suffering from hunger, living on under $1 per day, without access to safe drinking water and without access to basic sanitation
• environmental sustainability		
MDGs should ensure the basic rights of:		
• health	• shelter	• eliminate gender disparity in education
• education	• education	

Figure 21.8

Living in extreme poverty

Millennium Development Goals on water

As shown in Figure 21.7, a lack of clean water is one of six features that characterises living in extreme poverty, and two of the MDGs were to reduce by half by 2015 the number who in 1990 lived without access to safe water and without access to basic sanitation. An earlier attempt by the UN to provide water and sanitation for all by 1990 was the International Drinking Water Supply and Sanitation Decade launched in 1980. This ambitious target was never reached. The year 2008 was designated the International Year of Sanitation. Will this attempt be more successful? It was also in 2008 that the UN claimed a

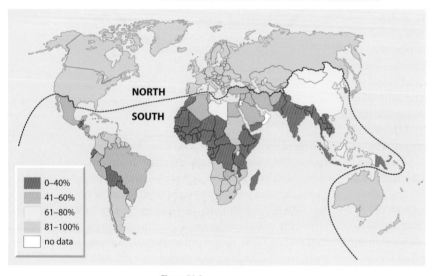

Figure 21.9

Percentage of the population with access to safe water

number of facts:

■ It would take an extra US$10 billion per annum to achieve the MDGs by 2015.

■ 1.1 billion people – 1 in 6 of the world's population – did not have access to safe water (Figure 21.9).

■ 2.6 billion people – more than 2 in 6 of the world's population – did not have adequate sanitation.

■ If all the Earth's water was poured into a bucket then, as 97.5 per cent of it is saltwater, the fresh water available for drinking (the remaining 2.5 per cent) would be the equivalent of one teaspoonful (and that assumes it is not polluted).

■ At any given time, almost half the total population of the developing countries is suffering from one or more of the main diseases such as diarrhoea, cholera, typhoid and bilharzia (Figure 21.27) that result from the inadequate

provision of safe water and sanitation, and half of the hospital beds in the developing world are occupied by people with water-related illnesses.

■ Water-related disease is the second major cause of death for children, with a total of almost 2 million dying across the world each year and 5000 a day in developing countries.

■ In semi-arid areas, obtaining water is time consuming at the best – Figure 21.11 shows women carrying water, which could weigh 20 kg, on their heads and taking several hours to collect from a source several kilometres away. Such unreliable sources become life threatening during times of drought (Figure 16.5).

■ Whereas an average person living in Europe uses 200 litres of water a day – half that of someone living in the USA – a person living in a developing country may only have 10 litres for washing, cooking and drinking.

■ The demand for water in the 20th century increased by more than twice the rate of population growth and this demand is expected to rise by another 40 per cent by 2030.

■ Although safe water and adequate sanitation may be difficult to find in shanty settlements of cities in developing countries (pages 443 and 445), urban areas are usually much better off than more remote rural areas (Figure 21.10).

■ Increasing attention needs to be paid to **virtual water**. This is water that appears in food products or is needed to manufacture goods. Agriculture accounts for over 70 per cent of water consumption as it can take 1000 litres to produce 1 kg of potatoes, 1450 litres for 1 kg of wheat and 3450 litres for 1 kg of rice. A country consumes even more water if it imports fresh fruit and vegetables.

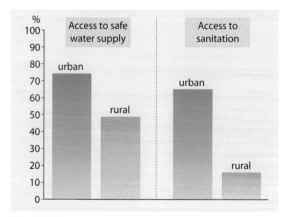

Figure 21.10

Safe water and sanitation: world total

Figure 21.11

Women carrying water

In early 2008, the Secretary-General of the UN expressed concern over the increasing number of global conflicts resulting from water shortages. As consumption increases and resources dwindle, conflicts over water are becoming more heated as people downriver find themselves at the mercy of those upriver, with irrigation and dam construction the major flashpoints. While water disputes may not be a single cause for warfare, they can inflame existing tensions.

Places 97 Malawi and Ethiopia: WaterAid

In 2006, WaterAid celebrated its twenty-fifth year and was credited to be Britain's most admired charity. To date, WaterAid has helped over 12 million people in developing countries to gain access to safe, clean water and to improved sanitation. Its aims are to help people in some of the poorest countries:

- to set up, operate and maintain their own safe domestic water and sanitation facilities
- to learn about safe hygiene practices so that they gain maximum health benefits.

It achieves these aims by helping local organisations to set up low-cost, sustainable projects that use appropriate technology and which can be managed by the community itself. WaterAid, which relies on donations, can provide safe water, sanitation and hygiene education for just £15 per person – basic services that are essential if vulnerable communities are to have any hope of escaping from the stranglehold of disease and poverty. It also lobbies governments and decision-makers to prioritise water and sanitation in their poverty reduction plans.

Malawi

Malawi is one of the world's poorest countries with 65 per cent living below the poverty line and a life expectancy of less than 40 years. Only 73 per cent of the 11.2 million inhabitants have access to safe water and only 61 per cent to sanitation. WaterAid began to work here in 1999 and now has four ongoing projects in rural areas and one in the capital of Lilongwe. Two schemes in rural areas include digging over 200 wells in the Salima District to reach clean supplies of underground water and then using modern pumps to raise this water to the surface where it is providing safe water for 26 000 people (Figure 21.12), and rehabilitating existing piped water systems in Machinga District to provide 15 000 people with safe water. One innovative approach encourages villagers to construct composting latrines in which human waste is mixed with soil and ash to form a rich compost. This could be significant in a country where most people depend on farming for their livelihood and where the soil is often infertile and fertiliser is both scarce and expensive. In low-income areas of Lilongwe, sustainable systems for managing water kiosks are being developed.

Ethiopia

The villages of Deyata Dodota and Dewaro in central Ethiopia are just 8 km from each other in distance but seem poles apart in their ways of life. Thanks to WaterAid, Deyata Dodota now has water piped to it, allowing villagers to grow vegetables in their front gardens. In Dewaro, villagers rely on crude, earth-banked dams that hold water for just six months a year, water which they not only use for drinking, washing and disposing of sewage, but which they share with their animals. For half the year they have a long trek for water. Deyata Dodota is essentially self-sufficient; Dewaro needs food aid and lives in the hope that £3000 will be found to extend the pipeline to them.

Figure 21.12

WaterAid pumps bring clean water to African villages

Scattergraphs

It was suggested on pages 606–608 that there was a correlation between certain criteria and the level of development. 'Correlation' in this sense is used to describe the degree of association between two sets of data. This relationship may be shown graphically by means of a **scattergraph**. This involves the drawing of two axes: the horizontal or *x* axis and the vertical or *y* axis. Usually one variable to be plotted is dependent upon the second variable. It is conventional to plot the **independent variable** on the *x* axis and the **dependent variable** on the *y* axis.

Figure 21.13 shows two relationships, one from physical geography and one from human geography. In the physical example, rainfall is the independent variable, with runoff being dependent upon it. The human example shows GDP as the independent variable and energy consumption per capita to be dependent upon this measure of a country's wealth.

The data are plotted against the scales of both axes. The degree of correlation is estimated by the closeness of these points to a **best-fit line**. This line is usually drawn by eye and shows any trend in the pattern indicated by the location of the various points. One or two points, or **residuals**, may lie well beyond the best-fit line and, being anomalous, may be ignored at this stage. (Later it may be relevant to try to account for these **anomalies** or exceptions.)

The best-fit line may be drawn as a straight line (on an arithmetic scale) or as a smooth curve (on log or semi-log scales). If all the points fit the best-fit line exactly, there is a **perfect correlation** between the two variables. However, most points at best will lie close to and on either side of the drawn line. A positive correlation is where both variables increase – i.e. the best-fit line rises from the bottom left towards the top right (Figure 21.14a and b). A **negative correlation** occurs where the independent variable increases as the dependent variable decreases – i.e. the best-fit line falls from the top left to the bottom right (Figure 21.14d and e). In some instances, the arrangement of the points makes it impossible to draw in a line, in which case the inference is that there is no correlation between the two sets of data chosen (Figure 21.14c). In the event of one, or both, of the variables having a wide range of values, it may be advisable to use a logarithmic scale (Figures 3.22 and 18.25).

If the scattergraph shows the **possibility** of a correlation between the two variables, then an appropriate statistical test should be used to see if there is indeed a correlation, and to quantify the relationship.

Figure 21.13

Plotting the dependent and independent variables

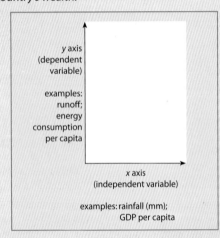

y axis (dependent variable)

examples: runoff; energy consumption per capita

x axis (independent variable)

examples: rainfall (mm); GDP per capita

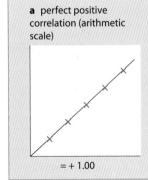

a perfect positive correlation (arithmetic scale)

= + 1.00

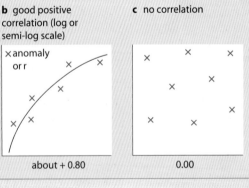

b good positive correlation (log or semi-log scale)

× anomaly or r

about + 0.80

c no correlation

0.00

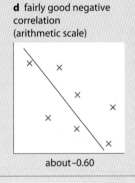

d fairly good negative correlation (arithmetic scale)

about –0.60

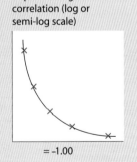

e perfect negative correlation (log or semi-log scale)

= –1.00

Figure 21.14

Types of correlation and their associated Spearman's rank coefficients

GDP per capita			Energy consumption per capita					Birth rate per 1000	Rank	d	d²
	US$	Rank	kg oil-equivalent	Rank	d	d²					
Norway	53 000	1	5284	2	1	1		12			
USA	45 800	2	8051	1	−1	1		14			
Switzerland	41 100	3	3622	6	3	9		10			
UK	35 100	4	3992	5	1	1		13			
Germany	34 200	5	4267	3	−2	4		8			
Japan	33 600	6	4058	4	−2	4		9			
Argentina	14 300	7	1653	8	1	1		19			
Malaysia	13 300	8	1950	7	−1	1		21			
Brazil	9700	9	1012	9	0	0		20			
Colombia	6700	10	799	11	1	1		20			
Egypt	5500	11	638	12	1	1		27			
China	5300	12	902	10	−2	4		12			
India	2700	13	476	13	0	0		24			
Kenya	1700	14	466	14	0	0		40			
Sierra Leone	700	15	230	15	0	0		48			
					$\sum d^2 = 28$					$\sum d^2 =$	

Figure 21.15

Ranked data for GDP, energy consumption and birth rates for selected countries, 2007

Spearman's rank correlation coefficient

This is a statistical measure to show the strength of a relationship between two variables. Figure 21.15 lists the GDP per capita for 15 selected countries. Fifteen is the minimum number needed in a sample for the Spearman's rank test to be valid.

The first stage is to see if there is any correlation between the GDP and the energy consumption per capita. This can be done using the following steps:

1 Rank both sets of data. This has already been done in Figure 21.15. Notice that the highest value is ranked first. Had there been two or three countries with the same value, they would have been given equal ranking, e.g. rank order: 1, 2, 3.5 , 3.5 (3.5 is the mean of 3 and 4) , 5, 7 , 7 , 7 (7 is the mean of 6, 7 and 8), 9, 10.

2 Calculate the difference, or d, between the two rankings. Note that it is possible to get negative answers.

3 Calculate d^2, to eliminate the negative values.

4 Add up ($\sum$) the d^2 values (in this example, the answer is 28).

5 You are now in a position to calculate the correlation coefficient, or r, by using the formula:

$$r = 1 - \frac{6\sum d^2}{n^3 - n}$$

where: d^2 is the sum of the squares of the differences in rank of the variables, and n is the number in the sample.

In our example it follows that:

$$r = 1 - \frac{6 \times 28}{3375 - 15}$$

$$= 1 - \frac{168}{3360}$$

= 1 − 0.05 (then do not forget the final subtraction)

= 0.95 (it is usual to give the answer correct to two decimal places).

In this example, there is a strong positive correlation (remember, a perfect positive correlation is 1.00) between GDP and energy consumption per capita.

Although the closer r is to +1 or −1 the stronger the likely correlation, there is a danger in jumping to quick conclusions. It is possible that the relationship described may have occurred by chance. The second stage is therefore to test the **significance** of the relationship. This is done by using the graph shown in Figure 21.16. Note that the correlation coefficient r is plotted on the y axis and the **degrees of freedom (df)** on the x axis. Degrees of freedom are the number of pairs in the sample minus two.

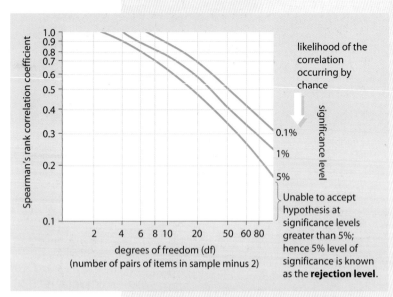

Figure 21.16

The significance of the Spearman's rank correlation coefficients and degrees of freedom

and 6 in area C. Had chance been the only factor affecting this distribution, then it might reasonably be expected that as area A covers 50 per cent of the total area, then half the villages would be located there. Similarly, areas B and C, each covering 20 per cent of the area, should both have 10 villages, leaving area D, with only 10 per cent of the area, with the remaining 5 villages. This means, as shown in Figure 21.18a, that we have two sets of data showing the **observed** (O) number and the **expected** (E) number of villages. In reality, however, Figure 21.17 shows that areas B and D have more villages than might be expected and A and C fewer than expected. It is tempting, therefore, to suggest that there could be a relationship between the observed and expected distributions and that this relationship is dependent upon the height of the land, whereas the difference may in fact be due entirely to chance factors. Chi-squared is used to estimate the probability that the differences are due to chance.

It is often best to begin with a null hypothesis, which in this case might be 'There is no significant relationship between the distribution of villages and the height of the land.' We can now use the formula for chi-squared, which is:

$$\chi^2 = \sum \frac{(O - E)^2}{E}$$

Using the correlation coefficient of GDP per capita and energy consumption per capita, which we have worked out to be 0.95, we can read off 0.95 on the vertical scale and 13 (i.e. 15 in the sample minus 2) on the horizontal. We can see that the reading lies above the 0.1 per cent significance level curve. This means that we can say with 99.9 per cent confidence that the correlation has not occurred by chance. The graph also shows that if the correlation falls below the 5 per cent significance level curve then we can only say with less than 95 per cent confidence that the correlation has not occurred by chance. Below this point, the correlation or hypothesis is rejected in terms of statistical significance – i.e. there is too great a likelihood that the correlation has occurred by chance for it to be meaningful. Even if there is a significant correlation, the result does not prove that there is necessarily a *causal* relationship between variables. It cannot be assumed that a change in A causes a change in B. Further investigation is necessary to establish this.

Chi-squared

Whereas Spearman's rank seeks **associations** between x and y values, chi-squared looks for **differences** between groups (or areas). The symbol for chi-squared (*chi* is a Greek letter pronounced 'ky') is χ. Figure 21.17 shows the hypothetical distribution of villages over an area of land consisting of four contrasting categories of height, i.e. frequencies of 0–50 m, 51–100 m, 101–150 m and over 150 m (it could have been different types of soil, or rock type, etc.). Of the 50 villages located here, 20 are in area A, 12 in each of areas B and D,

AREA B

20% of area

12 villages

AREA A

50% of total area

20 of total villages

AREA C

20% of area

6 villages

AREA D   10% of area   12 villages

Figure 21.17

Chi-squared: observed and expected villages

Figure 21.18b shows how to use the formula and, in this example, how we obtain a calculated value of chi-squared of 12.8. We can now, by using Figure 21.19, test for the significance of this value and determine the probability that the distribution was due to chance. Notice that, as in Spearman's rank (Figure 21.16), the horizontal axis is labelled 'degrees of freedom' (df). We read the degrees of freedom by subtracting 1 from the total number of distributions (areas), in this case $4 - 1 = 3$. Using our two coordinates ($\chi^2 = 12.8$ and df = 3) we can obtain a location on the graph which is just above the 1 chance in 100 curve, i.e. our distribution is only likely to occur by chance once in every 100 situations. We can assume, therefore, that there is a possible connection between the distribution of villages and height of the land and so we can start looking for causes (had the location on the graph been below the 5 chances in 100 curve, then we could assume that there was no connection between village distribution and height of the land and therefore we need not spend time seeking reasons).

Figure 21.18

A worked chi-squared example

a

Area	A	B	C	D	Total
O (Observed)	20	12	6	12	50
E (Expected)	25	10	10	5	50

Using chi-squared

b

(i)	$(O - E)$	– 5	+2	–4	+7				
(ii)	$(O - E)^2$	25	4	16	49				
(iii)	$\dfrac{(O - E)^2}{E}$	1.0	0.4	1.6	9.8				
(iv) Σ (sum of)	$\dfrac{(O - E)^2}{E}$	1.0	+	0.4	+	1.6	+	9.8	= 12.8

$\therefore \chi^2 = 12.8$

Figure 21.19

The significance of chi-squared and degrees of freedom

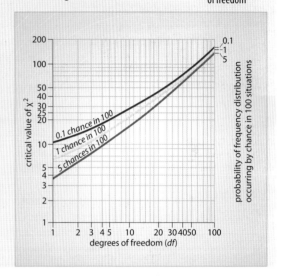

Stages in economic growth

The Rostow model

Various models, with a wide range of criteria, have been suggested when trying to account for differences in world development. These include those based on capitalist and Marxist systems as well as those more concerned with wealth, social and cultural differences. One of the first models to account for economic growth, and probably still the simplest, was that put forward by W.W. Rostow in 1960. Following a study of 15 countries, mainly in Europe, he suggested that all countries had the potential to break the cycle of poverty and to develop through five linear stages (Figure 21.20).

Figure 21.20

Rostow's model of economic growth

level of development →

5 high mass consumption

4 the drive to the maturity

3 take-off

2 preconditions for take-off

1 the traditional society

time →

Approximate date of reaching a new stage of development

Country \ Stage	2	3	4	5
UK	1750	1820	1850	1940
USA	1800	1850	1920	1930
Japan	1880	1900	1930	1950
Venezuela	1920	1950	1970	—
India	1950	1980	—	—
Ethiopia	—	—	—	—

Figure 21.21

Changes in employment structure based on Rostow's model

	1 Primary	2 Secondary	3 Tertiary (services)
Stage 1	vast majority	very few	very few
Stage 2	vast majority	few	very few
Stage 3	declining	rapid growth	few
Stage 4	few	stable	growing rapidly
Stage 5	very few	declining	growing rapidly

Stage 1: Traditional society A subsistence economy based mainly on farming with very limited technology or capital to process raw materials or develop industries and services (Figure 21.21).

Stage 2: Preconditions for take-off A country often needs an injection of external help to move into this stage. Extractive industries develop. Agriculture is more commercialised and becomes mechanised. There are some technological improvements and a growth of infrastructure. The development of a transport system encourages trade. A single industry (often textiles) begins to dominate. Investment is about 5 per cent of GDP.

Stage 3: Take-off Manufacturing industries grow rapidly. Airports, roads and railways are built. Political and social adjustments are necessary to adapt to the new way of life. Growth is usually limited to one or two parts of the country (**growth poles** – page 569) and to one or two industries (**magnets**). Numbers in agriculture decline. Investment increases to 10–15 per cent of GDP, or capital is borrowed from wealthier nations.

Stage 4: The drive to maturity By now, growth should be self-sustaining. Economic growth spreads to all parts of the country and leads to an increase in the number and types of industry (the **multiplier effect**, page 569). More complex transport systems develop and manufacturing expands as technology improves. Some early industries may decline. There is rapid urbanisation.

Stage 5: The age of high mass consumption Rapid expansion of tertiary industries and welfare facilities. Employment in service industries grows but declines in manufacturing. Industry shifts to the production of durable consumer goods.

Criticisms of Rostow's model

Rostow's model, put forward in 1960, suffers the same criticisms as several other models, of being both outdated and oversimplified (Framework 12, page 352), although, as one critic concedes, 'the alternatives are just too difficult to explain and to apply'. You should be aware, however, of such valid criticisms:

- The model assumes, incorrectly, that all countries start off at the same level.

- While capital was needed to advance a country from its traditional society, often the injection of aid has been dwarfed by debt repayments which delayed, and has even prevented some countries (especially in Africa), from reaching the 'take-off' stage.

- The model underestimates the extent to which the development of some countries in the past was at the expense of others, e.g. through colonialism and imperialism.

- It predicts too short a timescale between the beginning of growth and the time when a country becomes self-sustaining. It over-emphasises the effect of the **learning curve**, i.e. the time taken for a country to develop diminishes as countries learn from others that are already developed. While the emergence of the NICs (page 578) in the late 20th century and of Russia, India and China in the early 2000s seem to support Rostow's claim, he was, like most people, to underestimate the effects of globalisation.

- The model has not seen a universal sequence and is, according to Barke and O'Hare among others, too Eurocentric.

Barke and O'Hare's model for West Africa

Barke and O'Hare (*The Third World*, 1984) claimed that although developed industrial countries may have moved through Rostow's five stages, it seems increasingly unlikely that countries that have yet to develop economically will follow the same pattern. This may be because capital alone is insufficient to promote take-off. Perhaps what is needed is a fundamental structural change in society which encourages people to save and invest and to develop an entrepreneurial, business class, as was the case in Hong Kong. Possibly the process which allows transition from traditional agriculture to advanced industry is a relict one, being applicable only to the early industrialised countries which had unlimited use of the world's resources and markets. Barke and O'Hare have suggested a four-stage model for industrial growth in developing countries, pointing out that elements from different stages often exist side by side, providing a 'dual economy'.

Stage 1: Traditional craft industries These were in existence before European colonisation, e.g. cloth weaving, iron working, wood carving and leather goods in northern Nigeria (Kano).

Stage 2: Colonialism and the processing of primary products Raw materials were initially exported in an unprocessed form (cocoa and palm oil) while the chief imports (textiles and machinery) came from the colonial power and, being cheaper, destroyed many local craft industries. Later, some processing took place, usually in ports or the primate city (page 405), if it reduced the weight for export (vegetable oils and sugar), if it was too bulky to import (cement), or if there was a large local market (textiles). To help obtain raw materials from their colonies, the European powers built ports (Accra and Lagos), but railways were only constructed if there were sufficient local resources to make them profitable. Education, along with the development of industrial and management skills, was neglected.

Stage 3: Import substitution During the Second World War and, later, following their independence, countries had to replace the import of textiles, furniture, hardware and simple machinery with their own manufactured goods. Production was in small units with limited capital and technology.

Stage 4: Manufacture of capital, goods and consumer durables As standards of living rose in several countries (notably in the NICs in Latin America and South-east Asia), there was an increased demand for heavier industry and 'Western'-style durable consumer goods. These industries, often because of the investment and skills needed, were developed by transnational companies wishing to take advantage of cheap labour, tax concessions and entry to a large local market (page 573). The American Valco company, for example, in the mid-1950s constructed a dam on Ghana's River Volta, a hydro-electric power station at Akosombo, and an aluminium smelter at Tema, in return for duty and tax exemptions on the import of bauxite and the export of aluminium, and the purchase of cheap electricity. Projects developed by transnationals are usually prestigious, of limited value to the country, and may be withdrawn (Volkswagen have stopped operating in Nigeria) should world sales drop. In other cases, where private capital was not forthcoming or where the dominance of transnational corporations was felt to be undesirable, as in China and India, large-scale industrial development was promoted through five-year national plans for economic development (Case Study 19).

Core–periphery model

Economic growth and development are rarely even. We have already seen how Myrdal (page 569) identified 'growth poles' which, he claimed, developed into core regions; how in the 19th century it was the coalfields that formed Britain's major industrial areas; and, since 1980, how

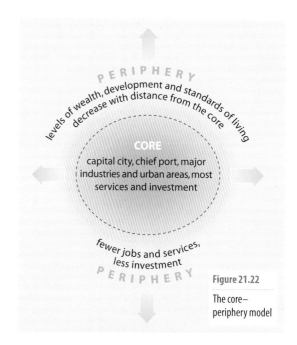

Figure 21.22

The core–periphery model

the artificially created Special Economic Zones became growth centres in China (Case Study 19). Economic activity, including the level of industrialisation and intensity of farming, decreases rapidly with distance from the core regions and towards the periphery – as shown in the **core– periphery model** (Figure 21.22).

The **core** forms the most prosperous and developed part of a country, or region. It is likely to contain the capital city (with its administration and financial functions), the chief port (if the country has a coastline) and the major urbanised and industrial areas. Usually, levels of wealth, economic activity and development decrease with distance from the core so that places towards the **periphery** become increasingly poorer.

As a country develops economically, one of two processes is likely to occur:

1 Economic activity in the core continues to grow as it attracts new industries and services (banking, insurance, government offices). As levels of capital and technology increase, the region will be able to afford schools, hospitals, shopping centres, good housing and a modern transport system. These 'pull' factors encourage rural in-migration (page 366). Meanwhile, in the periphery jobs will be relatively few, low-paid, unskilled and mainly in the primary sector, while services and government investment will be limited. These 'push' factors (page 366) force people to migrate towards the core. This process still seems to operate in the NICs and in many of the economically less developed countries (Kenya, Peru). Barke and O'Hare have suggested that 'just as it is possible to conceive of cores (MDCs) and peripheries (LDCs) on a

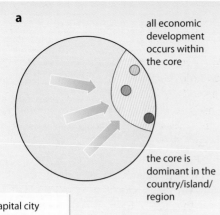

a

all economic development occurs within the core

the core is dominant in the country/island/region

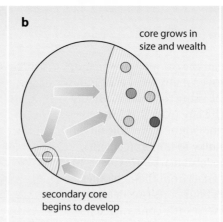

b

core grows in size and wealth

secondary core begins to develop

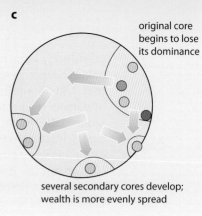

c

original core begins to lose its dominance

several secondary cores develop; wealth is more evenly spread

- ⬤ capital city
- ◯ important city
- ⬤ main port
- ⬭ primary core
- ⬭ secondary cores
- ⬭ periphery
- ➡ migration of people

Figure 21.23

The hoped-for economic growth in a country

global scale, it can be acknowledged that colonialism inspired cores (enclave economies) and peripheries (rural subsistence sector) within Third World countries themselves'.

2 Industry and wealth begin to spread out more evenly. Initially, a second core region will develop followed by several secondary regions (Figure 21.23). This can result in the decline in the dominance of the original core. Even so, there will still be peripheral areas that are less well off. This process has occurred in many of the economically more developed countries, e.g. USA and Japan (Figure 19.20) and, more recently, the emerging China (Places 98).

Places 98 China: core–periphery

Economic development has, until very recently, been severely restricted in China partly due to the country's vast size and partly due to physical barriers such as mountains and deserts. In the early 20th century most of China's limited commercial activity was concentrated around three core regions (Figure 21.24a). These were Beijing, the capital, in the north; Shanghai, the only international port and city, near to the mouth of the Yangtze River in the centre; and Canton (modern Guangzhou) and the Pearl River estuary in the south (adjacent Hong Kong was then a British colony). In the 1950s Mao Zedong attempted to industrialise China but his efforts only further impoverished an already economically poor country that had virtually isolated itself from the rest of world.

Real progress only took place after his death in 1976 when China slowly began to open its doors to outsiders. In 1980 five Special Economic Zones (Figure 19.42) were established, creating a new industrial core along parts of the south-east coast. About that time 14 'open cities', or ports, were designated at intervals along the entire coastline with the aim of encouraging overseas trade (Figure 21.24b). Even so, apart from the heavy industrial region between Shenyang and Harbin in the north-east and around Chongqing far up the Yangtze River, economic development did not spread far into the huge periphery.

Yet within the last two or three decades, China has developed to such an extent that it is expected, in the next few years, to become the world's third largest economy and its increasing wealth, albeit from a low base, is beginning to spread to even remote villages (Case Study 14B) and provinces (Figure 21.24c). Even so, most development has been, and still is, in the coastal provinces and the Yangtze Basin where 94 per cent of the population now live. The Yangtze Basin, where the Three Gorges Dam (page 545) provides electricity for new high-tech industries and the lake behind it has improved river navigation as far as Chongqing (Figure 21.24c), is the only large core region to have developed far inland.

Figure 21.24

Core and periphery in China (see Figure 21.23 for key)

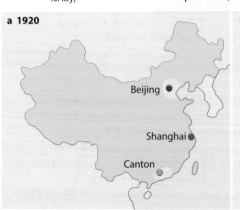

a 1920

Beijing

Shanghai

Canton

b 1980

Harbin

Shenyang

Beijing

Chongqing

Shanghai

Guangzhou

Hong Kong

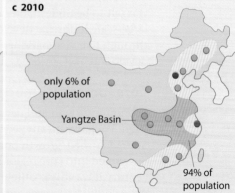

c 2010

only 6% of population

Yangtze Basin

94% of population

Health and development

Health, according to the UN Millennium Development Goals, is one of four basic human rights (Figure 21.7). This particular basic right is most likely to be denied to people living in extreme poverty, especially those in the least economically developed countries where there may be disease, hunger and a lack of safe water and adequate sanitation. Health is closely linked with economic development, and indeed several measures of development named on page 608 were birth and death rates, infant mortality, life expectancy, a balanced diet and the number of people per doctor or hospital bed. Good health, according to the World Health Organization (WHO), is 'a state of complete physical, mental and social well-being and not merely the absence of disease and infirmity' – a statement that implies complex interactions between humans and their various environments (Figure 21.25).

Bearing in mind Phillips' warning in Figure 21.25 concerning difficulties in trying to correlate health with economic development, there do appear to be marked differences in the types of illness (Figure 21.27) and in health care (Figure 21.26) between the more and the less economically developed countries. It has been suggested that, as a country develops, it is likely to pass through several stages of **epidemiological** or **health transition**.

Figure 21.25

The complex inter-relationship between health and development

'It has long been acknowledged that the health status of the population of any place or country influences development. It can be a limiting factor, as generally poor individual health can lower work capacity and productivity; in aggregate in a population, this can severely restrict the growth of economies. On the other hand, economic development can make it possible to finance good environmental health, sanitation and public health campaigns – education, immunisation, screening and health promotion – and to provide broader-based social care for needy groups. General social development, particularly education and literacy, has almost invariably been associated with improved health status via improved nutrition, hygiene and reproductive health. Socio-economic development, particularly if equitably spread through the population – although this is rarely the case – also enables housing and related services to improve. The classical cycle of poverty can be broken by development.

However, it is notoriously difficult to provide generalisations about the relationship between economic development and a population's health status. We can cite examples in which correlations between GNP and life expectancy are not straightforward. There are many examples to show how economic development has contributed to improving quality of life and health status, via indicators such as increased life expectancy, falling infant, child and maternal mortality and enhanced access to services. By contrast, there are examples in which economic development, infrastructure expansion and agricultural intensification do not always coincide with improved human well-being. There is, in fact, a growing realisation that macroeconomic changes may not always filter down to benefit all of the population, and many perhaps soundly based policies in economic terms can have devastating human effects in increasing poverty and maldistribution of resources.'

David Phillips and Yola Verhasselt

Figure 21.26

Differences in health care
a Cataract camp, Kolkata
b Intensive care unit, St Bartholomew's, London

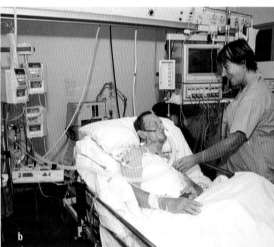

Figure 21.27

Differences in types of disease between less and more developed countries

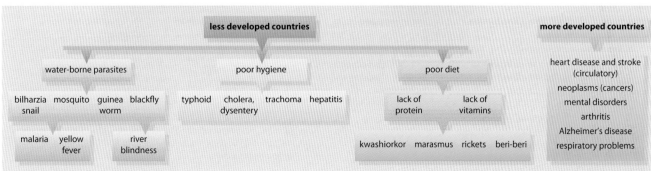

The epidemiological (health) transition

The demographic transition model (Figure 13.10) suggests that fertility (birth rate) declines appreciably, probably irreversibly, when traditional, mainly agrarian societies are transformed by modernisation, industrialisation and bureaucratic urban-oriented societies. This rather straightforward and simplistic demographic transition assumes that, for example, a simple industrial-economic modernisation will occur in societies accompanied by changes in lifestyles, living conditions and health levels. Of greater interest to epidemiologists, health planners and medical geographers is that with 'modernisation' and increasing affluence and life expectancy comes a very different disease or ailment profile in most countries from that which previously existed in a 'traditional' state or developing country (Figure 21.28). Figure 21.29 has been adapted from Omran's epidemiological transition. Initially, three stages of the transition were envisaged:

1   the age of pestilence and famine which gradually merges into ...
2   the age of receding pandemics (worldwide diseases), giving way to ...
3   the age of degenerative and human-induced diseases.

More recent studies have suggested the emergence of ...

4   the age of delayed degenerative diseases and, associated with a lengthening of life, poorer health.

Omran suggested that there were three variations in the basic model:

1   The classical or 'Western' model, which took place over a prolonged period (100 to 200 years).
2   The 'accelerated' model, which occurred in Japan after the Second World War, and more recently in Hong Kong (Places 99), Singapore and other NICs in South-east Asia. This showed rapid declines in mortality and fertility.
3   The 'delayed' model, which is common to many of today's less developed countries. It contains elements of morbidity and mortality from both degenerative and infectious diseases but, at the same time, lacks the marked reduction in fertility experienced in the 'Western' model.

Figure 21.28

A view on health transition

It has long been recognised that societies pass through various patterns of morbidity (illness and disease) and mortality (causes of death) during the development process, even if not all the stages and sequences are identical in every case. In general, health improves, morbidity and mortality fall and come from different causes, and life expectancy increases; this comprises the **'epidemiological transition'** [after Omran, 1971]. More recently the term **'health transition'** is being used, as it has a broader concept than epidemiological, i.e. it focuses on health rather than just on morbidity. These changes generally come with 'modernisation' and are indeed part and parcel of the process. They seem to occur at a different pace in varying countries and, in recent years, they are related to the application of modern medical techniques and technology as well as to changing standards of living, nutrition, housing and sanitation.

David Phillips,
The Epidemiological Transition in Hong Kong, 1988

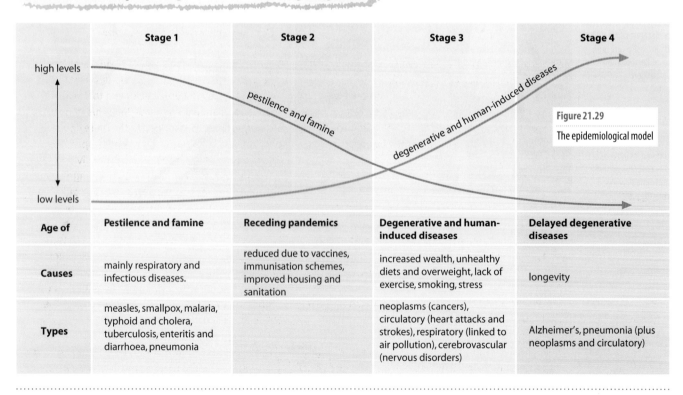

Figure 21.29

The epidemiological model

	Stage 1	Stage 2	Stage 3	Stage 4
Age of	Pestilence and famine	Receding pandemics	Degenerative and human-induced diseases	Delayed degenerative diseases
Causes	mainly respiratory and infectious diseases.	reduced due to vaccines, immunisation schemes, improved housing and sanitation	increased wealth, unhealthy diets and overweight, lack of exercise, smoking, stress	longevity
Types	measles, smallpox, malaria, typhoid and cholera, tuberculosis, enteritis and diarrhoea, pneumonia		neoplasms (cancers), circulatory (heart attacks and strokes), respiratory (linked to air pollution), cerebrovascular (nervous disorders)	Alzheimer's, pneumonia (plus neoplasms and circulatory)

Figure 21.30 shows the epidemiological changes for Hong Kong between 1951 and 2001. The graph illustrates three trends that closely match Oman's accelerated model:

1 A rapid decline in infective/parasitic diseases (due to improved standards of living, better housing conditions and improved medical care including immunisation) and digestive complaints (the result of improved health care and a better diet).

2 An initial drop in respiratory illnesses and pneumonia which has since been reversed (partly as a result of increased traffic emissions).

3 A rapid increase in deaths from 'Western' diseases, especially malignant neoplasms (cancers) and heart disease (due to overweight and an increase in stress).

Singapore shares these three characteristics with Hong Kong, probably because it too has a fairly homogenous ethnic mix living mainly in urban areas. Similar patterns showing changes in the cause of death can also be seen in other existing and emerging NICs in South-east Asia such as Malaysia, South Korea, Taiwan and, presumably in time, China. Where dissimilarities do appear, they may be credited to differences in wealth, social status, ethnic mix, religion and level of urbanisation, both within and between countries.

Figure 21.30

Epidemiological change in Hong Kong, 1951–2001

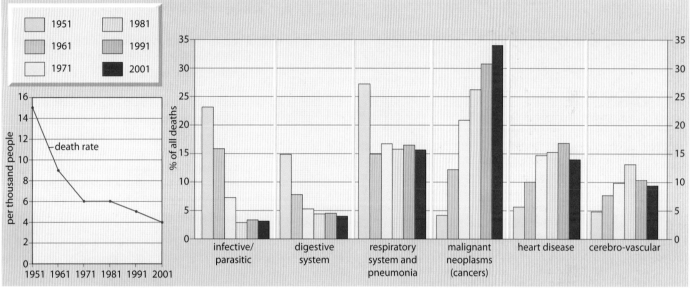

The value of the epidemiological (health) transition

Perhaps the most important role the epidemiological transition can play is to provide a formal framework within which to set health and health-care strategies over the medium to long term. As such, it could provide a major stimulus to future health-care needs both within countries when directed by governments, or globally through international health agencies. It could help health planners in places where change is very rapid (NICs), is varied between social groups (rich and poor communities in developing countries) and ethnic groups (South Africa), and where health care is expensive and finance is limited (the UK). It could also point out the growing needs for care from causes like mental illness, especially in developing countries, and Alzheimer's disease, in more developed countries, which are both considerably underestimated in much health sector planning.

The epidemiological transition is relevant for manufacturers and suppliers of medicines and health equipment, for researchers looking for new vaccines, and for governments trying to decide where best to allocate funds.

Finally, by identifying a fourth stage, that of the age of delayed degenerative diseases, the epidemiological transition draws attention to the world's ageing population (pages 359–360). This stage, although at present confined to the more developed and wealthy 'Western' countries (Japan, the UK), suggests a lengthy old-age potentially dogged with chronic, but non-fatal, ailments. Old age, faced by an ever-increasing proportion of the population and whose health and social needs are often greater than those in younger age groups, may not be attractive unless public and family support are forthcoming. Although developing countries are further from this stage, nevertheless many are experiencing a rapid increase in longevity, resulting in more people needing care as they live longer. Due to the increasing numbers of the elderly in many developing countries (China , Case Study 13; India), and due to the absolute totals, it is necessary to start planning now for their future health and social care.

HIV/AIDS

AIDS (acquired immune deficiency syndrome), first described in medical literature in 1981, had become **pandemic** (an epidemic that spreads over a wide geographical area) by the 1990s and remains one of the greatest threats to global public health. The three main means of transmitting **HIV (human immunodeficiency virus)** are by the exchange of body fluids during sexual intercourse (with greater efficiency from male to female than vice versa), through infected blood (sharing needles/syringes and by contaminated blood transfusions) and parentally from mother to child during pregnancy or birth. The dominant forms of transmission, and the way the virus spreads, vary worldwide (Figure 21.31).

UNAIDS and WHO announced in 2007 that 'HIV/AIDS continues to be a major development, global health and security challenge, especially in southern Africa. It reverses life expectancy gains, erodes productivity, decimates the workforce, consumes savings, and dilutes poverty efforts threatening the realisation of the Millennium Development Goals' (page 609). They also pointed out that this report reflected improved epidemiological data collection and analysis which resulted in substantial revisions of all previous estimates. This latest data suggested that the number of new HIV infections had begun to level off and the number of deaths attributable to AIDS had begun to fall. The pandemic, still dominant in sub-Saharan Africa, had also become increasingly infectious in Indonesia and Vietnam, followed by Eastern Europe and Central Asia (Figure 21.31).

The 2007 report claimed that:

- although all countries across the world were affected, HIV/AIDS was most prevalent in countries in sub-Saharan Africa (Figure 21.31 and Places 100) where 22.5 million people were affected, followed by South and South-east Asia with 4 million
- the percentage of people living with HIV worldwide, many of whom had been born with it, had declined from a peak of 38 million in 2003 to 33.2 million (although this in part may have been due to the improved method of data collection mentioned above)
- there was a decrease in the number of reported new infections, down from just over 3 million a year in the late 1990s and 5 million in 2003 to an estimated 2.5 million in 2007. Even so, worldwide that was an average of 6800 new infections per day
- in 2007, 2.1 million people died of AIDS – an average of 5700 each day
- life expectancy, especially in the worst-affected countries of sub-Saharan Africa, was continuing to fall although there were encouraging signs that, since 2005, antiretroviral therapy was beginning to prolong life even if, as yet, there was no known cure.

Figure 21.31

Estimated global distribution of HIV infections, 2007

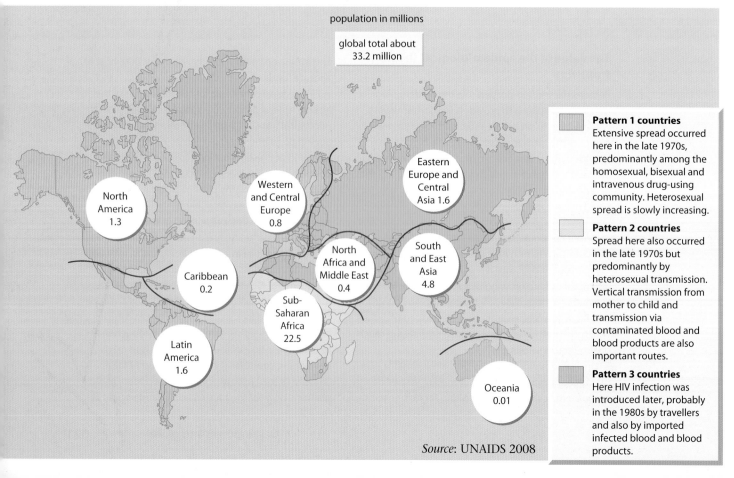

population in millions

global total about 33.2 million

North America 1.3

Western and Central Europe 0.8

Eastern Europe and Central Asia 1.6

Caribbean 0.2

North Africa and Middle East 0.4

South and East Asia 4.8

Sub-Saharan Africa 22.5

Latin America 1.6

Oceania 0.01

Pattern 1 countries
Extensive spread occurred here in the late 1970s, predominantly among the homosexual, bisexual and intravenous drug-using community. Heterosexual spread is slowly increasing.

Pattern 2 countries
Spread here also occurred in the late 1970s but predominantly by heterosexual transmission. Vertical transmission from mother to child and transmission via contaminated blood and blood products are also important routes.

Pattern 3 countries
Here HIV infection was introduced later, probably in the 1980s by travellers and also by imported infected blood and blood products.

Source: UNAIDS 2008

Sub-Saharan Africa remains the global epicentre of the epidemic. In 2007 there were an estimated 22.5 million infected people living in this region who had HIV, i.e. 68 per cent of those affected globally and 35 per cent of this region's total population. The region also contained 43 per cent of all children aged under 15, and 52 per cent of all women above the age of 15, who were affected across the world by the virus. Eight countries in southern Africa (Botswana, Lesotho, Malawi, Mozambique, South Africa, Swaziland, Zambia and Zimbabwe) accounted for almost one-third of all the new HIV infections and AIDS deaths across the world (Figure 21.32). Although the 1.7 million new infections in sub-Saharan Africa in 2007 was a significant reduction on previous years, it was still nearly 70 per cent of the world's total, while the 1.6 million deaths due to AIDS in this region was 76 per cent of the world's total.

One of the worst effects of HIV/AIDS has been a reduction in life expectancy. By 2005, in southern Africa it had, on average, fallen by 10 years since the pandemic was first recorded. In 2007 it still appeared to be falling, with the average age for the 10 countries in the world with the lowest life expectancy – all in

this region – now being only 42.5 years (Figure 21.32). Latest predictions for these countries is that by 2015 it is likely to be under 42 years – more than 20 years less than the 63 years it might have been had HIV/AIDS never occurred. In the worst-affected countries, such as Botswana, the pandemic is creating a 'chimney-shaped' population structure (Figure 21.33), which leaves fewer people in the economically active age group (page 354). It has also left an estimated 11.4 million children in the region as orphans – just over 1 in every 4 children. More recently, and resulting from the reduced effectiveness of people's immune system, the risk of tuberculosis (TB) has increased by 50 per cent and deaths from TB by 25 per cent. Of the 14 million people globally co-infected with TB and HIV, 10 million live in sub-Saharan Africa where treatment is both harder to get and less effective. As more people are weakened by HIV, there are fewer doctors and nurses to treat patients, fewer teachers to educate children about the causes and effects of the illness, and fewer healthy farmers to produce sufficient food (page 503).

Figure 21.33

Projected population structure for Botswana in 2020

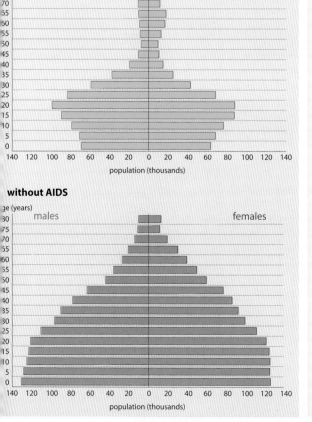

with AIDS

ge (years)

males   females

population (thousands)

without AIDS

ge (years)

males   females

population (thousands)

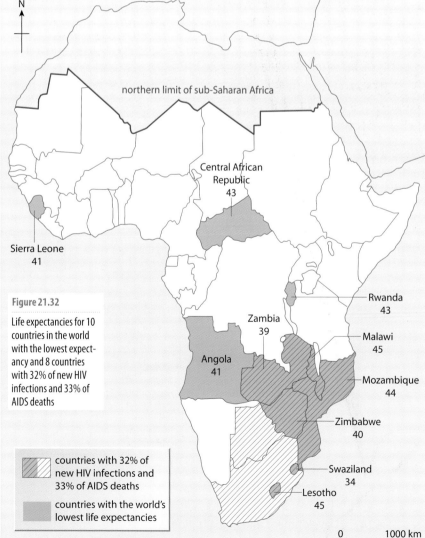

N

northern limit of sub-Saharan Africa

Central African Republic
43

Sierra Leone
41

Rwanda
43

Zambia
39

Malawi
45

Angola
41

Mozambique
44

Zimbabwe
40

Swaziland
34

Lesotho
45

Figure 21.32

Life expectancies for 10 countries in the world with the lowest expectancy and 8 countries with 32% of new HIV infections and 33% of AIDS deaths

countries with 32% of new HIV infections and 33% of AIDS deaths

countries with the world's lowest life expectancies

0   1000 km

International trade

Development of world trade

Trading results from the uneven distribution of raw materials over the Earth's surface. It plays a major role in the economy of all countries as none has an adequate supply of the full range of minerals, fuels and foods; of manufactured goods; or of services to make it self-sufficient. Countries that trade with other countries are said to be **interdependent**. During colonial times, several European countries began to use raw materials found in their colonies to develop their own domestic manufacturing industries. This saw the beginning of modern international trade between those countries that provided many of the relatively cheap raw materials and those that made a much greater profit by manufacturing or processing those raw materials. Later, in the 20th century, the more economically developed countries came to specialise in particular aspects of manufacturing, as this created greater benefits than in trying to compete with other countries that had equal, or better, opportunities. Even more recently, international trade has come to be dominated not just by a few wealthy countries but by an increasing number of large **transnational corporations** (page 573 and Places 101, page 630).

Balance of trade and balance of payments

The raw materials, goods and services bought by a country are called imports and those sold by a country are exports. The **balance of trade** for a country is the difference between the income it receives from its visible exports and the cost it incurs in paying for its visible imports. The **balance of payments** includes the balance of trade together with any invisible earnings or costs such as from banking and insurance, tourism, remittances from migrant workers abroad, professional advice and air/sea transport. Countries that earn more from their exports than they pay for their imports are said to have a **trade surplus** enabling them to become richer. Those countries that spend more on imports than they earn from their exports have a **trade deficit** and so become increasingly less well-off. It is this difference between the trade of countries that has largely been responsible for the creation, and widening, of the **development gap** (page 605).

Figure 21.34

Major global trading blocs, including associate members

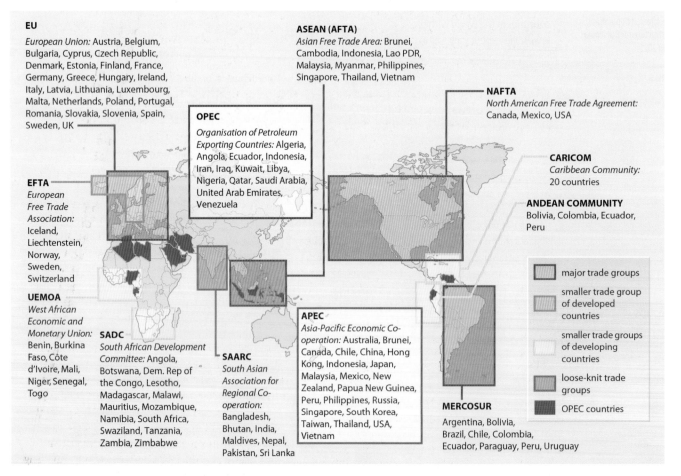

EU

European Union: Austria, Belgium, Bulgaria, Cyprus, Czech Republic, Denmark, Estonia, Finland, France, Germany, Greece, Hungary, Ireland, Italy, Latvia, Lithuania, Luxembourg, Malta, Netherlands, Poland, Portugal, Romania, Slovakia, Slovenia, Spain, Sweden, UK

EFTA

European Free Trade Association: Iceland, Liechtenstein, Norway, Sweden, Switzerland

UEMOA

West African Economic and Monetary Union: Benin, Burkina Faso, Côte d'Ivoire, Mali, Niger, Senegal, Togo

SADC

South African Development Committee: Angola, Botswana, Dem. Rep of the Congo, Lesotho, Madagascar, Malawi, Mauritius, Mozambique, Namibia, South Africa, Swaziland, Tanzania, Zambia, Zimbabwe

OPEC

Organisation of Petroleum Exporting Countries: Algeria, Angola, Ecuador, Indonesia, Iran, Iraq, Kuwait, Libya, Nigeria, Qatar, Saudi Arabia, United Arab Emirates, Venezuela

ASEAN (AFTA)

Asian Free Trade Area: Brunei, Cambodia, Indonesia, Lao PDR, Malaysia, Myanmar, Philippines, Singapore, Thailand, Vietnam

SAARC

South Asian Association for Regional Co-operation: Bangladesh, Bhutan, India, Maldives, Nepal, Pakistan, Sri Lanka

APEC

Asia-Pacific Economic Co-operation: Australia, Brunei, Canada, Chile, China, Hong Kong, Indonesia, Japan, Malaysia, Mexico, New Zealand, Papua New Guinea, Peru, Philippines, Russia, Singapore, South Korea, Taiwan, Thailand, USA, Vietnam

NAFTA

North American Free Trade Agreement: Canada, Mexico, USA

CARICOM

Caribbean Community: 20 countries

ANDEAN COMMUNITY

Bolivia, Colombia, Ecuador, Peru

MERCOSUR

Argentina, Bolivia, Brazil, Chile, Colombia, Ecuador, Paraguay, Peru, Uruguay

major trade groups

smaller trade group of developed countries

smaller trade groups of developing countries

loose-knit trade groups

OPEC countries

Trading blocs

During the latter part of the 20th century an increasing number of countries grouped together for the purpose of trying to increase the volume and value of their trade. Two of the earliest and largest **trading blocs** were the EU and NAFTA (Figure 21.34), each of which now has an internal market of around 500 million people. By creating trading blocs, countries can eliminate custom duties (tariffs) between member states which, in turn, will reduce the price of products sold between them. Although this made the EU, for example, more competitive against non-member countries or rival trading blocs such as Japan or NAFTA, it also created restrictions (trade barriers) between goods made in the EU and those of developing countries. This has meant that the LEDCs have found it increasingly difficult to sell their products to MEDCs, increasing further the trade and development gap.

Figure 21.35

Selected inter-regional and intra-regional trade flows, 2006

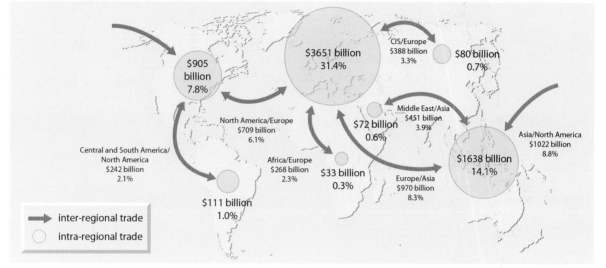

Selected inter-regional and intra-regional trade flows, 2006

- North America/Europe $709 billion 6.1%
- Central and South America/North America $242 billion 2.1%
- Africa/Europe $268 billion 2.3%
- CIS/Europe $388 billion 3.3%
- Middle East/Asia $451 billion 3.9%
- Europe/Asia $970 billion 8.3%
- Asia/North America $1022 billion 8.8%
- $905 billion 7.8%
- $3651 billion 31.4%
- $80 billion 0.7%
- $72 billion 0.6%
- $33 billion 0.3%
- $111 billion 1.0%
- $1638 billion 14.1%

→ inter-regional trade
○ intra-regional trade

The direction of world trade

Figure 21.35 shows the pattern of world trade, by value and including finance, that has taken place over the last few decades.

- Most of a nation's international trade is with one or more neighbouring countries, e.g. Canada with the USA, South Korea with Japan, the UK with countries in Western Europe.
- Most of the world's trade is between the advanced market economies of NAFTA, the EU and Japan, although their share fell from 72 per cent in 1990 to 68 per cent in 1998 and 58 per cent in 2007.
- The advanced market economies have had relatively little trade with the developing countries. Where they have – as was seen when accounting for the development gap (page 605) – they have generally exported high-value goods and imported low-value goods in return.
- There has been relatively little trade between the developing countries themselves. This is partly because many of them have had low rates of economic growth and partly because they have tended to produce similar, and limited, types of goods, i.e. the same one or two materials.
- Since the 1970s the advanced economies have faced increasing competition from the so-called newly industrialised countries (NICs) in Asia (Hong Kong, Singapore, South Korea and Taiwan, page 578) and in Latin America (Brazil and Mexico). Even more recently there has been, in terms of scale and speed, an unprecedented emergence of a new trading nation – China (Case Study 21).
- Today world trade is dominated not by countries but by large and powerful transnational corporations (TNCs, pages 573 and 630).

Trade links

Figure 21.36 gives an indication of the importance of trade for 12 selected countries that belong to different levels of economic activity. It also shows the three main groupings of agricultural products, fuels and minerals, and manufactured products, into which most items of world trade are manageably placed together with, as a measure of their development, the trade per capita. The advanced economies, the NICs and TNCs, and now the emerging markets, have manufactured goods accounting for a high proportion of their total exports. This has enabled them to accumulate the capital and technology needed to buy and process requisite raw materials such as fuels and minerals. In contrast, although most developing countries have some manufacturing, it is usually often only primary processing or is operated by TNCs taking advantage of their cheap labour (page 573).

	Advanced economies			NICs	
	USA	UK	Japan	Singapore	Malaysia
World rank – exports	2	7	4	14	19
World rank – imports	1	4	5	15	23
Exports a Type (agricultural / fuels and minerals / manufactured / others)	4, 9, 7, 80	5, 5, 13, 77	1, 3, 6, 90	2, 5, 14, 79	2, 10, 15, 73
b Value US$	1 038 278	448 291	649 931	271 772	160 676
c % world's exports	8.59	3.71	5.38	2.25	1.33
Imports a Type (agricultural / fuels and minerals / manufactured / others)	4, 5, 21, 70	13, 9, 12, 66	3, 11, 35, 51	3, 3, 21, 73	3, 7, 13, 77
b Value US$	1 919 420	619 385	579 574	238 652	131 152
c % world's imports	15.46	4.99	4.67	1.92	1.06
Trade per capita US$	10 864	21 389	10 112	124 769	11 603

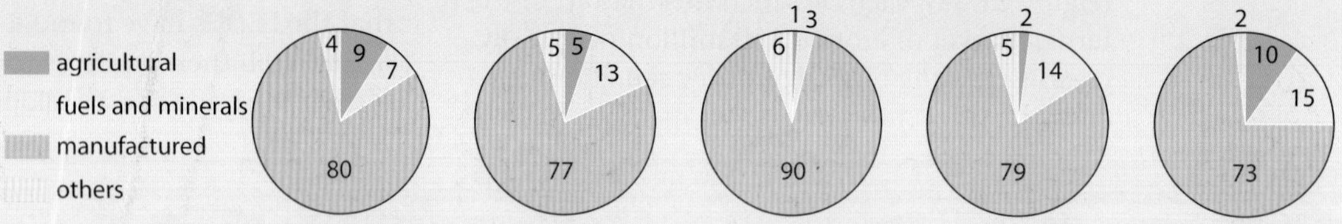

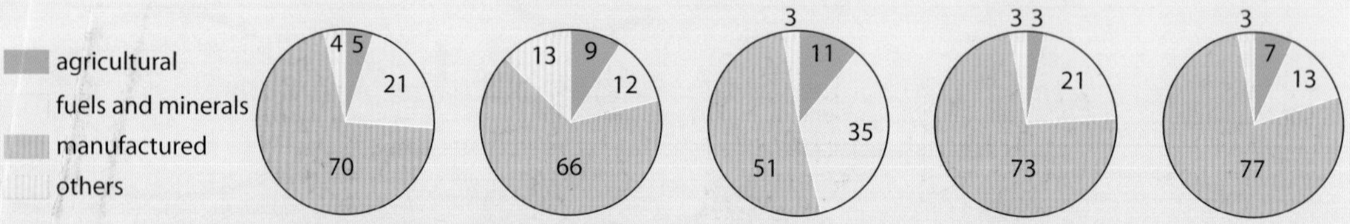

Figure 21.36

Selected exports, imports and trade per capita of selected countries

The world market in fuels, usually oil and natural gas, is dominated by the OPEC countries and, recently, Russia. Most is exported to fuel-short advanced economies in the EU and Japan, although the rapid increase in demand since about 2005 has come from China. The price of these fuels tends to be beyond the reach of developing countries, retarding their economic development even more. The pattern of mineral exports is less obvious, with both developed (Australia and Canada) and developing (Jamaica and Zambia) countries being major exporters. Again, however, it is the advanced economies, NICs and, most recently, China, that are the chief importers.

Agricultural products often account for over half of a developing country's exports, although an increasing number of African countries are now having to import cereals as their food production decreases (pages 503 and 629). While many of the more industrialised countries rely on imports of foodstuffs, some that have extensive (USA, Canada and Australia, page 486) or intensive (Netherlands, Denmark, page 487) farming systems, are net exporters.

For many years developing countries have made demands for a fairer trading system. One request is for higher or fixed prices for their primary products so as to limit the widening of the development gap; a second request is for better access to markets within the more well-off countries. There is still the tendency for some MEDCs to try to impose quotas, to add tariffs, to try to limit the quantity, or to raise the price, of goods imported from the LEDCs. Other demands have included changes in the international monetary system so as to eliminate fluctuations in currency exchange rates; encouraging MEDCs to share their technology; dissuading MEDCs from 'dumping' their unwanted, and sometimes untested, products cheaply; lowering interest rates; and an increase in aid free of economic and political strings (page 632).

The WTO report of 2008 confirmed that the growth of world trade had declined from 8.5 per cent in 2006 and 5.5 per cent in 2007 to a forecast of 4.5 per cent for 2008. This decline began with a slowdown in the North American economy which later spread to the EU and Japan, giving them average forecast growth of only 1.1 per cent in 2008. Figure 21.37 shows that, partly due to an increase in the price of raw materials, especially metals and fuels, and having to rely less on the advanced economies for trade, the emerging markets and developing countries had not, so far, been affected as much by this decline; this gave them a predicted growth of 5 per cent in 2008.

NICs		OPEC		Emerging markets		Developing economies	
Brazil	UAE	Nigeria	China	India	Kenya	Sierra Leone
24	23	43	3	28	107	170
28	27	60	3	17	87	180
2, 29, 50, 19	1, 2, 34, 63	1, 2, 5, 92	1, 3, 4, 92	1, 12, 19, 68	1, 18, 46, 35	13, 7, 80
137 470	139 353	52 000	968 936	120 254	3437	216
1.14	1.15	0.43	8.00	1.00	0.03	0.00
2, 6, 22, 70	4, 9, 6, 81	1, 14, 5, 80	1, 6, 20, 73	11, 4, 37, 48	5, 11, 29, 55	1, 29, 30, 40
95 886	97 754	21 809	791 461	174 845	7311	389
0.77	0.79	0.18	6.37	1.41	0.06	0.00
1234	39 288	447	1207	307	275	127

World Trade Organization (WTO)

A basic aim of the WTO is to bring together countries that belong to various customs unions, allowing them the opportunity to take decisions on multilateral trade agreements. It was established in 1995, replacing GATT (the General Agreement on Trade and Tariffs) which had been set up in 1948 to try to reduce tariffs (import duties) and to provide a forum for discussing problems of international trade. Although over 150 countries are members of the WTO, effectively most decisions are made by only eight – the so-called G8 of Canada, France, Germany, Italy, Japan, Russia, the UK and the USA – which, with the exception of Russia, also form the inner circle of the **Organisation for Economic Co-operation and Development (OECD)**. In contrast, the many developing countries, with their limited wealth, products and technology, have least say and find it difficult to obtain a fair share of the world's trade.

The first of many summit trade talks took place in 1986 when 65 developing countries and NICs met to discuss tariffs, subsidies and trade reform. Subsequent meetings, known as the Uruguay Round, followed. By 1995, some tariffs had been removed but generally only on industrial products that benefited the NICs. In contrast, mainly due to strong farming lobbies in the USA and the EU, there was little reform on agricultural products, much to the detriment of the developing countries.

Figure 21.37

Recent changes in world trade, 1999–2008

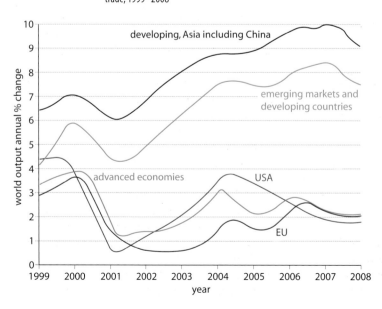

The Doha round of talks, named after the capital of Qatar where the first summit took place in 2001, initially had 101 developing countries attending. In 2002 the World Bank estimated that freeing international trade boundaries and subsidies could lift 320 million people above the $2 a day poverty line by 2015. However, after only little progress was made at Cancun (Mexico) in 2003, at the talks in Hong Kong in 2005 the MEDCs agreed to grant duty-free and quota-free market access for at least 97 per cent of tariff lines on products originating in the least developed countries. This decision, which addressed Millennium Development Goal 8: Aid, Trade, Growth and Global Partnership (page 609), still had the potential to lift millions of people out of poverty, but at the reduced figure of 75 million, not the previously hoped-for 95 million (and this assumed *all* tariffs, quotas and other obstacles to free trade would be removed – an assumption that in 2008 was seen to have been a fanciful scenario).

With agriculture dominating the poorest economies in Africa, Latin America and parts of Asia, much of the negotiations between 2001 and 2008 centred on proposals for lowering barriers to trade in farm products, and curtailing subsidies that richer nations pay their farmers to grow cotton, corn and other crops. Such subsidies can lead to gluts that depress world prices and put farmers in developing countries at a disadvantage. But not all developing countries have the same interests. While sweeping reforms of global farm policies could benefit places like Argentina and Brazil, they would make life even more difficult for the poorest countries that have to import food, especially when, in 2008, the price of cereals shot up.

The 2008 talks, attended by 153 nations, were held in Geneva but soon ran into difficulties (Figure 21.38). The talks were extended, allowing further discussions between the top trading nations of the EU, the USA, China, Japan, India, Brazil and Australia – leaving, as usual, the poorer nations to watch and wait. After nine days the

Doha Trade Talks Collapse

July 2008

THE Doha round of world trade talks has collapsed in what one former trade chief called the biggest blow to globalisation since the end of the Cold War.

Negotiators warned that there was now little or no chance of salvaging the talks, which promised to bring down trade tariffs, pull millions out of poverty and keep food and goods prices under control. It is the first time a major set of world trade talks has collapsed entirely, and insiders warned that the consequences would be weaker economic growth and a less globalised world.

Officials warned that there was now 'little or no appetite' to return to the round. Insiders said the talks had stumbled after the USA, China and India failed to compromise on the size of their agricultural tariffs. At the centre of the dispute were so-called 'safeguard clauses' which allowed developing nations to slap emergency tariffs on imports if they leaped to unmanageable levels. US negotiators apparently balked at Indian and Chinese proposals to trigger these safeguards on their cotton exports.

A WTO spokesperson said: 'We have missed the chance to seal the first global pact of a reshaped world order. We would all have been winners. Years of negotiation which were and are important for globalisation have been sacrificed by this failure.'

Figure 21.39

Collapse of the Doha trade talks

talks collapsed (Figure 21.39), with neutrals blaming the USA, China and India. It will be interesting to see what the situation will be in, say, 2010 or even 2015.

Food shortages: a global issue

In mid-2008, the UN called for action to tackle hunger and malnutrition in a world of rising food prices, claiming that 'they have become the forgotten Millennium Development Goal [page 609]. This goal has received less attention, but increased food prices and their threat, not only to people but to political stability, have made it a matter of urgency to give it the attention it needs [Figure 21.41].'

While headline news about high food prices is a relatively new phenomenon, they have been rising since 2001 after half a century of being depressed (Figure 21.40a). Imagine a low-income family in a developing country earning less than $1 a day who might have paid 20 cents for a kilogramme of wheat one year and had to pay 30 cents the next. For people in poverty spending over half their income on food in order to survive, price rises of staples can be devastating.

Figure 21.38

The hopes and problems at Geneva, 2008

Hopes

- Farm tariffs could be reduced to 30 per cent.
- A reduction in money for subsidies on farm products by 60 per cent or even 70 per cent.
- A resultant benefit in trade could increase income for developed and developing countries.
- Reduced prices for consumers in the advanced economies and fairer prices for farmers in emerging economies.
- Millions of people could be pulled out of poverty.

Problems

- The USA, EU and Japan insisted that the larger trading nations of the emerging economies – Brazil, China and India – open their markets to Western manufactured goods.
- The emerging nations insisted on large cuts in farm subsidies and tariffs paid to farmers in the USA and the EU.

The root causes of these unprecedented rises have been the large increases in energy (especially oil which is needed for machinery and transport) and fertiliser costs, the demand for food crops in biofuel production, and a record low level in cereal stocks. The price of oil appears likely to remain high and the demand for biofuels to increase further. In 2007, one-quarter of the US maize crop (11 per cent of the global total) went into biofuel production when, previously, the USA had supplied over 60 per cent of the world's exports. Other factors include: a higher demand for grain to feed livestock in China, where increasing affluence means more people are eating meat (50 kg per capita in 2007 compared with 20 kg in 1990); a four-year drought in Australia which, instead of being a major exporter (page 485), has had to import wheat itself; water shortages in general when, as seen on page 610, over 70 per cent of water supply goes to agriculture; and a global reduction in the area under cereals from a peak in 1980 (Figure 21.40b). This includes the Commonwealth of Independent States (CIS) where, according to a Moscow bank, only 43 per cent of arable land in the world's largest cereal grower is still under cultivation, and the EU with, until 2008, its set-a-side land policy.

According to the FAO, in 2008 there were 36 countries in crisis as a result of higher food prices, and in need of external assistance (aid). Of these, 21 were in Africa, 10 in Asia and 5 in Latin America. In many of these places, food shortages have been worsened by internal conflicts and extreme weather – both floods and drought (Places 75, page 503).

Responding to this crisis, the UN Secretary-General listed, at a G8 meeting prior to the Doha round of trade talks in 2008 (page 628), the following needs which he said could only be met with global co-operation:

- Ensure vulnerable populations are given urgent help by scaling up food assistance, giving financial support for food aid and exempting relief food from export restrictions and taxes.
- Boost agricultural production by giving seeds and fertiliser to up to 450 million of the world's small-scale farmers and for the G8 leaders to give more development assistance to agriculture.
- Improve fair trade by reducing agricultural subsides in G8 countries (page 631).
- Increase significantly investment in farming, agricultural research and rural development.
- Strengthen global food commodity markets and provide an aid package on trade for LEDCs.
- The G8 countries and their partners to reassess subsidies and tariffs on biofuel production.

Figure 21.40

World cereal prices and production
Sources: World Bank, FAO

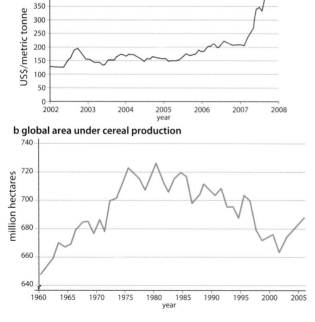

a wheat prices (2002–08)

b global area under cereal production

Figure 21.41

Predicted impact of food price rises on trade balances
Source: World Bank

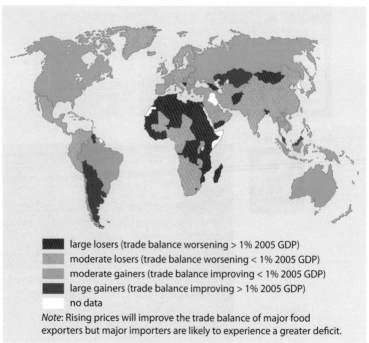

- large losers (trade balance worsening > 1% 2005 GDP)
- moderate losers (trade balance worsening < 1% 2005 GDP)
- moderate gainers (trade balance improving < 1% 2005 GDP)
- large gainers (trade balance improving > 1% 2005 GDP)
- no data

Note: Rising prices will improve the trade balance of major food exporters but major importers are likely to experience a greater deficit.

TNCs and world trade

It is argued that globalisation is similar to the colonial period except that it is large transnational corporations, not countries, that are increasing their wealth and dominating world trade. Certainly in the last century, TNCs – usually with their headquarters in the advanced economies or in the NICs – located most of their factories in developing countries as these could provide both raw materials and the cheap labour needed to produce goods that were to be sold in developed countries. Yet, given the chance, many developing countries welcomed the presence of TNCs, seeing them as an opportunity to obtain investment and to create employment.

Places 101 South Korea: Samsung – a TNC

Figure 21.42
Samsung welcomes visitors

South Korea's tenth president, elected in 2007, had always been involved with giant corporations, in his case Hyundai. Hyundai is one of many similar family-run businesses that have become TNCs, and which are collectively known as *chaebols*. The growth of these chaebols, unique to South Korea, in the 1970s–1980s made them leading world TNCs in shipbuilding, steel, cars, construction, computers and electronics, and made South Korea one of Asia's four 'tiger economies' (page 578). The largest TNC is Samsung (Figure 21.43).

Figure 21.43

The Samsung factory at Suwon, south of Seoul

The organisation was set up as a family trading company in 1938 and was to benefit after the Korean War by supplying UN forces. In 1969 it opened a factory in conjunction with the Japanese firm Sanyo, to make black-and-white televisions and with a workforce of 36 employees. Today, the site of that factory covers an area the size of over 200 football pitches (Figure 21.43) and employs 22 000 workers, nearly all in Research and Development (one in eight has either an MA or a PhD). The corporation now has 124 offices in 56 countries, 16 overseas production factories of which 13 are in China and the others elsewhere in South-east Asia, and a global workforce of 154 000. Samsung is composed of numerous businesses, the three largest being Samsung Electronics, the world's biggest electronics company, Samsung Heavy Industries, one of the world's biggest shipbuilders, and Samsung Construction and Engineering. The three businesses reflect the meaning of the Korean word *samsung*, meaning 'three stars'. With over 20 per cent of the nation's exports, Samsung has a powerful influence on the country's economic development, politics, media and culture and has become a role-model for national pride.

It is the world's leader in LCD and flat-screen TVs, is second (to Nokia) in the production of mobile phones, and is a major producer of laptops, cameras and printers as well as air conditioners, fridges, washing machines, microwaves and vacuum cleaners. It also sponsors an English Premier League football team – another example of globalisation.

FAIRTRADE ®

Guarantees a better deal for Third World Producers
The Fairtrade Mark guarantees:
- farmers get a fair and stable price for their products
- farmers and workers get the opportunity to improve their lives
- greater respect for the environment
- a stronger position for farmers in world markets
- closer links between shoppers and producers
- investment in local community projects.

Figure 21.44

The Fairtrade Mark

Fairtrade

For many years developing countries have made demands for a fairer trading system (page 626). **Fairtrade** in the UK was established in the early 1990s as a strategy for poverty alleviation and sustainable development aimed at small-scale, disadvantaged farmers in some of the world's poorest countries. Fairtrade guarantees a fair price to farmers for their produce, and providing decent working conditions and improvements in local community amenities such as schools and health centres (Figure 21.44).

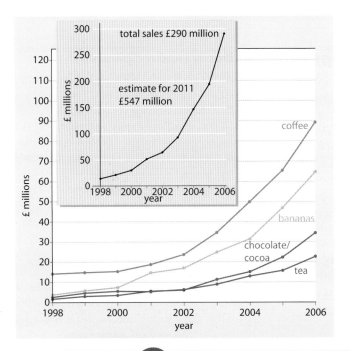

total sales £290 million

estimate for 2011 £547 million

Figure 21.45

Fairtrade sales, 1998–2006

More than 4000 Fairtrade products have been licensed for sale in the UK. Shoppers can choose wine, cotton products, flowers and sports balls as well as food and soft drinks carrying the Fairtrade Mark. In 2006 alone, sales of Fairtrade products increased by 46 per cent (Figure 21.45), providing further evidence of the growth of **ethical consumerism**. This is when an increasing number of shoppers are prepared to pay more for products if they feel it will help provide jobs and lift people out of extreme poverty. Large TNCs such as Nestlé (coffee in El Salvador) and Tate & Lyle (sugar cane in Belize), together with superstores such as Asda, Sainsbury's, Tesco and Marks and Spencer, are being encouraged by shoppers to stock and support Fairtrade products.

Places 102 Ghana: Fairtrade

In 1993, a group of cocoa farmers in Ghana, together with Twin Trading (a UK trading association), set up their own Kuapa Kokoo co-operative on Fairtrade terms. Their aim was to create an organisation with farmers' welfare at its heart and with a reputation for quality and efficiency. Once the co-operative members had harvested the cocoa pods, split them open with a machete and dried the beans found inside (Figure 21.46), they were able to sell their produce to the co-operative and enjoy the benefits of selling to the Fairtrade market: prompt payment, a regular bonus, democratic rights and community improvements funded by Fairtrade income. Kuapa Kokoo, which means 'good cocoa farmers', then weighed the bags and sold the cocoa to the government cocoa board, which then sold it on all over the world. In 2008 – and still the only farmer-owned company in Ghana – the co-operative had 45 000 members (28 per cent of whom were women) in 1200 small villages which produced 5 per cent of the country's cocoa (Ghana is the world's second largest cocoa grower).

In 1997 the members of Kuapa Kokoo voted to set up their own chocolate company, and with the help of Twin Trading, the Body Shop, Christian Aid and Comic Relief, and with a loan guaranteed by DFID (the UK's Department for International Development), Divine Chocolate was born (Figure 21.47). Today Divine Chocolate is the leading Fairtrade chocolate company in the UK, and after the Body Shop kindly donated its shares to Kuapa Kokoo, the co-operative now owns 45 per cent of the business. This means that as well

Figure 21.46

Splitting open the cocoa pods

as receiving the Fairtrade minimum price and the Fairtrade social premium, the co-operative also shares the profits and has a real say in how its products are produced and marketed. In 2007, Divine Chocolate Inc, also co-owned by Kuapa Kokoo, was established in the USA and with all debts paid off Divine Chocolate delivered the first dividend to Kuapa Kokoo.

Fairtrade has transformed the lives of many villagers in Ghana, delivering fundamental improvements in living and working conditions, and enabling participation in an organisation that values women, education and the needs of the farmer. As one teenager whose family was a member of Kuapa Kokoo said: 'We sell cocoa for the Divine bar getting a fairer price for our beans. My family now earn enough for me to stay at school and to buy for ourselves better machinery while the profits and end-of-year bonus have enabled the village to construct a well, which now gives us a clean water supply (Places 97), a new school and a mobile health centre.' It has also enhanced the status of women.

Figure 21.47

The Divine chocolate bar

Overseas aid and development

Overseas aid is the transfer of resources at non-commercial rates by one country (the donor) or an organisation, to another country (the recipient). The resource may be in the form of:

1 money, as grants or loans, which has to be repaid, even at low interest rates
2 goods, food, machinery and technology
3 know-how and people (teachers, nurses).

The basic aim in giving aid is to help poorer countries develop their economies and to improve services in order to raise their standard of living and quality of life. In reality, the giving of aid is far more complex and controversial as it does not always benefit the recipient.

Types of aid

Basically, there are two main types of aid: **official** and **voluntary**. The differences in their purposes and aims are summarised in Figure 21.48.

Donors and recipients

Although it is the advanced economies that are the largest donors in terms of US dollars, the amount that each country gives as a proportion of its own GDP is small – certainly well below the 0.7 per cent recommended by the UN. Indeed it is often only the Scandinavian countries which, while giving less in total amounts, achieve the UN figure. As for the recipients, while the two-thirds of the world's lowest-income countries located in sub-Saharan Africa do receive most of the overseas aid, there is no simple correlation between the level of poverty and the amount of aid received. Donor countries are just as likely to give aid to those countries that have supported them in times of war or provide land for military bases, possess a valuable raw material or have strong historic ties as to countries that are the least well-off. Some organisations such as the International Monetary Fund (IMF) also aim to help the poorest countries while others, such as the World Bank, lend capital for specific projects.

Figure 21.48

Official and voluntary aid

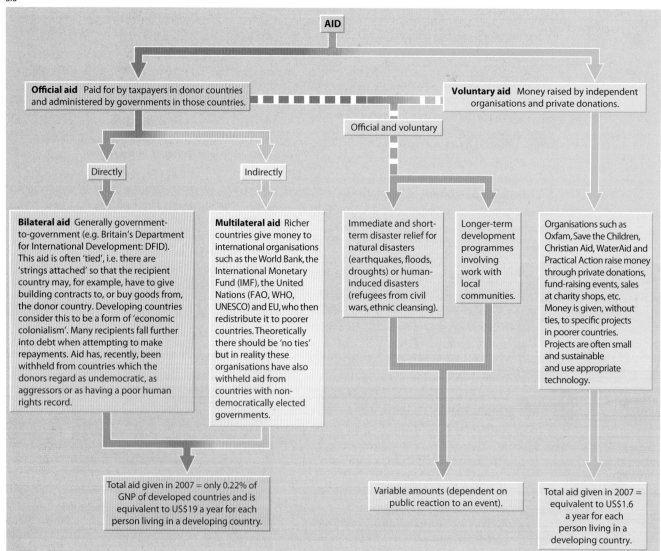

Figure 21.49

Arguments for and against the giving of aid

For	Against
• Response to emergencies, both natural and human-induced.	• Aid is a conscience-salver for the rich and former colonial powers.
• Helps in the development of raw materials and energy supplies.	• Better to use money on the poor living in the donor countries.
• Encourages, and helps to implement, appropriate technology schemes.	• An exploitation of physical and human resources.
• Provides work in new factories and reduces the need to import certain goods.	• Used to exert political and economic pressure on poorer countries.
• Helps to increase yields of local crops (green revolution) to feed rapidly growing local populations.	• Increases the recipient country's external debt.
• Provides primary health care, e.g. vaccines, immunisation schemes, nurses.	• Often only goes to the rich and the urban dwellers in recipient countries, rather than to the real poor.
• Helps to educate people about, and to implement, family planning schemes.	• Encourages corruption among officials in donor and recipient countries.
• Grants to students to study in overseas countries.	• Undermines local activities, e.g. farming.
• Can improve human rights.	• Does not encourage self-reliance of recipient countries.
	• Often not given appropriate technology.

Is aid good or bad?

While few people would argue against emergency aid, except to say that it is often 'too little and too late', other forms of aid are more controversial. Some consider that no non-emergency aid should be granted, especially as it is usually given in the political, industrial or commercial interests of the donor without concern for the environment or the long-term improvement in the quality of life of the recipient. Too often, aid tends to address the symptoms of poverty rather than its causes. Others feel that aid can make important contributions to the economy of many of the least well-off countries and to the welfare of some of their poorest communities. Some of the arguments of the pro-aid and anti-aid groups are listed in Figure 21.49.

Places 103 Sri Lanka: aid after the 2004 tsunami

One major effect of globalisation is the speed at which news is flashed around the world. In some cases, like the Indian Ocean tsunami in 2004 (Places 4) or the Chinese earthquake in 2008 (Places 2), people across the globe feel as if they themselves are involved in the event and consequently are anxious to help in whatever way, however small, they can.

In Sri Lanka, a place known by overseas tourists, the tsunami left almost 40 000 dead, 575 000 homeless and 16 000 seriously injured. Hospitals, schools, homes, hotels, roads and the mainline railway between Colombo and Gale were destroyed. Aid came from three main sources:

- **Emergency aid** came from voluntary international relief organisations who are used to responding rapidly to any global disaster – although they admitted never one so great as this. Initially they help to locate possible survivors and treat the injured. They then seek to satisfy the urgent needs of the survivors which, these organisations claim, is always for shelter, clothing, food, toilets, clean water and medical supplies.
- **Short-term aid** is provided partly by the voluntary relief organisations and partly by ordinary people. After the tsunami and following appeals in British newspapers (Figure 21.50) and on television, people began phoning, using the Internet or sending cheques to organisations such as Oxfam, Christian Aid and CAFOD. Within a few days over £100 million had been donated and when the Disaster Appeal closed after two months, £300 million had been raised. People in many other countries did the same.
- **Long-term aid** is provided by governments which, in this case, pledged £3700 million – easily a world record. This money was used to rebuild communications, hospitals, schools, houses and in trying to recreate jobs.

Two years later, the Sri Lankan Reconstruction and Development Agency (RADA) announced that nearly 90 per cent of the pledged money had been received – a remarkably high figure as often governments, agencies and people fail to meet their promises as their memory of an event fades – and that 1020 projects had been either completed or started.

Figure 21.50

Tsunami appeal advert

TSUNAMI
EARTHQUAKE

Hundreds and thousands of people across a dozen countries have been affected by the major disaster and devastation caused by the earthquake in the Indian Ocean and the Tsunamis that followed.

Aid agencies are working to provide emergency relief and need your support. DEC members are ActionAid, British Red Cross, CAFOD, Care International UK, Christian Aid, Concern, Help the Aged, Merlin, Oxfam, Save the Children, Tearfund, World Vision.

Give now to the Disasters Emergency Committee.

0870 60 60 900

or www.dec.org.uk

Or by cheque to PO Box 999, London EC3A 3AA, payable to DEC Tsunami Earthquake Appeal

World transport

Transport is referred to several times in this book:
- It can be viewed as an indicator of wealth and economic development, e.g. as measured by the number of cars per 1000 people. While the more developed countries have less than one-fifth of the world's population, they have over three-quarters of its cars and lorries.
- It is essential in linking people, resources and activities; in increasing personal mobility; and for the exchange of goods (trade) and ideas (information).
- It was considered a major factor in industrial location (Weber, page 557) and in determining agricultural (von Thünen, page 471) and urban (page 425) land use. The relative decrease in transport costs since the 1950s has made this a less significant location factor.
- In early economic/geographical theory, costs were thought to be proportional to distance (von Thünen's central market and Christaller's central place), especially on a flat plain where transport costs were equally easy and cheap in all directions. Later, costs were regarded to be a function of a raw material's weight and the distance it had to be moved (Weber).

- Improvements in transport resulting from **space-shrinking technologies** include containerisation, Airbus A380 and the Internet. These increase speed and ease, and all contribute to globalisation.

Characteristics of modern transport systems

A comparison of the characteristics of the major forms of present-day transport – canal, ocean shipping, rail, road, air and pipeline – is given in Figure 21.53, with each type having its advantages and disadvantages over rival forms of transport. Figure 21.53 also refers to terminal and haulage costs. **Terminal costs** are fixed regardless of the length of time of journey and are highest for ocean shipping and lowest for road transport. **Haulage costs**, which increase with distance but decrease with the number of passengers carried or the amount of cargo handled, are lowest for water transport and highest for air (Figure 21.51). It is now accepted that, as transport costs comprise terminal costs plus haulage costs, then the cost per tonne/km declines with distance. Figure 21.52 shows the changes in passenger and freight traffic in the UK in the last 50 years.

Figure 21.51

Transport costs

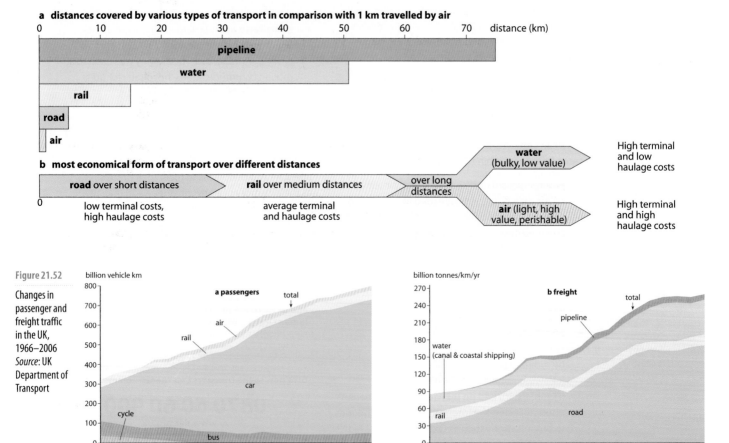

Figure 21.52

Changes in passenger and freight traffic in the UK, 1966–2006
Source: UK Department of Transport

		Canals and rivers	Ocean transport and deep-sea ports	Rail	Road	Air	Pipelines
Physical	**Weather**	Canals can freeze in winter. Drought/heavy rains make rivers unnavigable.	Storms, fog. Icebergs in North Atlantic.	Very cold (frozen points). Heavy snow (blocks line). Heavy rain can cause landslides, heat can buckle lines.	Fog and ice both can cause accidents/pile-ups. Cross-winds for big lorries; snow blocks routes; sun can dazzle.	Fog, icing and snow: less since planes have had automatic pilots. Airports better if sheltered from wind and away from hills and areas of low cloud.	Not greatly affected.
	Relief	Width of channels. Need flat land or gentle gradients. Soft rock/soil for digging, problems with deltas. Rivers must be slow-flowing, have a constant discharge and have no rapids.	Harbours need to be deep, wide and sheltered. Tidal problems.	Cannot negotiate steep gradients so have to avoid hills. Estuaries can be obstacles. Flooding in valleys.	Avoids/takes detours around high land. Valleys may flood. May go around estuaries if no bridges.	Large areas of flat land for runways, terminal buildings and warehousing. Firm foundations. Ideally, cheap farmland or land needing reclamation. Relief not a barrier.	Difficult to lay, then relief is not a problem.
Economic	**Speed/time**	Slowest form of transport. Long detours and possible delays at locks.	Slow form of transport, yet most economical.	Fast over medium-length distances.	Fast over short distances and on motorways. Urban delays.	Fastest over long distances, not over short ones due to delays getting to and passing airport security.	Very fast as continuous flow.
	Running or haulage costs (wages and fuel): increase with distance	Often family barge. Limited fuel use means the cheapest form of transport over lengthy journeys.	Expense (oil used as fuel) increases with distance.	Relatively cheap over medium-length journeys. Fuel costs and wages rising.	Cheapest over shorter distances. Haulage costs increase with distance. Recent rise in cost of petrol.	Very expensive, yet speed makes it competitive over very long distances.	Cheapest as no labour is involved (provided diameter is large).
	Terminal costs (loading and unloading costs and dues): no change with distance	Canals expensive to build and to maintain, unless natural waterways used.	Ports expensive – harbour dues/taxes. Expensive to build specialised ships. Less since containers. Cheapest over long distances.	Building and maintenance of track/stations/signalling/rolling stock are very expensive.	Expensive building and maintenance costs, especially motorways. Car tax instead of dues, but roads built from taxation therefore lower overheads. Congestion charges.	Very expensive to build and maintain airports. High airport dues. Planes expensive to purchase and maintain.	Very expensive to build. Need surveillance.
	Number of routes	Relatively few. Inflexible.	Relatively few ports, inflexible due to increased specialisation of ships. Links to hinterland. Coastal shipping.	Not very flexible. Recent increase in urban rail and new high-speed intercity routes.	Many and at different grades. Great flexibility, most in urban and industrial areas.	Often only a few internal and international airports/routes. Not very flexible because of safety.	Limited to key routes. Inflexible and one-way flows.
	Goods and/or passengers carried	Heavy, bulky, non-perishable, low-value goods. Present-day tourists.	Heavy, bulky, non-perishable low-value goods. Cruise passengers. Goods carried in containers.	Intercity passengers. Heavy, bulky (chemicals, coal) and rapid (mail) goods. Can carry several hundred passengers. Dependable and safe.	Many passengers. Perishable, smaller loads by lorry. Relatively few people carried by one bus or car.	Mainly passengers. Freight is light (mail), perishable (fruit) or high-value (watches).	Bulk liquid (oil, gas, slurry, liquid cement, water).
	Congestion	Very little except at locks.	Increasing delay and congestion in many deep-sea ports.	Considerable congestion on intercity and commuter routes.	Congestion heavy in urban areas, at peak times and in holiday periods.	Heavy at large airports and at peak holiday times.	None.
	Convenience and comfort	Neither very convenient, unless for leisure/relaxation, nor very comfortable.	Not very convenient. Cruise liners very comfortable.	Commuter routes uncomfortable. Some intercity routes better.	Door-to-door (except for some city centre destinations): most convenient and flexible. Safety is questionable; strain for drivers, but independent.	Country to country. Jet lag if more than three time zones crossed. Cramped, dehydrating and tiring over longer journeys.	Raw material or port to industry.
Environmental	**Environmental problems**	Some oil discharged, but relatively few problems.	Tankers discharging oil. Much land needed for ports, hard-standing and warehousing.	Noise and visual pollution limited to narrow belts. Noise decreases with welded rails, increases with high-speed trains. Electric trains cause less pollution.	A major cause of noise and air pollution. Effect on ozone layer, acid rain, and global warming (greenhouse effect). Uses up land, especially farmland. Structural damage caused by vibrations.	High noise levels. Some air pollution. Uses up much land for airports.	Few are buried underground. Eyesore on surface.

Figure 21.53

Comparable characteristics of transport systems

Ocean shipping

Many ports in Western Europe developed either by trading with their former colonies or across the Atlantic to the Americas. In turn, large ports were created within the colonies to export raw materials or acting first as entrepot ports and now as **free-ports**. A freeport is an area of land exempt from taxes paid by the rest of the country in which it is located. As such, it can attract imports that can be manufactured into goods that are then exported without having to pay duties or tax, e.g. Singapore with, amongst other industries, its oil-refining (Places 104). Just as ocean shipping continues to grow in quantity, so too have ships increased both in size and in specialisation, e.g. oil tankers and bulk iron ore carriers. This in turn has meant that it is the wider, deeper estuaries that have seen the most concentrated growth in the world trade by sea, a trade that has been increasing steadily for several decades and which has, since 2000, grown enormously since China began exporting its wide range of cheaply manufactured goods. Most of the world's trade is moved by water.

A ship berthed at a quayside is not only not earning money, it is having to pay out harbour dues. Two innovations have enabled the turn-round time (the time it takes to unload and load cargo) to be shortened:

1 The development of **roll on/roll off (Ro-Ro)** methods whereby lorries can drive straight on to ships, reducing the need for cranes and, indeed, dock workers.

2 The introduction of **containerisation** in which goods are packed into containers of a specific size at, for example, a factory and taken by train or lorry to the container port where they are easily and quickly loaded onto ships using specialised equipment (Places 104). Containerisation is considered to have been one of the major driving forces in the process of globalisation.

The *Emma Maersk* is the world's largest container vessel (capable of carrying over 11 000 containers) and longest ship (at 397 m). Its regular run is between China and Western Europe.

Places 104 Singapore: an ocean port

On founding the port of Singapore in 1819, Sir Stamford Raffles decreed that it was open to all maritime nations. Today over 400 shipping lines with links to more than 600 ports worldwide have taken advantage of that decree and since 1986 Singapore has been the world's busiest port in terms of shipping tonnage, and its main bunkering port (i.e. fuel container). At any given time, over 800 ships are likely to be in port, with a new one arriving or weighing anchor every seven minutes (128 568 vessels in 2007 compared with 81 000 in 1992). To save time, harbour pilots are flown out by helicopter to meet incoming vessels. With its modern handling equipment, it takes less than a second to move 1 tonne of cargo. Warehouses are also automated and computerised. Vessels vary from modern supertankers, bulk carriers and container ships to the more traditional bumboats and barges (Figure 21.54). In 2007 the port also handled 27.9 million containers making it, along with Shanghai, the world's busiest container port (Figure 21.55).

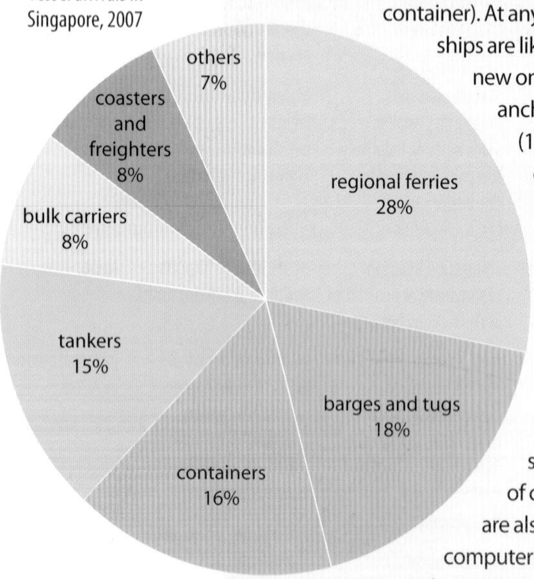

In 2007 Singapore was voted – for the twentieth time since 1987 – the best port for its cost competitiveness, container shipping-friendly regime, adequacy of investment in port infrastructure, and visionary developments. Singapore is a freeport, still open to all countries, with seven free trade zones of which six are for seaborne cargo and one is at nearby Changi international airport. Goods can be made or assembled in these zones without payment of import or export duties and profits can be sent back to the parent company without being taxed. Many high-tech TNCs assemble their goods here before selling them at competitive prices. However, the port's largest money-earner is oil, a resource that the country does not possess. This is because Singapore imports crude oil from the Middle East, Indonesia and Malaysia, refines it in the freeport and then exports a range of oil products, making it the world's third largest oil-refining centre.

Figure 21.54

Vessel arrivals in Singapore, 2007

- others 7%
- coasters and freighters 8%
- bulk carriers 8%
- tankers 15%
- containers 16%
- barges and tugs 18%
- regional ferries 28%

Figure 21.55

The port of Singapore

Air transport

Air transport has the highest terminal charges, high haulage costs (aviation fuel) and affects large numbers of people living on flightpaths near to airports. Its advantages (Figure 21.53) include speed over long distances both for passengers such as tourist and business people, and for freight especially if it is of high value (watches, diamonds), light in weight (mobile phones) or perishable (fruit). Apart from employing large numbers of people at airports, air transport is important to countries that are of considerable size (Brazil), where ground terrain is difficult (Sahara Desert, the Alps), when crossing stretches of sea (London to Belfast), or when relief aid is essential following a human (Rwanda) or natural (earthquake) disaster or international conflict (Afghanistan).

Since deregulation in the EU in 1993, there has been increased competition between existing airlines, a wider availability of routes and the advent of low-budget airlines with their reduced fares. This led to an increase in the number of flights, passengers and freight, with congestion at airports and competition for airspace. This increase in demand, especially during holiday periods and at 'hub' locations, has resulted in the building of more and larger airports.

Beijing's third terminal, opened in time for the 2008 Olympics, is 2.9 km from end to end and is larger than all five Heathrow terminals put together (Figure 21.56). It will increase Beijing's passenger capacity from 35 million to 85 million. China plans to build another 97 airports by 2020, bringing the country's total by then to 239. National passengers have grown from 7 million in the mid-1980s to 185 million by 2007, in response to China's rapid economic growth.

This, and other world airport planned development, was before the surge in oil prices in 2008, which left airlines in a state of uncertainty, not knowing whether fuel costs will remain high, go higher or even fall, and air travel was included in carbon-credit trading.

Figure 21.56

Beijing's new Terminal 3

Places 105 London Heathrow and Dubai: 'hub' international airports

Figure 21.57

Numbers of people passing through Heathrow

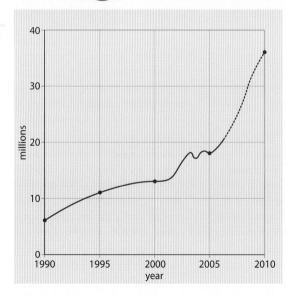

Heathrow

Heathrow is the leading European 'hub' airport for international flights and is said to be Britain's main gateway to the global economy. A 'hub' is when, instead of small planes flying relatively short journeys between many cities, large planes fly between the biggest cities with feeder flights (Figure 21.57). Part of Heathrow's importance stems from the fact that 21 per cent of passengers arriving at the airport are 'in transit', just stopping long enough to change flights. This causes congestion in the airport and little income for the UK but is essential for filling seats on British Airways flights and maintaining the airport in pole position. However, to maintain this position it is argued that a third runway will be needed by 2015 and the two existing ones need to be used more. Events leading to the final decision will provoke a major economic, social and environmental debate. Some of the advantages and disadvantages of the proposed expansion are summarised in Figure 21.58.

Figure 21.58

Arguments for and against expansion

For expansion	Against expansion
• The prestige of being Europe's major 'hub' airport and the world's busiest.	• Aviation is the fastest-growing source of avoidable carbon emissions, and must be curtailed.
• Heathrow is vital to the British economy with 170 000 jobs dependent on it.	• Residents in the south-east will experience an increase in noise, congestion and pollution; some 700 existing homes will have to be demolished, and a further 150 000 people will be under the new flightpaths.
• If Heathrow does not expand, flights and jobs will go to rival airports in Paris, Frankfurt and Amsterdam.	• The new runway is expected to cater more for short-haul flights for which there are less damaging alternatives.
• In 1991, 16 per cent of the total arrivals were passing through 'in transit'; by 2006 this was 21 per cent and by 2010 it is predicted to be 31 per cent (Figure 21.57). These are essential for filling, and maintaining, BA flights (40 per cent of Heathrow's total).	• The vast number of the present 18 million 'in transit' passengers spend virtually no money as they pass through the airport, contributing little to Britain's balance of trade.
• The environmental damage is exaggerated – aircraft only contribute 6 per cent of Britain's total carbon emissions, far less than cars and coal-fired power stations.	• The airport already has a reputation for congestion, long delays and lost luggage.

Dubai

Dubai has made itself the new 'hub' for air transport in the Middle East and beyond. It is a time-zone bridge between the Far East and Europe on the east–west axis and between the CIS and Africa on the north–south axis. A third terminal was opened in 2008 to relieve pressure created by the 34 million passengers and 260 000 flights that used the airport in 2007. It has been constructed to take the new Airbus A380 which has 525 seats. Dubai's success as a 'hub' has been its linking together of seemingly unlikely pairs of cities, e.g. Nagoya and São Paulo, Moscow and Cape Town, Guangzhou and Dar es Salaam (Figure 21.59). Emirates airline also uses Dubai airport to link smaller cities with major world centres, for example passengers from Newcastle can fly

Figure 21.59

Dubai as an air transport 'hub'

to Dubai and have a night's rest before travelling on to places in Japan, China and Oceania.

Transport, carbon trading and international agreements

After power stations and industry, transport is the major cause of carbon release into the atmosphere. The effect of cars and other road vehicles emitting carbon dioxide, a greenhouse gas, on global warming, have been known for some time. It is only more recently that the increase in air traffic has been seen as a further factor in climatic change. What is still to be broadly accepted is the effect of ocean transport which handles most of the world's trade, and of an increasing number of cruise liners. The UK government, as just one example, claims that it has reduced carbon emissions in the last decade but, as environmentalists point out, it has ignored both ocean and air transport in its calculations. If these emissions were included, it would mean that Britain had an overall increase in carbon emissions.

Under the Kyoto Protocol – which was drawn up in 1992, adopted in 1997, came into force in 2005 and is due to expire in 2012 – industrialised countries were meant to cut greenhouse gas emissions by an average of 5.2 per cent. Since Kyoto, total global emissions have in fact soared; the economies of China and India have boomed at a rate that was not predicted and the world's population has grown by about 1 billion. At present it is the industrialised countries that emit most carbon (Figure 21.60) while the poorest nations often emit so little that any cutbacks by them would have minimal effect on a global scale (Figure 21.60). As with development, there is a wide gap between the high-emitting rich countries and the low-emitting poor countries. One suggested solution is **carbon trading**. The EU already has an emissions trading mechanism in operation, together with voluntary offset schemes.

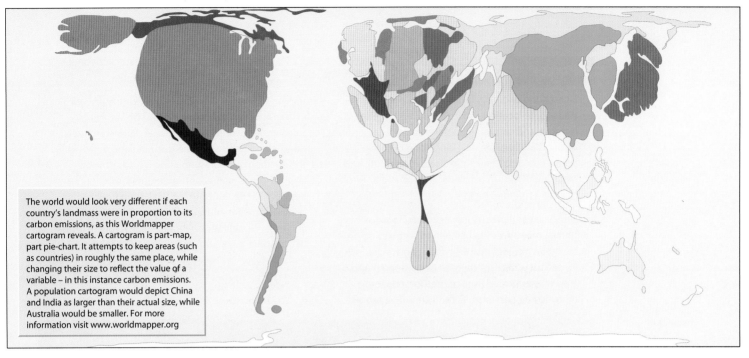

The world would look very different if each country's landmass were in proportion to its carbon emissions, as this Worldmapper cartogram reveals. A cartogram is part-map, part-pie-chart. It attempts to keep areas (such as countries) in roughly the same place, while changing their size to reflect the value of a variable – in this instance carbon emissions. A population cartogram would depict China and India as larger than their actual size, while Australia would be smaller. For more information visit www.worldmapper.org

Figure 21.60

Cartogram to show contribution to carbon emissions by different parts of the world
Source: © 2006 SASI Group (University of Sheffield) and Mark Newman (University of Michigan)

Carbon trading is when each country is given a quota for its emissions. Those countries that emit most would be able to buy from countries that do not use their full quota, allowing those that emit less than their quota to earn money by selling their surplus. While this may be a way for the poorest countries to earn extra income, it hardly solves the global problem as rich countries will presumably buy extra credits rather than reduce their own emissions. Problems relating to international trade and transport would remain. Take two examples:

1   A country in the EU buys bananas, even through Fairtrade, from a country in the Caribbean. Which country is liable for the carbon transport emissions – the exporter or the importer?
2   Another country in the EU, or a TNC based there, orders goods to be made in China where they can be produced more cheaply. Is it the country/TNC that orders and sells the goods that is responsible for the transport emissions, or China where the goods were manufactured?

Carbon trading can only work through international co-operation but getting 200 countries with a wide divergence of interests to agree is a different matter. These interests include the following:

■ The USA fears that a reduction in its emissions would mean job losses and a possible fall in the country's standard of living. It agreed, for the first time in 2008, to talk about emissions at the 2009 Copenhagen conference.
■ The EU countries argue for a 30 per cent reduction but are finding it hard to achieve.
■ Emissions of emerging countries, such as China and India, are surging and these countries are under no pressure to cut back. China, where numerous new coal-powered stations are being built, claims that it needs this energy to create jobs, while India says it needs the extra energy just to improve, or even to maintain, the standard of living of its rapidly growing population.
■ Developing countries do not see why they should help solve a problem that was not of their making, and to do so would mean their being given money and technology by the developed countries.

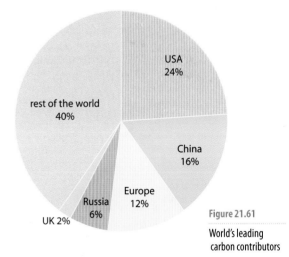

Figure 21.61

World's leading carbon contributors

Integrated transport systems

Although most long-distance transport is either by ship (freight), plane (passengers) or pipeline (oil and natural gas), both road and rail can be used to cross continents such as North America, or to travel from Western Europe to the former Eastern bloc countries. In an ideal world, there would be a stronger link between these various types, whereas in fact integrated systems tend to be limited to regions and large urban areas (Places 106) than being on a global scale.

Hong Kong originally grew as a result of its strategic trade route location and its large, deep, sheltered harbour, and continued to develop partly as a result of later industrialisation. Hong Kong became one of South-east Asia's four 'little tigers' (page 578), and trade with China in particular and the Pacific Rim in general expanded rapidly.

Early transport was mainly restricted to water due to the limited amount of flat land. As building on the steep hillsides proved difficult and hazardous (Case Study 2B), especially on Hong Kong Island, land was reclaimed from the sea for industry, housing and transport. Three forms of transport

used at the beginning of the 20th century are still in operation today (Figure 21.62). The Star Ferry transfers large numbers of people daily from Hong Kong Island to Kowloon on the mainland; trams link the northern part of Hong Kong Island (although land reclamation means their routes are no longer adjacent to the sea); and the Peak Tram funicular railway carries wealthy commuters and tourists to and from Victoria Peak (Figure 21.63). A fourth form of transport, the Kowloon Railway, linked the colony with the New Territories and the Chinese cities of Guangzhou and Shenzhen (page 581).

Figure 21.62

Hong Kong's Star Ferry, funicular railway and tram

Figure 21.63

The development of transport in Hong Kong before 1992

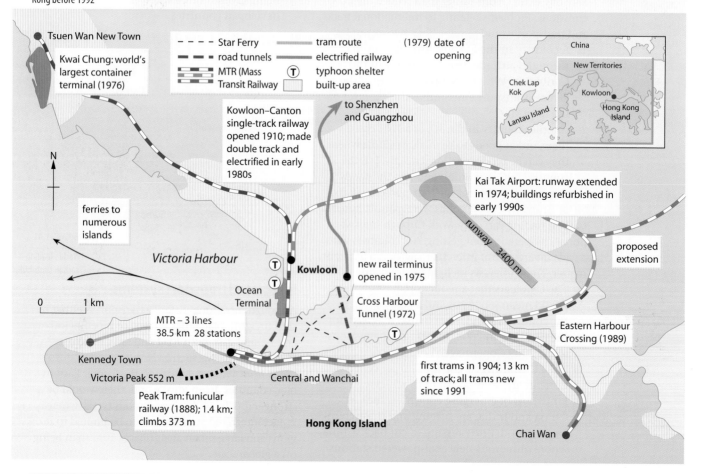

Figure 21.64

The Tsing Ma bridge

Transport since 1997

In 1997, the British handed Hong Kong back to China and the former colony became a Special Administrative Region. By 2008, the following additions and changes had been made to the transport system.

The largest development was the construction of a new international airport at Chek Lap Kok (Figure 21.65). The decision to re-locate the airport here was made in 1989 as part of a comprehensive plan to incorporate air, road, rail and port developments. The airport itself was opened in 1998 with a second terminal nine years later. By that time it was handling 47 million passengers a year. The airport is connected to Tung Chung (a new town on Lantau Island), Kowloon and Hong Kong Island by a 27 km expressway that includes two bridges (Figure 21.64) connecting islands west of Kowloon, and a new tunnel under Victoria Harbour. Adjacent to the expressway is the Airport Express (AEL) whose trains cover the 35.3 km to Hong Kong Island in 24 minutes. Of two new MTR lines, one connects with Disneyland on Lantau Island and the other was built between eastern Hong Kong Island and eastern Kowloon using yet another new under-harbour tunnel. At present the MTR tracks cover 91 km and have 53 stations. The Kowloon–Canton Railway (KCR) has extended its east coast line (2004), and opened a new west coast route (2003) between northern Kowloon and the new town of Tuen Mun (Figure 21.65). These two routes will themselves be linked in 2009. The east coast route of the KCR now provides a high-speed direct link with Shanghai and Beijing. The port of Hong Kong received 39 000 vessels in 2006 while the twin container terminal of Kwai Chung and Tsing Yi remains one of Asia's largest although it has now been overtaken by Singapore (Places 104) and Shanghai (Case Study 15B).

Each day, about 11 million passenger journeys are made including over 4 million by bus, 3.8 million by MTR, 1.4 million by rail, 240 000 by tram, 155 000 by ferry and 28 000 by AEL.

Figure 21.65

Transport developments since 1997

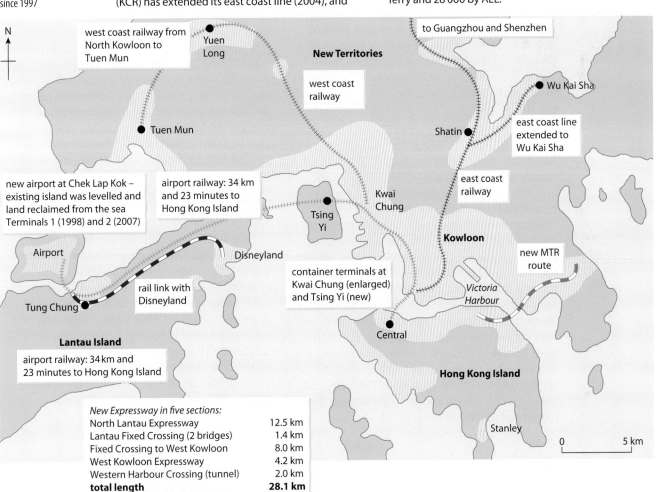

west coast railway from North Kowloon to Tuen Mun

to Guangzhou and Shenzhen

New Territories

Yuen Long

west coast railway

Wu Kai Sha

Shatin

east coast line extended to Wu Kai Sha

Tuen Mun

new airport at Chek Lap Kok – existing island was levelled and land reclaimed from the sea Terminals 1 (1998) and 2 (2007)

airport railway: 34 km and 23 minutes to Hong Kong Island

east coast railway

Tsing Yi

Kwai Chung

Kowloon

Airport

Disneyland

new MTR route

rail link with Disneyland

container terminals at Kwai Chung (enlarged) and Tsing Yi (new)

Victoria Harbour

Tung Chung

Lantau Island

airport railway: 34 km and 23 minutes to Hong Kong Island

Central

Hong Kong Island

Stanley

0   5 km

New Expressway in five sections:

North Lantau Expressway	12.5 km
Lantau Fixed Crossing (2 bridges)	1.4 km
Fixed Crossing to West Kowloon	8.0 km
West Kowloon Expressway	4.2 km
Western Harbour Crossing (tunnel)	2.0 km
total length	**28.1 km**

Information and communications technology (ICT)

Since the mid-1990s, the telecommunications/ICT sector has undergone major changes. Indeed when a previous edition of this book was being written in 1998 and when the latest figures available would have been for 1996, only 12.9 per cent of the world's population possessed a fixed or landline telephone; only 2.5 per cent owned a mobile cellular telephone; and only 1.4 per cent had access to the Internet (Figure 21.66). Consequently this section did not appear in that book. Within a decade these **space-shrinking technologies** have become a major reason for the growth and spread of globalisation.

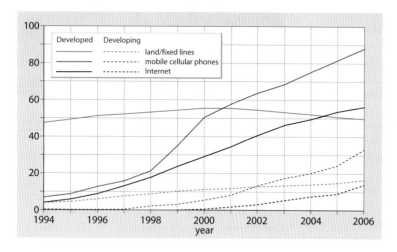

ICT	Developed countries %		Developing countries %	
	1994	2006	1994	2006
Fixed lines	48.8	51.5	4.4	13.9
Mobiles	5.2	90.9	0.2	32.4
Internet users	2.2	58.6	0.03	10.2

Figure 21.66

Fixed telephone lines, mobile cellular phones and Internet users per 100 inhabitants, 1994–2006

Fixed telephone lines (landlines)

Fixed lines were easily the most available of the three forms of ICT in the 1990s with a frequency ten times greater in developed than in developing countries. Although the number of landlines in developed countries peaked at 56.8 per cent in 2000 – a time of rapid growth of both mobile phones and the Internet – it has slowly continued to increase in developing countries although in Africa there are, on average, only 3 fixed lines per 100 people.

Mobile cellular telephones

Latest data suggest that the global number of mobile telephones surpassed 3 billion in 2007 and that by 2008 over half of the total population would own their own mobile and could collectively be sending up to 300 000 texts each minute. By 2007, the number of subscribers in developed countries exceeded 90 per cent whereas the number for developed countries was still in the mid-30s. Even so, despite this large difference, mobile phones have been critical in enhancing access to telecommunications in many developing countries, and especially in rural areas where fixed lines remain limited or are non-existent. The 13 per cent of the world's population that live in the G8 countries (Canada, France, Germany, Italy, Japan, Russia, the UK and the USA) account for 30 per cent of mobile ownership.

Internet

Although access to the Internet has also been growing rapidly, the number of users in developing countries again remains limited with only just over 10 per cent compared with almost 60 per cent in developed countries. There are also major discrepancies in international Internet bandwidth – the critical infrastructure that dictates the speed at which websites in other countries can be accessed. Other constraints for developing countries include the high cost of international bandwidth (they often have to pay the full cost of a link to a hub in a developed country), literacy and a lack of electricity. At present over 40 per cent of the world's Internet users live in the G8 countries whereas in as many as 30 developing countries, Internet users number less than 1 per cent of the total population.

The ICT 'revolution' has seen the speeding up of the globalisation process and is contributing to the disintegration of national economies (page 605). It is, arguably, the resultant flow of data, finance and migrant remittances that forms the most indicative feature of globalisation. ICT has also allowed industries and services, from large-scale TNCs and international banks down to self-employed individuals, a freer choice of location for their site or place of work (post-Fordism, page 561).

The global value chain

The **value chain**, a later development of the **commodity chain**, is a connected group of activities that are required to see a product through a series of stages from concept-design to marketing-distribution (Gereffi 1994). The process of globalisation has promoted two types of chain:

1 The **producer-driven chain** is characteristic of capital- and technology-intensive industries (automobiles, computers and other high-technology activities) where the system is controlled by large TNCs.

2 The **buyer-driven chain** is typical of labour-intensive activities such as the fashion industry (Case Study 21), retailing (Wal-Mart – Figure 19.46 – and Ikea) and merchandising (Nike and Adidas) which involves the setting up of a global network and which increasingly depends on access to, and advances in, ICT.

The value chain involves dividing the industry into several components, each of which may be located in a country that offers the lowest-cost factors. For example:

■ The head office for administration and from where raw materials may be ordered and designs for the product drawn is in the USA with its available finance.

■ Research and development (R&D) into improved methods of processing is carried out in the UK with its skilled technical labour force.

■ Processing/manufacturing of parts is done in China where labour is plentiful and cheap (Case Study 21), although the final assembly may be in an NIC such as Malaysia.

■ Marketing and distribution are carried out in North America and the EU with their large and wealthy consumer markets.

■ After-sales and customer services operate through a call centre located in India (Places 107), taking advantage of its low-priced but skilled labour.

Each year, the value chain becomes more complex, dynamic and service industry oriented. The issue is where and when can value to a product be added and how far can factories and locations at the lower end of the chain manage to upgrade (Places 102 on chocolate in Ghana). For example, in the garment industry (Case Study 21), where design adds considerable value, sub-contractors in places such as Turkey and India used only to make the clothes designed elsewhere but now, increasingly, they design their own.

Places 107 India: call centres

The rapid growth in **call centres** is one consequence of space-shrinking technologies. Call centres represent a company–customer relationship in which a wide range of support services, including after-sales advice, marketing, technical support, claims enquiries, seat reservations and data provision, are provided over the telephone from dedicated centres to a widely dispersed customer base by firms such as American Express, Bank of America, BT (British Telecom) and Dell. Call centres provide information and advice for existing customers as well as trying to attract new ones. In the last decade, globalisation has seen India, with the city of Bangalore in particular, specialising and becoming a world hub in this sector of business process outsourcing, which is the final link in the value chain. India has become such a prime location for call centres and offshore services for firms based in the USA and Britain that it is in danger of becoming stereotyped for providing that specialised type of service, rather than being known for its wider economic development. It is the world's leading exporter of ICT services and its volume of outsourcing is doubling every three years, to the detriment of similar jobs back in the USA and the UK. Some of India's call centres are adding value by moving up into business service provision.

To many American and British employers, India has a stable democracy, a huge English-speaking population and a sound education system that turns out more than a million graduates a year, all of whom are looking for well-paid jobs. But 'well-paid' is a relative term. The average income per capita in India is under $1000 a year; for a person working in a call centre in that country it is between $15 000 and $25 000 a year; to pay someone in America or the UK to do the same job is likely to cost $70 000–$90 000 a year. So large firms in the Western world are moving their call centres to India in order to reduce their financial costs in an attempt to remain competitive in today's world.

But it is not an easy life for Indian call centre workers (Figure 21.67). Due to the time difference – Bangalore is 11 hours behind New York – the manning of phones has to be done throughout the night. By day the agents, as the call centre telephonists are known, have typical Indian names but by night they take on names that sound like the boy or girl next door in America or Britain. In their training they are taught to identify different 'Western' accents and to use those accents whenever possible themselves in order to make them sound more friendly and helpful to the caller. By the end of their training, only 5 out of every 100 of the original applicants are likely to get a job.

Figure 21.67

A call centre in Bangalore

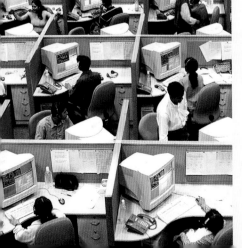

In the early 19th century, Britain was a leading producer of textiles. At that time, it imported silk from China and some cotton from India. Later, Britain began exporting textile machinery to countries such as India which were then able to export manufactured textiles back to the UK. This, on hindsight, was the beginning of globalisation in industry and the creation of a relatively simple value chain.

Today, the mention of globalisation can provoke extreme opinions. Certainly its impact on the peoples and economies of both China and India has been considerable – sometimes for the good, sometimes for the worse. But this impact of globalisation has not just been one-way. The growth of the Chinese and Indian economies has affected many people across the world, again sometimes for their benefit, sometimes to their detriment. The textile and fashion industry provides a good example of how a global value chain affects people and where, as so often is the case, some are winners and some are losers:

garment design (sourcing) ⟶ production ⟶ retail (sales) ⟶ supply (wearing)

China

Many designer clothes, including sportswear, trainers and jeans, are produced on a global scale by large TNCs which have located their main production factories in developing countries, especially if, like Mexico and Turkey, these countries are near to the market for mid-range products. As designs and styles of clothing are constantly changing, then it is quicker, easier and cheaper to get employees to adapt to these changes than it is to replace expensive machinery geared to specific garments. This means that the TNCs locate their main factories in countries like China where labour costs are still low, although in China's case many garment factories were initially financed from Hong Kong.

Many people living near to new textile and fashion factories, which have modern machinery, have **benefited from globalisation**. They are likely to get the better jobs and, should two or three members of the family also be employed, may earn enough to build a new house for themselves (Figure 21.68a). Unlike the house they will have left, this will be larger, lighter and cleaner; it will have electricity, running water and sewerage; and the new owners can probably afford a washing machine, TV, fridge and computer.

However, the number of new factories that have opened has greatly exceeded the supply of local labour. This has led to thousands of people from the surrounding poorer, rural areas being attracted to the large cities, creating a scale of rural to urban migration never before seen anywhere in the world (page 366). As is so often the case, the reality of urban factory life is far from the migrants' perception and so China's 150 million migrant workers, many of whom are women, have **benefited less from globalisation**. They are likely to get the worst jobs, may have to work more than 12 hours a day for at least six days a week, and earn under £100 a month (£4 a day). The worst factories have been described as 'sweatshops' as working conditions are often cramped and sometimes unhealthy and the jobs repetitive and boring (Figure 21.69). Accommodation may be in single-sex dormitories (Figure 21.68b), sharing a room with up to 12 other workers. There is little space or privacy, and washing facilities may have to be shared with up to 50 people. Most of their wages will be sent home as remittances but the workers can rarely afford to return to their villages themselves.

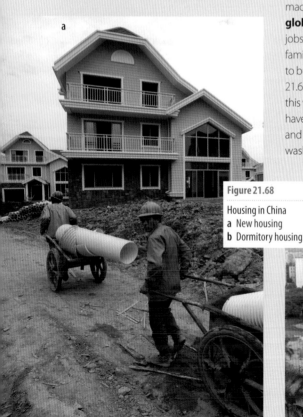
a

b

Figure 21.68

Housing in China
a New housing
b Dormitory housing

Figure 21.69

Sweatshop conditions

Figure 21.70

Women sewing in a small shop/workshop

There is another group of people who certainly have **not benefited from globalisation**. Tang Lee's family have been making children's clothes in Beijing for five generations and then selling them in their small shop in a quiet back street (Figure 21.70). Now his business is failing in the face of globalisation. As China becomes richer, more of its people can afford the brand-named fashion products that are pouring into or being made in China and which are available for sale in the hundreds of new, large department stores. Added to this, the increase in the number of foreign television programmes has made the Chinese, especially the younger ones, more aware of 'Western-style' designer products. As Tang Lee said: 'People want whatever they see on television but it will mean the end of small clothes makers, small shops and the traditional Chinese culture'.

India

In 2005, India's newspapers reported that, due to the end of textile quotas and Chinese trade disputes, 'exports to the USA had increased by 36 per cent and the textile boom has given jobs to India's poor' and a government minister predicted that 'as many as 12 million jobs could be created in the textile sector over the next few years'.

The globalisation policy of the government has seen the construction of modern capital-intensive spinning mills that use modern technology (Figure 21.71); the introduction of a promotion and marketing strategy aimed at capturing both the urban and rural market; and a diversification in the range of the products which are aimed to be low-price and high-quality. The industry has also invested heavily in acquiring sophisticated high-technology equipment and tools from overseas countries and introduced production and marketing collaboration with foreign manufacturers. The high-volume production of quality synthetic and cotton items, which has benefited so many people as well as the national economy, has, however, given it a competitive advantage over traditional handloom products.

The government has recognised the impact that the entry of global competitors is having on the handloom sector (Figure 21.73) which is mainly located in rural areas and operated by women working in weaver communities (Figure 21.72). The weakening of the handloom sector is posing a serious threat to the socio-economic life and livelihood of traditional weaver communities in general and the socio-economic status of rural women working in those communities in particular. As rural women constitute a major segment of the labour force in the handloom industry, it will have a far-reaching effect on the government's drive for rural poverty alleviation and economic empowerment for women.

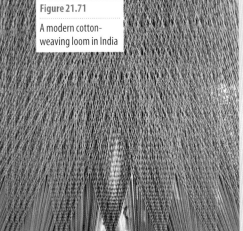

Figure 21.71

A modern cotton-weaving loom in India

Figure 21.72

Hand spinning yarn – in a weaving community

Although Bangalore has become the global hub for call centres (page 643), it is still important for textiles. Figure 21.74 describes how the less wealthy, less educated members of the community can also benefit from globalisation.

With increasing globalisation, the degree of competition for marketing textile items has intensified with the entry of foreign suppliers and foreign brands. The position and share of handloom products has been suffering by the entry of major competitors. It is imperative that the handloom industry sector, with its distinct and unique features, prepares and strengthens itself to meet the challenges and intensity of competition in the global and internal market.

Source: Ministry of Human Resource Development, government of India

Figure 21.73

Extract from a government report

Champa Kala does not have the English nor the computer skills needed to find work in one of the many call centres located in the skyscrapers of central Bangalore, nor with one of the software firms that have transformed the region into a high-tech hub. Instead she works in an industrial suburb as a seamstress in a new garment export factory which, since the expiration six months earlier of a 30-year-old global system of textile quotas and the end of a long USA–EU trade dispute with China, is part of India's booming textile sector. Naturally she does not earn the wages nor work in the air-conditioned atmosphere of a call centre but she is happy enough simply to have found a job that pays around $1200 a year as she helps produce jackets for Gap Inc.

Many economists believe that it is new factories like this that typify the low-end, labour-intensive manufacturing sector that India needs if it is to improve the standard of living of its 400 million low-skilled, poverty-stricken citizens who live on less than $1 a day and who have been largely by-passed by the country's high-end job growth. A director for the garment factory claimed that it was providing jobs for the illiterate and semi-illiterate classes by taking up to 300 people a week straight from villages and farmland and, within a month in their training centre, giving them the skills to work the machines. The garment firm opened this factory in 2004 and within 12 months employed 1600 people. It has since opened several more in the region. In 2006 the textile sector, which nationally employed 35 million people and generated $14 billion in exports, had raised the hopes for sustained job creation, especially if India's share of the global textile market rises from the 4 per cent of 2004 (China had 20 per cent) to a predicted 15 per cent by 2010.

Figure 21.74

Textiles in Bangalore

China, the EU and North America: the quotas row, 2005

This crisis had its origins in the scrapping, at the beginning of the year, of the Multi-Fibre Arrangement (MFA) which set quotas on how many garments could be imported from individual countries into the EU and North America.

Cheaper manufacturing costs in China mean that it can undercut other countries by up to 25 per cent and so hundreds of retailers switched production there (as manufacturing costs in China are only 4 per cent those of the USA and the EU, then production, packaging, shipment to and then distribution in the EU and the USA can all be paid and still leave a decent profit). As a result, imports from China soared by up to 1200 per cent and, by the middle of the year, several *billion* more garments were en route to European markets.

Within months, at least 50 000 jobs were lost in traditional textile countries in Southeast Asia as factories closed down, and by the end of the year over 1 million jobs were to be lost in Bangladesh, Sri Lanka (still recovering from the previous year's tsunami), Cambodia and the Philippines (Figure 21.75). According to the UN, these countries lost 10 per cent of their export earnings in eight months. However, it was only when firms in Italy (Europe's leading garment manufacturer), France and the UK began closing that the EU acted. In July it imposed quotas on ten categories of garments coming from China in order to protect its own domestic market from a deluge of cheap goods – but by then more than 80 million items were already made up and on their way!

Retailers in the EU were unhappy, claiming that the quotas inhibited free trade and that consumers would be hit through price rises and shortages of jumpers, jeans, trousers and lingerie. Campaigners said that Western demand for cut-price clothes was fuelling a vicious circle of supply-chain switches, rapid wage reductions in the poorest countries and worsening labour relations globally. They argued that the introduction of quotas was protectionism at its worst and that while the EU had been preaching to the developing countries about the need to open up their markets, the EU then imposed restrictions to protect their own. The general secretary of the International Textile, Garment and Leather Workers' Federation said: 'Our concern is that countries like Bangladesh and Sri Lanka are being forced to try to undercut China and each other. They can only do this by increasing their already long working hours and reducing their already low wages. Garment manufacturing provides one of the few economic opportunities for poorer countries to raise their incomes. Now people working in textile factories in those countries are having to live at a subsistence level, and undercutting means that these nations cannot lift themselves out of poverty.'

Although the crisis was eventually brought to a conclusion, it did not prevent further factory closures and job losses in both developing countries and the EU as China continues to dominate the world's garment trade (Figure 21.76).

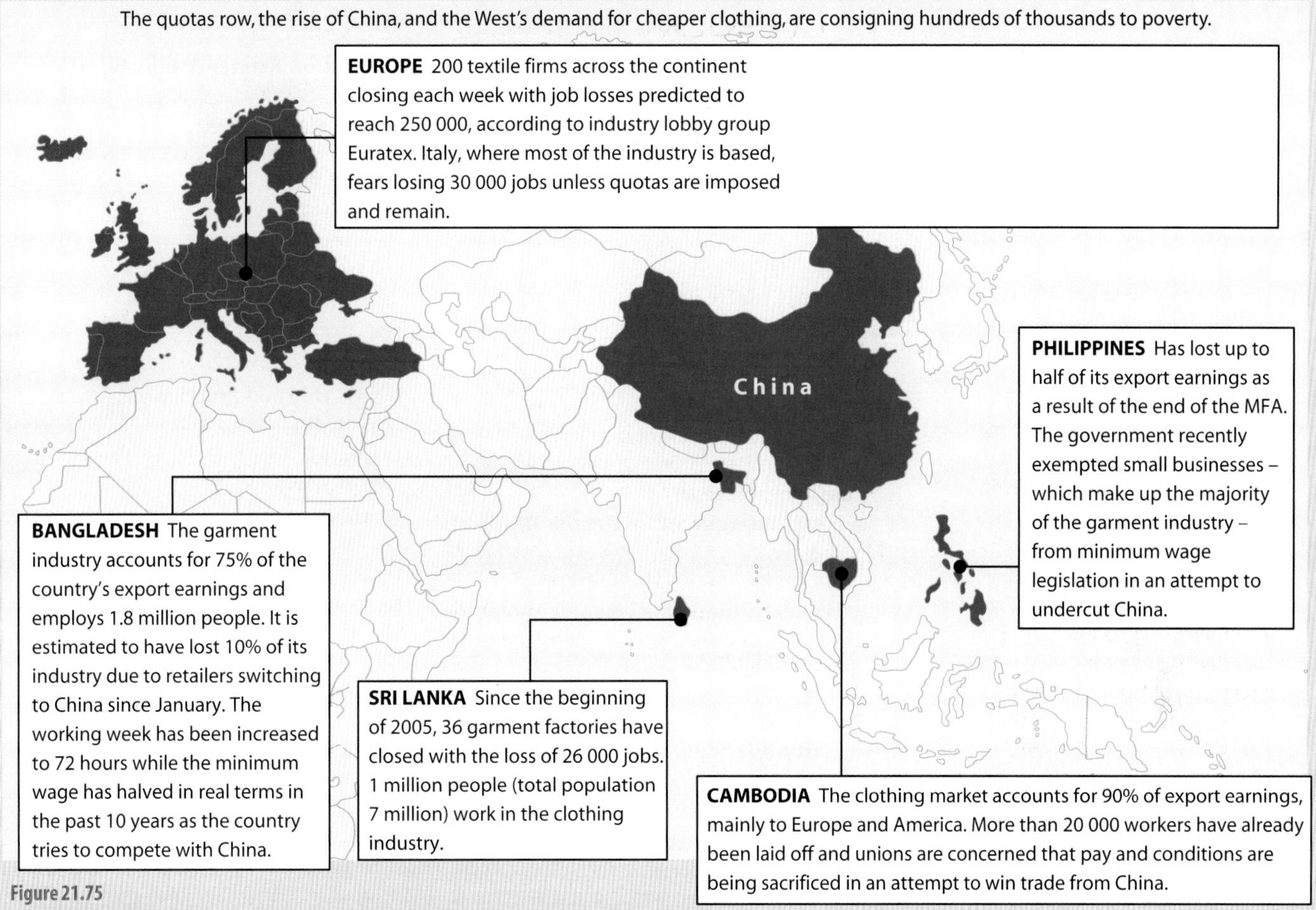

The quotas row, the rise of China, and the West's demand for cheaper clothing, are consigning hundreds of thousands to poverty.

EUROPE 200 textile firms across the continent closing each week with job losses predicted to reach 250 000, according to industry lobby group Euratex. Italy, where most of the industry is based, fears losing 30 000 jobs unless quotas are imposed and remain.

PHILIPPINES Has lost up to half of its export earnings as a result of the end of the MFA. The government recently exempted small businesses – which make up the majority of the garment industry – from minimum wage legislation in an attempt to undercut China.

BANGLADESH The garment industry accounts for 75% of the country's export earnings and employs 1.8 million people. It is estimated to have lost 10% of its industry due to retailers switching to China since January. The working week has been increased to 72 hours while the minimum wage has halved in real terms in the past 10 years as the country tries to compete with China.

SRI LANKA Since the beginning of 2005, 36 garment factories have closed with the loss of 26 000 jobs. 1 million people (total population 7 million) work in the clothing industry.

CAMBODIA The clothing market accounts for 90% of export earnings, mainly to Europe and America. More than 20 000 workers have already been laid off and unions are concerned that pay and conditions are being sacrificed in an attempt to win trade from China.

Figure 21.75

The quotas row: the situation in August 2005

Figure 21.76

From the *Newcastle Journal*, February 2008

Slow death of region's textile industry

THE North-East clothing and textile industry has been in decline for two decades, when customers such as Marks & Spencer – which once prided itself on selling British-made clothes – began to source garments from cheaper foreign suppliers.

Dewhirst, once one of M&S's largest suppliers, employed up to 20,000 people in the UK at one stage. Now, the business has around 1,500 British staff, mainly in design,

sampling and office-based roles. In the region, it retains a menswear manufacturing site in Sunderland and a plant in Peterlee.

Much of the work at these two sites is alterations to clothing that is made abroad. The region's textile industry has been hit hard by cheap imported goods in recent years. Since the late 1990s, more than 5,000 clothing jobs have been lost in the North-East.

The North-East's textile industry is made up of around 600 firms – mainly working in the areas of design, laundry and distribution – the majority of which employ fewer than 20 staff.

The manufacturing side of the business tends to concentrate on quality, expensive goods such as the garments made by Barbour in South Shields. Beau Brummell in Seaham makes blazers and other school clothing.

Fred Kirkland, from the North-East Textile Network and Skillfast – the UK sector skills council for fashion and textiles – said: 'Clothing and textiles as an industry has changed and moved on. What we do retain is the design and technical aspect. This is the high value end of the industry. It is very important we keep these skills in order to compete with companies abroad.'

Further reference

Barke, M. and O'Hare, G. (1991) *The Third World*, Oliver & Boyd.

Bek, D. and Binns, T. (2000) 'Putting ethics on the table', *Geography Review* Vol 21 No 2 (November).

Black, W.R. (2003) *Transportation: A geographical analysis New York*, The Guildford Press.

Bumpus, A. (2008) 'Energy matters: carbon offsets', *Geography Review* Vol 21 No 4 (April).

China Business Handbook 2008.

China Economic Review 2008.

Coe, N.M., Kelly P.F. and Yeung, H.W.C. (2007) *Economic Geography*, Blackwell.

Dicken, P. (2007) *Global Shift*, Sage Publications Ltd.

Digby, B. (2004) 'The changing geography of HIV/AIDS', *Geography Review*, Vol 18 No 1 (September).

Global Civil Society Yearbooks.

Held, D. *et al.* (1999) *Global Transformations*, Polity Press.

Phillips, D.R. and Verhasselt, Y. (eds) (1994) *Health and Development*, Routledge.

Singapore Yearbook of Statistics 2007, Singapore Ministry of Trade & Industry.

Statistical Yearbook for Asia and the Pacific 2007, Economic and Social Commission for Asia and the Pacific.

Thottathil, S. (2008) 'Energy matters: fair trade or food miles', *Geography Review* Vol 21 No 3 (February).

Canadian International Development Agency, Virtual Library on International Development:
www.acdi-cida.gc.ca/cidaweb/acdicida.nsf/En/Home

Centre for Health Protection, Department of Health, Government of the Hong Kong Special Administrative Region:
www.chp.gov.hk

CIA World Fact Book:
www.cia.gov/library/publications/the-world-factbook/

Food and Agriculture Organisation (FAO):
www.fao.org/index_en.htm

HDI data, human development reports:
http://hdr.undp.org/en/reports/global/hdr2007-2008/

Human Development Index, life expectancy tables:
http://hdrstats.undp.org/indicators/2.

International Energy Agency:
www.worldenergyoutlook.org/html

International Labour Organisation (ILO), child labour:
www.ilo.org

Population Reference Bureau:
www.prb.org

United Nations (UN):
www.un.org/Pubs/CyberSchoolBus/index.shtml

UN AIDS:
www.unaids.org

UN Development Programme (UNDP):
www.undp.org/

UN Water for Life:
www.un.org/waterforlifedecade/

UN World Food Programme:
www.wfp.org/english/

Wateraid:
www.wateraid.org/

World Bank:
www.worldbank.org/

World Bank, development indicators, annual GNP: (by country and the world)
www.worldbank.org/data/

World Energy Council:
www.worldenergy.org/

World Health Organisation (WHO):
www.wto.int/

World Trade Organisation:
www.wto.org/

Questions & Activities

Activities

1 **a i** What is meant by 'gross domestic product' (GDP) per capita? *(2 marks)*

    ii Why is this often chosen as a useful indicator of a country's level of development? *(2 marks)*

    iii Sometimes the Human Development Index (HDI) is used to indicate level of development, rather than using GDP/capita. What are the advantages of using the HDI? *(4 marks)*

  b Study Figure 21.3 on page 606.

  To what extent does this map support the view that the old division of the world into the 'rich north' and the 'poor south' is no longer very useful? *(7 marks)*

  c Choose one of the following sets of statistics that can also be used to show development:

- energy consumption/person
- number of doctors/thousand people
- level of education of females.

Explain why your chosen set of statistics is a good indicator of a country's level of development. *(10 marks)*

2 **a** The Rostow model shows the economy of a country going through five stages:
- traditional society
- preconditions for take-off
- take-off
- the drive to maturity
- high mass consumption.

Describe the characteristics of each stage. *(10 marks)*

b In Myrdal's core–periphery model, why does population often move from the periphery towards the core? *(5 marks)*

c Name a country that shows evidence of having a core and a periphery. Explain how Myrdal's model helps you to understand the distribution of economic development in that country. *(10 marks)*

Exam practice: basic structured questions

3 Study Figure 21.36 on pages 626–627.

a Describe the main features of the imports and exports of:
  i the developed countries (USA, UK and Japan) *(3 marks)*
  ii the OPEC countries (UAE and Nigeria). *(3 marks)*

b Choose **one** of the emerging market countries (China or India) and explain how that country has succeeded in developing its economy in recent years. *(9 marks)*

c Referring to **two** developing economies, for example Kenya and Sierra Leone, explain how changes in the world trade system might help their process of development. *(10 marks)*

4 **a** **i** What is a 'hub' airport? *(2 marks)*
  ii Name **one** international hub airport and explain why it has become important on a world scale. *(3 marks)*

b Discuss the economic and environmental arguments for and against an increase in the number of aircraft flights around the world. *(10 marks)*

c Name **one** city with an integrated transport system. Outline the main components of that system and explain the social, economic and environmental benefits of the integrated system. *(10 marks)*

Exam practice: structured questions

5 **a** Choose **one** of the MDGs in Figure 21.77 numbered 2, 3, 4 or 5. Explain why your chosen MDG can make an important contribution to the development of poor countries. *(5 marks)*

b Many people think that HIV/AIDS is a disease that is particularly damaging to the development process in many poor countries.
Suggest reasons for this view. *(8 marks)*

c With reference to **one or more** case studies, show how the process of economic development can take place whilst also ensuring environmental sustainability. *(12 marks)*

6 **a** Explain the importance of capital investment in Rostow's model of industrial development. *(4 marks)*

b Barke and O'Hare developed a different model to help explain the way many African countries were developing. Explain the importance of transnational corporations in their model. *(5 marks)*

c Name a country where a clear core–periphery relationship exists. Explain why the core developed much more than the periphery and discuss whether the difference between the core and the periphery is likely to be reduced in the future. *(15 marks)*

1	Eradicate extreme poverty and hunger
2	Achieve universal primary education
3	Promote gender equality and empower women
4	Reduce child mortality
5	Improve maternal health
6	Combat HIV/AIDS, malaria, and other diseases
7	Ensure environmental sustainability
8	Develop a global partnership for development

Figure 21.77

The eight Millennium Development Goals (MDGs)

7 **a** Study Figure 21.33 on page 623.
Discuss the economic and social significance of the prevalence of HIV infection in the different age cohorts in Botswana, and in similar countries of southern Africa. *(12 marks)*

b Outline the main features of the epidemiological transition model, and explain how a study of the model can help with an understanding of the process of economic and social development. *(13 marks)*

Exam practice: essays

8 With reference to **one** country where there are marked differences between the level of development in the core region and the periphery:
- explain why the different levels of economic development have arisen
- explain what the government is doing to try to reduce the differences between the core and the periphery. *(25 marks)*

9 With reference to countries at different stages of economic development, discuss how globalisation has affected *either* the textile and clothing industry *or* the ICT industry. *(25 marks)*

10 'Free Trade is more important than Fair Trade in encouraging the economic development of the poor countries of Africa, South America and Asia.'
Discuss this statement. *(25 marks)*

Index

rotational Slumping (handwritten)